The Greatest Lie on Earth

Proof That Our World Is Not a Moving Globe

EXPANDED EDITION

Edward Hendrie

Say among the heathen that the LORD reigneth: the world also shall be established that it shall not be moved: he shall judge the people righteously. (Psalms 96:10)

Keep that which is committed to thy trust, avoiding profane and vain babblings, and oppositions of science falsely so called: Which some professing have erred concerning the faith. Grace be with thee. Amen. (1 Timothy 6:20-21)

GREAT MOUNTAIN PUBLISHING

Eleventh Edition (Expanded)

Hardcover Book ISBN: 978-1-943056-05-7
Paperback Book ISBN: 978-1-943056-03-3
Ebook ISBN: 978-1-943056-04-0
EPUB Ebook ISBN: 978-1-943056-13-2
EdwardHendrie@gmail.com
Other books from Great Mountain Publishing®

- The Sphere of Influence
- 9/11-Enemies Foreign and Domestic
- Solving the Mystery of BABYLON THE GREAT
- The Anti-Gospel
- Bloody Zion
- What Shall I Do to Inherit Eternal Life?
- Murder, Rape, and Torture in a Catholic Nunnery
- Rome's Responsibility for the Assassination of Abraham Lincoln
- Antichrist: The Beast Revealed

Available at:
www.antichristconspiracy.com
https://greatmountainpublishing.com
www.911enemies.com
www.mysterybabylonthegreat.net
www.antigospel.com
https://play.google.com
www.barnesandnoble.com
www.amazon.com

Edward Hendrie rests on the authority of the Holy Bible alone for doctrine. He considers the Holy Bible to be the inspired and inerrant word of God. Favorable citation by Edward Hendrie to an authority outside the Holy Bible on a particular issue should not be interpreted to mean that he agrees with all of the doctrines and beliefs of the cited authority. All Scripture references are to the Authorized (King James) Version of the Holy Bible, unless otherwise indicated.

Table of Contents

Introduction

This book reveals the mother of all conspiracies. It will set forth biblical proof and irrefutable evidence that will cause the scales to fall from your eyes and reveal the truth that the world you thought existed is a myth. The book addresses hard facts, not fanciful theories.

An honest man who is shown to be wrong must either admit his error or cease to be honest. There are many so-called scientists and theologians who have made accommodations for falsehood. In order to keep their station in life, they have compromised their integrity and knelt in fealty to mammon and the praise of men.

George Orwell stated that "in a time of universal deceit - telling the truth is a revolutionary act." We live in such a time. The most universally accepted deceit today is the ingrained teaching that the earth is a globe spinning at a speed of approximately 1,000 miles per hour at the equator, while at the same time orbiting the sun at approximately 66,600 miles per hour.[1] While the earth is orbiting and spinning, the sun, in turn, is supposed to be hurtling through the Milky Way galaxy at approximately 500,000 miles per hour. The Milky Way galaxy is, itself, alleged to be racing through space at a speed ranging from 300,000 to 1,340,000 miles per hour.

What most people are not told is that all of this spinning,

orbiting, and speeding through space, has never been proven. Indeed, those hypothesized movements and speeds are completely made-up. In fact, every scientific experiment that has ever been performed to determine the motion of the earth has proven that the earth is completely stationary. Yet, scientific textbooks ignore the scientific proof that contradicts the myth of an orbiting and spinning Earth.

The "Christian" churches have gone along with the deception and teach in their "Christian" schools a heliocentric solar system, with a spinning and orbiting earth, even though the Bible clearly states that the earth does not move and is not a globe. How could this deception have been so complete as to include both the scientific world and the "Christian" churches? Why would the scientific world go along with a falsehood that has been proven to be wrong? Why would the "Christian" churches go along with a myth that is contrary to the Bible? This book will explain why and how it happened.

People have been conditioned to have a visceral reaction to any evidence that contradicts the heliocentric model. Such rejection of hard evidence, without a fair hearing, brings folly and shame. "He that answereth a matter before he heareth it, it is folly and shame unto him." (Proverbs 18:13) This folly and shame is manifested in devilish philosophies that permeate society. The deception of a spherical, spinning earth is the foundation for Darwinian evolution, Freudian psychoanalysis, and Marxist communism. Indeed, the progressive emergence of the sodomite subculture into a government protected privileged class is the direct result of the prevailing theory of heliocentrism (the sun at the center of a solar system). "How so?" you might ask. It is quite simple. To remove the earth as the center of God's creation, and accept in its place an earth that is just one of millions of wandering planets in the universe, removes man as God's unique creation, made in his image. Once the centrality of the earth in God's creation is removed, it is only a small half-step further to remove the existence of God himself from the minds of men.

Once God is removed from man's consciousness, then also is removed the authority of God's word and his law. Man is then enthroned, being a law unto himself. The common thread running through heliocentricity, evolution, psychology, and communism is that there is no God. Indeed, atheism is logically a necessary element for each of those man-made philosophies to stand.

The sodomite privileged class springs from a godless generation, with no fear of God. Indeed, they must necessarily reject the God of the Bible, because within the Bible is found God's condemnation of sodomy. "Thou shalt not lie with mankind, as with womankind: it is abomination." (Leviticus 18:22) That sin is so abhorrent to God that he rained fire and brimstone upon Sodom and Gomorrah as punishment for that filthy sin. "Then the Lord rained upon Sodom and upon Gomorrah brimstone and fire from the Lord out of heaven." Genesis 19:24. Nowadays, however, sodomy is viewed as a protected lifestyle, with the U.S. Supreme Court judging sodomy a good thing, even to the degree of creating a right for same-sex couples to get married. God curses those who call the sin of sodomy good. "Woe unto them that call evil good, and good evil; that put darkness for light, and light for darkness; that put bitter for sweet, and sweet for bitter!" (Isaiah 5:20)

The State and Federal governments in the United States have used the pro-sodomy laws to create a protected class with special government-granted, group privileges, which governments call "civil rights." "Woe unto them that decree unrighteous decrees, and that write grievousness which they have prescribed." (Isaiah 10:1) Those government-granted, group privileges will be used to strike at the God-given, individual rights of Christians. There is no new thing under the sun. God has spoken against such laws contrived to persecute his people. "Shall the throne of iniquity have fellowship with thee, which frameth mischief by a law? They gather themselves together against the soul of the righteous, and condemn the innocent blood." (Psalms 94:20-21)

One might ask, how can granting special privileges to a class of sinners infringe on the God-given, individual rights of Christians? Let us look at the example of the Oregon Equality Act of 2007, which is a law that grants special state privileges to sodomites. In 2013, Christians Aaron and Melissa Klein were fined $135,000 by the Oregon Bureau of Labor and Industries (BOLI) for declining to bake a cake for a lesbian wedding ceremony in violation of the Oregon Equality Act of 2007.[2] The fine was levied after the lesbian couple, Rachel Cryer and Laurel Bowman, filed a civil rights complaint with BOLI against the bakery, "Sweet Cakes by Melissa," for "emotional, mental, and physical suffering."[3] Tod Starnes reported for Fox News how the couple's small bakery was crushed by the sodomite protests, and as a result they were forced to close their bakery:

> The backlash against Aaron and Melissa Klein, owners of the bakery, was severe. Gay rights groups launched protests and pickets outside the family's store. They threatened wedding vendors who did business with the bakery. And, Klein told me, the family's children were the targets of death threats. The family eventually had to close their retail shop and now operate the bakery out of their home. They posted a message vowing to stand firm in their faith.[4]

Melissa Klein stated: "(I) honestly did not mean to hurt anybody, didn't mean to make anybody upset, (it's) just something I believe in very strongly."[5] After the Kleins were forced to shut down their bakery, Aaron Klein had to find employment as a trash collector to make financial ends meet. The State of Oregon "frameth mischief by a law" against the soul of the righteous. The Klein case in Oregon is not an anomaly.

Christian liberties are being crushed all over the country.[6] New Mexico's Supreme Court ruled unanimously that two Christian photographers, Elaine and Jonathan Huguenin, who declined to photograph a same-sex union violated the state's Human Rights Act.[7] The Attorney General of the State of Washington filed a lawsuit against a florist who refused to provide flowers for a same-sex couple's wedding.[8] Jack Phillips, the owner of Masterpiece Cakeshop in Colorado was ordered by a judge either to serve sodomite couples or face fines.[9] "When the righteous are in authority, the people rejoice: but when the wicked beareth rule, the people mourn." (Proverbs 29:2)

Phillips appealed his persecution all the way to the U.S. Supreme Court, which ruled that the Colorado Civil Rights Commission did not treat the baker fairly and that unfair treatment violated his constitutional rights of free speech and religion.[10] The decision by the U.S. Supreme Court was not a complete victory. The Court did not rule that the Colorado law itself that protected sodomites was an infringement on the baker's constitutional rights to free speech and religion, it was only the unfair way in which the Colorado Civil Rights Commission enforced the law that was unconstitutional.

The *Masterpiece Cakeshop* decision is only the beginning of a long fight. The decision is one step back for the sodomites that will be followed by two steps forward. Justice Thomas pointed that fact out in his concurring opinion. He stated that the *Masterpiece Cakeshop* case is the very type of case that he predicted would surface as a result of the 2015 U.S. Supreme Court decision in *Obergefell v. Hodges*[11] that granted sodomites the constitutional right to marry. That right to marry granted to sodomites, he said, will be used as a lever to persecute Christians and "stamp out every vestige of dissent and vilify Americans who are unwilling to assent to the new orthodoxy."[12] After the *Masterpiece Cakeshop* decision by the U.S. Supreme Court, the agencies of government won't be as overtly evil and stupid as was the Colorado Civil Rights Commission against Phillips. They will try to conceal their animus against Christians, while subtly turning the screws of government against them. As Justice Thomas states, Christians have only lived to fight another day. But it will be a death by a thousand cuts.

The Colorado Civil Rights Commission saw the U.S. Supreme Court decision for what is was: only one step back, which will be followed by two steps forward. Within just two weeks after the U.S. Supreme Court issued its ruling in favor of Phillips, the Colorado Civil Rights Commission was at it again with another prosecution against Phillips for his refusal to make a cake for a transgender customer. The commission alleged that Phillips discriminated against a customer based upon gender identity in violation of Colorado law. No doubt, the commission is going to try to be more subtle in how it handles Phillips this time. But make no mistake about it, their aim is to trample his religious freedom by providing special protections for sodomites.

What did Philips do that violated the laws of Colorado? He refused to make a custom cake with a blue exterior and a pink interior to celebrate the fact that the customer transitioned from male-to-female and had come out as transgender. That transgender customer was a lawyer who had made many previous obscene requests for custom cakes that were refused by Philips. One request by that customer was for Phillips to design a birthday cake for Satan that would feature an image of Satan smoking marijuana. Rod Dreher described the actions of the Colorado Civil Rights Commission in prosecuting Phillips again: "It's as if the state is giving a finger to the Supreme Court."[13]

The promotion and protection of sin acts to degenerate society. One sin begets another sin, as the perversions spiral downward. On July 19, 2016, the U.S. Government announced a new policy of allowing transgender persons to serve in the U.S. Military. A new sexual identity military manual states that "[e]ffective immediately, no otherwise qualified service member may be involuntarily separated, discharged or denied re-enlistment or continuation of service, solely on the basis of their gender identity."[14] The manual goes on to spell out the procedures for service members to obtain sex-change operations at government expense.

The Persecution of Jack Phillips illustrates the militancy of the transgender perversion. The militant transgender movement has infiltrated government agencies, and it will use the force of government to push its Satanic agenda. For example, on the night of January 28, 2018, a drunk and bloodied man, who self-identified as a transgender woman (biologically, he was a man), was ejected from a nearby homeless shelter for fighting. The drunk and bloodied man, wearing a woman's nightgown, sought to stay at Hope Center, which is a shelter for battered women and women escaping from sex-trafficking. He was refused entry, because he was a man, and all of the women in the shelter sleep in a common area. He was given cab fare by the staff at Hope Center so he could go to the emergency room for treatment.

The man returned the next day seeking to stay there, but again he was refused. The man responded by filing a complaint with the Anchorage Equal Rights Commission, under a statute prohibiting discrimination in public accommodations.[15] Essentially, the man (who self-identified as a transgender woman) was suing the women's shelter, claiming it was legally required to allow him (a drunk and violent man) to sleep in a common area among the women it housed, who were fleeing from violent and abusive circumstances. Incidently, in the past, the shelter had provided its services to that very man by serving him meals and allowing him to shower by himself — he simply was not allowed to sleep there. Howard Slugh Aylana Meisel, in an article for the National Review, reported:

> Many of the women had been abused by men and

> would not feel safe or comfortable sleeping so near a biological man. The lawyer noted that allowing a man to sleep in the room would "traumatize and present unreasonable safety risks for the abused and battered women who are admitted for overnight shelter." If this case alone wasn't disturbing enough, the Anchorage Equal Rights Commission sued Hope's lawyer after he made comments to a local reporter defending the shelter.[16]

The lesbian, gay, bisexual, transgender (a/k/a LGBT) communities have an even more nefarious agenda, which they keep secret from the public at large. Michael Swift's 1987 Gay Manifesto reveals the dirty secret of the LGBT rights movement. The LGBT community wants the right to sodomize children. The Gay Manifesto brazenly states:

> We shall sodomize your sons, emblems of your feeble masculinity, of your shallow dreams and vulgar lies. We shall seduce them in your schools, in your dormitories, in your gymnasiums, in your locker rooms, in your sports arenas, in your seminaries, in your youth groups, in your movie theater bathrooms, in your army bunkhouses, in your truck stops, in your all male clubs, in your houses of Congress, wherever men are with men together. Your sons shall become our minions and do our bidding. They will be recast in our image. They will come to crave and adore us.[17]

Former California State Assemblyman Steve Baldwin researched the connection between homosexuality and pederasty. He published his findings in the Spring 2002 Regent University Law Review. Baldwin found that "[s]cientific studies confirm a strong pedophilic predisposition among homosexuals."[18] One 1988 study published in the *Archives of Sexual Behavior*, reported that 86% of pedophiles who victimized boys "described themselves as homosexual or bisexual."[19] Research statistics further show that homosexuals, as a population, molest children at a rate that is ten to twenty times greater than heterosexuals.[20] Those facts are well known in the homosexual community. San Francisco's leading homosexual newspaper, *The Sentinel*, bluntly states that "[t]he love between men and boys is at the foundation of homosexuality."[21]

Homosexual publications openly promote pederasty and are often populated with travel ads for sex tours to Burma, the Philippines, Sri Lanka, Thailand, and other countries infamous for boy prostitution. Baldwin reveals that "[t]he most popular travel guide for homosexuals, Spartacus Gay Guides, is replete with information about where to find boys for sex and, as a friendly warning, lists penalties in various countries for sodomy with boys if caught."

Baldwin found that "the mainstream homosexual culture commonly promotes sex with children. Homosexual leaders repeatedly argue for the freedom to engage in consensual sex with children."[22] He determined that one of the principal aims of the LGBT rights movement is the legalization and promotion of child molestation. Mainstream LGBT organizations such as the International Lesbian and Gay Association (ILGA) and the National Coalition of Gay Organizations have passed many organizational resolutions calling for lowering or eliminating age of sexual consent laws, as a way to legalize pedophilia.[23]

The ILGA represents more than 620 member organizations. In 1995, the UN decertified the consultative status of the ILGA as a Non-Government Organization (NGO), because of political pressure from the U.S. Congress, when it was revealed the several member organizations of the ILGA promoted pedophilia. In an effort to get reinstated, ILGA severed ties with the North American Man Boy Love Association (NAMBLA), whose members practice pederasty. Severing ties with NAMBLA was apparently window dressing, since ILGA refused to expel a half-dozen more powerful (and discreet) pro-pedophile groups.[24] The UN has maintained its ban on the consultative status for ILGA.

The rampant pederasty among sodomites is nothing new. And it is the result of the evil spiritual influence that drives the sodomite perversion. For example, Aleister Crowley was a high Freemason and Satanist who styled himself as "The Beast" and proudly proclaimed himself to be "the wickedest man on earth." Crowley practiced ancient tantric sex magic (a/k/a magick). The core of tantric sex magic is to engage in increasingly cruel and perverse sex acts, gravitating to homosexual acts, and culminating in the rape of a child. The dark masonic secret, known only to the very highest adepts of Freemasonry, is that their way to immortality is to sexually violate a child.[25]

That is the purpose behind the granting of government protected privileges to sodomites. That government-protected privilege is only the first step toward the degenerative goal of the evil occultists who control the governments of the world, which is to legalize sex with children. The powerful elite of the world today secretly engage in all manner of pederasty, which this author details in his book, *Antichrist: The Beast Revealed.*

Indeed, the elite have plans to normalize pedophilia. The first step toward that goal is to convince the populace that pedophilia is simply a sexual preference with which a person is born.[26] Marjan Heine advanced that argument during a 2018 TEDx Talk at the University of Wurzburg. Heine's presentation was titled: *Why Our Perception of Pedophilia Has to Change.*[27] Heine defined pedophilia as a person's inborn sexual preference for pre-adolescent children. She argued that nobody is responsible for their feelings, and so pedophiles are not responsible for their feelings of sexual attraction toward pre-adolescent children.

Heine claimed that it is social isolation that drives pedophiles to sexually abuse children. Heine suggested that if the shame of pedophilia was removed, and pedophiles were welcomed into society, they would not act out their impulses to sexually abuse children. Heine argues that society must overcome its negative feelings toward pedophiles and treat pedophiles with respect, because, after all, they did not choose their sexuality. Pedophiles should not be blamed for their sexual preference, they should be welcomed for who they are: pedophiles.

Heine uses the evil sophistry that a person's proclivities are dissociated from their actions. She reasons that a pedophile is not necessarily a child molester and a child molester is not necessarily a pedophile. Heine ignores the reality that a person is identified by what he does, not by what he feels or thinks. For example, a football player is a football player because he plays football. He is identified by what he actually does. A person may feel like playing football, but until he actually plays football, he is not a football player. A thief is a thief, because he steals. He is identified as a thief, because of what he does. He is not a thief if he only desires to have someone else's property, but does not take it. A pedophile is a pedophile because he sexually abuses children. He is identified as a pedophile because that is what he does. Heine ignores that reality.

To make a distinction between pedophiles and child molesters is to make a distinction without a difference. It is a false dichotomy. Heine thinks she can get away with making that misleading distinction because most people do not know that recidivism among pedophiles is very high. Show me a pedophile who denies he molests children and I will show you a liar. Research has proven that pedophiles do not control their impulses to sexually molest children. For example, an Emory University Study conducted by a leading child abuse researcher, Dr. Gene

Abel, found that the average child molester (i.e., pedophile) claims 380 victims in a lifetime.[28]

The high recidivism of molestation among pedophiles impeaches Heine's statement made during her presentation that "scientific studies indicate that only 20% to 30% of all child molesters are pedophiles."[29] No, 100% of child molesters are pedophiles; just as 100% of those who steal are thieves. All pedophiles sexually abuse children; that is why they are called pedophiles. Pedophiles act on their perverted impulses. The suggestion that they don't is a devilish perversion in-and-of itself. The most disturbing aspect of Heine's speech was that at its conclusion she received applause from the audience in the packed auditorium.

Sin begets sin. Once a sin is legitimized and protected by the force of government, it begins a downward trajectory, where more sin is legitimized and protected. Such sin creates a violent and chaotic society. This chaos gives rise to new restrictive laws, which ultimately brings about a police state to reestablish order. Under the new police powers granted to the state, the exercise of God-given rights is suppressed by the government.

That is not hyperbole. Former U.S. House of Representatives Majority Leader Tom DeLay revealed in an interview on The Steve Malzberg Show that the U.S. Department of Justice plans to expand the "civil rights" granted by the U.S. Government to include a raft of new perversions, including, but not limited to, bestiality, polygamy, and pedophilia. Those abominable "civil rights" to rape children, have polygamous marriages, and have sex with animals will be used as a lever to persecute Christians who exercise their God-given rights, object to those sins, and seek to protect the children. The U.S. Justice Department has been busy strategizing on how to use the force of government to persecute recalcitrant Christians.

Former U.S. House Majority Leader Tom DeLay claims the Justice Department has drafted a memo that spells out a dozen "perversions," including bestiality and pedophilia, that it wants legalized.

"We've ... found a secret memo coming out of the Justice Department. They're now going to go after 12 new perversions. Things like bestiality, polygamy, having sex with little boys and making that legal," DeLay said Tuesday on "The Steve Malzberg Show" on Newsmax TV.

"Not only that, but they have a whole list of strategies to go after the churches, the pastors and any businesses that try to assert their religious liberty. This is coming and it's coming like a tidal wave."

The Texas Republican's bombshell claim comes four days after the Supreme Court ruled that same-sex marriage is now legal in all 50 states — a landmark decision DeLay strongly opposes.

When Steve Malzberg repeated to DeLay his assertions that the Justice Department seeks "to legitimatize or legalize" practices such as bestiality — defined as sex acts between humans and animals — DeLay responded:

"That's correct, that's correct. They're coming down with 12 new perversions ... LGBT [short for lesbian gay, bisexual and transgender] is only the beginning. They're going to start expanding it to the other perversions."[30] (bracketed parenthetical in original article)

Satan knows that if man cuts loose his passions without regard to the laws of God, the sinful conduct will necessitate a tyrannical government. How so? The escalating sin is in essence an encroachment upon the God-given rights of another. Jesus explained the concept:

> Jesus said unto him, Thou shalt love the Lord thy God with all thy heart, and with all thy soul, and with all thy mind. This is the first and great commandment. And the second is like unto it, Thou shalt love thy neighbour as thyself. On these two commandments hang all the law and the prophets. (Matthew 22:37-40)

The love that Jesus speaks of is not an emotional feeling toward another as Hollywood movies would have you believe. Emotions come and go. Love is not an emotion, it is an action. Jesus explained: that "as ye would that men should do to you, do ye also to them likewise." (Luke 6:31). Indeed, doing good for another person has nothing at all to do with how you feel about that person. Jesus stated that you should do good to even your enemies who have treated you badly. Luke 6:27-36. God's word has a built in dictionary where words are defined by the use of parallelism. One can see in Luke chapter 6 at verses 32 and 33 that God uses parallelism to define love, not as a feeling, but rather as an action of "doing good" to another person.

"For if ye **love** them which **love** you, what thank have ye? for sinners also **love** those that **love** them." (Luke 6:32)

"And if ye **do good** to them which **do good** to you, what thank have ye? for sinners also do even **the same**." (Luke 6:33)

Doing good to another is not just a suggestion, it is a command from God. Jesus states without equivocation that all of the commands of God are subsumed in the commands to love God and love your neighbor. Love has nothing to do with feelings. You should do nice things for others, whether you "feel" like it or not, because by doing so you are obeying God's command to do good to others. If one loves his neighbor, he will not violate the command, "thou shalt not steal." Exodus 20:15. While the command not to steal is a prohibition, Jesus takes it a step further and commands that we are to affirmatively do good to others (i.e., love).

Doing good often involves putting our needs second to the needs of another. If, however, a person does not regard God and seeks to satisfy the pleasure of his fleshly desires, without regard to the detrimental effect of his conduct on others, it will cause the government to step in to protect the aggrieved party. Sinful behavior brings about more crime in society, which is then used as a justification to bring about more government regulation and control of the masses. As explained by Edmond Burke: "Men are qualified for civil liberty in exact proportion to their disposition to put moral chains on their own appetites. Society cannot exist unless a controlling power upon will and appetite be placed somewhere, and the less of it there is within, the more there is without. It is ordained in the eternal constitution of things that men of intemperate minds cannot be free. Their passions forge their fetters."

That is why Satan wants to drive any thought of God from the minds of men. Indeed, by doing so, a satanically controlled government can impose its tyranny on the people, without much objection from the people, because the people are ignorant that the fetters applied by the government violate the laws of God. For example, the communist collective ideal is based upon larceny. It uses the government to redistribute wealth. The redistribution of wealth is always justified by some purported need, and it usually involves helping children; it is always cloaked in nice sounding labels, like "welfare." There are always powerful corporate interests that are lined up to be on the receiving end of the government forced transaction.

Rather than charity coming willingly from one individual to another, the government uses its force, through taxation, to take money from one group and give it to another. Taking money from one group and giving it to another group cannot be done if the government accepted the existence of God. If God were to be accepted, then necessarily God's commands must be then obeyed. The government would have a hard time justifying taking money from one group of persons for the benefit of another group of persons, when faced with God's command: "Thou shalt not steal." (Exodus 20:15) Using the government as an intermediary does not make the stealing any more justifiable.

Like communism, Darwinian evolution is necessarily atheistic and cannot stand if the authority of the Bible is accepted. The Bible makes it clear that God created man in his image, not after the image of an ape. "So God created man in his own image, in the image of God created he him; male and female created he them." (Genesis 1:27)

Heliocentrism destroyed the landmark of geocentrism. Heliocentrism laid the groundwork for Darwinism. Once Darwinian evolution settled in, it was time for the planting of Marxist communism. In 1860 Karl Marx stated the following about Darwin's book, *On the Origin of Species*, which announced the theory of evolution: "Although it is developed in the crude English style, this is a book which contains the basis of natural history for our views."[31]

After allowing the theory of evolution to ferment in his mind for the next 7 years, in 1867, Marx brought forth to the world his economic philosophy, based upon godless Darwinism, titled *Das Kapital*. Marx sent a copy of his new book to Charles Darwin inscribed with the following: "To Charles Darwin from a devoted admirer, Karl Marx, 1873."

In October, 1873, Charles Darwin wrote to Karl Marx the following letter of thank you for the inscribed book sent to him by Marx:

Dear Sir:

I thank you for the honour which you have done me by sending me your great work on Capital; & I heartily wish that I was more worthy to receive it, by understanding more of the deep and important subject of political Economy. Though our studies have been so different, I believe that we both earnestly desire the extension of Knowledge, & that this is in the long run sure to add to the happiness of Mankind.

I remain, Dear Sir
Yours faithfully,

Charles Darwin

The synergism between Darwin's godless evolution and Marx's godless communism is evidenced by the fact that in 1959 the communist regime in the Soviet Union struck a coin and printed a stamp[32] in honor of the 150th anniversary of Darwin's birth (1809).

Sigmund Freud based his theories of psychoanalysis on the premise that there is no God. In his *The Future of an Illusion*, he described belief in God as a collective neurosis: he called it "longing for a father."[33] Psychology is essentially religion without God. Like all man-made religions, there are schisms. There are more than 200 different theories of psychology (Freudian psychoanalysis being only one of them).

Heliocentrism is truly the mother of all conspiracies. None of the other Satan-inspired, man-made philosophies would ever dare be presented without the geocentric landmark first being removed by the general acceptance of heliocentrism. Heliocentrism is the keystone to the deception; remove that keystone, and the entire edifice of deception will fall. With that keystone in place, the satanic source of the deception is inscrutable, and the facade of scientific authority for the satanic philosophies remains believable.

The educational authorities use a tried and true method of conditioning students to reject true scientific evidence and instead accept the myth of heliocentricity. William H. Poole explained this stratagem. "There is a principle which is a bar against all information, which is proof against all argument, and which cannot fail to keep a man in everlasting ignorance. This principle is, contempt prior to examination."[34] Satan has used this method to keep people in a state of nescience about God's creation. He has created a hive mentality instilled at the earliest stages of education that protects the myth of heliocentricity. Anyone who questions the legitimacy of heliocentricity has his character attacked by the hive as being ignorant. The evidence for a stationary, flat earth, no matter its validity, will be dismissed without examination. Thus the hive is kept in "everlasting ignorance."

The importance of heliocentrism to Satan's plan for world domination is why it is introduced at the earliest stage possible in the education of students. From the first day in government schools, the foundational scientific truth presented to a student is a globe of the earth that can spin on a pedestal. Heliocentricm is repeatedly reinforced throughout a student's scientific education. Heliocentrism is the most glaring example of mass mind control. To control the masses, heliocentrism must be introduced early and often, so that it becomes a deeply ingrained belief. While at the same time, belief in a stationary, flat earth is always portrayed by the educational authorities as evidence of abject ignorance.

Each of the satanic philosophies of the world (heliocentrism, communism, evolution, psychology) are necessary cogs in the New World Order, which is an ancient trans-generational conspiracy. There is nothing new about the New World Order. The New World Order conspiracy is a conspiracy against God and man; it began in heaven and continues today on earth.

> And there was war in heaven: Michael and his angels fought against the dragon; and the dragon fought and his angels, And prevailed not; neither was their place found any more in heaven. And the great dragon was cast out, that old serpent, called the Devil, and Satan, which deceiveth the whole world: he was cast out into the earth, and his angels were cast out with him." (Revelation 12:7-9)

Satan was cast out of heaven. He then gathered with his minions on earth and began his war against God and man anew.

> And when the dragon saw that he was cast unto the earth, he persecuted the woman which brought forth the man child. ... And the dragon was wroth with the woman, and went to make war with the remnant of her seed, which keep the commandments of God, and have the testimony of Jesus Christ. (Revelation 12:13,17)

The target of Satan's malevolence is the true church of Jesus Christ. Satan has a secret army of devil possessed adversaries of Christ who do Satan's bidding in working to suppress the gospel of Jesus Christ and enslave the world. In order to achieve those ends, Satan must control the minds of the masses. He must enslave our minds, before he can enslave our bodies. Heliocentrism is the foundational belief for enslaving the minds of men.

Satan's goal is a world government under his control; his minions call it a New World Order. One of Satan's minions is President George H.W. Bush. In his January 29, 1991, State of the Union speech before a Joint Session of Congress, President Bush brazenly announced his efforts toward achieving a New World Order. "What is at stake is more than one small country, it is a big idea, a New World Order, where diverse nations are drawn together in common cause."[35]

Isa Blagden stated that "if a lie is only printed often enough, it becomes a quasi-truth, and if such a truth is repeated often enough, it becomes an article of belief, a dogma, and men will die for it."[36] Heliocentrism has become the dogma of our world today. The purpose of the heliocentric model is to hide the existence of God. This book will not only prove that heliocentrism is false; it will go further and prove that God created a stationary, flat earth that is the center of his creation, just as stated in the Bible. You will come to understand that God is real, and he is watching.

1 Samuel Rowbotham Proved the Earth Is Flat

Evidence that the earth is flat is all around us. For example, if the earth were a sphere, then the curvature of the earth would be manifested in the physical landscape. But we do not see any curvature of the earth. The earth's circumference is purported to be 24,901 miles at the equator. That means that the horizon should drop from sight at the rate of 8 inches per mile distance squared. Samuel Birley Rowbotham (1816–1884), who went by the *nom de plume* of "Parallax," in 1881 explained the principle in his book, *Zetetic Astronomy, Earth Not a Globe*,

> If the earth is a globe, and is 25,000 English statute miles in circumference, the surface of all standing water must have a certain degree of convexity--every part must be an arc of a circle. From the summit of any such arc there will exist a curvature or declination of 8 inches in the first statute mile. In the second mile the fall will be 32 inches; in the third mile, 72 inches, or 6 feet, as shown in the following diagram:[37]

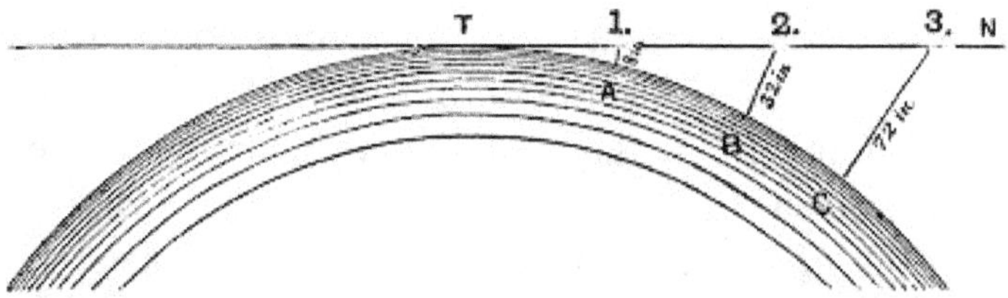

In Cambridge County, England there is an artificial canal, called the "Old Bedford." It is approximately twenty miles in length. Rowbotham arranged to have a boat, with a flag on it that was exactly 5 feet above the surface of the water. The boat was rowed to Welney Bridge, which was exactly six statute miles in a straight line from Rowbotham, who had waded into the water. Rowbotham stood in the middle of the canal with a telescope exactly 8 inches above the surface of the water. In looking through the eyepiece of the telescope and observed the receding boat during the whole period required to sail to Welney Bridge. "The flag and the boat were distinctly visible throughout the whole distance!"[38]

If the earth were a sphere, as it is supposed, the flag should not have been visible to Rowbotham. The boat and flag should have been 11 feet 8 inches below the horizon as depicted in the diagram below.

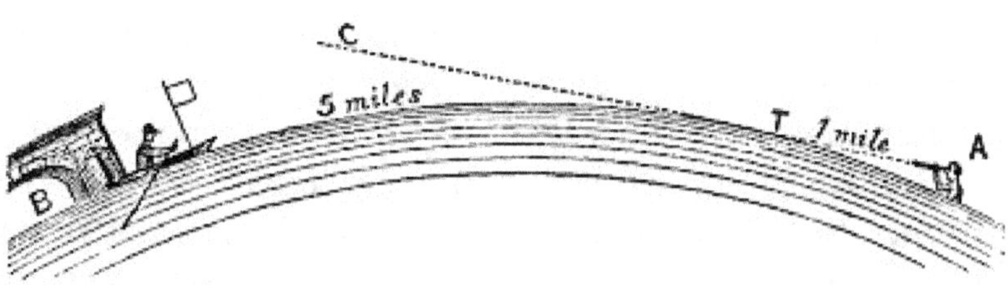

Rowbotham explains the depiction in the diagram above: "Let A B represent the arc of water 6 miles long, and A C the line of sight. The point of contact with the arc would be at T, a distance of one mile from the observer at A. From T to the bridge at B would be 5 miles, and the curvature from T to B would be 16 feet 8 inches. The top of the flag on the boat (which was 5 feet high) would have been 11 feet 8 inches below the horizon T, and altogether out of sight. Such a condition was not observed."[39]

Rowbotham was able to see the flag and boat for the entire 6 mile journey in the canal, all the way to the bridge, as depicted in the diagram below.

The fact that Rowbotham was able to see the flag and boat for the entire 6 miles, proves that the earth is flat. Rowbotham did many similar experiments at the canal that all proved that there was no curvature of the earth.

The formula for the expected drop per mile is $M^2 \times 8 = D$, where "M" is the number of miles from the observer at ground level, "8" is the number of inches dropped per mile squared, and "D" is the distance of the drop over the horizon. The above equation assumes an earth with a circumference of 24,901 miles. J. Clendenning, the former British Surveyor General of the Gold Coast (Ghana), in his book, *The Principles of Surveying,* confirmed that "as regards curvature, a level line parallel to the earth's surface ... falls away from a plane tangential to the earth at any point by about 8 in. in a distance of a mile, and this difference increases with the square of the distance."[40]

The U.S. Army Topographic Surveying Technical Manual uses a different equation for calculating the supposed curvature of the earth. The U.S. Army manual states that the earth's curvature is "equal to the square of the distance to the point divided by the earth's diameter."[41] The results using the U.S. Army manual equation or $M^2 \times 8 = D$ are the same.

The validity of the formula $M^2 \times 8 = D$ can be verified by using the Pythagorean theorem. Under the Pythagorean theorem the length of any side of a right triangle can be determined if one knows the length of the other two sides. The formula for the Pythagorean theorem is $a^2+b^2=c^2$, where a, b, and c are the three sides of a right triangle.

In the graphic below, this author used the Pythagorean theorem to determine the drop below the horizon at a distance of 50 miles. The answer using the Pythagorean theorem is exactly the same answer through the formula of $M^2 \times 8 = D$. To find the drop from the horizon at a distance of 50 miles we would take the miles squared multiplied times 8 inches per our formula above, $M^2 \times 8 = D$. Assuming a 50 mile distance, we get $50^2 \times 8 = 20{,}000$ inches (1,666 feet). Converting the 20,000 inches (1,666 feet) to miles, we get a drop of .32 miles.

The radius of the earth is determined to be 3,963 miles. The radius of the earth is determined through the formula of $R = C \div (2 \times \pi)$ where "π" is 3.14159, "R" is the radius, and "C" is the circumference. The number "π" is a mathematical constant that is based upon the ratio of a circle's circumference to its diameter. To arrive at π, simply divide a circle's circumference by its diameter. All circles, no matter their size, have the same ratio of their circumference to their diameter of 3.14159. π is an irrational number that never ends. 3.14159 only carries π out to the fifth decimal place, but it in fact continues indefinitely.

If the circumference of the earth at the equator is taken to be as alleged, 24,901 miles, that would mean that the earth's radius is $24{,}901 \div (3.14159 \times 2) = 3{,}963$ miles. For some reason, the alleged radius of the earth is published by many authorities to be 3,959 miles. That, however, cannot be correct, according to the geometric equation for determining the radius of a circle. This author will use the figure of 3,963 miles obtained from the equation for the radius of the earth. The difference of four miles between the two figures does not make a material difference in our conclusions.

Using the Pythagorean theorem ($a^2+b^2=c^2$) we can easily determine the length of the hypotenuse (side "c"). Side "a" is the radius of the earth (3,963 miles), side "b" is 50 miles, which means that side "c" is 3,963.32 miles. If we subtract the length of the hypotenuse (side "c") from the known radius of the earth, we can confirm that the drop below the horizon at a distance of 50 miles from a supposedly globular earth is .32 miles. That is the exact same figure we obtained using the formula $M^2 \times 8 = D$, thus confirming the validity of that formula.

Using two separate formulae to compute the distance of drop below the horizon at 50 miles, assuming a round earth with a circumference of 24,901 miles and a radius of 3,963 miles.

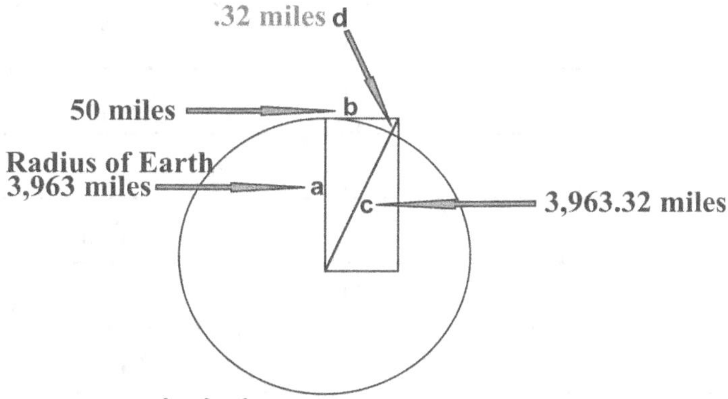

$a^2+b^2=c^2$

$\sqrt{a^2+b^2} = c$

$\sqrt{(3{,}963 \times 3963) + (50 \times 50)} = 3{,}963.32$

c - radius = d

3,963.32 - 3,963 = .32 miles

Miles^2 x 8 inches = drop

50 miles x 50 miles x 8 inches = 20,000 inches = .32 miles

Below is a chart that gives the results of the equation ($M^2 \times 8 = D$) for the expected drop per mile assuming an earth with a circumference of 24,901 miles. "M" is the number of miles from the observer at ground level, "8" is the number of inches dropped per mile squared, and "D" is the distance of the drop over the horizon.

Statute Miles	Drop Below Horizon Due to Earth's Supposed Curvature
1	8 inches
2	32 inches
3	6 feet
4	10 feet
5	16 feet
6	24 feet
7	32 feet
8	42 feet
9	54 feet
10	66 feet
20	266 feet
30	600 feet
40	1,066 feet
50	1,666 feet
60	2,400 feet
70	3,266 feet
80	4,266 feet
90	5,400 feet (1.02 miles)
100	6,666 feet (1.26 miles)
120	9,600 feet (1.82 miles)

The shortcut formula ($M^2 \times 8 = D$) is accurate up to the limits of vision through the atmosphere for calculating the supposed drop on the mythical, spherical earth. But there is a slight error that creeps into the results of the calculated drop over the mythical sphere of the earth at extremely long distances using the shortcut formula ($M^2 \times 8 = D$) when comparing it to the more precise formula using the Pythagorean theorem. For example, at a distance of five hundred (500) miles on the mythical, spherical earth, there is a one-half percent (.5%) difference between the results using the shortcut formula (M2 × 8 = D) (calculated to be a supposed 31.57 mile drop) and the results using the Pythagorean theorem (calculated to be a supposed 31.42 mile drop). At a distance of one thousand (1,000) miles there is a one and one half percent (1.5%) difference between the results using the shortcut formula (M2 × 8 = D) (calculated to be a supposed 126.26 mile drop) and the results using the Pythagorean theorem (calculated to be a supposed 124.22 mile drop). As the above calculations demonstrate, the slight error using the shortcut formula (M2 × 8 = D) is not manifested to any measurable degree until extremely long distances are reached. The slight differences in the outcome at very long distances over the mythical, spherical earth are not material, as the differences do not affect the fact that the earth cannot possibly be a sphere, but only affect the precision (and then only at extreme distances) of the proof that the earth cannot possibly be a sphere. This does not in any way undermine the validity of the formula nor the overwhelming evidence that proves the earth is flat.

2 Visibility of Distant Objects Over Water

If, in fact, the earth were a round ball as is supposed, there is simply no way to explain how lighthouses can be seen from such great distances at sea. This phenomenon can be experienced by modern mariners today. The only interpretation of the phenomenon of being able to see lighthouses and other structures many miles out at sea is that the earth is flat. Rowbotham gave a series of examples of lighthouses that could, in fact, be seen at great distances. Those lighthouses should have been well below the level curvature of the earth and thus out sight if the earth were a sphere.

> The Egerö Light, on west point of Island, south coast of Norway, is fitted up with the first order of the dioptric lights, is visible 28 statute miles, and the altitude above high water is 154 feet. On making the proper calculation it will be found that this light ought to be sunk below the horizon 230 feet.
>
> The Dunkerque Light, on the south coast of France, is 194 feet high, and is visible 28 statute

> miles. The ordinary calculation shows that it ought to be 190 feet below the horizon.
>
> The Cordonan Light, on the River Gironde, west coast of France, is visible 31 statute miles, and its altitude is 207 feet, which would give its depression below the horizon as nearly 280 feet.
>
> The Light at Madras, on the Esplanade, is 132 feet high, and is visible 28 statute miles, at which distance it ought to be beneath the horizon more than 250 feet.
>
> The Port Nicholson Light, in New Zealand (erected in 1859), is visible 35 statute miles, the altitude being 420 feet above high water. If the water is convex it ought to be 220 feet below the horizon.
>
> The Light on Cape Bonavista, Newfoundland, is 150 feet above high water, and is visible 35 statute miles. These figures will give, on calculating for the earth's rotundity, 491 feet as the distance it should be sunk below the sea horizon.[42]

Rowbotham used standard, readily available authorities to prove that the earth is a flat plane. For example, he cites many instances in the authoritative text of its time on lighthouses, *Lighthouses of the World,* by Alexander G. Findlay (1861). Findlay was an internationally renowned geographer and hydrographer, who wrote many reference guides, which were invaluable to mariners throughout the world. Findlay had unquestioned competence in hydrography. Mariners relied upon the accuracy of his *Lighthouses of the World* reference guide. Indeed, it was imperative that the entries in that reference book be precise, as the very lives of the mariners rested on the accuracy of the information in that text. While the text was certainly accurate,

Rowbotham demonstrated that many of the lighthouses should not have been visible from the distances listed in that text if the earth were a globe. The only rational conclusion is that the earth is a flat plane. One example Rowbotham cites from page 39 of *Lighthouses of the World* is the Bidston Lighthouse.

> By the same authority, at page 39, the Bidston Hill Lighthouse, near Liverpool, is 228 feet above high water, one bright fixed light, visible 23 nautical or very nearly 27 statute miles. Deducting 4 miles for the height of the observer [10 feet above the water], squaring the remaining 23 miles and multiplying that product by 8 inches we have a downward curvature of 352 feet; from this deduct the altitude of the light, 228 feet, and there remains 124 feet as the distance which the light should be below the horizon![43]

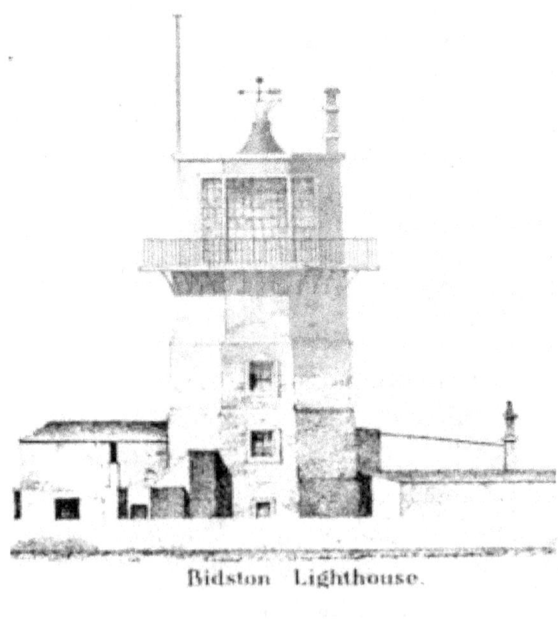

Bidston Lighthouse.

Above is a sketch of Bidston Lighthouse, from *Sailing*

Directions from Point Lynas to Liverpool. The Bidston Lighthouse was replaced in 1873 with a new lighthouse built nearby. Below is a photocopy extract of the page in the book, *Lighthouses of the World*, which Rowbotham cited.

ENGLAND. LIGHTHOUSES. **West Coast.** 39

Name and Character of Light.	Lat. N. Long. W. ° ′	Description, &c.	Description of Apparatus	Height above H. W.	Visible in Miles.	Year established.
HOYLAKE Two br. fixed lights	53 23.7 3 10.7	In one, S.W. by S., 1,200 feet apart, near the Church	●	55 31	13 11	1763
BIDSTON One bright fixed light	53 24. 3 4.4	A stone tower, 68 feet high, on the hill	●	228	23	1771
LEASOWE One bright fixed light	53 24.8 3 7.5	On the shore, between the Mersey and Dee................	●	94	14	1763
BLACK ROCK One rev. lt. 1 minute br. twice; *red* once	53 26.6 3 2.	A white tower, 94 ft. high. Also a *fixed light*, while 11 ft., down Rock Channel and up Mersey	●	61	14	1830

Please note that Findlay explained that the distance at which a lighthouse could be seen given in his book as "the minimum distance to which the light can be seen, in clear weather, from a height of 10 feet above the sea level."[44] Since Findlay is giving the minimum distance in clear weather, that means that in clear weather the lighthouse could potentially be seen from further than the listed distance. Furthermore, the listed distances assume clear weather. If the weather was not clear, due to humidity or other causes, then the distance from which a lighthouse could be seen, would, of course, be shorter.

Rowbotham illustrated an example where passengers on a ship in St. George's Channel could see the lighthouses on each shore at the same time from the middle of the channel. The lighthouses were separated by a total distance of 60 miles, which put each lighthouse 30 miles from the ship going down the center of the channel. Such a phenomenon would be impossible if the earth were a globe. Rowbotham explains:

The distance across St. George's Channel, between

Holyhead and Kingstown Harbour, near Dublin, is at least 60 statute miles. It is not an uncommon thing for passengers to notice, when in, and for a considerable distance beyond the centre of the Channel, the Light on Holyhead Pier, and the Poolbeg Light in Dublin Bay, [as shown in the diagram below].[45]

If the earth were a globe, however, the lighthouses could not have been seen at all. Indeed, due to the curvature of the supposed globular earth, at the distance of 30 miles each from the ship, and even subtracting for the height of the observer above the water, each lighthouse would be over 300 feet below the horizon. Rowbotham explains:

> The Lighthouse on Holyhead Pier shows a red light at an elevation of 44 feet above high water; and the Poolbeg Lighthouse exhibits two bright lights at an altitude of 68 feet; so that a vessel in the middle of the Channel would be 30 miles from each light; and allowing the observer to be on deck, and 24 feet above the water, the horizon on a globe would be 6 miles away. Deducting 6 miles from 30, the distance from the horizon to Holyhead, on the one hand, and to Dublin Bay on the other, would be 24 miles. The square of 24, multiplied by 8 inches, shows a declination of 384 feet. The altitude of the lights in Poolbeg Lighthouse is 68 feet; and of the red light on Holyhead Pier, 44 feet. Hence, if the earth were a globe, the former would always be 316 feet and the latter 340 feet below the horizon,

as seen in the following diagram .... The line of sight H, S, would be a tangent touching the horizon at H, and passing more than 300 feet over the top of each lighthouse.[46]

Evolutionist Robert Schadewald, who believes that the world is a sphere, tried to refute Rowbotham's findings. Schadewald quoted Rowbotham's book, *Zetetic Astronomy*, regarding his calculations for the Ryde Pier Lighthouse:

> This conclusion [that the Earth is flat] is greatly confirmed by the experience of mariners in regard to certain lighthouses. Where the light is fixed and very brilliant it can be seen at a distance, which the present doctrine of the Earth's rotundity would render altogether impossible. For instance, at page 35 of "Lighthouses of the World," the Ryde Pier Light, erected in 1852, is described as a bright fixed light, 21 feet above high water, and visible from an altitude of 10 feet at the distance of 12 nautical or 14 statute miles. The altitude of 10 feet would place the horizon at the distance of 4 statute miles from the observer. The square of the remaining 10 statute miles will give a fall or curvature downwards from the horizon of 66 feet. Deduct from this 21 feet, the altitude of the light, and we have 45 feet as the amount which the light ought to be below the horizon![47]

What did the skeptic, Schadewald, conclude regarding

Rowbotham's findings? He examined Rowbotham's calculations and found that Rowbotham's was absolutely correct. Shadewald tried to introduce refraction of light as the explanation for the visibility of the Ryde Pier Lighthouse. However, his refraction theory did not explain what was observed regarding the visibility of the lighthouse that should have been below the horizon on a spherical earth. Schadewald was left scratching his head.

> Rowbotham's calculation is correct, although he made no allowance for atmospheric refraction. But even deducting 1/7 of the dip for refraction (a generous correction) does not solve the problem for sphericity.[48]

Schadewald scrutinized Rowbotham's findings regarding the Bidston Hill Lighthouse, discussed above. Once again, Schadewald had to admit that Rowbotham was correct. Schadewald's attempt to explain Rowbotham's results by introducing atmospheric refraction again failed to account for the visibility of the lighthouses.

> Again, Rowbotham's arithmetic is correct, and even a generous correction for atmospheric refraction cannot solve the problem for sphericity. Rowbotham gave about 20 such examples in *Zetetic Astronomy*, and he averred that "many other cases could be given from the same work, shewing that the practical observations of mariners, engineers, and surveyors entirely ignore the doctrine that the earth is a globe."
>
> Suspicious reader that you are, you probably wonder whether Rowbotham cited *Lighthouses of the World* correctly. Bresher wondered, too. When he consulted the work, however, he found that the published numbers were exactly as Rowbotham

stated them.[49]

Although the visibility of the lighthouses proved the earth was flat, Schadewald still did not accept Rowbotham's conclusion. Schadewald was left arguing that Rowbotham cherry picked lighthouses from Findlay's *Lighthouses of the World* that could be seen, which otherwise should have been below the horizon and invisible. He argued that most of the lighthouses could be seen both on a spherical and a flat earth. He, therefore, dismissed those lighthouses that should have been out of sight, below the horizon on a spherical earth, as "anomalies." Shadewald stated:

> This technique is the common property of those determined to convince others of their position by whatever methods they find expedient. Thus, many creationist evangelists comb the scientific literature trying to find things that don't seem to fit the conventional view. Then they present these anomalies to the public as representative, just as Rowbotham presented his anomalous lighthouses. Looking for lighthouses is, of course, easier than trying to construct a creation model. Despite assertions to the contrary, no predictive "creation model" of the biological world exists.[50]

Schadewald criticizes Rowbotham for pointing to proof of a flat earth through lighthouses that should be invisible on a globular earth, by arguing that he could only cite 20 examples. Schadewald misses the point. Those 20 lighthouses existed and Schadewald does not dispute that they were visible, just as claimed by Rowbotham. How does he address those facts? He doesn't. His explanation is a shrug of the shoulders: "Beats me." He then engages in conjecture that "perhaps" the lighthouses were observed under unusual conditions, or "perhaps" new observations may not confirm the published observations. That is not science, that is conjecture. Indeed, Shadewald knows that neither of those things could be true, because at the time Rowbotham published his book, there were many trying to refute his claims. His opponents could easily have established whether the lighthouses were not visible at the listed distances. Schadewald even cites to one such man, M. R. Bresher, who in fact had to admit to the accuracy of Rowbotham's lighthouse data. Schadewald's conclusion sounds like a man who has stumbled across the truth, yet gets up, dusts himself off, and continues on his journey, without a thought of what just happened.

> And what about Rowbotham's anomalous lighthouses? Beats me. Perhaps the reported observations were made under unusual conditions. Perhaps, for those lighthouses still operating, new observations would not confirm the reported anomalies. By now, however, some of Rowbotham's lighthouses presumably have been closed, torn down, or destroyed by the elements. For these, we will never know. One thing is certain; those who seek only anomalous lighthouses will never find light.[51]

While many of the lighthouses have been closed down, there are many other lighthouses that still stand today as silent sentinels proving that the earth is flat. For example, the Bell Rock Lighthouse is still in use today.[52] It is the oldest lighthouse in the British Isles and is an engineering marvel that has saved countless ships and lives. The lighthouse is built on a submerged reef, approximately 10 miles out to sea, off the east coast of Scotland. The difficulty in constructing a lighthouse in an area of such stormy seas moved the BBC to describe it as one of the seven wonders of the industrial age. It was first lit in 1811. It is the oldest surviving sea washed lighthouse in the world.

According to Alexander Findlay's authoritative reference guide, *Lighthouses of the World*[53], the light on the Bell Rock

Lighthouse is 90 feet above the high water mark of the sea and it can be seen at sea by ships for a distance of 14 nautical miles (which is a little more than 16 statute miles). Please note, again, that Findlay listed "the minimum distance" at which the lighthouse could be seen, in clear weather, from a height of 10 feet above the water.[54]

If the earth were a globe, a mariner on a ship at sea would not be able to see that light from a distance of 16 statute miles. Squaring the miles and multiplying by 8 inches for the curve of the earth, and making an allowance for the fact that a mariner would be approximately 10 feet above the water, we find that the light at the top of the lighthouse would be approximately 6 feet below the horizon if the earth were a globe. We know, however, that mariners can see the light in clear weather from a "minimum" distance of 16 statute miles. Only on a flat earth, could a mariner on the deck of a ship see the light on the Bell Rock lighthouse from 16 statute miles away. That minimum viewing distance in clear weather for the Bell Rock Lighthouse is unattainable on a globular earth.

Figure 13: Bell Rock Lighthouse

Another example of a functioning lighthouse that stands as a sentinel proving that the earth is flat is the Fourteen Foot Bank Lighthouse. That lighthouse is built 12 miles off of Bowers Beach in the middle of Delaware Bay.[55] The lighthouse is 59 feet above the sea and can be seen by ships at sea from a distance of 14 statute miles (12 nautical miles).[56] If the earth were a globe, a mariner on a ship at sea would not be able to see that light from a distance of 14 statute miles. Squaring the statute miles and multiplying by 8 inches for the curve of the earth, and making an allowance for the fact that a mariner would be approximately 10 feet above the water, we find that the light would be approximately 8 feet below the horizon if the earth were a globe. We know, however, that mariners can see the lighthouse from a distance of 14 statute miles. The New Jersey Lighthouse Society explains that "[i]t is 59 feet to the focal plane of the light. The lighthouse is still in operation and described on current nautical charts as 'GP Fl (2) 20 sec. 50 ft. [sic] 12 m Horn,' which indicate[s] to mariners that the light is a group flashing light (2), every 20 seconds on a 59 foot tower, visible 12 nautical miles [14 statute miles] at sea and that the station is equipped with a fog horn."[57] Only on a flat earth, could a mariner on the deck of a ship see the Fourteen Foot Bank Lighthouse 14 statute miles away.

The evidence of a flat earth is everywhere to be seen. Rowbotham's experiments can be replicated very easily today. For example, in the photograph below, the Chicago skyline can be seen clearly. The picture was taken by Joshua Nowicki as he stood at Grand Mere Park, Michigan, which is approximately 57 miles away, across lake Michigan from Chicago.[58]

Figure 14: Photograph of Chicago taken by Joshua Nowicki, as he stood at Grand Mere Park, Michigan, 57 miles away.

If the earth were a globe, Chicago would be below the horizon. The only way that Chicago could be seen from the western shore of Michigan is if the world is flat. Below is a map showing the 57 mile distance across lake Michigan from Grand Mere Park, Michigan to Chicago, Illinois.

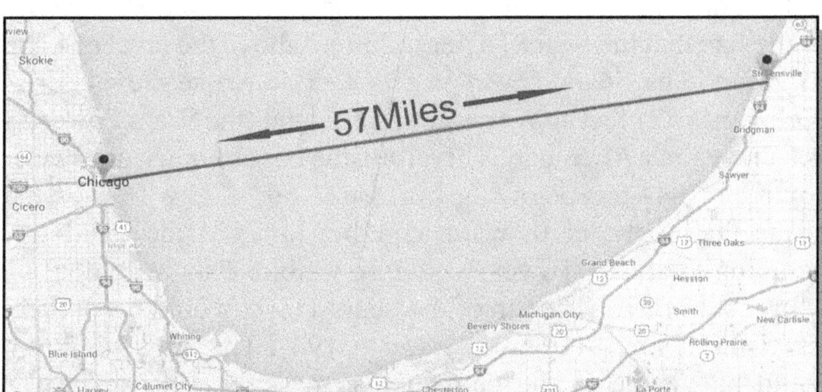

Let us calculate the position of Chicago in relation to Grand Mere Park, assuming that the earth were a globe. Grand Mere Park is 600 feet above sea level. Lake Michigan is 577 feet above sea level. Therefore, Grand Mere Park is 23 feet above Lake Michigan (600 - 577 = 23). Let's assume that the photographer was at the highest point in Grand Mere Park. We will add six feet for the height of the photographer. We come up with an estimate that the camera was, at most, 29 feet above the level of the water on Lake Michigan (23 + 6 = 29). We will subtract seven miles from the distance of 57 miles to account for the 29 foot elevation of the camera above Lake Michigan (29 foot drop to horizon = approximately 6.6 miles, which is rounded up to 7 miles) (29 feet = $6.6^2 \times 8$ inches).

Calculating the curvature of the earth using the 50 mile distance, we find that the street level of Chicago should be 1,644 feet below the horizon. The tallest building in the picture is the Sears Tower (it has been recently renamed the Willis Tower). The Sears Tower was, from 1974 to 1998, the tallest building in the world. It stands 1,450 feet above the street. However, the antennae on top of the tower brings the total height to 1,729 feet above the street level.

In coming up with the 1,644 feet drop below the horizon, it must be understood that it was necessary to subtract 23 feet from the total drop of 1,667 feet ($50^2 \times 8$ inches = 1,667 feet) to account for the fact that the Sears Tower is 23 feet above the level of Lake Michigan. The Sears Tower is 595 feet above sea level. Lake Michigan is 577 feet above sea level. That puts the Sears Tower 23 feet above Lake Michigan. Therefore, the base of the Sears Tower would be 1,644 feet below the horizon (1,667 - 23 = 1644). That means that if the earth were a globe, none of the buildings, including the Sears Tower, would be visible. They would all be below the horizon. The top of the Sears Tower would be 194 feet below the horizon (1,644 - 1,450 = 194). The only thing that would be visible in the entire Chicago skyline would be the uppermost 85 feet of the antennae on the top of the Sears Tower.

In fact, however, the entire Sears Tower and all of the other buildings along the shore in Chicago can be seen in the Nowicki photograph. That picture by Joshua Nowicki of the Chicago skyline proves that the earth is flat.

Figure 16
Sears Tower, indicating the height of the tower and where it would be in relation to the horizon when viewed from 57 miles away across Lake Michigan if the earth were a globe.

Robert Sungenis, who is a geocentrist, but believes that the earth is a globe, tried to refute the flat earth explanation for the visibility of the Chicago skyline in the Joshua Nowicki photograph taken from Grand Mere Park. But Sungenis had no way to explain the visibility of the Chicago skyline on a globular earth, so he simply dismissed it. Sungenis explained in his book, *Flat Earth Flat Wrong*:

> Unfortunately, the above clip that Nowicki took of the Michigan skyline on April 28, 2015 from Grand Mere State Park suffers from this lack of information. Although we know the distance from the Park to Chicago is 60 miles, only Nowicki knows the height from which he took the film. ... Since Nowicki, to my knowledge, has not posted the height from which he took the film on April 28, 2015, the film becomes rather useless in the debate over whether the Earth is flat.[59]

Sungenis is being disingenuous. He knows full well that he can easily determine the elevation of Grand Mere State Park, which is 23 feet above Lake Michigan. Why did he not do that? Because the only reasonable explanation for Nowicki's photograph is that the earth is flat. So, how did Sungenis address Nowicki's photo from Grand Mere Park? Sungenis began by first moving on to a different photograph taken by Nowicki from Warren Dunes Park, which is approximately 10 miles south of Grand Mere Park. Why did he do that? Because Warren Dunes Park is closer to Chicago (53 miles away) and the elevation is approximately 250 feet above Lake Michigan. Sungenis stated: "We are also told that one can go to a height of 250 feet on the Michigan shore to take the shot."[60]

Why would Sungenis be so quick to dismiss the picture taken from Grand Mere Park and move on to a different picture taken from Warren Dunes Park? Because Sungenis stated without equivocation: "this author's purpose is to discredit the flat Earth theory."[61] He cannot do that with the Nowicki photo from Grand Mere Park. So he moved on to a different Nowicki photo. Sungenis opined that Nowicki was at an elevation of 200 feet when he took the Warren Dunes Park picture. Thus, with the elevated position of the photographer at Warren Dunes Park at a closer distance of 53 miles, Sungenis can use sophistry and obfuscation to explain how from that elevated position the Chicago skyline could be seen on the allegedly globular earth. From the 200 foot elevation of the photographer, the drop below the horizon on the hypothetical spherical earth over a distance of 53 miles from Warren Dunes Park would be 864 feet, rather than the 1,667 foot drop expected from Grand Mere Park with the photographer at an elevation of 29 feet (23 feet + 6 feet height of photographer = 29 feet) 57 miles away.

Sungenis was able to shave off a pesky 819 feet (1,667−848=819) to arrive at a 848 foot drop and begin his argument for the visibility of the 1,450 foot tall Willis (Sears) Tower being plausible on a spherical earth. He knows that his argument would lack merit if he were to use the Nowicki photograph from Grand Mere Park with the required horizon drop of 1,667 feet. And so he simply dismissed that photograph as being "rather useless in the debate over whether the Earth is flat"[62] and moved down the Michigan shore to a photograph from Warren Dunes Park that was more helpful to his argument. Indeed, Sungenis admitted as much. "That 9 miles makes a big difference, as we shall see later, especially when the height of the snapshot is included, as well as the atmospheric conditions and the sea level height."[63]

Later in his book, Sungenis came full circle and reentered the discussion of the Grand Mere Park photograph. He did so by posing a hypothetical; but not before reiterating that the Grand Mere Park photograph "is really of no use to anyone in the flat-earth debate"[64]

> Again, since Nowicki does not reveal the height from which he took the clip, it is really of no use to anyone in the flat-earth debate. Unless we know the height, we cannot explain what will be seen of the Chicago skyline. But let's say that Nowicki was at the same height at Grand Mere State Park, 60 miles away, as he was at Warren Dunes State Park, 53 miles away, when he took the April 28, 2015 photo.[65]

Sungenis begins his argument by assuming that Nowicki was at an elevation of 200 feet above Lake Michigan, but not at Warren Dunes. He, instead, adds the 200 feet to Nowicki's elevation at Grand Mere Park. Sungenis goes one better. He arbitrarily adds another 25 feet to the 200 foot elevation. Sungenis then applies that 225 foot elevation to his calculations used to argue that the Willis (Sears) Tower can be seen on a spherical earth from the Grand Mere Park location. Sungenis fudged his numbers to get to the outcome he wanted.

Sungenis is essentially posing a hypothetical wherein he assumes that the photographer was 225 feet above Lake Michigan at Grand Mere Park, when the highest elevation of Grand Mere Park is only 23 feet above the lake. Why would he do that? Because he wants to shave off hundreds of feet from the drop required if the earth were a globe. He is determined to defend the globe model by hook or by crook. He has no real evidence, so he just makes things up. He is putting Nowicki at an elevation of 225 feet above Lake Michigan at Grand Mere Park, which he knows did not happen, and indeed such an elevation is impossible from that location. Sungenis then begins his argument at that point of fiction and constructs his dissimulation for how the Willis (Sears) Tower could be visible on a spherical earth from Grand Mere Park.

Joshua Nowicki's photograph created such a sensation that

it was necessary for the media powers to explain how Chicago could be seen from Grand Mere Park on the Michigan shore, 57 miles away, which is an impossibility if the earth were a globe. Tom Coomes, a weatherman for ABC News argued that it was impossible to actually see Chicago from 57 miles away. Coomes stated that "Chicago is beyond the horizon; you should not be able to see it."[66] However, it can be seen; it is clearly revealed in Joshua Nowicki's photograph. How does Coomes explain this apparent impossibility? Coomes explained away Nowicki's photograph by averring that the Chicago skyline wasn't really there at all.[67] He claimed that Nowicki's picture was depicting what is called a "superior mirage." Coomes showed an amazing display of self control, as he kept a straight face throughout his ridiculous explanation.

Figure 17: Tom Coomes explaining that Joshua Nowicki's photo of the Chicago skyline was not really there at all; it was a mirage.

The problem with Coomes' explanation is that a superior

mirage is a mirage of an object that is usually inverted above the actual object. Sjaak Slanina explains that "[a] superior mirage occurs when an image of an object appears above the actual object."[68] The drawing below is from the National Snow and Ice Data Center (NSIDC) and illustrates the inversion of a superior mirage.

Figure 18: NSIDC Explanatory Diagram of a Superior Mirage

The picture below is an actual superior mirage appearing above a freighter. Notice that the mirage is inverted and appears above the actual freighter. Both the freighter and its mirage are in view. This particular mirage is unusual because of the amount of detail it displays, yet there is still extreme distortion of the mirage; the mirage fades toward the stern of the ship, without the rear cabin being depicted in the mirage at all.

Is it possible to see the Chicago skyline in a superior mirage? The answer is yes. Below is a picture of a superior mirage of Chicago taken from western shore of Michigan, in St. Joseph by J. Michael Hall on April 18, 2015.[69] Notice that the Chicago towers are quite indistinct and

distorted. Most notably, observe that the towers are upside-down. That is quite different from what is seen in Nowicki's photograph from Grand Mere Park.

Figure 20: Superior mirage of Chicago skyline taken from St. Joseph, Michigan.

The picture by Joshua Nowicki of the Chicago skyline looks nothing like a superior mirage. Nowicki's photograph is detailed and shows the skyline right-side-up. If Nowicki's picture were a superior mirage, the Chicago skyline would most likely be upside-down and extremely distorted. There is a refraction of light caused by the moisture in the air, which is separate and distinct from unusual occurrence of a superior mirage.[70] A consequence of the moisture in the air is that it causes a lensing effect of the atmosphere. This effect magnifies distant objects and causes the bottoms of those distant objects to be cut off from view.[71] Many misinterpret the missing bottoms of buildings and ships to be caused by the earth's curvature, not realizing it is due to the lensing effect of the atmosphere. The proof that the missing bottoms are due to the atmospheric lensing effect is in the fact that often the object observed should not be able to be seen at all if the earth were a globe, as it should be completely below the supposedly curved horizon.

The Chicago skyline seen in the Nowicki photograph is most definitely not a mirage. That was confirmed by Rob Skiba and Rick Hummer. On June 24, 2016, Rob Skiba and Rick Hummer chartered a boat, from which they were able to see the Chicago skyline from the mouth of the New Buffalo Harbor on the Michigan shore, a distance of approximately 42 statute miles.[72] More importantly, they confirmed that the Chicago skyline was not a mirage by traveling almost the entire width of Lake Michigan from the Michigan shore at New Buffalo Harbor to within approximately 9 statute miles (approximately 8 nautical miles) of Chicago. The entire time during the trip across Lake Michigan the skyline of Chicago was upright and in view.[73] That confirmed that the skyline that they saw from approximately 42 miles away, and was kept in view the entire length of travel across Lake Michigan, most definitely was not a mirage.

Below is a picture of the Isle of Man taken from across the Irish Sea at Rossall Beach, England, on August 10, 2012.[74] The distance between Rossall Beach and Port Soderick, on the east coast of the Ilse of Man, is approximately 63 miles.[75] Rossall beach is at an elevation of 20 feet.[76] Adding 6 feet to that for the height of the photographer, we arrive at a height of the photographer above the water of 26 feet. Assuming a curved earth, at 26 feet above the water the horizon will not begin to drop below sight from the photographer until a distance of approximately 6 miles. Subtracting 6 miles from the 63 mile distance, between the Isle of Man, we get 57 miles. We must account for the distance of 57 miles of curvature of the earth between the photographer and the shore of the Isle of Man. Using the equation, $\text{miles}^2 \times 8$ inches = distance below horizon, we get 2,166 feet of drop below the horizon of the Isle of Man ($57^2 \times 8$ inches = 2,166 feet) from

Rossall Beach. The Isle of Man has a maximum elevation of 2,037 feet, which means that if the earth were a globe, the top most peak of the Isle of Man would be 129 feet below the horizon.

Figure 21: Isle of Man seen from Rossall Beach, 63 miles away.

Since it is the highest point, and not the shoreline on the Isle of Man, that is at issue, one might wonder what is the distance from Rossall Beach to the highest peak on the Isle of Man. The highest point on the Isle of Man is Snaefell, at 2,307 feet above the Irish Sea. The distance across the Irish sea to Snaefell is also approximately 63 miles.[77]

The windmills that are seen in the photograph are approximately 9 miles offshore from Rossall Beach. Because the windmills are much closer to the photographer than the Isle of Man they look larger in relation to the Isle of Man than they really are. The windmills have a maximum height of 425 feet above the water to the tip of the rotors turned at 12 o'clock.

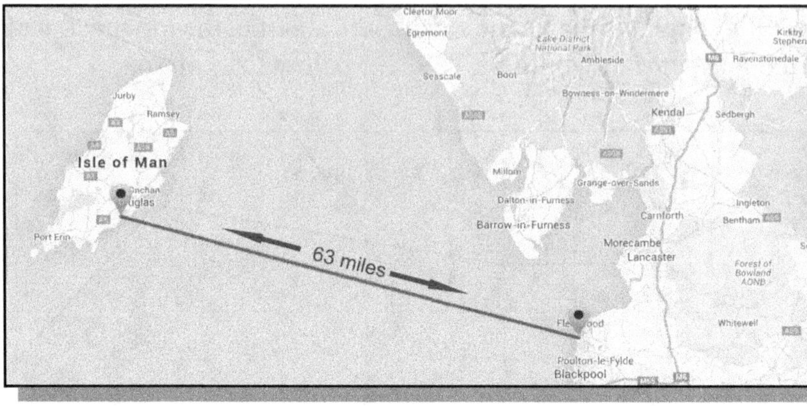

Figure 22: Map showing distance from Rossall Beach to Port Soderick on the east coast of the Isle of Man.

There is no way that the Isle of Man could be seen from Rossall Beach, unless the earth was flat. The above photograph of the Isle of Man from Rossall Beach is stark proof that the earth is in fact flat. The photograph was taken from Rossall Beach and posted on the Rossall Beach Residents and Community Group website to show the beautiful sunset.[78] The Rossall Beach Residents and Community had no idea that the photograph shows much more than a beautiful sunset; it is evidence of a flat earth.

The picture below is a frame from a video. A video camera with a telephoto lens was set up approximately 15 feet 11 inches off the water on Fisherman's Pier in Edmonds, Washington.[79] The camera is looking North-Northeast across Puget Sound toward a ferry that crosses from Mukilteo, Washington to Clinton, Washington. In the picture you will see the ferry about to arrive at Clinton, Washington. The distance from Fisherman's pier to Mukilteo directly across the water is 10.34 miles. As the ferry travels from Mukilteo to Clinton, the distance from the camera gets greater. At the halfway point of travel from Mukilteo to Clinton the distance between the camera at Fisherman's Pier to the lengthens to 10.81 miles. The distance from the Fisherman's Pier to just prior to arrival at Clinton is approximately 11.3 miles. The distance to Clinton, itself, in a straight line from Edmonds as reported on www.FreeMapTools.com is 11.65 miles. For simplicity, as the ferry in the picture is about to arrive at Clinton, we will round down the distance to 11 miles. If the earth were spherical, as is commonly believed, the entire ferry should not be able to be seen from Fisherman's Pier. At a distance of 11 miles, and adjusting for the height of the camera off the dock, (for simplicity, we will round up to 16 feet), a globular earth would put the ferry below the horizon by approximately 25 feet. That is, the bottom 25 feet of the ferry should not be visible at all. In the picture, however, the black topsides of the hull can be seen against the white deck cabins. The first deck cabin is a car-deck, with a clearance of 16 feet. It is visible in the picture, but it should be below the horizon if the earth were a globe. However, we can see the entire ferry, including the black topside of the hull on the waterline, from the bow to the stern. Very simply, the only way the entire ferry could be seen in the picture below is if the earth is flat.

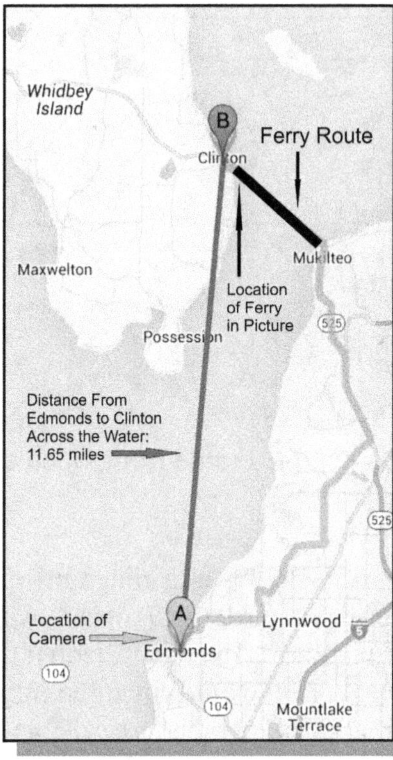

Figure 24: Frame from video of ferry taken from 11 miles away.

Below is a diagram of the actual ferry depicted in the above picture, the Toitea.[80] Approximately 25 feet should have been blocked by the horizon if the earth were a globe. You can see that 25 feet would cover the entire black topside of the hull and the first deck, which are visible in the photograph above. The manufacturer of the ship states that the black topside is 8 feet off the water.[81] The green stripe is approximately 3 feet wide. The black topsides and the green stripe, which amount to 11 feet total, should be completely below the curvature of the earth, if the earth were a globe. However, you can see clearly the black topside of the ferry against the waterline.

Robert Sungenis, in his book *Flat Earth Flat Wrong*, published the series of photos on the following pages as a way to prove that the earth is a sphere because they depict a ship disappearing below the horizon, hull first.[82] He argues that phenomenon proves that the earth is a sphere, because it demonstrates that the ship is dropping over the horizon as it gets further from the horizon.

The problem with the video from which Sungenis has taken the photo array is that it is a fraud. The videographer subtly moves the camera lower, which has the effect of raising the horizon and cutting off the hull of the ship. This trickery gives the false impression that the ship is dropping below the horizon on a globular earth. This author tracked down the original University of

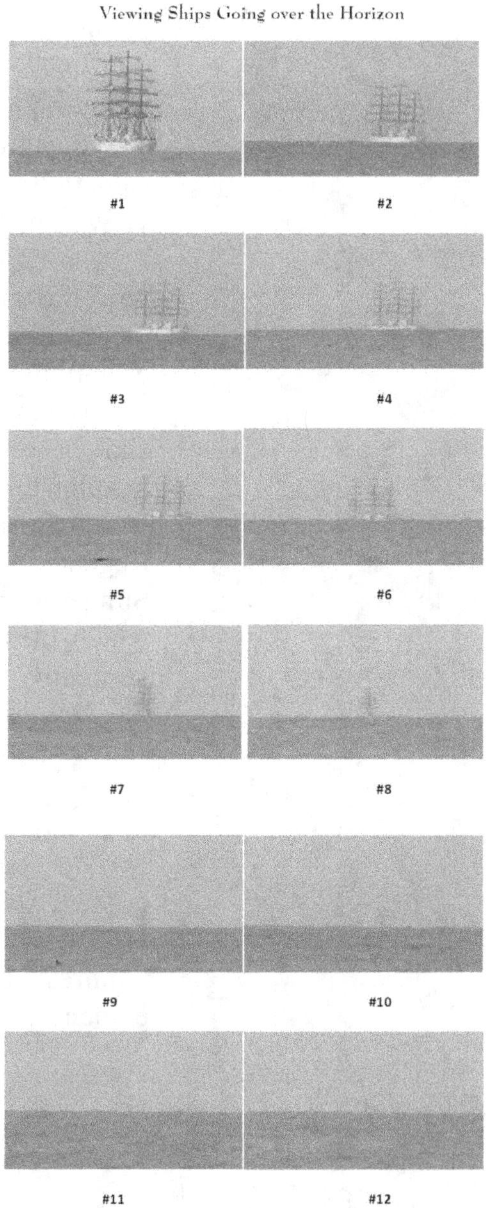

Figure 26: Array of Pictures from Robert Sungenis' Book, *Flat Earth Flat Wrong*

California video and aligned frames from left to right.[83] When that is done, it becomes obvious what the videographer is lowering the camera, which causes the horizon to rise in the camera frame and cut off the ship from view from the bottom-up. Proof that the frames depicted by this author have been lined up accurately can be seen in the small white text on the bottom of each frame from the original video. The text is imprinted in the original video and is positioned on at the same location throughout the video. That text is: "www.fossweb.com © 2010 UC Regents." You will notice in the Sungenis frames the text is missing, which means that his frames are not vertically aligned from the bottom of the frames. Thus the raising of the camera is not detectable in his frames.

If the earth is truly a globe, why is it necessary to use video trickery to prove it? The earth is flat. The evidence proves it. To try to show otherwise, requires the fabrication of false evidence.

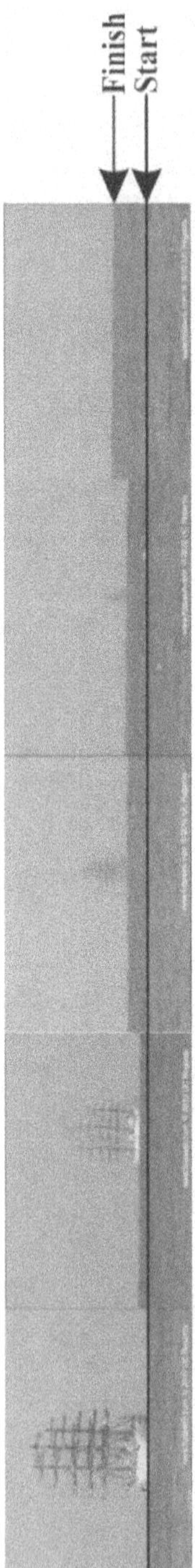

In the time lapse video when the frames are laid side by side from left to right, one can see that the ship is being cut off from the bottom up because the camera is being lowered in stages to elevate the horizon. That is being done to give the false impression that the ship is dropping below the horizon on the supposed curved earth. In reality, the ship is disappearing over the flat earth due to perspective and refraction.

Below is a photograph taken from the Promenade of Nevi.[84] The picture was taken on December 4, 2007.[85] The picture shows the Island of Corsica in the distance. The tallest mountain on Corsica is Monte Cinto, which rises 8,878 feet above sea level. Monte Cinto is 139 miles from Nevi. Nevi has a maximum elevation of 82 feet above sea level. Taking into account the elevation of the photographer at Nevi (82 feet above sea level + 6 feet in height = 88 feet), we subtract 12 miles from the 139 mile distance. If the earth were a globe, there would be 127 miles of curvature between the photographer at Nevi and the highest point on Corsica, Monte Cinto. That supposed curvature would create a drop of 10,753 feet in the 127 mile distance. That means that if the earth were a globe, the top of Monte Cinto should be completely out of view 1,875 feet below the horizon (10,753 - 8,878 = 1,875). That, however, is not the case. The Island of Corsica, all the way to the top of Monte Cinto, is in view in the photograph, which means that the earth is not a globe. The photograph of Corsica, taken from Nevi, could only happen on a flat earth.

Figure 28: Picture of the Island of Corsica, taken from Nevi, 139 miles away from the tallest peak on Corsica, Monte Cinto. If the earth were a globe, the peak on Monte Cinto should be 1,875 feet below the horizon.

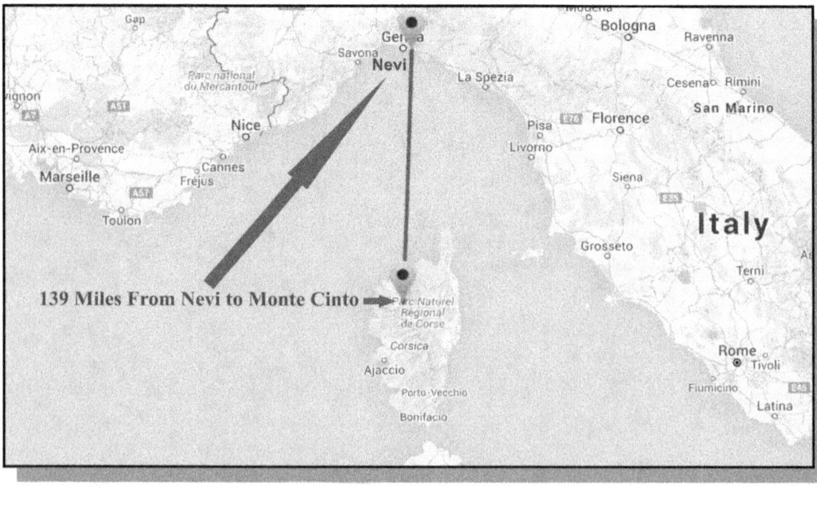

The United States Coast Guard has provided a chart of navigational lights and buoys.[86] The chart proves that the earth is flat. For example, the U.S. Coast Guard lists the focal plane of the Cape Canaveral Lighthouse as being 137 feet above the mean high water level. The nominal range of the light is listed as 24 nautical miles (approximately 28 statute miles).[87] The nominal range is the "maximum" distance at which a light can be seen in clear weather, which is defined as conditions that allow for the visibility of 10 nautical miles.

Assuming the earth is a globe and the sailor observing the Cape Canaveral Lighthouse is 10 feet above the level of the sea when observing the Cape Canaveral Lighthouse, the curvature of the earth would put the light on the lighthouse 247 feet below the horizon at the distance of 28 statute miles from the observer. To account for the height of the observer above the level of the sea (10 feet), we take the 28 statute miles and subtract 4 statute miles (which gives us 24 miles). We then take the square of the 24 miles (576) and multiply that number by 8 inches and we get 4,608 inches (384 feet) drop below the horizon on a supposed curved globe. We then subtract the height of the focal plane of the lighthouse (137 feet) from the horizon drop (384 feet) and we find the light on the Cape Canaveral Lighthouse would be 247 feet below the horizon assuming that the observer was 10 feet above the sea level at a distance of 28 miles from lighthouse. However, the lighthouse is not 247 feet below the horizon. That can mean only one thing: the earth is not a globe. The ability of mariners to see the Cape Canaveral Lighthouse from a distance of 28 statute miles is proof that the earth is in fact flat.

The above cited publication from the United States Coast Guard posts a Geographic Range Table that gives "the approximate geographic range of visibility for an object which may be seen by an observer at sea level."[88] The table can be seen below. The objects at the given height could not be seen at the distances listed in the table if the earth were a globe. For example the table establishes that an object that is 200 feet above sea level can be seen by an observer at sea level from a distance of 16.5 nautical miles (19 statute miles). If the earth were a globe, the top of that 200 foot object would be 40 feet below the horizon from 19 statue miles away (19^2 x 8 inches = 240 feet). The range table is accurate, as it is based upon real world testing and experience. The objects listed in the range table could only be seen from the distances on the chart if the earth is flat. The range table presents official evidence from the United States Coast Guard proving that the earth is flat.

GEOGRAPHIC RANGE TABLE

The following table gives the approximate geographic range d visibility for an object which may be seen by an observer at sea level. It is necessary to add to the distance for the height of any objea the distance corresponding to the height of the observer's eye above sea level.

Height Feet / Meters	Distance Nautical Miles (NM)	Height Feet / Meters	Distance Nautical Miles (NM)	Height Feet / Meters	Distance Nautical Miles (NM)
5/1.5	2.6	70/21.3	9.8	250/76.2	18.5
10/3.1	3.7	75/22.9	10.1	300/91.4	20.3
15/4.6	4.5	80/24.4	10.5	350/106.7	21.9
20/6.1	5.2	85/25.9	10.8	400/121.9	23.4
25/7.6	5.9	90/27.4	11.1	450/137.2	24.8
30/9.1	6.4	95/29.0	11.4	500/152.4	26.2
35/10.7	6.9	100/30.5	11.7	550/167.6	27.4
40/12.2	7.4	110/33.5	12.3	600/182.9	28.7
45/13.7	7.8	120/36.6	12.8	650/198.1	29.8
50/15.2	8.3	130/39.6	13.3	700/213.4	31.0
55/16.8	8.7	140/42.7	13.8	800/243.8	33.1
60/18.3	9.1	150/45.7	14.3	900/274.3	35.1
65/19.8	9.4	200/61.0	16.5	1000/304.8	37.0

Example: Determine the geographic visibility of an object, with a height above water of 65 feet, for an observer with a height of eyeof 35 feet.

Enter above table;
Height of object 65 feet= 9.4 NM
Height of observer 35 feet= 6.9 NM
Computed geographic visibility= 16.3 NM

Figure 30: United States Coast Guard Geographic Range Table

On August 8, 2018, Darrell Marble conducted an experiment where he placed a laser light seven (7) feet above the water on Owen Beach at Point Defiance Park in Tacoma, Washington. He shined the laser 10.3 miles across Puget Sound to a camera that was 3 feet off the water on the opposite shore at Saltwater State Park in De Moines, Washington.[89] Marble documented that from his position on Saltwater Park he was able to see the laser light's point of origin at Owen Beach, 10.3 miles away. Assuming, arguendo, that the earth is a globe, and accounting for the fact that the laser is seven (7) feet above the water and the target camera was three (3) feet above the water, the laser's point of origin should have been approximately 28 feet below the horizon as viewed from the camera. Being able to see a laser light's point of origin from the opposite shore across 10.3 miles of water, requires the water, and thus the earth, to be flat.

An organization of field engineers known as FECORE conducted a series of experiments using lasers to determine whether there was any curvature over water. In one experiment, on February 26, 2018, over Lake Balaton in Hungary, the FECORE engineers determined that there was no curvature.[90] A laser that was 2.2 meters (7.2 feet) above the water was shined across the lake. A target camera was set up 12 kilometers (7.46 miles) across the lake on an opposite shore at a height of 1.6 meters (5.25 feet) above the water. The camera (Nikon P900) was able to photograph the the source of the laser beam on the opposite shore, 7.46 miles away. If the earth were a sphere, the camera should never have been able to see the source of the laser beam. Indeed, at 5.25 feet above the water, the laser beam should have been completely out of sight, 2.21 meters (7.26 feet) below the horizon with an impenetrable hump of water blocking the beam from its position on the other side of the lake. But the laser beam was not blocked. The source of the laser beam was detected at the other end of the lake, 12 kilometers (7.46 miles) away. That experiment proves that the water over the lake was flat, and thus the earth is flat.

In April 2018, The FECORE team traveled to Lake Ijssel in the Netherlands to conduct the same experiment over a longer distance. On April 8, 2018, the FECORE team set up a laser on the shore of Lake Ijssel, 2.85 meters (9.35 feet) above the water. They shined the laser across Lake Ijssel to the opposite shore, 28.68 kilometers (17.82 miles) away.[91] A camera (Nikon P900) set up at a height of .85 meters (2.79 feet) above the water was able to photograph the source of the laser beam on the opposite shore, 17.82 miles across the lake. If the earth were a globe, the laser would not have been visible 17.82 miles away. From a vantage point of 2.79 feet above the water, the spherical earth model requires the laser light on the opposite shore to be more than 122 feet below the horizon. That would mean that the beam would be blocked by a massive mound of water. But it was not blocked. It was visible. That means the 17.82 miles of water over which the laser beam passed must necessarily be flat. That refutes the globe model and is proof that the earth is flat.

3 Reflections on Refraction

Many heliocentric scientists point to atmospheric refraction to explain the visibility of distant objects over the alleged curvature of the supposed spherical earth. The problem with their claim is that refraction simply does not, and cannot, explain the phenomenon.[92] A 2006 scientific study of atmospheric refraction was performed by Diego Gutierrez, Francisco J. Seron , Adolfo Munoz , and Oscar Anson at the Department of Computer Science and Systems Engineering, University of Zaragoza. The scientists, in pertinent part, created atmospheric computer modeling of superior and inferior mirages, Fata Morgana mirage, and the Novaya-Zemlya mirage. A significant premise for their computer modeling study was explained by the authors:

> Under standard circumstances, the effects are negligible. It is under non-standard circumstances when we really see some amazing atmospheric phenomena. Probably the best known is the inferior mirage, which occurs when lower regions of the atmosphere are much hotter than higher regions, as in deserts.[93]

That statement by the scientists is significant because it

points out that the best known atmospheric refraction is the inferior mirage, which is caused by the common (and indeed usual) occurrence of warm air below colder air. That atmospheric configuration in fact causes the apparent horizon and distant objects to drop in elevation. That drop in the apparent height of distant landmarks is the opposite from what is promoted by geodesists.

Before discussing the specifics of how light refraction causes inferior mirages, it would be best for us to look at the science of light refraction. We will rely on one of the foremost experts in the world on light refraction, Michael Davidson (1950-2015). Davidson was head of the National High Magnetic Field Laboratory Optical Microscopy Program at Florida State University. Davidson was known worldwide for his groundbreaking research. Some of Davidson's work with refraction can be found on the Florida State University National High Magnetic Field Laboratory Optical Microscopy Program website.

Davidson's models for refraction are based upon the established work of the Dutch mathematician Willebrord Snell (1580-1626) and another Dutch scientist, Christiaan Huygens (1629-1695). The combination of their work gave rise to what is known today as Snell's law. That law is the basis for predicting the refraction of light as it travels from one medium to another. The study of light refraction has given rise to a refraction index, where different materials are given a value depending on their ability to refract light. As Davidson explains "light would travel more slowly through materials having a greater refractive index. Stated another way, the velocity of light through a material is inversely proportional to its refractive index."[94] That means the higher the index number, the lower the speed of light as it travels through the medium.

Ken Lee Wheeler, who is an expert in light, photography,

and magnetism and author of *Uncovering the Missing Secrets of Magnetism*,[95] disputes the theory that refraction of light has to do with the variability of the speed of light. Wheeler bases his theory of light on the brilliant work of Nikola Tesla. Wheeler makes a compelling argument that if the speed of light is in fact affected by the medium through which it travels, then it would violate the physical law of conservation of energy for light to increase in speed when traveling from a higher refractive index medium to a lower refractive index medium. Thus, according to the law of the conservation of energy, it would be impossible for light to increase in speed when entering a medium with a lower refractive index. Wheeler, instead, views light as being a field effect, where the light is induced through a medium and the refraction is a function of the difference in the permeability of the distinct mediums.[96] Wheeler views light refraction as the light's rate of induction changing from one medium to another.[97] Resolving the differences between Richardson's orthodox theory and Wheeler's Teslian theory of light is beyond the scope of this book, and it is not necessary, because the theory of light travel does not change the observed reality regarding light refraction. Whichever theory you accept, the fact remains that light is refracted at the boundary of two mediums with different refractive indices.

The convention in the scientific community is to assign a refractive index for a vacuum of 1.0. Since light is thought to travel fastest through a vacuum, all other substances necessarily have a refractive index greater than 1.0. The refractive index of air is 1.000292.[98] The refractive index for water is 1.333.[99] To arrive at the refractive index number (n) the speed of light through a vacuum (c) is divided by the speed of light through water (v). Because light is thought to travel through water at a slower speed than through a vacuum it has a higher light refraction index number. Davidson explains what happens with light as it travels from air through water:

> When a light wave passes from a less refractive

> medium (such as air) to a more refractive medium (such as water), the velocity of the wave decreases. Conversely, when light passes from a more refractive medium (water) to a less refractive medium (air), the velocity of the wave increases. The normal is defined as a line perpendicular to the boundary, or interface, between two substances. The angle of incidence in the first medium, with respect to the normal, and the angle of refraction in the second medium (also with respect to the normal), will differ in proportion to the difference in indices of refraction between the two substances. If a light wave passes from a medium of lower refractive index to one of higher refractive index, it is bent toward the normal. However, if the wave travels from a medium of higher refractive index to a medium of lower refractive index, it is bent away from the normal.[100]

There is an equation born of Snell's law that explains the refraction of light when traveling from air to water and vice versa: n1 x sin(q1) = n2 x sin(q2). Davidson explains that "in Snell's equation, the variable n1 represents the refractive index of the medium in which the incident ray travels, while n2 is the refractive index of the medium through which the refracted ray travels. The value q1 represents the angle (with respect to the normal) at which the incident ray strikes the boundary, and q2 is the angle at which the refracted ray travels."[101]

Applying Snell's law to refraction of light as it travels from air to water we find that light traveling at a 60° angle if incidence (q1) is refracted to an angle of 40.5°(q2). The angle of incidence and the angle of refraction are both measured from an imaginary line, called the normal line, that is perpendicular to the border between the air and water. In the graphic below we see the effect of light refraction by water according to Snell's Law.

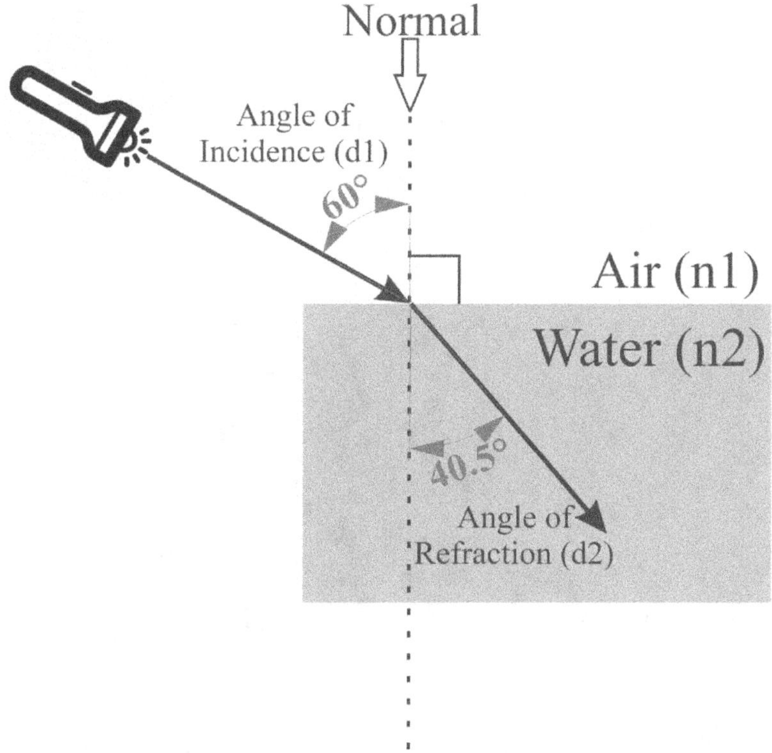

Light is refracted toward the normal line as it travels from air to water.

If we reversed the mediums and had the light traveling from water with a refractory index of 1.333 to air with a refractory index of 1.000292, then the light refraction would be in the opposite direction. For example, if the light were traveling at an angle of incidence to the normal line of 45° through water, then upon entering the air, the angle of refraction at the boundary between the water and air would be away from the normal line at an angle of 70° as it passed through the air. That greater angle of refraction away from the normal line is because the light is traveling from a high refractory index medium (water) to a lower refractory index medium (air). The equation, n1 x sin(q1) = n2 x sin(q2), gives a resulting angle of refraction of 70°with light exiting water into air, assuming an angle of incidence of 45 ° going from water to air.

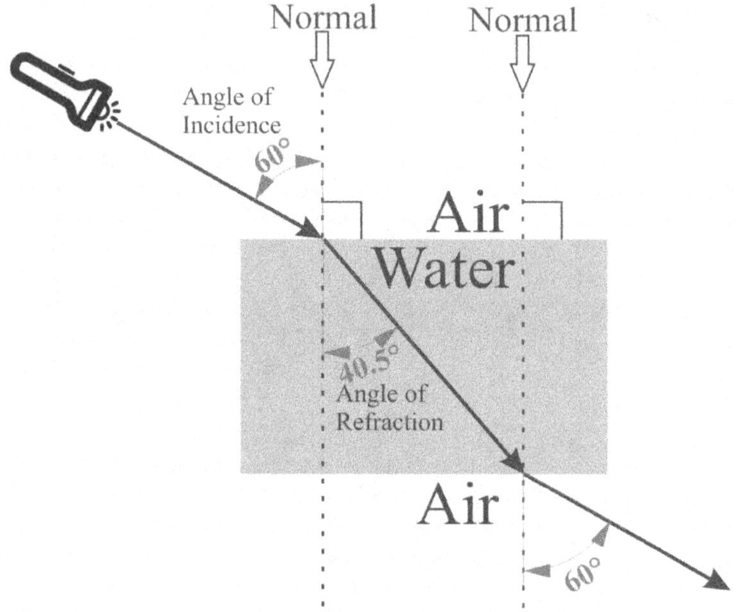

Light traveling from air to water is refracted by water toward the normal line. But upon exiting the water and entering the air again, that light is refracted away from the normal line and returns to its original angle of incidence.

Light can be refracted both when entering and when exiting a medium. Michael Davidson explains the effect of light entering water and then exiting again from water into air. Davidson explains that a light beam is refracted upon entry into the water, and again upon exiting the water. But the refraction upon exiting the water is reversed from the entry path.[102] A light beam will exit the water into the air at the same angle it had upon its entry into the water. This causes the point of exit to shift laterally along the boundary due to the different angular path taken by the light while it traversed the higher refractory index water.[103] The graphic above illustrates the principle.

A little known fact is that there is a distinction between the physical densities of different mediums and their optical densities. A medium's greater physical density often means that it will have a greater optical density, but that is not always the case. For example, water is more physically dense than ethyl alcohol. But ethyl alcohol, with a refraction index of 1.36, is more optically dense than water, which has a lower refraction index of 1.333. As explained by the Physics Classroom:

> The speed of the wave depends upon the optical density of that material. The optical density of a medium is not the same as its physical density. The physical density of a material refers to the mass/volume ratio. The optical density of a material relates to the sluggish tendency of the atoms of a material to maintain the absorbed energy of an electromagnetic wave in the form of vibrating electrons before reemitting it as a new electromagnetic disturbance. The more optically dense that a material is, the slower that a wave will move through the material.[104]

Cold air is optically more dense than warm air, and cold air is also physically more dense than warm air when they are both at the same atmospheric pressure. But keep in mind that air pressure progressively increases in the atmosphere as altitude decreases.[105] As air pressure increases, assuming all other variables remain equal, the refractive index also increases. The physical air density increases at lower altitudes. Most of the air and mass of Earth's atmosphere lies in the troposphere, which is the air mass covering an altitude within approximately seven (7) miles of the earth's surface. At sea level the air is physically densest because it is compressed by the weight of the air above it. As Dr. Dale Ward of the University of Arizona explains:

> The more air above a level (and hence the more

weight of air above a level), the greater the squeezing effect (or compression). Since air density is the number of air molecules in a given space (volume), air density is typically greatest at the surface or sea level (where it is squeezed by the weight of the entire atmosphere above) and decreases as we move up in the atmosphere because the weight of air above becomes less and hence there is less of a squeezing effect.[106]

Above the troposphere is the stratosphere; after that comes the mesosphere, and then the thermosphere.[107] While there are fluctuations in temperatures in the upper atmosphere, the warmest temperatures are found closest to the surface of the earth, which are only exceeded by the temperatures in the upper band of the thermosphere. According to the Lyndon State College, Atmospheric Sciences Department, in the troposphere the air progressively gets warmer at an approximate rate of 6.5°Celsius (11.7°F) per kilometer as you descend in altitude.[108]

Warm air has a lower optical density and therefore also a lower refractive index than cold air. Dr. Danny R. Faulkner of Answers in Genesis believes that the earth is a spinning sphere. But he acknowledges that "[s]imply put, the speed of light is slightly greater in warmer air than it is in cooler air. ... Since the speed of light is greater in warmer air, warmer air has a lower index of refraction than cooler air."[109]

Dr. Faulkner is correct that at a higher temperature, the index of refraction is lower. For example, the National Institute of Science (NIST) calculated that index of refraction through air is 1.000243285 for a light with a wavelength of 633 nm, traveling through the air with a temperature of 50°C (102°F), with no humidity. But when the temperature drops to 5°C (41°F), with no change in the other variables, NIST determined that the index of refraction, increases to 1.000282756.[110] Another variable for the index of refraction through air is the air pressure. The greater the air pressure, the greater the index of refraction.[111] But if you assume that the air pressure remains the same, the refractive index is inversely related to the temperature.

The troposphere contains 90% of the atmospheric mass, and it also contains 99% of the water vapor in the atmosphere.[112] The more water vapor in the air, the lower the refractive index. It sounds backwards that atmosphere with more water vapor would have a lower refractive index, since water vapor contains water and thus one would think that water (which has a higher refractive index than air) mixed in air as a vapor would have a higher refractive index. But that is not the case. Kaye & Laby, National Physical Laboratory, explains that "[t]he refractivity of water vapour is less than that of air, so that if the air is moist its refractive index will be smaller than the value calculated for dry air."[113] The refractive index of dry air is 1.000292, but the refractive index for water vapor is 1.000256.[114] Water vapor is concentrated in the troposphere.

As the humidity increases in the air, the refractive index for the air is lowered. For example, the refractive index is 1.000267805 for a light with a wavelength of 633 nm traveling through air with 0% humidity at 20°C (68°F). But when the humidity is increased to 100% at 20°C (68°F) the refractive index of the air drops to 1.000267394.[115]

The fact that humid air has a lower refractive index than dry air is not generally understood, because it is counterintuitive that humid air would have a lower refractive index than dry air. Some of those who promote the spherical earth use this to their advantage to pull the wool over the eyes of the general public. For example, Robert Sungenis, Ph.D., who is a geocenrist and the author of more than 20 books on science and religion, is adamant that the earth is a globe. Sungenis haughtily claims that his scientific judgments are based upon rigorous study.[116] With all of his rigorous study and background in science, Sungenis nonetheless wrongly states that humid air has a higher refractive index than dry air. It is illustrative of the depth of the misconception regarding light refraction, when an accomplished science author with a Ph.D. can be wrong about it. In his book, *Flat Earth Flat Wrong*, Sungenis tries to explain that the Chicago skyline was visible from 57 miles across Lake Michigan when it should have been below the curve of the supposed spherical earth because the light was refracted upward over the curve of the horizon:

> Be that as it may, normally the refraction due to air is minimal. Air's refractive index is calculated to be 1.000293, which means that except for a 0.000293 bending, a light beam going through air will travel straight. Conversely, a light beam going through water will be bent by 1.333, which means that it will be bent by one third. Since there is water in the air in the form of moisture (humidity), we can safely say that the bending of the light beam across Lake Michigan due to water content alone will be somewhere between 1.000293 and 1.333. As noted, the general humidity in Chicago on April 28, 2015 fluctuated from 74% in the early morning to 54% in early evening but the humidity just above the lake is not known.[117]

Sungenis theorized that the light across Lake Michigan was refracted to cause the city of Chicago skyline to be visible above the supposed curve of the earth in the famous photograph by Joshua Nowicki. Sungenis' argument actually works in the opposite direction. The humid air over the lake has a lower refractive index, not a higher refractive index as Sungenis argued. That lower refractive index would cause the city skyline to be refracted downward, not upward as Sungenis argued. Sungenis compounds his error by considering light refraction as though it is a constant and not accounting for the fact that the refraction of light happens at the boundary between two mediums of different optical density and that the direction of refraction is dependent on the direction in which the light is passing through the different media.

Sungenis also misapplies the refractive index. Sungenis states that "a light beam going through water will be bent by 1.333, which means that it will be bent by one third." That is not true. The refractive index of 1.333 means that the light traveling through water will travel slower than light traveling through a vacuum. To arrive at the refractive index number of 1.333 (n) the speed of light through a vacuum (c) is divided by the speed of light through water (v). The refractive index of a vacuum is 1.0, whereas the refractive index of water is 1.333. The refractive index of 1.333 for water does not mean that a light beam will be bent by one third, as alleged by Sungenis. To determine the angle of refraction when traveling from one medium to another, we must solve for that angle using Snell's Law in the following equation: n1 x sin(q1) = n2 x sin(q2).

As we have seen, although it is counterintuitive, in fact it is true that dry air is optically more dense than humid air. Since water is physically more dense than air, one would also logically think that humid air would be physically more dense than dry air. But that is not the case. Dry air is in fact physically more dense than humid air (assuming equivalent atmospheric pressure). That is because 99% of the air's gases are hydrogen and oxygen. Nitrogen in the air has a molecular weight of 28 and the oxygen molecules have a molecular weight of 32.[118] Compare those weights with the weight of water vapor, which has a molecular weight of only 18.[119] The water vapor must displace the surrounding air containing the heavier hydrogen and oxygen, thus having the effect of reducing the physical density of the air. We witness proof that moist air is less dense than dry air every day when we see clouds, which contain water vapor, floating overhead.

The U.S. Geological Survey explains that because the water vapor in a cloud is less dense than the air beneath it, "the lesser density of the cloud allows it to float on the dryer and more-dense air."[120] Thus, dry air is both optically and physically more dense than humid air.

The Institute of Physics is a leading worldwide scientific society made up of members who are at the "top of their fields in academia, business, education and government."[121] The institute explains on their physics website that the illusion of a mirage of what appears to be water in a desert is due to the refraction of light as it passes from the higher cold air to the lower hot air.

> The illusion results from the way in which light is refracted (bent) through air at different temperatures. Cold air is denser than warm air, and therefore has a greater refractive index. This means that as light passes down from cool to hot air, it gets bent upwards towards the denser air and away from the ground. To your eyes, these distorted rays seem to be coming from the ground, so you perceive a refracted image of the sky on the ground. This looks just like a reflection on the surface of a pool of water.[122]

The point in citing to the Institute of Physics article on mirages is that the very phenomenon that causes the sky to be refracted to the ground (hot air beneath cold air) is the status quo for the atmosphere (with the warm air of the troposphere being beneath the cold air of the stratosphere). The same principle is at work with the heat gradient between the stratosphere and the troposphere.

There are gradients of moisture in the air, particularly at the boundary layer of the atmosphere. The constant movement of the air can cause distortions and rippling of objects viewed through the atmosphere as their images are refracted. The boundary layer is the layer of the atmosphere that composes an area within approximately two kilometers of the surface of the earth.

Frédéric Fabry, from the Department of Atmospheric and Oceanic Sciences, McGill University, Montreal, Quebec, Canada, explains that "[m]oisture variability naturally arises in the convective boundary layer."[123] His research has shown that there is continual variability in moisture in the air. "Near the surface, the air is warmed and moistened by surface fluxes. It then rises to the top of the boundary layer where it mixes with air from the capping stable layer that is generally drier, causing the growth of the boundary layer. Finally, it sinks back toward the surface."[124] Fabry has found that "[m]oisture variability in the top half of the boundary layer is typically larger than at the surface, especially at smaller scales. This is because moisture variability in the higher part of the boundary layer arises from the contrast between updrafts bringing moist surface air and downdrafts entraining drier capping-layer air with them."[125]

Experiments conducted by scientists from the Department of Applied Mechanics and Engineering Sciences at the University of California have confirmed that in the boundary layer over water "humidity fluctuations can contribute significantly to refractive-index fluctuations."[126] The scientists also determined that temperature variance contributed to variations in humidity to affect the refractive index of air in the boundary layer over the water.[127]

There is a significant lensing effect of moisture and heat in the atmosphere. This can be seen when viewing objects transversely through the boundary layer of the atmosphere. The objects are thus seen to appear lower on the horizon than their actual elevation. We experience differing moisture gradients in extreme circumstances when approaching a fog bank and when exiting a fog bank. Of course, that extreme moisture creates a translucent effect in the atmosphere that completely obscures distant objects from view. As we will see below, the effect of moisture and heat in the boundary layer of the atmosphere causes both the objects in the distance and the horizon to shift downward.

Nonetheless, many who try to explain the visibility of objects that should be below the horizon on a supposedly spherical earth argue that they are superior mirages.[128] Superior mirages are due to a temperature inversion, where warmer air exists in a layer above significantly cooler air. It is the more usual case to have warmer air closer to the surface of the earth. The more unusual temperature inversion with warm air above the cold air often causes a concomitant mirage inversion and severe distortion of objects viewed through the thermal inversion. Temperature inversions that cause upside down superior mirages simply cannot explain how right-side up landmarks can be seen long distances away, when they should be below the supposed curve of the earth.

Those pushing the spherical earth model ignore the much more prevalent lensing effect of heat and moisture in the atmosphere causing inferior mirages and a drop in the apparent height of landmarks and the horizon. Instead, they propagandize a superior mirage theory. But heat and moisture layers in the atmosphere weigh against the spherical earth model, because the resulting refraction of the light makes distant objects more difficult to see. It is during periods of clear weather when distant landmarks become more visible from long distances over water. The mirage argument used to refute a flat earth is truly a *Fata Morgana*.

Olympus Corporation is one of the world's leading manufacturers of professional opto-digital products for medicine, science and industry.[129] Olympus Corporation has a unique expertise in light refraction, and it makes products that work in the real world of optics. Olympus explains the effect that causes mirages. They point out what is often glossed over and not explained by geodesists. That glossed over fact is that "[t]he actual

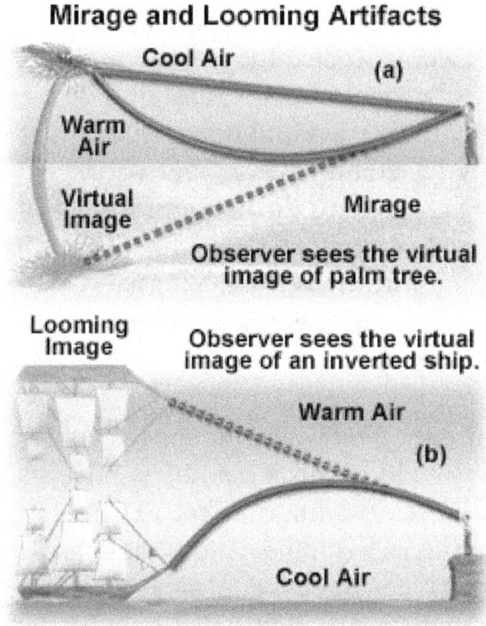

Figure 7

mirage effect that is visualized depends upon whether cooler air overlies warmer air, or vice versa."[130] Olympus Corporation explains that where air layers have "significantly different temperatures, overlapping layers of air in the atmosphere are responsible for producing what are often termed mirages, a phenomenon in which the virtual image of an object is observed to be positioned **either above or below** the actual object."[131] The graphic posted on the Olympus corporate website (Figure 7) illustrates the effect of the atmospheric layering.

Note that the mirage images are upside-down. Notice also that the most common occurrence is where warm air that is closer to the surface; this causes an inferior mirage. The inferior mirage is not actually lower, it only appears lower. The inferior mirage is labeled in Figure 7a as "Virtual Image." Such refracted images are commonly called apparent objects, to distinguish them from the actual objects. Refraction takes place at the boundary between the cool air and the warm air. The most significant point is that the warm air causes refraction of light which lowers the apparent image (Figure 7a). It is much more common to have warm air below cold air. In that more common circumstance, where warm air is below cold air, the apparent object (what is seen by the observer) is lowered. Olympus explains:

One type of mirage appears as an upside-down

virtual image directly beneath the real object, and occurs when a layer of warm air near the ground or water surface is trapped by denser, cooler air lying above. Light from the object traveling downward into the warmer air adjacent to the ground (or water) is refracted upward toward the horizon. At some point the light reaches a critical angle for the warm air, and is bent upward by total internal reflection, **resulting in the virtual image appearing below the object.**[132]

This more common mirage effect where the apparent object is dropped lower from view is seldom discussed by geodesists. They focus only upon the superior mirage effect. There is a refractive effect related to an inferior mirage called "sinking."[133] The sinking effect is manifested when there is cool air above warm air, which is the usual atmospheric condition. Assuming an equivalent atmospheric pressure, warm air is both physically less dense and optically less dense than cold air. Thus, the apparent image of a distant object seen through the border of the warm air is refracted and caused to sink. In sinking, the apparent object is lower than the actual object, but there is no inversion, as is the case with an inferior mirage.

The observed image in a superior mirage is seen to hover upside-down above the actual object. There is a related effect, similar to a superior mirage, called "looming."[134] This rare looming effect does not invert the image, but as with a superior mirage, it happens where cool air is below warm air.

Whether there is a superior mirage or an inferior mirage, the effect is an inversion of the object that causes extreme distortion. The superior mirage effect is not what is seen by observers of distant objects that are supposed to be over the horizon on a purported spherical earth.

The reason that the geodesists shy away from discussing inferior mirages or sinking is that there cannot be an inferior mirage or sinking of an object that is below the horizon on a supposedly spherical earth. That is because in order for an inferior mirage or sinking to manifest, the actual object must be in view. Because it is most common to have warm air below cold air in the troposphere, the recurring visibility of landmarks that should be below the horizon cannot be explained as mirages, since most mirages are inferior and inferior mirages by definition appear below the actual object. An inferior mirage requires the actual object to be above the horizon in order for the apparent object to be in view below the actual object. If the actual object is below the horizon on the mythical ball earth, an inferior mirage of that object would necessarily be below the actual object and, of course, also below the horizon.

Furthermore, the inferior mirage causes a drop in elevation (and inversion) of the object. With a sinking object there is no inversion. If there is an inferior mirage or sinking seen from an object that is supposed to be below the horizon on the purportedly curved earth, that is proof that the earth cannot be curved and must be flat. That is because such an inferior mirage or sinking would require the actual object to be visible above the horizon in order to refract the light to cause the lower apparent image. Thus, light refraction actually works against the spherical earth theory.

Refraction through the warm air causes the apparent object to appear below the actual object. In addition to refraction in the atmosphere from temperature, there is also light refraction in the atmosphere caused by moisture. That moisture in the atmosphere acts as a lens that affects the appearance of objects that are long distances away from the observer. The orthodox view that refraction of light in the atmosphere always causes objects to shift in appearance upward in elevation is wrong. The reality is just the opposite. Light refraction in the atmosphere more often makes distant objects drop lower from view, making them more difficult to see, not easier to see. That is what is defined as sinking.[135]

Thus, light refraction does not enhance the argument for a spherical earth; it works against it. Light refraction does not explain how distant objects that are supposed to be below the horizon on a supposedly spherical earth can still be seen. Indeed, in the usual environment, light refraction makes distant objects more difficult to see. So, we are left with the only logical explanation for the visibility of distant objects that should be below the horizon if the earth were a globe. The earth cannot be a globe; it must be flat.

Supposed authorities on refraction seem to focus on the variability of the physical density in the atmosphere, without accounting for the variability in its optical density. For example, W.J. Humphreys, C.E, Ph.D., former Meteorologic Physicist for the U.S. Weather Bureau, in his book, *Physics of the Air*, based his view that light is refracted in such a way in the atmosphere under normal circumstances to make the apparent height of an object higher than the actual height on the fact that the air becomes more physically dense as one descends in altitude.[136] The greater air density at lower altitudes is due to atmospheric pressure. It is true that the air has a greater physical density closer to the ground. Indeed, increased atmospheric pressure at lower altitudes increases the index of refraction.[137] But that is abated by the fact that air at the lower altitude, being warmer and containing more moisture, also has a lower optical density. That is important because the issue of the effect of atmospheric refraction on distant landmarks involves a viewer who is looking through the boundary layer of the atmosphere at an acute angle to the horizontal plane. The variability of temperature, humidity, and density in the boundary layer refracts the light in such a way that the apparent horizon and the apparent height of the distant landmarks drop in height. That is the opposite effect to that which is alleged by Humphreys and other geodesists.

It is notable that neither Humphreys nor any geodetic surveying authorities provide photographic evidence for their theory of light refraction causing the elevation of distant landmarks. Their books are full of diagrams and equations, but contain no photographs proving their claims. Why not? That is because their theory is a contrived deception, designed to support the orthodox, but illusory, position that the earth is a sphere. Indeed, if geodetic surveyors tried to document through photographs atmospheric refraction, they would end up with pictures showing an apparent drop in elevation of distant landmarks.

The fact that atmospheric refraction actually causes distant landmarks to appear lower (and not higher) impeaches the only argument that geodesists have to explain how distant objects can be seen when they should be below the curve of the supposed spherical earth. With the theory that atmospheric refraction causes distant objects to appear elevated being shown to be bogus, the only logical explanation that remains standing to explain the visibility of such distant objects is that the earth is flat. Geodetic surveyors and orthodox scientists (so-called) simply cannot allow that to be known. And so the fallacious explanations of atmospheric refraction in geodesy books are limited to equations and sketches, with no real photographic evidence presented.

Humphreys, of course, accepts that the earth is a sphere. He combines his misunderstanding of the earth's shape with his misunderstanding of the optical refraction of the atmosphere to conclude that it is theoretically possible for the density gradient to create symmetry in the elevation of the horizon such that "it is conceivable, therefore, that the size of a planet and the vertical density gradient of its atmosphere might be such that one's horizon on it would include the entire surface that he could look all the way round, and, as someone has said, see his own back."[138] That is just silly; it is mythology; but that is the science (or rather religion) of the globe earth.

Humphreys' absurd statement is not unlike the rhetorical question often asked by geodesists: "if the earth is flat, why can't we see the continents across the ocean?" It seems that common sense is not a common trait among the geodesists. Indeed, that is the whole point of heliocentricity: to drive from the populace any ability for independent, critical thinking. Heliocentrists seem unaware of the common observation that the ability to see a long distance through the atmosphere is attenuated by temperature, moisture, and particles in the air. In astronomy, this is known as atmospheric extinction. But when heliocentrists argue against the flat earth, they ignore the reality that there are limits to visibility through the atmosphere. Atmospheric translucence varies depending on the consistency of the atmosphere and the degree to which it absorbs or scatters light.[139] Atmospheric attenuation is compounded by the limitations of visual perspective and the distortion of light refraction. That is why objects disappear as they travel away in the distance on the level sea.

Refraction happens when the medium surrounding the observer is different to that in which the object is being observed.[140] It is possible for the atmosphere to have different temperatures, air pressure, and humidity at two locations; some degree of refraction would then result.[141] The little known fact is that because refraction in the atmosphere is usually due to the lower optical density of warm humid air between the observer and the object, the effect is to cause the apparent image of the distant landmark to be blurred and drop in elevation. Thus refraction does not enhance the observation of distant objects, but rather works against it.

As Frédéric Fabry, from the Department of Atmospheric and Oceanic Sciences, McGill University, explained above, the atmosphere is dynamic in the boundary layer. There is constant flow up and down of warm moist air. This dynamic atmosphere causes a concomitant fluctuation in the lensing effect. The lensing effect magnifies distant objects and causes the bottoms of those distant objects to be cut off from view.[142] As the moisture and temperature in the air fluctuate so also does the magnification effect of the atmosphere on distant objects. The objects can be seen to fluctuate in size and even disappear from sight depending on the temperature and amount of moisture that is in the air at a given time.[143] Often, landmarks and buildings will appear on a coastline with very little distortion, but if the moisture and temperature in the air fluctuate, those same buildings and landmarks may take on a fuzzy and squashed down appearance. In those circumstances, the bottoms of the buildings and landmarks along the shoreline become invisible and the tops are reduced in elevation.[144]

Many misinterpret the missing bottoms of buildings and ships to be caused by the earth's curvature, not realizing it is due to the lensing effect of the atmosphere.[145] The proof that the missing bottoms are due to the atmospheric lensing effect is in the fact that often the object observed should not be able to be seen at all if the earth were a globe, as it should be completely below the supposedly curved horizon.

This lensing effect can be seen with changes in the temperature and humidity in the air over time. Light refraction due to temperature and humidity in the atmosphere is well known by scientists and surveyors. Indeed, atmospheric light refraction was observed in experiments in the early 19th century.[146] In those early experiments, the atmospheric refraction of light was attributed, for the most part, to humidity.[147]

The problem with most research regarding light refraction is that it almost always begins with the false premise that the earth is spherical. When a drop in the apparent elevation of a distant landmark is seen, but the drop is much less than would be expected by the alleged curvature of the earth, it is assumed by geodetic surveyors that the lesser drop is due to atmospheric light refraction that has caused the distant object to be elevated. This leads to the false conclusion that the refraction of light in the atmosphere causes distant objects to appear higher than they really are. What is actually being seen is a drop in the apparent elevation of the distant object on a flat earth that is due to atmospheric light refraction (combined with, as we will see later, lens refraction).

The light refraction causes a drop in the apparent elevation of distant landmarks that is much less than would be expected if the drop in elevation was due to earth being a globe. The fact remains, however, that the apparent drop in elevation of distant landmarks is due to light refraction and not due to the mythical curvature of the earth.

The orthodox view of geodetic surveyors that the atmosphere refracts the light to make distant objects to appear elevated is backwards. It is based on the false premise that the earth is a globe. As we will see later, the upward refraction theory of geodetic surveyors is refuted by real science and actual observations. In fact, except in unusual circumstances, heat and humidity in the atmosphere cause the apparent horizon and the apparent height of distant landmarks to drop in elevation. Atmospheric refraction actually makes distant landmarks more difficult to see over the flat earth. The ineluctable conclusion, therefore, is that when distant landmarks can be seen, when they should be below the horizon if the earth were a globe, is that they can be seen only because the earth is flat.

In the two pictures below we see the same rock depicted in pictures taken by a camera from the same location from a stationary camera, fixed in place, with no adjustments made to the camera between shots.[148] The rock is 1.54 kilometers from the camera. The photographer used a 25-minute time-lapse series of photos. The photos show the effect of atmospheric refraction causing both the horizon and the rock depicted in the picture on the viewer's left to drop.

Above are two pictures from a 25 minute time lapse series of photos taken from the same location from a stationary camera, locked in place, with no adjustments made to the camera during the entire 25 minutes. Atmospheric refraction is causing both the horizon and the rock on the viewer's left to drop.

As the distance increases, the lensing effect of the atmosphere becomes more dramatic, with not only the horizon and height of objects dropping, but also the bottom of distant objects disappearing. In the pictures below we see those effects manifested with three oil platforms off the coast of Santa Barbara, California.[149]

The oil platforms are from the viewer's left to right: Platform Gail, Platform Habitat, and Platform Grace. The cameraman was shooting the pictures using Nikon P900 at an elevation of 76 feet above sea level, from One Thousand Steps Beach, in Santa Barbara, California. Platform Habitat (the middle platform) is closest to the camera and is 90 feet above sea level and approximately 10 miles away. Platform Grace (the platform on the viewer's right) is 122 feet above sea level and is approximately 20 miles from the camera. Platform Gail (the platform to the viewer's far left) is the furthest from the camera; it is 150 feet tall and approximately 25 miles from the camera. The pictures below are screen shots from two separate videos taken on two separate days. The picture on the left was from a video taken on February 23, 2017, at 9:13 a.m.[150] The picture on the right was from a video taken on January 30, 2017, at 8:42 a.m.[151]

The videos were taken from shore at the exact same

location from the same 76-foot elevation above sea level. The picture on the left shows the effects of light refraction on distant objects. Notice that the lensing effect of the atmosphere is to drop both the horizon and the distant platforms. Light refraction causes the apparent location of the platforms to drop below the actual location of the platforms. The actual effect of light refraction (which is to drop the apparent location of distant objects) is the opposite of what the "scientific" community states.

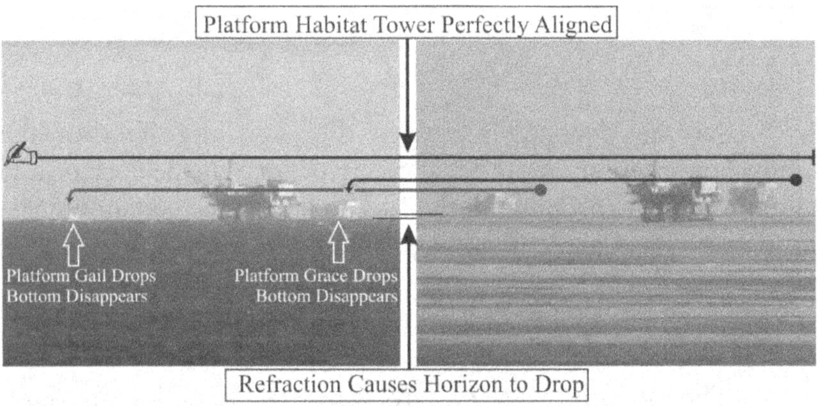

The light refraction in the atmosphere is seen in two pictures taken from the exact same location on different days. Refraction causes the horizon to drop, the bottom of distant objects to disappear, and a drop in elevation of the distant objects.

Temperatures at the times the pictures were taken were roughly the same. The temperature was between 51 °and 54° at the time the picture was taken on February 23, 2017,[152] and it was approximattely 55 °F at the time the picture was taken on January 30, 2017.[153] Oddly, the relative humidity was higher the day in the picture on the right. How can it be explained that the higher humidity weather is clear, when it otherwise should be expected to create greater atmospheric refraction? Be mindful that light refraction only happens at the boundary between layers of differing atmospheric humidity and warmth. If the atmosphere is calm and uniform in density there may be less light refraction than would otherwise be the case in high humidity level.

The interaction between warm humid air and cold dry air can be a complicated puzzle. While one can make generalities about the isotropic refraction of light passing from one uniform medium to another, the earth's atmosphere is not uniform. Light refracts in the atmosphere in an anisotropic fashion. When a variable like wind is introduced, it creates a more dynamic environment for air movement. In the picture on the right, it was a dead calm day with no wind, whereas in the picture on the left there was a steady wind of between three (3) and nine (9) miles per hour. It is not clear if that movement created a more dynamic movement of the air and the boundaries between relatively warm moist air and dry cold air. Whatever the cause, the atmospheric refraction can be demonstrated to cause a drop in the apparent elevation of landmarks.

Those who adhere to the globe earth model, may be suspicious that the lensing effect seen above is the reverse of what is depicted. They may suspect that the oil platforms on the right are seen over the alleged curve of the earth because of the refraction of light. That suspicion can be dispensed with by a simple experiment. In the pictures below a magnifying lens is used to simulate the refraction of the atmosphere.[154] As the magnifying lens is drawn toward the camera (in the picture to the reader's right), the greater refraction caused by the lens, in turn, causes the apparent ruler to drop below the height of the actual ruler, the horizon drops, and the bottom of the ruler becomes cut off from view.[155] This is precisely what we see in the oil platform photographs, where the oil platforms are dropped from view due to light refraction.

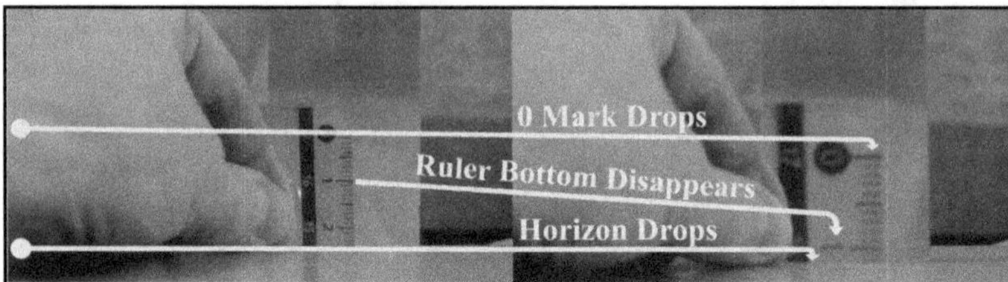

In this single motion shot with a fixed camera, the magnifying lens is drawn toward the camera in the frame on the viewer's right. Light refraction causes the ruler to squash outward, the horizon to drop, the bottom to disappear, and the top 0 mark of the ruler to drop down.

In the next picture, the dropped horizon after it has been refracted is raised up to the same level as the pre-refraction horizon. We see that when that is done, the 0 mark on the ruler has still dropped in height. The refraction causes the ruler to become squashed down. Again, that is what we see in the oil platform photographs of the atmospheric refraction, where the distant oil platforms take on a squashed look as they are refracted downward.

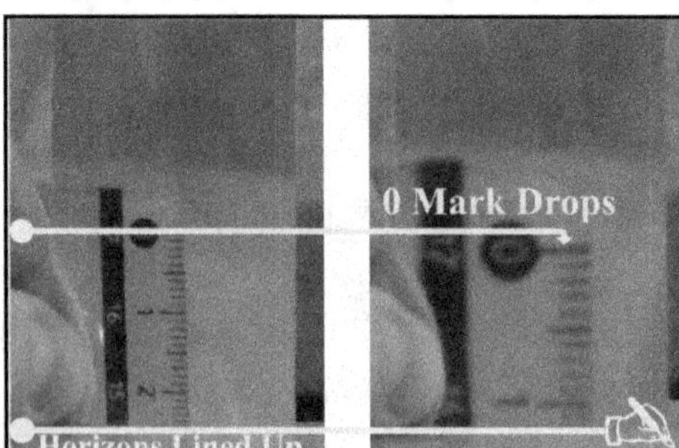

Even by lining up the pictures of the ruler at the horizons, and thus not including the horizon drop, we find that after refraction the 0 mark on the ruler has dropped in relation to the horizon. The ruler has been magnified along its width, but the top has dropped down in relation to the horizon. Thus the ruler takes on a squashed down appearance after refraction.

The refraction of light that is seen in the photographs of the ruler experiment using a magnifying lens confirm the atmospheric refraction in the oil platform photographs. That confirmation proves that the effect of refraction is to obscure distant objects by causing the horizon to drop, cutting off the bottoms of distant objects, and lowering the height of those distant objects.

Notice the blurring effect on the ruler as the lense is drawn toward the camera. We can see that same blurring in the picture taken from 4.87 miles away of the buildings on Santa Claus Lane on the right, below.[156] The lensing effect of the atmosphere causes extreme distortion, compression of the buildings, and a drop their apparent height. The picture depicts the combined effect of both shrinking and an inferior mirage. The combined effect is that the apparent height of the landmarks drop in elevation. This is yet more evidence that light refraction makes distant objects more difficult to see. The argument of geodesists that the distant landmarks can be seen below the supposed curvature of the earth due to light refraction is a fallacy. The landmarks can be seen beyond the drop, because the drop is not there. The earth is flat.

Both of the above pictures were taken of buildings on Santa Claus Lane using a zoom lens at different times from the same location across a body of water to buildings 4.87 miles away. In the picture on the left the atmosphere is clear, and there is little atmospheric refraction. The atmospheric refraction in the picture at right is causing blurring and distortion of the buildings, including a shrinking effect with a drop in height. Indeed, the buildings are squashed by the lensing effect of the atmosphere. By aligning the pictures at the border of the rock shore and buildings, it is apparent that there is an inferior mirage of the shoreline, perhaps caused by the radiant heat and evaporation of moisture from the water's surface that gives the appearance of the horizon dropping lower. The lensing effect in the atmosphere is seen here to drop distant landmarks lower and makes them more difficult to see. This reality is the opposite of the outcome from atmospheric refraction algorithms used by geodesists.

When we are looking at light refraction in the atmosphere

due to the lensing effect it is because of the interaction between warm moist air and cold dry air. In absolute terms, warm air can hold more moisture than cold air. For example, at 100% relative humidity at 0°C/32°F, the atmosphere has the capacity to hold approximately 5g of water vapor per cubic meter (5g/m3). But if we raise the temperature to 30°C/86°F and assume 100% relative humidity, the air can hold approximately 30g of water vapor per cubic meter (30g/m3). Warmer moist air has a lower refractive index than cold dry air. Their dynamic interaction causes the lensing effect in the atmosphere. The refraction takes place at the boundary between the two atmospheric mediums of different temperatures and humidity.

In the pictures of Santa Claus Lane above, the picture on the left was taken on June 12, 2016 at 4:43 p.m.[157] The picture on the right was taken on November 22, 2015 at 4:10 p.m.[158] The refraction in the picture on the right seems to be based upon the heat and perhaps the evaporation of moisture from the water due to the heat. The temperature for the Nearest city, Santa Barbara, California, was recorded to be 84°F, sunny, with a relative humidity of only 9% at 2:53 p.m. on November 22, 2105.[159] Whereas the temperature when the picture on the left was taken was only 68°F, mostly cloudy, with a relative humidity of 57% at 4:53 p.m. on June 12, 2016.[160]

Although the humidity was higher in the picture on the left, the heat over the water on the sunny day of November 22nd in the picture on the right was higher. The difference in air temperature has a greater effect on light refraction than does humidity. Nonetheless, water vapor also seems to have played a role, since the warmth of the sun would cause evaporation of the water and create more actual water vapor in the air close to the water in the November 22nd picture on the reader's right. The synergism between the heat and evaporation created the refraction even though the surrounding atmosphere had less humidity. The refraction takes place at the boundary of different mediums. It seems with the surrounding air having very low humidity, there was a more severe change and thus refraction. Indeed, the inferior mirage of the rocks seen in the picture on the right is exactly what would be expected by the warm rising air. It is notable that there was no wind at 2:53 p.m. on November 22, 2015. That seems to have had the effect of allowing the warm air and water vapor to sit in place, even as the temperature cooled thereafter. By 4:53 p.m., the temperature had dropped to 70° (there was no reported data for 3:53 p.m. on November 22, 2015).

In the picture of the buildings on Santa Claus Lane on the left, however, the humidity was higher, but the temperature was lower, and it was mostly cloudy. That created an environment with a more homogenous atmosphere that was not impacted by evaporation from the water. There was not the severe change in mediums through which the light would pass to be refracted. Please note that these hypotheses are based upon the general weather conditions reported for Santa Barbara, California. As all sailors know, there can be local variations in weather conditions. Andrew Young, an astronomer writing for San Diego State University, explains that "the refraction near the horizon depends mostly on the local temperature gradient, which is much more important than the local temperature itself."[161] Young states that the variable density gradient of the atmosphere results in the concomitant distortion of objects.[162] That is what we see with the Santa Claus Lane picture on the right, above, where both the height of the buildings shrunk and at the same time there was an inferior mirage that mirrored the shoreline boulders downward.

The effects seen in the picture of the buildings on Santa Claus Lane on the right, above, are the opposite of what mainstream scientists, like W.J. Humphreys, C.E, Ph.D., aver atmospheric refraction should cause. The orthodox view is that atmospheric refraction will cause the elevation of the apparent height of a landmark, not a lowering of its height, as is depicted in the picture. Indeed, the alleged elevation of the apparent height of a distant object due to atmospheric refraction has become such a dogma, that geodetic surveyors claim to automatically deduct an arbitrary refraction adjustment of one-seventh that they deduct from the calculation of the alleged drop due the curvature of the earth. They do that to, in effect, flatten out the supposed curvature of the earth as a way to explain why distant objects that should be out of view below the horizon can be seen above the horizon. In the face of contrary photographic evidence, like the pictures of the distant buildings on Santa Claus Lane, that refraction works in the opposite direction, scientists are left to argue, as did Chico Marx: "Who ya gonna believe, me or your own eyes?"[163]

The picture below is a screen shot of a zoom lens video of a distant boat. While videoing the distant boat a fish was seen jumping out of the water.[164] Because the light had refracted the apparent horizon downward, the fish seemingly appeared out of midair. In fact, the fish was jumping out of the water at the level of the actual horizon. Mirror images of both the boat and the fish can be seen in inferior mirages below the actual horizon.

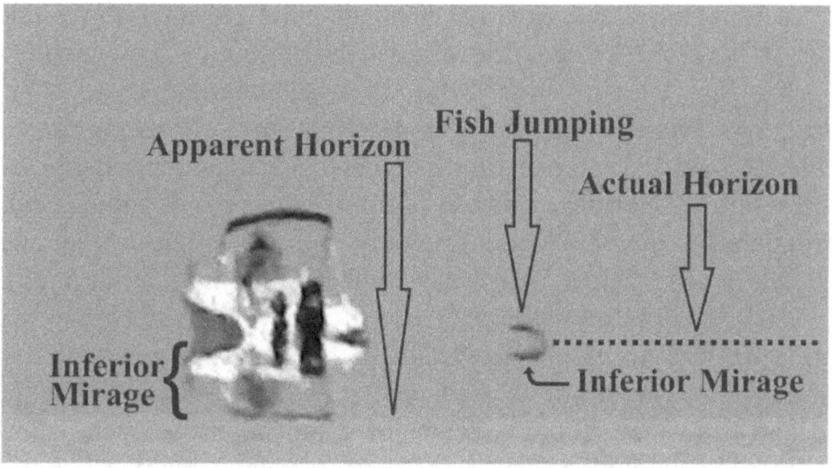

The atmospheric refraction causes the apparent horizon to drop. The actual horizon is above the apparent horizon. The atmosphere causes the boat to be refracted downward in an inferior mirage. A fish can be seen appearing seemingly out of mid air, but he is actually jumping approximately 6 inches out of the water and thus 6 inches above the actual horizon. An inferior mirage of the fish can also be seen below the actual fish.

The drop in the horizon due to atmospheric refraction is ordinarily not detected, because the atmosphere above the waterline is homogeneous and the water below the waterline is also homogeneous. Thus, the refraction typically goes unnoticed. It is only when there are objects on the horizon, such as the boat and the jumping fish in the picture above that the downward refraction of the apparent horizon can be seen. The uniformity of the atmosphere causes the seam to go unseen between the actual horizon and the refraction of the atmosphere in an inferior mirage. This atmospheric refraction causes the horizon to appear lower.

As the above picture proves, light refraction causes the apparent horizon to drop in elevation below the actual horizon. Yet, the orthodox view is just the opposite from reality: that light refraction causes the apparent horizon to be elevated above the actual horizon. Indeed, the alleged elevation of the apparent horizon has become such a dogma in science that there is an arbitrary one-seventh deduction from the earth curvature calculations used by geodetic surveyors to account for atmospheric refraction. The arbitrary one-seventh rule for refraction has been around since the 1800's, but it is only in modern times that it has allegedly been factored in to surveying calculations. For example, in 1881 Samuel Rowbotham cited to the Encyclopædia Britannica on "levelling," in which was explained that while unequal densities of air will cause refraction, such refraction is so slight and variable that no allowance should be made for it. Indeed, the Encyclopædia Britannica concluded that "though the refraction may at a mean compensate for about one-seventh of the curvature of the earth, it sometimes exceeds one-fifth, and at other times does not amount to one-fifteenth. We have, therefore, made no allowance for refraction in the foregone formulæ."[165]

Rowbotham cited to the Encyclopædia Britannica because he assumed that the "experts" on light refraction of his day were being honest about the refraction causing the apparent height of an object to be elevated. Rowbotham cited to the Encyclopædia Britannica to make the point that the light refraction was insignificant. Rowbotham did not realize that the claims of geodesists that light refraction elevates the height of an apparent image were just as false as their ultimate claim that the earth is a globe. Rowbotham apparently did not know that light refraction in most cases actually lowers the apparent height of a distant landmark.

It should be noted that the light refraction that the Encyclopedia Britannica says should be ignored is claimed today by geodetic surveyors to supposedly always be factored in surveys at the arbitrary rate of one-seventh, under all conditions, and always in the direction of a deduction from the supposed curvature of earth. But that light refraction rate is being considered by geodetic surveyors in the wrong direction. The pictures of Santa Claus Lane are exhibits A and B of that fact. Indeed, in all of the research this author has done, I was not presented with a single actual photograph in any research paper or treatise proving the direction of refraction alleged by geodetic surveyors. One will find lots of calculations and diagrams presented by geodetic surveyors, but there seems to be no actual documentation of their claimed direction of refraction. Indeed, there can be no actual evidence, because the reality is contrary to their postulation.

Indeed, in one of the few examples where a scientist tried to show some evidence to support the geodetic refraction algorithms, the photographic evidence actually refuted the algorithm. Jurgen Giesen has created a light refraction simulation, which includes charts and algorithms, based on Jean Meeus' writings on atmospheric refraction.[166] The charts and algorithms indicated that atmospheric refraction causes an apparent object to be elevated. But the photograph of the sun that Giesen depicted on the website, which Geisen posted to display the actual effect of atmospheric refraction, demonstrates that the apparent sun is actually compressed (not elevated) and the apparent horizon actually drops (it does not rise). Those actual effects depicted in the photograph on his website are the opposite of what the equations and charts he displayed indicate should happen. This author sent Giesen an email asking how he could explain the contradiction, but I did not receive a response from him.[167]

Harald Edens, Ph.D., is a professor of physics at New Mexico Institute of Mining and Technology and specializes in atmospheric physics.[168] He has witnessed the flattening effect of the sun at the horizon caused by atmospheric refraction. He presented a series of photographs on his website illustrating the compressed sun and drop in the apparent horizon. The photographs are similar to the picture of the sun posted in this chapter below and also similar to the photograph on Giesen's website. Edens explains that the effect is caused by atmospheric refraction that bends the light rays.

> The temperature (and pressure) gradient in the atmosphere causes bending of light rays. This is called atmospheric refraction. Light rays from the low sun or moon will refract more closer [*sic*] to the horizon. This is why the setting or rising sun (and moon) appears flattened: the light rays from the lower part of the sun's disk refract more than the rays emerging from the top, and the vertical angle over which you see the sun is decreased - it is flattened.[169]

Dr. Edens states that because of atmospheric refraction, "the vertical angle over which you see the sun is decreased - it is flattened." Dr. Edens is explaining what his photographs actually illustrate. Dr. Edens is able to prove, with his photographs, the very opposite of the claim by geodists in their algorithms. Dr. Edens' photographs contradict the geodists claim that the vertical angle at which you see the sun through the atmosphere should increase, thus resulting in an elevated apparent sun. The reality illustrated in Dr. Edens' photographs is that the atmosphere actually causes the apparent horizon and the apparent height of distant objects to drop in elevation. The most strikingly obvious fact in his photographs is that atmospheric refraction causes the sun to appear flattened. Dr. Edens' photographic proof impeaches the deceptive algorithms from geodesists that they present to argue that the apparent horizon and distant objects are elevated in height by atmospheric refraction.

Incidently, this author sent an email to Dr. Edens to ask him about the apparent conflict between what is observed in his photographs and the geodetic algorithms.[170] I attached the above pictures from Santa Claus Lane and the compressed sun photograph posted below as examples of how actual atmospheric refraction differs from the predicted refraction using geodetic algorithms. I received no response from Dr. Edens.

Plane surveying assumes that the earth is a flat plane (hence, the name plane surveying); it is a valid discipline, involving precise measurement. Geodetic surveying, supposedly takes into account the purported curvature of the earth; it is an elaborate scam. Geodetic surveyors admit that in the end, they factor out the refraction/curvature calculation when recording survey results. They go through a ritual of taking two equidistant measurements in opposite directions and then simply cancel out the supposed adjustment made for curvature and light refraction.[171] It is a meaningless exercise. Geodetic surveyors in reality only pretend to consider light refraction and the earth's curvature.

In the picture below we see a view from a camera along the pavement over a bridge.[172] The heat and humidity in the air causes the apparent horizon to drop in elevation. The cause of the horizon drop is atmospheric refraction, creating an inferior mirage. That causes the scene above the actual horizon to be mirrored below the actual horizon. This effect is a regular occurrence. Over water, however, this effect is not detected, because of the uniformity of the air and water. But the effect can be seen in the picture below due to the fact that the guard rails are also being refracted downward. The downward refraction of both the horizon and the guard rails demonstrates the fact that atmospheric refraction causes both the horizon and the height of distant objects to drop in elevation.

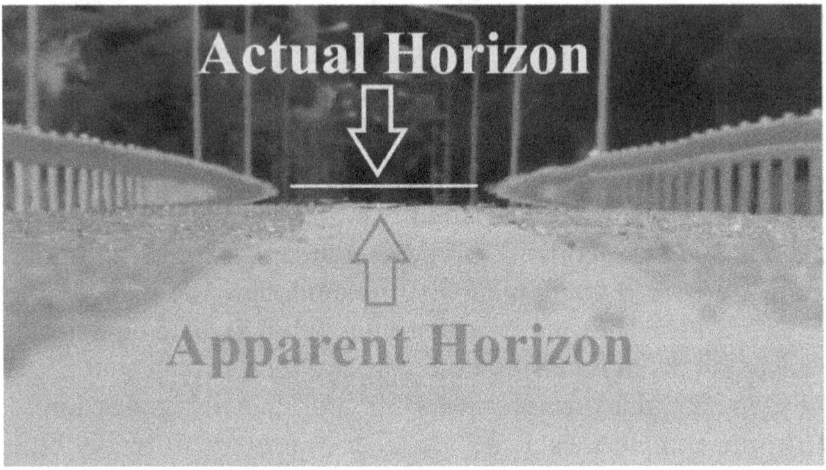

The picture above shows a view across a flat bridge. The atmospheric refraction has caused the apparent horizon to drop below the actual horizon. This is usually not perceived in sky/water horizons because of the uniformity of the sky and water. But here we can see the mirroring effect of the inferior mirage on the straight guard rails that appear bent downward as they are refracted in the inferior mirage.

It is difficult to perceive the horizon drop due to atmospheric refraction when looking over water. But it can sometimes be detected during sunset. The picture below was taken from the west coast of Florida looking over the Gulf of Mexico.[173] As the sun sets its light makes contact with the refracted atmosphere and an inferior mirage of the sun is manifested. That inferior mirage is a mirror image of the sun. The inferior mirage marks out the area between the actual horizon and the apparent horizon. The picture depicts the apparent horizon below the actual horizon. Notice also that the atmospheric refraction shrinks the elevation of the sun, giving it a squashed down oblate appearance. A perfect circle has been positioned at the bottom of the apparent sun, which reveals that the oblate sun is actually the shrinking of the elevation of the apparent sun. Modern astronomers assume that "refraction raises both the Sun and the apparent horizon above their geometric positions."[174] As we see with the picture below, that assumption is wrong.

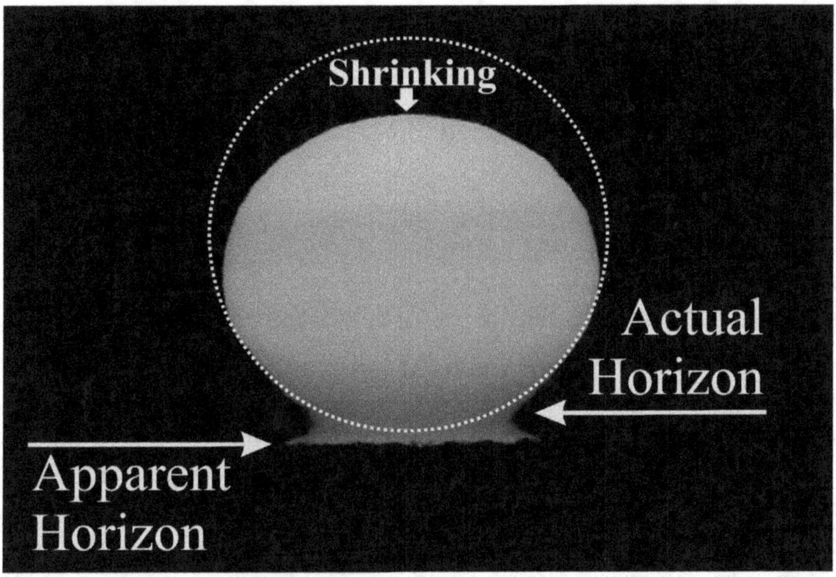

As the sun sets, atmospheric refraction causes the sun to shrink in height, giving it an oblate appearance. As the sun gets closer to the horizon, an inferior mirage of the sun is mirrored beneath it, which reveals the apparent horizon that is below the actual horizon.

The regular atmospheric refraction is in the opposite direction from what is represented by modern science. The dropping of the horizon due to atmospheric refraction makes landmarks more difficult to see. Visible landmarks that would otherwise not be seen over the horizon of the supposedly spherical earth can be seen because the earth is flat. The argument by geodesists that light refraction is the cause for the visibility of distant objects is wrong. Light refraction actually makes distant objects more difficult to see.

As the sun in the above picture later sets over the ocean, it can be seen in the picture below through a zoom lens to disappear while it is still above the horizon.[175] It seems that the impression that the sun is dropping below the horizon is caused by perspective and atmospheric refraction.[176] Added to those phenomena is the extinction of light as it travels through the atmosphere. Atmospheric extinction is the absorption of the light by the atmosphere as the light gets more distant from the observer. Andrew T. Young, PhD, adjunct professor of astronomy at San Diego State University, and an expert in atmospheric refraction, explains that "extinction is responsible for the dimness of the setting Sun."[177] Thus what we see is that the sun is simply disappearing into the distance as it gets further away. Perspective, atmospheric refraction, and atmospheric extinction give the appearance that the sun is dropping below the horizon. The picture of the sun below was taken just prior to the sun disappearing from view. It is a far-off oblong light that is being squashed by atmospheric refraction. When the oblate light is zoomed in, it shows that the sun is actually traveling away from the viewer with half of the sun above the actual horizon and the other half being duplicated in an inferior mirage below the actual horizon as it disappears from sight.[178]

This is observable evidence that the sun is traveling in a circuit over the earth. The heliocentrists argue that light refraction causes the apparent sun to appear above the actual sun.[179] They must maintain that position in order to explain how distant objects, which should be below the supposed curve of the globular earth, can still be seen above the horizon. But, as we can see, the heliocentric light refraction argument is backwards. In reality, the actual sun is above the apparent sun. The apparent sun is manifested in an inferior mirage below the actual sun. The disappearance of the sun below the horizon is caused by the combination of: 1) perspective, which causes distant objects to drop; 2) light refraction, which causes distant objects to shrink; and 3) atmospheric extinction[180], which is the absorption of the light by the atmosphere as the light gets more distant from the observer. We can observe these phenomena with our own eyes in the photographs above and below. The heliocentric explanation for the setting sun, on the other hand, is not based upon observable and verifiable science; it based upon lies.

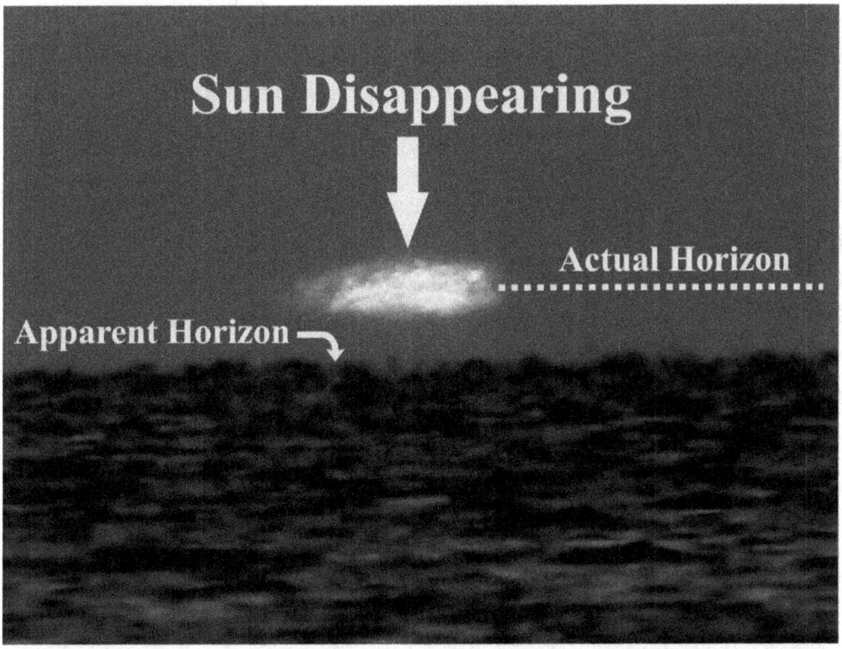

The sun is traveling in a circuit over the earth. In this screen shot from a zoom lens video of the sun just prior to disappearing from view, we can see that the sun is shrinking in size above the horizon as it disappears into the distance due to a combination of perspective, atmospheric refraction, and atmospheric extinction.

The bottom line (pun intended) is that atmospheric refraction does not explain how distant objects can be seen over a supposedly curved earth. That is because atmospheric refraction usually makes distant objects less visible, not more visible as would be required under the spherical earth theory. Atmospheric refraction actually works against the argument for a spherical earth. Atmospheric refraction, which is the only argument that heliocentric scientists have to explain the visibility of distant objects that should be below the horizon on a supposedly spherical earth, is thus turned on its head. Atmospheric refraction usually has the opposite effect from what is portrayed by orthodox science. Atmospheric refraction actually supports the evidence for a flat earth.

When buildings that should be below the supposed curvature of the earth can be seen above the horizon, it is because the earth is flat. The reason the far off buildings are visible is because the atmosphere is clear and there is less atmospheric refraction. When there is more atmospheric refraction, the distant landmarks will be obscured and harder to see. But when there is less atmospheric refraction, the landmarks become visible. That can mean only one thing; the earth is flat.

What is the dynamic for the atmospheric refraction to cause the height of landmarks to appear lower to an observer? The three primary factors affecting atmospheric refraction are temperature, humidity, and atmospheric pressure. The effect of atmospheric pressure on light refraction can sometimes be the most significant factor. For every .9 kilopascal (kPa) (.13 psi) increase in atmospheric pressure, there is a 1% increase in the optical density of the air.[181] At 26.67°C (80°F) the atmospheric pressure at sea level is 101.35 kPa. At 26.67°C (80°F), the atmospheric pressure at 1,000 feet is 97.90 kPa.[182] That is a difference of approximately 3.5 kPa. That means that the percentage increase in the refractive index due to barometric pressure when light travels from 1,000 feet to sea level is approximately 3.88% (3.5 ÷ .9).

Temperature and humidity also have an effect on the optical density of the air, but they are inversely related to optical density. That means that the higher the humidity, the lower the optical density (refractive index), and also the higher the temperature, the lower the optical density. That is the opposite of the effect of atmospheric pressure, where the higher the air pressure, the higher is the refractive index. Barometric pressure and thus the refractive index increases at lower elevations. But humidity and temperature also increase at lower altitudes, but the increased temperature and humidity each lower the refractive index at lower altitudes. For example, for every 3 C (1.8 F) increase in temperature there is a 1% decrease in optical density. When considering the fact that temperature decreases 3 C (5.4°F) for every 1,000 feet you increase in elevation, we find that there is a 1% decrease in the refractive index due to temperature when light travels from 1,000 feet to sea level.

The effect of humidity is more difficult to quantify, but the greater the humidity, the lower the optical density of the air. Humidity does not have as great an impact on refraction as does temperature and barometric pressure. There is only a .13% reduction in the refractive index when going from 0% humidity to 100% humidity (assuming a temperature of 20°C (68°F) and a light beam with a wavelength of 633nm).[183]

The net effect of barometric pressure, humidity, and temperature is that light traveling from 1,000 feet will be refracted toward the normal. With that predicate in mind, one would think that the much more significant effect that barometric pressure has on light refraction would cause distant objects to be elevated, just as argued by orthodox geodesists. But that is not what we see in reality, as the above photographs testify. So, what is going on? Why does light refraction cause the apparent height of distant landmarks to drop, when both the physical density and the optical density of air increases with increased air pressure at lower altitudes? The apparent height of distant landmarks drop because they are being viewed from an acute angle to the horizontal plane through the atmosphere. That acute viewing angle through the atmosphere means that the light is traveling through the air to the viewers eye at a somewhat consistent elevation. That means that the barometric pressure would have little variability along the light's horizontal path of travel. The acute horizontal viewing angle nullifies the effect of the barometric pressure on light refraction. That leaves the humidity and warmer air temperature at the lower altitudes to create an optically less dense environment that refracts the light away from the normal and causes distant landmarks to drop in height.

Initially, the light is refracted toward the normal as a result of the light traveling downward through the increasingly more optically dense atmosphere created by the steady increase in the barometric pressure as the light gets closer to the surface of the earth. But when the light from the distant landmark reaches the lower elevations, the atmospheric refraction is reversed away from the normal as it travels to the viewer's eyes at an acute angle through the optically less dense, warm, humid air close to the surface of the earth.

At the lower elevations, the acute viewing angle (which is virtually horizontal) causes the light to be refracted almost completely by the warm, humid air near the surface of the earth. The barometric pressure is more constant along that acute viewing angle. Thus, the light is refracted away from the normal line by the warm, humid air at a constant barometric pressure, which has a lower refractive index. This refraction away from the normal line causes distant landmarks to drop in height. That fact is concealed by geodesists.

Geodesists have oversimplified the phenomenon of light refraction. They focus on the greater physical (and optical) density of the air with the higher barometric pressure at lower altitudes. But they ignore the fact that the viewer, who is at that lower elevation, is looking at an acute angle to the horizontal plane through warm, humid air, which has a lower optical density. This causes the apparent height of distant landmarks to drop.

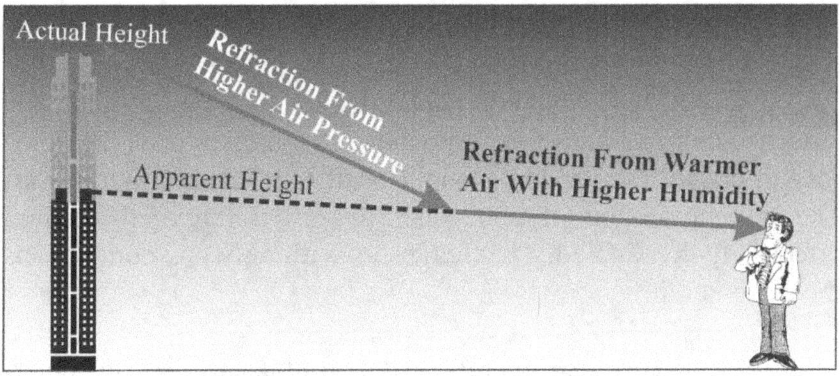

The light is initially refracted toward the normal line as it travels from the lighter air with lower atmospheric pressure (lower refractive index) to denser air with higher atmospheric pressure (higher refractive index). When the viewer looks out at an acute angle to the horizontal plane, the light is being refracted in the other direction, away from the normal line, by the warmer, more humid air, which has a lower refractive index. The net effect of the atmospheric light refraction, first toward and later away from the normal line, causes the apparent height of the distant building as seen by the observer to drop.

The above diagram is a simplified rendering, but it illustrates the interplay of the variables that cause atmospheric refraction. There are likely many layers of atmosphere with different optical densities between the observer and the landmark, with a combination of effects. Scientists have confirmed that there is great variation in warmth and humidity that affect the refractive index of air in the boundary layer.[184] But it is clear that the net effect of the variable refraction in the atmosphere is that generally both the apparent horizon and the apparent height of distant landmarks drop. The photographic evidence, samples of which are provided above, proves that fact. That actuality, in turn, refutes the argument of geodesists that light refraction causes landmarks to be elevated and visible when they should be below the horizon of the alleged spherical earth. Geodesists provide no real evidence for their claim. They only provide one-size-fits-all algorithms. But the real, observable evidence contradicts their claim. The photographic evidence demonstrates that atmospheric refraction goes the other way; it causes the apparent height of distant landmarks to drop. The only argument of geodesists against a flat earth is thus effectively invalidated. That leaves us with only one conclusion: the earth is flat.

Steve Torrence entered the data for the earth's atmosphere into 3D software (Cinema 4D®). The software showed that the apparent sun drops lower in the sky than the actual sun.[185] If the apparent sun drops in elevation (as we witness actually happening in the above photographs) then the same would be true for distant landmarks (which we also witness actually happening in the above photographs).

That drop in apparent height of the sun is the opposite of the theory of orthodox science, so-called, which posits an erroneous hypothesis of the apparent sun being higher in the sky than the actual sun.[186] Orthodox "science" must create a model of the earth that has light necessarily following the curvature of the earth in order to explain how distant landmarks are visible when they should be below the horizon of the supposed spherical earth. But that supposed elevation of the height of distant objects is not what is replicated in experimental models, nor is it what we actually see. Indeed, we find the evidence to be the very opposite; the atmosphere refracts the light to cause the apparent height of distant landmarks (and the apparent horizon) to drop in height.

A fellow researcher, who goes by the *nom de plume* of Dr. Zack, verified Torrence's results, using AutoCad software.[187] Dr. Zack next went a step further by taking a glass bowl full of water and holding it under a light.[188] Dr. Zack was able to determine that when he held the water under the light, the shadow cast by the light striking the vertical implements on a table grew longer than they were previously without the light passing through the bowl of water. The fact that the shadows became longer when the light shined through water indicated that the apparent position of the light was caused to drop by passing through the water.

Dr. Zack's glass bowl containing water simulated the effect of the atmosphere causing the apparent sun in the sky to drop lower than the height of the actual sun. The water simulated the double refraction of light traveling through the increasing density of the earth's atmosphere as the barometric pressure increases at lower altitudes. The drop in height of the apparent light is based upon the light being initially refracted toward the normal line as it entered the water-filled bowl; the light was then refracted away from the normal line upon exiting the water-filled bowl. That refraction away from the normal line is what happens when looking at an acute angle through the lower refractive index atmosphere as the sun approaches he horizon.

Dr. Zack, in collaboration with Dr. Anggoro Aji, Ibam Archerstone, and Ayat Hadiyat performed a much more elaborate test using a bi-concave lens. He was once again able to establish that a light refracted by the lens caused the apparent light to drop in elevation in relation to its actual height.[189] Those experimental results offer an explanation for what we witness happening with the drop in the apparent height of distant landmarks.

The drop in height of the apparent sun as we look at it approaching the horizon, like the drop in the apparent height of landmarks as we look at them on the horizon, and the drop in the apparent height of the horizon itself, is what we witness in the above photographs in this book. Geodesists want the public to ignore that empirical evidence. They, instead, try to bedazzle the public with long algorithms that have no relationship to reality. Those algorithms are smokescreens to conceal the reality of the flat earth.

4 Refraction Skulduggery

Another example where atmospheric refraction can be seen over water is in the picture below of electrical towers spanning Lake Pontchartrain. That particular picture is being pushed by spherical earth advocates, who claim that it proves the curvature of the earth, because it gives the appearance of curved water and concomitant curved electrical towers.[190] Closer examination, however, reveals that picture does not prove convex water. The picture instead illustrates the combined effects of perspective and atmospheric refraction. One researcher (Dr. Zack) has also found that the power lines in the picture are not in a straight line.[191] They are of equal height, but they are not laterally in a straight line. That lateral offset enhances the curved drop effect.[192] Those factors combine to create the optical illusion of curved power lines as they get more distant from the observer. The primary cause for the curved appearance is light refraction. It seems that the gallows upon which the spherical earth Haman intended to hang the truth of the flat earth has become the instrument that instead hangs the spherical earth lie. See Esther 7:9-10.

Above is a picture used by spherical earth advocates to prove the earth is a ball. But what the picture actually demonstrates is the effect of atmospheric refraction, which causes the horizon to drop and distant objects to shrink in height.

The key indicator that it is primarily atmospheric refraction, and not curvature, that is causing the bending effect in the picture is the inferior mirage evident at the horizon point. That inferior mirage is proof that the picture is not depicting curved water, but rather atmospheric refraction that is merely giving the appearance of curved water. Adding to refraction is the visual limitation of perspective. The curving effect in the picture is due to the combined effects of atmospheric refraction, the limitations of perspective, and the fact that the towers are not in a perfectly straight line.[193] If the camera were to move to the side profile of the towers over Lake Pontchartrain the towers and water would be seen to be perfectly flat and level for the entire length of the lake.[194] Indeed, a researcher did just that, not with the electrical towers, but with the Lake Pontchartrain bridge.[195] And, sure enough, the bridge was level and spanning a perfectly flat level lake; there was not an iota of curvature.[196] But that should not be a surprise to anyone. Water is always flat and level. In the real world, standing water is perfectly flat and is not ever convex. That fact proves that the earth is flat, since more than 70% of the earth's surface is covered by water. If water is flat, the earth must necessarily also be flat.

According to the photographer of the above picture, who is a spherical earth advocate, there are 80 towers that span the 16-mile width of Lake Pontchartrain at that particular point in the lake. The picture depicts approximately 30 towers before we reach the apparent horizon. That puts the horizon at approximately 6 miles in the distance. At that point, there is noticeable atmospheric refraction that can be seen as it causes an inferior mirage of the cement pylons. The pylons are clearly being refracted downward, which gives the appearance of bending in a curve as the tops of the pylons are mirrored upside-down next to the water in the inferior mirage. This atmospheric refraction has the effect of cutting off all subsequent pylons from view. The missing pylons in the distance illustrate a consequence of perspective and light refraction, which is to cut off the bottoms of distant landmarks. At the same time, the apparent horizon drops in elevation. This is the very same effect that is depicted in the previous bridge picture of the guard rails over the bridge.

At the point where the pylons are being mirrored in an inferior mirage and the apparent horizon is dropping, the tops of the towers are also bending downward as they shrink in height. This illustrates the effect of atmospheric refraction on distant objects. When combined with the limitations of perspective, the effect is that the apparent height of distant landmarks drop in elevation and they are more difficult to see. Once it is understood what the picture actually depicts, we can see that it refutes the orthodox view of geodetic surveyors that the light always refracts in such a way to make the apparent height of objects to be elevated. The picture impeaches the claim by spherical earth advocates that landmarks that should be invisible below the horizon become visible due to light refraction. The result we see in the picture is from the combined effects of atmospheric refraction and the limitations of perspective, which always cause the apparent height of distant landmarks to drop.

Geodetic surveyor Jesse Kozlowski engaged a plane surveyor, Chris Van Matre, in a discussion of the evidence for a flat earth. Jesse Kozlowski shared a screen shot of a theodolite image showing the horizontal crosshair on his theodolite above the horizon.[197] Kozlowski represented the picture as being a picture through a perfectly horizontal theodolite looking over the ocean. His argument was that the horizon of the ocean being below the theodolite crosshairs shows the drop due to the curvature of the earth.

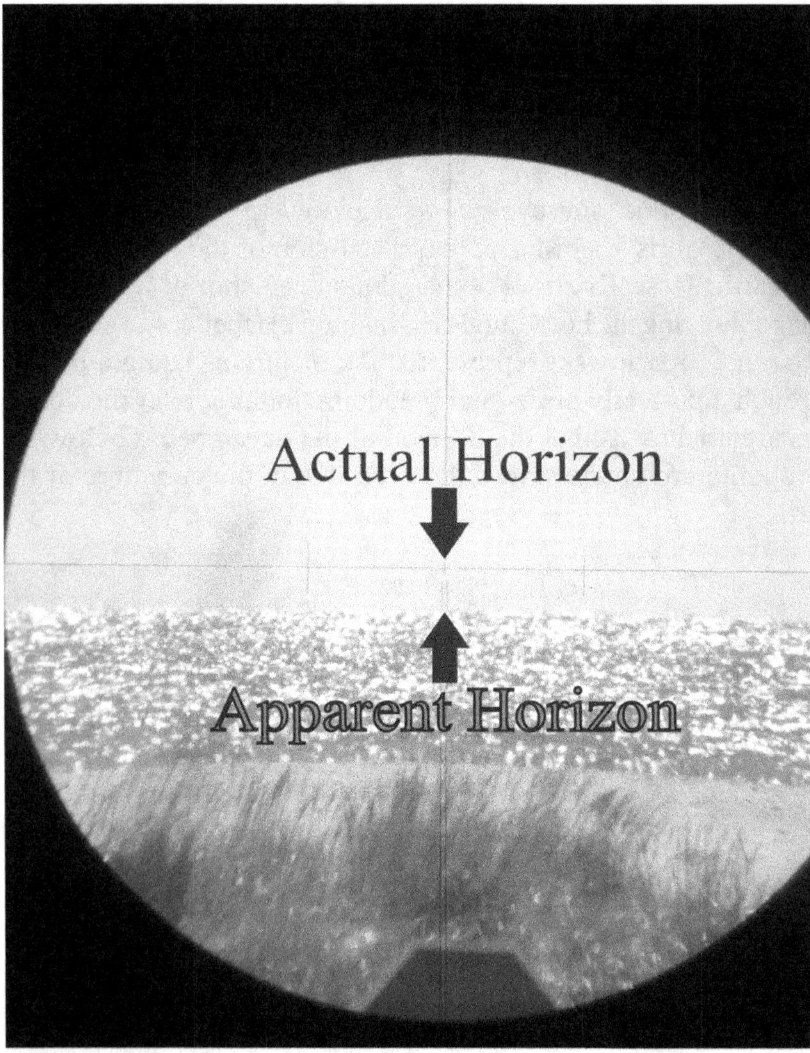

This picture was taken by Geodetic Surveyor Jesse Kozlowski, which he stated was of the ocean horizon from a horizontally level theodolite. Kozlowski alleges that the drop in the horizon below the horizontal cross-hairs is because the earth is a sphere. It is actually due to a combination of atmospheric refraction and theodolite lens refraction. Indeed, you can see the bowed beach shore caused by the theodolite lens refraction. The horizontal line on the theodolite shows where the actual horizon is located.

Samuel Rowbotham, in his day, was confronted by surveyors who would present similar evidence as presented by Kozlowski to prove the rotundity of the earth. Rowbotham discovered that the lenses in theodolites create a refraction of light. This refraction causes the apparent horizon to drop in relation to the actual horizon. Rowbotham explains:

> But in Ireland, in Scotland, and in several parts of England, he has been challenged by surveyors to make use of the theodolite, or ordinary "spirit level," to test this appearance of the horizon. It was affirmed that, through this instrument, when "levelled," the horizon always appeared below the cross-hair, as shown in figure 30--C, C, the cross-hair, and H, H, the horizon. In every instance when the experiment was tried, this appearance was found to exist; but it was noticed that different instruments gave different degrees of horizontal depression below the cross-hair. The author saw at once that this peculiarity depended upon the construction of the instruments. He ascertained that in those of the very best construction, and of the most perfect adjustment, there existed a certain degree of refraction, or, as it is called technically, "collimation," or a slight divergence of the rays of light from the axis of the eye, on passing through the several glasses of the theodolite.[198]

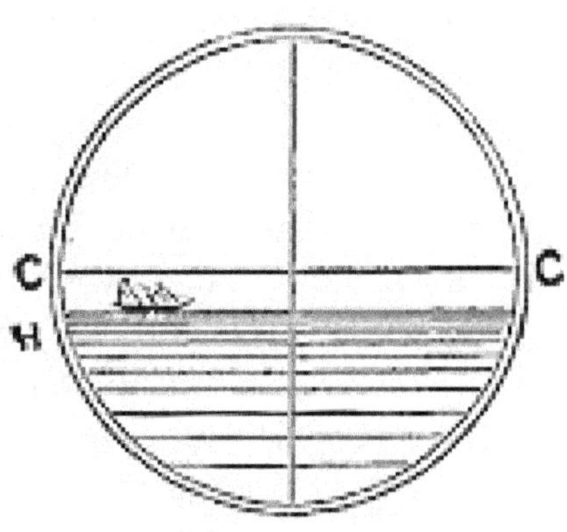

Figure 30

Van Matre explained to Kozlowski that Kozlowski has misinterpreted the theodolite --

image. Van Matre told Kozlowski that the actual horizon is located at the horizontal crosshairs and the apparent horizon has dropped in elevation due to mirroring from atmospheric refraction. The drop in the apparent horizon has nothing to do with the curvature of the earth. It is due to a combination of refraction by the theodolite lenses and atmospheric refraction. This causes the apparent horizon to drop.

Notice that the refraction caused by the lenses in the theodolite as observed by Rowbotham in 1881 is the same phenomenon that we see in the picture of the theodolite view presented by Kozlowski. Rowbotham explains why this occurs:

> A lens is a magnifying glass because it dilates, or spreads out from its centre, the objects seen through it. The infinitesimal or mathematical point actually in the centre is, of course, not visibly influenced, being in the very centre or on the true axis of the eye, but any part in the minutest degree out of that abstract centre is dilated, or diverged, or thrown further away from it than it would be to the naked eye; hence its apparent enlargement or expansion. Whatever, therefore, is magnified, is really so because thrown more or less out of the centre, and the more or less magnifying power of the lens is really the more or less divergence of the pencils of light on passing through the substance of which it is composed. In the telescope of a theodolite, or spirit-level, the spider's web of which the cross hair is made is placed in the actual centre; hence, in an observation, the point absolutely opposite to it is not seen, but only some other point minutely distant from it, but the distance of which is increased by the divergence caused by the lenses; and this divergence is what is called the "magnifying power." This is the source of those

> peculiarities which have been so very illogically considered to be proofs of the earth's rotundity. It is from this peculiarity that several gentlemen prematurely concluded that the water in the Bedford Canal was convex.[199]

Kozlowski feigned ignorance of any light refraction in his ocean horizon photograph. He stated that he could not perceive any light refraction in the photograph and argued, therefore, there must not be any. Kozlowski knows all about theodolite lens and atmospheric refraction. Kozlowski is an expert geodetic surveyor with many decades of surveying experience. He is a New Jersey state licensed surveyor, Deputy Surveyor General of West Jersey, and Surveying Technology Manager at Maser Consulting. He also holds a U.S. patent (#7,748,129) for an apparatus and method for measuring the trigonometric cosine and trigonometric sine of an angle.[200] In his capacity as Surveying Technology Manager at Maser Consulting, Kozlowski is responsible for "[d]irecting surveying standards and specifications related to project control and quality assurance, [and] providing training, technical support and guidance to project teams of field crews and office staff."[201] With all of that expertise, Kozlowski, nonetheless, claimed not to be aware of light refraction affecting his theodolite photo of the ocean horizon. His claim of ignorance simply does not pass the smell test.

Kozlowski should know that atmospheric refraction is undetectable at the boundary line between air and water, because of the uniformity of color and character of the air and water. The mirroring of the atmosphere goes undetected in such circumstances. The only time the atmospheric refraction can be perceived is when there is an object that meets the horizon, as we have seen in the drilling platforms, buildings, the sun, power line towers, and bridge depicted in the previous photographs.

It should be noted that Kozlowski did not contest what Van

Matre said about the effect of the refraction; that it caused a drop in the horizon. Kozlowski only contested whether there was refraction at all. His failure to object to Van Matre's conclusion regarding the direction of the refraction is notable, since geodetic surveyors publically proclaim that the light refraction actually acts in the reverse direction; they claim it causes an elevation in the apparent horizon.

The dissimulation by Kozlowski is apparent when considered in view of the fact that geodetic surveyors claim to always consider light refraction over long distances and deduct that refraction from the supposed curvature of the earth. Geodetic surveyors account for atmospheric refraction in the wrong direction, claiming that it causes the apparent horizon to be raised in elevation. Yet when presented with an explanation of refraction for the drop in the apparent horizon over his long distance ocean sighting, Kozlowski seems to have forgotten his geodetic script; he did not argue that the atmosphere refracted the horizon upward, instead he denied there was any atmospheric refraction at all.

Kozlowski inadvertently confirmed that his feigned ignorance of the light refraction in the theodolite picture was a false pretense when an hour later he suggested to Van Matre that he avoid light refraction by taking a short survey of 1,000 feet. Kozlowski explained that he recommended a short distance because he stated that "there's no doubt about it" that light refraction would otherwise affect his survey results. Kozlowski stated:

> You know why I'm keeping it short Chris is to eliminate the whole thing about refraction. Look if we say, yeah, you know what, **the further you go you are going to have refraction involved. And there's no doubt about it, you will.** So let's try to eliminate that from the test. And say we are going to minimize that to an inconsiderable amount. And

> that's why I'm suggesting about 1,000 feet. Because at 1,000 feet you can see your target pretty well, pick good atmospheric conditions, afternoon is best. ... So let's try to reduce the refraction so it becomes not a consideration.[202] (emphasis added)

Kozlowski's statement that "there's no doubt about it" that the further you look, the more the atmosphere and the theodolite lenses distort what you see, contradicted his earlier phony claim that there was no light refraction in his theodolite image that looked out into the infinite sky over the horizon. His statement indicates that he knows full well that as the distance increases there is light refraction. Indeed, there is "no doubt" that light refraction is a certainty when looking off into an infinite sky over the horizon, as Kozlowski did in his theodolite photograph. The many other photographs above demonstrate that atmospheric refraction has the effect of lowering the apparent horizon. The refraction caused by the theodolite lenses does the same thing.

Kozlowski claims that his zenith angle in one direction is greater than 90°when he lines up his theodolite horizontal crosshair on a distant horizon.[203] A zenith angle is measured from the top vertical plumb downward. Kozlowski documented on his blog that the mean zenith angle was 90° 05' 28" when he dropped the crosshairs down to the horizon in the ocean theodolite view depicted above.[204] The effect of light refraction in causing a drop in the horizon over long distances explains why Kozlowski obtained that result. Kozlowski knows this, but at no time on his blog where he posted the above picture and explains his zenith angle results does he even mention light refraction.[205]

Kozlowski engaged Jeran Campanella in a discussion during a live chat on You Tube. During the chat Kozlowski admitted that refraction affects observations through a theodolite and "can tend to" cause distant objects to drop down lower. Kozlowski, however, did not want to go down that line of discussion. He was determined to discuss the topic of horizontal plane, without reference to the effects of refraction on observations through at theodolite. He is trying to pull the wool over the eyes of the uninformed, because he knows that refraction explains why the objects in the distance drop in view through his theodolite, but Campanella correctly focused on the effects of refraction.

> Kozlowski: "Well there is absolutely refraction is [*sic*] involved in this photograph, no doubt!
>
> Campanella: "And there will always be involved when looking in the distance."
>
> ***
>
> Campanella: "Anything you are looking at in the distance will be down lower than it is in reality, period."
>
> Kozlowski: **"Well, I wouldn't say period, I would say: "can tend to."** It depends on the time of day and so forth. You can watch refraction. If I video recorded this all day long, you would see this moving. It would be moving. **But I am not here to talk about refraction right now.** We can get to refraction. I really want to stick with the concept right now of establishing the horizontal reference. This is the part that gets left off most flat earth discussions about the whole eight inches per mile squared."[206]

Kozlowski tried his best to stay away from any discussion of refraction. But later in the disussion, Campanella forced the topic again on Kozlowski and asked him a direct question about the effects of refraction. Kozlowski at first tried to claim that the atmosphere causes objects to be elevated in appearance. He prefaced his statement by saying "when refraction is worst." Campanella immediately realized what Kozlowski was trying to do by that qualification, and asked him what was the general effect of atmospheric refraction. Kozlowski had to admit that the atmospheric refraction generally causes objects in the distance to drop from view.

> Kozlowski: "You've got 32 miles of atmosphere."
>
> Campanella: "When we know that light bends due to going into a heavier medium is the light going to bend down or up for the most part in general for 32 miles away from you."
>
> Kozlowski: "When refraction is worst, you could see things coming up."
>
> Campanella: "No. I'm not talking about a rare thing happening. I'm not talking about a Fata Morgana. I am talking about, in general, when you are looking at 32 miles of distance and there is going to be a compaction of atmosphere, between you and the distance. ... **In the distance over 32 miles would things generally drop or rise?**"
>
> Kozlowski: **"Drop."**
>
> Campanella: "Okay."
>
> Kozlowski: "But Jeran ... You said 'admit,' **I've already said there is definitely refraction here.** And what's the magnitude of it?"[207]

Kozlowski tried to argue that refraction was only a small part of the downward appearance of objects in the theodolite. Campanella disagreed and maintained that every bit of the downward appearance of distant objects through a level theodolite is due to refraction. Kozlowski realized quickly that he had a losing argument. The first chance Kozlowski got, he jettisoned his mitigation of drop from downward refraction argument and instead switched to an argument that light causes an elevation of distant objects. He did that because Campanella painted Kozlowski in a corner when Campanella objected to a depiction of a computer model of the flat earth to which Kozlowski was referring.

Campanella objected because the computer model made no allowance for refraction. Kozlowski told Campanella that he could apply refraction to the model, but Campanella pointed out that the computer model drafter in other models had inaccurately applied refraction in the wrong direction. At that point, Kozlowski decided that he had to defend the misleading computer model by stating that refraction causes distant objects to rise up. That statement was a direct contradiction of his earlier acknowledgment that refraction causes distant objects to drop.

> Campanella: It's a flat earth model with no refraction, and no nothing at all that happens in real life. You're right it's a great drawing, great.
>
> Kozlowski: Yeah, but you can apply refraction to this.
>
> Campanella: What kind of refraction? They applied reverse refraction. They make things rise up in the distance.
>
> Kozlowski: That's what refraction does. It pulls things up over. It helps the flat earth.
>
> Campanella: It pulls things up, huh?
>
> Kozlowski: Yeah. It does.

> Campanella: This is why you guys are so confused. The last time we talked, you heard Steve McCray say that you can't see boats that are beyond the curve because their photons wouldn't hit your eyes. Yet, you do believe that things that are behind the curve can be lifted up by refraction and then the photons are shown to you. So, it doesn't work when the flat earth wants to employ the same thing. But you guys don't understand what you have. You've been given a blue print of how the deception works, and you guys go around repeating the blue print.[208]

Campanella hit the nail on the head. Geodesy requires that the atmosphere refracts objects upward, not because that is what happens, but because otherwise they would not be able to explain the visibility of distant objects that should be hidden by the curve of the earth. The upward refraction of distant objects has become the dogma of geodesy. Kozlowski, inadvertently let the cat out of the bag when previously, in a rare fit of honesty, he stated that light refraction generally causes distant objects to drop in elevation.

Kozlowski initially tried to explain that the drop from refraction was only a small part of the drop seen in distant objects, but he had no substance to support his argument. Indeed, if any part of the drop is due to refraction, it would impeach all of the surveying studies that he presented to prove the earth is a sphere; because he never did any correction for refraction. Kozlowski twice surveyed a mile span of New Jersey Route 206, in order to try to prove that the earth is a sphere. He also surveyed Cooper River Lake, which he also alleged proved the earth's curvature. It is notable that in reporting his surveys he never mentioned refraction as playing any role in his results.

Kozlowski's admission to Campenella that the atmosphere

refracts light in such a way to make distant objects drop in elevation impeaches all of his survey results and his credibility. Because to admit that light refraction plays a role and then not to mention light refraction when reporting his results, suggests purposeful dissimulation. This is especially clear when viewed in light of all of the steps he took to document his supposed careful precision. Kozlowski must have realized that and so when he was pressed on the matter by Campanella he returned to the deceptive geodetic script that falsely portrays light refraction causing distant objects to rise in elevation over the supposed curve of the earth.

Kozlowski initially admitted in his conversation with Campanella that atmospheric refraction generally causes distant objects to drop in elevation. And Kozlowski's statement to Van Matre reveals that he knows that a survey of greater than 1,000 feet creates an environment conducive to light refraction. How could he not know? He is an expert surveyor. With that in mind, Kozlowski set about to prove that the earth is a sphere by surveying a distance of 5,280 feet (one statute mile) on NJ Route 206. Being an expert surveyor who is on record advising another surveyor (Chris Van Matre) to keep his survey length to no more than 1,000 feet to avoid the effects of light refraction, brings into bold relief his deception in trying to prove a spherical earth by surveying a 5,280-foot (one statute mile) road span.[209]

Kozlowski did not follow his own advice. He told Van Matre to keep his observations to 1,000 feet to avoid the effects of light refraction. But Kozlowski increased the length of his observation fivefold, to 5,280 feet. Why would he do that? The answer is obvious. Kozlowski purposely surveyed a long distance in order to ensure that there would be plenty of atmosphere between his theodolite and his target to refract the target lower. That way he could deceive people into believing that the drop in his survey target was due to the curvature of the earth.

Where did Kozlowski get his guidance to keep the survey

length to within 1,000 feet? Kozlowski certainly has read, because he has recommended, the publication, *Propagation of Refraction Errors In Trigonometric Height Traversing and Geodetic Levelling* by G. A. Kharaghani from the Department of Surveying Engineering at the University of New Brunswick. That research study explains how light refraction error is accounted for in geodesy.[210] That research report states that "[r]efraction error is less easily controlled and is more complex because, in addition to height difference, it is a function of temperature gradients and the square of the sight length (Vanicek et al., 1980)."[211] The report further states that "for sighting distances greater than 100 m [328 feet], a very accurate temperature gradient along the line of sight is required to compute the refraction correction."[212]

The problem is that it is very difficult to obtain an accurate temperature gradient. Because of that difficulty, the reality is that light refraction cannot be accurately computed. Consequently, the final recommendation of the report is to keep surveys of vertical elevation to no longer than 150 meters (492 feet) for the leap frog method of surveying. In leap frog surveying the instrument is set up midway between two target-reflector stations. When conducting reciprocal surveying, the recommendation is to keep the sight length to 250 meters (820 feet).

> Computation of the refraction effect based on the meteorological approach is neither accurate enough nor practical for long lines of sight in trigonometric height traversing, since the profile of the line must be known in addition to the temperature gradient (measured or modelled). A more appropriate solution is to limit the sight length to 150 m for the leap-frog and 250 m for the reciprocal methods and keep the clearance at more than 1 m; at the same time one should avoid unfavourable conditions as examined in Chapter 5.[213]

That guidance from Kharaghani to keep the survey length no longer than 850 feet is from a publication that Kozlowski has personally recommended as an authoritative source on how to address light refraction when surveying. Yet, Kozlowski expanded the length of his survey of vertical elevation at NJ Route 206 to more than 6 times the length advised in the very publication he cites as giving authoritative guidance on how to address light refraction when surveying. Kozlowski knew all about the recommendation in Kharaghani's publication on light refraction, and that is why he recommended that Van Matre keep his sightings to within 1,000 feet. Kozlowski knew what he was doing in his NJ Route 206 survey. He purposely stretched the length of his survey to a full mile in order to obtain the greatest light refraction possible.

Below is a picture Kozlowski posted on the internet that shows the view in 2016 through his theodolite located at mile marker 27 at his target located one mile in the distance at mile marker 28 on New Jersey Route 206.[214] Notice the distortion of the target and all other images in the photograph caused by atmospheric refraction. The blob to the viewer's right of the tripod is actually an orange road safety cone. The rectangular street signs adjacent to the tripod target, which are on a straight post, appear misshapen and the post appears crooked. The straight road markings are misshapen from the atmospheric refraction.

The atmospheric refraction has distorted the distant objects. That atmospheric refraction has also caused a drop in the apparent height of the target. The atomospheric refraction is obvious, and it confirms what Chris Van Matre, who has approximately 30 years of experience as a surveyor, stated. Van Matre averred that readings in theodolites at long distances are affected a great deal by atomospheric distortion from the thermal heat waves.[215] Notably, Kozlowski never mentioned light refraction as playing any role in his results.

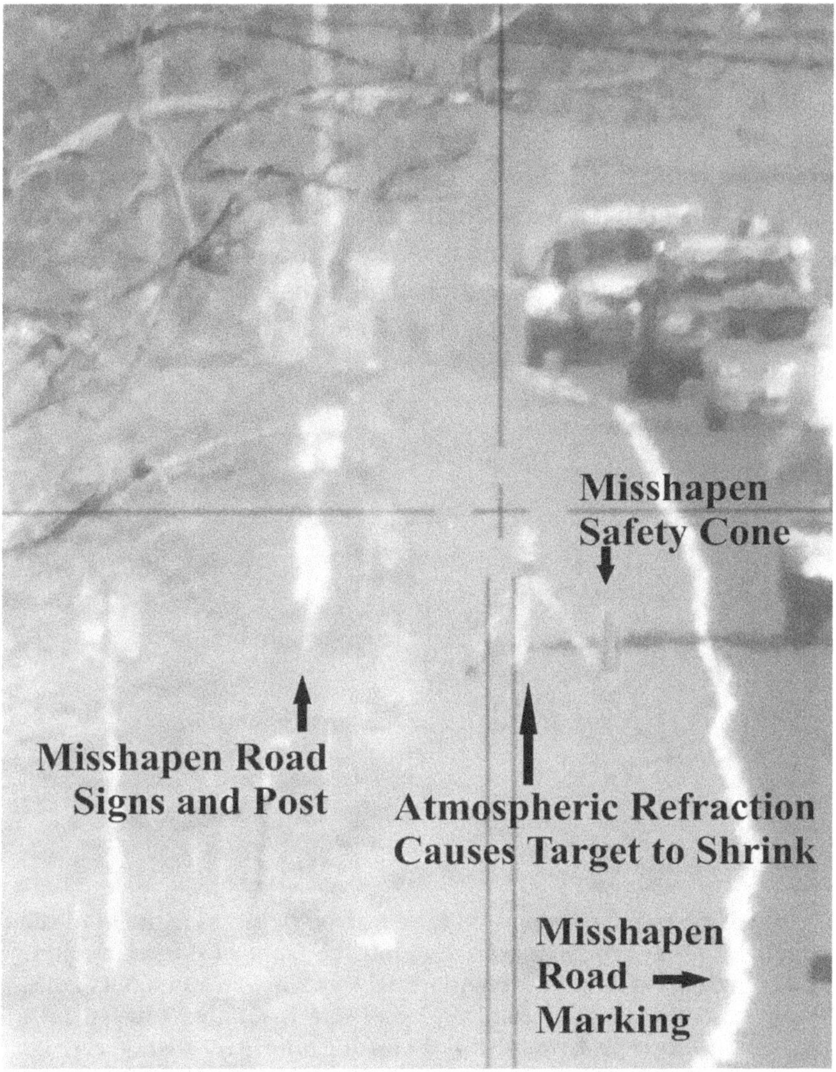

This is Kozlowski's picture of the view through his theodolite of the target located at mile marker 28, one mile in the distance. There is distinct blurring due to atmospheric refraction. The atmospheric refraction is most apparent in the misshapen road signs, post, road marking, and road safety cone. The refraction has caused the apparent height of the target to drop in elevation. This is known as shrinking. The horizontal line on the theodolite lens crosses above the target, not because the earth is a sphere, but rather because the target is being refracted downward.

Below is a screen frame from a video taken by Kozlowski through his windshield of his tripod target at mile marker 28.[216] He is behind the target and facing toward mile marker 27 where he has set up his theodolite. Notice that the traffic safety cone is perfectly conical and the traffic signs are rectangular and are on a straight post. That is not what we see in the photo taken from his theodolite that is one mile away at mile marker 27. Looking through a mile of atmosphere caused refraction which distorted the shapes and dimensions of the distant objects and caused the target to shrink in height.

Frame of a video by Kozlowski taken through his windshield of the tripod target at mile mark 28 showing the front of the rectangular road signs on a straight post, and perfectly conical traffic safety cone. Notice also the straight road marking. He is looking toward mile marker 27 where his theodolite is set up one mile in the distance.

Theodolites should be calibrated annually or biannually depending on the manufacturer's recommendation or industry standard. Testing and calibration of a theodolite for precision on the horizontal plane is relatively simple because there are angle reference standards that can be used. But reference standards for testing and calibration of theodolite for vertical angle accuracy cannot practically be used and so calibration of a theodolite for vertical angles is consequently more complicated.[217] The particular theodolite that Kozlowski used for his surveying is made by Wild Heerbrugg (Model T-2). A controlled laboratory test was performed by P.T. Kuduntwane & O.A. Kruger, from the National Metrology Institute of South Africa (NMISA), of two Wild Heerbrugg theodolites. The tests by Kuduntwane and Kruger showed that after a series of tests the average values for vertical angles measured by the Wild Heerbrugg theodolites were 3.5° in error from the nominal angles.[218] That degree of error being discovered in controlled tests of theodolites, should give one an idea of how temperamental and potentially inaccurate theodolites can be for measuring vertical angles if they are not properly calibrated. It should give one pause before accepting on blind faith Kozlowski's results as being as precise as he alleges, when his measurements assume a precise and calibrated theodolite.

It all comes down to trusting that the theodolite is accurate and the surveyor properly operated it without bias. Kozlowski is not operating the theodolite to obtain objective measurements without regard to the implications of those measurements. He is, instead, taking sightings with his theodolite to prove a point. He has a bias in favor of a particular outcome. He is reporting his theodolite measurements in order to prove that the earth is a sphere. He has an interest in the results showing that the earth is spherical. Indeed, that was his purpose in conducting the survey. That bias must be considered when deciding his credibility and the reliability of his reported measurements.

Surveyor Chris Van Matre reveals how an error in a theodolite of one hundredth of a foot (less than 1/8th of an inch) from level, will be exponentially exaggerated as the distance increases. The error would be massive at a distance of one mile.

> If you're just slightly off level, let's say that it's just a hundredth of a foot [less than 1/8th of an inch] off level where you're standing, so the further

> you project that out, the bigger the gap is going to be. So, I don't think, in my own personal opinion, that a theodolite is accurate enough to do the measurements that they're doing, the geodesists. And then I believe that geodesists kind of fall in the same category with astrophysicists. They're handed their mathematics and they go out and they make observations, and their math concludes that there's actual earth curvature.[219]

Are Kozlowski's 2016 NJ Route 206 measurements off? Yes. Kozlowski inadvertently documented his inaccurate measurements. The photograph below of the target taken by Kozlowski reveals the height of his target at 4.78 feet. Notice that the ruler is tilted. The ruler is not being held vertically plumb to the ground. A measurement at an angle that deviates from a 90° vertical angle will necessarily show a longer measurement than the true measurement. That means that the actual height of the target was set slightly lower than the 4.78-foot height reported by Kozlowski. Van Matre's statement regarded a theodolite being off level by one hundredth of a foot, not being off in height by that much. Because Kozlowski's inaccuracy involves target height and not theodolite horizontal level inaccuracy, it is not enough to make a material difference. But his sloppiness does raise an issue regarding his portrayed precision of his measurements.

Kozlowski's mismeasurement of the target height reveals that Kozlowski's alleged precision is nothing of the sort. He stated that his target was set "at a measured height of 4.78 feet." But it is more accurate to say that the target was set at a close approximation of 4.78 feet. Kozlowski tries to represent his results as based upon strict exactness, but his method of target measurement demonstrates that he can be careless. His portrayal of meticulous measurement is theater to make people believe his assertion that the earth is a sphere. Kozlowski's surveying on New Jersey Route 206 was a sham, which is demonstrated by the fact that Kozlowski's exactitude in his reported results in the drop of the target is impossible because he cannot see the center of the target a mile in the distance.

When looking at the target through Kozlowski's theodolite, one cannot see any centering markings on the target. The head of the target and the prism in its center are completely obscured by atmospheric refraction. Yet, Kozlowski is reporting his measurements to the nearest arcsecond. He is giving the misleading impression of precise measurement, when in fact the supposed precision he is reporting cannot be accomplished with the blurred target. The target is so misshapen from atmospheric refraction that the only thing that can be seen in the distance is a yellow blob at the top of the tripod.

The Oregon Department of Transportation explains that "differential leveling is the process used to determine a difference in elevation between two points."[220] Kozlowski's measurements of the vertical angles on New Jersey Route 206 are similar in concept to differential leveling. Kozlowski is a New Jersey state licensed surveyor and Deputy Surveyor General of West Jersey. So we know that he is familiar with the New Jersey Surveying Manual. Section 3.7.6.2.1 B 3 of that manual requires that when conducting differential leveling "[s]ight distances should never exceed 75 meters (250 feet)."[221]

There are a number of reasons for limiting the sight distance to 250 feet. One reason is that most surveyors performing differential leveling have difficulty reading the rod accurately at distances longer than 250 feet. Indeed, while the State of New Jersey states that the maximum distance should be 250 feet, some surveyors try to limit their sighting distances in one direction to 150 feet. For example, surveyor Brent MacDonald explains "that we can really only have a maximum distance of 300 feet from our back sight to our foresight. Because we can't exceed the 150 feet from the center of this instrument to the rod. Because we just frankly can't read accurately over that distance. So it all depends on what instrumentation you have, but generally speaking the 150-foot rule is a good rule of thumb."[222]

One might argue that in differential leveling the surveyor must be able to see the numbers on the rod, whereas Kozlowski did not need to see the numbers. He only needed to see the center of the target. But that argument misses the point. Not being able to clearly see the readings on the rod when conducting differential leveling is similar to the problem created by Kozlowski with his one mile sight distance. The center of his target is obscured and misshapen due to light refraction. Indeed, there are supposed to be black marks surrounding and pointing to the center prism, but those marks cannot be seen at all. It is obvious when looking at the blurred picture above that there is no way that Kozlowski could accurately determine the center of his target. Without being able to see the center of the target, how could Kozlowski measure the supposed drop of the target, down to a single arcsecond, as he claimed he did?

Kozlowski rigged the results. He put one mile of atmosphere between his theodolite and the target. He knew full well that the combination of lens refraction and atmospheric refraction at that distance would cause the target to drop in elevation. That long

Picture taken by Kozlowski of the target at mile marker 28 in his New Jersey Route 206 survey. He portrayed it as being precisely set at "a measured height of 4.78 feet." When viewed through his theodolite from a mile away, the precisely set target markings could not be seen.

distance sighting also caused distortion of his target. But he did not care that the distortion would make a precise sighting impossible, because precision was not his real objective. His true purpose was to give the impression of precision, without actual precision. That is why he reported the measured height of the target at precisely 4.78 feet, which is an accuracy of within less than 1/8th of an inch. And that is also why he reported his sightings down to an arcsecond. There are 60 arcminutes per degree, and sixty arcseconds per minute.

Kozlowski claimed his 24 measurements of the alleged zenith angle drop of the target at mile marker 28 averaged 35.83 arcseconds. His reported standard deviation over the 24 measurements was only 3.36 arcseconds. That level of precision is simply not possible with only the distant, misshapen target blob as the basis for his sightings. His claimed precision is an artifice.

Kozlowski's readings, in reality, were not precise, because, as we can see in the blurred picture above, he could not possibly find the center of the target with the precision he claimed. The target is supposed to have centering black marks pointing toward a center prism. Yet, all he could see was a yellow blob. He claimed to set his target a precisely measured height of 4.78 feet. What is the point of such precision in setting the center of the target when he could not possibly see the center of the target? Kozlowski is presenting supposedly precise results, when it is clear that he could not see the center of the target. He could only center his theodolite where he guessed the center should be located. His reported results are just window dressing to make it appear that his conclusion that the earth was a sphere is authoritative. In fact, his survey is a charade.

The drop that Kozlowski alleges is due to the sphericity of the earth can be explained by the combined effect of light refraction caused by the lenses of the theodolite and atmospheric refraction. But Kozlowski presents the evidence that is suggestive of light refraction as instead evidence of a spherical earth.

The above measurements at mile markers 27 and 28 on NJ Route 206 were performed in 2016. Kozlowski portrays his survey as proof the earth is a sphere. But Kozlowski actually did the same survey a year earlier, in 2015. He used the same equipment and the same methods and protocol. But his 2015 survey impeaches the alleged accuracy of his 2016 survey.

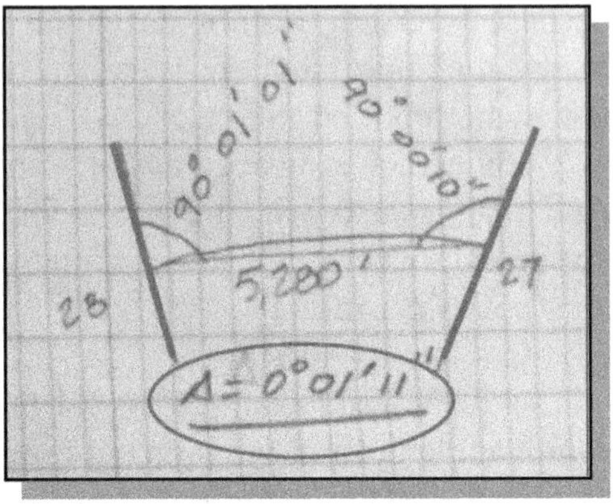

Figure 50: Kozlowski's sketch of his 2015 survey of mile markers 27 to 28 on NJ Route 206. The red annotations are his. The zenith angle at mile marker 28 was more than six (6) times greater than at mile marker 27. This could only happen from lens and atmospheric light refraction.

In Kozlowski's 2015 survey he determined from a total of 12 measurements (6 in each direction) that the sum of his zenith angle deltas totaled 0° 01' 11". The problem with his conclusion is that the zenith angle at mile marker 28 was a whopping 90° 01' 11", whereas the zenith angle at mile marker 27 was only 90° 00' 10". That is a six-fold difference. Kozlowski's hypothesis is that the earth is a uniform gradient ellipsoid. His survey should have returned equivalent readings in each direction, just as he allegedly did in his 2016 survey.

The six-fold disparity in the zenith angles in opposite directions in his 2015 results is evidence of error. Indeed, his results showed that in one direction, toward mile marker 27, that the earth was only slightly curved. In the other direction, toward mile marker 28, the curvature is much greater than the reported curvature of the alleged spherical earth. Atmospheric refraction is dynamic because atmospheric conditions can change within minutes.[223] Assuming that there is no operator error, the only other variables that could account for a six-fold difference in arc measurement, showing the earth curved much less than the orthodox opinion in one direction, while it is much greater than the supposed curvature in the other direction, would be from lens and atmospheric refraction.

Kozlowski tried to cover over the disparity in zenith angles in his 2015 survey by simply adding them together and anouncing that "the sum of the two angles is greater than 180 degrees."[224] And explaining that those results "were within 19 seconds of the estimation."[225] The orthodox view is that the sum of the delta of the zenith angles should equal 52 arcseconds. Kozlowski came up with a combined delta figure of 0° 01' 11" (0° 00' 10"+0° 01' 01"=00° 01' 11").

Kozlowski then puts in a dig, which is the entire reason for his 2015 NJ Route 206 survey: "How do these results fit the flat earth model?"[226] Well, the measurement of the mile marker 27 drop of only 10 arcseconds certainly does not confirm the standard model of a spherical earth. And the 61 arcsecond reading in the other direction far exceeded the orthodox curvature model.

The disparate delta results were not missed on Kozlowski, but he was hoping that people would not pay that close attention to the details of his survey; so he just added together his zenith angle deltas and announced that figure as proof that the earth is a sphere. But that is not proper. Kozlowski is trying to cover up his erroneous results by adding his two errors together. That may be okay to do in geodesy, but it is considered cheating in the real world.

If the earth were a sphere, as alleged by Kozlowski, his

zenith angles should have been equivalent at both mile markers 27 and 28. The massive error in zenith angles in his 2015 survey impeaches his alleged proof. Very simply, if the earth is a sphere, and it is thus curved in one direction, it should also be curved to the same degree in the other direction. Kozlowski's 2015 survey results could only mean that there is an error in the measurement.

Indeed, the disparity in zenith angles in the 2015 survey explains why Kozlowski decided to repeat his survey between mile markers 27 and 28 on NJ Route 206 again in 2016. He portrayed the 2105 results as being "within 19 seconds of the estimation." That sounds pretty convincing, but he obviously did not believe his own claim. If the results were so conclusive, as Kozlowski claimed, there would be no reason to redo the survey in 2016. There was a difference of 61 seconds (71" - 10"=61") between his 2015 reciprocal survey readings. Kozlowski was able to even that out to roughly a two (2) second difference in the zenith angle deltas at mile markers 27 and 28 in his 2016 survey.

He knew that the results of the 2015 survey were invalid. So, he returned to the scene of the crime with the same equipment and used the same protocols, but he ended up with a more uniform result. Was it only a coincidence? The sum of the delta of his zenith angles was 0° 1' 10" in his 2016 survey. The sum of the delta of his zenith angles in his 2015 survey was 0° 1' 11". There was only one (1) second difference in the sum of the delta of his zenith angles between the 2015 and 2016 surveys.

But he was able to accomplish his goal of evening out the zenith angles. In his 2016 survey, he reported the delta at mile marker 27 as 0° 0' 36" and the delta at mile marker 28 as 0° 0' 34". Seemingly, that solves Kozlowsi's 2015 survey problem. But he cannot erase the fact that his 2015 results impeach the legitimacy of his conclusion that the earth is a sphere. That is because if the earth were truly a sphere, the degree of curve would be identical in all directions. The 2015 NJ Route 206 disparate readings in each direction are evidence of error caused by light refraction. Although he tried, he cannot unring that bell. The light refraction did not go away in 2016. His 2016 results were just as much a result of light refraction as his 2015 survey. The only thing he accomplished in his 2016 survey was to even out the sight readings in each direction.

All of Kozlowski's measurements, photographs, and documentation are a smokescreen. He expected that most people would simply accept his conclusion that his survey proves that the earth is a sphere. He did not intend that anybody would actually examine his figures to see if they supported his conclusion. But the devil is in the details. When his figures are scrutinized they do not support his claim that the earth is a sphere.

STANDPOINT	ELEV	FOREPOINT	ELEV	DELTA ELEV	OBS Z ANGLE***	RECIPROCAL SUM	COMPUTED DELTA ARC	MILES	CURVE DROP*	TRIG**
27	59.30	28	59.30	0.00	90° 00' 36"		0° 0' 50"	0.96	-0.62	-0.88
28	59.30	27	59.30	0.00	90° 00' 34"	0° 01' 10"	0° 0' 50"	0.96	-0.62	-0.88

* COMPUTED RESULTS ARE DECIMAL FEET (8 INCHES PER MILES^2) DOWN FROM HORIZONTAL PLANE AT STANDPOINT.
** COMPUTED RESULTS ARE DECIMAL FEET (COS OBS Z ANG * 5047.29 FEET) DOWN FROM HORIZONTAL PLANE AT STANDPOINT.
*** MEAN RESULTS OF 6 SETS OF 4 F1&F2 POINTINGS ROUNDED TO NEAREST SECOND.

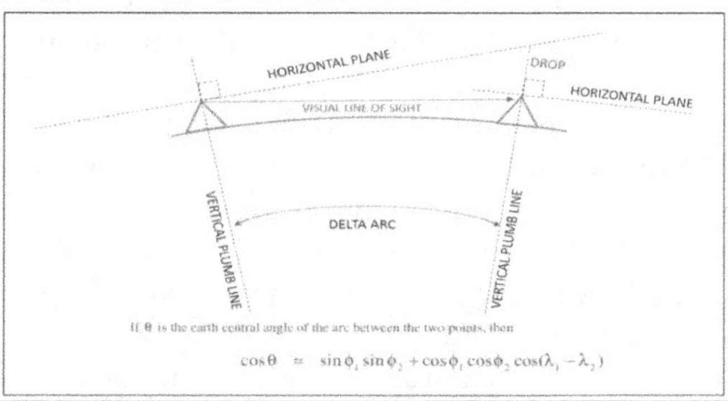

Figure 51: Chart of the 2016 NJ Route 206 survey posted by Kozlowski showing his final calculations and his conclusion that the earth is a sphere. This illustrates the obvious fact that uniformity in curvature is required by the globular earth model. It further demonstrates that the delta in the zenith angles must also necessarily be the equivalent; thus proving that his previous 2015 survey was erroneous, because the measurements of the curvature drop at each end of the 2015 survey were not equivalent.

The New Jersey Route 206 road surveys were not Kozlowski's only attempts to use light refraction to falsely portray a spherical earth. Kozlowski also went about surveying the length of Cooper River Lake in New Jersey, which is 7,311 feet (1.39 miles). He surveyed Cooper River Lake in order to show that the water on the lake has a convex bow, as a way to prove that the earth is a sphere.[227] He published his data on the internet. The data showed that his theodolite averaged in elevation, for twelve readings (six in each direction), a 90.009 degree vertical angle (zenith angle) (90.009° = 90° 00' 33.17"). His argument is that if the earth were flat, then his zenith angle measurements in one direction should be precisely 90° and 180° for two readings in opposite directions (2 × 90° = 180°). But since his zenith angle measurements were greater than 90° he asserts that must mean that the lake is convex and thus the earth is a globe.

Assuming the earth is a sphere, there should be a measurable angle of curvature of one arcminute per nautical mile (1 arcminute = .0166°). That means that there should be .02° (1.2 arcminutes) of curvature over a length of 7,311 feet (1.39 statute miles or 1.21 nautical miles). In order to obtain the angle of curvature, we must multiply Kozlowski's average zenith angle by two. When doing that we come up with a .018° angle of curvature. That is very close to the expected .02° if the earth were a globe. Does that mean the earth is a globe? No, it does not. What Kozlowski was detecting and measuring was a combination of theodolite lens refraction and atmospheric refraction. Cooper River Lake is not Convex. Indeed, no body of water on the earth is convex.

Kozlowski's claim of finding convex water across Cooper River Lake is easy to refute. For example, one researcher was able to use a telephoto lens to film a spotlight from across Utah Lake, Utah, that was lowered to water level on the opposite bank, 7.53 miles away.[228] His Nikon P-900 camera was mounted on a tripod, 30.5 inches off the surface of the lake on the opposite shore. The filming was performed at night, with the temperature ranging between 19°F and 25°F. That ensured that there would be little light refraction from the heat of the sun and evaporation from the surface of the water. If the earth were a globe, the spotlight held at water level on the opposite shoreline should have been completely out of sight, almost 21 feet below the horizon. But the fact that the researcher was able to see the light and film it using a telephoto lens at the other end of the lake refutes Kozlowski's claim that water is convex. Indeed, being able to video the light on the opposite shore of the lake is evidence that the water on the lake is level and the earth is flat.

It is notable that at no time in Kozlowski's blog, where he posted his data and conclusions of his Cooper River Lake survey, did he ever mention theodolite lens or atmospheric refraction. That is not an inadvertent oversight, it is purposeful. Kozlowski cunningly set up a test spanning 7,311 feet (1.39 statute miles), knowing full well that the theodolite lenses and atmosphere would refract the light over the lake. Without mentioning the theodolite lens and atmospheric refraction, he then falsely portrayed his results as proving that the water is curved. Kozlowski admitted to Jeran Campanella that he knew that light refraction definitely causes distant objects to appear to drop in elevation. Kozlowski advised Chris Van Matre to keep his surveying sighting length to 1,000 feet to avoid light refraction. Yet, Kozlowski took a survey sighting that was seven-times longer. He did that to ensure that the light refraction would be manifested. But Kozlowski did not breathe a word about light refraction when explaining his Cooper River Lake surveying results, so he could portray the false impression that the earth is a globe.

The drop in elevation of the target as Kozlowski sighted through his theodolite was due to a combination of theodolite lens refraction and atmospheric refraction. That light refraction was expected by Kozlowski, and that is why he selected the 7,311-foot (1.39 mile) long Cooper River Lake. It is the same light refraction that Chris Van Matre revealed was depicted in Kozlowski's theodolite view over the ocean. Kozlowski is using the general public's ignorance about light refraction to pull off a type of parlor trick to convince them that the earth is a globe. Indeed, once one realizes the trick Kozlowski is pulling, it becomes clear that what he is actually proving is that light refraction causes the apparent height of distant objects to drop in elevation. By doing that, he is impeaching the legitimacy of geodesy, which claims the opposite effect.

Some have tried to use atmospheric refraction to explain how landmarks that should be below the supposed curve of the alleged spherical earth can be seen. But as we have learned, refraction cannot explain that phenomenon. Indeed, atmospheric refraction actually makes objects less visible over long distances, not more visible. The only explanation for how landmarks can be seen beyond the supposed curve of the earth is that there is no curve to the earth; the earth is flat.

Since atmospheric refraction actually has the effect of obscuring distant landmarks, some charlatans try to use that obscuring effect as a way to deceive people into believing that the earth is a sphere. They will point to the missing portion of landmarks as evidence that the landmark is below the curvature of the earth and thus the earth is a sphere. When in fact the missing portion of the distant landmark is being obscured by atmospheric refraction.

One such effort to deceive is geodetic surveyor Jesse Kozlowski's use of a theodolite aimed across Lake Pontchartrain in Louisiana. Kozlowski tried to prove that the earth is a sphere, in a video he made of his Lake Pontchartrain survey.[229] He asserted that because the Marriott hotel at the other end of Lake Pontchartrain was partially hidden from his view as he looked across Lake Pontchartrain through his theodolite, the earth must be a sphere. As we will see, this was just a parlor trick by Kozlowski.

Refraction of light is caused by the gradient of warm moist air in the boundary layer of the atmosphere.[230] A consequence of the warm moist air is that it causes a lensing effect of the atmosphere. This effect magnifies distant objects and causes the bottoms of those distant objects to be cut off from view.[231] Many misinterpret the missing bottoms of buildings and ships to be caused by the earth's curvature, not realizing it is due to the lensing effect of the atmosphere. The proof that the missing bottoms are due to the atmospheric lensing effect is in the fact that often the object observed should not be able to be seen at all if the earth were a globe, as it should be completely below the supposedly curved horizon.

The video by Kozlowski of the Marriott hotel from across Lake Pontchartrain illustrates the point that atmospheric refraction causes the bottoms of buildings to be cut off from view. The more humidity in the air, the more the air refracts the light. That is why Kozlowski traveled to Lake Pontchartrain to take his theodolite shot to try to prove the earth is a sphere. Lake Pontchartrain abuts New Orleans. Liz Osborn of the weather and science website Current Results explains that "New Orleans has the highest relative humidity among big US cities."[232]

Kozlowski took his sighting across Lake Pontchartrain on December 10, 2017. He documents that he was at Lake Pontchartrain from early morning and throughout the entire day. The purpose for being there in the first place was to take a sighting across Lake Pontchartrain to prove the earth is a ball. One would think that he would want to take the citing while there was plenty of daylight. But he dilly-dallied by taking irrelevant sightings. It seemed that he was purposely wasting time. He waited until after the sun had gone down, during twilight, to finally take his theodolite sighting across Lake Pontchartrain. Indeed, in the photograph you can see that the lights in the buildings are illuminated, indicating it was twilight when the picture was taken.

Only Kozlowski knows if he was being dilatory on purpose, to ensure that he took his sightings at twilight. But the effect of his delay was to take his sightings over Lake Pontchartrain when the relative humidity was elevated.

Kozlowski knows full well the effects of light refraction on theodolite sightings because he specifically told Chris Van Matre, a plane surveyor, that the best time to take a theodolite sighting to avoid light refraction is in the afternoon.[233] Van Matre is a pilot and a surveyor with 30 years of experience who is convinced, based upon his experience and empirical studies, that the earth is flat. Kozlowski explained to Van Matre the long distance surveying that was done over Lake Pontchartrain prior to building the Lake Pontchartrain Bridge. Van Matre interjected that once you get beyond one to one and a half miles, even in the dry climate of Colorado [let alone, the high humidity over Lake Pontchartrain], the atmospheric distortion from heat and humidity makes it difficult to even see the target through an ordinary theodolite.[234]

Kozlowski proceeded to explain to Van Matre that the historical surveying of Lake Pontchartrain, which spanned from 6 to 20 miles, required the building of high observation towers, more than a hundred feet tall, and the use of massive telescopic theodolites [that collimate the light to mitigate refraction] as they survey over long distances at bright lights at the tops of other tall towers.[235] Kozlowski explained that they did that in order to ensure "they were way above the heat waves."[236] By making that statement, Kozlowski inadvertently revealed his awareness of the reason that the long distance surveyors used high towers was to rise above the ground layer heat gradient in order to reduce the effects of that heat gradient in refracting the light. That means that Kozlowski knows that the heat over Lake Pontchartrain causes light refraction that affects the height of the target when viewed through an ordinary theodolite near ground level. Yet, he misrepresented the picture he took through his theodolite of the Marriot Hotel as being an unrefracted image that shows the supposed curvature of the earth.

The relative humidity in the air increases in the evening. The steady increase in relative humidity as the evening hours advance is a common weather phenomenon. As evidenced by Kozlowski's advice to take theodolite sightings in the afternoon in order to reduce the effects of atmospheric refraction, Kozlowski is fully aware of the increasing humidity as the evening advances. That seems to explain why Kozlowski dilly-dallied taking irrelevant sightings until dusk. According to the Time and Date weather statistics for New Orleans, on December 10, 2017, the relative humidity in New Orleans increased from 51% at 1:53 p.m. to 67% at 4:53 p.m.[237] The pictures indicate that they were taken in early twilight, which began at sunset at 5:01 p.m. on December 10, 2017, in New Orleans.[238]

On December 10, 2017, the temperature at 1:53 p.m. was 57°F, and the barometric pressure was 30.32"Hg. That means that, with a relative humidity of 51% at 1:53 p.m., the absolute humidity in the air was 6.1 grams per cubic meter (6.1g/m3). By 4:53 p.m. the temperature had only dropped one degree to 56°F, and the barometric pressure remained at 30.32"Hg. That means that with a relative humidity of 67%, the absolute humidity at 4:53 p.m. had risen to 7.2 grams per cubic meter (7.2g/m3).

Kozlowski took his readings and pictures at the time that there was the most humidity in the air. Added to the generally high humidity in the New Orleans area is the fact that the water vapor close to surface of water bodies creates an even greater atmospheric lensing effect.[239] Research has shown that the water vapor in the air within 20 feet of the surface of water is significantly greater than the amount of water vapor above that level.[240] Kozlowski had his theodolite approximately 12.13 feet off the surface of the lake and so his view would be greatly affected by the water vapor in the atmosphere close to the water.

The high humidity over Lake Pontchartrain would be ideal to create a lensing effect that would cause the horizon to drop, the bottom of the Marriott hotel to disappear, and the elevation of the hotel to drop. Indeed, we can see the squashed down appearance of both buildings in his theodolite photograph.

The proof that refraction explains the horizon drop, the disappearance of the lower parts of the buildings, and their elevation to drop is the fact that the building next to the Marriott, Two Lakeway Center, is still in view. If the earth were a globe, Two Lakeway Center should have dropped below the horizon due to the supposed curvature of the earth. Kozlowski did not realize that his supposed proof of a spherical earth actually proves that the earth must be flat. That is the only way that Two Lakeway Center could be in view.

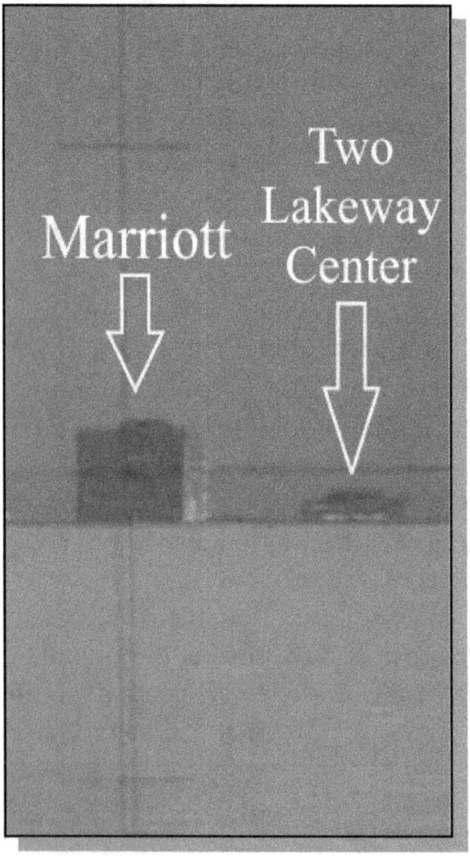

Photograph of image in Kozlowski's theodolite pointed across Lake Pontchartrain. If the earth were a globe, Two Lakeway Center would be completely out of view, below the horizon.

The shorter building just to the west (the viewer's right) of the Marriott is Two Lakeway Center. As Kozlowski documented, he was 127,375 feet (24.12 miles) across Lake Pontchartrain from the Marriott Hotel. Two

Lakeway Center is catacorner next door and roughly the same distance away from Kozlowski as was the Marriott. When we deduct the 4 miles from the 24-mile distance for Kozlowski's 12.13 feet in height to account for the supposed spherical earth horizon being 4 miles away, we come to a distance of 20 miles past the supposed spherical earth horizon for Two Lakeway center. If the earth were a globe, there should be a 266-foot drop below the supposed spherical earth horizon 20 miles in distance past that horizon. Two Lakeway Center is a 19-story, 259-foot tall building. That would put the Two Lakeway Center completely out of view, 7 feet below the horizon, if the earth were a globe. Yet, we can clearly see Two Lakeway Center. We could only see Lakeway Center if the earth is flat.

Geodetic surveyors, like Kozlowski, aver that they deduct for atmospheric refraction from the 8 inches per mile squared by arbitrarily subtracting one-seventh of the supposed drop (thus ending up with a 6.86 per mile squared drop figure).[241] For example, the U.S. Army Topographic Surveying Technical Manual states that "[t]he effect of curvature on the apparent elevation of a distant point is equal to the square of the distance to the point divided by the earth's diameter [that is equivalent to 8 inches per mile squared] ... To allow for refraction, this result must be diminished by about one-seventh."[242] I guess in the unreal geodetic surveying world, light is always refracted in the atmosphere at a uniform one seventh of the curvature drop. One would think that surveyors would be more precise than to arbitrarily use the same refraction deduction for all conditions. But that is what you do, when you are just making things up.

As we have seen, atmospheric refraction works in the opposite direction as is supposed by geodetic surveyors, and furthermore, the degree of refraction is not uniform. So the deduction for atmospheric refraction by geodetic surveyors is meaningless. Their entire calculation for the supposed curvature (deducting for refraction) is pointless, because the earth is in fact flat. Their claim to account for the curvature and refraction of the earth is a fiction.

In any event, after adjusting (in the wrong direction) for light refraction, Kozlowski calculated the drop using the figure of .57 feet (6.86 inches) per mile squared. He came up with a figure for the expected drop in his theodolite image of 332 feet. He actually came up with a figure that would put the 259-foot Two Lakeway Center even further out of sight below the horizon. He made a mistake by using the entire 24.12 mile distance between his theodolite and the Marriott hotel. He did not deduct for the height of his theodolite, which accounts for his error.

Kozlowski's supposed proof of a spherical earth actually proves that the earth could not be a sphere. Kozlowski is an expert who claims that what he saw proved that the earth was a globe. Yet, his observations prove just the opposite.

This author asked Kozlowski in an email to explain in detail how the earth could be a globe when he was able to see Two Lakeway Center above the horizon. He did not answer that question. Of course, how could he? The only truthful answer would be because the earth is flat. And he was not going to admit that.

5 The “Plane” Truth About Geodetic Surveying

In plane surveying the surveyor makes no accommodation for the alleged curvature of the earth. All theodolite measurements assume a flat earth. The United States Defense Mapping Agency explains that surveys of small areas do not require any accommodation for the supposed curvature of the earth. What does the U.S. Defense Mapping Agency consider a “small area?” It gives a city as one example. It is acceptable for a surveyor to consider the earth as being flat when surveying short distances, such as the length and breadth of a city. It seems then that for most purposes geodetic surveying is irrelevant. And the U.S. Defense Mapping Agency pretty much says so:

> [T]he sphere is a close approximation of the true figure of the earth. ... **The idea of flat earth, however, is still acceptable for surveys of small areas.** Plane-table surveys are made for relatively small areas and no account is taken of the curvature of the earth. **A survey of a city would likely be computed as though the earth were a plane surface the size of the city.** For such small areas, exact positions can be determined relative to each

> other without considering the size and shape of the total earth.[243]

The U.S. Defense Mapping Agency, of course, portrays a requirement for accommodating the supposed curvature of the earth when measuring long distances, which it describes as areas spanning continents and oceans. This accommodation is called geodetic surveying. It is notable that the U.S. Defense Mapping Agency uses imprecise language to indicate when geodetic surveying is required. The agency claims that it is not required for a "city" but it is required for distances spanning continents and oceans. There is no standard sized city. The largest city by land area in the United States is Sitka, Alaska (2,870.3 square miles). The 150th ranked city by land area in the United states is Presque Isle, Maine (75.8 square miles). There is a significant range in area between those two cities. And there is a wide gulf in area between even a large city and continents and oceans. The North American Continent covers an area of approximately nine (9) million square miles. When government agencies just make things up, they know it is best to be imprecise.

That imprecision runs throughout the discipline of geodesy. For all of its lip-service about precision, it is not at all precise when it comes to determining when to introduce the alleged geodetic calculations. The cut-off point between plane surveying and geodetic surveying is usually a vague distance like "long." When figures are given in geodetic surveying publications, they vary widely. The point at which geospatial dimensions must be factored in to surveying calculations can range from approximately 100 square miles to 1,000 square miles. Professor John Clayton Tracy of the Structural Engineering Sheffield Scientific School at Yale University, in his text book on surveying, explained that plane surveys are accurate up to an area of 100 square miles.

> In *plane* surveying the curvature of the earth's surface is not taken into account, in *geodetic*

> surveying it is; hence the necessity for the two terms *plane* and *geodetic*. For the ordinary surveys extending over small areas, the methods of plane surveying involve no appreciable errors. The limit is roughly placed at **about 100 square miles**, although plane surveys of much greater extent are often made when great accuracy is not required.[244] (bold emphasis added, italics in original)

J. Clendinning, the former British Surveyor General of the Gold Coast (Ghana), in his book, *The Principles of Surveying,* on the other hand, stated that geodesy does not come into play until a survey covers 1,000 square miles or more.

> When surveys cover only a limited area, they can be regarded as having been made on a plane surface, so that ordinary theorems in plane geometry and plane trigonometry are applicable, but when very extensive areas, covering, say, **1,000 square miles or more**, are involved, it is necessary to take the spheroidal shape of the earth into account. Nearly all surveys made for purely engineering purposes cover only a limited area and can therefore be treated as plane surveys.[245] (emphasis added)

According to Professor Tracy, although the earth is supposed to be a globe, a surveyor can ignore the gradual and uniform 16 foot bulge rising in the middle of the 100 square mile survey area. All of the border elevations would be off by 16 feet compared to the center elevation. Surveyor General Clendinning is more forgiving. Under his guidance, surveyors can ignore the supposed gradual and uniform 166 foot bulge rising in the center of the 999 square mile survey area that would necessarily have to be there if the earth were a globe. All of the border elevations would be off by 166 feet compared to the center elevation. If the earth were a globe, surveys that ignore that supposed fact would necessarily be inaccurate. The bulges in the supposed curved earth could not be ignored if they were there, particularly when building water projects, such as water reservoirs. The slope of the ground must be represented accurately, because water cannot flow uphill. Surveying requires precision, and surveying is in fact precise. We, therefore, know that no earth curve bulges are there to be ignored. If there is no earth curvature that can only mean that the earth is flat. And surveyors consider it as such when conducting plane surveying. Geodetic surveying is an artifice to bamboozle the public.

Incidentally, the U.S. Defense Mapping Agency not only measures the nonexistent curve of the earth; it also has an instrument to measure non-existent gravity. "Modern relative gravity measurements are made with small, very portable, and easily used instruments known as gravimeters (gravity meters)."[246]

What is this sophisticated gravimeter? The U.S. Defense Mapping Agency reveals that "[t]he heart of all modern gravimeters consists of a weight suspended by a very sensitive spring. Changes in length of the spring due to changes in the acceleration of gravity as the gravimeter is moved from place to place are translated by the mechanisms of the gravimeter into 'dial reading' differences which are proportional to gravity differences."[247]

According to the description by the U.S. Defense Mapping Agency, a gravimeter is essentially a spring scale. It simply measures the weight of a known object. To suggest that things have a different weight, depending on the location on the earth would throw off all measurement and indeed render international commerce almost impossible. Can you imagine a gold transaction, where gold would have a different weight depending on where on the earth it was weighed? There would be long negotiations with each party arguing for a different location for weighing the gold.

That does not happen. By the way, there is an infinitesimal difference in weight based, not on differences in gravity, but atmospheric pressure. But the difference is so slight, it is almost unmeasurable and completely irrelevant in commerce. The fact that no adjustment is made for any difference in weight at different locations on earth is strong evidence that weight is generally uniform all over the earth and the "gravimeter" (as is gravity that it purports to measure) is yet another "scientific" myth.

There is no such thing as gravity; gravity is not necessary on a flat earth. It is density that keeps objects from floating off the surface of the earth. People and objects are heavier than the air and therefore do not float off the ground. Gravity does not exist.

The contrivance of geodetic surveying was required in order to overcome the logical implication that the earth is flat, which is raised by the practice in plane surveying of assuming a flat earth. Geodetic surveying (as distinct from plane surveying) is supposed to be a special type of surveying over large areas, which is alleged to make adjustments for the curvature of the earth and for atmospheric light refraction. But when reading closely what is actually done during geodetic surveying, it seems that the adjustments made for the curvature of the earth (and refraction of light) are actually ignored in the results of a geodetic survey. Curvature and light refraction corrections for geodetic surveying are alleged to exist, but in practice they are not actually made. Instead, the recorded results are plane survey results.[248]

Surveying equipment measures distances along a horizontal line of sight. According to the geodetic survey theory, an error is manifested in measurements due to the spherical shape of the earth. But instructions for geodetic surveying point out that the error due to the curvature of the earth is nominal when surveying short distances.[249] An accurate solution, therefore, can still be found by excluding any accommodation for the curvature of the earth. But geodetic surveying requires an adjustment for the supposed curvature of the earth at very long distances. Geodetic surveyors also allege that adjustments must also be made for light refraction.

In a geodetic surveying instructional video from California State University, the instructor explains that if two equidistant measurements are taken, then no adjustment needs to be taken for the curvature of the earth or refraction of light.[250] Former Surveyor General J. Clendinning confirms this methodology in his book *The Principles of Surveying*. Clendinning stated that "[i]n all cases in trigonometrical levelling it is advisable to observe vertical angles in both directions, simultaneously if possible, as this tends to eliminate the effects of curvature and vertical refraction."[251]

That makes no sense if the earth were in fact a globe. The instructor in the California State University video explains that if there is a measurement taken from the midpoint between a known benchmark at a backsight and then to an equidistant point to a temporary foresight benchmark, then no adjustment calculation is necessary for the curvature of the earth or light refraction. If an arc in fact exists, then it does not suddenly disappear because you are measuring from its midpoint. The only reason to disregard the existence of an arc when surveying is if the arc does not in fact exist.

Geodetic surveyors deduct from the supposed curvature of the earth by a factor of one-seventh due to atmospheric refraction, even though geodetic surveyors acknowledge that atmospheric refraction is not a constant. In addition, they always make the adjustment in one direction (to deduct from the alleged curvature rate). They don't care that the adjustment is arbitrary, always in one direction (the wrong direction) and may have no correlation to reality, because that adjustment is designed to deceive people. It is not a real adjustment. The point of the adjustment is to convince people that the supposed curvature is observed to be less than the actual curvature, in order to explain why distant objects that should be out of view below the horizon can be seen above the horizon.

The scam of adjusting survey results for the earth's supposed curvature and the refraction of light can be seen in the equation that is used. The geodetic theory is that refraction (r) should be subtracted from the supposed curvature rate of the earth (C) at an arbitrary rate of one-seventh of the earth's supposed curvature rate. The equation should be C−r. Yet when we look at the actual equation, it is given as c+r. The instructor in the geodetic survey video explains that C−r is changed into a variable. The equation (C−r) is represented in geodetic surveying by the variable (c+r).

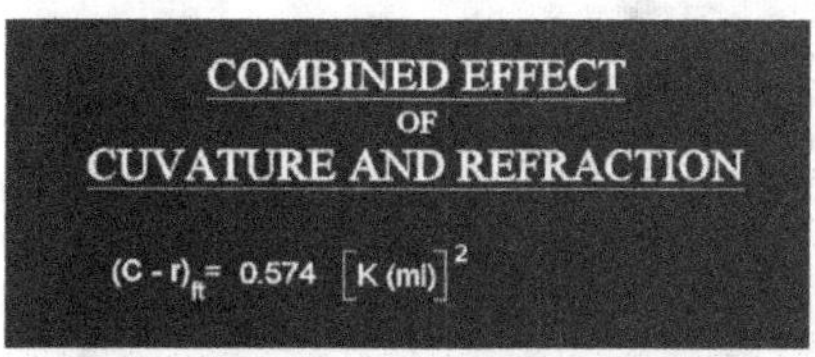

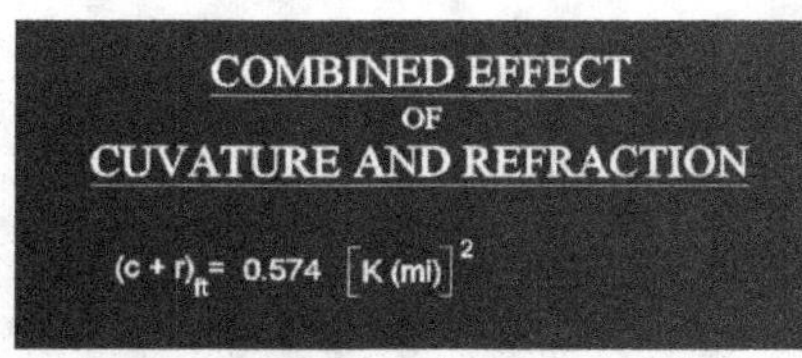

Above are screen shots from the California State University geodetic surveying instructional video showing the transformation of the equation for curvature and refraction (C−r) into the variable (c+r). In the equation and the variable both the capital "C" and small "c" stand for curvature and the letter "r" stands for refraction.

It makes no sense to change an equation (C−r), which is easily understandable, into a variable (c+r) that means the very opposite of what it is supposed to represent. The only difference is that the capital "C" representing curvature in the equation is changed to a small "c" in the variable. But reversing the equation from subtraction to addition creates nothing but confusion.

Causing confusion is the whole idea behind the supposed geodetic accommodation for curvature and refraction. Geodetic surveying is based upon a netherworld of obfuscatory math equations designed to beguile, not to inform. To the geodetic surveyors it matters not how their accommodation for curvature and refraction is represented, because in the end, the geodetic surveyors simply discard the variable (c+r). The obfuscatory variable (c+r) is discarded, since it represents something that is not real. It is part of an elaborate deception.

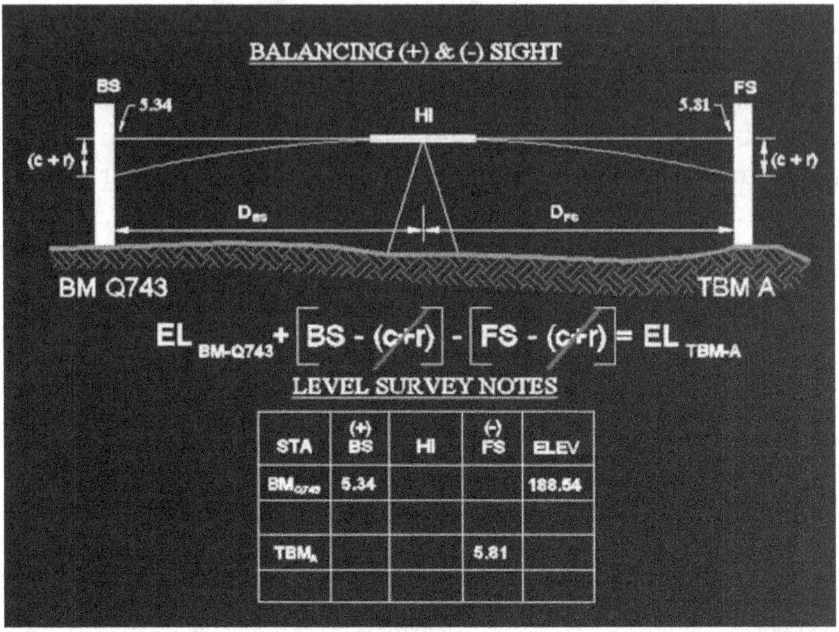

The above chart, from a California State University geodetic surveying instructional video, illustrates that after all of the adjustments are made for the supposed curvature of the earth and light refraction, those adjustments are factored back out at the end. The instructor explains that if two equidistant measurements are taken, then no adjustment needs to be taken for the curvature of the earth or refraction of light. The curvature minus refraction equation (C−r) is needlessly changed to a variable (c+r) that portrays the opposite meaning of the actual equation it is supposed to represent. The variable (c+r) is really meaningless. In the end, it is deleted.

A mathematical formula may be used to prove something that is imaginary and not real. That is why the equation (C−r) is replaced by a variable (c+r) that means the opposite of the equation it is supposed to represent. Dr. Herbert Dingle states that "in the language of mathematics we can tell lies as well as truths, and within the scope of mathematics itself there is no possible way of telling one from the other."[252] Geodetic surveyors are telling lies with their mathematical formulae, where they purport to account for the curvature of the earth and atmospheric refraction, but in reality they do neither.

In the real world, the only way that (C−r) could equal (c+r) would be if both "c" and "r" are always 0. Perhaps that is the reason for changing (C−r) into the variable (c+r). It may be an occult way for initiates into the geodetic secrets to reveal that both "c" and "r" are always zero (0).

If the issue is that the survey must accurately record the distance along the supposed curved surface of the globe and there must also be an adjustment for the refraction of light, then the calculation adjustments should necessarily be done, regardless of the number and types of measurements. Indeed, if a surveyor is measuring to only one point at a foresight, he is measuring from a known benchmark, which is the elevation of his theodolite. So, the known backsight benchmark would seem to add nothing new to allow for the disregarding of either the curvature of the earth or the refraction of light. If the earth is a globe, there is still the same arc of a sphere that is present, for which an adjustment should be made.

If no adjustment is made for curvature that necessarily means that there must not be curvature. Regarding refraction, as we saw above, light refraction is apparent at very long distances. But the fact that the adjustment for refraction is to be ignored in surveying seems to be handy way to cover up the fact that the purported refraction adjustment in geodetic surveying is being done in the wrong direction. The survey results are deemed accurate, because both refraction and the alleged curvature have been ignored. It seems that all of the talk about geodetic surveying requiring accommodating curvature and refraction is a facade. In the end, no adjustments are made for curvature or refraction. The ineluctable conclusion is that accommodation for curvature is not needed because the earth is flat and any light refraction is negligible at typical surveying distances. Plane surveying is legitimate. Geodetic surveying, on the other hand, is not legitimate; it appears to be theater to beguile the uninformed masses.

This author shared the above findings on geodetic surveying with Christopher Van Matre, who is a land surveyor with more than 30 years of experience. Van Matre concurred with this author's assessment of geodetic surveying.[253] The very existence of the discipline of land surveying, where the methods and instruments assume a flat earth, was a stark impeachment of the spherical earth myth. Surveyors throughout history had obtained verifiably precise results by assuming the earth is a flat plane. That simple fact had to be concealed.

Most think math algorithms are ways to honestly solve problems. But, as the above geodetic surveying algorithms demonstrate, math can be used to deceive. This fact is confirmed by Professor Herbert Dingle, an eminent physicist and former president of the Royal Astronomical Society. Professor Dingle reveals that "in the language of mathematics we can tell lies as well as truths, and within the scope of mathematics itself there is no possible way of telling one from the other."[254] How can that be? Professor Dingle explains that "the symbols that compose a mathematical expression may, with equal mathematical correctness, correspond both to that which is observable and that which is purely imaginary or even unimaginable."[255] The algorithms used in geodetic surveying are just such mathematical deceptions.

George Orwell was a communist insider who tried to warn the world about the communist menace by leaking the secrets behind the Satanic conspiracy against God and man in his dystopian novel, *1984*. His novel divulged the actual methods used by the elite communists to program obedience in the population.

In his revelatory novel, Orwell alludes to mathematicians who are used to condition the masses to believe lies. Geodetic surveyors are just such mathematicians. They do their part in an intergenerational Orwellian conspiracy to deceive the masses with their obfuscatory algorithms.

In Orwell's novel, there is a character, O'Brien, who is an Inner Party member and the principal antagonist in the novel. O'Brien explains to the protagonist, Winston Smith, that "[w]e control matter because we control the mind."[256] Winston disputes O'Brien's claim. Winston argues that the Inner Party does not control all matter and gives an example of the stars being beyond the Inner Party's control; some stars being, as he believes, a million light-years away. Winston misunderstands O'Brien's point. By way of explaining his point, O'Brien reveals to Winston that Winston's conception of the universe is contrary to reality. Winston had been convinced by the state propaganda that the stars are light-years away. But O'Brien explains that the stars are in reality only kilometers away, and that the sun and stars circle the earth, which is the center of the universe. O'Brien implies that Winston's misconception of the universe is the result of the clever work of mathematicians. "The stars can be near or distant, according as we need them. **Do you suppose our mathematicians are unequal to that?** Have you forgotten doublethink?"[257]

O'Brien revealed to Winston that if the people can be deluded by state propaganda to think that the world is much different from how it really is, it is as though the state has changed reality; in a sense, it controls matter by controlling what people think. That is what O'Brien meant when he said that "[w]e control matter because we control the mind."[258] The state propaganda worked on Winston, as he was convinced by that propaganda that the stars were light-years away, when in fact they are much closer. Almost all people today are like Winston. They believe the earth is a moving globe, because that is what has been inculcated in them by the state-controlled propaganda; they do not realize that the moving globe earth is an unreal fantasy that they have been programmed to believe. It is not real. Geodetic surveyors are Orwellian mathematicians, using their bewildering algorithms, as they ply their doublethink trade to create in the minds of men the thought that the earth is a sphere. They do that to conceal the reality that the earth is flat.

Heliocentrists could not refute the results obtained from plane surveying, so they did the next best thing. The scientific priesthood created another "scientific" denomination called geodetic surveying. The first alleged geodetic survey of any notoriety was performed in 1669, in France.[259] Great Britain did not began geodetic surveys until 1784.[260] The U.S. Government department of the National Geodetic Survey was established in 1807.[261] They created a make-believe schism between "plane surveying" and "geodetic surveying." The schism in fact does not truly exist. It is a facade. Accounting for the curvature of the earth in geodetic surveying is all an elaborate game of pretend. That is evidenced by the fact that, in the end, geodetic surveyors factor out all of the calculations for the curvature of the earth. Geodetic surveyors end up with the same measurements as plane surveyors. The sole purpose for which geodetic surveying exists is to enchant the masses to maintain the belief in a globular earth.

Geodetic surveyors get away with fibbing about light refraction and a globular earth by, in the end, factoring out both curvature and light refraction. The convoluted calculations of geodetic surveyors serve to obfuscate the reality of the flat earth and give the false impression that the earth is a globe. The U.S. Military does not have that luxury. The U.S. Military cannot base its missile guidance calculations on myths. And so the U.S. Military assumes that the earth is flat, because it really counts, and it cannot afford to be wrong.[262] For example, in *Technical Memorandum, No. 104*, which is a 1953 declassified official government document from the U.S. Air Force Flight Determination Laboratory, the U.S. Air Force not only assumes that the earth is flat when calculating how light refraction affects the accuracy of guided missiles, but it verified that fact.[263] Indeed, it began its analysis of light refraction over the flat earth by validating that the earth is flat under the heading: "Validity of Flat Earth Assumption for Atmospheric Calculations."[264]

Many surveyors know that the earth is flat, but they keep quiet about it. One exception is retired United States Department of Agriculture (USDA) Surveyor Ray Goodwin, who has 27 years of experience as a plane surveyor. Goodwin is convinced by the evidence that he has seen over his 27 years as a surveyor that the earth is flat. Goodwin is so certain that the earth is flat that he has laid down a $100,00 winner-take-all challenge to any surveyor in the world to prove that the earth is a sphere.[265] He and the challenger would each put up $100,000. Whoever proves his case wins the other's $100,000. Goodwin made clear that it is a bona fide challenge that he will honor.

Goodwin's challenge is simple. Goodwin suggested surveying a 60 mile long stretch of public beach at Myrtle Beach, South Carolina. Each surveyor will bring one other person, two ATVs will be rented, two survey rods will be used, one calibrated and certified theodolite transit, and one video camera. The benchmark will be set, and foresights and back sights will be taken along the waterline down the beach. If the earth is a sphere, there should be a drop of 2,400 feet over the 60 mile distance. If the earth is flat, there will be no measurable drop.

Geodetic surveyor Jesse Kozlowski was notified about Goodwin's challenge. Kozlowski refused to accept the challenge. Kozlowski gave excuses for why he refused to accept the challenge.[266] Kozlowski can spend days-on-end traveling the country and conducting surveys and documenting those surveys in his effort to refute the claim that the earth is flat, but oddly, he states that he would not waste his time with Goodwin's challenge. Goodwin's challenge would take no more than a day to perform.

And if Kozlowski is so sure of his belief that the earth is a sphere, he would make a cool $100,000 for a single day's work. The downside for Kozlowski is that if he is wrong, he will lose $100,000. And that is why Kozlowski will have none of it.

Kozlowski refused to even watch the video that lays down the specifics of Goodwin's challenge. Kozlowski, nonetheless, asked questions about the challenge, all of which would have been answered for him had he watched the video of the challenge. His questions were obviously a smokescreen for his refusal to accept the challenge. Why has Kozlowski refused to accept Goodwin's challenge? Because Kozlowski knows he cannot win. The earth is flat, and he knows it. Indeed, no surveyor has accepted Goodwin's challenge. Kozlowski has blocked Goodwin from being able to comment on Kozlowski's videos that Kozlowski has posted on the internet, which he alleges prove that the earth is a sphere.[267] Kozlowski apparently will not allow someone who knows surveying to expose his sophistry, obfuscation, and erroneous methodology, so Kozlowski has blocked Goodwin from commenting.

Update: On April 6, 2020, Jesse Kozlowski emailed this author and requested that I send him a PDF copy of my book so he could offer corrections and additional information. I sent him a PDF copy of the book, but he did not respond as promised. I emailed him on August 26, 2020, and asked for his opinion of the book. He emailed me back that day and said he would try to read it that weekend and get back to me. As of the date of this writing (November 19, 2022), it has been more than two years, and this author has not heard back from Kozlowski. He initially reached out to this author because he was eager to offer corrections to this book, but after more than two years, Kozlowski apparently could find nothing to correct.

6 Water Not Convex Proves Earth Is Not a Globe

Water always seeks its own level, because water is always level. Water cannot be anything but perfectly flat. There is never any convexity to water, as would be required by a globular earth. All oceans, seas, and lakes are perfectly level. Indeed, we use the term "water level" to denote the status of the flat and level water. The very terms we use are based upon water always being flat and level across its length and breadth.

Figure 55: Water is always perfectly level, and thus the horizon on water is always perfectly flat, with no curvature.

No matter what configuration you pour water into, it is always level. It is an irrefutable physical reality that can be replicated by anyone. The image below illustrates that reality.[268]

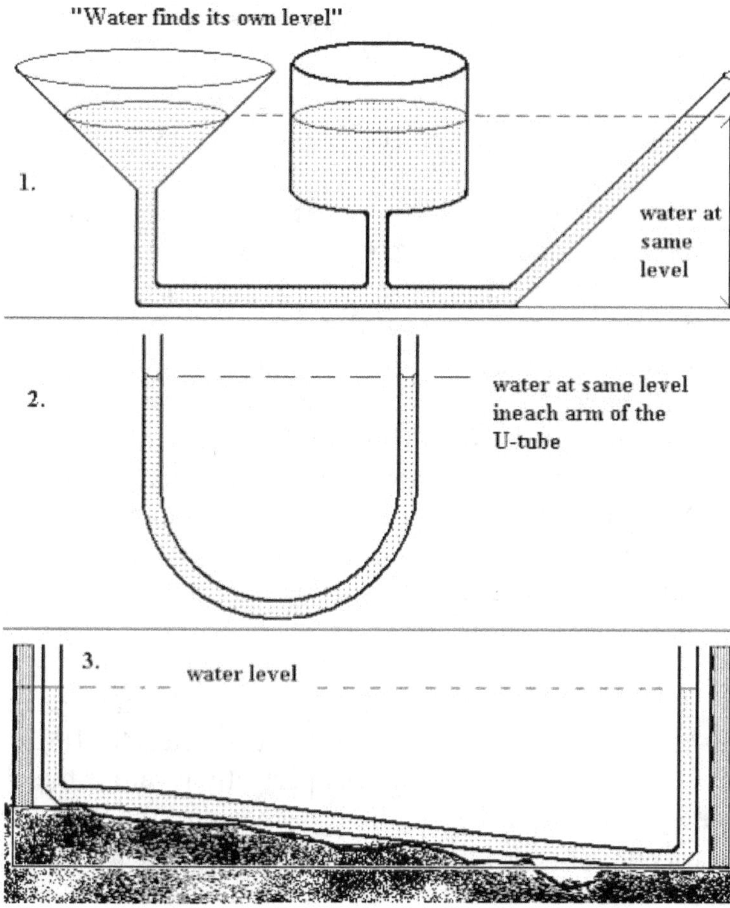

Eric Dubay, in his book, *The Flat Earth Conspiracy*, quotes an engineer, whose job it was to construct canals and railways. The engineer explained that no allowance is made for the supposed curvature of the earth. Why not? Because the earth is flat.

> Engineer, W. Winckler, wrote into the Earth Review October 1893 regarding the Earth's supposed curvature, stating, "As an engineer of

> many years standing, I saw that this absurd allowance is only permitted in school books. No engineer would dream of allowing anything of the kind. I have projected many miles of railways and many more of canals and the allowance has not even been thought of, much less allowed for. This allowance for curvature means this – that it is 8" for the first mile of a canal, and increasing at the ratio by the square of the distance in miles; thus a small navigable canal for boats, say 30 miles long, will have, by the above rule an allowance for curvature of 600 feet. Think of that and then please credit engineers as not being quite such fools. Nothing of the sort is allowed. We no more think of allowing 600 feet for a line of 30 miles of railway or canal, than of wasting our time trying to square the circle."[269]

The Suez Canal is just one example that stands as irrefutable proof that the earth is flat. The canal runs for 100 miles. There are no locks in the canal; it is perfectly level for the entire 100 miles. If the earth were a globe, each end of the canal would necessarily be over one mile below the horizon from the other end. That would mean that the middle of the canal would be elevated in a round hump 1,666 feet in the mid-point of the canal. However, that is not the case. David Wardlow Scott, in his book, *Terra Firma*, explained:

> The distance between the Red Sea at Suez and the Mediterranean Sea is 100 statute miles, the datum line of the Canal being 26 feet below the level of the Mediterranean, and is continued horizontally the whole way from sea to sea, there not being a single lock on the Canal, the surface of the water being parallel with the datum line. It is thus clear that there is no curvature or globularity for the

> whole hundred miles between the Mediterranean and the Red Sea; had there been, according to the Astronomic theory, the middle of the Canal would have been 1,666 feet higher than at either end, whereas the Canal is perfectly horizontal for the whole distance.[270]

The Suez Canal offers such stark proof that the earth is flat that a more detailed study of that canal is warranted. Samuel Rowbotham provides us that explanation, complete with actual illustrations from the authorities. The official illustrations reveal a flat earth.

> The completion of the great ship canal, which connects the Mediterranean Sea with the Gulf of Suez, on the Red Sea, furnishes another instance of entire discrepancy between the theory of the earth's rotundity and the results of practical engineering. The canal is 100 English statute miles in length, and is entirely without locks; so that the water within it is really a continuation of the Mediterranean Sea to the Red Sea. "The average level of the Mediterranean is 6 inches above the Red Sea; but the flood tides in the Red Sea rise 4 feet above the highest, and its ebbs fall nearly 3 feet below the lowest in the Mediterranean." The datum line is 26 feet below the level of the Mediterranean, and is continued horizontally from one sea to the other; and throughout the whole length of the work, the surface of the water runs parallel with this datum, as shown in the following section, fig. 38, published by the authorities.[271]

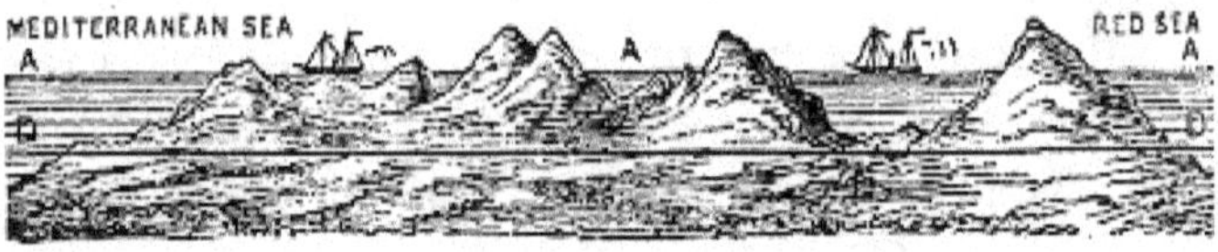

fig. 38

> A, A, A, is the surface of the canal, passing through several lakes, from one sea to the other; D, D, the bed of the canal, or horizontal datum line to which the various elevations of land, &c., are referred, but parallel to which stands the surface of the water throughout the entire length of the canal; thus proving that the half-tide level of the Red Sea, the 100 miles of water in the canal, and the surface of the Mediterranean Sea, are a continuation of one and the same horizontal line.[272]

The actual survey of the Suez Canal proves that the surface of the water is flat for 100 miles from one end of the canal to the other. That flat span of water could only be possible if the earth is in fact flat. Rowbotham modifies the official illustration of the Suez Canal to show a fictional depiction of how the Suez Canal would appear if the earth were a globe. The unreal depiction (which would be required if the earth were a globe) is not at all what is seen and measured at the Suez Canal. The convex bow of water (which is a physical impossibility) illustrates the impossibility that the earth is a globe.

fig. 39

> If the earth is globular, the water in the centre of the canal, being 50 miles from each end, would be the summit of an arc of a circle, and would stand at more than 1600 feet above the Mediterranean and Red Seas (50^2 x 8 inches = 1666 feet 8 inches), as shown in diagram, fig. 39.[273] A, the Mediterranean Sea; B, the Red Sea; and A, C, B, the arc of water connecting them; D, D, the horizontal datum, which, if the earth is globular, would really be the chord of the arc, A, C, B.[274]

Rowbotham continues with his proof of the flat earth by explaining the survey of a body of water that is exponentially larger than the Suez Canal. Rowbotham reveals the actual survey measurements of the floor of the Atlantic Ocean from Valencia, Ireland, to Newfoundland, Canada, prior to laying the transatlantic cable. That survey proves beyond any doubt that the earth is flat. Rowbotham describes the depth of the Atlantic Ocean in fathoms; one fathom equals six feet, which is two yards. Please note that Rowbotham discusses the distance across the Atlantic Ocean in nautical miles, whereas his discussion of the span of the Suez Canal is in statute miles. A nautical mile is equal to approximately 1.15 statute miles. In the equation for the supposed drop below the horizon expected if the earth were a globe (M^2 x 8 inches = Drop), the miles (M) are always in statute miles.

> The bed of the Atlantic Ocean, from Valencia (western coast of Ireland) to Trinity Bay, Newfoundland, as surveyed for the laying of the cable, is another illustration or proof that the surface of the great waters of the earth is horizontal, and not convex, as will be seen by the following diagram [fig. 40], contracted from the section, published October 8, 1869, by the Admiralty. C, D, is the horizontal datum line, and A, B, the surface of the water, for a distance of

> 1665 nautical, or 1942 statute miles. At about one-third the distance from A, Newfoundland, the greatest depth is found--2424 fathoms; the next deepest part is 2400 fathoms; at about two-thirds the distance from A, towards B, Ireland, while in the centre, the depth is less than 1600 fathoms.[275]

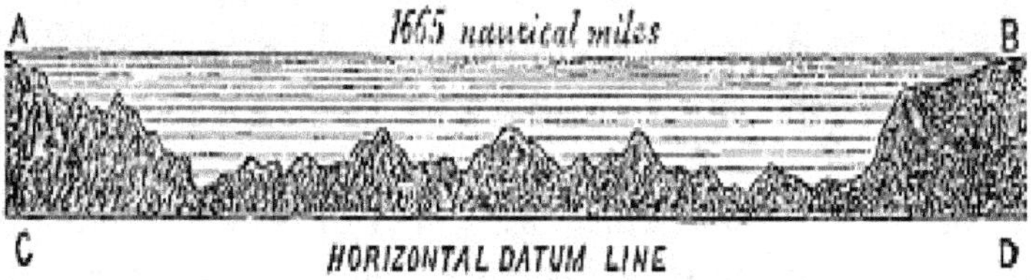

fig. 40

The actual depiction of the Atlantic Ocean is a large flat body of water spanning 1,665 nautical miles from Newfoundland to Ireland. Such would be the case only if the earth is in fact flat. Rowbotham, again, modifies the official illustration to create a fictional depiction of the span across the Atlantic Ocean as it would necessarily appear if the earth were a globe (fig. 41). The unreal depiction is not at all what is found when surveying the Atlantic Ocean.

> [W]hereas, if the water of the Atlantic is convex, the centre would stand 628,560 feet, or nearly 120 miles, higher than the two stations, Trinity Bay and Valencia; and the greatest depth would be in the centre of the Atlantic Ocean, where it would be 106,310 fathoms, instead of 1550 fathoms, which it is proved to be by actual soundings. Fig. 41 shows the arc of water which would exist, in relation to the horizontal datum line, between Ireland and Newfoundland, if the earth is a globe. Again, if the water in the Atlantic Ocean is convex--a part of a great sphere of 25,000 miles

circumference--the horizontal datum line would be a chord to the great arc of water above it; and the distance across the bed of the Atlantic would therefore be considerably less than the distance over the surface. The length of the cable which was laid in 1866, notwithstanding the known irregularities of the bed of the Ocean, would be less than the distance sailed by the paying-out vessel, the "Great Eastern;" whereas, according to the published report, the distance run by the steamer was 1665 miles, while the length of cable payed out was 1852 miles.[276]

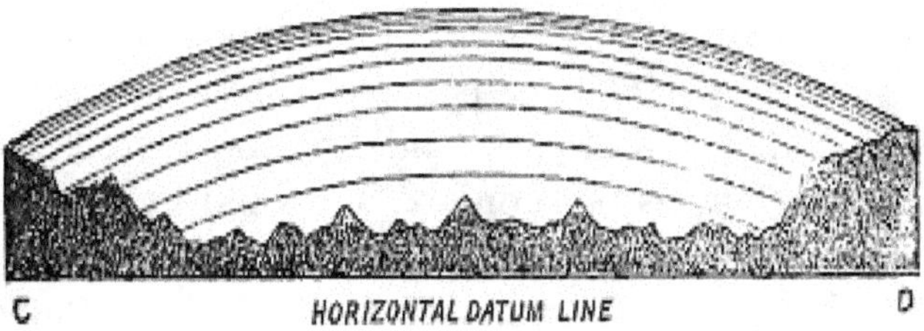

fig. 41

Rowbotham puts the final nail in the coffin of the globular earth fiction by pointing out that the distance that the steamer traveled across the Atlantic Ocean as it laid the cable was approximately 1,665 miles, which is exactly what would be expected if the vessel was traveling in a straight line across the Atlantic Ocean on a flat body of water.

All water is flat and level. Indeed, in common parlance, the height of the surface of a body of water is called the water "level." The oceans cover 71% of the earth's surface. It is impossible for the earth to be a sphere when 71% of its surface is covered by water, which necessarily must be flat and level. The very existence of the oceans proves that the earth must necessarily be flat.

7 The Earth Is Flatter Than a Pancake

Of course, land, with its mountains and valleys, is not perfectly level, as is water. However, when large expansive areas of land are considered, it can be determined that land is not convex, as would be expected on a spherical earth. Land is generally flat. Scientific research proves that fact. For example, Mark Fonstad, Ph.D., William Pugatch, and Brandon Vogt, Ph.D., used data from the United States Geological Survey to determine that, on scale, the State of Kansas is literally flatter than a pancake.[277]

The researchers used what is referred to as the flattening ratio to compare Kansas to a pancake. The flattening ratio is f = a-b/a. What that means is that the length of an area (a) is measured and its deviation in height (b) is subtracted from that length (a). That figure is then divided by the length (a) to come up with the flattening ration (f). If the area is perfectly flat, the result should be one (1.000).[278] On a scale where one (1) is perfectly flat, the geographers used a confocal laser to determine that a pancake had a measured flatness of .957. The State of Kansas was scaled down using a 1:250,000 scale digital elevation model (DEM). Kansas was found to have a measured flatness of .9997. That means that Kansas is literally flatter than a pancake.

Jerome Dobson, President of the American Geographical Society and Professor of Geography at the University of Kansas and Joshua Campbell, geographer and GIS architect in the Office of the Geographer and Global Issues at the U.S. Department of State, came to the defense of the State of Kansas. They did not want people to think that Kansas was flat and boring. Dobson and Campbell analyzed the research study of Dr. Fonstad, et al.

Dobson and Campbell used a different measure. Instead of the flattening ratio used by Fonstad, et al., Dobson and Campbell used a simpler equation that we will call a relief quotient to analyze the research study of Dr. Fonstad, et al. In their study, Fonstad, et al., compared transections of a pancake and the east-west profile of merged relief data from the State of Kansas. The pancake used by Fonstad, et al., was obtained from an IHOP restaurant. Its relief was measured over a 2 millimeter sample strip taken from across the 130 millimeter diameter of the pancake.[279] Relief means the quantitative measurement of vertical elevation change in a landscape over a given area. For an area of land, the relief can be obtained by subtracting its highest point in elevation from its lowest point.

The highest point in the State of Kansas is 4,039 feet above sea level, and the lowest point is 679 feet above sea level. Subtracting those figures from one another, we arrive at 3,360 feet (.64 miles). The relief for Kansas, therefore, is 3,360 feet (.64 miles). The east-west transection of Kansas is approximately 400 miles across, resulting in an approximate relief quotient of .0016 (.64 ÷ 400 = .0016).

Dobson and Campbell accepted the results of Fonstad's study as accurate, but were surprised to find that the relief quotient comparison revealed that every state in the United States is flatter than a pancake:

The pancake measured in the article was 130

> millimeters, and its surface relief was 2 millimeters. Apply that ratio to the east-west dimension of Kansas, approximately 644 kilometers [400 miles], and the state would need a mountain (2/130 x 664,000 meters) 9,908 meters [32,507 feet] tall in order not to be flatter than a pancake. Since the highest mountain in the world is 8,848 meters [29,029 feet] tall, every state in the U.S. is flatter than a pancake.[280]

It turns out that Dobson and Campbell miscalculated the relief quotient for the pancake. They misinterpreted the 2 millimeter sample strip from which the relief was measured as meaning that the relief of the pancake was 2 mm. The relief of the pancake should have been 5.59 mm. How was that figure determined? Working from the reported flattening ratio of .947 arrived at by Fonstad, et al. in their study for a 130 mm diameter pancake, one can simply takes the flattening ratio equation (f = a−b/a) and solve for the deviation in elevation (b). When doing that, we arrive at an average relief of 5.59 mm (not the 2 mm reported by Fonstad and Campbell). That gives us a relief quotient for the pancake of .043 (5.59 ÷ 130 = .043). Dobson and Campbell came up with an incorrect relief quotient for the pancake of .015 (2 ÷ 130 = .015).

Using the more accurate relief quotient of .043 we find that for Kansas not to be flatter than a pancake would require a mountain that is 90,816 feet (17.2 miles) high (.043 × 400 miles). If the earth were a globe, Kansas would have a bulging arc more than 26,666 feet (5 miles) above sea level. Interestingly, that would give Kansas a relief quotient of .0125, which would still make Kansas flatter than a pancake, which has a relief quotient of .043. Does that mean that the earth is a globe? No, it does not. Because, the maximum actual relief in Kansas is only 3,360 feet, giving it an actual relief quotient of .0016, which means that there is no such five mile bulging arc, as would be required on a globe.

Furthermore, the larger the land mass the greater is the arc that would be expected in the hypothesized spherical earth. That means that Florida, which was determined by Dobson and Campbell to be flatter than Kansas, could only be flatter than Kansas if the earth was flat. Florida is 447 miles across from north to south, compared to Kansas at only 400 miles at its longest from east to west. If the earth were a sphere, that would give Florida a bulging arc of 33,153 feet (6.27 miles) and a relief quotient of .014 over its 447 mile length. That means that if the earth were a globe, Kansas, with a globular relief quotient of .0125 on a supposed globular earth, would be flatter than Florida. But in actuality, we find that Florida is flatter than Kansas. Which would be impossible on a spherical earth. Indeed, Florida is the flattest state in the United States. The fact that Florida was determined by actual measurements to in fact be flatter than Kansas, means that the earth cannot be a globe; it must be flat.

In general, the larger the area measured on earth, the flatter is the actual relief quotient, which is the very opposite of what would be expected on a spherical earth. That fact alone is proof that the earth is flat. Indeed, when the relief quotient calculations are applied across a large expanse of land, for thousands of miles, the difference in relief quotients between the figures for a globular hypothesis and the actual relief quotients are stark. The larger the expanse measured, the lower the relief quotient. That could only happen on a flat earth.

For example, the continental United States is approximately 2,800 miles across. If the earth were a globe, the continental United States would have a terrain with a bulged arc approximately 2,613,333 feet (495 miles) above sea level across it. No such topographical bulge exists. If the earth were a globe, the continental United States should have a relief quotient of 0.177 (495 ÷ 2,800 = 0.177). The actual relief quotient of the continental United States, however, does not come close to the relief quotient of 0.177) that would be expected on a spherical earth.

The highest point in the continental United States is 14,494 feet above sea level, and the lowest point is 282 feet below sea level. The relief across the 2,800 mile breadth of the continental United States is therefore 14,776 feet (2.8 miles) (14,494 + 282 = 14,776) (14,776 feet = 2.8 miles). 282 feet is added to the 14,494 feet because the 282-foot elevation is below sea level. Dividing 2.8 miles by the 2,800-mile breadth of the continental United States gives us a relief quotient of .001 (2.8 ÷ 2,800 = .001).

The actual relief quotient for the continental United States of .001 means that the earth cannot be a sphere. If the earth were a sphere, the relief quotient for the continental United States would be exponentially greater (.177). The actual relief quotient of the continental United States is 177 times less than it would be if the earth were a globe. That means that the earth must be flat.

Using a pancake as a gauge of flatness, we find that the terrain of the continental United States is, on scale, significantly flatter than a pancake. It is not even close. The relief quotient of a pancake is approximately .043, which is exponentially greater than the .001 relief quotient of the continental United States. The actual relief quotient for the continental United States is 43 times less than a pancake. That means that the continental United States is flat, which, in turn, means that the earth is flat.

Simple calculations, as were done above with the continental United States, prove that the earth is not a sphere but is, in fact, flat. The study by Dr. Fonstad, et al., has far reaching implications, not lost on geographers. Lee Allison, the Director of the Kansas Geological Survey, concluded from that research study that "everything on Earth is flatter than the pancake as they measured it."[281]

Dr. Dobson, performed additional research on the issue of the flatness of Kansas.[282] Dr. Dobson was joined in his research by Joshua Campbell. Dobson and Campbell used a different methodology than did Dr. Fonstad, et al., but their research confirmed the results of Dr. Fonstad, et al.

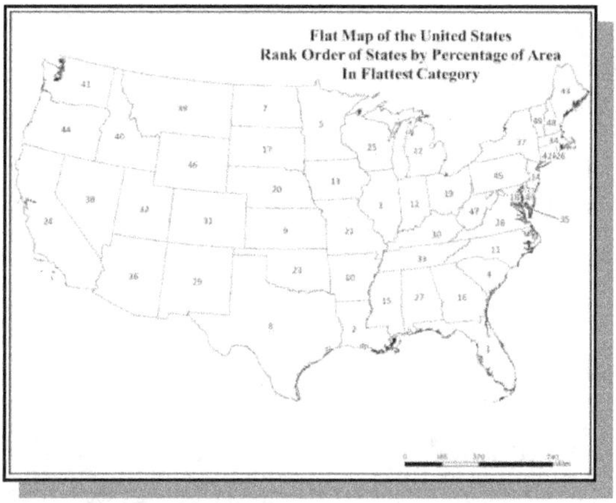

Figure 61: **Flatness ranking map for states prepared by Jerome Dobson and Joshua Campbell as published in *The Atlantic*.**

Most notably, Dobson and Campbell found that the entire United States was flatter than a pancake. Dobson and Campbell further discovered that Florida, Illinois, North Dakota, Louisiana, Minnesota and Delaware were all flatter than Kansas.[283]

Dr. Dobson extrapolated from his own confirmatory research that the entire world is flatter than a pancake. Dr. Dobson had this to say about the research study by Dr. Fonstad, et al.: "Our own findings did not refute their conclusion about Kansas but rather proved that their conclusion applies to the whole world."[284] Dr. Dobson's research was published in the Geographical Review, a peer-reviewed journal published by the American Geographical Society. Neither Dobson and Campbell's findings, nor those of Dr. Fonstad, et al., have ever been refuted or even challenged. For the United States, on scale, to be flatter than a pancake, necessarily means that the earth must be flat. The research of Dobson, Campbell, and Fontad, et. al., proves that to be the case.

8 Mountains of Proof

Jon McIntyre was troubled by the evidence of the flat earth and simply could not bring himself to accept that such a massive conspiracy to hide the very nature of the earth could exist. He came up with an ingenious way to determine, once and for all, whether the earth was a sphere or flat. He concluded that if the earth were flat then two equally high mountains separated by many miles would appear to the observer to be the same height if the observer was stationed at a vantage point that was equal in height to the peak of the two mountain tops.

All he needed to do was to move perpendicular to the alignment of the mountains, thus creating a parallax between the mountains, and he could then see both mountains side-by-side. If, however, the earth were a globe, then the more distant mountain top would drop below the height of the nearer mountain top by the distance in miles squared multiplied by eight inches (miles^2 × 8 inches = distance of drop on the supposed spherical earth).

McIntyre searched and found three mountains that met his criteria.[285] The three mountains were found in the Black Mountain Range in North Carolina. He positioned himself with his camera at an elevation of 5,385 feet on the ridge of Tennent Mountain. He then trained his camera at Fryingpan Mountain, six (6) miles in the distance from his position on the ridge at Tennent Mountain. The peak of Fryingpan Mountain has an elevation of approximately 5,380 feet above sea level. Thirty four (34) miles beyond Fryingpan Mountain was Graybeard Mountain, with an elevation of approximately 5,395 feet.

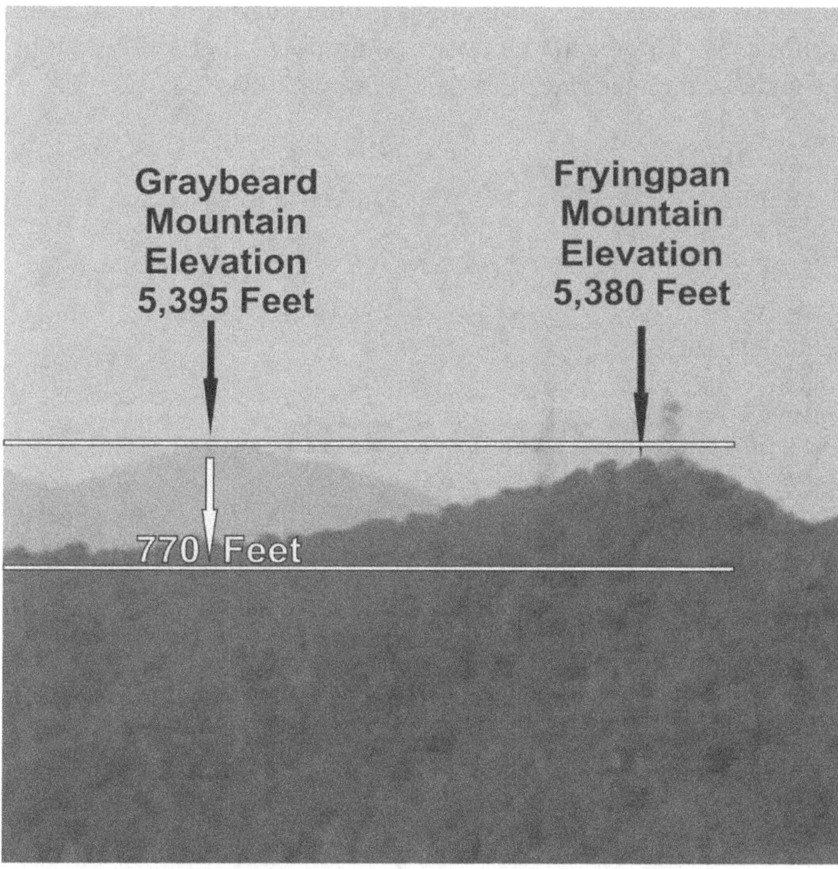

Graybeard Mountain, which is 34 miles behind Fryingpan Mountain, would have dropped with the curvature of the earth 770 feet below the horizon and out of sight behind Fryingpan Mountain if the earth were a globe.

McIntyre's position was to the side of Fryingpan Mountain, which created a parallax between it and Graybeard Mountain. This allowed him to view both the mountains juxtaposed to one another, although they were separated by 34 miles. He discovered that in fact the mountain tops were almost the same height, just as indicated by their official reported elevations.

If the earth were a globe, then the curvature of the earth would cause Graybeard Mountain to drop 770 feet and be out of the sight of the observer. But that is not what we see.

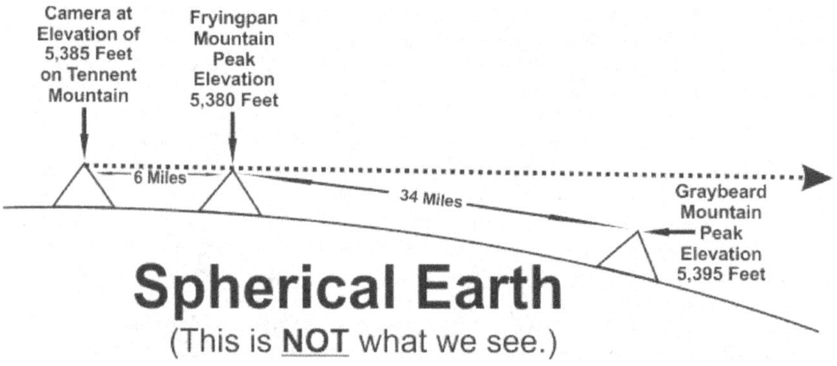

The picture taken by McIntyre clearly shows Graybeard Mountain and Fryingpan Mountain at their reported heights juxtaposed to one another.

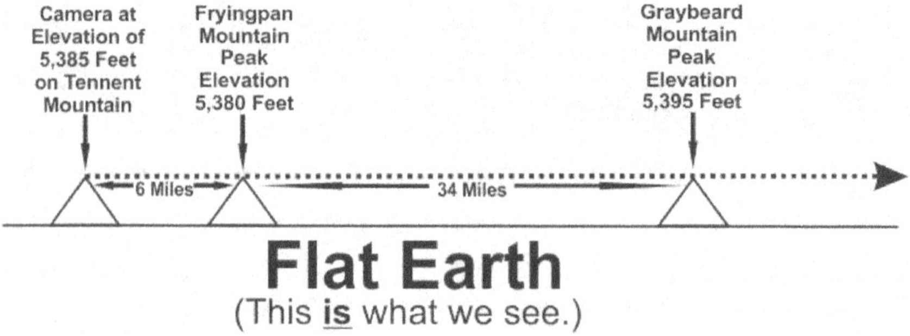

Such a configuration as depicted in McIntyre's photograph would be impossible on a spherical earth. His picture is compelling evidence that the earth is flat.

It should be noted that McIntyre is not a believer in the flat earth. The most he can bring himself to say on the issue of whether the earth is flat is: "I don't know." But he cannot otherwise explain the phenomenon of the mountain peaks being the same height in relation to one another as officially reported, although they are separated by 34 miles. He states that he can "find no way to interpret this data other than to say that it clearly supports the conclusion that the earth is flat." Yet, he cannot bring himself to believe it. The implications for him are just too disturbing.

> This shot I find to be completely bizarre. Do I believe that the earth is flat? That's a big leap to make. A really big leap to actually believe in a conspiracy theory of that size. It is craziness! But what am I supposed to do with this evidence? I don't really know. Honestly, I don't really know. But I felt compelled to make this video. I felt compelled to collect irrefutable evidence one way or the other, so I could end the internal debate I was having about it and answer the questions I was having regarding flat earth, because I did find the debate interesting.[286]

It seems that McIntyre finds the evidence in his experiment that proves that the earth is flat to be compelling. He put a lot of work and study into conducting his experiment, but when the results pointed clearly and irrefutably toward a flat earth, McIntyre could not fully believe that the earth is flat. To do so would be to make "a really big leap to actually believe in a conspiracy theory of that size." He has been thoroughly conditioned to ignore his senses and view the results of his test as "craziness" no matter how convincing is the evidence. McIntyre admits that, although the evidence in his experiment is compelling proof that the earth is flat, he cannot bring himself to accept it, because he cannot overcome his conditioning from "a lifetime of being told that I live on a globe flying through space."

> My love of truth and love of the pursuit of truth compelled me to conduct this experiment and produce this video. I find the evidence I collected to be very convincing. In fact I find no way to interpret this data other than to say that it clearly supports the conclusion that the earth is flat. Yet if you were to ask me if I believe the earth is flat my answer would be "I don't know" because at this point I truly do not know what I believe. I am fully aware of what this evidence is pointing to and yet my mind seems unable to firmly settle on the belief that the earth is flat. Maybe this is due to a lifetime of being told that I live on a globe flying through space.[287]

9 Trying to Make a Molehill Out of a Mountain

McIntyre's video and photographic proof of a flat earth has been attacked by deceptive debunkers. They use sophistry, obfuscation, and lies to confuse readers and conceal the truth of the flat earth. One example is the purported debunking website, Metabunk.com, which was so full of errors and misrepresentations that McIntyre called the deceptive posters: "liars." The whole purpose of the heliocentric posters is to deceive the unwary and ignorant who the posting charlatans know will not truly dig into the details. It is an age-old strategy of befuddling people with misleading details so they will just throw up their hands in frustration and reject the otherwise convincing evidence that would upset the mythology of the status quo.

Below is a quote from McIntyre about the supposed debunking of his video on the forum at Metabunk.com. Keep in mind when reading McIntyre's reaction to the deceptive posts that he, to this day, does not accept that the earth is flat. That means he agrees with the viewpoint of the debunkers that the earth is a sphere, but he nonetheless calls them liars. He calls them liars, because he knows his facts and understands how they are twisting the evidence to mislead others. McIntyre will not tolerate misrepresentations of the evidence. McIntyre strives to find and accurately report the truth and will not allow his work to be maligned by liars who have an agenda to deceive people into believing that the earth is a sphere; even though he, himself, believes the earth is a sphere. Indeed, he pulls no punches and describes the so-called debunkers as "liars."[288] McIntyre states:

> Look, this is exactly what I'm not going to deal with. I spent many, many days hiking around those mountains trying to locate the perfect position to get the shot. The whole entire time I was intermittently seeing Fryingpan and Graybeard. And I've been up there probably more than twenty times since and I still see them all of the time. I hike over Tennent regularly. Guess what? I see Fryingpan and Graybeard and they look exactly the same as in my shot because guess what??? They are the same mountains. How in the world do I go up there and spend all of this time up there and hike all of these mountains. I've been to the top of all of these peaks. I know these peaks. I see them all the time. And then somehow when I need to find a shot on my line of sight and at the right elevation I suddenly forget what they look like.
>
> I quite seriously do not appreciate this. Not you bringing this to my attention but a liar lying and then trying to force me in some way to pay attention to these lies as though they are worthy of attention. The assertions by that page you sent me are.... I don't know... absurd, outright lies, inanity. I verified those peaks. I live near those peaks. I hike those peaks. I was directly on that line which is obvious to anyone who can think because it looks exactly like the image on Peakfinder.

Furthermore, just take a look at the mountain range in the distance that includes Graybeard. LOOK AT IT. To the left of Bald Knob you see steady incline all the way up to the hogback of Mount Mitchell. It's obvious. Look at it. Then look at Peakfinder and examine the way it looks. That ridge of mountains is unique and completely identifiable to anyone willing to think.

I hike up there all the time. I drive up the Blue Ridge Parkway all of the time. And I've hiked down the ridge to Graybeard from the road. When I'm driving up the Blueridge I have Mt. Mitchell on my left and I'm looking to the right and guess what I see... to the furthest left: Graybeard, slightly closer: Pinnacle, then even closer: Bald Knob. I've sat on top of those peaks. I know them. I've hiked that trail and been up there. I've used my topo maps for navigation. I've seen them from multiple angles and multiple mountain tops.

... I hate liars. I've seen those peaks from many angles and and made sure to be thorough and yet I have some guy making up lies about me and my test. Look, you are free to read that trash and believe it. You are being lied to. [If] you bring it to my video again, sincere or not, I [will] delete [your] comments and I [will] block you.

Why? Very simple. I will not allow lies and deception to be promoted along with my video on my channel. I won't have it. I'm done. Bring legitimate, honest debunking to my comments or don't bring anything. Stop being lazy.

Go to Peakfinder. Take a look at how that

> mountain range looks and then watch my video. It is obviously the same mountain range. I should not have to spend my time explaining something to you that is so completely obvious. Instead of checking with me to see if some debunker has made a point just go spend an hour and look for yourself. It is easy. And you will see they are lying.[289]

McIntyre's tirade against the liars on Metabunk would also apply to Jesse Kozlowski, who portrays himself as an expert in geodetic land surveying. Kozlowski states that he has been land surveying for many decades. Jesse Kozlowski has posted videos and other information on the internet that he alleges proves that McIntyre was wrong and that the earth is a globe. Kozlowski was first introduced to this author by Dr. Jim Fetzer. Dr. Fetzer is professor emeritus of the philosophy of science at the University of Minnesota Duluth, who believes the earth is a spinning sphere. Dr. Fetzer engaged in an email discussion with True Ott, Alex Studer, and this author about flat earth. Dr. Fetzer explained his view of the flat earth thusly:

> As a professional philosopher, I offered courses in epistemology and the philosophy of science as well as in logic and critical thinking. My first book, SCIENTIFIC KNOWLEDGE (1981), focuses on the nature of scientific knowledge. The belief in a flat Earth is a manifest absurdity in contradiction to thousands of years of scientific research.[290]

Dr. Fetzer followed up that email by calling upon Jesse Kozlowski as one of his experts to address the evidence with which Ott, Studer, and this author had presented Fetzer that proved the earth was flat. Dr. Fetzer introduced him as "the brilliant Jesse Kozlowski."[291] So, here we have an eminent professor in turn bringing forth his "brilliant" expert to once-and-for-all put to rest the flat earth nonsense. Let us examine the effort closely, because the refutation of their evidence does nothing but prove the earth is flat.

Sometimes the most convincing proof of the flat earth actually comes from those who promote the heliocentric model. "How so?" you might ask. In the effort of heliocentric "scientists" to attack the evidence of the flat earth, they necessarily must engage in all manner of sophistry, obfuscation, and deception. When that deception is revealed, the logical question in the minds of men of reason is "why engage in such deception if the earth were truly a moving globe?" The ineluctable conclusion is that there would be no reason for deception if the earth were as the "scientists" portray it. The earth, therefore, must be flat.

Jesse Kozlowski weighed in as Fetzer's champion by first referencing us to his analysis of Jon McIntyre's flat earth findings. Kozlowski has posted a video on the internet that purports to impeach McIntyre's findings by, in principal part, alleging that Jon McIntyre could not know if he was looking on a horizontal plane when looking across Fryingpan Mountain and Graybeard Mountain.[292]

Kozlowski is simply wrong, and his inaccurate statement seems to be a purposeful effort to deceive. Because the evidence of McIntyre's angle of view is memorized by pictures taken with his camera, the issue comes down to whether McIntyre held his camera level. All of the altitudes for the mountains are known, although there is some variation within a couple of feet, depending on the authority that is referenced. The distance between the mountains is known. The elevations of the mountains were measured from sea level. The altitude is measured vertically plumb from sea level to the highest point. So all references to altitudes begin from a level plane, sea level.

McIntyre was on a ridge on Tennent Mountain at an elevation of 5,385 feet. Fryingpan Mountain (5,380 feet) was 6 miles away, and Graybeard Mountain (5,395 feet) was 34 miles beyond Fryingpan Mountain. So all three of the mountain tops are roughly the same elevation. When McIntyre shot his picture, he framed the mountain peaks in the vertical center of the lens in his camera. Assuming that he was at the reported height, he necessarily had to hold the camera relatively level in order to frame the mountain peaks in the center of the frame. That is simple geometry.

Picture two chairs of equal height on a level floor. Now place a straight board across the gap between the seats of the chairs. That board would be level to the ground. Once it is established that the floor is level and the chair seats are at equal heights, it is a geometric fact that the horizontal board would also be level. That is the same situation with which McIntyre was presented. He was on a ridge at Tennent Mountain at 5,385 feet in altitude measured up from sea level, and he was looking across 6 miles to the peak of Fryingpan Mountain that was 5, 380 feet in altitude, which was also measured up from sea level. The picture that McIntyre took was in the vertical center of his camera lens, which means that he necessarily had to be looking on a horizontal level plane across the peak of Fryingpan Mountain.

Take, for example, identical twins of equal height and have them look each other in the eye on level ground. They would be looking at each other on a horizontal plane. Now, that is what McIntyre was doing. He was at a given height looking across six miles at the top of a mountain that was at the same elevation as was he. It is simple geometry that he necessarily is pointing his camera horizontally.

Suppose McIntyre held the camera at a slight angle upward. That would only move the mountain tops lower in the frame, but it would not cause them to change positions in relation to one another. If he tilted the camera down, it would cause the tops of the mountains to move up in the frame. But it would not cause the mountains to change position in relation to one another. The only way that the camera could change the relative locations of the mountains would be if he raised or lowered the camera in altitude. But he did not do that.

For McIntyre to take a picture across six miles to the top of Fryingpan Mountain, he must point the camera along a straight level horizontal plane. That is because each end of that plane, where he is on Tennent Mountain and the peak of Fryingpan Mountain, must have vertical coordinates (at their base) that are at a right angle to sea level. In the case of the flat earth he necessarily must be looking out at roughly a right angle to a vertical line from sea level as though forming a table top.

All of the above assumes a flat earth. Let's now assume a spherical earth. Assuming a spherical earth, we would have the same three mountains as configured above only on the arc of a sphere. The first thing that must be realized is that on a sphere, once you rise in altitude above the earth the viewer must look down in order to see the horizon. Indeed, that alone is proof that the earth is flat, because the horizon is always at eye level no matter how high one rises in altitude above the earth. The fact that no one in real life who rises above the horizon must look down at an acute angle to keep the horizon in view is proof that the earth cannot be a sphere.

Looking at an elevated object from the same elevation in the distance on a sphere, a cameraman would need to tilt his camera downward. That means that if McIntyre were on a spherical earth he would necessarily need to point his camera downward at an acute angle measured up from the plumb vertical line to keep the mountains on the arc of the curve in view.

If we assume a flat earth, we know McIntyre must have held his camera horizontally level. If we assume, a spherical earth, McIntyre must hold his camera at an acute angle. But McIntyre did not point his camera downward on an acute angle, because he stated that he pointed his camera horizontal. Again, more evidence of a flat earth.

Regardless, the tilt of the camera is irrelevant when one is at a known height and the objects you have lined up are at known equal heights to the viewing platform. That holds true regardless of whether the earth is a sphere or flat. In the case of a sphere, McIntyre would be holding his camera on a flat plane from the ridge on Tennent Mountain viewing the top of Fryingpan Mountain at the same elevation. Once that flat viewing plane is established, on a spherical earth all distant mountain tops beyond Fryingpan Mountain (e.g., Graybeard Mountain) should drop from view at the rate of the supposed curvature of the earth. But there is no drop. Thus, there is no curvature.

Whether you accept that the earth is a sphere or flat, for McIntyre to take a picture across six miles to the peak of Fryingpan Mountain, he must point the camera along a straight plane across the peak of Fryingpan Mountain. In the case of a spherical earth McIntyre must be looking out at an acute angle to a vertical line to sea level. On a spherical earth, McIntyre would still be looking out over a flat plane, at the peak of Fyringpan Mountain, where it and Tennent Mountain would be angled inward as though table legs bent inward. But that would not change the fact that whether on a flat earth or a spherical earth McIntyre would be looking out on a flat plane toward the peak of Fyringpan Mountain. If the earth were a sphere, Graybeard Mountain should have dropped along the curved surface of the earth 770 feet below the ridge of Fyringpan Mountain. But that is not what we see. There is no drop in Graybeard Mountain.

The fact that McIntyre could see Graybeard Mountain as he looked on a straight plane across the top of Fryingpan Mountain while he was on the ridge of Tennent Mountain at the same height as Fryingpan Mountain means that the earth must necessarily be flat. There is no getting around that fact.

On a flat earth, each end of that plane must have vertical coordinates (at their base) that are at a right angle to sea level. In the case of the flat earth McIntyre necessarily must be looking out at roughly a right angle to a vertical plumb line from sea level as though forming a table top. Once McIntyre sets up the plane along Tennent Mountain and the peak of Fyringpan Mountain, then Graybeard Mountain must line up, if the earth is flat. And that is what we see in the photograph. If the earth were a globe, then Graybeard Mountain should not be in view, as it should have been 770 feet lower and out of sight below Fyringpan Mountain's ridge on the left. But Graybeard Mountain does not drop from view. Graybeard Mountain is viewed precisely at its reported height equivalent to Graybeard Mountain as expected on a flat earth.

Once MacIntyre's camera shoots across the same altitude at Fryingpan Mountain, he necessarily is pointing his camera on a level plane. The fact that Graybeard Mountain can be seen 34 miles past Fyringpan Mountain can only mean one thing. The earth is flat.

10 The Testimony of Buildings and Bridges

Parallel lines are two lines that are always the same distance apart and never touch. The fact that walls on buildings and columns on bridges are parallel, testify to the fact that the earth is flat. Indeed, no wall on any building on a spherical earth could be parallel if they are constructed using plumb bobs or spirit levels. Rob Alphanostrum illustrates in the diagram below how it can be verified that the earth is in fact flat by using weighted plumb bobs.[293]

If two plumb bobs are hung from a bridge, or any high platform, the lines attached to the plumb bobs will be the same distance apart at the top as at the bottom. The lines will always be parallel no matter how far apart they are hung. That parallel phenomenon could only happen on a flat earth. If the earth were a sphere, the spherical earth theory requires that the distance between the plumb bobs at the top would be greater than at the bottom. That is because on a spherical earth, the theory of gravity requires that a plumb bob, wherever it is on earth, must be pulled to the center of the earth. The vectors of the force of gravity would be like the spokes on a wheel splaying outward from the center of the wheel. The plumb bobs would follow the force vectors of gravity, if the earth were a globe. But that is not what we see. What we see is that the lines holding the two adjacent plumb bobs will always be parallel no matter how far apart they hang. That could only mean that the earth is flat.

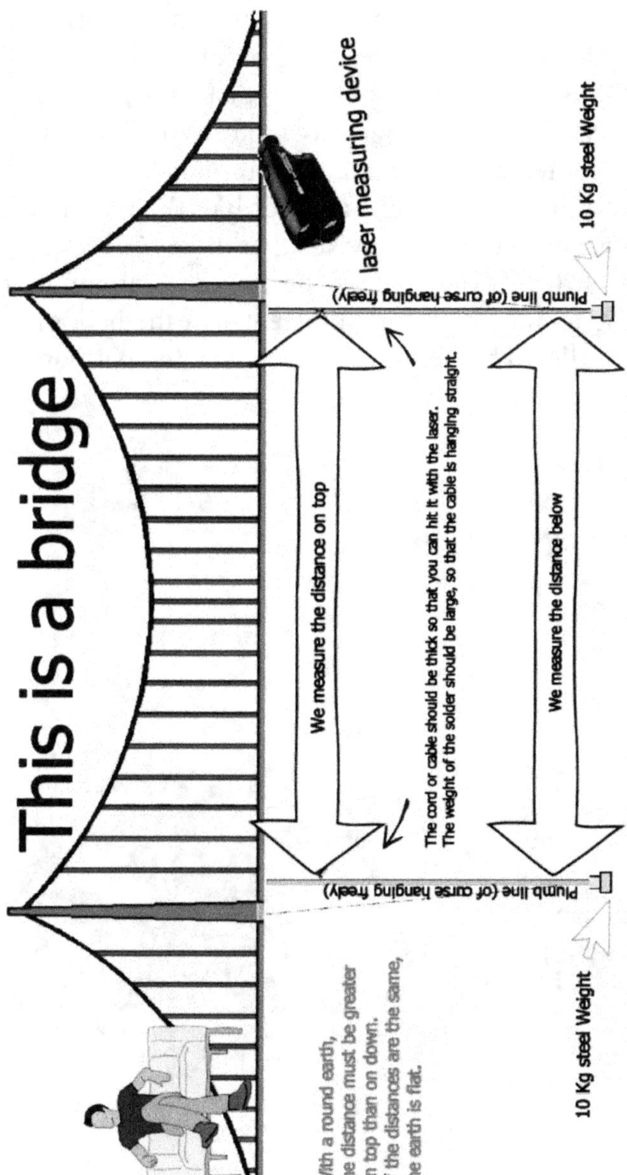

The fact that plumb bobs are always parallel, proves the earth is flat. This phenomenon can be seen in hot air balloons floating over the earth. The hot air balloons in the photograph below span a distance of approximately one mile on the vertical plane over St. Louis.[294] The photograph was taken during the 2012 Great Forest Park Balloon Race. The gondolas hanging beneath the hot air balloons act as giant plumb bobs that hang perfectly vertical to the ground. The theory of gravity requires the balloons to hang toward the center of the globular earth. That would cause the balloons to splay out from one another like the ribs in a Chinese fan. But close measurement reveals that there is no splaying. The vertical lines drawn over all of the balloons running from the center of the gondolas through the center of the balloon tops are perfectly parallel. That parallelism between the balloons is proof that the earth is flat.

Photograph with vertical lines drawn over the hot air balloons showing how the gondolas hanging underneath the balloons act as giant plumb bobs. The vertical plumb lines are perfectly vertical and are also perfectly parallel to one another as they span a vertical plane of approximately one mile over St. Louis. The balloons could only create vertical plumb lines perfectly parallel to each other if the earth was flat. If the earth were a globe, the theory of gravity requires the balloons to be splayed out over the curved earth like the ribs in a Chinese fan.

One could argue that a one mile arc on a spherical earth with an alleged radius of 3,963 miles is so slight that the divergence from parallel of plumb bobs would be imperceptible. Indeed, on the supposed spherical earth the angle of divergence would only be approximately .015°, which is a little less than one arc minute. Assuming the hot air balloons that are more than one mile apart in the photograph above are at a height of 2,000 feet, as they appear to be, the line of divergence from parallel a mile apart would result in only about a 6-inch spread by the time the vertical lines reached from the surface of the earth to the balloons. Accepting those facts, as we must when assuming a spherical earth, there would seem to be merit to the argument that the perceptible divergence of the vertical lines is slight. But there is a big difference between perceptible divergence and measurable divergence.

The divergence from parallel of plumb bobs expected on a globular earth, although arguably not perceptible, would in fact be measurable, if it existed. But in the history of the world, no such divergence from parallel has ever been measured in plumb bobs on the earth. Why not? Because the earth is flat. The measurement proposed by Rob Alphanostrum using plumb bobs suspended from a bridge would prove the earth is flat.[295] It would matter not whether the divergence on the supposed spherical earth would be perceptible to the observer; the issue is whether the divergence would be measurable. A laser measurement device would be precise enough to measure any deviation from parallel.

It is not necessary that a bridge be used to verify that all plumb bobs are parallel on the flat earth. Any span, or even any separate buildings, where the plumb bobs could be suspended with an unobstructed view between them would suffice. But we do not need to even perform the measurement proposed by Alphanostrum. The fact that all rectangular buildings have parallel vertical walls is all the proof we need. All rectangular buildings are built with walls using spirit levels or plumb bobs. All such buildings have parallel walls. Deviation from parallel, as would be required on a spherical earth, would be minute, but it would nonetheless be measurable if it existed, particularly with uniform steel construction. But the walls of such buildings are perfectly parallel. Those buildings, with their parallel walls, stand as silent monuments proving that the earth is flat.

The measurability of the splaying is the important fact that proves that the earth is flat. It would be particularly difficult, nay, impossible, to build skyscrapers with plumb and parallel walls on a spherical earth. We will see that fact proven when we discuss the construction of the World Trade Center later in this chapter. For now, we will view the graphics below, which illustrate the point. Assuming a height of 2,000 feet for the hot air balloons on the far right and far left separated by approximately one mile on the vertical plane in the St. Louis hot air balloon photograph, there would be a six-inch splay between the plumb lines of those balloons if the earth were a globe. The advocates for a spherical earth would certainly argue that such a splay is imperceptible. But, again, it is not the perceptibility that is at issue, it is the measurability. And if such splaying existed, it would be measurable.

Proof of the measurability of the plumb bob splaying required on the spherical earth can be seen in the graphic below. Graphics software was used to graph a 90.015° splayed vertical line abutting the base of the 90° vertical plumb line running through the hot air balloon on the viewer's far left in the St. Louis hot air balloon photograph above. When the line is magnified 300x we can see the measurable difference in the lines at the balloon height of 2,000 feet depicted in the photograph. The graphic below shows the six inch spread between the lines that would be expected on a spherical earth as the plumb lines are splayed from one another. But there is, in reality, no such splaying.

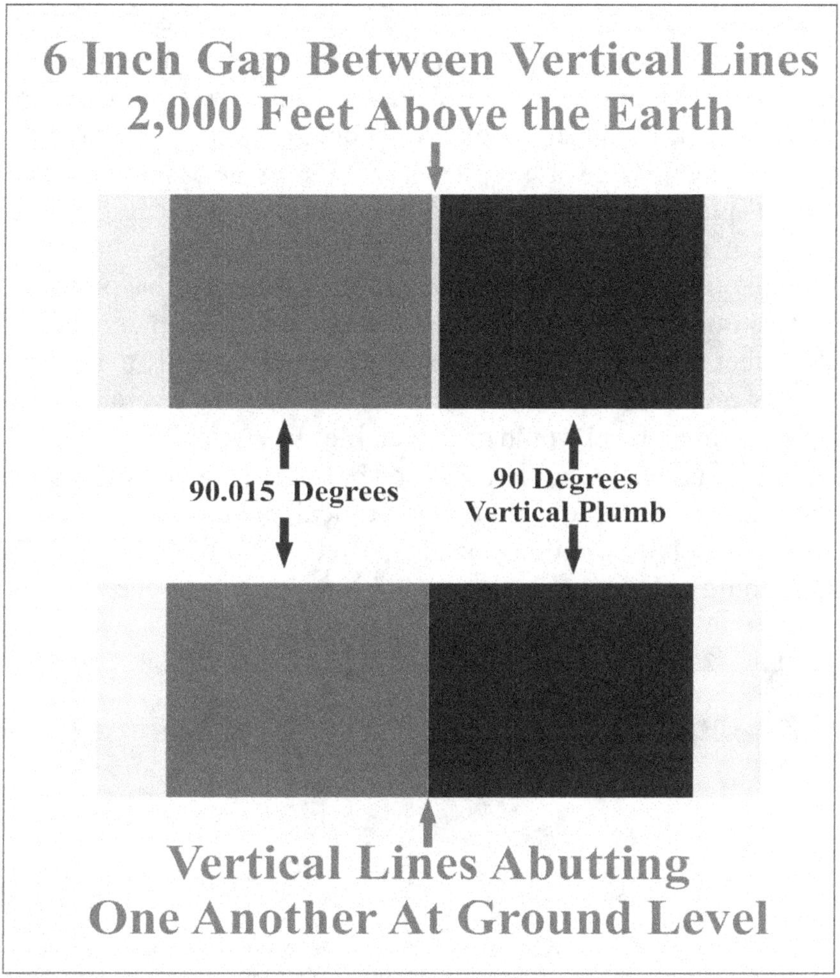

The dark line above on the viewer's right is a screen shot at 300x magnification of the 90 degree plumb line drawn over the hot air balloon on the viewer's far left floating over St. Louis depicted in the previous photograph. The lighter line above on the viewer's left is a screen shot at 300x magnification of a line set at a 90.015 degree angle. The edges of the lines above were drawn with them abutting one another at the bottom. The 300x magnification reveals the measurable 6 inch gap that would occur between two plumb lines offset by a .015 degree angle a mile apart at a height of 2,000 feet, as would be required if the earth were a globe. There is no such gap in reality. All vertical plumb lines everywhere in the world are in fact perfectly parallel, with the distance between them being equal at all points, because the earth is flat.

The graphic below depicts a 300x magnification of the ends of a horizontal line that abutted the vertical plumb lines over the far right and far left hot air balloons that were approximately one mile apart in the above St. Louis hot air balloon photograph. That horizontal line was set precisely abutting the bases of the two vertical plumb lines. That same line was then moved up to the height of approximately 2,000 feet, which was just above the balloon on the viewer's extreme left. The horizontal line perfectly abutted both vertical plumb lines without any overlap or space at 2,000 feet. The light horizontal line was set graphically to be in an overlay on top of the black vertical line so that any overlap in the lines at ground level would be detectable. The perfect abutment of the horizontal line at both the top and bottom of the vertical plumb lines proves that the 2,000 foot vertical plumb lines that are approximately one mile apart are perfectly parallel. The plumb lines could only be perfectly parallel if the earth was flat.

The top graphics show 300x magnified screen shots of a horizontal line abutting the left and right vertical plumb lines drawn over the hot air balloon on the far left and the hot air balloon at the far right. The horizontal line above was drawn at approximately the 2,000 foot level. The bottom graphics show 300x magnified screen shots of the very same line at ground level abutting the far left and far right plumb lines. The hot air balloons are floating perfectly plumb to the earth and at the same time are perfectly parallel. This could only happen if the earth was flat. If the earth were a globe, the vertical plumb lines would be splayed out and there would be a noticeable gap between the horizontal line and the vertical lines at the 2,000 foot level.

If the earth were a globe, the vertical plumb lines a mile

apart would splay out by six inches at the 2,000 foot level. The horizontal line in the above picture would not touch the left and right most vertical plumb lines as it is moved from the ground to a height of 2,000 feet, if the earth were a globe. There would instead be a six-inch gap at the 2,000 foot height, just as depicted in the above graphic with the 90.015° vertical line running next to the 90° vertical line. But, as the horizontal line graphic shows, the horizontal line perfectly abutted both vertical plumb lines at the ground level and at 2,000 feet. There was no six-inch gap as would be required if the earth were a globe. The missing six-inch gap means that the earth cannot be a globe and must be flat.

If the earth were in fact a globe, measuring the angle of splay of distant plumb bobs, which is required under the spherical earth theory, would seem to be a very easy and precise way to verify the circumference of the supposed globular earth. Why don't the spherical earth advocates just do that and put an end to the argument? Because there is no angular splay between distant plumb bobs on earth. All plumb bobs on earth hang perfectly parallel. The above graphics illustrate that truth.

All measurements purporting to gauge the curvature of the earth actutally refute the globular earth mythology and prove the earth is flat. The spherical earth advocates argue that the curvature of the globular earth is so gradual as to be almost imperceptible on buildings and that is the reason that the earth appears flat. But they miss the point, which is that if the earth were a globe, its alleged sphericity would be measurable. That is what the graphics above show. The problem with the globular earth theory is that all measurements performed do not show any curvature; on the contrary, they show that the earth is flat.

Indeed, there is no measurable splaying of any walls on any rectangular buildings in the world. Of course, there are exceptions, but those exceptions are due to poor construction and other anomalies, not design. All rectangular buildings are designed to have vertically plumb and parallel walls. They are precisely measured and constructed to be so. That would not, and could not, be done on a spherical earth. The graphic above demonstrates that fact.

Malcolm Bowden is a retired civil and structural engineer, and a well respected geocentrist. While Bowden is an ardent believer that the earth is the center of the universe, he also believes that the earth is a sphere, and he is very much against any notion of a flat earth. He has created a video titled, *Flat Earth Stupidity Exposed.*[296] In his video he presented an explanation of the construction of a bridge which showed the columns of the bridge, where each column is required **not** to be parallel, because he believes the earth is spherical. He claims that is how bridges are built on the supposed spherical earth. In his video, he showed how a spirit level is used on theodolites to construct bridges. The two bridge depictions below are from Bowden's video. His annotations as they appeared in the video are the small black typeface. My annotations are the larger gray typeface framed by brackets. The top depiction is a bridge on a flat earth that Bowden falsely claims is wrong. The bottom depiction is a bridge on a spherical earth that Bowden falsely claims is correct.

[Bowden falsely claims that this is wrong.]

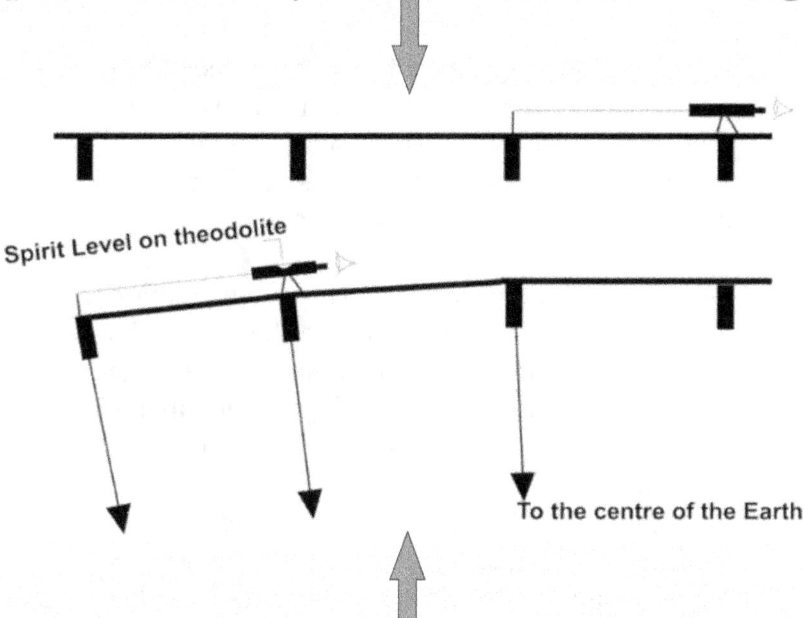

[Bowden falsely claims that this is correct.]

Bowden claims that the spirit level creates bridge columns that point toward the center of a spherical earth. He is correct that theodolites with spirit levels are used to build bridges, but they do not create columns that point to the center of a spherical earth, because the earth is not a sphere. The spirit level ensures that the theodolite is precisely level to the earth, thus ensuring that the bridge columns are vertically plumb. The resulting columns in the real world would be perpendicular to the level water and perfectly parallel to each other, just as depicted in the top bridge depiction. Bowden's bowed bridge in the lower depiction is a myth, created entirely in his mind. The fact that bridge columns in the real world are in fact parallel proves that the earth is flat.

Bowden's video is provably false. I posted a comment under the video explaining how his argument regarding bridge construction in fact proved the very opposite of his claim. His argument proved that the earth is flat. Below is my posted comment:

> **Edward Hendrie**: The entire video is chock full of erroneous information, but for brevity, I will focus on Bowden's single bridge explanation, because in his explanation he has in fact proven that the spherical earth is impossible.
>
> On one point, he is correct, that all construction of buildings and bridges use a level. As he indicated, all vertical support columns on a bridge are plumbed straight. That single truth in his bridge explanation actually proves the earth is flat. That is because if the earth were a globe as he suggests, and bridges are thus curved on a flat earth, as he claims, the columns on the bridges would splay out from one another. Indeed, his video illustrates that point. None of the columns on his mythical bridge are parallel. However, all support columns on real bridges in the real world are parallel, which can only mean that the earth is flat.
>
> Buildings are built upon the same principle as bridges, with the walls plumbed level to the surface of the earth. If the earth were a sphere, as Bowden claims, the walls would splayed outward from one another on all buildings. On a sphere, no walls in any building could ever be parallel. Every floor on a high rise building would be progressively larger as you got higher and higher. High rise building are built using uniform spans of steel. That uniformity would be impossible on a spherical earth, as each span of the higher floors would be progressively longer by a small fraction

and the walls could never be parallel. As Bowden inadvertently proved, you cannot have parallel building walls on a spherical earth. For example, the World Trade Center Towers, with their parallel and square walls offered silent testimony that the earth is flat. Such a buildings could not be built on a spherical earth. Real buildings in the real world have level floors with parallel walls, thus proving that the earth is flat.

The curved bridge illustrated by Bowden exists only in his imagination. There is no such bridge in the real world. How would Bowden explain the flat and level, 23 mile-long, Lake Pontchartrain Causeway? It is not a bridge of his imagination on a mythical spherical earth. It is a real bridge on a real flat earth with real support columns that are perfectly parallel to one another.

How would Bowden explain the Suez Canal? The Suez Canal is just one example that stands as irrefutable proof that the earth is flat. The canal runs for 100 miles. There are no locks in the canal; it is perfectly level for the entire 100 miles. If the earth were a globe, each end of the canal would necessarily be over one mile below the horizon from the other end. That would mean that the middle of the canal would be elevated in a round hump 1,666 feet in the mid-point of the canal. The actual survey of the Suez Canal proves that the surface of the water is flat for 100 miles from one end of the canal to the other. That flat span of water could only be possible if the earth is in fact flat.

Edward Hendrie

Author of The Greatest Lie on Earth: Proof That Our World Is Not A Moving Globe[297]

Malcom Bowden replied to my posted comment as follows:

> **Malcolm Bowden**: REPLY-BRIDGE COLUMNS. His assertion that they are "columns that are perfectly parallel to one another." is NOT based on evidence but upon his THEORETICAL ASSUMPTION of a flat earth!. If he were to measure the two distances from the top of the first and last columns and their bottoms he would find that they are fractionally longer at the top than at the bottom. This is extremely difficult to do so he has NO evidence that I am wrong. Suez Canal - The survey would have been carried out using theodolites which are automatically corrected for the curvature of the earth due to their leveling bubble - AS WITH THE CONTRUCTION OF BRIDGES.
>
> MB[298]

Bowden's posted reply alleged that my statement that in the real world bridge columns are perfectly parallel to one another was based entirely on a theoretical assumption of a flat earth. No. My statement was based upon reality. Bridge columns, and indeed building walls, are designed and plumbed to be 90° vertical. The walls and columns are parallel to one another in the real world. Bowden seems to live in a fantasy world, where buildings have walls that splay out from one another and can never be parallel.

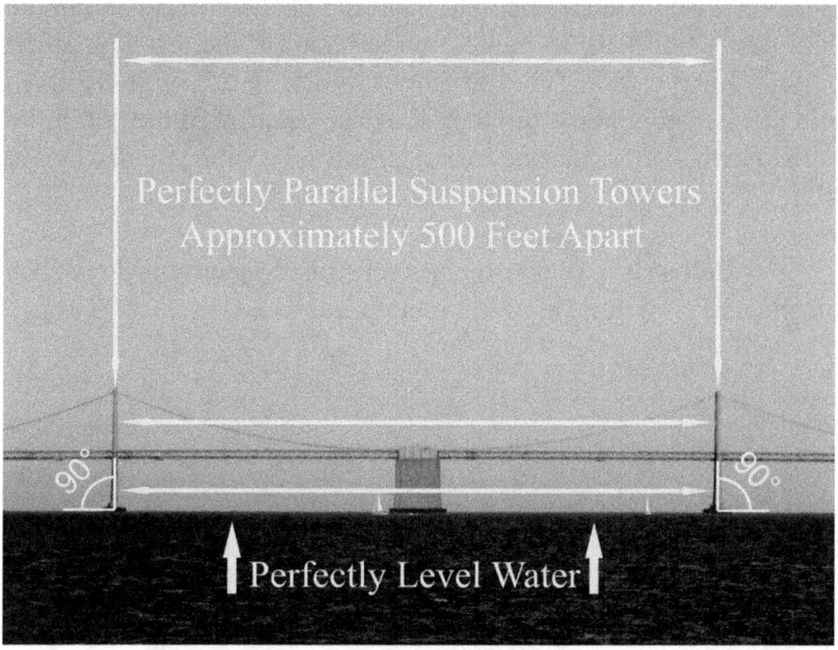

Above is a photograph of a portion of the western span of the San Francisco Bay Bridge. The suspension towers of the bridge, which are approximately 500 feet apart, are perfectly parallel as they rise 90° vertically above the perfectly level water. There is no measurable indication of any splaying from the surface of the water as Bowden alleges, and indeed would be required if the earth were a sphere. The San Francisco Bay Bridge is what bridges look like in the real world. This bridge, and indeed all bridges in the world, with their parallel, 90° plumbed, vertical suspension towers, testify that the earth is flat.

Nick Berry was "educated as a rocket scientist and aircraft designer, graduating with a Masters Degree in Aeronautical and Astronautical Engineering."[299] He has an internet blog called DataGenetics in which he claims that "because of the curvature of the Earth, the towers [on the Golden Gate Bridge] are a little wider apart at the top than at the bottom." Below is the image Berry posted showing his concept of the Golden Gate Bridge on a spherical earth. Berry calculated that the bridge towers should be 46.2 mm, (1 and 13/16 inches) wider apart at the top than at the bottom. Berry states: "The top of the Golden Gate Bridge is almost two inches wider at the top than the base because of the curvature of the Earth! Tell everyone in the car this next time you drive over it!"[300]

Below is a diagram depicting his conception. $l = 1{,}280$ meters and $l + x = 1{,}280.0462$ meters.

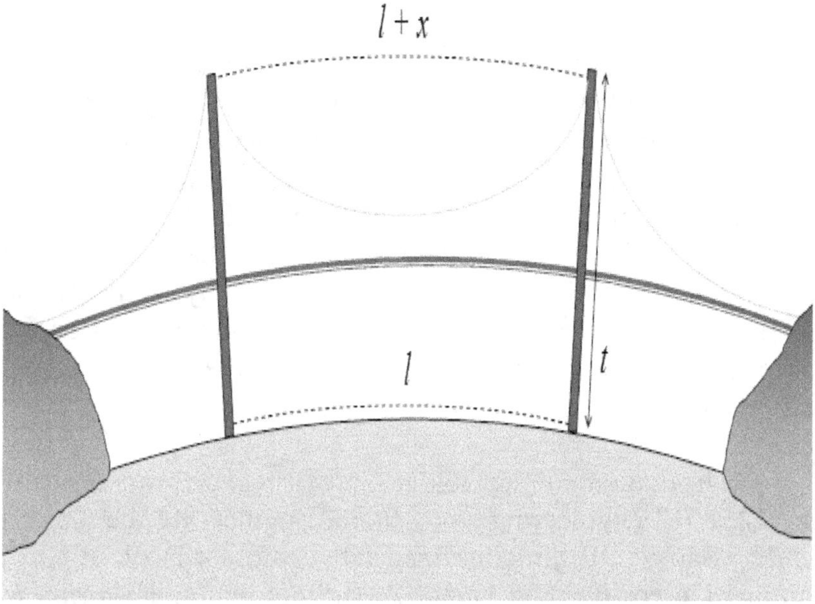

Figure 71: Nick Berry's depiction of his misconception of the splaying out of the Gold Gate Bridge Towers.

The suspension towers on the Golden Gate Bridge are precisely 4,200 feet apart, which is the equivalent of approximately 1,280 meters. Berry's calculations are based upon the false premise that the earth is a sphere (he clarifies that he thinks that the earth is actually an oblate spheroid). I checked Berry's trigonometric calculations and obtained a slightly different answer of 1.55 inches (approximately 1 and 9/16 inches) for the added distance between the top of the towers and the bottom. The difference in our calculations may be due to the fact that I assumed a radius of the earth of 3,963 miles, whereas Berry assumed a radius of 3,959 miles (6,371 kilometers). It makes no difference anyway, since the earth is not a globe, and thus both figures are mythical.

As with Bowden, Berry seems to be living in a fantasy world. In the real world, the real Golden Gate Bridge does not have suspension towers that splay out. The suspension towers on the Golden Gate Bridge that exist in the real world are perfectly parallel. Real engineers who were responsible for building the real Golden Gate Bridge created diagrams that clearly depict parallel suspension towers. The real engineers made no adjustments for a mythical, spherical earth. Indeed, the official website of the Golden Gate Bridge Highway & Transportation District depicts the spans of the Golden Gate Bridge at the top as being 4,200 feet (1,280 meters) apart.[301] There is no indication of any compensation for a spherical earth. Indeed, the engineering blue prints depict perfectly parallel suspension towers constructed 90° vertical from their bases.

The diagram of the Golden Gate Bridge below is from the June 22, 2012 engineering study titled, Engineering the Golden Gate, by Sarah L. Billington, Associate Professor Dept. of CEE, Stanford University, and David P. Billington, Jr., Historian & Independent Scholar.[302] The diagram was created by the Board of Engineers for the Golden Gate Bridge project, Leon Moisseiff, Joseph Strauss, Othmar Ammann, and Charles Derleth Jr.

Figure 72: Diagram of the real Golden Gate Bridge over perfectly level water on a flat earth.

Notice that the diagram depicts a perfectly level mean high water mark. It also depicts a slight ten foot bow upward in the center of the bridge as measured against the perfectly level mean high water mark. If there were any bow in the earth and resulting splaying in the towers, the drafters would have indicated that, just as they indicated the bowed roadway on the bridge. The suspension towers are perfectly perpendicular to the level water. The towers are perfectly plumb and parallel to each other. Notice also that the vertical support cables are also perfectly vertical and perfectly parallel with one another.

Nick Berry's spherical earth construct cannot hold up to scrutiny. The towers on the Golden Gate Bridge are diagramed by the design engineers as parallel to one another. If one reads any treatise on bridge building, one finds that all bridges across water are constructed assuming that the water is perfectly flat and that the bridge columns will rise precisely plumb and perpendicular to the water.[303] The towers are always diagramed as being parallel to one another. That is because that is how bridges are in fact built on a flat earth.

Below is a diagram from the Akashi Kaikyo Bridge, which is the world's longest suspension bridge. There are 1,991 meters (6,532 feet or 1.24 miles) between the suspension towers. The massive towers stand 282.8 meters (928 feet) in height over the water. Notice the measurement given for the distance between the towers in the diagrams is given from the top of the towers. That indicates that the engineers expected that there would be no deviation from the distance at the surface of the water, from which they would begin the construction of the towers. The engineers planned for the towers to be vertically plumb and parallel

Berry's point in calculating the supposed difference in the distances between to the tops of the towers on the Golden Gate Bridge was to argue that the difference is so slight as to be insignificant. What may seem slight can be quite significant when put under the immense stress carried by the bridge suspension towers. For example, the builders of the Akashi Kaikyo Bridge built the bridge suspension towers perfectly vertical with exacting tolerances. They had to set the base caissons within 10 centimeters (3.9 inches) of perfectly level. The immense stresses placed upon the towers required that the towers be built vertically plumb. The Sumitomo Group Public Affairs Committee website explains just how carefully the builders were in ensuring that each tower was built perfectly vertical. "If the caissons were placed even slightly off level, the towers would not stand vertically; so the foundations had to be set on the seabed at a depth of 60 meters, horizontal to less than a 10-centimeter margin of error."[304]

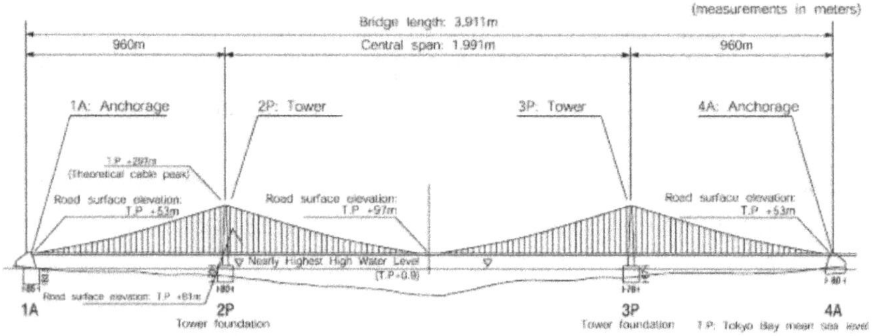

Figure 73: Diagram of the Akashi Kaikyo Bridge showing perfectly parallel suspension towers perpendicular over flat level water.

The seabed had to be leveled and the caissons for the towers had to be laid horizontal with the level seabed, with less than10 centimeters (3.9 inches) as the margin of error. Otherwise, the bridge would be out of specification. Apparently, the towers could only be adjusted 3.9 inches at their base to make up for any slight error in laying the caissons.

Assuming the earth is a globe, we find by using trigonometry that the Akashi Kaikyo Bridge towers would be 3.17 inches further apart at the top than at the bottom. Suspension bridge towers are designed to withstand massive downward compression loads, but they are not designed for massive lateral loads. If the towers are not vertically plumb and parallel, it would put added lateral stress on the towers.

If the earth were a sphere, no walls on any buildings or columns on any bridges would be parallel. The fact that all walls on square buildings and all columns on bridges built in the real world are designed and built to be parallel is proof that the earth is flat.

Malcolm Bowden presents his theoretical assumption that bridge columns (and by necessity all building walls) are wider apart at the top than they are at the bottom, due to the curvature of the earth. Being a civil engineer, Bowden knows that the splaying out of the columns needs to be true if the world is a sphere. He also knows that if it is not true, then the earth must be flat. He simply cannot accept that the earth is flat, and so he must necessarily accept the fiction that there are not, and indeed cannot be, any parallel bridge columns or building walls. The fact that his belief is not supported by the real world evidence proves his spherical earth construct is wrong.

Bowden's artifice is unsupportable by observable evidence. For example, below is a picture of the New York skyline. It is clear to see that the skyline is perfectly level. This is confirmed by the fact that the vertical walls on the buildings that are more than a mile apart are parallel to one another. The walls and floors of all buildings are built using levels. Levels create perfectly plumb vertical walls. The fact that the plumb vertical walls are also perfectly parallel means that the earth cannot be a sphere.

Vertical walls of buildings more than a mile apart are perfectly parallel.

In the above picture of the New York skyline, the lines placed along the side of the vertical walls of the buildings more than a mile apart have been extended past the height of the antenna at the top of the Freedom Tower. The tip of the antenna is 1,792 feet in height above the ground. If the earth were a globe, the vertical lines running from the walls would not be parallel (as they in fact are), but would be splayed apart at an angle of approximately .015°, which is approximately one minute of arc. If the earth were a sphere, the extended vertical lines would be splayed out a distance of approximately 5.43 inches further apart at the height of the antenna than at ground level. But no such splaying can be measured; the vertical lines remain perfectly parallel. Indeed, the vertical lines could be extended to infinity and they would remain perfectly parallel, thus indicating that the earth is flat.

Even in the unusual case when a building is designed with angled walls (like the Freedom Tower depicted in the photograph above), the floors on such buildings are always horizontally level and parallel to the ground, as well as being horizontally parallel to all floors on other buildings.

With some exceptions (like the Freedom Tower), almost all buildings have walls that are perpendicular (90°vertical) to level ground. The fact that almost all buildings are designed with walls that are plumbed vertical, and that age old construction method results in the opposing walls being perfectly parallel to one another, can mean only one thing: the earth is flat.

The Freedom Tower, which is the tallest building depicted in the picture above, was built to replace the destroyed World Trade Center Towers. The original blueprints of the North Tower of the World Trade Center were leaked to the media in 2007 by a whistle-blower.[305] Those blueprints prove both that the earth is flat and that the 2002 FEMA Performance Study[306] and the 2005 NIST Report[307] on the collapse of the twin towers were deceptive cover-ups. Read this author's book, *9/11 - Enemies Foreign and Domestic*, for more details on the who, how, and why of the 9/11 attacks. 9/11 attack researchers Jim Hoffman, Gregg Roberts, and Jan Hoyer explain that "[t]he blueprints to the Twin Towers and Building 7 remained off-limits to the public for more than five years after the attack [until they were leaked], despite the fact that the buildings were built with public money and that the engineering drawings of public buildings are supposed to be public information."[308]

Hoffman, Roberts, and Hoyer explain that the blueprint drawings contain a wealth of detail about the buildings, including the dimensions of the structural members. The original blueprints of the North Tower reveal that the FEMA and NIST reports were misleading regarding the massive core columns. The FEMA and NIST reports gave the false impression that the twin towers were weakly constructed.

> Both of the government-sponsored engineering studies of the Twin Towers' "collapses" -- FEMA's and NIST's -- are highly misleading about the core structures. Neither FEMA's Study nor NIST's Report discloses dimensions for core columns -- dimensions that are clearly evident in the architectural drawings. Both Reports use a variety of techniques seemingly designed to minimize the strength of the cores or to conceal their structural role entirely. So effective was FEMA at concealing the nature of the cores that the 9/11 Commission Report, citing the FEMA Report, denied the very existence of the core columns.[309]

The blueprints of the WTC North Tower also prove that the earth cannot be a sphere and must be flat. How could blueprints of a building prove that the earth is flat? Because the blueprints show that the North Tower was built as a perfectly square cuboid (with each corner chamfered 6 feet 11 inches). The blueprints show that the horizontal width for each side of the building was precisely 207 feet, 2 inches. That dimension ran from the bottom to the top of the building.[310]

Figure 75: World Trade Center Towers. The North Tower is on the left with the antenna.

The North and South World Trade Center Towers were sister towers, with exactly the same exterior dimensions. The North and South Tower buildings were both 110 stories tall. The North Tower rose 1,368 feet from the ground to its roof. Other than their corner chamfers, the Twin Towers were perfectly square cuboids. The North Tower had exactly the same dimensions, with each side of the building being 207 feet, 2 inches wide, at the 110th floor, 1,368 feet above the ground, as it did at the first floor at ground level. Such an occurrence would not be possible if the earth were a globe.

Using trigonometry, we find that if the earth were a sphere the 110th floor should have been splayed out from the base by .16 inches (almost 3/16 of an inch). One might think that 3/16th of an inch is not very much of a difference in width between the first floor and the 110th floor. But when one realizes that the tower was built using massive steel girders that were prefabricated offsite in uniform dimensions with preset holes, it becomes clear that the tower could not be built with any deviation in floor dimension. That is because the holes for the bolts and rivets must line up perfectly in order for the structure to be assembled. The holes on the girders would line up perfectly on the ground floor, but they would be almost 3/16 of an inch off center by the time the steel workers reached the 110th floor. Indeed, this problem would gradually manifest itself long before the steel workers ever reached the 110th floor. The girders could not be assembled, because the holes for the bolts and rivets would not line up and steel girders do not stretch. Indeed, this would be an issue for all four walls of the building, as the splaying outward of the walls and corners would be in all directions on each floor.

The 110 story World Trade Center Towers, with their perfectly parallel, vertically plumb walls, could not be built on a spherical earth. The architectural diagrams for the North Tower indicate that it was designed to have each side of each successive floor measure horizontally precisely 207 feet, 2 inches. There is no indication in the architectural drawings for a 3/16 inch allowance for the splaying out of the walls as the tower rose from the ground of a supposedly spherical earth to the 110th floor. The blueprints list the dimensions of every single one of the 110 floors as having a horizontal measure of precisely 207 feet, 2 inches. Such could only be the case on a flat earth.

Engineer W. Winckler stated that no construction engineer would dream of allowing for the curvature of the earth.[311] For example, Brian Mullin is a professional civil and structural engineer, who has been practicing for ten years and is licensed in four states.[312] Mullin stated in July 2016 that "[s]ome people think it [the idea of a flat earth] is ridiculous, but a lot of engineers and surveyors out there have really started to ask questions. They have started to realize that there aren't answers to these things. That we ignore the alleged curvature and rotation of the earth all the time and we never have any error because of it."[313]

What Mullin says is true. No engineer takes into account curvature of the earth, because to do so would be to take into account something that does not exist. Yolanda Whittle responded to Mullin's comment by stating, that during her two years as a surveying crew chief "we NEVER took curvature into effect."[314] (emphasis in original) Whittle stated that she is licensed in 3 states as a professional civil engineer. She further stated that "[w]e did make adjustments for the lens refraction on long jobs, but doing many traverse surveys over long roads, I never saw any curvature or deducted for curvature. When I started, I even asked the head licensed surveyor [with] over 25 yrs [experience], and he said that it's never used. That should tell you everything you need to know!"[315]

David Barnard Steinman, Ph.D., was a brilliant structural engineer who designed and built bridges in the United States, Thailand, England, Portugal, Italy, Brazil, Haiti, Puerto Rico, Canada, Korea, Iraq and Pakistan. He was awarded the Franklin Institute's Louis E. Levy Medal in 1957 for his design of the Mackinac Bridge.[316] Dr. Steinman wrote an authoritative book on constructing bridges titled, *A Practical Treatise on Suspension Bridges*.[317] Nowhere in his painstakingly thorough and precise treatise on bridge design and construction does he ever mention making an allowance for the supposed curvature of the earth.

In the fantasy world in which Malcom Bowden and Rick Berry live, they can make things up and say that an allowance must be made for the earth's curvature. They can get away with it, because they do not actually design or build anything. Engineers who design real buildings can't just make things up. In the real world, engineers who design structures that are actually built, must be accurate and precise. There is not a construction blueprint for any building or bridge, no matter how large the structure, that makes any allowance for (or even any mention of) the supposed curvature of the earth. That is because the earth is flat.

11 Airplane Level Flight Proves the Earth is Flat

If the earth were a globe, airliners would not be able to fly on a flat and level path. Any pilot will tell you that once a plane gets to its cruising altitude, the pilot "levels off" and flies in a straight and level path. Indeed, the pilot uses the horizon to ensure that the plane is flying level. When visibility is poor, the pilot will use the instruments and fly a level heading, using an artificial horizon. Either way, at all times the plane flies level, once the pilot reaches cruising altitude. If the earth were a globe, the pilot would have to constantly adjust the heading of the plane and dip its nose down to keep a constant altitude. However, the earth is in fact flat, so if a pilot dipped the nose of the plane down, to adjust for the supposed curvature of the earth, he would find that he is losing altitude. If the pilot kept on that downward trajectory, the plane would crash into the earth. The fact that no such downward adjustment is made by pilots for the supposed curvature of the earth is proof that the earth is flat. Eric Dubay explains:

> If the Earth were a sphere, airplane pilots would have to constantly correct their altitudes downwards so as to not fly straight off into "outer space!" If the Earth were truly a sphere 25,000

miles in circumference curveting 8 inches per mile squared, a pilot wishing to simply maintain their altitude at a typical cruising speed of 500 mph, would have to constantly dip their nose downwards and descend ... Otherwise, without compensation, in one hour's time the pilot would find themselves 166,666 feet (31.5 miles) higher than expected! A plane flying at a typical 35,000 feet wishing to maintain that altitude at the upper-rim of the so-called "Troposphere" in one hour would find themselves over 200,000 feet high into the "Mesosphere" with a steadily raising trajectory the longer they go. I have talked to several pilots, and no such compensation for the Earth's supposed curvature is ever made. When pilots set an altitude, their artificial horizon gauge remains level and so does their course.[318]

Indeed, the Federal Aviation Administration (FAA)

assumes a flat earth when training pilots and air traffic controllers using their Target Generation Facility (TGF). The TGF consists of several software programs that control simulation scenarios, using simulated aircraft. The TGF computer simulator drives almost all of the air traffic control laboratories at the FAA William J. Hughes Technical Center.[319] The FAA states: "Our lab has worked closely with the TGF group to have aircraft perform the way air traffic controllers would expect them to behave in the real National Airspace System (NAS). The simulated aircraft in TGF are quite realistic representations of their real-life counterparts."[320]

The software used in the TGF must be accurate in order to properly train pilots and air traffic controllers on the behavior of aircraft in flight. The assumption of a flat earth can be seen in a publication from the FAA, *The Engineering Analysis and Design of the Aircraft Dynamics Model For the FAA Target Generation Facility [TGF]*. That FAA publication explains the software that is used by the TGF to accurately simulate the behavior of aircraft over the earth.

> Specifically, this document discusses the detailed engineering design and software implementation of an Aircraft Dynamics Model (ADM) suitable for incorporation into the FAA TGF simulations at the FAA William J. Hughes Technical Center, Atlantic City, NJ. The model is designed to be implemented on computers located within the facility, and to work in conjunction with software models of radar, data links, and other Air Traffic Management (ATM) equipment to provide realtime simulation of aircraft operating within the National Airspace System (NAS).[321]

The FAA publication explains that in order for the software to accurately replicate the behavior of aircraft during flight over the earth, the software running the TGF simulator assumes that the earth is flat.

> The observant reader will notice that the aircraft equations of motion were calculated **assuming a flat Earth** and that we here assume the development frame was the NorthEast-Down frame. This implies necessarily that **earth rotation and the variation of the gravity vector with position over the earth were ignored** in developing the aircraft equations of motion. This simplification limits our mathematical model to the flight of aircraft only. The model will not properly handle the flight of sub-orbital craft and spacecraft such as intercontinental ballistic missiles, satellites, or the space shuttle. The model is adequate for all vehicles traveling under Mach 3.[322] (emphasis added)

Notice that the FAA document both assumes a flat earth and ignores the rotation of the earth and gravity. The FAA maintains that the computer model assumes a stationary, flat earth and at the same time remains precise and accurate regarding the flight of aircraft. The only way that a stationary, flat earth can be the basis for a model, and that model still be accurate, is if the earth is in fact flat and stationary. That means that the ignored gravity and rotating, globular earth do not affect the accuracy of the model, because they do not in fact exist in reality.

The proviso in the last sentence that the computer model does not address spacecraft or satellites is meaningless surplusage. The statement that the model is adequate for all vehicles traveling under Mach 3 is also meaningless, since it does not specify that it would be inaccurate for vehicles traveling over Mach 3.

The entire purpose of the TGF simulation is to create as real a simulation of actual aircraft flight over the earth as is possible. The only way to do that is to simulate what is real. That is why the TGF software assumes a flat earth. The TGF is designed to have "high fidelity" to reality. The FAA publication states:

> Currently, the aircraft dynamics incorporated in the TGF are based on the first principles of physics and aeronautics. The models provide the performance characteristics needed to support **high fidelity simulations**. The TGF incorporates fuel burn models environmental (weather) effects. Additionally, the modeled aircraft are representative of commercial air traffic in the US National Air Space (NAS). As future simulations are developed or brought to the Technical Center, **higher fidelity will be required to identify NAS operational safety and performance issues**. The TGF is prepared to increase its fidelity and operational connectivity required to meet the demands by the other FAA programs and simulators. **The goal of this project has been to develop and maintain a high-fidelity simulation** capability to meet the needs of the FAA in operating, testing, and evaluating its NAS.[323] (emphasis added)

The reality to which the TGF software is faithful is that the earth is stationary and flat. If the TGF simulator based its model on the false premise of a globular, spinning earth, it would create a safety issue for aircraft. The FAA simply had to assume a flat earth, because that is reality. If the FAA had assumed a spinning, spherical earth model, the pilots and air traffic controllers would be trained improperly and create the messy and embarrassing result of planes crashing with regularity.

Indeed, since safety is paramount, it was necessary for the FAA to verify, with real-life testing, the algorithms in the TGA simulator that assume a flat, stationary earth. The FAA publication states that "[t]he document concludes with a section on verification and validation, the process by which the various features of the simulation are tested and verified."[324] What did the section on verification reveal? The FAA publication reports that "[t]he testing that was done to verify and validate the TGF simulation gives us a high degree of confidence that the models contained herein have sufficient fidelity for use as a target generating tool."[325] The models that the testing validated as being faithful to reality to a high degree of confidence assumed a flat, stationary earth. That means that the FAA verified, through actual testing with aircraft, with a high degree of confidence, that the earth is flat and stationary.

The U.S. Government implicitly acknowledged the reality of the flat, stationary earth in another official publication. In NASA's 1988 publication (#1207) titled *Derivation and Definition of a Linear Aircraft Model*, it states that the generally accepted linear aircraft model is based upon "a rigid aircraft of constant mass flying over a flat, nonrotating earth."[326]

The U.S. Air Force is also on record implicitly admitting that the earth is flat. The Systems Engineering Branch, Flight Determination Laboratory of the United States Air Force prepared a technical memorandum in 1953 titled: *Atmospheric Refraction Errors for Optical Instrumentation, Preliminary Report, Technical Memorandum, No. 104*. That technical memorandum, which has since been declassified and posted on the U.S. Department of Defense Technical Information Center website, contains the following warning:

> This document contains information affecting the National Defense of the United States within the meaning of the Espionage Laws, Title 18, U. S. C., Sections 793 and 794. Its transmission or the revelation of its contents in any manner to an

unauthorized person is prohibited by law.[327]

What is so secret about atmospheric refraction that revelation would be a national security concern? The technical memorandum was written to provide technical information for the design of guided missiles. It is required to be accurate, as the missiles must be able to precisely hit their targets. Notably, the formulae in the technical memorandum were constructed assuming the validity of a flat Earth. Indeed, the assumption of a flat earth was the foundational premise for the atmospheric refraction formulae.[328] It makes perfect sense to assume a flat earth, because that is the actual environment in which the guided missiles will be deployed. With that in mind, the technical memorandum begins with the following passage:

> II. VALIDITY OF FLAT EARTH ASSUMPTION FOR ATMOSPHERIC CALCULATIONS
>
> The relative mass of the atmosphere at any elevation angle is given approximately by the cosecant of the elevation angle. **This relationship is correct for a flat earth and a flat atmosphere.**[329] (bold emphasis added)

In a footnote to the above passage there is mention of the fact that a formula is also given for a spherical earth, but that no account is given for the curved path that the spherical earth is supposed to present. "In the present comparison of the relative air masses at various elevation angles for flat and spheroid earth, no account is taken of the slightly curved path followed by radiation arriving at a given angle."[330] Why would the publication present two different formulae, one for the flat earth another for a spherical earth, and then disregard the spherical earth formula? Because the spherical earth formula does not comport with reality. It is simply cover language to suggest that the earth is a sphere and thus hide the reality of the flat earth.

Throughout the technical memorandum, the military scientists repeatedly reaffirmed the validity of their assumption of a flat earth. For example on page 5 of the memorandum, it states that "[a] flat earth assumption, then, may be safely used for total angular refraction down to about 7° elevation."[331] When it really counts, as when designing missiles, the military assumes that the earth is flat.

In March 2010, a research memorandum was published from the U.S. Army Research Laboratory on the issue of six-degree-of-freedom (6-DOF) projectile flight dynamic modeling. That esoteric topic simply means that the research memorandum outlines a model to predict the effect of liquid payloads on the trajectories of projectiles. The memorandum is talking about lethal bombs, and so the predictive model must be precise. The research memorandum states at the outset that "[t]hese equations assume a flat Earth."[332] Again, when precision is required in calculations, it is necessary for the U.S. Army to assume the reality of a flat earth.

A 2001 report from the U.S. Army Research Laboratory was titled: *Electromagnetic Fields Over Flat Earth*.[333] And sure enough, the U.S. Army Research Laboratory "assumed that the transmitting antenna and the target (or receiver) are located above, but near the surface of a flat idealized earth."[334]

In a research project on atmospheric oscillations conducted for NASA by the Georgia Institute of Technology, the researchers acknowledged that a flat, nonrotating earth is a common model used by scientists. "A model frequently used is that of a flat, nonrotating earth. ... The most one can profitably simplify the problem is to consider an isothermal atmosphere, plane level surfaces, and a nonrotating earth."[335] That is a secret that not expounded in the classroom; but it is freely shared with NASA.

In a 253 page Master of Science thesis paper published by the Massachusetts Institute of Technology (MIT) Department of Aeronautics and Astronautics, about targeting systems, "[t]hree targeting methods were considered: assuming a flat Earth."[336] What is significant about the research paper is that it addresses precision guided munitions targeting. The military strike must be pinpoint accurate to be effective. It states that "small changes in elevation can lead to drastic errors due to the orientation of the camera."[337] Precision guidance requires that the targeting system in the unmanned aerial vehicle (UAV) have a correct model of the earth, because it must have precise data about its elevation and orientation above the earth and the elevation of the target. That is why the paper assumes the reality of a flat earth.

The research paper states that the target will be identified for a strike when the UAV is within 1,200 feet of the target. But the UAV must travel many miles over the flat earth at an elevation of between 500 to 1,000 feet to locate the target. The General Atomics MQ-1 Predator UAV, which is similar to the UAVs depicted in the diagrams in the research paper, has a range of 770 miles.[338] The assumption of a flat earth would not work for the targeting system if the earth were a globe, because on the UAV proposed in the research paper "the camera is fixed at a 65.38° angle from the vertical (looking straight down)."[339] If the earth were a globe, the assumption of a flat earth would cause the targeting camera to quickly be out of calibration with the curve of the earth during level flight over the hypothetic globular earth as the UAV searched for a target. It, therefore, would make no sense to design a precision targeting drone assuming a flat earth if the earth were a globe. The necessary accuracy of the targeting UAV to locate a target and precisely guide a strike on that target after assuming a flat earth is evidence that the earth must be flat.

If the paper did not accurately describe the proper method for precise targeting, it would be worthless. If it were worthless, it would never be published by MIT. In order to be certified for publication by MIT the research paper had to be reviewed and approved by three experts in Aeronautics. But it's not worthless. It's accurate. It's accurate because it assumes the proper positions of targeted vehicles on a flat earth. The illustrations below are from the research paper and depict targeting systems being implemented assuming a flat earth.[340]

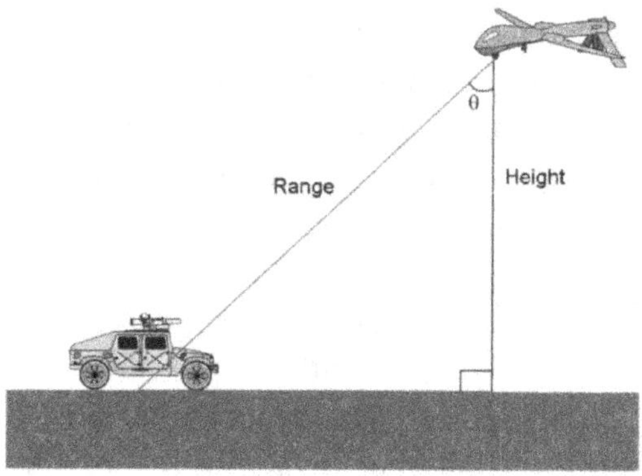

Figure 4-17: Example of Flat Earth method. The range (hypotenuse) and horizontal leg are in the xy direction of the camera's LOS vector, which is the red vector (heading vector) in Figure 4-16.

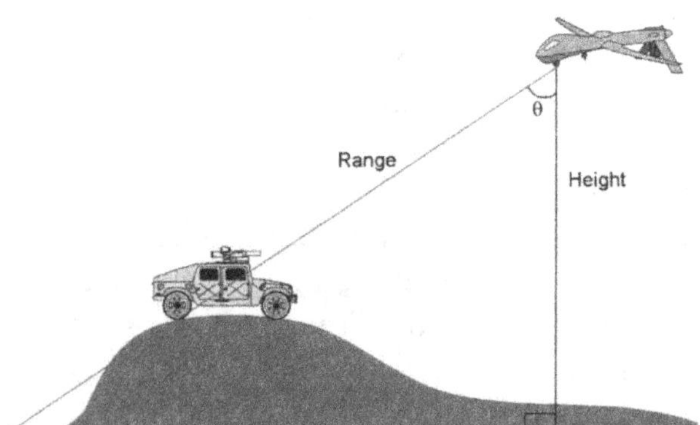

Figure 4-18: Example of changing terrain while using Flat Earth method. Again, red vectors along the heading vector, as in Figure 4-16.

Figure 77: Illustrations and annotations from the original MIT article on targeting systems that assume a flat earth.

In an article from the NASA Langley Research Center titled: *General Equations of Motion for a Damaged Asymmetric Aircraft*, the research paper states that "in this paper, the rigid body equations of motion over a flat non-rotating earth are developed that are not necessarily referenced to the body's center of mass."[341] Note that the equations in the paper were based upon motion of a damaged aircraft over a flat non-rotating earth. Why is that important? Because the paper states that "[i]n order to analyze the dynamics of damaged aircraft the dynamic equations of motion must properly reflect the underlying physics."[342] The underlying physics must be accurate, which is why the equations in the paper were based upon the flat non-rotating earth.

In March 1972, NASA prepared a technical memorandum at their Langley Research Center titled *Determination of Angles of Attack and Sideslip from Radar Data and a Roll-Stabilized Platform*.[343] Incidently, Langley, Virginia, is also home to the CIA. In that technical memorandum NASA assumed a "flat, nonrotating earth."[344]

Of course, no reference to the flat earth by NASA would be complete without some qualification. The 1972 NASA report states that "[i]t is assumed herein that the earth is represented as a flat, nonrotating reference frame. (This assumption limits the analysis to short-range, short-duration flights.)"[345] (parenthetical in original) What is the point in assuming facts, if the conclusions drawn are virtually useless in real life application? The assumed facts must be true for the conclusion of the study to have any validity. The stated qualification in the study that the flat earth assumption makes the data only germane to short-range, short-duration flights is a lie. It is put there to cover up the reality that the earth is flat. Indeed, the NASA researchers had to assume a flat earth, because otherwise, their study would not make sense. That is because they used gyroscopes in their study. Whether the flights are long range or short range, gyroscopes are only useful on a flat earth. Using a gyroscope is virtually worthless, even dangerous, on a spherical earth.

Every commercial and military aircraft has onboard an instrument that can only work on a flat earth. That instrument is the attitude indicator, which is also known as an artificial horizon. That artificial horizon instrument has a display, that shows the pilot the attitude of the plane to the actual horizon, even if he cannot see the horizon because it is obscured by inclement weather or darkness. The pilot can determine the roll of the aircraft, (i.e., if he is flying level) and the pitch of the aircraft (i.e., if the nose of he aircraft is pointed below or above the horizon). That attitude indicator works by means of a spinning gyroscope mounted on a gimbal.

A gyroscope has two important properties: 1) rigidity in space, and 2) precession. All practical applications of gyroscopes are based on these two properties. Precession means that the gyro will resist any force that attempts to change its plane of rotation. A force applied to the gyro will result in a movement of the gyro, but not in the direction of the applied force. The gyro will instead move at right angles to the direction of the applied force. The rate of precession is in direct proportion to the applied force.

The most important property of a gyroscope for a pilot is rigidity in space. That means that a gyroscope will retain its horizontal attitude parallel to the direction of the spin of its rotor and retain its vertical attitude in relation to its rotor's axis. The rotor's axis around which it spins is the y axis (vertical); the horizontal plane of the rotor is on the x (lateral) axis and the z (depth) axis. Thus, if the spinning gyroscope is on a gimbal that allows for rotation along three axes (x, y, and z), the gyroscope will retain its original attitude in relation to the x, y, and z axes, and the gimbal will move around the fixed gyroscope. The x, y, and z axes of the gyroscope will remain rigid in space, offering a fixed matrix upon which the airplane can move. Any difference between the fixed axes of the gyroscope and the moving airplane will show up in the coordinates on the attitude indicator instrument. That is what the gimbal attached to the aircraft in the instrument bay allows, as illustrated in the following frames from a U.S. Navy training film produced in the 1960s, explaining how gyroscopic flight instruments work.

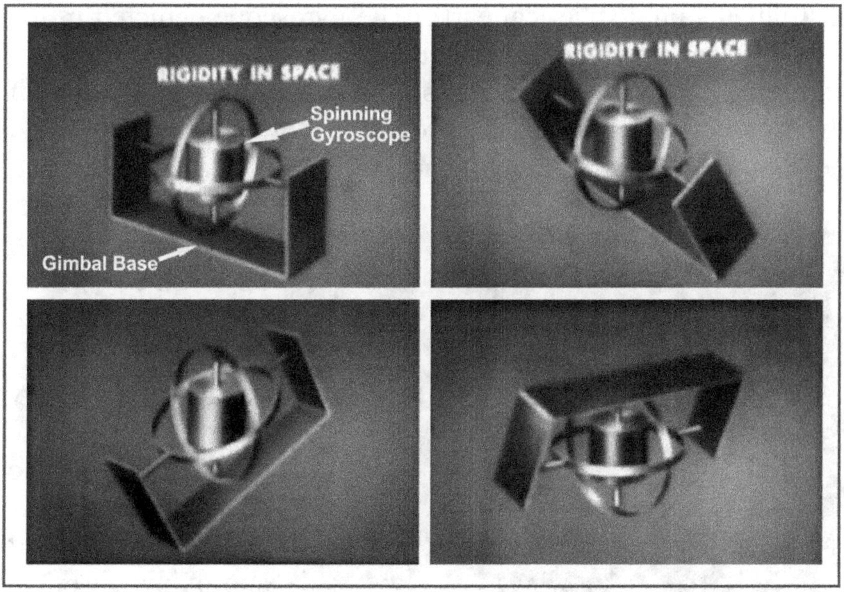

The spinning gyroscope in the attitude indicator stays fixed and rigid in space. The gimbal in the attitude indicator instrument allows for the aircraft to move in relation to the fixed axes (x, y, and z) of the spinning gyroscope. Thus the pilot can see by looking at the attitude indicator instrument the exact attitude of his plane in relation to the horizon.

The problem with the spherical earth model is that a plane with a attitude indicator showing straight and level flight will cause the plane to fly off into the upper atmosphere as the curve of the globe falls away from the aircraft. The only way that a pilot on an aircraft can fly a straight and level path using his attitude indicator is on a flat earth.

Below is a frame from a United States Navy training film, circa 1960.[346] The frame illustrates the inherent impossibility of maintaining level flight for an aircraft using gyroscopic instruments on a spherical earth. The gyroscope depicted in the frame is seen to maintain its rigidity in space as it hypothetically circumnavigates the globe. As is obvious by the rigid attitude of the gyroscope, a plane with a gyroscopic attitude indicator would be gradually heading out toward the upper atmosphere as it traversed the globe. Indeed, as indicated in the film illustration, the plane would be in a vertical trajectory after flying one quarter of the circumference of the earth, a distance of approximately 6,250 miles.

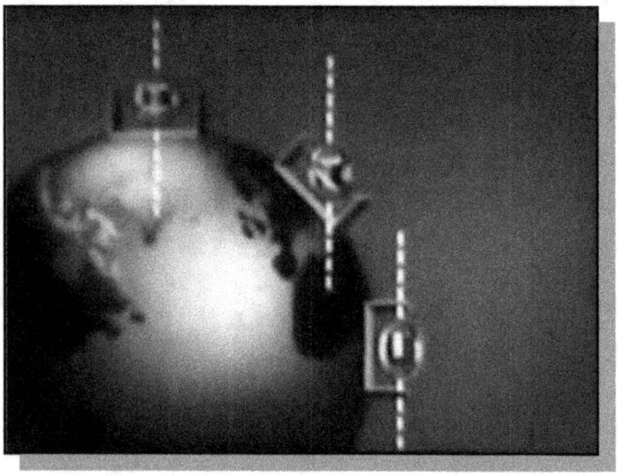

Incidently, the training film narrator tried to explain away this apparent problem by stating that there is a continuous adjustment to the attitude of the gyroscope built into the attitude indicator instrument. However, the mythical, on-the-fly adjustment to the gyroscope is impossible, as such an adjustment would render the gyroscope useless. Recall that the fundamental property of a gyroscope is that it maintains its rigidity in space. If the fundamental property of a gyroscope (rigidity) can be so easily altered, that means that rigidity is not truly a fundamental property of a gyroscope. An adjustment to the fundamental characteristic of rigidity in space of the gyroscope, if it were possible, would defeat the whole reason to use the gyroscope. It would make no sense to use an instrument with a gimbaled gyroscope that would require an additional feature to neutralize the primary property of the gyroscope (rigidity). The supposed neutralizing feature of the instrument would need to detect the precise curvature of the supposedly ball earth and maintain a true course over its surface. If it could do all of that, why add the complication of a gimbaled gyroscope? An attitude indicator with gimbaled gyroscope would only be a hindrance on a ball earth.

Rigidity in space is exactly the opposite of what is called for in an attitude indicator on a spherical earth. A gyroscopic attitude indicator instrument is necessary on a flat earth, but that same instrument would be pointless, and even dangerous, on a spherical earth.

A gyroscopic attitude indicator keeps its attitude precisely perpendicular on its axis, on a fixed plane to the spin of its rotor, and perfectly level in relation to the surface of the earth, because the earth is flat. The rigidity in space of a moving gyroscope that maintains its attitude is only useful in an attitude indicator on a flat earth. A gyroscopic attitude indicator simply will not work on a globe.

One interesting thing revealed in the above illustration from the United States Navy Training Film is the fact that if the earth were a spinning globe, a gyroscope in a gimbal placed on a level table should show the spin of the earth. Thus, if the earth were spinning, the gimbal surrounding the gyroscope would show a movement of 1.25 degrees within 5 minutes and 15 degrees in one hour. As the United States Navy Training Film illustrates, after 6 hours, the gimbal should turn 90 degrees underneath the gyroscope on a spinning globe. Such an experiment using a gimbaled gyroscope has been performed. The experiment showed no movement whatsoever of the gimbal around the gyroscope, thus proving that the earth is not a spinning globe.[347]

12 Can You Hear Me Now?

Exalt Wireless, which bills itself as a leading supplier of next-generation wireless connectivity systems for Internet infrastructures, announced in a February 8, 2016, news release that its ExploreAir LR 7 GHz microwave system was able to establish a reliable, high-speed link spanning what it believes is a record-breaking 235 kilometers (146 miles) over water.[348] The notable thing about the accomplishment is that microwave transmissions, as with all electromagnetic waves, travel in a straight line. An engineering and technology website explains:

> A clear path is critical to the microwave link's success. Since microwaves travel in essentially straight lines, man-made obstacles (including possible future construction) that might block the signal must either be overcome by tall antenna structures or avoided altogether.[349]

The supposed record claimed by Exalt is dubious, since Andy Singer reported that in 1979 there was an even longer successful microwave transmission.

> We did some research and perhaps not surprisingly the record for the world's longest

> microwave link was a hop commissioned decades ago in 1979. The laws of physics have not changed. This record setting link was 360 km [224 miles] long and crossed the Red Sea over a good part of its path connecting Jebel Ebra, Sudan with Jabal Dakka, Saudi Arabia.[350]

Exalt Wireless was reorganized due to financial insolvency in 2014.[351] Their corporate website presently makes no mention of their 2016 alleged 235 kilometer microwave transmission.[352] Singer's revelation of a 1979 record was reported by him on March 23, 2016, one month after the Exalt press release. This author could not verify even the existence of a place called Jebel Ebra, Sudan, which was reported by Singer to be one end of the 1979 world record microwave transmission. It could be that Jebel Ebra existed in 1979, but it no longer exists due to the political instability in Sudan over the years.

The odd thing about the supposed microwave signal records is that they are not really that far. Ordinary radio signals have been sent with regularity exponentially further for more than a hundred years. Is it possible that the record transmissions of microwave signals are in fact disinformation to give the false impression that radio signals cannot travel very far? Why would someone do that? To conceal the flat earth, which would be required for long range microwave transmission.

It is possible on a supposed spherical earth for a transmission station to be high enough over sea level to explain how the radio signal could pass above the intervening curved earth for a very long distance. In any event, the above reported cases of record microwave transmissions are of little help in proving whether or not the earth is flat, since neither Singer nor Exalt reported the height of the transmission towers. That critical missing information seems purposeful. There is an unverified claim that the Exalt towers were 50 feet in height, but that information comes from an anonymous source and thus cannot be verified.[353]

The information from Exalt and Singer is not completely irrelevant, because it reinforces the fact that all microwave transmissions are line of sight; that is, they travel in a straight line. Microwave transmissions are simply electromagnetic transmissions in the microwave spectrum. A fundamental characteristic of all electromagnetic waves is that they travel in a straight line.[354] At the beginning of the development of worldwide radio communications, the developers of radio technology let the cat out of the bag. They revealed that transmissions of electromagnetic waves can be sent thousands of miles. That distance far exceeds what is claimed today. How was it possible to send radio communications thousands of miles 100 years ago, and yet today it is claimed that microwave communications can only travel a little more than 100 miles? The proof that radio waves can be sent thousands of miles, necessarily proves that the earth is flat.

Radio waves are electromagnetic waves at a longer wavelength than microwaves. Both microwaves and radio waves are electromagnetic waves that require line of sight for communication. While electromagnetic waves can travel through some surfaces, like walls, it is generally thought that "earth's curvature is a direct block to line-of-sight communication. When enough distance separates the two radio stations so that their antennas fall behind the curvature, the Earth itself blocks the transmitted signals from the receiver."[355] This fact confused the father of radio, Guglielmo Marconi. On 12 December 1901 he was able to make the first long distance Morse code wireless communication between St. Johns, Newfoundland, Canada, and the Poldhu Wireless Station, Cornwall, England.[356] That radio signal was transmitted a distance of more than 2,000 miles across the Atlantic ocean. Such a transmission is an impossibility on a spherical earth. Indeed, it would be blocked by a 126 mile high bulge, if the earth were a sphere. Because Marconi thought the world was a sphere, and he understood that radio waves travel in a straight line, he was at a loss to explain how the radio waves traveled more than 2,000 miles on a globular earth.

Ham radio operators today can communicate with people all over the earth. This author remembers talking with a ham radio operator in the early 1970's. He had a massive radio antenna in his backyard. He explained that he had records of contact with ham radio operators in every country in the world, except Russia. How could he communicate with fellow ham radio operators all over the world if the earth were a sphere?

Modern science has come up with all sorts of convoluted theories to explain such long distance radio transmissions, because they cannot allow it to be known that the earth is flat. The most prevalent theory is the ionosphere bounce theory.[357] Under that mythology, radio operators can talk to people on the other side of the supposed spherical earth, not because the earth is flat, but because their radio signal bounces off the thin upper atmosphere, called the ionosphere.

Think about it logically. Under the ionosphere bounce stratagem, radio waves travel through the atmosphere until the atmosphere gets extremely thin (it is then called the ionosphere), and when those radio waves reach that thinnest part of the atmosphere, they bounce off and return to earth. Now, the modern scientists are not sure where the bouncing takes place, because the height of the ionosphere ranges from 50 miles to 600 miles in altitude. Does that make sense? No.

Radio waves, like light waves, are electromagnetic waves, and thus follow similar rules of refraction as they travel through the atmosphere. In most cases, the atmosphere refracts electromagnetic waves, it does not bounce (i.e., reflect) them. But reflection of radio waves as an explanation for long distance radio communication is not out of the question. The issue is what is reflecting the waves. It certainly cannot be reflected by a thin upper atmosphere. There must be something with much more physical density.

The modern myth of the ionosphere-bounce theory is completely impeached by the practice of what is known as moon-bounce, or earth-moon-earth (EME) communication.[358] Moon-bounce communication or EME is where radio operators, including amateur radio (ham) operators, bounce radio signals off the moon.[359] The moon is supposed to be 238,900 miles from earth. Think about this logically; if the ionosphere, which is the atmosphere that is at a height of 50 to 600 miles from the earth's surface is supposed to reflect radio signals back to earth, how can those same radio signals pass through the ionosphere and travel all the way to the moon, which is supposed to be 238,900 miles from earth?

Obviously, the ionosphere bounce theory has a problem. Scientists cannot have a theory whereby radio waves bounce off the ionosphere, but also have those same radio waves pass through the ionosphere to bounce off the distant moon supposedly located beyond the ionosphere in outer space. The moon must be closer than the alleged 238,900 miles from earth and it cannot be above what scientists call the ionosphere. The only way the same radio signals that are supposed to bounce off the ionosphere can also bounce off the moon would be if both the moon and ionosphere were in the same place. The ineluctable conclusion is that the radio signals bounce off the firmament and the scientists are deceptively calling that phenomenon an ionosphere bounce. It also means that the moon must necessarily be in the firmament. That perfectly explains how radio signals can be reflected from both the moon and every other area in the sky where the moon is not located. Indeed, the evidence of radio signal reflection from both the sky and the moon confirms God's word; the moon is in the firmament.

And God said, **Let there be lights in the**

> **firmament** of the heaven to divide the day from the night; and let them be for signs, and for seasons, and for days, and years: And let them be for **lights in the firmament** of the heaven to give light upon the earth: and it was so. And God made two great lights; the greater light to rule the day, and the lesser light to rule the night: he made the stars also. And **God set them in the firmament of the heaven** to give light upon the earth, And to rule over the day and over the night, and to divide the light from the darkness: and God saw that it was good. And the evening and the morning were the fourth day. (Genesis 1:14-19)

Elizabeth Howell, writing for Space.com, explains the generally accepted scientific theory that "Earth's magnetic field and atmosphere shields the planet from 99.9 percent of the radiation from space."[360] That scientific theory poses a problem for the mythology of the moon being 238,900 miles away in outer space. Modern scientific theory has the earth's atmosphere acting as a giant Faraday cage to protect the earth from the allegedly dangerous radiation made up of gamma rays, x-rays, and cosmic rays. There is not agreement among scientists whether cosmic radiation is made up of electromagnetic waves or subatomic particles.[361] NASA categorizes cosmic rays as part of the electromagnetic spectrum.[362] Just below cosmic rays on the electromagnetic spectrum are gamma rays and x-rays. Radio waves are also considered electromagnetic waves.

God is above us in heaven, watching over man. (Job 22:12; Ecclesiastes 24:5; Psalms 14:2 & 33:13) God states that "heaven is my throne." (Acts 7:49) But Satan has used "science falsely so called" to remove God in the minds of men from his throne and replace the glorious abode of God with a dark and hostile empty vacuum, full of deadly x-rays, gamma rays, and cosmic rays.

Outer space is always portrayed as a black void. Apollo 12 Astronaut Pete Conrad described space as the blackest black he has ever seen.[363] Apollo 17 Astronaut Eugene Cernan described the earth from his alleged Apollo spacecraft as "surrounded by the blackest black you can conceive in your mind."[364] Astronaut Leroy Chiao stated that "when you look out into space away from the sun, it's the darkest black you can imagine."[365] Neil Armstrong, the alleged first man to walk on the moon said "the sky is a deep black when viewed from the moon as it is when viewed from ... the space between the earth and the moon."[366]

Why do astronauts describe the mythical outer space as the blackest black? Because "God is light, and in him is no darkness at all." (1 John 1:5) Satan must remove any hint that God might be in heaven and so outer space must be void of all light. Because, in God there is no darkness at all, and outer space, which is supposed to be where the heaven of God is located is full of darkness. NASA, therefore, implants in the minds of men that the heaven of the Bible is a myth and there is no God.

Indeed, they have turned the glorious abode of God in heaven into hell. Jesus Christ describes hell as **"outer darkness."** Which is exactly how outer space is described by astronauts. "And cast ye the unprofitable servant into **outer darkness**: there shall be weeping and gnashing of teeth." (Matthew 25:30) That "outer darkness" is the everlasting fire of hell. Jesus explained that fact 11 verses after describing "outer darkness" when he said: "Then shall he say also unto them on the left hand, Depart from me, ye cursed, into **everlasting fire**, prepared for the devil and his angels:" (Matthew 25:41) Jesus explained that the fire of hell is the everlasting punishment due for those for their sins who have not been elected to salvation. "And these shall go away into **everlasting punishment**: but the righteous into life eternal." (Matthew 25:46)

Outer space, as portrayed by modern science, is a subtle and blasphemous attack on God. Only the chosen high priests of science (called astronauts), can venture into the devilish black void of space. They must wear special ceremonial protective suits before they enter into the godless void of space. The chosen astronauts come back to earth to report that instead of finding God, they found an empty, hostile, and blackest of black voids. Of course, outer space as portrayed by NASA is not real; it is religious mythology. Indeed, NASA has never explained how the special ceremonial garb, called space suits, that have no lead linings, can protect their astronauts (priests) from the dangerous x-rays, gamma rays, and cosmic rays that are alleged to exist in outer space. The dangerous environment of outer space seems very much like the perilous "everlasting fire" of hell. But making that point is impolite and marks a person as a lunatic (i.e., heretic) against science (i.e., religion).

The ingrained societal acceptance of the orthodox view that the earth is a sphere, usually brings with it the conditioned deprecatory charge that anyone who believes that the earth is flat must be crazy. Such a crazy person must be ostracized and excluded (excommunicated) from the scientific community. That marginalization is part of the conditioning process and is systemic in the hierarchy of the heliocentric scientific system. If ostracization is determined to be insufficient, the dissenter is marked as insane (apostate); such a person must then be subjected to psychological (ecclesiastical) treatment to cure him of his insanity (apostasy). They have established a special class of inquisitional priests (psychologists) who wear white priestly garb to administer the necessary treatment. Psychology is a religion without God.

Everybody who has had medical or dental x-rays knows that they are dangerous, and typically vital organs are covered by a lead shield. The x-ray technician will usually stand behind a lead lined wall when activating the x-ray machine. That is done because it is understood that the atmosphere is no barrier to x-rays. But modern scientists would have us believe that the same atmosphere that offers no barrier to x-rays in a medical facility, magically becomes an impenetrable barrier when the x-rays are trying to reach earth from outer space.[367] It is impossible for the atmosphere to block x-rays. That means that there cannot be any x-rays coming from outer space. Indeed, x-rays cannot emanate from outer space, because there is no outer space.

If Earth's atmosphere is actually able to block dangerous gamma rays, x-rays, and cosmic rays supposedly bombarding earth from outer space, it can certainly block radio waves. How could astronauts supposedly in outer space, and even on the moon, use electromagnetic radio waves to communicate through that purportedly impenetrable atmospheric shield to NASA mission control back on earth? Indeed, how can the alleged protection offered by the atmosphere in blocking out the supposedly powerful and dangerous electromagnetic waves, at the same time allow exponentially weaker electromagnetic transmissions from amateur radio operators to pass back and forth through that supposedly impassable atmosphere as they bounce off the distant moon?

The scientific community simply states that the atmosphere is able to block gamma, x-rays, and cosmic rays, but it cannot block radio waves. But they contradict themselves when they say the atmosphere (ionosphere) can in fact block and reflect radio waves when they travel upward from earth. Why have such illogical and opposing theories? Because the high priests of science need the ionosphere bounce phenomenon to explain long distance radio communications on the supposed spherical earth. And they, at the same time, need to allow for radio communications from earth with astronauts who are supposed to be in outer space.

Obviously, the ionosphere bounce theory poses a problem for outer space radio communications. There is simply no way for the radio signals to reach outer space if they purportedly bounce off the ionosphere. For radio signals to reach outer space would require the ionosphere to have an on/off switch that turns off the ionosphere reflector and allow waves through it when NASA wants to communicate with astronauts in space and on the moon. The on/off switch would also come in handy in allowing radio signals through the ionosphere when ham radio operators want to bounce their signals off the purportedly distant moon.

It sounds ridiculous, but it is a fact that the modern scientific theories have the ionosphere being both a reflector and a conductor of radio waves. Modern "science" uses the ionosphere bounce theory to explain the long distance radio communication that otherwise would be impossible on a globular earth. And then, presto-chango, the formerly impenetrable ionosphere reflector suddenly allows the radio waves to pass right through when they need to explain radio communication to and from the mythical vacuum of space.

That is obvious nonsense. It is "science falsely so called." 1 Timothy 6:20. It is satanic religious dogma, masquerading as science, being promoted by evil men trying to keep a lid on the true nature of God's creation. There is, in fact, no vacuum of space. The moon is close; it is in the firmament. When radio waves are being reflected back down to earth, they are being bounced off the firmament.

The radio signal sent by Marconi was Morse code. On September 29, 1915, the U.S. Navy, in cooperation with AT&T and Western Electric, was able to communicate by voice over radio waves from their naval radio station in Arlington, Virginia, near Washington, D.C., to their station in Honolulu, Hawaii. That milestone in radio communication was reported in the November 1915 issue of The Electrical Experimenter:

> The most wonderful feat in wireless telephony

> was accomplished on the night of Sept. 29 last, when the human voice was projected through the ether from Washington, D.C., to Honolulu, a distance of 4,900 miles![368]

Notice that it was reported that the radio waves traveled through the ether. You will read in subsequent chapters how the existence of the ether has been hidden from the masses. The notable thing about the September 29, 1915, broadcast from Arlington to Honolulu is that radio waves trying to reach a distance of 4,900 miles on a supposed spherical earth would find an impassable hump of earth and water 758 miles high. Modern science gets around that obstacle by postulating the ionosphere-bounce theory.

The successful long distance radio communication was not a fluke. It was the culmination of years of experimentation and product development. The transmission was made in 1915, at a time before any relay stations were constructed and before any alleged space satellites were launched. It was so early in the development of radio communication that there were no means for those at the naval radio station in Honolulu to respond to the signal. "Owing to the lack of adequate sending apparatus at Honolulu it was impossible to communicate back to Washington, but a cablegram announced the fact that the message had been received and distinctly heard."[369]

The 4,900-mile Arlington/Honolulu radio transmission happened on the same day that a radio transmission was made 2,000 miles across the Atlantic Ocean from Arlington, Virginia to Paris, France. On the 100th anniversary of the event, Julia Zorthian, writing for Time Magazine explained:

> A hundred years ago, the word "hello" spoken in Arlington, VA was heard in Paris. The word originated from the vibrations in the vocal chords

of Mr. B. B. Webb, an engineer at the Arlington radio station. That sound passed through Webb's lips, crossed Virginia airspace, entered a radio mouthpiece. There it was converted into electromagnetic waves, and in that moment on Oct. 21, 1915, human speech did something it had never done before: crossed the Atlantic.

On the European end, an antenna atop the Eiffel Tower picked up the radio waves. Two engineers with the American Telephone and Telegraph Company (now AT&T) stationed at the receiver, which demodulated information from the signals into audio, heard Webb's voice on the Paris side.

***

That night, the American Telephone and Telegraph Company president Theo Vail sent a dispatch that said:

"The talk from Washington this morning, heard both in Paris and Honolulu...establishes as a fact that under favorable atmospheric and electrical conditions, with proper equipment which the engineers of the American Telephone and Telegraph Company have developed, we will be able to carry on conversation between New York and European points as well as to the western coast and points across the Pacific Ocean."[370]

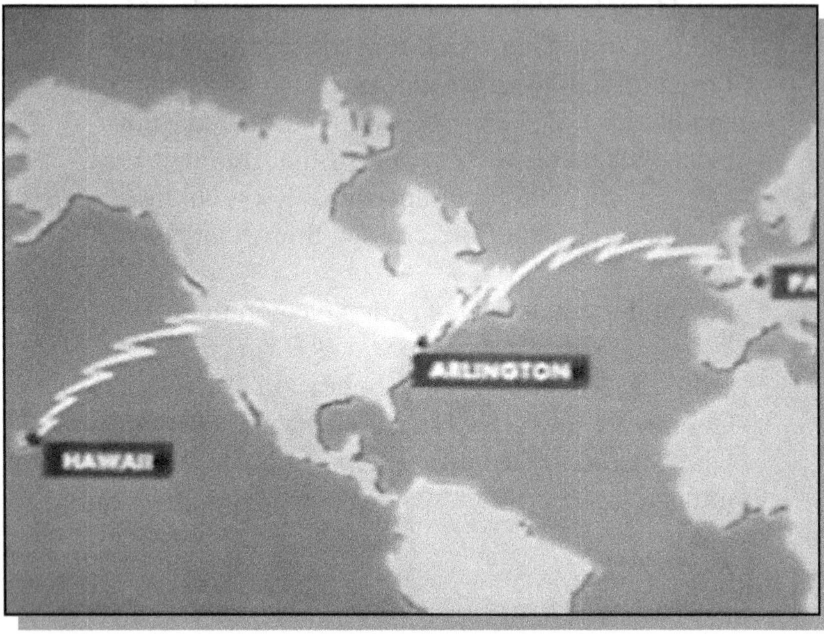

Figure 80: Screen shot from a 1953 documentary film about transistors, which illustrates the 1915 radio transmissions from Arlington, VA, across the world to Hawaii and Paris. Such transmissions would be impossible if the earth were a sphere.

The radio transmission from Arlington, Virginia to Paris, France, spanned 2,000 miles. That transmission to Paris would be impossible if the earth were a sphere. There would be a 126 mile high hump of water blocking the transmission. The fact that the radio transmission was successful is yet more confirmation that the earth is flat.

What is suspicious about Julia Zorthian's Time Magazine commemorative article is that Zorthian only mentions the 2,000-mile Arlington-to-Paris transmission. The more significant radio communication from Arlington to Honolulu, at a distance of 4,900 miles, was not mentioned, other than the passing reference to Hawaii in the quote from AT&T President Theo Vail.

It seems that Time Magazine was trying to play down the distance of the 1915 radio transmissions by focusing on the shorter 2,000-mile Arlington-to-Paris radio transmission and omitting mention of the much longer 4,900-mile Arlington–to-Honolulu radio communication. That makes no sense if the objective of the article was to explain the monumental achievement on that day one hundred years earlier. But it makes perfect sense to play down the significance of the 1915 achievement if the objective is to hide proof that the earth is flat.

The inherent illogic and contradiction in the ionosphere-bounce theory make it a flawed and inconvenient explanation for long-range radio transmissions. But it is all modern science has, so they just go with it. A straight line transmission on a spherical earth from the Arlington Naval Radio Station to Honolulu (or Paris for that matter) would be impossible. Perhaps Time Magazine did not want to reveal or explain the longer 4,900 mile Arlington-to-Honolulu radio transmission because it reveals too much; i.e., it is stronger evidence that the earth is flat.

The Arlington Naval Radio Station had two 450-foot antenna towers and one 600-foot antenna tower.[371] Arlington, Virginia, is 272 feet above sea level. That means that the radio signal would be broadcast from a maximum height of 872 feet above sea level. The problem is that the radio signal would have to travel in a straight course across the continental United States. That poses a problem even on the flat earth. That is because the radio signal would traverse over land that is on average more than 2,000 feet above sea level. While 2,000 feet is the average elevation, there would be areas much higher than that average. For example, the radio signal would have to travel over southern Colorado and northern New Mexico. The topography in that region has an impassible barrier of mountains rising to more than 6,000 feet above sea level. There is simply no possible way that a radio signal emitted at 872 feet above sea level could travel in a straight path across the continental United States and reach Hawaii with all of that mass of earth in the way. The radio signals would be stopped by that impenetrable land barrier.

We have established that the radio signal could not travel in a straight line of sight path to Hawaii, even on the flat earth, and the ionosphere bounce is an impossible myth. So, how was the 1915 radio transmission made from Arlington to Honolulu? The radio waves were bounced off the firmament that covers the flat earth. God describes the firmament as "strong, and as a molten looking glass." Job 37:18. The radio signals that reached Honolulu were reflected off that strong firmament.

That is, perhaps, why the Time Magazine article did not mention the Arlington-to-Honolulu transmission. Any close examination of the details would cause people to ponder how the transmission was accomplished. That, in turn, would lead to the conclusion that the means of reflection could not be the thin upper atmosphere (ionosphere), it must be something that is "strong, and as a molten looking glass." That would lead to an acceptance of the biblical model of a flat earth with a tent-like firmament over it. That could simply not be allowed. And, so, the decision was made to not mention the more significant achievement of the 4,900-mile Arlington-to-Honolulu radio communication and instead only mention the less significant 2,000-mile Arlington-to-Paris radio communication.

The radio signals that absolutely prove that the earth is flat are those of the LORAN system. LORAN, which is an acronym for Long Range Navigation, was a radio navigation system developed in the United States during World War II. It involved radio signals sent straight out to sea from shoreline antennae. The ships at sea would be able to mark their location on charts by LORAN instruments that compared the arrival time of the radio signals from different shoreline radio towers. The very existence of the LORAN system proves that the earth is flat, because the system would not work if the earth were a globe.

The Encyclopedia Britannica gives a succinct description of LORAN:

> Loran, abbreviation of long-range navigation, land-based system of radio navigation, first developed at the Massachusetts Institute of Technology during World War II for military ships and aircraft located within 600 miles (about 970 km) of the American coast. In the 1950s a more accurate (within 0.3 mile [0.5 km]), longer-range system (over 2,000 miles [3,200 km]), known as Loran-C, operating in the 90–110 kilohertz range, was developed for civilian use, and the original loran (renamed Loran-A) was phased out.[372]

LORAN-C was terminated as a service in the U.S. by the U.S. Coast Guard on 8 February 2010.[373] The tallest radio tower used by the Loran-C system was the 1,350 foot tower located at the LORAN Station at Port Clarence in Alaska.[374] Port Clarence is 16 feet above sea level, which would put the top of the radio tower at a height of 1,366 feet above sea level.

What is notable is that the LORAN system required that the radio signal be sent straight out to sea to ships on the water. There could be no bouncing of the signals off the firmament. The water, being flat, offered no obstacle to the LORAN radio signals. The system required that the radio signals be sent horizontal to the water. For the system to work, the ships needed to receive at least two intersecting radio signals from two different towers. From those two signals they obtained their first line of position on their shipboard LORAN instrument. The shipboard LORAN instrument then needed to receive two separate signals from two other towers to obtain a second line of position. Where those two lines of position intersected gave the ship a fix on its position at sea. The captain would then plot his position on his shipboard LORAN chart.[375]

The LORAN-C system had the capacity to send signals more than 2,000 miles out to sea. Assuming towers at a height of 1,366 feet above sea level, if the earth were a globe, the LORAN-C system would only be able to send radio signals 45 miles out to sea, before having the signals blocked by the earth's curvature. And that 45 mile distance assumes that the signals are being sent out from the tallest tower in the United States. That means that if the earth were a globe, a ship 2,000 miles away from the towers would have the LORAN-C signals blocked by an impenetrable 1,955 foot long, 115-mile high, hump of water rising above the height of the LORAN-C radio signals.

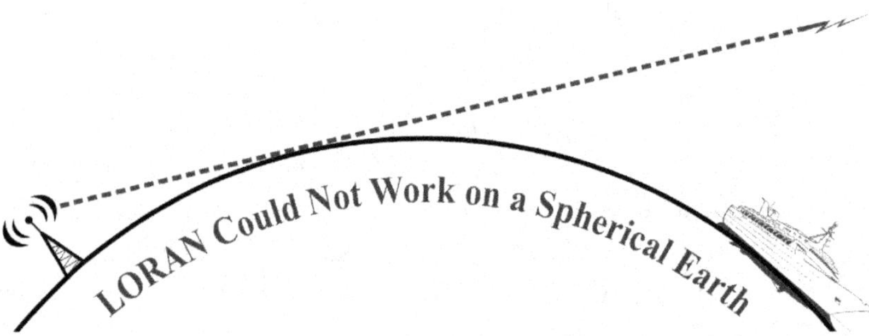

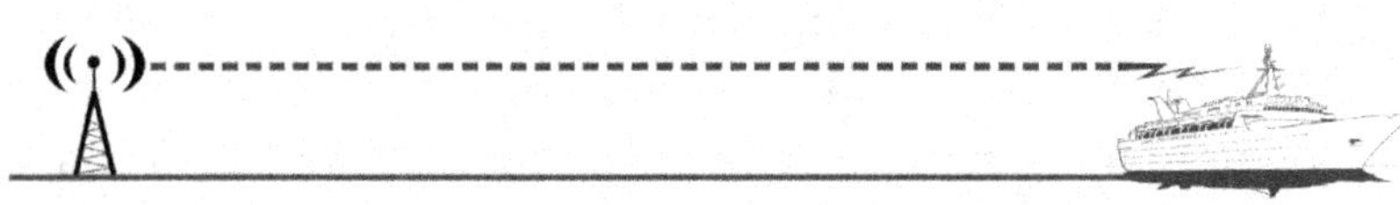

LORAN Could Only Work on a Flat Earth

Ships navigated across the vast oceans for more than 60 years in reliance on the LORAN radio signals. The LORAN system worked. It could not work if the earth were a globe. It could only work on a flat earth. The effectiveness of the LORAN radio navigation system is proof that the earth is flat.

Some have alleged (with good reason) that LORAN still exists; that it has just been revamped with improved technology and then rebranded as the Global Positioning System (GPS). GPS is being portrayed as a satellite-based navigational system. Updating and rebranding LORAN as GPS is the perfect scheme to keep a lid on the flat earth and convince the world that the earth is a spinning globe.

It seems that all supposed satellite signals are actually signals from land-based towers. For example, this author's wife has a "satellite" radio in her car. We have noticed that when traveling over a mountain road that rises in elevation approximately 2,000 feet above the foot of the mountain, we always lose the "satellite" reception on the radio as we reach the top of the mountain. It seems odd to lose reception from an alleged "satellite" that is supposed to be overhead as one rises in elevation. The radio receiver should be closer to the supposed "satellite" and thus receive a stronger signal the higher we rise in elevation. We have noticed that at the same moment that we lose satellite reception, we also lose cell phone tower reception. I infer, therefore, from the simultaneous loss of cellular phone reception and "satellite" radio reception that the "satellite" radio transmissions are not actually broadcast from satellites at all, but are broadcast via cellular towers.

There may very well be objects in the atmosphere that perform the functions of what are today called satellites. But these satellites are not in outer space; they are actually tethered beneath massive helium balloons, floating very high in the atmosphere.[376] Indeed, a little-known fact is that NASA annually consumes millions of cubic feet of gaseous helium that it uses to fill balloons to carry satellites into the upper atmosphere.[377]

In 2014, NASA contracted with Air Products and Chemicals, Inc. of Allentown, Pennsylvania, to supply "128.6 million cubic feet of gaseous helium to support operations at 13

NASA locations"[378] over the next five years, from 2015 to 2019. Yet in a report to Congress by the Secretary of the Interior, it was revealed that NASA needed a total of 401,535 million cubic feet of gaseous helium from 2015 to 2019.[379] The 2015 report states that "[t]he two major Federal users of helium are NASA and the Defense Logistics Agency (DLA)."[380] The report reveals that NASA consumes more than twice the helium consumption of the entire U.S. Military. It is not clear why the contract to supply NASA's helium needs understates the actual needs of NASA by almost threefold.

Of course, NASA has a ready cover story for this massive helium consumption by what is supposed to be a space agency. In its publically posted 2014 document awarding the helium supply contract, NASA reported its reason for needing helium.

> Helium is used throughout NASA as a cryogenic agent for cooling various materials and in precision welding applications, as well as lab use. Helium also is used as an inert purge gas for hydrogen systems and as a pressurizing agent for ground and flight fluid systems of space vehicles.[381]

Notice that there is no mention of using helium for balloon projects. But in the 2015 report to Congress, the Secretary of the Interior explained in detail one of the reasons NASA needed helium. It was a reason that NASA did not mention in its public notice of the helium contract. The 2015 Congressional report states that "[h]elium is used at all NASA Field Centers, as well as at NASA's Columbia Scientific Balloon Facility (CSBF) for their balloon campaigns in Antarctica and in the United States."[382] The real reason NASA needs helium is to fill balloons for floating objects in the upper atmosphere that they falsely portray as satellites located in outer space.

Who knew that NASA had such an extensive balloon program? NASA seems to be keeping the wraps on its balloon program. No doubt, they are concerned that people will add 2+2 together and realize that the NASA balloon program is actually the NASA satellite launch program, which is why they did not mention it in their publically posted helium contract. All of the gee-whiz rocket launches of alleged satellites are just theater to be shown on the nightly news to convince the public that satellites are being launched into space. There have been hundreds of NASA rocket launches of satellites reported in the news. But there is never any mention of NASA balloon launches.

Google, which is a company with a market capitalization of more than 700 billion dollars, with deep ties to the CIA, is certainly capable of launching space satellites, if space were real. Google has, instead, decided to forgo the facade of launching space satellites. Instead, Google has embarked on a program to launch atmospheric satellites tethered to helium balloons to provide worldwide internet service. Google explains its balloon strategy, which it calls Project Loon:

> We aim to launch and maintain a fleet of balloons to provide Internet coverage to users on the ground, with our Autolaunchers capable of safely and consistently launching a new balloon every 30 minutes. We have flown over 25 million km of test flights to date since the project began - with one of our record-breaking balloons surviving for 190 days aloft in the stratosphere.[383]

According to Google, "transceivers transmit connectivity from ground stations, across balloons, and back down to users' LTE phones."[384] And just like the satellites that are supposed to be traveling in mythical outer space, "solar panels power the equipment during the day and charge an onboard battery to allow for nighttime operation."[385] The technology for the Loon balloon satellites is essentially that of cell towers. Google explains that "Project Loon has taken the most essential components of a cell tower and redesigned them to be light enough and durable enough to be carried by a balloon 20 km up in the stratosphere."[386]

Notice that there would be a regular need for balloon assisted satellites, as each atmospheric satellite could only, at most, last 190 days. Under that program, there would be a constant need for replacement balloon satellites. Google is one of the few companies to publically announce a balloon satellite program, but it is not the only player in the balloon satellite game. Satellite company OneWeb has entered into a joint venture with Airbus to construct the manufacturing capacity to build up to 15 satellites per week. "The initial production of 900 satellites is planned for launch into low Earth orbit beginning in 2018, to deliver affordable Internet access globally."[387] OneWeb portrays the satellites as being placed in low Earth orbit in space; but it is incredible to suggest that they would have 900 rocket launches in 2018. It is quite clear that OneWeb's planned manufacturing capacity is for low atmospheric satellites tethered on helium balloons.

This author personally attended the Satellite 2018 conference.[388] Oddly, it was a satellite conference with no actual satellites on display. That is not an exaggeration. I saw a total of four small models of satellites, two hanging by strings (approximately 3 feet in breadth from end to end, including the solar panels), one in a display case and one on a table (each about 12 inches in breadth from end to end, including the solar panels), but there were no actual satellites on display. There were all kinds of massive ground-based parabolic antennae, mock-ups of ground communication centers, and other large telecommunications equipment on display, but no actual satellites.

Imagine attending a car show that, instead of automobiles themselves, had booths dedicated to displaying the wonders of gasoline stations and car washes. Suppose that the only cars on display were a handful of small plastic models. Such an occurrence would be surreal. The Satellite 2018 conference was that kind of surreal experience. It was a satellite show with no real satellites on display.

It is hard to put something on display that does not actually exist. It is easier to create a false reality using CGI, or small plastic models that are portrayed as miniature replicas, than it is to create a convincing physical construction of what is supposed to be a real satellite. What if the materials used in the CGI cannot be convincingly fabricated? You can imagine the difficulty in creating an object that is supposed to work in space, but when put on display may be subjected to scrutiny that would likely reveal that it could not possibly work as designed. That is a risk that the devil and his minions could not take. And so, there were no actual satellites on display at the Satellite 2018 conference.

One of the small satellite models was by OneWeb on a display podium with a graphic that warned not to take pictures of it. Why would there be a warning not to take a picture? It looked no more sophisticated than a fun-meal toy. There was nothing about the model that was materially different from the many graphics displayed in the videos being projected and the publications handed out at the conference. Could it be that OneWeb did not want pictures taken because the display was so pathetic that it would undermine its credibility as a legitimate satellite manufacturer? It must be worrisome to portray yourself as a manufacturer of something that doesn't actually exist.

This author walked throughout the entire Satellite 2018 conference area and only saw four booths from companies that claimed that they made things for satellites (not including the few other companies like OneWeb, Airbus, and Boeing who claim to build entire satellites). Every other booth addressed ground signal reception and dissemination of alleged satellite signals. This author spoke with one engineer who worked for a company that claimed to make thrusters for satellites in space. The engineer admitted that the thrusters his company made require that they expel a mass in order to work in space. He explained that is necessary because in the vacuum of space the simple thrust from a rocket does not work because there is no atmosphere to push against. So the thruster must supply the mass necessary to push against when in space. He stated that all satellites must have their own onboard fuel and that the fuel will run out within 15 years. As my questions became more searching, the company officials became very interested in who I was and who I represented.

When we look into the space satellite companies, we find some very colorful characters, who have a history of questionable business practices. Can one expect a businessman to be honest when his business is based upon pretending to make something that does not truly exist (outer space satellites) that are designed to go to a place that also does not truly exist (outer space)? For example, the founder and executive chairman of OneWeb is Greg Wyler. Eric Berger reveals the following information about Wyler:

> More than 15 years ago, his [Wyler's] company, Terracom, sought to bring the Internet to Rwanda through a contract to run fiber optic cables across the country. A few years later, after Terracom's targets to connect schools to the Internet were not met and amid questions about the company's business practices, Rwanda fined Terracom, and Wyler was out as its leader.[389]

Ron Nixon reported about the sordid mess left by Wyler in Rwanda:

> [A]fter nearly four years, most of the benefits hailed by him [Wyler] and his company have

> failed to materialize, Rwandan officials say. "The bottom line is that he promised many things and didn't deliver," said Albert Butare, the country's telecommunications minister.[390]

According to Nixon, Wyler's company, Terracom, ended up owning "99 percent of the shares in Rwandatel, the country's national telecommunications company."[391] Rwandan officials state that was where Wyler's true interest was all along. "Government officials say [Wyler's] company has spent more time marketing and signing up cell phone customers than on expanding Internet service."[392] It seems that the promise of linking 300 schools with internet service was merely a ploy to obtain the more lucrative cell phone business in Rwanda.

Wyler then tried to pull a fast one by cashing out Terracom's shares in Rwandatel. Nixon reveals that "government officials contended that Terracom secretly tried to trade its shares in the Rwandan telecom to GV Telecom, a regional African telecommunications company incorporated in the British Virgin Islands. Rwandan officials were furious, saying this was a violation of the contract signed by the two parties. The plan was scrapped and Mr. Wyler was widely criticized."[393]

13 Horizon Is Always Flat

The evidence of a flat earth is easy to discern, once a person perceives what he is seeing around him. The horizon is always seen to be flat over water. That proves that the earth is itself flat, otherwise the horizon would be bowed. B. Charles Brough explains:

> The marine horizon, from whatever position it is viewed, always appears to be, and is, in fact, a perfectly level line, and since this appearance is the same in all parts of the world, its surface must be level; and therefore the Earth is a Plane. This may be proved to be the case, by erecting at a suitable elevation on the sea shore, a duly-levelled board, or a string - at right angles to a plumb-line - tightly stretched between two vertical poles. On looking towards the sea, the horizontal line for a distance of 20 miles may be easily observed, and throughout its entire length it will be found to coincide with the straight-edge, or string: but if the earth were a globe, the horizontal line would form an arc of twenty miles in length, curveting both ways from the center, at the rate of eight inches, multiplied by the square

> of the distance. Hence the horizontal line at either end of the distance ought to be depressed some 66 feet below the horizon in the center. But as no such appearance is ever presented, it necessarily follows that the earth cannot be a globe, or other than a plane.[394]

Samuel Rowbotham proved the correctness of Brough's statement. Rowbotham stated:

> It is known that the horizon at sea, whatever distance it may extend to the right and left of the observer on land, always appears as a straight line. The following experiment has been tried in various parts of the country. At Brighton, on a rising ground near the race course, two poles were fixed in the earth six yards apart, and directly opposite the sea. Between these poles a line was tightly stretched parallel to the horizon. From the center of the line the view embraced not less than 20 miles on each side making a distance of 40 miles. A vessel was observed sailing directly westwards; the line cut the rigging a little above the bulwarks, which it did for several hours or until the vessel had sailed the whole distance of 40 miles. The ship coming into view from the east would have to ascend an inclined plane for 20 miles until it arrived at the center of the arc, whence it would have to descend for the same distance. The square of 20 miles multiplied by 8 inches gives 266 feet as the amount the vessel would be below the line at the beginning and at the end of the 40 miles.[395]

Indeed, Rowbotham presents graphic proof that the earth is flat, simply by looking at the sea horizon. His proof can be replicated by anyone.

> The sea horizon, to whatever distance it extends to the right and left of an observer on land, always appears as a perfectly straight line, as represented by H, H [in the figure below].
>
> 
>
>
> Not only does it appear to be straight as far as it extends, but it may be proved to be so by the following simple experiment. At any altitude above the sea-level, fix a long board--say from 6 to 12 or more feet in length--edgewise upon tripods, as shown in [the figure below].
>
> 
>
>
> Let the upper edge be smooth, and perfectly levelled. On placing the eye behind and about the centre of the board B, B, and looking over it towards the sea, the distant horizon will be observed to run perfectly parallel with its upper edge. If the eye be now directed in an angular direction to the left and to the right, there will be no difficulty in observing a length of ten to twenty miles, according to the altitude of the position; and this whole distance of twenty miles of sea horizon will be seen as a perfectly straight line. This would be impossible if the earth were a globe, and the water of the sea convex. Ten miles on each side would give a curvature of 66 feet (10^2 x 8 = 66 feet 8 inches), and instead of

> the horizon touching the board along its whole length, it would be seen to gradually decline from the centre C, and to be over 66 feet below the two extremities B, B, as shown in fig. 18. Any vessel approaching from the left would be seen to ascend the inclined plane H, B, C, and on passing the centre would descend from C towards the curvating horizon at H. Such a phenomenon is never observed, and it may be fairly concluded that such convexity or curvature does not exist.[396]

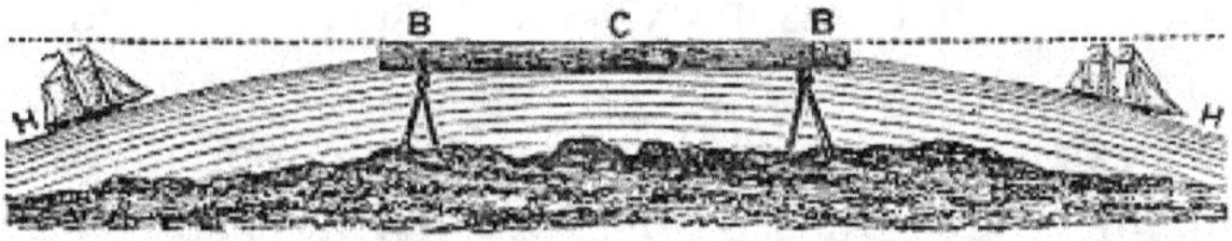

Rowbotham presents further evidence that the earth is flat by observation of a specific coastline in Great Britain. The observations he made regarding the flatness of the coastline of North Wales can be replicated today. Anyone today can view the coastline of North Wales from the Isle of Man and confirm that the earth is flat, just as Rowbotham did in 1881.

> From the high land near Douglas Harbour, Isle of Man, the whole length of the coast of North Wales is often plainly visible to the naked eye, a distance extending from the point of Ayr, at the mouth of the River Dee, towards Holyhead, not less than fifty miles. Whatever test has been employed, the line, where the sea and the land appear to join, is always found to be perfectly horizontal, as shown in the following diagram; fig. 21, whereas, if the earth is spherical, and therefore the surface of all water convex, such an appearance could not exist. It would of necessity appear as shown in fig. 22. A line stretched

horizontally before the observer would not only show the various elevations of the land, but would also show the declination of the horizon H, H, below the cross-line S, S. The fifty miles length of the Welsh coast seen along the horizon in Liverpool Bay, would have a declination from the centre of at least 416 feet (25^2 x .8 inches = 416 feet 8 inches). But as such declination, or downward curvation, cannot be detected, the conclusion is logically inevitable that it has no existence. Let the reader seriously ask whether any and what reason exists in Nature to prevent the fall of more than 400 feet being visible to the eye, or incapable of detection by any optical or mathematical means whatever. This question is especially important when it is considered that at the same distance, and on the upper outline of the same land, changes of level of only a few yards extent are quickly and unmistakably perceptible. If he is guided by evidence and reason, and influenced by a love of truth and consistency, he cannot longer maintain that the earth is a globe. He must feel that to do so is to war with the evidence of his senses, to deny that any importance attaches to fact and experiment, to ignore entirely the value of logical process, and to cease to rely upon practical induction.[397]

Fig. 21: The flat coastline of North Wales as viewed from the high land near Douglas Harbour, on the Isle of Man. The coastline of North Wales is approximately 50 miles in length and approximately 70 miles distance from the Isle of Man.

Fig. 22: The coastline of North Wales from the high land near Douglas Harbour, on the Isle of Man, as it would appear if the earth were a globe. Most of the coastline would be below the horizon (points S, S), and further the coastline would be curved laterally, with a 416 foot, 8 inch, bow at the middle, from the two points H, H. This, however, is not what is seen, thus proving that the earth is not a globe.

14 Flat Horizon is Always At Eye Level

Not only is the horizon always flat to the observer, but it always rises to eye level with the observer who takes flight, no matter how high the observer ascends. If the earth were a globe, the lateral horizon would be bowed and it would drop below the eye level of the observer as he ascended above the earth. Eric Dubay explains:

> Whether at sea-level, the top of Mount Everest, or flying over a hundred thousand feet in the air, the always horizontal horizon line always rises up to meet the eye-level of the observer and remains perfectly flat. You can test for yourself on a beach or hilltop, in a large field or desert, aboard a hot-air balloon or helicopter; you will see the panoramic horizon ascend with you and remain completely level all around. If the Earth were actually a big ball, however, the horizon should sink as you ascend, not rise to your eye-level, and it would dip at each end of your periphery, not remain flat all around. Standing in a rising balloon, you would have to look downwards to the horizon; the highest point of the ball-Earth

> would be directly beneath you and declining on each side. ... J. Glaisher wrote in his, "Travels in the Air," that "On looking over the top of the car, the horizon appeared to be on a level with the eye, and taking a grand view of the whole visible area beneath, I was struck with its great regularity; all was dwarfed to one plane; it seemed too flat."[398]

Samuel Rowbotham illustrates that if the earth's surface were convex, as it would need to be on a globe, a balloonist, instead of seeing the horizon remain at eye level as the balloon ascends, the horizon would drop down, requiring the balloonist to look down to see the horizon.[399] However, that is not what the balloonist sees. The balloonist sees the horizon always at eye level.

Rowbotham illustrates in the image below that "[i]n the case of the balloon at an altitude of two miles, the horizon would have been 127 miles away, and more than 10,000 feet below the summit of the arc of water underneath the balloon, and over 20,000 feet below the line of sight A, B, as shown in [the below figure]; and the dip C, H, from C, B, to the horizon H, would be so great that the aëronaut could not fail to observe it; instead of which he always sees it 'on a level with his eye,' 'rising as he rises,' and 'at the highest elevation, seeming to close with the sky.'"[400]

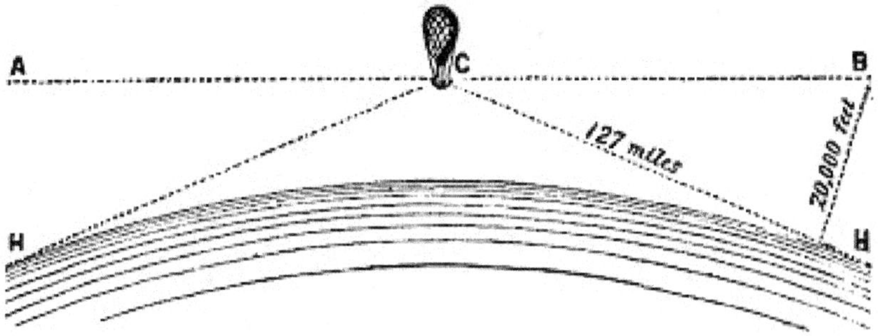

Today, any passenger on a commercial airplane can see the very same phenomenon as a balloonist. The horizon of the earth remains at eye level from the point of the plane's takeoff to its cruising altitude of 30,000 feet. If the earth were a globe, the horizon would drop below the view of the observer on the airplane. The fact that the passenger sees the horizon always at eye level is proof positive that the earth is a flat plane.

David James used a theodolite application on an Apple iPhone 7 to demonstrate that the horizon always rises to eye level. He verified the accuracy of the inclinometer on the iPhone and found that it read within a tolerance of ±00.6°.[401] He used the theodolite in an airplane at an altitude of 39,401 feet. He pointed the theodolite at a horizontal angle (-00.3°).[402] In doing so, he was able to put the cross hairs of the theodolite on the horizon. If the earth were a globe, he would have needed to point his theodolite down from horizontal at an angle of approximately 3.46°. Below is a screen shot from James' theodolite showing where the horizon should have been located if the earth were a globe. Notice also that the horizon is perfectly flat, as would be expected on a flat earth.

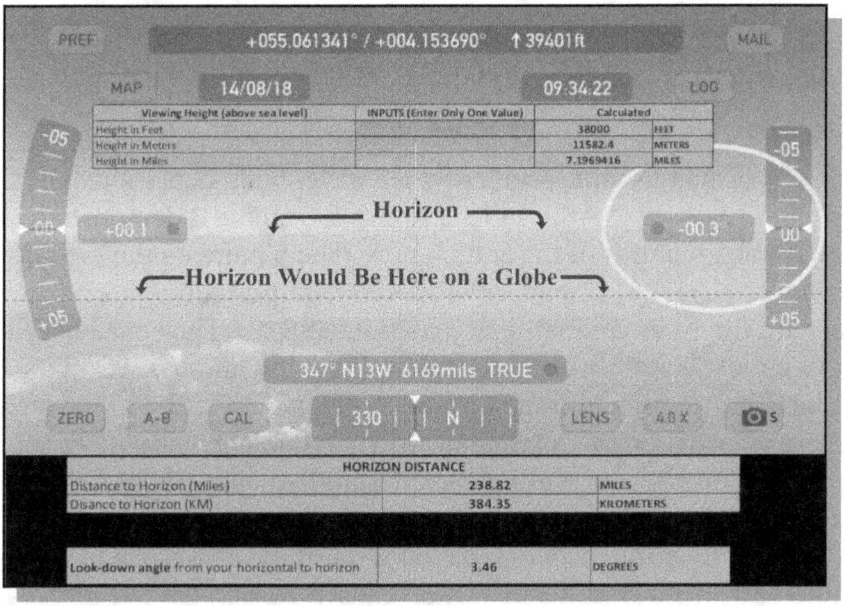

Figure 88: Screen shot of theodolite application showing the flat horizon at an altitude of 39,401 feet is at eye level. If the earth were a globe the horizon would have dropped to the dotted line.

Eric Dubay explains the recent appearance of photographic trickery used in producing curved earth depictions in NASA and other high altitude photographs.

> Amateurs have sent balloons to heights of over 121,000 feet and you can watch video online of the horizon rising with the camera-level and remaining perfectly flat 360 degrees around. NASA videos and other "official" sources, however, such as the recent Red Bull skydive at 128,000 feet have been caught adding fake curvature to the Earth via wide-angle lenses and post-production work. Panoramic photos atop Mount Everest also often claim to be displaying

> Earth's curvature, but this is simply the result of distortions and limitations inherent in wide-angle lenses.[403]

Dubay's allegation of the misleading depictions of a globular earth by NASA and the controlled media is demonstrated by the two pictures of the earth below. The picture on the top was taken by NASA Commander Scott Kelly, purportedly while he was aboard the International Space Station (ISS).[404] The space-based ISS almost certainly does not exist; the below picture was likely taken from a high altitude aircraft. The camera and wide-angle, fish eye lens used by Commander Kelly is depicted inset in the top photo; Commander Kelly himself supplied that picture of the camera. Adobe Photoshop has a feature that allows one to input the kind of camera and type of lens, the software will then automatically adjust the distortion caused by the fish eye lens. A photographic expert did that very thing, and the result is a perfectly flat horizon, as revealed in the bottom picture, below. Notice the hot spot caused by the sun, which indicates that the sun is close overhead and much smaller than the earth.

Dubay's allegation that the depictions of a spherical earth by NASA and the controlled media are misleading can be further verified by watching the Red Bull sponsored world record skydive of Felix Baumgartner. Before Baumgartner exited his capsule suspended by a balloon at approximately 128,000 feet above the earth, a camera with an ordinary lens shows the flat, eye level horizon of the earth. The picture states a height of 127,518 feet, because it was taken a few minutes before the jump; the balloon continued to ascend to the jump height of over 128,000 feet. If the earth were a globe, the horizon should not have been visible at eye level at 127,518 feet above the earth.

When Baumgartner steps out of the balloon's capsule, and the video is switched to the exterior GoPro camera, suddenly the flat horizon is transformed into a curved horizon. The curved horizon is depicted, because the exterior GoPro camera used a wide angle, fish eye lens, which caused the earth's horizon to appear bowed.

Below is a picture shortly after Baumgartner jumps from the capsule. Notice the fish eye lens on the GoPro camera shows a concave earth. The fish eye lens will show the earth's horizon either concave or convex, depending on the camera's orientation to the horizon. In the picture below the earth is at the outer circumference of the fish eye lens, causing the flat horizon to appear concave. In the picture above, however, the sky is at the outer circumference of the fish eye lens, thus causing the earth's flat horizon to appear convex. In fact, the earth's horizon is neither convex nor concave, it is flat.

Using a fisheye lens to give the earth a curved appearance is a trick commonly used by the mass media to inculcate people to accept the globular earth mythology as fact. For example, Adam Savage, who is an actor (and a self-described atheist) on the Discovery Channel television science show *Mythbusters*, filmed a flight aboard the U-2 spy plane. The stated reason for the flight was to determine whether the U-2 spy plane is the most difficult plane in the world to fly. How was Savage supposed to determine how difficult the U-2 is to fly when he is only going to be a passenger and they were not going to compare it to other planes? Indeed, at the end of the show, they acknowledged that being a passenger in the plane and not testing other planes made it impossible to determine if the U-2 was the most difficult plane in the world to fly. They simply concluded that it was plausible to say it is. But they knew that going in. It seems that the given reason for the U-2 flight was just a cover story for the real reason for the flight, which was to demonstrate that the earth is a globe by filming the supposed curvature of the earth from a high altitude by using fisheye lens trickery. Not just anybody can be a passenger in a U-2 spy plane. Some very powerful people pulled some strings to make the flight happen. The episode was appropriately titled: *Flights of Fantasy*.

During the U-2 flight, the fisheye lens camera filmed what appeared to be the curvature of the earth. Savage narrated at that moment: "From this height, 67,500 feet, our planet has a definitive curve to it. It is more three dimensional than I have ever seen it."[405] Savage's narration was a provable deception. Within a second of Savage describing the curved earth, the video cut to a second cockpit camera where the horizon was inadvertently centered in the lens. Even with a fisheye lens, there is no distortion when the object is horizontal to the center of the lens. It is only when an object in the fisheye lens is off-center that it takes on the curved distortion of the lens. Consequently, when the video cut to Savage in the cockpit, the horizon behind him can be seen to be perfectly flat.

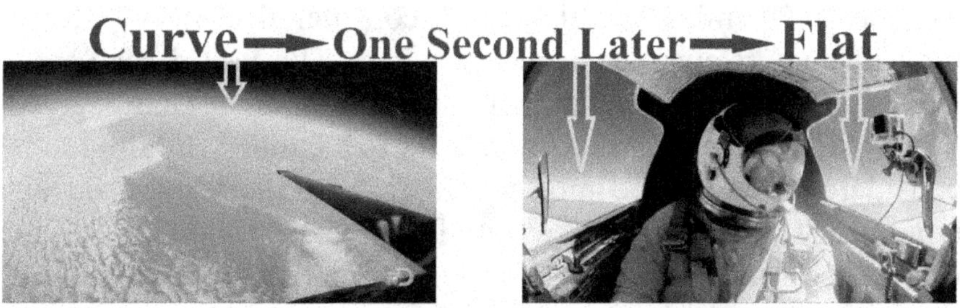

Adam Savage states: "from this height, 67,500 feet, our planet has a definitive curve to it," as the fisheye lens camera showed the curved earth, as depicted in the screen shot from the video in the picture on the left. Within a second after he makes that statement, the video cuts to a camera inside the cockpit of the plane as depicted in the screen shot of the video on the right above. Directly behind Savage, the flat horizon can be seen. Clearly, Savage was lying about what he saw. He is a special effects expert and knew full well that the fisheye lens would make the flat earth appear to be curved. He narrated the video, not according to the reality of what he saw, but according to the distortion he knew would be depicted by the fisheye lens. You can see the fisheye lens camera to Savage's left attached to the cockpit window that was used to film the curved earth picture.

It is impossible for the earth to go from being curved to being flat within one second. The Mythbusters have been busted trying to support the mythical globe. Obviously, Savage was lying about what he was seeing. He is a special effects expert and knew full well that the fisheye lens would give the flat horizon the appearance of a curved horizon. Savage's dissimulation was a preplanned deception by Savage, the Discovery Channel, and the U.S. Military to sell the lie of a globular earth and conceal the truth of the flat earth. They have been caught selling a myth on, of all shows, *Mythbusters*.

Below is a picture taken from a camera aboard a seized German V-2 rocket launched from the White Sands Missile Range in New Mexico, on October 24, 1946.[406] The photograph was taken from a reported altitude of 65 miles (343,200 feet). The picture from the upper atmosphere is hailed as the first photograph ever taken of the earth from what scientists call "space." Today, pilots exceeding 50 miles in altitude are considered to have entered "space" and are awarded astronaut wings. The World Air Sports Federation (Fédération Aéronautique Internationale) has decreed that a pilot exceeding 100 kilometers (62.1 miles) in altitude is considered to have gone into "space." It seems that "space" occurs at arbitrary altitudes, in nice round numbers; in the U.S. it's 50 miles, in Europe it's 100 kilometers. That doesn't seem very scientific.

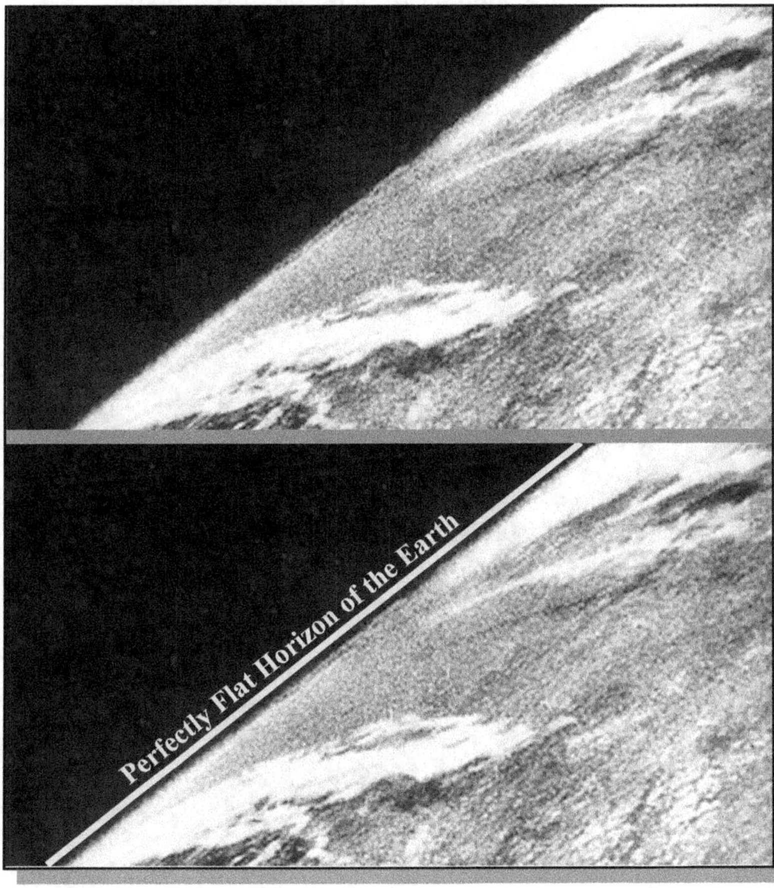

Figure 94: First Photograph taken from what scientists call "space," at an altitude of 65 miles (343,200 feet), on October 24, 1946.

The picture taken from the unmanned V-2 rocket was taken almost three times the distance in altitude from which Baumgartner jumped. Notice that the earth's horizon is perfectly flat. One would think that if the earth were a globe, the curvature of that globe would be more pronounced the higher in the sky from which the earth is viewed. Again, this is evidence that the Baumgartner photograph, supposedly showing the curvature of the earth, is in fact distorted by the fish eye lens on the Go-Pro camera used to take the picture. The earth is in fact a flat plane.

Below is a screenshot from a video taken from a camera aboard the V-2 launched on October 24, 1946. It shows a little bit longer view of the horizon. The published picture above, was cropped to show a shorter horizon than is depicted in the video.

NASA has cropped the originally published photo of the horizon even further in its version of the V-2 high altitude picture posted on its website.[407]

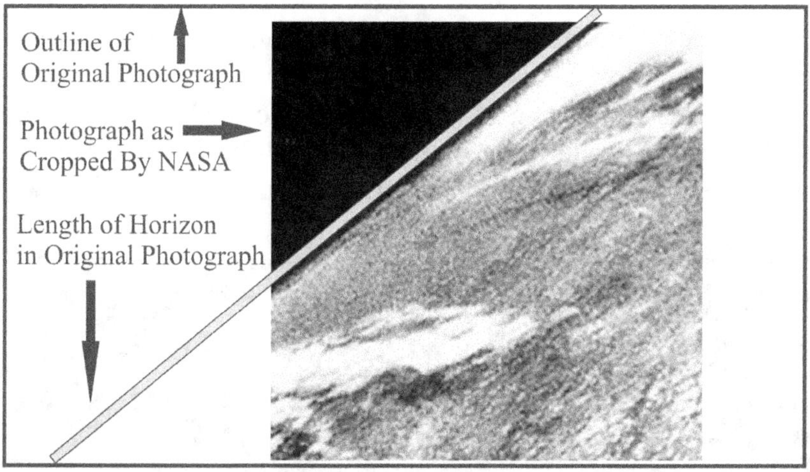

NASA displayed the cropped photo behind a circular frame on its website, which provided additional cropping of the horizon. The circular frame was used by NASA, presumably to suggest to the viewer a spherical earth.

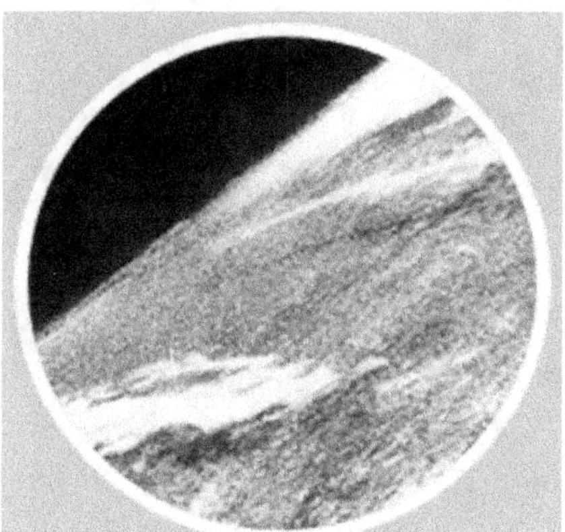

Figure 97: Horizon from V-2 photograph as it appeared on the NASA website

The NASA website shows a very limited, double-cropped view of the photograph that was originally published as the first photograph from space. There seems to be an effort by the government to conceal the flat horizon. One would think that NASA would want to present the most expansive view of the earth possible. Yet, NASA does just the opposite. There is no rational explanation for cropping an already small grainy photograph of earth from the upper atmosphere, other than to conceal from the public the flat horizon of the earth depicted in that photograph.

Below is yet another screenshot taken of a video from an altitude of approximately 65 miles (343,200 feet) above the earth using a camera aboard a separate V-2 rocket launched from the White Sands Missile Range in the late 1940's.[408] It shows yet a longer view of the perfectly flat horizon of the earth.

Below is a frame from a video taken in the 1960s at a reported altitude of 317,000 feet (60 miles) from a camera aboard an X-15 hypersonic experimental aircraft in the 1960s.[409] Note the flat horizon.

Below is a picture taken from a camera attached to a balloon floating at approximately 80,000 feet in altitude. Notice the flat, eye level horizon. Notice also, the hot spot on the clouds cast from the sun. The only way that the sun could cast a hot spot on the clouds would be if the sun was in fact directly over the surface of the earth. It is impossible for a sun that is 93 million miles away, as alleged by heliocentric scientists, to cast a hot spot on the clouds, as is revealed in the photograph.

Figure 100: Flat horizon at eye level from an altitude of 80,000 feet. Note the hot spot on the clouds directly under the sun, which indicates that the sun is much smaller than the earth and close overhead.

15 No Coriolis Effect Proves a Stationary Earth

One principle of movement on a spinning globe, is that the spinning will necessarily produce what is known as a Coriolis effect. The Coriolis effect was first postulated by Gustave-Gaspard Coriolis, a French engineer, mathematician, and physicist who was born on May 21, 1792 and died on Sept. 19, 1843. The *Encyclopedia Britannica* states that the Coriolis force is "an effect of motion on a rotating body, of paramount importance to meteorology, ballistics, and oceanography."[410]

The Encyclopedia Britannica further explains the Coriolis force as it pertains to the supposedly spinning spherical earth:

> In 1835 he [Coriolis] published a paper, "Sur les équations du mouvement relatif des systèmes de corps" ("On the Equations of Relative Motion of Systems of Bodies"), in which he showed that on a rotating surface, in addition to the ordinary effects of motion of a body, there is an inertial force acting on the body at right angles to its direction of motion. This force results in a curved path for a body that would otherwise travel in a

straight line. The Coriolis force on Earth determines the general wind directions and is responsible for the rotation of hurricanes and tornadoes.[411]

The Encyclopedia Britannica provided a graphic to explain the Coriolis force on the supposedly spinning globular earth. The graphic is reproduced below.

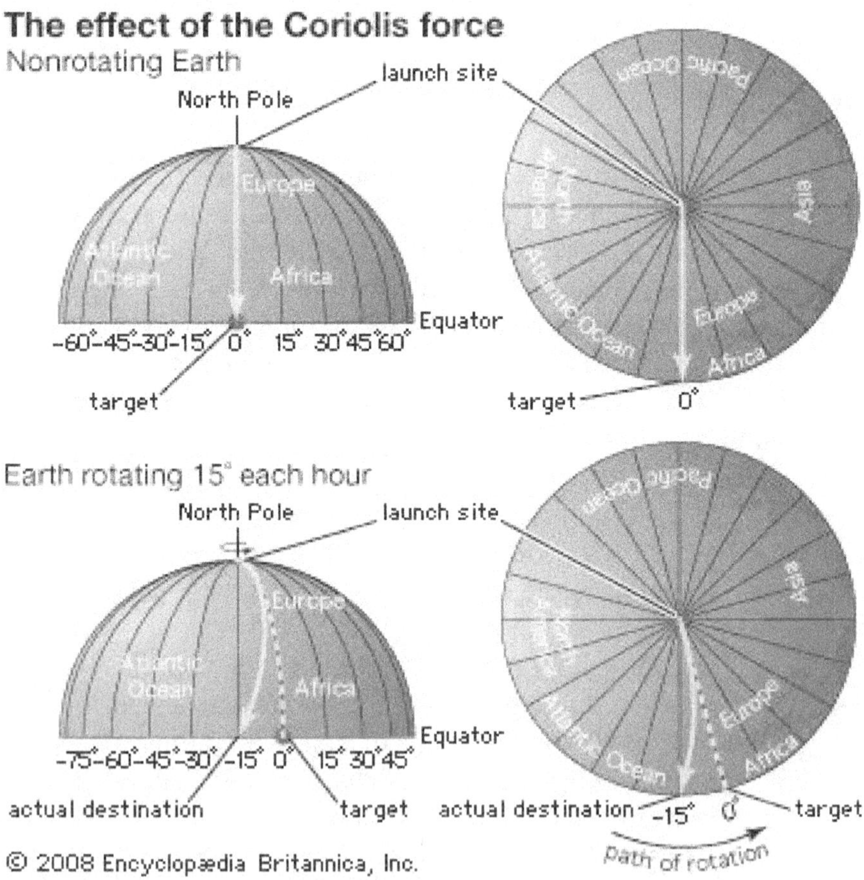

The problem with the above illustration from the Encyclopedia Britannica, is that it has no basis in fact. The Coriolis Force is very real. If the earth were in fact a spinning globe, the Coriolis effect would be manifested. The problem is that there is no such Coriolis effect taking place on earth. Which means that the earth is not spinning. The different directions of rotation of hurricanes in northern and southern latitudes has nothing to do with the claimed Coriolis effect of the spinning earth.

Airplanes that fly north and south do not adjust their flight paths to account for any Coriolis effect. For example, assuming the heliocentric model with the earth traveling at more than 1,000 mph at the equator, the Coriolis effect would cause a plane flying from Buffalo, New York to Miami, Florida to fly off course in a westerly direction due to the supposed faster spin of the earth as the airplane approaches the wider circumference of the earth at the latitude of Miami, Florida. Yet, in reality, the flight arrives in Miami on time and without the pilot having to adjust for any Coriolis effect due to the rotation of the earth.

Indeed, if there were a Coriolis effect it would be nearly impossible to land a plane on a runway. A runway that runs north and south would be careening at approximately 1,000 miles per hour across the path of the airplane, which would make it impossible to line up the airplane for a landing.

The Coriolis effect for spinning objects is real. Modern scientists must sell the myth that there is a Coriolis effect manifested on the earth, in order to make the spinning earth seem real. The fact that there is no Coriolis effect on the earth, creates a real problem for "scientists." Their solution to that little problem is to lie. They claim that there is a Coriolis effect, when there is not. The following from the National Geographic is an example of the modern explanation of the Coriolis effect that is supposed to be manifested on earth, but is, in fact, completely absent.

> Let's pretend you're standing at the Equator and you want to throw a ball to your friend in the middle of North America. If you throw the ball in a straight line, it will appear to land to the right of your friend because he's moving slower and has not caught up.
>
> Now let's pretend you're standing at the North Pole. When you throw the ball to your friend, it will again appear to land to the right of him. But this time, it's because he's moving faster than you are and has moved ahead of the ball.
>
> This apparent deflection is the Coriolis effect. ...
>
> Fast-moving objects such as airplanes and rockets are influenced by the Coriolis effect. Pilots must take the Earth's rotation into account when charting flights over long distances. This means most planes are not flown in straight lines, even if the airports are directly across the continent from each other. The line between Portland, Maine, and Portland, Oregon, for instance, is very long, and fairly straight. However, a plane flying from Portland, Oregon, could not fly in a straight line and land in Portland, Maine. Flying east, the Coriolis effect seems to bend to the right, in a southerly direction. If the Oregon pilot flew in a straight line, the plane would end up near New York or Pennsylvania.
>
> Military aircraft and missile-control technology must calculate the Coriolis effect for similar reasons. The target of an air raid could be missed

> entirely, and innocent people and civilian structures could be damaged. ...
>
> The Coriolis force applies to movement on rotating objects. It is determined by the mass of the object and the object's rate of rotation. The Coriolis force is perpendicular to the object's axis. The Earth spins on its axis from west to east. The Coriolis force, therefore, acts in a north-south direction. The Coriolis force is zero at the Equator.
>
> Though the Coriolis force is useful in mathematical equations, there is actually no physical force involved. Instead, it is just the ground moving at a different speed than an object in the air.[412]

The National Geographic is just one of many examples of a massive deception. The earth is supposed to be spinning at approximately 1,000 miles per hour at the equator. Because the circumference of a ball is smaller north and south of the equator, the earth does not spin at as great a speed at higher and lower latitudes. Portland, Oregon, is at 45 degrees north latitude from the equator and the purported spin of the earth at that latitude is approximately 700 miles per hour. Portland, Maine, is at 44 degrees North latitude with the spin of the earth only a tiny bit faster than 700 miles per hour. The Coriolis effect is supposed to put the airplane in New York if the pilot simply tried to fly the airplane straight and level toward Portland, Maine. That is simply not true. The pilot sets his heading toward Portland, Maine and accounts only for wind conditions. The pilot makes no accommodation whatsoever for a Coriolis effect, because the earth is not spinning; there is no Coriolis effect to calculate.

The national Geographic alleges that "military aircraft and

missile-control technology must calculate the Coriolis effect." The National Geographic cites to no authority for its statement, for the simple reason that no authority exists. No authority exists because it is not true. The National Geographic is simply making things up to fool the gullible public into believing that the earth is spinning at an incredible speed.

The theory of the Coriolis effect in the National Geographic example is that the eastbound airplane would be able to keep up with the speed of the allegedly spinning earth, because the airplane at take-off would be adding its speed to the 700 mph speed of the runway in Portland, Oregon. The problem with that argument is that it assumes that the runway is lined up due-east, and the airplane is taking off from the runway in a due-east direction.

The Coriolis effect is supposed to be based upon the spin of the earth and the fact that objects in motion over the spinning earth are moving independent of the spin of the earth once they are in motion. If there truly were a Coriolis effect on earth, it would pose a real problem for plane flights. If a plane were to take off from an airport in Portland, Oregon, in a North/South runway and turned east to fly to Portland, Maine, the airplane would never make it to Portland, Maine. That is because, the airplane would be traveling at approximately 560 miles per hour once it reached cruising altitude. The earth, however, would be spinning at 700 miles per hour eastbound beneath the airplane. The airplane would never be able to catch up with the speed of the earth's spin. The airplane would be constantly losing distance over the ground at the rate of 140 miles per hour. Essentially the airplane would be moving backwards over the ground.

What is found is exactly the opposite. An airplane traveling eastbound from Portland, Oregon, to Portland, Maine, would in fact have a shorter flight time than an airplane traveling westbound from Portland, Maine, to Portland, Oregon. The reason has nothing to do with the spin of the earth. The high velocity eastbound winds at high altitude, known as the jet stream, carry the airplanes along and allow eastbound flights to have a faster ground speed. The jet stream can have wind speeds ranging from 60 miles per hour to over 250 miles per hour. That same jet stream is a hindrance to westbound flights. It was reported in January, 2015, that a British Airways Boeing 777-200 Jet was able to travel at ground speeds in excess of 745 miles per hour as it traveled in the eastbound jet stream of approximately 250 miles per hour.[413] Incidently, the speed of sound is 760 miles per hour at sea level.

David Wardlaw Scott explains experiments done in England at the turn of the 20th century, using a cannon, which showed that there was no Coriolis effect whatsoever manifested on earth, thus indicating that the earth is motionless. In the experiments, a cannon was fixed firmly on the ground in a precisely vertical position. The cannon was fired. The cannon ball ascended for 14 seconds vertically and it took 14 seconds for the cannon ball to fall back to earth, for a total of 28 second aloft. If the earth were traveling eastward at 600 miles per hour at the latitude in England, the cannon ball would be expected to land almost 5 miles to the west of the cannon. However, that did not happen. The cannon ball fell generally within 2 feet of the cannon. In a couple of instances, the cannon ball actually returned to the cannon's mouth.[414]

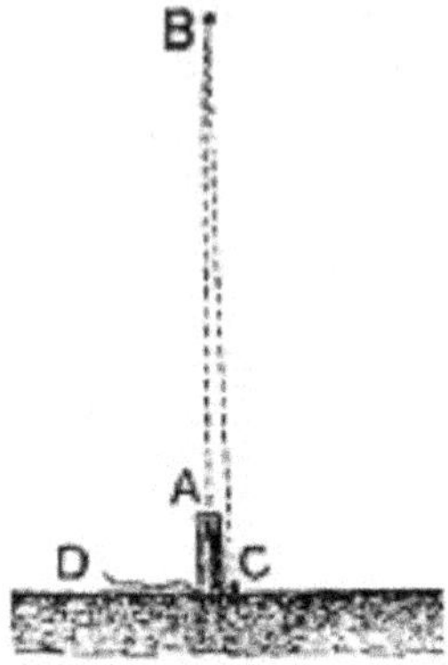

Figure 102: Cannon shot vertically and returning to ground near cannon, proving no Coriolis force on Earth.

If there were any spin to the earth causing a Coriolis effect, it would have been discovered by now. Yet every single experiment ever performed to detect the motion of the earth has returned a null result (e.g., the Michelson/Morley experiment). Real-life experiments have proven that there is no Coriolis effect on Earth, because the earth is stationary. Governmental educational systems, however, persist in pushing the myth of a Coriolis effect on a supposedly spinning Earth. Indeed, they must argue that there is a Coriolis effect, because if they let it be known that there is no Coriolis effect, that would let the cat out of the bag that the earth is stationary. For example, below is an illustration from the Coastal Practice Network, which is funded by the European Union Regional Development Fund.[415]

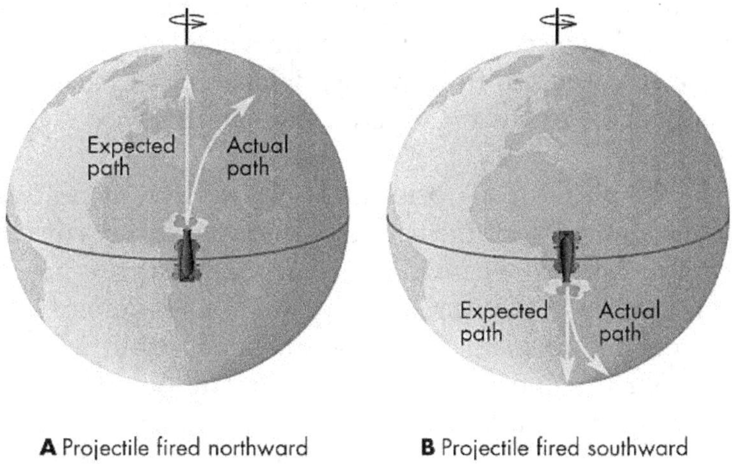

Figure 103: Coastal Practice Network showing their depiction of Coriolis effect of a cannon shot on Earth due to spin of earth. However, no such effect actually happens, because the earth is not a globe and does not spin.

Indeed, if there were truly a Coriolis effect on earth as depicted in the above illustration, then the artillery officers and snipers would be trained to consider the spin of the earth in making their calculations for accurate firing. Yet, you will look in vain for any mention of Coriolis effect in any military artillery[416] or sniper instruction manual.[417] In all of the wars fought throughout history, no soldier has ever been instructed to consider the Coriolis effect of a spinning Earth when sighting in a target with his artillery piece or other weapon.[418] The folly of adjusting for a mythical Coriolis effect would be immediately apparent, as the soldier's round would travel off course and miss the target.

This author is a former federal firearms instructor. I have seen many thousands of rounds shot from all kinds of firearms at various distances. I have never witnessed any round fired ever be affected by the alleged Coriolis effect. I have never read the Coriolis effect discussed in any firearms manual; nor have I ever heard it be discussed by any firearms instructor as something to consider in the use of any firearm, from any distance.

If, in the real world, those whose lives depend on the accuracy of their performance using weapons do not consider the Coriolis effect, that is convincing evidence that there is no Coriolis effect on earth. That means that the earth is not spinning. All of the pretty colored diagrams showing a spinning earth with cannon balls going off-course due to the supposed Coriolis effect depict an unreal myth. The earth is fixed and does not move.

16 Emergency Landing Evidence of a Flat Earth

On October 7, 2015, a Chinese Airlines (CAL) flight made an emergency landing at Ted Stevens Anchorage Airport in Alaska.[419] The emergency landing was necessary, because a woman had gone into labor and subsequently gave birth during the flight. The flight originated from the Taiwan Taoyuan International Airport and was scheduled to land in Los Angeles, California. The pilot decided to divert the flight to Anchorage Airport, six hours after taking off, after he was informed that the woman's water broke.

Below is the route charted on a globe model of the earth. The dashed line below depicts the scheduled route from Taiwan to Los Angeles, with a solid line showing the emergency diversion route to Anchorage, Alaska, and then on to Los Angeles on a globular earth. The emergency landing in Anchorage, Alaska, makes little sense if the earth were a globe, because the flight diversion means that the airline decided to divert a flight from its scheduled destination of Los Angeles, which is a city of almost 4 million people that offers some of the best medical care in the world, to a far-off city less than one-tenth its size, in a wilderness state, and saving very little time in the process. If the earth were a globe, the pilot could have just as easily diverted the flight to

Hawaii. But he did not do that. Why not? As we will see on the map of the flat earth diversion of the flight to Hawaii would not have made sense. Why did the pilot divert the flight to Anchorage, when he seemingly did not save very much time and went way out of his way? Again, the flat earth map explains that fact as well.

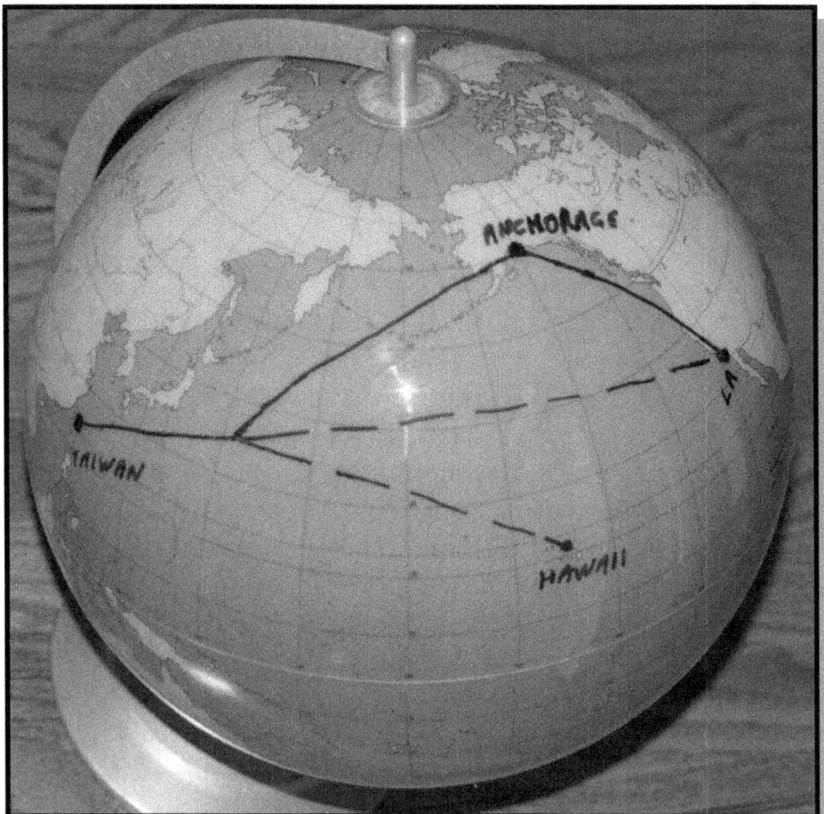

According to CAL, the "flight took off from Taiwan Taoyuan International Airport at 11:50 p.m. Wednesday and was scheduled to arrive in the U.S. at 8:35 p.m., Wednesday, Los Angeles time [12:35 p.m., Thursday, Taipei time]."[420] Due to the flight diversion, the airplane landed at Anchorage at 9:06 a.m. Thursday (Taipei time).[421] That means that the diversion to Anchorage, Alaska, caused the airplane to land in Anchorage 3 hours and 29 minutes earlier than the scheduled arrival time in Los Angeles. That 3 hour and 29 minute difference would be impossible if the earth were a globe, because on a globe, the distance to Anchorage is supposed to be only a little less than the distance to Los Angeles at the point of diversion, not a full 3 hours and 29 minutes.

The anomaly of the diverted airplane arriving in Anchorage 3 hours and 29 minutes ahead of its scheduled arrival time in Los Angeles is explained by the fact that the earth is not a globe. Below is a graphic that shows the reality of the flat earth and perfectly explains the decision to divert the flight to Anchorage, Alaska. It illustrates how the flight arrived in Anchorage 3 hours and 29 minutes before the scheduled arrival time in Los Angeles. Assuming the typical cruising speed for an airliner of 560 miles per hour, that airplane would be expected to travel from the airspace over Anchorage, Alaska, to Los Angeles, California, a little over 2,300 miles away, in approximately 4 hours. That seems to explain the 3 hour and 29 minute early arrival in Anchorage.

On the actual flat earth, Anchorage is a destination that is almost directly along a straight flight path from Taiwan to Los Angeles. On that flight path, Anchorage is more than 2,300 miles northwest of Los Angeles, making it the logical place for an emergency landing. That is the reason the pilot chose to stop in Anchorage, rather than continue to fly to Los Angeles. As you can see in the polar azimuthal equidistant flat earth map below, the diversion of the flight to Anchorage, Alaska, makes perfect sense on the flat earth. The flat earth map shows that Anchorage is on the way to Los Angeles from Taiwan and therefore was the logical place to stop to see that the woman received medical attention.

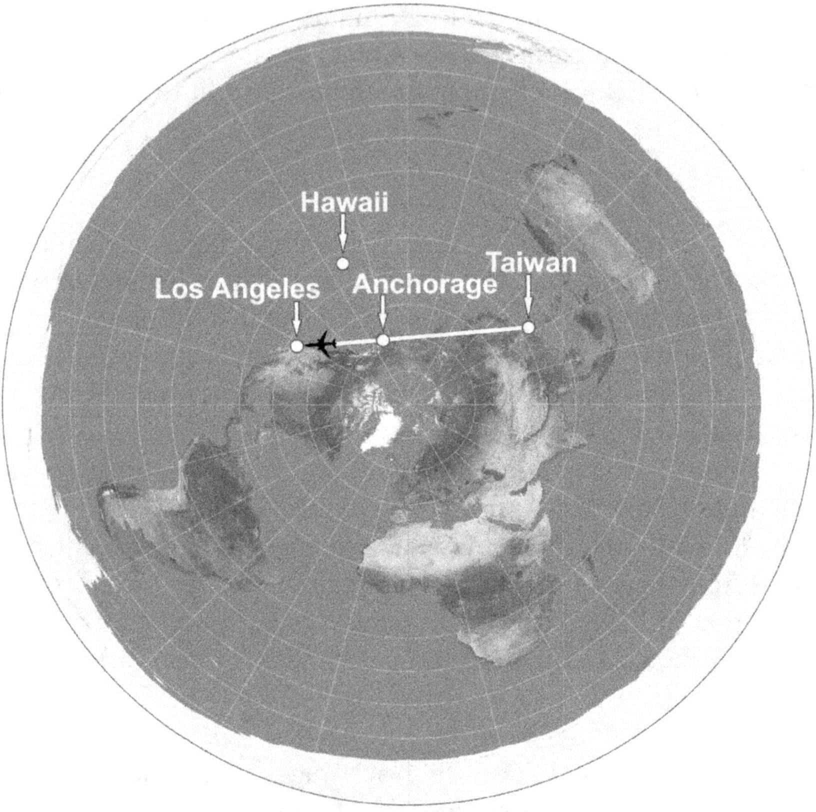

Figure 105: Flight path of the CAL emergency landing in Anchorage as it actually happened on the flat earth.

The flat earth map illustrates why the pilot did not divert the flight to Hawaii. In reality, flying to Hawaii would take the airplane on a course that would be thousands of miles out of the way and create many hours of delay. The fact that the pilot diverted his Los Angeles bound flight to Anchorage is evidence that Anchorage was along the flight route from Taiwan to Los

Angeles on a flat earth.

If the earth were a globe, the pilot never would have chosen to divert the flight to Anchorage, since the flight distance to Los Angeles appears only slightly longer on a globe. On a flat earth it only made sense to divert the flight, as the pilot did, from Los Angeles and stop on the way at Anchorage, Alaska. The pilot made the rational decision to land in Anchorage, because Anchorage was along the actual flight route on a flat earth. The emergency landing at Anchorage reveals the reality that the earth is flat.

17 Alaskan Detour

The picture below was taken of the flight path of an airplane as it traveled from Washington, D.C., to Tokyo, Japan. This author took that picture of the flight path of the airplane as depicted on the seat back screen while aboard United Airlines Flight Number UA7941, which was actually flown by ANA (All Nippon Airways), which was the contract carrier for United under the United Star Alliance. The 14 hour non-stop flight had a scheduled departure at 12:20 p.m. EST (GMT−4) on October 16, 2017, from Washington/Dulles International Airport (IAD) and a scheduled arrival at 3:25 p.m. Tokyo time (GMT+9) on October 17, 2017, at Tokyo/Narita International Airport.

Yet, we see in the picture of the airplane's path provided by the airline on the seat back screen that the airplane is traveling an elongated arc that takes the airplane in a northwest direction above the north coast of Alaska, before it dips back down in its arc of travel in a southwest direction toward Tokyo. The arc is caused by the distortion of the Mercator projection of the supposed spherical earth that is used in the graphics for the map.

Flight path of Airliner depicted on seat back screen showing it traveling northwest from Washington, D.C., in an elongated arc above the northern coast of Alaska, before it dips back down and travels in a southwest direction toward Tokyo.

Tokyo, Japan, is located at 35.7° North Latitude. Washington, D.C., is located at 38.9° North Latitude. That means that Tokyo is a little more than 3° South in latitude in relation to Washington, D.C. The map depicted on the screen on the seat back is the Mercator projection. The Mercator projection purports to convert the curvature of the earth to a flat map configuration. It presents perfectly horizontal latitude and perfectly vertical longitude lines intersecting one another at right angles.

Below is a clearer picture of the flight path of United Airlines Flight 7941 from the wesbsite, FlightAware[422], which tracks the coordinates of specific flights and plots them on a Mercator map projection of a supposed spherical earth. Indeed, FlightAware has a minute-by-minute flight log of the specific coordinates of United Airlines Flight 7941.[423] This author emailed a copy of the seat back photo to several people, including a pastor, who is a former NASA engineer. The point this author was making was that the flight route proves that the earth is flat, as the route displayed on the seat back screen makes little sense on a globular earth. But when the flight route is plotted on a flat earth it makes perfect sense. The flight route on a flat earth is seen to be a direct straight path from Washington, D.C. passing above the north coast of Alaska and then on to Tokyo, Japan.

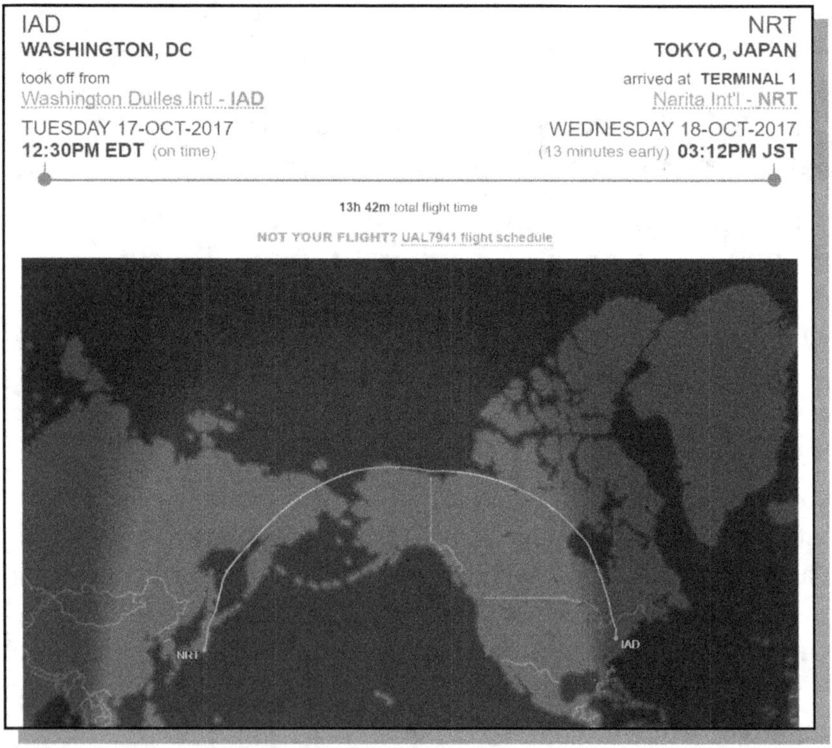

Figure 107: Screen shot of graphic from the FlightAware website, which plotted the United Airlines Flight 7941 flight coordinates on a Mercator projection of a supposed spherical earth.

The former NASA engineer, to whom this author sent the information, adamantly believes that the earth is a sphere. He sent me the following photo explaining how the arched flight path from Washington, D.C., to Tokyo displayed on the Mercator projection on the back seat screen in no way indicates that the earth is flat. He averred that it is perfectly explainable on a spherical earth. Keep in mind, this former NASA engineer plotted the shortest route on a globe to prove to me that the seemingly roundabout route on the Mercator projection is actually the shortest route on the spherical earth.

He stated that "the only map that is accurate anywhere on the Earth is a globe because it reflects the shape of the Earth, a sphere." He further stated: "On a globe, the shortest distance between two points is a Great Circle. I took out my globe and drew a 'straight' line between Washington and Anchorage and then Anchorage to Tokyo. Here is the picture I took in my office a few minutes ago:"

Figure 108: **Depiction by former NASA engineer and spherical earth advocate showing the shortest flight path from Washington, D.C., to Tokyo, Japan, on a spherical earth.**

When one examines the former NASA engineer's depiction of the shortest route, it becomes clear that it is not the route actually flown by the UAL 7941 pilot. The route actually flown by the pilot took the airplane along the northern coast of Alaska, whereas the route from the former NASA engineer takes the flight to the southern coast of Alaska. The former NASA engineer's mythical flight path takes the airplane fully 10 degrees (approximately 700 miles) south of the actual course of the airplane. Whereas, the airplane on which this author flew in fact passed across the north coast of Alaska. The back seat screen listed the coordinates on the screen as 70°43' North Latitude, 164°20' West Longitude. When the shortest route on a spherical earth and the actual route that was flown are both plotted on a flat earth, it becomes clear why the pilot flew along the north coast of Alaska. In reality, it is a shorter distance. That is because in reality, the earth is flat.

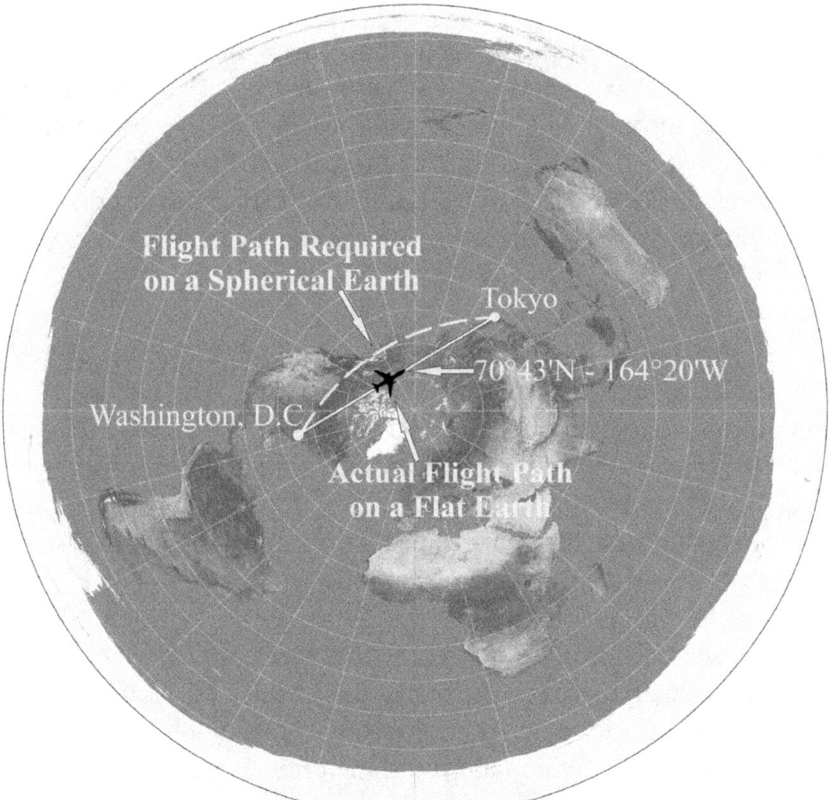

The solid line depicts the actual straight flight path from Washington, D.C., to Tokyo, Japan, on the flat earth. It perfectly explains why the plane traveled northwest through the coordinates of 70°43'N - 164°20'W off the north coast of Alaska before traveling southwest toward Tokyo. The dashed line shows a flight path along the southern coast of Alaska required on a spherical earth.

This author pointed out to the former NASA engineer that his plotted course proved that the earth could not be a globe, because the supposed flight path he plotted was not the flight path actually taken by the pilot. The flight path actually taken by the pilot would not make sense on a globular earth, since it would cause the pilot to fly 700 miles further north than necessary if the earth were in fact a globe.

When confronted with that fact, the former NASA engineer tried to backtrack by claiming that he plotted the coordinates through Anchorage by choice and not by necessity. He claimed that he plotted the path through Anchorage as a "stopover." He claimed that if he had plotted the shortest distance on a nonstop flight it would have passed over the north coast of Alaska.

In this author's response email to the former NASA engineer I asked him: "Well then why did you not plot the non-stop from Washington, D.C. through 70°43' North Latitude, 164°20' West Longitude and then on to Tokyo? Why did you fabricate a completely different route?" I then challenged him: "Let's see your new plot with the actual route." He did not answer my questions or respond to my challenge.

The dissimilation of the former NASA engineer's statement can be proven in many ways: 1) This author never indicated any stopover in Anchorage or anywhere else, and in fact the graphics I sent (which are the same ones listed above) clearly indicate that UAL 7941 was a nonstop flight. 2) This author specifically told him that the flight path passed through 70°43' North Latitude, 164°20' West Longitude, which is along the north coast of Alaska, thus making a stopover in Anchorage impossible. 3) The former NASA engineer specifically stated that his drawn course "closely matches the route shown," which indicates that he tried to draw a straight course that matched the actual course but failed. 4) Indeed, there was no reason to plot a flight path south of the actual flight path, unless he had no choice. 5) He stated that the close match between the course he drew and the actual one flown by the pilot was "proof that the globe is true;" but it was not a close match, his route took the airplane on a course that was 700 miles south of the actual route. 6) His entire reason for plotting the route on the globe was to prove that "on a globe, the shortest distance between two points is a Great Circle." 7) He described his "great circle" route as in fact being a "'**straight**' line between Washington and Anchorage and then Anchorage to Tokyo." 8) The picture he sent clearly shows a **straight** course on a globe from Washington, D.C., to Tokyo, Japan that passes through the southern part of Alaska at Anchorage. 9) It is not true that the shortest route on a globe would pass north of Alaska. As his picture clearly illustrates, the shortest route between Washington, D.C., and Tokyo, Japan, on a globe passes over the southern coast of Alaska.

If the former NASA engineer tried to work the coordinates for UAL 7941 into his globe, setting a course from Washington, D.C. through 70°43' North Latitude, 164°20' West Longitude and then on to Tokyo, he would find that he could not have a straight line but will have to create an arc. Although it would be a more gradual and less severe arc on the globe than on the Mercator projection, it would be an arc nonetheless.

Why is that important? Because when the actual flight path of UAL 7941, which passed over the north coast of Alaska, is plotted on the flat earth map we see that it is a perfectly straight path. The shortest route between Washington, D.C., and Tokyo Japan on the flat earth is the shortest route that was in fact taken by the pilot of UAL 7941. Whereas the shortest route on the supposed globe was not taken by the pilot. That is evidence that the earth is flat.

The former NASA engineer's globe picture proves that the earth cannot be a globe. He has proven that the path that was in fact flown is slightly longer on a globular earth. His effort has only confirmed that the earth is flat. Indeed, if the path of the airplane was over the southern coast of Alaska, then that would indicate the earth is not flat, as that would not be a straight course on the flat earth. But, the flight path actually flown by the pilot of UAL 7941 passed over the north coast of Alaska, thus indicating a flat earth and disproving the globe model.

The depiction of the flight path on the flat earth in the north polar azimuthal equidistant map perfectly explains the flight path going northwest and passing along the northern coast of Alaska, and then after passing its most northern point, finishing in a southwest trajectory as the airplane approaches Tokyo. That flight path, being a straight line, is in fact the shortest distance between Washington, D.C., and Tokyo, Japan, thus indicating that the earth is flat.

The flight path passing over northern coast of Alaska is only slightly shorter in distance than the flight path over the southern coast of Alaska. Focusing on the fact that the difference in the distance of the routes is slight misses the point. The proof that the earth is flat is found not in the magnitude of the difference in the distances between the optimal flight paths on the flat earth and spherical earth models, but in the precision of the navigation equipment on modern aircraft. The fact that the modern aircraft (Boeing 777-300ER) for UAL Flight 7941 with its precise, advanced avionics equipment would travel the slightly shorter northern route is evidence that the earth is flat. If the earth were a sphere, the advanced avionics of the aircraft would have taken the aircraft on a route over the southern coast of Alaska; but it did not do that. The advanced avionics ensured that the airplane traveled a precise course on the shortest route, which was the northern route. It is axiomatic that the shortest distance between two points is a straight line. The advanced navigation equipment on the Boeing 777-300ER of UAL Flight 7941 was programmed to plot that straight course. The fact that the straight course actually flown by UAL 7941 between Washington, D.C., and Tokyo, Japan, took the aircraft over the northern coast of Alaska indicates that the earth is flat.

Because the Mercator projection depicts a spherical earth, it distorts the flight path as a long arc. Notice on the picture of the back seat screen it indicates the coordinates of the airplane at the point the picture was taken: 70°43' North Latitude, 164°20' West Longitude. Taking those exact coordinates and plotting them on the flat earth map, we see that the airplane is in actuality flying a straight line course on the flat earth from Washington, D.C., to Tokyo, Japan. The shortest distance between two points is a straight line. There is no way that any airline would purposely lengthen a flight, particularly when the destination is close to its maximum safe flying distance for the aircraft. The straight line course is depicted on the flat earth map. That is because the earth is flat and UAL 7941 flew a straight path from Washington, D.C., to Tokyo, Japan. The depiction of a flight path that is an arc as displayed on the seat back screen can only be explained by the fact that Mercator projection of a spherical earth is a distortion. The depiction of the straight line flight path using the same coordinates on the flat earth map is proof that the earth is flat.

The effort by the former NASA engineer to illustrate the shortest distance on a globe from Washington, D.C., to Tokyo, Japan, by flying along the southern coast of Alaska, only proves that the earth cannot be a globe. The flight path he drew was not the flight path plotted by the advanced avionics of the Boeing 777-300ER. Looking at the southern coast route plotted on the flat earth shows why the advanced avionics of the Boeing 777-300ER did not choose that route. It would have taken the flight in a needless roundabout route that would only serve to unnecessarily lengthen the flight. The pilot was flying the airplane on the real flat earth, not a mythical globular earth, and the flight path actually flown, as plotted by the advanced avionics of the Boeing 777-300ER, proves it.

18 Strange Connections

Other anomalies in airline flight paths that demonstrate the truth of a flat earth are the strange connecting flights from cities south of the equator to cities north of the equator, when the city of departure and final arrival are both south of the equator. For example, one flight itinerary for a traveler going from Sydney, Australia, to Johannesburg, South Africa (which according to Google Earth is a distance of 6,868 miles) involves taking Emirates Flight 415 north to Dubai, UAE, in order to catch a connecting flight to Johannesburg. Emirates Flight 415 is a 14 hour and 20 minute flight that traverses 7,490 miles. The passengers then have a 14 hour and 50 minute layover in Dubai. They then take a connecting 8 hour and 15 minute flight on Emirates Flight 761 from Dubai to Johannesburg. Google Earth puts the distance from Dubai to Johannesburg at 3,974 miles.

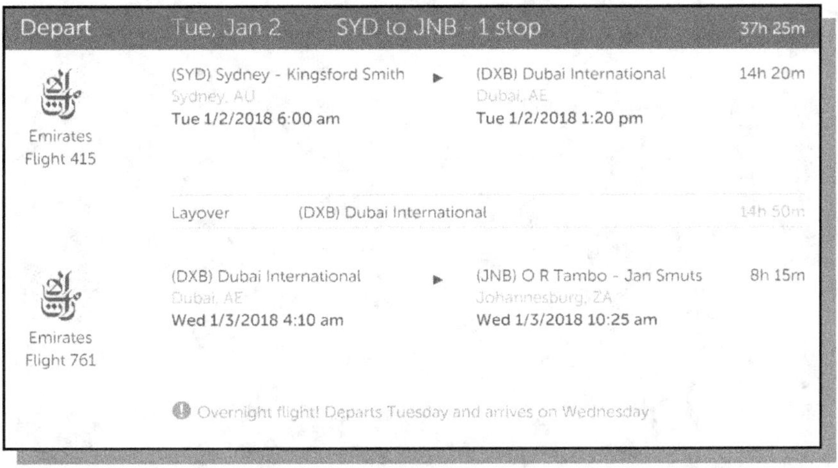

Figure 110: Screenshot of flight itinerary for flight from Sydney to Johannesburg, with a layover in Dubai

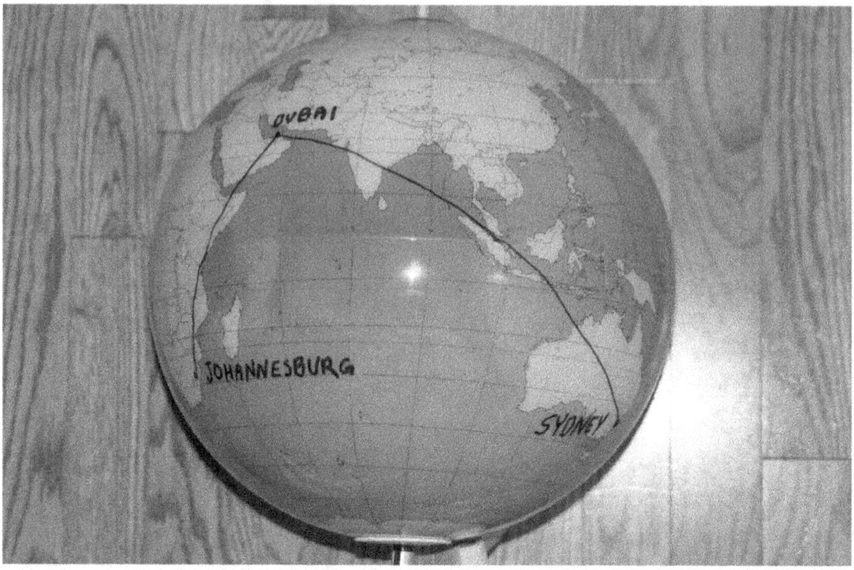

Having a connecting flight north of the equator in Dubai for a flight from Sydney to Johannesburg makes no sense on a spherical earth.

According to Google Earth, the distance from Sydney to Johannesburg is 6,868 miles. But the distance from Sydney to Dubai is 7,490 miles. How does it make sense to fly more than 600 miles further to Dubai, then take a connecting flight and spend another 8 hours and 15 minutes flying an additional 3,974 miles from Dubai to Johannesburg? It does not make any sense if we live on a globe and the figures provided by Google Earth are accurate. One might argue that the hub for Emirates Airline is in Dubai, and therefore it only stands to reason that Emirates would fly to its hub before flying on to Johannesburg. It is true that airline routes are based upon the hub location for an airline. That is how an international airlines select the routes they will fly. If a route is not economically feasible with the established hub, the airline will simply not fly that route.

The apparent roundabout route through Dubai as it appears on the spherical earth is misleading. There is an economically feasible explanation for the layover in Dubai. The itinerary with a layover in Dubai makes perfect sense if the earth is flat. As you can see on the diagram of the flight path on the flat earth, Dubai is a good location to stop on a long distance flight from Sydney to Johannesburg. It is because Emirates Airline has a hub in Dubai that it flies the Sydney to Johannesburg route.

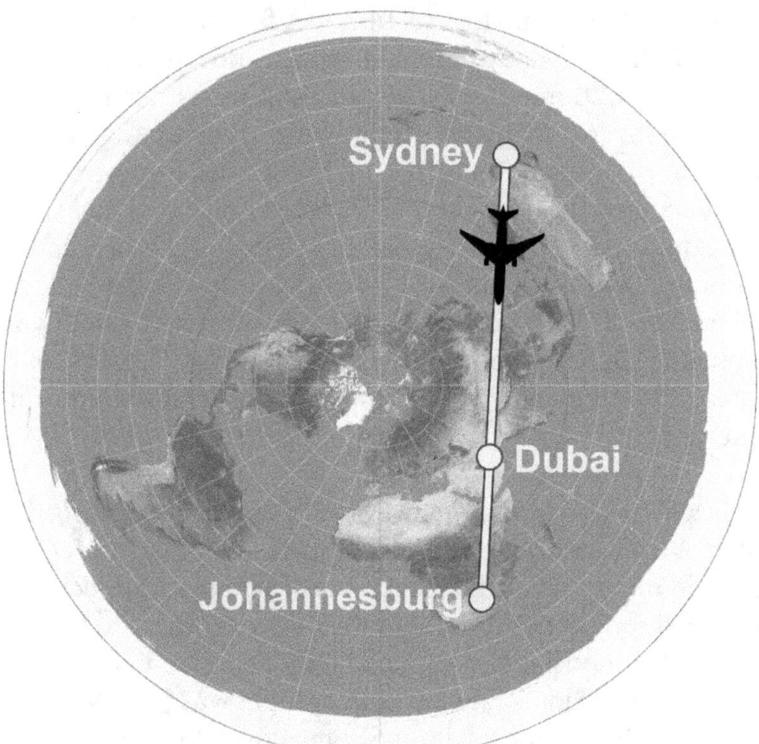

Having a connecting flight north of the equator in Dubai for a flight from Sydney to Johannesburg makes perfect sense on the flat earth.

Notably, there is a listed nonstop flight from Sydney to

Johannesburg, Qantas Flight 63. Qantas reports that flight takes 14 hours to fly the 6,869 miles purported by Google Earth to be the distance from Sydney to Johannesburg.

In view of the nonstop Qantas Flight 63, it would seem that the Emirates Flight 415 to Dubai, and then the connecting flight of Emirates Flight 761 to Johannesburg is significantly less preferable. The Emirates itinerary has a total flight time of 22 hours and 35 minutes (not including the 14 hour and 50 minute layover in Dubai) compared to 14 hours for the Qantas Flight 63 direct flight from Sydney to Johannesburg. Indeed, that first leg to Dubai alone is 20 minutes longer and 600 miles further than the Qantas Flight 63 nonstop flight from Sydney to the final destination of Johannesburg. Furthermore, the Emirates Airline itinerary requires a14 hour and 50 minute layover in Dubai and an extra 8 hours and 35 minutes in the air. That would be a grand total of 23 hours and 25 minutes of additional travel time. The total travel time for the direct flight is 14 hours, whereas the total travel time for the Dubai stopover flight itinerary is 37 hours and 25 minutes (37 hours and 25 minutes - 14 hours = 23 hours and 25 minutes).

There is one argument for taking the connecting flight through Dubai instead of the nonstop Qantas Flight 63. While keeping in mind that airline flight rates fluctuate from day to day (and indeed sometimes form hour to hour), according to the prices published for the flights when this author researched the flights, the price for the Qantas Flight 63 direct flight was $3,089, whereas the price for the itinerary with the connecting flight in Dubai was $2,456. That would be a $633 savings to opt for the Emirates Airline itinerary.

Apparently, the $633 savings is worth the extra 23 hours and 25 minutes of travel time for some travelers, but that is not the only consideration when analyzing these flights. Assuming the earth is a globe, Emirates Airline would have a hard time justifying such a convoluted flight itinerary as depicted on a globular earth of flying the airplane an extra 8 hours and 35 minutes over a further distance of 4,596 miles by stopping in Dubai. Indeed, it would be foolish for any airline to charge less money to fly a significantly further distance with an out of the way stopover between two cities, particularly, when the first leg of the trip (from Sydney to Dubai) is further than the Google Earth reported distance for a direct flight from Sydney to its final destination in Johannesburg. Even if Dubai is the hub for Emirates Airline, it would seem that a flight from Sydney to Johannesburg with a layover in Dubai would be uneconomical for Emirates Airline if the earth were a globe. That suggests that the flight exists as it does because the earth is in fact flat.

But the 14 hour Qantas Flight 63 nonstop flight seems to suggest that the earth could not be flat. How can a flight traverse the expanse between Australia and South Africa on a flat earth in only 14 hours? The nonstop flight time of Qantas Flight 63, assuming everything else is equal, should be roughly equivalent to the total flight time for the Emirates Airline itinerary through Dubai, if the earth is flat. But the nonstop Qantas Flight 63 is only 14 hours in flight time, which is 8 hours and 35 minutes shorter than the Emirates Airline itinerary.

If the earth is flat, how can the flight time of only 14 hours for Qantas Flight 63 be explained? The answer is found in the air. There are jet streams that flow at different altitudes that can speed an aircraft along at a much faster ground speed than its air speed. The jet stream can have wind speeds ranging from 60 miles per hour to more than 250 miles per hour. For example, in 2015, a Boeing 777-200 set a record crossing the Atlantic from New York to London as it reached a ground speed of 745mph.[424] The typical cruising speed for a Boeing 777-200 is 562 mph. While the airspeed for a cruising Boeing 777-200 is only 562 miles per hour, the jet stream through which the airplane traveled carried the jet at a much faster ground speed of 745 mph. For reference, at sea level, the speed of sound travels at 760 mph.

The New York to London ground speed record was north of the equator. It is notable that there is no similar news reporting any ground speed records being set by commercial jet aircraft traveling between cities across the ocean using the jet stream south of the equator. Why not? Because, the only way that they could report a record ground speed for a commercial airplane across the ocean south of the equator using the jet stream would be to reveal the truly large expanse of the area on the earth traversed by the aircraft that actually exists south of the equator. That would expose the truth that the earth is flat. In order to keep a lid on the flat earth, the fast ground speed traveled in the jet stream south of the equator must be concealed. The jets traveling nonstop flights across the ocean south of the equator must always be falsely portrayed as flying at the slower cruising speed for that aircraft over distorted shorter distances than they are really traveling, thus giving the fictitious impression that the earth is a sphere.

Jeran Campanella analyzed the jet streams and found that the jet streams have different speeds at different altitudes, and most notably, they also reverse direction at different altitudes.[425] For example, while a jet stream in the southern latitudes would be a westerly jet stream at 30,000 feet, once a plane climbs to a higher altitude, the jet stream will reverse in direction and become an easterly jet stream. Incidentally, after Campanella and others began publicizing their findings about the wind currents using the website https://earth.nullschool.net, and then basing their explanation of the time compression in the southern flight routes on that analysis, the website, without explanation, removed the alternative of allowing the viewing of wind currents on an Azimuthal Equidistant map. That removal prevents anyone from seeing how the jet stream flows between locations on the flat earth.

When checking flight data for previous Qantas 63 flights on FlightAware.com, a strange phenomenon presented itself. The flight data was reported for the period from takeoff until approximately 1 and ½ hours into the flight, and then suddenly there is a ribbon through the data explaining: "Gap in available data."[426] After reviewing several historical Qantas 63 Flights, it was found that the gap in data was typically more than 10 hours in length. What was happening during those gaps? It can reasonably be hypothesized that it was during the gap the airplane was catching a jet stream. That jet stream can then push the airplane along at a much faster ground speed than its typical cruising air speed. Thus, an airplane can make the trek across the expanse below the equator at a much faster speed, giving the impression that the earth is a sphere, when in fact, it is not. If Qantas Flight 63 is catching a jet steam, that means that the purported direct distance between Sydney and Johannesburg as reported by Google Earth must necessarily be inaccurate.

Time (EST)	Latitude	Longitude	Course	Direction	kts	mph	feet	Rate	Reporting Facility
Mon 09:40:07 PM	-42.0967	142.3782	217°	Southwest	477	549	30,000		Aus ATC (ADS-B)
Mon 09:40:17 PM	-42.1138	142.3599	218°	Southwest	477	549	30,000	11	Aus ATC (ADS-B)
		Gap in available data							
Tue 08:28:18 AM	-33.8787	31.0517	0°	North	470	541	37,000	11	FlightAware ADS-B (FADN)
Tue 08:28:48 AM	-33.8093	31.0522	0°	North	470	541	37,000		FlightAware ADS-B (FADN)

Cropped screen shot of the relevant portion of the Qantas Flight 63 flight data log from 10 December 2017 showing a 10.8 hour data gap during the approximately 14 hour total flight time.

The existence of the jet streams sets up the perfect situation for an airline to use the jet stream traveling westbound from Sydney to Johannesburg and then use a different jet stream blowing in the opposite direction when returning westbound from Johannesburg back to Sydney. Notably, the same data gap is found for the Qantas Flight 64 that flies from Johannesburg to Sydney as was found in the data for Qantas Flight 63 that flies from Sydney to Johannesburg.[427] Thus, there is no way to determine the ground speed of the jet aircraft during the more than10 hour data gap as they travel back and forth from Sydney to Johannesburg.

Assuming the Google Earth figures of 6,868 miles for the distance from Sydney to Dubai and 3,974 miles for the distance from Dubai to Johannesburg are accurate, that would bring the total distance from Sydney to Johannesburg as calculated by Google Earth to be 10,842 miles as graphed on the flat earth. To catch the jet stream Qantas Flight 63 would need to fly a slightly longer route. Because we don't know the deviation, for the sake of simplicity, we will use the 10,842 mile figure. For Qantas Flight 63 to fly that distance nonstop, puts the average ground speed for Qantas Flight 63 using the jet stream at 774 miles per hour.

But that is not all. Taking the flight data from a representative Qantas Flight 63 from December 10, 2017, we find that there was a 10.8 hour gap in flight data.[428] That leaves us with flight data for 3.2 hours out of the total of roughly 14 hours of flight time. If we approximate an average speed of 541 miles per hour, the airplane would have traveled 1,731 miles during the reported 3.2 hours. That leaves the aircraft with a distance of 9,111 miles, out of the total distance of 10,842 miles, yet to travel during the remaining 10.8 hours. In order to have traveled that distance in 10.8 hours the plane must have traveled at a ground speed of 844 miles per hour (9,111 miles ÷ 10.8 hours = 844 mph). In order for the aircraft to reach a ground speed of 844 mph would require it fly in a jet stream. That would be a ground speed record for a non-SST commercial aircraft. It would certainly explain why there is a more than10 hour data gap in all of the Qantas nonstop flights from Sydney to Johannesburg. Should one think that such a speed is improbable, consider that Egyptair Flight MS 648 Riyadh to Cairo was recorded on FlightRadar24 to be traveling at a ground speed of 803 mph.[429] In addition, Egyptair Flight MS 542 was recorded to be traveling from Cairo to Istanbul at a ground speed of 832 mph.

The most compelling evidence that Qantas Flight 63 is catching a jet steam is the longer flight time for Emirates Airline to fly essentially the same distance. Indeed, Quantas Flight 63 flies all the way to Johannesburg faster than Emirates Airline can fly to Dubai. The reason that the Emirates Airline itinerary on a straight line of travel to Johannesburg takes longer in flight time than Qantas Flight 63 is because Emirates Airline cannot ride the jet stream as can Quantas Flight 63. Emirates is locked into a stop at its hub in Dubai, which causes it to miss the strong jet streams that are found in the lower latitudes. While Emirates Airline is traveling a more direct route from Sydney to Johannesburg than is Qantas Flight 63, it cannot travel as fast because it is traveling without the benefit of a jet stream.

If the earth is a sphere, then the Emirates Airline flight itinerary is not a sound business decision for the Emirates Airline. The Dubai layover flight itinerary has Emirates Airline charging 20% less money to fly a passenger jet 67% further (an extra 4,596 miles). But if the earth is flat, that means that the Emirates Airline itinerary makes perfect sense, as the airline is not flying a further distance than the nonstop flights. The shorter travel time for the nonstop Qantas Flight 63 is explained by the jet stream. Qantas Flight 63 is traveling the same distance as the two legs of the Emirates Airline Sydney-Dubai-Johannesburg itinerary, it is just doing it faster due to the jet stream. Indeed, it is likely that the Qantas Flight 63 is traveling a slightly longer distance, because it must deviate from a perfectly straight course in order the catch the jet stream.

That means that the reported nonstop flight distance between Sydney and Johannesburg on Google Earth is inaccurate. It also means that the flight itinerary from Sydney, with a layover in Dubai, and then on to Johannesburg is not as it appears on a spherical earth. That flight itinerary only appears circuitous because the spherical earth does not portray an accurate depiction of the route. When looking at the flight itinerary on a flat earth, we see that Emirates Airline is in reality not being circuitous. They are not insane clowns on a global joyride. The earth is not a globe. The stopover in Dubai is perfectly explained by a flat earth. The hub for Emirates Airline in Dubai is directly along the flight path from Sydney to Johannesburg on the flat earth. That is why Emirates Airline flies that route.

The flight path deception has been going on for years. As soon as travel by aircraft emerged, efforts were made to conceal the flat earth. Indeed, it was understood immediately that with air travel, the airlines and military needed accurate maps of the flat earth to get where they were going. The Mercator projection simply would not do. While they show the Mercator projection today on the seat back screen to the ignorant masses, that is not the true configuration of the word. The world is flat.

Below is the World War II north polar azimuthal equidistant *Air Map of the World,* described as the "Global War Strategy Map" of the earth. The cartographer creators explain that the air routes depicted on the map approximate the actual military air routes taken during World War II.

The drafters cautioned that the air routes depicted on the map are only approximations and "are not intended to show air routes in actual operation. In times of war air routes are military secrets." It is notable that all air routes on that map, both north and south of the equator, are depicted as straight lines. It is axiomatic that the shortest distance between two points is a straight line. The only way that the routes on the map could approximate the actual military air routes is if the earth is actually flat. The map must be accurate in order for the military to know exactly where certain areas are located; for example, locating a small island in the middle of the Pacific Ocean. Furthermore, if the earth were a globe, the routes would need to be distorted arcs on the map, in order to account for the curvature of the earth.

How did the cartographers of this military map explain the straight lines, without any accommodation for the supposed spherical shape of the earth? They stated that straight lines (instead of arcs) were drawn "to avoid confusion." How can making routes inaccurately drawn on a map "avoid confusion?" It does not. If the earth were truly a globe, drawing straight lines would only add confusion. It is ridiculous to think that cartographers, whose very profession requires precision, and who can draw accurate shapes of continents and islands, cannot seem to draw simple arcs. That deceptive statement by the cartographers is an effort to keep the lid on the truth of the flat earth. When the governments of the world need to know precisely where things are located on the earth (as during a war) they refer to a north polar azimuthal equidistant map of the earth (i.e., a flat earth map). The Global War Strategy Map is evidence that the earth is flat.

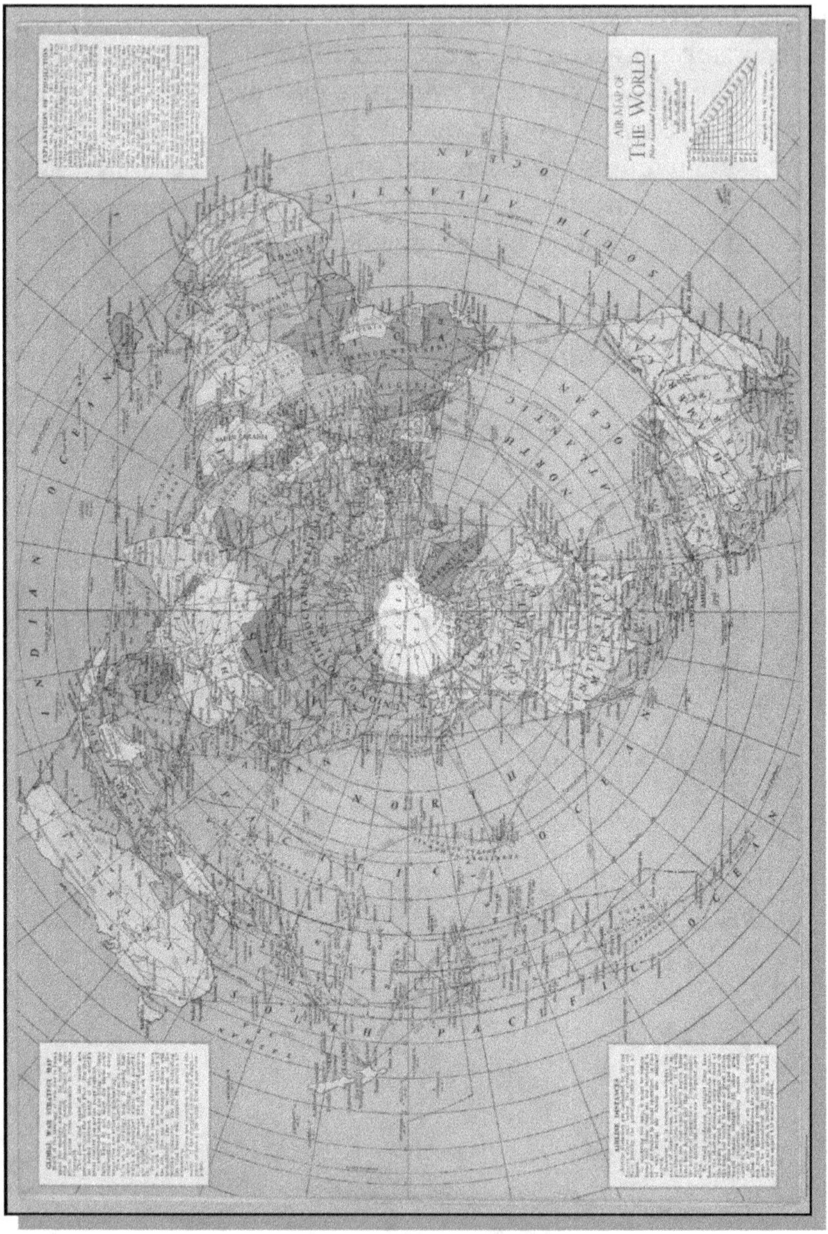

WW II *Air Map of the World* Global War Strategy Map

19 Simple Proof That Heliocentrism is Unfeasible

For most people it is a new concept to think that the sun might be much smaller than the earth and confined to the space directly above the plane of the earth. Before discussing the scientific proof for that, let us examine the standard theory of the heliocentric model. Under the heliocentric theory, the earth orbits the sun once every 365.25 days. Since we only have a 365 day calendar, every four years ($4 \times .25 = 1$ day) we have a leap year and add a 29th day to February to make up that one missing day. In those 365.25 days, the earth is supposed to orbit a full 360 degrees around the sun. At the same time the earth is orbiting the sun, it is completing one 360° rotation every 24 hours.

Under the heliocentric model, as the earth travels around the sun, each day the earth will be a little less than one degree further in its orbit from the previous day. That means that every 6 months the earth will be 180 degrees in its orbit around the sun and be opposite the sun from where it was 6 months earlier. There is a problem with this model. Let us assume that we begin our observation of the sun at 12:00 noon in New York on September 22nd. Each solar day is precisely 24 hours in length. According to the heliocentric model, every 24 hour day, the earth is supposed to do an exact 360 degree spin on its axis and end up at the same place. At the same time, each day the earth has traveled a little less than one degree of the 360° orbit around the sun. That means that after 6 months, 12:00 noon in New York will arrive during the middle of the dark night on March 21st.

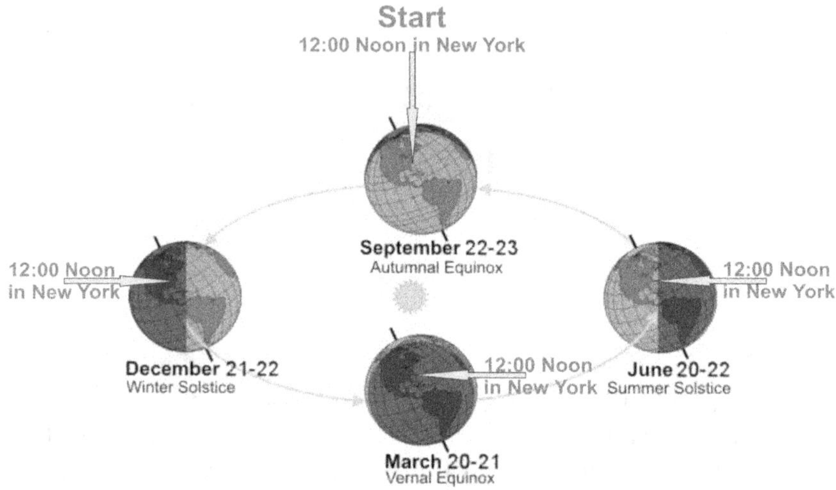

To recap, 12:00 noon arrives in the middle of the day in New York on our start day on September 22nd. The heliocentric theory requires the earth to be at the opposite side of the sun 6 months later. Consequently, New York will be on the dark side of the earth facing away from the sun at 12:00 noon on March 21st. In New York, therefore, on March 21st, 12:00 noon should arrive during the middle of the night. Under the heliocentric model, this occurrence should happen year-in and year-out, every year. We know, however, that does not happen, which means that the heliocentric model is wrong.

According to our 365 day calendar, we need a leap year once every 4 years to make up for a missing day. That means that after 6 months, the earth is supposed to have missed 1/8th of its orbit around the sun. However, that does not change the fact that under the heliocentric model, 12:00 noon in New York will progressively move a little less than one degree each day until it arrives during the middle of the dark night opposite the sun after six months. That is because each solar day is precisely 24 hours in length, which under the heliocentric model requires the earth to rotate exactly 360° on its axis in those 24 hours.

The Annenberg Foundation's teacher resource center, affiliated with Colorado State University, has a model of the globular spinning earth and explains that "[e]ach day, the earth rotates once on its axis, which equals 360 degrees."[430] The University of California at Berkeley's online self-study guide states that the earth spins 360 degrees every 24 hours, "Full turn = 360 degrees."[431] The University of Chicago Department of Astronomy and Astrophysics states that the earth rotates 360° "once every 24 hours."[432]

Professor of Astronomy Courtney Seligman, who has taught college astronomy for 39 years, states that "[t]o calculate how long it takes for the Earth to rotate through an angle of one degree, we divide the length of a day, 24 hours or 1440 minutes, by the 360 degrees it turns through during that rotation, obtaining a rotational speed of 4 minutes per degree."[433]

Professor Seligman created a diagram of the movement of the earth around the sun. He gave the following explanation of the diagram:

> In the diagram below the four blue dots on the right represent the position of the planet at four times, separated from each other by a third of a rotation period. The number of rotations that the planet has made is indicated by the numbers to the right of each dot. The white dot shows how the position of a specific place on the planet changes as the planet rotates to the east (counter-clockwise in this diagram), and the large

yellow dot on the far left represents the position of the Sun. The sizes of the Sun and planet and the angle that the planet moves through during one rotation have been exaggerated to make it easier to see what is happening. The movement of a planet during one rotation [*sic*]. The planet's movement around the Sun causes the Sun to appear to move around the sky. Each degree that the planet moves around the Sun causes the Sun to appear to move a degree around the planet.[434]

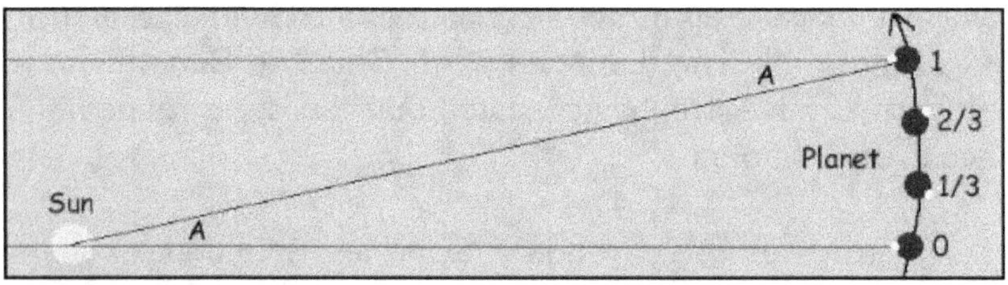

Figure 116: Diagram by Professor Seligman explaining the supposed 360° rotation of the earth as it orbits the sun.

Professor Seligman illustrates the standard model of heliocentricity with the movement of the earth around the sun as it rotates on its axis 360° each 24 hour day. His diagram confirms this author's diagram presented at the beginning of the chapter as the standard heliocentric construct. The depiction of the heliocentric orbit of the earth at the beginning of this chapter was from Penn State University John A. Dutton e-Education Institute, and it is the standard heliocentric model accepted by virtually all scientists today.[435] The diagram was annotated by this author with the captions pointing out where 12:00 noon in New York would be at each quarter of the earth's yearly orbit. Other than the annotations, the diagram appears as Penn State depicted it.

Professor Seligman has taught college astronomy for 39 years and has a full and complete understanding of heliocentrism.

His diagram illustrates that the heliocentric model does not comport with reality. If Professor Seligman's diagram was expanded to encompass a 180° travel of the earth around the sun, we would find that 12:00 noon would arrive in the middle of the night after six month's time, just as depicted in the diagram at the beginning of this chapter. We know that in fact every day 12:00 noon arrives in the middle of the day in New York. There has never been a day in the history of New York ever when 12:00 noon was in the dark of the night. That simple fact proves that the heliocentric model is wrong.

Professor Seligman, and many astronomers like him, are wrong that the earth orbits the sun. They are correct, however, about the relationship between the sun and the earth being one of 360° and 24 hours. The sun in fact does make a 360° circuit over the flat earth once every 24 hours.

The reason that the standard heliocentric model requires a 360° rotation in 24 hours is that it comports with the actual movement of the sun over the earth. The 360° path of the sun over the earth every 24 hours is observable and measurable.

All celestial navigation is based upon that truth. Celestial navigation is based upon the confirmable fact that the sun moves 15° each hour (360° ÷ 24 hours = 15°). Each hour can in turn be broken down to minutes per degrees by simply taking 60 minutes ÷ 15° = 4 minutes per degree.

James I. Sammons explains the basic concepts of celestial navigation. As with most people, he has been conditioned to believe that it is a rotating earth that causes the movement of the sun in its path. However, he is absolutely correct about the sun and the earth having a relationship of 360° in 24 hours. Indeed, that is the essential starting point for all celestial navigation. The fact that celestial navigation is based upon the sun's movement of 360° in 24 hours, and that it works, seemingly locks in the heliocentric model to the 360° spin every 24 hours. Sammons explains:

> Just as we saw in finding latitude, finding longitude is easy enough when we've learned three basic ideas. The first of these ideas is the relationship between time and the rotation of the Earth. It takes an average time of 24 hours for the Earth to rotate 360 degrees so that a spot on its surface will move from under the Sun and then just return to it's under-the-Sun starting position. In 12 hours, the Earth will turn half around. In 6 hours, a quarter. If you divide the number degrees in a circle by the number of hours in a day, we find that the Earth turns 15 degrees each hour. 360° ÷ 24 hours = 15° per hour. We can take this a step further and state that the Earth turns one degree in four minutes. 1 hour = 60 minutes ÷ 15° = 4 minutes per degree.[436]

Mariners have been using celestial navigation for hundreds of years to mark an accurate course. It is based upon the 360°movement of the sun over the earth in 24 hours. Of course modern mariners have been conditioned to believe that it is the globular earth that is spinning. Mariners, however, have often found that their charts based upon a globular earth were inaccurate (particularly in the southern latitudes) when they compared their dead reckoning using their charts with their more accurate celestial navigation.

But the purported travel of the earth in its orbit around the sun of approximately one degree (.986°) each day under the heliocentric model would throw off celestial navigation. The alleged orbit around the sun would change the sun's position in the sky by .986° each day. That could be disastrous for a ship trying to find an island or avoid a hazard. Under the heliocentric model, the sun will be seen in the sky to travel not 360° each day but rather 360.986° each day. And as we have seen, the .986° movement of the sun in the sky each day would cause noon on the allegedly spherical earth to be in complete darkness in the middle of the night after six months. Obviously, there is a big problem with the heliocentric model.

The high priests of heliocentricity have perceived the problem with their model, and they have adjusted it to accommodate its error. What did they do? They added .986° to the 360° rotation of the earth, to come up with the earth spinning 360.986° each day.[437] Where did they get the .986°? They simply made it up by dividing the 360° of orbit by 365.25 days in the year, to come up with .986°. With this little mathematical adjustment, their problem is seemingly solved. Because they claim the earth rotates 360.986° in 24 hours, they calculated that the period of time for the earth to rotate 360° is 23 hours, 56 minutes, 4.1 seconds.[438] They call this shorter day, a sidereal day. A sidereal day is supposed to be based upon the length of time which passes between the position of the stars in the night sky. A sidereal day is irrelevant to the actual movement of the sun.

Below is a diagram from Cornell University Department of Astronomy illustrating the contrived 360.986° rotation each day.[439]

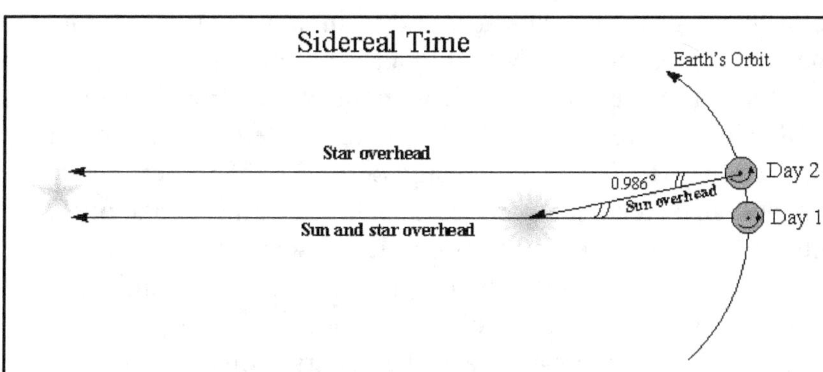

Sidereal literally means "relating to, or expressed in relation to stars or constellations."[440] A sidereal day is a measure of the movement of the stars. It is pure and simple deception to apply calculations for the movement of the stars to the movement of the sun. Never has there been celestial navigation involving the sun traveling 360.986° in 24 hours. But the concept so called scientists are trying to sell is that the extra .986° of spin cancels out the .986° change in position of the earth in its rotation around the sun each day. The argument is that while the earth orbits the sun .986° each day, the added rotation of the earth of .986° each day cancels out that .986° change in the earth's position in relation to the sun. They argue that with that cancellation the sun is seen to move over the earth 360° every 24 hours. But the .986° extra spin is a contrivance of the heliocentric priesthood. It is a superstitious belief that was conjured up in order to explain the otherwise impossible heliocentric model. The added .986° only has validity because the scientists say so. They don't explain how the earth's spin of an extra .986° happens to match precisely the .986 orbit of the earth around the sun. They don't need to because heliocentricity is more religion than science. You cannot think, and you cannot ask. You must believe the doctrine because the priests of science proclaim it to be so.

That .986° adjustment allows them to explain why high noon does not move each day and ultimately arrive at midnight once every six months. Max Planck was one of the most noted scientists of the last century. He won the Nobel Prize in Physics in 1918. He revealed the cult like belief system in the scientific community. He stated that "anybody who has been seriously engaged in scientific work of any kind realizes that over the entrance to the gates of the temple of science are written the words: Ye must have faith. It is a quality which the scientist cannot dispense with." So-called "scientists" today are more akin to witch doctors, who have mesmerized the superstitious tribe to believe their booga-booga nonsense.

20 The Testimony of the Stars

Polaris proves that the heliocentric model is pure fantasy. Polaris, the North Star, is fixed above the north pole of the earth. The North Star does not move; it has never moved from its fixed position over the north pole from the fourth day of creation (Genesis 1:16) until today.

The North Star is alleged by modern scientists to be 433.8 light years away from the earth. That means that a person would have to travel at the speed of light for 433.8 years before reaching the North Star. Light travels at approximately 186,282 miles per second, which is over 670 million miles per hour.

The diagram below proves the impossibility of the heliocentric theory. If the North Star is above the North Pole, it must be moving in precise and exact synchronization with the earth. The earth is supposed to be traveling at 67,062 miles per hour around the sun in a slightly elliptical path with an average radius of over 92 million miles (approximately 1.8 billion miles in circumference).

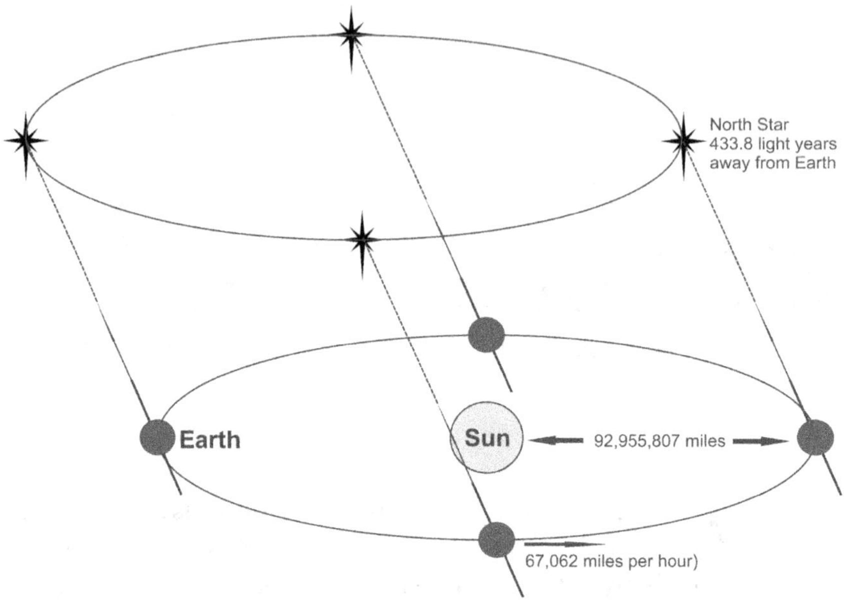

What makes the heliocentric model even more impossible is that while the North Star is supposed to be keeping itself in perfect synchronization with the orbit of the earth around the sun, the sun, itself, is supposed to be traveling at a speed of approximately 500,000 miles per hour around the Milky Way galaxy. Adding to the impossibility, the Milky Way galaxy is hypothesized by modern scientists to be racing through space at a speed ranging from 300,000 to 1,340,000 miles per hour. The heliocentric fairy tale gets even more fantastic; the perfect synchronization of the North Star with the earth finds its origin in a massive, chaotic explosion called the "big bang" that happened billions of years ago.

This insane fantasy (masquerading as science) of a spherical spinning and orbiting earth is so ingrained in the consciousness of the world, that any refutation of it is viewed by the world as itself evidence of ignorance and perhaps even insanity. Believing in a flat earth is immediately met with derision; the person is viewed as being an ignorant savage or part of a crazy fringe cult. President Barrack Obama recently tapped into that conditioned response, and used it against those who were impeding his plans for reducing the American standard of living by reducing our consumption of so-called fossil fuels. Obama stated in a speech that anyone who denied that global warming is real is an ignorant person who believes in a flat earth.

> President Obama angrily blasted climate change skeptics during his energy policy speech Tuesday at Georgetown University, saying he lacked "patience for anyone who denies that this problem is real." ... "We don't have time for a meeting of the flat-Earth society."[441]

How did scientists come up with the fantastic distance that the North Star is from the earth? Simple—they made it up. There is no evidence to support that distance. It is necessary in the heliocentric model that Polaris be 433.8 light years from Earth. Indeed, the fantastic distances in the heliocentric model were simply hypothesized to explain how Polaris can be fixed above a careening and spinning Earth and why all the stars move in unison, without any parallax between each other. Parallax is the movement of one star in relation to another star that would be expected to be seen if the earth were spinning and orbiting in space around the sun.

What is observed, however, is no parallax between the stars whatsoever. With the exception of a few wandering stars, which today are called the planets of the alleged solar system, the stars move in unison, as though they are part of one great single mass. That proved to be a problem for the heliocentric model. Indeed, such movement makes the heliocentric model impossible. In order to explain away the lack of parallax, the priests of heliocentrism simply expanded the universe in their minds and scientific papers and argued since the stars are so far away that the parallax is imperceptible to us on earth. The problem with that argument is that no matter how far away the stars are, parallax would be discernable if the earth were moving. There is no parallax, because the stars are in fact within the firmament (Genesis 1:14-18), moving in perfect unison over a flat, stationary earth.

21 Path of Midnight Sun Disproves Heliocentrism

Below is a 360° panorama time lapse photograph showing the circuit of the sun over the Island of Loppa in the northern part of Norway, between 1900 hours, July 21st and 1800 hours, July 22nd.[442] The Island of Loppa is at approximately 70° north latitude. The panorama photograph illustrates that in midsummer, north of the Arctic Circle, the sun stays above the horizon for 24 hours of the day. The photograph reveals the sun as it makes a complete circle over the head of the photographer as he took one picture each hour during a 24 hour period.[443] As you can see the sun is over a different area of the land at the time of each photograph. The picture was posted on a travel website in order to illustrate the beauty of the midnight sun in the north of Norway.

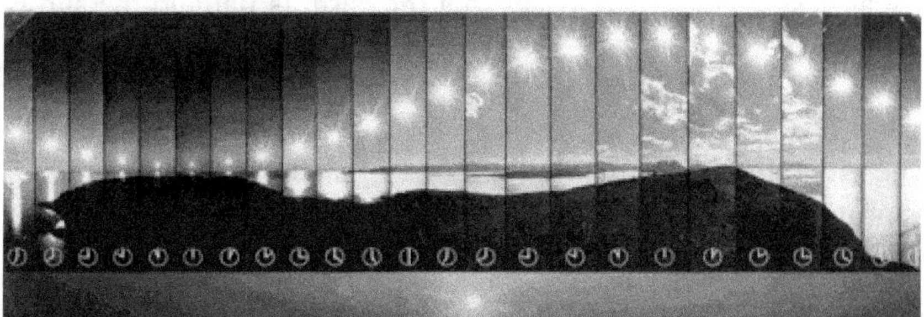

The above photograph is different from typical time lapse photographs, where the camera is left stationary. In the time lapse above, the photographer circles his camera a full 360° around his position. Each image of the sun is a separate photograph of the sun, with the background changing according to the movement of the camera as the photographer panned to shoot the sun once each hour.

The photographer has inadvertently proven that the sun circles over the flat earth. How so? The only way the sun would seem to circle the observer on a globular, spinning earth would be if the observer was standing directly over the North Pole. Any distance south of the north pole would create a non-uniform movement where the sun would speed up and slow down in the sky if the earth were spinning and orbiting a 93 million mile distant sun, as purported in the heliocentric model. That non-uniform motion would cause the topography to appear jagged at many of the seams when the photographic strips would be matched up, because the speed of the sun would not be constant relative to the photographer's uniform hourly time frames. The panorama photograph was taken above the Arctic Circle from approximately 1,400 miles south of the North Pole. The perfectly matched topography at each seam proves the sun's uniform speed.

The above time lapse photograph would be impossible if a spinning earth were orbiting around a distant sun, as is purported to be the case under the heliocentric model. The above time lapse photograph of the 360° circular path of the midnight sun showing a seamless topography proves that the earth is flat and the sun is circling overhead.

Below are two stills taken from a video by Rory Cooper.[444] The video illustrates that the only way that a time lapse series of photographs could depict a perfectly seamless topography is if the sun was circling 360° overhead north of the Arctic Circle within the boundaries of a flat earth.

Figure 120: Image showing the position of a cameraman standing south of the North Pole on a spinning earth with the sun millions of miles away. The midnight sun would be seen by the observer to speed up and slow down over the horizon.

Figure 121: Image showing the reality of the sun circling 360° at a uniform speed overhead within the boundaries of a flat earth. That is what is seen in all time lapse photographs of the midnight sun, north of the Arctic Circle.

22 Lunar Eclipses Explained

Many will point to a lunar eclipse to refute any notion of a flat earth. The argument is that both the sun and moon could not be smaller lights over a much larger earth, because an eclipse is argued to be the blocking of the sunlight on the moon by the intervention of the spherical earth. The problem with that postulation is that it assumes facts that have never been proven. That is, it assumes that a lunar eclipse is caused by the passing of a spherical earth between a distant sun and the moon. That supposed "scientific fact" can be proven to be a fallacy.

Samuel Rowbotham explains many historical instances where both the sun and moon were visible to an observer above the horizon during an eclipse of the moon. If the eclipse of the moon is caused by the intervention of the supposed shadow of a spherical earth between the sun and the moon, it would be impossible for an observer on the earth to see both the sun and the eclipsed moon above the horizon. Rowbotham cites an instance "on the 19th of July, 1750, when the moon appeared visibly eclipsed, while the sun was distinctly to be seen above the horizon."[445] He cites another example where "[o]n the 20th of April, 1837, the moon appeared to rise eclipsed before the sun had set. The same phenomenon was observed on the 20th of

September, 1717."[446] Those are not the only instances of this happening. Rowbotham continues:

> In the lunar eclipses of July 17th, 1590; November 3rd, 1648; June 16th, 1666; and May 26th, 1668; the moon rose eclipsed whilst the sun was still apparently above the horizon. Those horizontal eclipses were noticed as early as the time of Pliny.[447]

In a later 1881 edition of his book, Rowbotham cites to a Daily Telegraph report of the sun and eclipsed moon simultaneously appearing above the horizon. "On the 17th of January, 1870, a similar phenomenon occurred; and again in July of the same year."[448]

One might wonder if that same phenomenon is apparent today. In fact, it has been a regular occurrence throughout history. As reported by Joe Rao for Space.com, "On Oct. 8, [2014] interested skywatchers should attempt to see the total eclipse of the moon and the rising sun simultaneously. The little-used name for this effect is called a 'selenelion,' a phenomenon that celestial geometry says cannot happen."[449]

Notice that Rao explains that the phenomenon of the sun and the eclipsed moon appearing above the horizon at the same time is "a phenomenon that celestial geometry says cannot happen." Indeed, the very phenomenon is an impossibility under the heliocentric mythology. Modern scientists have even given a label to this impossibility: a "selenelion." Of course, modern cosmologists must now define their new dogma in scientific terms so that the masses won't discover that it is really a religious deception. How do they explain this phenomenon? Joe Rao presents the "scientific" explanation:

> And indeed, during a lunar eclipse, the sun and

> moon are exactly 180 degrees apart in the sky. In a perfect alignment like this (called a "syzygy"), such an observation would seem impossible. But thanks to Earth's atmosphere, the images of both the sun and moon are apparently lifted above the horizon by atmospheric refraction. This allows people on Earth to see the sun for several extra minutes before it actually has risen and the moon for several extra minutes after it has actually set. As a consequence of this atmospheric trick, for many localities east of the Mississippi River, watchers will have a chance to observe this unusual sight firsthand. Weather permitting, you could have a short window of roughly 2 to 9 minutes (depending on your location) with the possibility of simultaneously seeing the sun rising in the east while the eclipsed full moon is setting in the west.[450]

Rao would have the viewers of the phenomenon reject what their eyes see, and instead believe his "scientific" explanation that it is an "atmospheric trick." It is like the man caught by his wife in bed with another woman and explaining to his wife that she did not really see what she saw. He argues: "are you going to believe me or your lying eyes?" In like manner, the hapless eclipse observers must ignore what they see and yield their minds to the heliocentric dogma that they are not really seeing what is plainly before their eyes, the sun and eclipsed moon simultaneously above the horizon.

Figure 122: December 2011 Daytime Lunar Eclipse Viewed From New Mexico

The problem with Rao's explanation is that it cannot explain the observed event. To understand this, we must look at a time lapse film of a December 2011 daytime lunar eclipse taken from New Mexico.[451] In the time lapse, the moon is being eclipsed as the sun rises above the horizon. The dark eclipse of the moon does not appear progressing from the bottom of the moon and going up as expected by the "scientific" explanation of the selenelion as being due to atmospheric refraction of the light. The moon can be seen being eclipsed by a shadow that traveled over the moon from the top down. The atmospheric refraction explanation of the pseudo-scientists of modern heliocentricity requires the shadow to progress from under the moon. What actually happened, with the shadow progressing downward, refutes the light refraction explanation.

Below is a screen shot of a time lapse video of the December 2011 daytime lunar eclipse taken from Madison, Wisconsin.[452]

Figure 123: December 2011 Daytime Lunar Eclipse View from Madison, Wisconsin

Next we see a screen shot of a video taken of the December 2011 lunar eclipse viewed from East Saint Louis, Illinois, while both the sun and the moon were above the horizon.[453] The sun has risen behind the cameraman and the eclipse of the moon can clearly be seen with the shadow progressing from the top-down. That is just the opposite of what would be expected under the atmospheric refraction explanation given by the heliocentric pseudo-scientists.[454]

Figure 124: December 2011 Daytime Lunar Eclipse Viewed From East Saint Louis, Illinois

Early in the East Saint Louis eclipse video, the narrator is standing between the eclipsing moon and the rising sun. He states that "behind me is a full moon which will soon be going into a full lunar eclipse as it sets, kind of a rare experience, while at the same time behind us the sun is already beginning to rise."[455] The narrator then explains the unexpected occurrence of the shadow of the moon coming down across the moon. Rather than realize that such an occurrence is impossible if the earth were a spinning globe in space, he tries to reconcile the impossibility of what he is seeing with his ingrained heliocentric conditioning.

> What I find very interesting is that I was expecting the shadow to come upward across the moon ... and actually it is going the other way around ... the shadow is creeping down across the moon, even though the sun is coming up over here and that must be because of the moon's rotation around the earth, it's moving farther up, even though the sun's light is actually coming from over there, it's actually coming underneath the earth ... the moon must be rotating around even though we're turning so it looks like the moon is going down. It's kind of an optical illusion, pretty cool.[456]

His explanation makes no sense. He unconvincingly states that the light at the bottom of the moon is being caused by the light from the sun bending around the bottom of the earth and popping up on the other side to light the bottom of the moon. His explanation seems to be a variation on the atmospheric refraction argument, although it violates all known properties of light and refraction.

The narrator of the eclipse video explains that the sunrise is scheduled for 7:08 a.m. When 7:08 a.m. arrives, the narrator explains that the sun could be seen "peaking" above the horizon. He then states that "exactly 180 degrees in this direction and 180 degrees in the other direction is where the moon is. So there in a perfect line, the moon being this way and the sun being that way. If we drew a line from it [the sun] to the earth to the moon, it would be a nice perfect straight line, of course if it wasn't we wouldn't be having a lunar eclipse."[457] It is astounding that someone could be standing and looking at the sun above the horizon and in a straight line directly 180° across from him see the eclipsing moon and conclude that the earth he is standing upon is casting a shadow and causing the lunar eclipse. You would think that it would dawn on him (pun intended) that such an occurrence cannot be explained by the heliocentric model.

Daytime Lunar Eclipse
Darkness Moves From Top Down

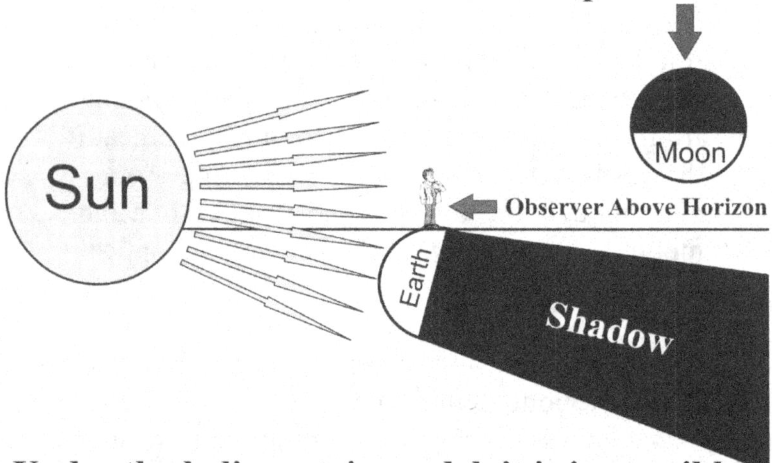

Under the heliocentric model, it is impossible for a daytime lunar eclipse to be caused by the earth's shadow.

The diagram above illustrates the impossibility of the earth's shadow causing a daytime lunar eclipse. Indeed, the progression of the shadow over the moon from the top down renders preposterous the explanation of heliocentric "scientists" that the phenomenon of a selenelion daytime lunar eclipse is due to atmospheric refraction. For the absurd atmospheric refraction theory to have any viability, the shadow would need to travel from the bottom up, not the top down, as actually occurs.

If the blocking of the sun by the earth is not the cause of the lunar eclipse, then what is the cause? Rowbotham opines that there is a semi-transparent body passing in front of the moon.

> We have seen that, during a lunar eclipse, the moon's self-luminous surface is covered by a semi-transparent something; that this "something" is a definite mass, because it has a distinct and circular outline, as seen during its first and last contact with the moon. As a solar eclipse occurs from the moon passing before the sun, so, from the evidence above collected, it is evident that a lunar eclipse arises from a similar cause--a body semi-transparent and well-defined passing before the moon; or between the moon's surface and the observer on the surface of the earth.[458]

Rowbotham based his conclusion on his observations and the findings of the astronomers of his day.

> That many such bodies exist in the firmament is almost a matter of certainty; and that one such as that which eclipses the moon exists at no great distance above the earth's surface, is a matter admitted by many of the leading astronomers of the day. In the report of the council of the Royal Astronomical Society, for June 1850, it is said:–
>
> "We may well doubt whether that body which we call the moon is the only satellite of the earth."
>
> In the report of the Academy of Sciences for October 12th, 1846, and again for August, 1847,

the director of one of the French observatories gives a number of observations and calculations which have led him to conclude that,–

"There is at least one non-luminous body of considerable magnitude which is attached as a satellite to this earth."

Sir John Herschel admits that:– "Invisible moons exist in the firmament."[459]

Sir John Lubbock is of the same opinion, and gives rules and formulæ for calculating their distances, periods.[460]

At the meeting of the British Association for the Advancement of Science, in 1850, the president stated that,—

"The opinion was gaining ground, that many of the fixed stars were accompanied by companions emitting no light." "The 'changeable stars' which disappear for a time, or are eclipsed, have been supposed to have very large opaque bodies revolving about or near to them, so as to obscure them when they come in conjunction with us."[461]
...

We have now seen that the existence of dark bodies revolving about the luminous objects in the firmament has been admitted by practical observers from the earliest ages; and that in our own day such a mass of evidence has accumulated on the subject, that astronomers are compelled to admit that not only dark bodies which occasionally obscure the luminous stars when in conjunction, but that cosmical bodies of large size exist, and that "one at least is attached as a satellite to this earth." It is this dark or "non-luminous satellite," which when in conjunction, or in a line with the moon and an observer on earth, is the immediate cause of a lunar eclipse.[462]

23 The Cold Light of the Moon

We need not rely on the opinion of Rowbotham or the eminent astronomers of his day. The very nature of the moonlight proves that the eclipse is almost certainly caused by a dark non-luminous body moving over the moon. The earth shadow as a hypothesis to explain an eclipse is refuted by the fact that at the middle point in the eclipse the moon does not go dark, as would be expected if sunlight were being blocked by a spherical earth. The moon, in fact, glows blood red at the apex of the eclipse. The moon turns a deep red hue, much like red-hot copper.[463] The moon is actually glowing red in the midst of what is supposed be the dark umbra of the earth's shadow.

That fact alone indicates that the moon is not a reflector of the sun's light, but rather is a light-source in-and-of itself. That confirms what God states in Genesis. God refers to both the sun and the moon as "lights." God did not call the moon a reflector of light, but rather a light in-and-of itself.

> And God said, Let there be **lights** in the firmament of the heaven to divide the day from the night; and let them be for signs, and for seasons, and for days, and years: And let them be

> for **lights** in the firmament of the heaven to give light upon the earth: and it was so. And God made two great **lights**; the greater **light** to rule the day, and the lesser **light** to rule the night: he made the stars also. And God set them in the firmament of the heaven to give light upon the earth, And to rule over the day and over the night, and to divide the light from the darkness: and God saw that it was good. (Genesis 1:14-18)

Genesis 1:15-16 states clearly that there are two lights (not one), "the greater light to rule the day, and the lesser light to rule the night." God states that they are to be "lights" in order "to give light upon the earth." God did not say that there would be one light and one reflector. Furthermore, God stated that the greater light rules the day. He said nothing about the greater light ruling the night, which would necessarily have to be the case if the moon (the lesser light) was not actually a light, but a reflector. The orthodox view that has the moon reflecting the sunlight effectively requires the greater light (the sun) to rule both the day and the night. But that cannot be, because God states clearly that the lesser light (the moon) rules the night. In order to rule the night, the moon must have its own light.

Indeed, Jesus stated, without equivocation, that the moon is not a reflector, but gives off its own light. "But in those days, after that tribulation, the sun shall be darkened, and **the moon shall not give her light.**" (Mark 13:24) The sun shall be darkened, and the corollary is that the moon shall not give her light. Obviously, that means that ordinarily the moon gives off its own light. *See also* Matthew 24:29; Ezekiel 32:7; Isaiah 13:10.

The truth of the Bible is proven by true scientific observation. Moonlight can be proven **not** to be the reflection of the light from the Sun. The moon in fact gives off its own light. And thus an eclipse can only be explained by the intervention of a non-luminous body in front of the self-luminous moon. Rowbotham explains the unique characteristic of moonlight, which absolutely destroys the theory that the moon reflects the light from the sun.

> A reflector will not throw off cold when heat is placed before it; nor heat when cold is presented. If a red light is received, red light will be returned, not blue or yellow. If the note C is sounded upon any musical instrument, a reflector will not return the note D or G, but precisely the same note, altered only in degree or intensity.
>
> If the moon is a reflector of the sun's light, she could not radiate or throw down upon the earth any other light than such as she first receives from the sun. No difference could exist in the quality or character of the light; and it could not possibly differ in any other respect than that of intensity or quantity. ...
>
> We have then, in order to know whether the moon is a reflector, merely to ascertain whether the light which we receive from her is, or is not the same, in character as that received from the sun.
>
> 1st. The sun's light is generally, and in an ordinary state of the atmosphere, of an oppressive, fierce, semi-golden, pyro-phosphorescent character; while that of the moon is pale, silvery and gentle; and when shining most brightly is mild and non-pyrotic.
>
> 2nd. The sun's light is warm, drying, and preservative, or antiseptic; animal and vegetable

> substances exposed to it soon dry, coagulate, shrink, and lose their tendency to decompose and become putrid. Hence grapes and other fruits by long exposure to sunlight become solid, and partially candied and preserved; as instanced in raisins, prunes, dates, and ordinary grocers' currants. Hence, too, fish and flesh by similar exposure lose their gaseous and other volatile constituents and by coagulation of their albuminous and other compounds become firm and dry, and less liable to decay; in this way various kinds of fish and flesh well known to travellers are preserved for use.
>
> The light of the moon is damp, cold, and powerfully septic; and animal and nitrogenous vegetable substances exposed to it soon show symptoms of putrefaction. Even living creatures by long exposure to the moon's rays, become morbidly affected. It is a common thing on board vessels going through tropical regions, for written or printed notices to be issued, prohibiting persons from sleeping on deck exposed to full moonlight, experience having proved that such exposure is often followed by injurious consequences.[464]

Can Rowbotham's claims regarding the different nature of moonlight and sunlight be proven? In fact, it has been proven. Rowbotham reveals that "[i]n the 'Lancet' (Medical Journal), for March 14th, 1856, particulars are given of several experiments which proved that the moon's rays when concentrated, actually reduced the temperature upon a thermometer more than eight degrees."[465] It is common knowledge that the magnification of the sun's rays increases the temperature at the point at which the magnified beam is cast. The moon, with its cold magnified light, cannot therefore be reflecting the warm light from the sun.

Experiments done as recently as 2016 confirm the results of the Lancet experiments. Magnifying the cold light of the moon resulted in demonstrated measurable cold temperature caused by the moonlight. One experimenter used a strong magnifying Fresnel Lens (although he did not indicate its magnification strength). He detected a difference of approximately 8.1° Celsius (14.6° Fahrenheit) between the surface upon which magnified moonlight was cast, which measured 4° Celsius (39.2° Fahrenheit) and the adjacent surface area shaded from moonlight, which measured 12.1° Celsius (53.8° Fahrenheit).[466]

Is magnification of the moonlight necessary to measure the coolness of moonlight? It is not. Experimenters have found that objects in moonlight are measurably colder than those shaded from the moonlight, without any magnification of that moonlight. One experimenter measured an approximate difference of 7.2°F between a white ceramic plate placed in moonlight (44.2°F) and a white ceramic plate shaded from the moonlight (51.4°F).[467] Another experimenter measured a difference of approximately 2°F between the portion of a metal ramp placed in moonlight (54°F) and the area of the same metal ramp that was shaded from the moonlight (56°F).[468] Yet, another experimenter found that the portion of a wallet left outside on the ground that was in the moonlight was 4°F colder than the portion of that same wallet that was shaded from the moonlight.[469]

This author has used a dual laser infrared thermometer [470] and detected that material in moonlight is indeed colder than material shaded from the moonlight. Depending on the material, the difference in temperature ranged from approximately 2°F to approximately 6°F colder for material in moonlight than for material shaded from the moonlight. Of course, as expected, the very opposite was the case for material in sunlight versus material shaded from the sun. This author found that material in sunlight was detected to be approximately 20°F to 25°F warmer than the same material shaded from the sunlight.

Moonlight has a unique characteristic (it is cold) that is different from sunlight (which is warm). That proves that moonlight cannot be the reflected light from the sun. If it were, the reflected light would have to be warm. As we have seen, moonlight shone on objects make them colder as compared to objects protected from the moonlight by shade, whereas an object in sunlight is warmer than an object shaded from the sun. There is another difference in the two lights. While magnification of sunlight causes magnification of the heat from the sun, magnification of the moonlight does not cause magnification of the cold. Even extreme magnification of moonlight does not cause significant reduced temperature. That is unlike extreme magnification of sunlight, which causes significant increased temperature. Dr. Henry Noad in his *Lectures on Chemistry* reveals results of experiments with magnified moonlight:

> The light of the moon, though concentrated by the most powerful burning-glass, is incapable of raising the temperature of the most delicate thermometer. M. De La Hire collected the rays of the full moon when on the meridian, by means of a burning-glass 35 inches in diameter, and made them fall on the bulb of a delicate air-thermometer. No effect was produced though the lunar rays by this glass were concentrated 300 times. Professor Forbes concentrated the moon's light by a lens 30 inches in diameter, its focal distance being about 41 inches, and having a power of concentration exceeding 6000 times. The image of the moon, which was only 18 hours past full, and less than two hours from the meridian, was brilliantly thrown by this lens on the extremity of a commodious thermopile.

Although the observations were made in the most unexceptional manner, and (supposing that half the rays were reflected, dispersed and absorbed), though the light of the moon was concentrated 3000 times, not the slightest thermo effect was produced.[471]

24 Let There Be Light

Modern scientists not only conceal the truth of the emitted light of the moon, but they conceal the fact that there is a light that shines during the day that is separate and distinct from the sun. Secular science has rejected the Genesis account of creation and instead placed the sun far away from the earth in an alleged vacuum of space. The scientific community is in a conspiracy against God and man and is trying to conceal the truth of God's creation. This author will take a different approach from that which is taken by modern scientists. I will not ignore God's word, but rather start with the premise that God's word is pure truth, and then try to explain observations in light of that truth.

Those Christians who think that the Bible should not be used as the basis for ascertaining the scientific nature of God's creation should consider what Jesus stated about that matter: "If I have told you earthly things, and ye believe not, how shall ye believe, if I tell you of heavenly things?" (John 3:12) That truth from Jesus is the reason that the "scientific" community pooh-poohs any reference to biblical authority. Indeed, the model of a spinning, spherical earth in a heliocentric system is propaganda framed for the very purpose of calling into question the authority of the Bible about "earthly things," with the ultimate objective of undermining its authority about "heavenly things." Heliocentricity contradicts the Bible at every turn (pun intended). Heliocentricity is what God calls the "opposition of science falsely so called." 1 Timothy 6:20.

Satan has an objective in mind by convincing people the Bible is scientifically wrong about "earthly things." His strategy is to get people to think that since the Bible is wrong about "earthly things," it must also be wrong about "heavenly things." Satan wants to drive from the minds of men the fact that God's "word is true from the beginning." (Psalms 119:160) Satan knows that if people believed that, "the word of the LORD is right; and all his works are done in truth" (Psalms 33:4), Satan's world conspiracy against God and man would collapse.

If we read the book of Genesis, we find that God created light on the first day and divided the light from the darkness. This formed our days and nights.

> And God said, Let there be light: and there was light. And God saw the light, that it was good: and God divided the light from the darkness. And God called the light Day, and the darkness he called Night. And the evening and the morning were the first day. (Genesis 1:3-5)

Notice that day and night began after God created light and divided it from the darkness. The sun and moon were not created until the fourth day; God did so after he created the firmament that divided the waters. God set the sun and moon as lights in the firmament.

> And God said, Let there be lights in the firmament of the heaven to divide the day from the night; and let them be for signs, and for seasons, and for days, and years: And let them be

> for lights in the firmament of the heaven to give light upon the earth: and it was so. And God made two great lights; the greater light to rule the day, and the lesser light to rule the night: he made the stars also. (Genesis 1:14-16)

From God's account of creation in Genesis we know that day and night, with the day being light and the night being darkness, were created by God on the first day. It was not until the fourth day that the sun and moon were created. There was daylight and night darkness for three days and nights before God created the sun and moon. That means that there is another source of light for our days that is separate from the sun. The sun rules the day, but daylight itself exists without the sun.

Some have opined that the source of the light for day is God himself, since "God is light, and in him is no darkness at all." 1 John 1:5. While it is true that God is light, it is also true that God is the creator of light. God states: "I form the light, and create darkness: I make peace, and create evil: I the LORD do all these things." Isaiah 45:7. When God stated "let there be light" in Genesis 1:3, the result was that "there was light." That indicates that God created the light, and that light is the light of day. The bottom line is that there is a source of light created by God for our days that is separate from the sun.

Rory Cooper, who is an artist, has illustrated in an instructional video that if the sun were the only source of light our days would not look at as they do.[472] Cooper opines, based upon his acceptance of the truth of God's word and his training and experience as an artist, that it would be impossible for our light during the day to appear as it does without another source of light. The fact that there is light that surrounds even the sides of objects that are faced away from the sun indicates that there another source of light that is backlighting everything.

We can see this second light source at dawn and at dusk. In this twilight stage, before the sun rises above the horizon and again after the sun sinks below the horizon, the sky is illuminated by a light that has a source which is separate from the sun. Indeed, this light could not come from the sun, since the only shadows that are cast are underneath parked cars that block light streaming straight down. This author is often outside at dawn and cannot see shadows anywhere, except those shadows underneath parked cars. That indicates that the light is being cast vertically straight down. Thus, the light could not be coming from the sun, which has not yet made its appearance above the horizon. If the sun were the source of the shadows they would be long and horizontal. Indeed, later, after the sun rises one will see that there are long shadows that are cast, but that is only after the sun is above the horizon.

The predawn and post-sunset light is called twilight, which is defined as "the soft, diffused light from the sky when the sun is below the horizon, either from daybreak to sunrise or, more commonly, from sunset to nightfall."[473] Mary Magdalene and the other Mary came to the sepulchre in which Jesus was buried "as it began to dawn toward the first day of the week." Matthew 28:1. Dawn means "to begin to grow light in the morning."[474] It describes "the first appearance of light in the sky before sunrise."[475] God describes Mary Magdalene's arrival at the sepulchre as being in the morning of the first day of the week. Luke 24:1. So we see that God marks the morning, not at sunrise, but rather at dawn, which is the first light, known as twilight. This is in accord with how God marked each day in Genesis, where he stated that each day was marked by evening and morning for 3 days before he created the sun and moon. Genesis 1:1-19. Indeed, twilight can be quite subdued, because when Mary Madgelene came early in the morning on the first day of the week, it was dark. John 20:1.

The orthodox view of twilight is that it is "produced by diffusion of sunlight through the atmosphere."[476] But there is another explanation that is completely ignored by the scientific community. It is more plausible that the twilight is coming from an entirely different source from the sun. This is supported by Genesis where God stated he created light and separated it from darkness to establish the day and night before he created the sun and moon.

The fact that there are no shadows except for those cast underneath cars, indicates that twilight is emanating from directly above the earth at a vertical angle that is perfectly perpendicular to the ground. This happens day in and day out, without variation. There are no conventional laws of light refraction or diffraction that can explain how the sun, being below the horizon, can be the source of the vertically descending light rays seen during twilight.

One could theorize that it is light reflection, and not refraction or diffraction, that is the cause of twilight. One candidate for a reflector would be the firmament, which God describes in Genesis 1:6. Indeed, God describes the sky as being a strong, molten looking glass. Job 37:18. That passage in Job suggests that at the outer reaches of the firmament it may be viscous and could act as a reflector. Arguably, sun dogs and rainbows are examples of the firmament acting as a reflector.

The problem with the firmament reflector theory for twilight is that the sun itself is "in" the firmament. Genesis 1:14-19. Since the sun is in the firmament, and it is beyond the horizon during twilight, the firmament would seem not to be a good candidate for reflecting sunlight perfectly perpendicular to the ground. Even if we ignored that fact, there is still no proof that the firmament would reflect the sunlight at a precisely perpendicular angle without variation every day, because, as we will see below, light is continuously refracted at different angles as it travels through the atmosphere. So a perfectly perpendicular light from a reflection without variation each day would be highly improbable.

The uniformity of the descending light rays during twilight suggests that the source of the light is not a reflection but rather an emanation. And that emanation must necessarily be independent of the sun. Twilight, being an independent emanation, explains why its light is cast at a vertical angle as it goes through the different layers of the atmosphere. While light ordinarily refracts when traveling through the different densities in the atmosphere, it does not refract if its angle of incidence is precisely perpendicular to each new layer of different density. Light is refracted only if it has an angle of incidence that deviates from perpendicular (i.e., normal). That would explain why the twilight is seen to be perfectly perpendicular to the ground.

The conclusion of an independent light source is helped along by the inerrant word of God, which states that indeed God created a source of light to illuminate the day, which he created before the sun, and that light is thus necessarily independent of the sun. “God called the light Day, and the darkness he called Night.” Genesis 1:5.

If the earth were a globe and the sun were the only source of light, such twilight could not exist. So, it is necessary for the scientific community to claim that the twilight is caused by the bending of light around the supposed curve of the earth. They simply cannot allow any discussion of another separate source of light that would confirm the creation account by God in the book of Genesis. Indeed, as will be explained below, atmospheric refraction cannot possibly explain the presence of twilight, because atmospheric refraction actually works the opposite of what is portrayed by "science falsely so called."

This author has noticed another phenomenon that further indicates a second source of light. Shortly after the sun rises above the horizon, the sun causes objects illuminated by it to cast long shadows. Automobiles that are in the direct sunlight from the risen sun do not have the shadows directly underneath them that were formerly there during twilight. Instead, they have long shadows cast from the sun that has just risen above the horizon. The sun washes out the shadows that were formerly underneath the cars during twilight. The sun is the "greater light" that God set up to "rule the day."

After the sun rises above the horizon, houses cast long shadows. It is no surprise that there are no other horizontal shadows within the borders of those long house shadows. The only exception is under automobiles that are parked in front of those houses. The automobiles, covered by the shadows of the houses, still have shadows of their own cast directly underneath them. Those cars are engulfed in the shadow of the houses. Consequently, they do not cast long shadows of their own, as do the cars in direct sunlight. But they still cast shadows directly underneath them as though there is a light directly above them. This shadow within a shadow effect indicates that there is a light source emanating from directly above that is separate from the sun.

In addition, when there are thick clouds, as during a rainstorm, there are no shadows from vertical objects that would ordinarily be cast by the sunlight. When those thick clouds block the sun, shadows can, however, be seen underneath automobiles. The shadows under cars could not be due to light refraction or diffraction, since no matter what time of day, the position of the sun in the sky over the clouds, or the altitude of the clouds, the light streams straight down, as though from a source independent of the sun. The greater light of the sun (Genesis 1:16) overwhelms the softer first light of the day and washes out the shadows from that softer first light (Genesis 1:3-5). It is only when the sun is not above the horizon, the sunlight is blocked by thick clouds, or the sunlight is blocked by some other obstruction that the shadows from the downward softer light of the day can be seen underneath cars.

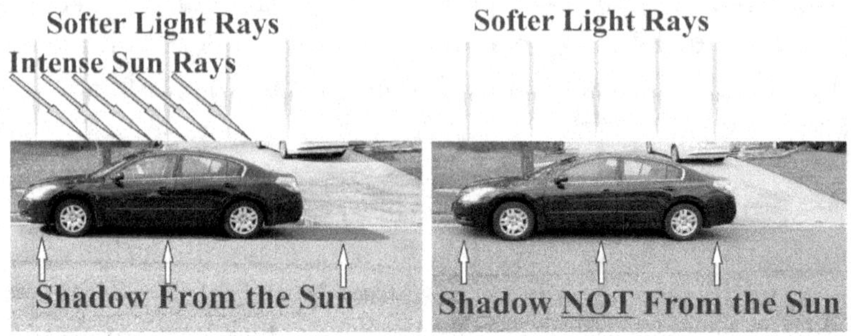

The above pictures were taken within one minute of one another. In the picture on the left the sun was shining. Within a minute of that picture being taken, thick dark clouds had completely obscured the intense sunlight, causing the long shadow under the car to disappear, being replaced by a shadow directly underneath the car from the remaining softer downward light, as depicted in the picture on the right. Indeed, after the clouds obscured the sun, all shadows on vertical objects disappeared. The only shadows remaining were directly underneath cars from the remaining softer downward light. That indicates that the source of the light in the picture on the right is shining straight down. That light is not coming from the sun.

If thick clouds blocking the sun reveal a source for daylight in addition to the sun, then certainly this second light source should be manifested during a total eclipse of the sun. An eclipse is where an object, believed to be the moon, blocks the sun. A total eclipse can be viewed from a location that is in line with the sun and the shadow of the moon that is cast when it blocks the sun. If the sun were the only source of light, one would expect that the area covered by the total eclipse would be plunged into total darkness. But that is not what happens. Instead, we find that the sky remains illuminated in a state of twilight during a total solar eclipse.

According to expert eclipse observers interviewed by Calla Cofield, senior writer for Space.com, "[a]t the edge of the path of totality, where the sun's disk is 99 percent covered by the moon, observers might notice a slight dimming of the light. But for the most part, this effect is visible only where the moon entirely covers the face of the sun."[477] It is interesting that when the sun is 99 percent covered by the moon there is only a slight dimming of the light, and that dimming is almost unnoticeable until the moon entirely eclipses the sun. What happens when the entire sun is blocked by the moon (assuming that it is the moon that is blocking the sun)? Cofield states that "[w]hen the moon completely blocks out the sun during a total solar eclipse, daylight suddenly turns into twilight."[478] This twilight is manifested during a total eclipse in very much the same way as before daybreak and after sunset each morning and evening.[479]

St. Joseph, Missouri, was dead center in the middle of the 70-mile wide path of totality for the eclipse that traversed the United States on August 21, 2017. Below is a screen shot from a video taken by Brent Butler during the total eclipse of the sun.[480] At the moment of totality, St. Joseph should have been plunged into complete darkness. But notice that the sky is still illuminated in twilight, while the sun is completely eclipsed.

Some will argue that the twilight is due to the diffusion of light in the atmosphere from the sun. That is certainly plausible. It is worth considering, as long as it is studied in view of God's account of creation in Genesis. The problem though with any argument that ignores God's account of his creation, is that such an argument is based upon willful ignorance. Conclusions from such an anti-biblical agenda are likely to be wrong. It is more appropriate to look to all theories for the illuminated sky in light of God's account of his creation. The twilight during the eclipse is best explained as from a source other than the sun, just as described in the book of Genesis.

This picture is a contrast enhanced screen shot from a video taken by Brent Butler during the totality of the August 21, 2017, eclipse. The video was taken at St. Joseph, Missouri, which was dead center in the 70 mile wide shadow band of eclipse totality. Notice that the sky is illuminated in twilight, despite the fact that at that moment the sun was totally eclipsed. Notice also that, with the sun eclipsed, there are no shadows, except directly underneath the cars. That is caused by a source of light coming straight down.

25 Attack From a Charlatan

The Bible is clear that the sun is in the firmament. But Steven Anderson, pastor of Faithful Word Baptist Church, attacks the notion that the firmament is in any way firm. Anderson tried to redefine the firmament during a strident rant, in which he condemned the truth of the flat earth.[481] Anderson describes belief in a flat earth as "foolishness" and "of the devil." Anderson calls those who believe in a flat earth, "feebleminded" "idiots." Anderson begins his attack on the verity of the flat earth by citing Genesis 1:20, which states: "And God said, Let the waters bring forth abundantly the moving creature that hath life, and fowl that may fly above the earth in the open firmament of heaven." (Genesis 1:20) Andserson opines that since the birds can fly in the open firmament that the firmament is the "atmosphere" and therefore cannot be firm.[482]

On the surface, Anderson's argument seems to have merit. But closer examination reveals that it is, in fact, sophistry. The Bible describes the firmament as being that which divides the waters beneath from the waters above. "And God said, Let there be a firmament in the midst of the waters, and let it divide the waters from the waters." (Genesis 1:6) In Job, we read that the firmament is like a molten looking glass. "Hast thou with him spread out the sky, which is strong, and as a molten looking glass?" (Job 37:18) Ezekiel 1:26 describes God's throne above the firmament.[483]

> And above the firmament that was over their heads was the likeness of a throne, as the appearance of a sapphire stone: and upon the likeness of the throne was the likeness as the appearance of a man above upon it. (Ezekiel 1:26 AV)

In Revelation 4:6 that area above the firmament, upon which is God's throne, is a sea of crystal glass. "And before the throne there was a sea of glass like unto crystal. (Revelation 4:6)

So, when we break it down we have the firmament, which is a firm barrier that has the consistency of a molten looking glass like that of crystal. The firmament separates the waters above the firmament from the waters beneath the firmament. God's throne is above the firmament. That separation has between it what we call the "atmosphere" (more accurately called the "atmosplane"). The reference to the birds in Genesis 1:20 flying states that they fly in the "open firmament of heaven." Notice that the reference is not to the firmament, but rather to the "open firmament." That is the area between the waters separated by the firmament.

Suppose you trapped a butterfly beneath a glass jar after it lands on a table outdoors. If you said, "the butterfly is in the glass jar," that would be a correct statement. You are expressing the fact that the butterfly is under the firm glass jar surrounding the butterfly. You are not suggesting that the butterfly is inside the hard glass itself. That is the same thing that God is expressing when he states that the fowl may fly "in" the "open firmament of heaven." God is describing the birds being in the open area surrounded by the firmament. God means that the birds, like the butterfly "in" the glass jar, are "in" the "open firmament of heaven."

Furthermore, Pastor Anderson expressly accepts as true the orthodox scientific position that the sun and moon are located in "outer space." The problem with Anderson's construct is that it contradicts the word of God. The Bible states clearly that the sun and moon are "in the firmament of the heaven." Genesis 1:14. Yet, Anderson contradicts God's word and places the sun and moon in a mythical place that so-called scientists call "outer space." That so-called outer space is necessarily outside the firmament (which Anderson states can only be the "atmosphere"). Anderson and, indeed, all "Christian" pastors who accept the heliocentric model, which places the sun and moon outside the firmament, implicitly reject God's account of his creation, which places the sun and moon "in" the firmament. Can such pastors be truly preaching the gospel when they contradict God's very word? "If I have told you earthly things, and ye believe not, how shall ye believe, if I tell you of heavenly things?" (John 3:12)

It gets worse. Anderson's defense of heliocentricity is downright blasphemy. God is in heaven, watching over man. Job 22:12; Ecclesiastes 24:5; Psalms 14:2 & 33:13. Heaven is God's throne. Acts 7:49. God's throne is above the firmament. Ezekiel 1:26; Job 37:18. To say that the area above the firmament is, instead, an empty and hostile vacuum of space is to dethrone God. Either the Bible is correct that God rules from his throne above the firmament, or science is correct that above the "atmosphere" is nothing but empty space. Both cannot be true. The heliocentric model is irreconcilable with the Christian gospel. Heliocentrism is not just unchristian; it is antichristian. Heliocentrism gives succor to both the atheist and the Satanist. The Satanists desire victory over God, and atheists claim God does not exist. The empty vacuum of space, which precludes God on his throne, is just fine for both of them. The devil just sits back and snickers at the "Christian" pastors, like Steve Anderson, who do his dirty work. Some pastors are ignorant dupes, others are willing agents of the devil.

There is an “open firmament” below the more viscous (molten) firmament that is a barrier to the waters above the firmament. The fact remains that the sun and moon are within the firmament of heaven and the fowl fly in the open firmament. Above the firmament is water, as God stated that he placed the firmament between the waters beneath and the waters above. "And God made the firmament, and divided the waters which were under the firmament from the waters which were above the firmament: and it was so." (Genesis 1:7) That seems to explain the blue sky during the day.

26 NASA Pictures of a Spherical Earth

On or before 1931, Wilbur Glenn Voliva offered $5,000 to anyone who could prove that the earth is a spinning sphere, floating in space, and orbiting the sun. Many tried to meet the challenge and collect the reward, but all failed. In1931, a Modern Mechanix magazine publicized the challenge. The article ended with the statement that "[t]he Voliva prize probably will remain uncollected unless some future space traveler some day anchors his ship a few thousand miles out in space and takes a movie of a globular world turning on its axis. That seems to be the only way the $5,000 can ever be collected."[484] The National Aeronautics and Space Administration (NASA) claims to have done just that.

Indeed, the alleged photographic proof from NASA showing a globular earth is the principal evidence cited by many scientists as proof that the earth is a spinning sphere in space. Bill Nye is an example of such a "scientist." Bill Nye is a famous television personality and science author, who is known as "The Science Guy." His television science program has been recognized for its educational excellence with many awards. Bill Nye has a bachelor of science degree in engineering. But Bill Nye is more an actor than he is a scientist. He began his career as an actor playing different roles. His persona eventually morphed into his own show "Bill Nye the Science Guy," where he portrays himself as a scientist. Nye recently appeared in a scripted video where he is asked if he can prove that the earth is a sphere. One of his replies was "if you'd like, look at pictures from space, where you see the earth as a sphere."[485]

In the video, Nye vouches for the authenticity of the NASA pictures of the spherical earth by saying "those pictures are not faked, if nothing else, here's why you can tell they are not faked."[486] This is the big moment now. Bill Nye "The Science Guy," award winning scientist and author, is now going to tell everybody how we can tell the pictures from NASA of the spherical earth are not fake. Bill Nye then gives his authoritative explanation, here it is: "Just to create the paperwork that NASA has created ... to send anything out into space, to send people into orbit or send them to the moon, that amount of paperwork would make faking it prohibitively expensive. No one could afford to generate that much documentation."[487]

Please, dear reader, reflect on the illogic of Bill Nye's explanation. He is stating that his scientific opinion is that the paperwork that NASA has in fact generated to send people into space would make it prohibitively expensive to fake the photographs. Huh? His argument is laughable when one considers that NASA's yearly budget is approximately 18 billion dollars.

But that is not the worst part. His explanation is illogical. He acknowledges that NASA has in fact generated the necessary paperwork to supposedly send people into space. Yet in the next breath he states that the paperwork to do so is so prohibitively expensive that NASA could not fake earth pictures. How does the paperwork expense of going into space even relate to faking pictures of earth? He is mixing the concept of the expense of the paperwork to actually go into space and the expense to fake the spherical earth photos. What he seems to be suggesting in his statement is that the paperwork to send someone into space to take fake pictures would be prohibitively expensive. But that is not the argument of those alleging that the pictures are fake. The argument of those alleging that the pictures are fake is that NASA did not go into space to take fake pictures, but that NASA faked the pictures using computer graphics and graphic artists in a studio on earth.

Nye does not realize it, but his argument implies that NASA did not actually go into space, because the massive expense of the paperwork to go into space makes it impossible to do so, because "no one could afford to generate that much documentation."[488] Bill Nye is not a true scientist. He is a propagandist. He has gotten so used to telling fantastic lies to keep the spherical earth myth going that he has lost touch with reality. He has gotten accustomed to people believing his lies, no matter how outlandish they are. He doesn't seem to realize that his lie about the prohibitive expense of the paperwork needed to go into space actually contradicts his argument that NASA really went to space. Nye just babbles nonsense, because he knows that most people will believe whatever he says, since he is the authoritative "Science Guy." To even suggest that NASA, with a yearly budget of approximately 18 billion dollars, does not have sufficient money to fake pictures of the earth is quite simply absurd.

The most telling thing is that is the best argument that Bill Nye could come up with to prove the authenticity of the NASA photographs. This was not an off-the-cuff statement by Nye, where he was caught unprepared for an unexpected question. This was a scripted studio produced video, where he prepared his answer ahead of time. That was the best answer he could come up with.

Another problem with Nye's explanation is that it is contradicted by NASA itself. NASA has admitted that many of its purported earth "photographs" are not truly photographs at all but are rather renderings using computer software. You have read that correctly, NASA has admitted that some of its photographs are fake. For example, NASA has gone on record acknowledging that the famous 2002 blue marble picture of earth that became the home screen for the 2007 Apple iPhone is a composite. NASA explains:

> In 2002, NASA scientists and visualizers stitched together strips of brand new data, in natural color, collected over four months from the Moderate Resolution Imaging Spectroradiometer, or MODIS, instrument aboard Terra. They added a layer of clouds to create this composite Blue Marble that became one of the most iconic Earth images of the new century when Apple selected it as their default background for the iPhone in 2007. A version of the MODIS Blue Marble is now used as the base layer in many visualizations of NASA Earth science data.[489]

Indeed, when examining the blue marble composite, one can see duplicate clouds on its surface, which is a dead giveaway that it is not truly a photograph from space.

Figure 128: NASA's 2002 Blue Marble composite of the earth showing duplicate clouds.

Indeed, the 2002 Blue Marble image is a piece of artwork. Robert Simmon created the 2002 Blue Marble earth image that graced the screens of the first iPhones in 2007. At the time Simmon created the Blue Marble earth image, he was the Senior Program Analyst, who worked as the Lead Data Visualizer and Information Designer, Code 613, Climate and Radiation Branch, Earth Sciences Division at the Sciences and Exploration Directorate.[490] That is quite a title to hold; it even comes with a code; that sounds very secretive. Interestingly, Simmon admitted in an interview that the 2002 blue marble was created by him using Adobe Photoshop software. Simmon stated: "It is Photoshopped but it has to be."[491] Simmon stated that he took data images from a NASA satellite to create the earth as "what I imagine it to be."[492]

If the earth is truly floating in an empty vacuum of space, a photograph of the earth does not need to be Photoshopped, as claimed by Simmon. NASA could simply send out a rocket with a camera into outer space and take a picture of the earth. NASA allegedly had no problem doing just that in 1972. However, it seems that with 30 years of advancement in technology NASA lost the ability to take pictures from space. The real reason that NASA must resort to Photoshop images of the earth from space is that there is no outer space and the earth is not a sphere floating in that non-existent space.

In a 2012 interview for an article for NASA, Simmon explains the following:

> The last time anyone took a photograph from above low Earth orbit that showed an entire hemisphere (one side of a globe) was in 1972 during Apollo 17. NASA's Earth Observing System (EOS) satellites were designed to give a check-up of Earth's health. By 2002, we finally had enough data to make a snap shot of the entire Earth. So we did. The hard part was creating a flat map of the Earth's surface with four months' of satellite data. Reto Stockli, now at the Swiss Federal Office of Meteorology and Climatology, did much of this work. Then we wrapped the flat map around a ball. My part was integrating the surface, clouds, and oceans to match people's expectations of how Earth looks from space. That ball became the famous Blue Marble.

> I was happy with it but had no idea how widespread it would become. We never thought it would become an icon. I certainly never thought that I would become "Mr. Blue Marble."[493]

Simmon reveals a couple of key facts: 1) Since the 1972 Apollo 17 moon mission and up until 2012, NASA had not taken a photograph of the entire hemisphere of the earth from space; 2) Simmon created the 2002 Blue Marble image of the earth by wrapping "the flat map around a ball;" and 3) Simmon's job was to create an image of the earth "to match people's expectations of how Earth looks from space."

Basically, the 2002 Blue Marble is not an actual image of the earth as seen from "space," but is rather Simmon's rendition of what he imagined would be an earth "to match people's expectations of how Earth looks from space."

Robert Sungenis, Ph.D., in his book, *Flat Earth Flat Wrong*, tries to refute the evidence of a flat earth. As part of his mission, he comes to the defense of NASA and tries to explain the duplicate clouds. He uses two mutually exclusive arguments to explain the duplicate clouds in the NASA Earth photographs. First, he argues that identical cloud formations happen naturally, and second, he argues that NASA was justified in adding CGI images of clouds. In support of his argument that the duplicate clouds were naturally occurring, Sungenis states:

> It is also true, however, that very similar cloud formations can stand right next to each other due to the fact that the same wind and pressure conditions that made the first cloud formation will also make the second cloud formation adjacent to it. Sometimes almost identical cloud formations can appear over and over again in the

> same photo, and most of these are caused by normal and natural occurrences.[494]

Sungenis subtly changes the evidence from obviously duplicate identical clouds to duplicate "almost identical" clouds. Why such a qualification by Sungenis? Because duplicate identical clouds simply do not happen naturally and everyone knows it. So in order for Sungenis to argue that the duplicate clouds happened naturally, he must argue that they are not identical, but they are rather "very similar" or "almost identical." Armed with his new "very similar" clouds argument, Sungenis doubles-down on his absurd NASA defense.

> Hence whatever pressure, wind, moisture and temperature made the first wave can also made [*sic*] the second, and so on. It is obvious, then, that similar cloud formations cannot prove whether an image is CGI or not. In fact, some of the very photos that flat-earthers claim are CGI because the cloud formations are repeated are, upon closer examination, not identical repetitions, and thus cannot be CGI images but rather proof that very similar cloud formations often conglomerate in certain areas of the Earth's atmosphere, all conditions permitting.[495]

Sungenis denies the duplicate clouds are identical, even though it is obvious to anyone looking at the duplicate clouds that they are in fact identical. That reality, causes Sungenis to switch arguments and admit that the clouds in the NASA earth photographs are computer generated images (CGI) after all. But he continues to cling to his first argument that the clouds are not CGI, but rather naturally occurring clouds. In Sungenis' mind the same clouds are both CGI and real.

> In any case, it is reported that NASA says that

> future photos of Earth will not have any photo-shopped cloud formations, since their technology has improved. As a result, flat-earthers lose this argument, not only because they fail to see that similarity in cloud formations does not mean the formations are identical and thus are not CGI images, but also because, in this case, NASA has a good reason for using some CGI images in order to provide something close to what Earth actually is instead of something distorted by the inadequacies of cameras and the altitude from which the pictures are shot.[496]

Sungenis argues that NASA pasted the duplicate CGI clouds on the alleged earth photograph to provide a realistic looking earth. He argues that was done because the alleged satellites, with their small lens cameras, could only take a picture of a narrow strip of the earth during each purported pass over the earth. After that admission, Sungenis promises that NASA will be good and not ever do it again, because their technology has improved. At the same time, he maintains that those same clouds he acknowledges are identical CGI clouds are not really identical after all, but only "very similar" clouds that naturally occur. Sungenis' arguments are inconsistent: 1) the clouds form "almost" identically in nature, and 2) those same "very similar" clouds are really identical clouds after all that were photo-shopped onto the earth image. It is like the defendant who lies by arguing that 1) he killed the man in self-defense, and 2) he could not be the killer, because he was not there. Why would Sungenis make such foolishly inconsistent arguments? Because his purpose is not to inform, but to obfuscate. He is throwing up a smokescreen of deception. If the reader does not believe that identical clouds naturally occur, perhaps he will believe that NASA had an innocent reason for pasting the CGI clouds on the globe image. Sungenis' hubris blinds him to the fact that the two arguments are mutually exclusive and cannot both be true at the same time. He, nonetheless, takes that nonsense position.

There are a very few photographs of the earth coming from NASA that NASA claims are actual photographs from space depicting the globular earth. One of those pictures is the famous picture allegedly taken by astronauts during the Apollo 16 moon mission. Below is that picture.

Figure 129: Picture of Earth allegedly taken by the Apollo astronauts on April 16, 1972, using a hand-held Hasselblad camera, after leaving their alleged orbit of Earth.

NASA provided the following description of the picture on the NASA website, where it is posted, as follows:

Apollo 16 view of the Earth taken with a

> hand-held Hasselblad camera about an hour and a half after translunar injection on 16 April, 1972. Most of the United States is visible at right center, including Lake Michigan and Lake Superior and Florida. Mexico and part of Central America are also visible. The slightly different shade of blue below Florida is the Bahama Banks. The Earth is 12,740 km in diameter and north is at roughly 1:00. (Apollo 16, AS16-118-18885).[497]

The problem with the picture is that it is a hoax, and the description provided by NASA is a complete fabrication. Below is that same picture after it has been adjusted on Adobe Photoshop by increasing the brightness and lowering the contrast.[498] You can see clearly that the picture is manufactured. When the contrast is lowered and the brightness is increased, it reveals the area from which the Apollo 16 image of earth was cropped by the creators. After the Photoshop adjustments are made, the area from which the image of earth was cropped shows up black against a very dark green background. The black area outside the cropped image of the earth is distinct on a computer screen, but it does not show up well in print. A white line has been drawn around the black area outside the cropped image so that it can be seen better in print. The crop line indicates that the image was composed using graphic cropping techniques. It is not a photograph from the Apollo capsule. Once NASA realized that their hoaxed photo of the spherical earth was detectable in Adobe Photoshop, they replaced the posted photo with a different one that would not show the cropping line evidence of photo manipulation.

Figure 130: Image of Earth alleged to be from Apollo 16 with brightness raised and contrast lowered, revealing the area from which the image of earth was cropped. The white line has been added so that the crop line can be seen better in print.

There are other NASA photographs showing the earth to be a globe. Every one of them is a fabrication. Take for instance the image below, which was allegedly taken by the Apollo 17 crew.

Figure 131: Alleged photograph taken during Apollo 17 mission to the moon.

Below is a description of the above photograph as it appears on the NASA website.

> Apollo 17 hand-held Hasselblad picture of the full Earth. This picture was taken on 7 December 1972, as the spacecraft traveled to the moon, the last of the Apollo missions. A remarkably cloud-free Africa is at upper left, stretching down to the center of the image. Saudi Arabia is visible at the top of the disk and Antarctica and the south pole are at the bottom. Asia is on the horizon is [*sic*] at upper right. The Earth is 12,740 km in diameter. (Apollo 17, AS17-148-22725)[499]

The depiction of the land masses in the picture is wrong.

Below is a comparison of the above Earth image with the image provided to Google using the official data from the U.S. Government.[500] The data used by Google in constructing its image of Earth comes from a number of organizations, including the National Geospatial-Intelligence Agency (NGA), the National Oceanic and Atmospheric Administration (NOAA), and the U.S. Navy, all of which are U.S. Government agencies.

The NGA, one of the agencies supplying data to Google, is an intelligence agency within the U.S. military. The NGA describes its mission as follows:

> The National Geospatial-Intelligence Agency has a responsibility to provide the products and services that decision makers, warfighters, and first responders need, when they need it most. As a member of the Intelligence Community and the Department of Defense, NGA supports a unique mission set. We are committed to acquiring, developing and maintaining the proper technology, people and processes that will enable overall mission success.
>
> Geospatial intelligence, or GEOINT is the exploitation and analysis of imagery and geospatial information to describe, assess and visually depict physical features and geographically referenced activities on the Earth. GEOINT consists of imagery, imagery intelligence and geospatial information.[501]

The two side-by-side globes below show a comparison between the Google Earth globe (on the right) and the picture from Apollo 17 (on the left). The two globes were lined up precisely in the same attitude. As you can see, however, the depictions in the globes of the land masses are quite different. For example, the upper horizontal line, which is drawn contiguous to the southern-most tip of Africa on the Apollo 17 image, shows that the southern tip of Africa in the Google image falls much further south of the southern tip of Africa in the Apollo 17 image. Also, the African land mass is depicted much smaller in the Apollo 17 image. In addition, the lower horizontal line, which is drawn contiguous to the southern-most tip of Madagascar in the Apollo 17 image, shows that its southern tip of Madagascar falls further south of the horizontal line in the Google image. Also, the land mass of Madagascar is much larger in the Google image.

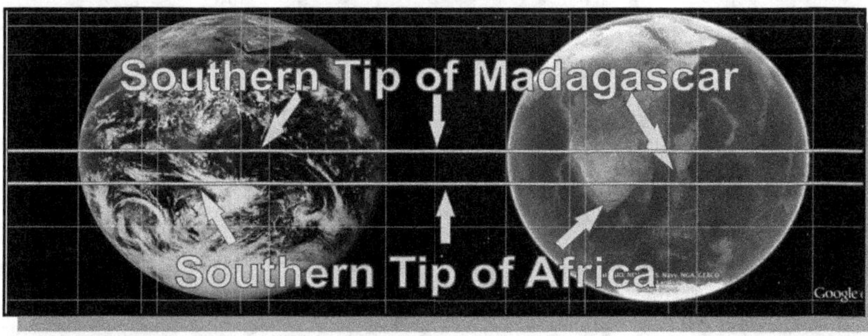

Figure 132: Comparison of the Apollo 17 Earth photograph and the image from Google that is based upon official U.S. government geographical data, showing inaccuracy in the Apollo 17 alleged photograph.

In the image below, you can see that the horizontal dimensions of the land masses on each of the two globes do not match up either. The Google image (top image) has the African continent much larger than the Apollo 17 image (bottom image). You can see the vertical line that is placed contiguous to the western most point of the African continent in the Apollo 17 image. Follow that line up and you can see how much larger is the Google depiction of the African continent by the amount of land that is west of that line.

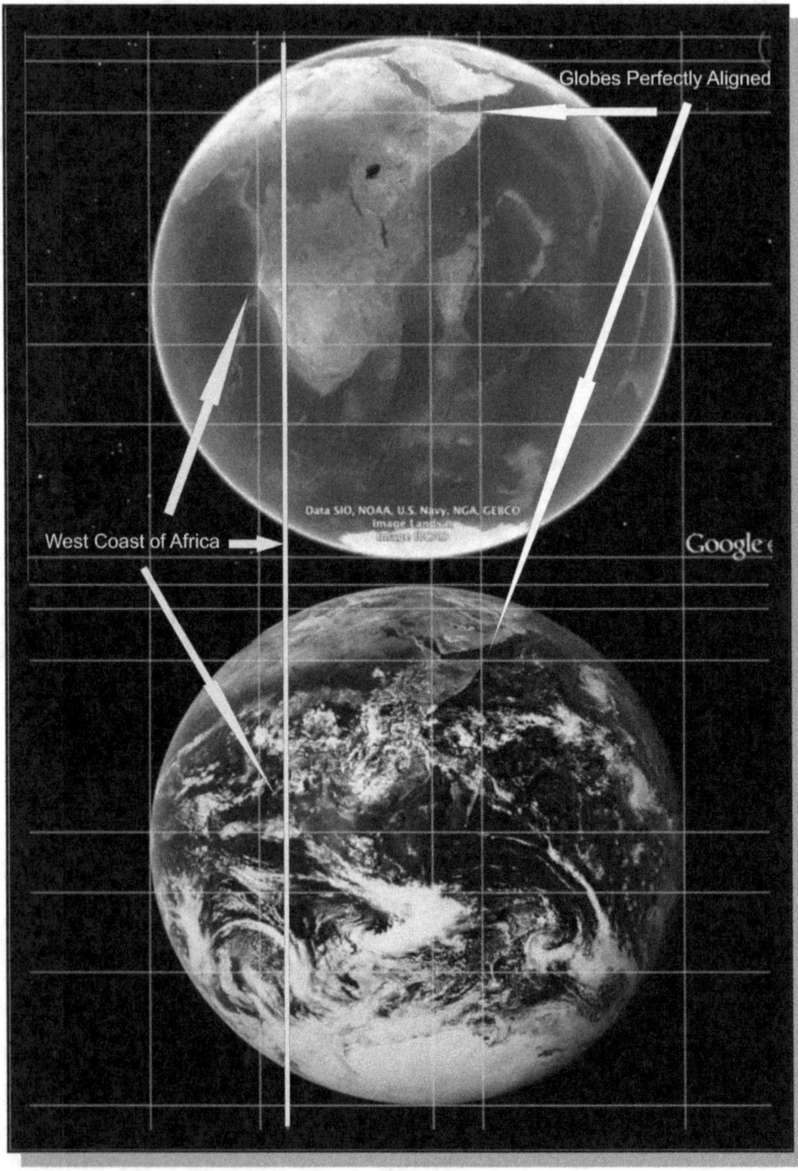

Figure 133: Comparison of the alleged Apollo 17 Earth photograph and the image of the earth from Google that is based upon official U.S. government geographical data, showing inaccuracy in the alleged Apollo 17 photograph.

Clearly, the alleged photographs provided by NASA of the earth are not real. NASA seems to be running a type of disinformation intelligence operation, whose purpose, in large part, is to convince a gullible public that the earth is a globe. Google is part of the disinformation campaign to conceal the flat earth.

It has become the custom for the public and even NASA to call all of the alleged full earth NASA photographs of the earth Blue Marbles. The Apollo 17 astronauts allegedly took their Blue Marble snapshot of the earth on December 7, 1972.[502] The Apollo 17 Blue Marble depicted above in this book (AS17-148-22725) is actually one of three separate and virtually identical Blue Marble photographs attributed to the Apollo 17 crew. All three of the photographs were allegedly taken between 4:59:05 and 5:08:14 after launch.[503] The launch of Apollo 17 was at 12:33 a.m. EST (5:33 a.m. UTC/GMT). That would mean that the three Blue Marble photographs depicting the earth were allegedly taken between 5:32 a.m. and 5:41 a.m. EST (between 10:32 a.m. and 10:41 a.m. UTC/GMT).[504] Two of the three photographs (AS17-148-22726 and AS17-148-22727) have been described by NASA as being taken at approximately 5 hours and 6 minutes after launch.[505] That would put the time of those two pictures at 5:39 a.m. EST (10:39 a.m. UTC/GMT).[506]

Since all three Blue Marble photographs of the earth show identical cloud formations and attitude of the earth, certainly NASA would have to claim that they were taken in close succession to one another. The nomenclature used by NASA is sequential. We know, therefore that AS17-148-227225 was the first picture of the three Apollo 17 Blue Marble Earth photographs taken. Since NASA must claim that it was taken close in time to the other two pictures, because they are virtually identical, we can infer that the time AS17-148-227225 was allegedly taken as being no more than one minute prior to the other two pictures. That would put the time that AS17-148-227225 was taken at approximately 5:38 a.m. EST (10:38 a.m. UTC/GMT). In any event, a few minutes difference either way will not effect our analysis.

Why is the approximate time of the Apollo 17 Blue Marble photograph important? Because the photograph depicts a fully lit earth. Since we can see the entire sunlit half of the earth, we know that the part of the earth that is not seen is in darkness. That fact proves that the Apollo 17 Blue Marble photograph is a fraud. How so? Because, Antonio Subirats discovered that on December 7, 1972, at approximately 5:38 a.m. EST (10:38 a.m. UTC/GMT), India, Europe, and Western Russia, and Western China were in complete daylight.[507] In addition, almost all of South America (except the northwest coast, made up of the western parts of Ecuador, Columbia, and Venezuela) and the western most part of Australia were in complete daylight. Yet, in the Apollo 17 Blue Marble photograph, we can see that none of Europe, Western Russia, Western China, South American, or Australia are visible; they are all on the dark side of the Blue Marble. A viewer looking straight on to a fully illuminated 180° of a globe is able to see every part of the half that is illuminated. The other half of the globe is out of view and must necessarily be in darkness.

Indeed, the actual daylight that is experienced each day on the earth proves that all depictions of the spherical earth, including Google Earth, are false. If you look at Google's depiction of Earth you will see that at the same attitude as depicted in the Blue Marble, neither Europe, Western Russia, Western China, South America, nor Western Australia are depicted on the Google Earth depiction of the globe. Yet, at that time and date all of those areas should be receiving the sun's light. Indeed, when you compare the sun cast for any day with any correlated attitude of the spherical earth, you will find it impossible to match the sun's actual illumination of the earth with the illumination theorized by the spherical model of the earth. The cast of the sunlight as actually experienced each day can only be manifested on a flat earth. The spherical earth is thus shown to be false.

In the picture below we see the circle indicating the dimensions of the Apollo 17 Blue Marble superimposed on a map showing the position of the sun and moon on December 7, 1972, at 10:38 a.m. UTC/GMT.[508]

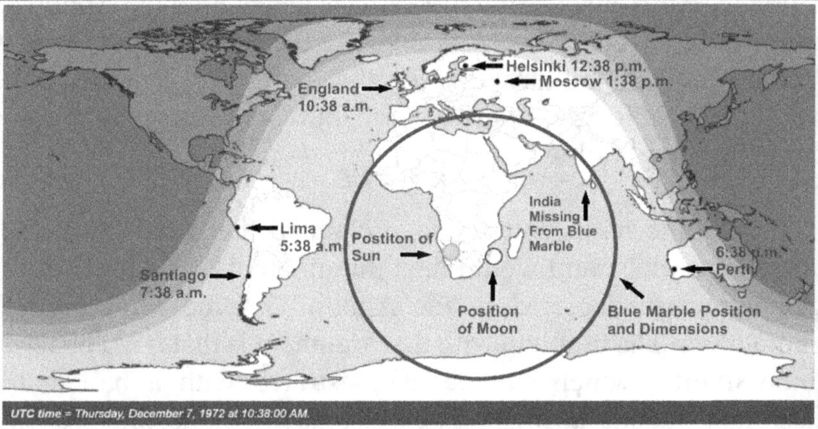

Map showing the position of the sun and moon and the area of sunlight on December 7, 1972, at 10:38 a.m. UTC/GMT, with an outline of the area of the NASA Apollo 17 alleged photograph of the Blue Marble Earth superimposed on it.

The fact that England cannot be seen on the illuminated side of the Blue Marble proves the fallacy of the globular earth model. England is on the dark side of the earth on the Apollo 17 Blue Marble Photograph. However, it was 10:38 a.m. UTC/GMT in Greenwich, London, England, at the time of the alleged Blue Marble photograph. Obviously, it was daylight in Greenwich, London, England at 10:38 a.m. UTC/GMT on December 7, 1972, when the Apollo 17 Blue Marble photograph was allegedly taken. Furthermore, when the Blue Marble picture was supposed to have been taken, it was 12:38 p.m. in Helsinki, Finland. Clearly, at 38 minutes after noon in Helsinki, it was daytime. Yet, according to the Blue Marble, Helsinki, Finland, is supposed to be in darkness. At the time the alleged photograph of the Bluc Marble was taken, it was 1:38 p.m. in Moscow. It was certainly daylight at that time in Moscow. The Blue Marble, however, has Moscow in complete darkness at 1:38 in the afternoon. Obviously the Apollo 17 Blue Marble is simply wrong in its depiction of the earth.

If the earth were a sphere, the position of the sun on December 7, 1972, at 10:38 a.m. UTC/GMT, should have created a gibbous earth, and not the full earth as depicted in the Apollo 17 Blue Marble. Additionally, the Blue Marble graphic artist failed to include India in his artwork. At that particular attitude, India should be visible on a spherical earth, but it is completely missing from the Blue Marble.

Some heliocentrists have claimed that the light passes around the globe and would thus illuminate the back side of the globular earth depicted in the Apollo 17 Blue Marble Earth photograph. The problem with that argument is that it is a physical impossibility, which can be demonstrated with a ball and a flashlight outside at night. The existence of atmosphere on the earth would not change this fact. NASA knows this and factors it into its construction of computer generated images (CGI) of the spinning earth. NASA and space agencies from other countries always portray the earth's terminator as being a boundary line at a 90° orientation to the sun, because that is how it must be for all spheres facing a light source. The backside of the globe is 180° in darkness, with the boundary marked by the terminator. The sunlit side, in contrast, is covered by 180° of sunlight that ends at the terminator. This fact is illustrated in the alleged photograph supposedly taken by the Himawari 8 geostationary satellite. The photograph below is a screen-shot from an alleged time lapse, which is taken at the point where the purported sun has illuminated 180°of the earth and the terminator is seen at a 90° angle to the position of the sun.

The theory that the sun's light can wrap around the globe on either side of the terminator beyond 180° is impeached by the physical laws, which the Himawari 8 animators seem to acknowledge in their CGI of the earth. Even gibbous depictions of

View of illuminated quarter earth allegedly taken from the Himawari 8 geostationary satellite showing that light cannot pass beyond the terminator line that is 90 degrees to the sun.

the earth require a 180° split between light and darkness. The gibbous depictions of the earth are all a matter of perspective created by CGI. Incidently, in the Himawari 8 alleged photograph there are no lights in the cities in eastern Australia. Don't Sydney and Melbourne have electricity? Obviously, the Himawari 8 imagineers forgot to include city lights in their CGI of the earth. I guess one cannot think of everything when constructing an earth CGI.

The fraud of the Apollo 17 Blue Marble can be clearly understood when one considers just how far from the terminator on the Blue Marble were the cities that were supposed to be in darkness on the Blue Marble but in fact enjoyed daylight on December 7, 1972, at 10:38 a.m. UTC/GMT. Assuming the earth is a sphere as depicted by the Blue Marble, on December 7, 1972, at 10:38 a.m. UTC/GMT, Moscow was supposed to be more than 1,000 miles distance from the terminator of the spherical Blue Marble earth and in darkness, dropping more than 126 miles on a curve below the horizon from the illuminated portion of the Blue Marble. On that date and time, Helsinki was supposed to be more than 1,300 miles distance from the terminator of the Blue Marble and in darkness, dropping more than 213 miles on a curve below the horizon from the illuminated portion of the Blue Marble. London was supposed to be more than 1,100 miles distance from the terminator of the Blue Marble and in darkness, dropping more than 152 miles on a curve below the horizon from the illuminated portion of the Blue Marble. Perth was supposed to be more than

1,200 miles distance from the terminator of the Blue Marble and in darkness, dropping more than 181 miles on a curve below the horizon from the illuminated portion of the Blue Marble. Both Lima and Santiago were supposed to be more than 3,000 miles distance from the terminator of the Blue Marble and in darkness, dropping more than 1,135 miles on a curve below the horizon from the illuminated portion of the Blue Marble.

These fantastic distances from the Apollo 17 Blue Marble's terminator would have all of these cities plunged into darkness, yet each and every one of them was enjoying daylight on December 7, 1972 at 10:38 a.m. UTC/GMT. That proves that the Apollo 17 Blue Marble is a NASA hoax. The only way that these cities could have been enjoying daylight at that time and date is if the world is in fact flat.

Both the Google Earth globe and the Apollo 17 Blue Marble have similar discrepancies regarding their terminators and the real area covered by the sun. The dimensions of the Google Earth (or any globe for that matter) cannot account for the area of daylight that is experienced on Earth each day. Only a flat Earth accurately explains the area illuminated by the sun.

In the 2001 movie, *A Funny Thing Happened on the Way To the Moon,* Bart Sibrel presents compelling evidence that the Apollo Moon landing missions that took place between 1969 and 1972 were elaborate hoaxes. One of the most compelling pieces of evidence is a piece of videotape footage that was inadvertently sent to Sibrel by NASA. The videotape shows Apollo astronauts Neil Armstrong, Buzz Aldrin, and Michael Collins allegedly in the Apollo space capsule narrating their view of the moon at a time that they claimed that they were 130,000 miles from Earth and on their way to the moon. The problem is that the videotape, which was never supposed to be seen by the public, shows the astronauts faking the pictures of Earth, by using some sort of transparency against the cabin window.

Evidence that the earth images from NASA are manufactured in the time-lapse pictures of the earth is obvious when looking at the clouds, which remain static without changing their form or position during hours of time lapse.[509] Such an occurrence would be impossible in the real world.

Below is a purported picture of the earth taken on July 6, 2015, and subsequently posted on the NASA website. It is described by NASA as follows:

> A NASA camera on the Deep Space Climate Observatory satellite has returned its first view of the entire sunlit side of Earth from one million miles away. This color image of Earth was taken by NASA's Earth Polychromatic Imaging Camera (EPIC), a four megapixel CCD camera and telescope.[510]

NASA can waste millions on their space launches, yet they use a pathetically anemic 4 megapixel CCD camera. One can search online and find for sale to the general public a 16 megapixel CCD camera for less than $10,000. Yet, NASA sends their alleged Deep Space Climate Observatory satellite into space with a camera having a mere 4 megapixel resolution. It seems that NASA is bending over backwards to provide the lowest resolution picture of earth with which they can get away. Why would they do that? Because it is easier to pull off the spherical earth myth with low resolution photographs. The higher the resolution, the more difficult it is to create a realistic globular earth. Higher resolution will show more detail and hence show signs of artificiality. The picture below is not actually a picture of earth from space, as alleged by NASA.

Figure 136: Photograph of earth allegedly taken from a camera aboard the NASA Deep Space Climate Observatory, from a distance of one million miles.

NASA states on its website that the image is made up of three layers of pictures taken using the alleged satellite's digital camera:

> The image was generated by combining three separate images to create a photographic-quality image. The camera takes a series of 10 images using different narrowband filters -- from ultraviolet to near infrared -- to produce a variety of science products. The red, green and blue

channel images are used in these color images.[511]

NASA admits that the image is a composite, but maintains that it is a composite of three photographic images, all taken from a million miles away with a digital camera on the Deep Space Climate Observatory (DSCOVR) satellite. Why was it necessary to make a composite image at all? Why not just take a picture of earth? Because, there is no Deep Space Climate Observatory satellite and the earth is not a globe. The image is a piece of very clever artwork. NASA knows that the layers in the image would likely be discovered in the future and decided to admit it is a layered image in order to have a plausible reason for the layers.

In any event, the claim by NASA that the image is an accurate photograph of earth from space, using its "Earth Polychromatic Imaging Camera (EPIC)," is a demonstrable deception. NASA claims that the image is a composite of three photographs of earth taken from space. In actuality, it is a composite of computer and human generated imagery. What proof is there of that? A professional photographer and designer, with over 20 years experience using Adobe Photoshop software, discovered that the NASA photograph was created using Adobe Photoshop software.[512] The expert inverted the image, using Photoshop, and discovered that whoever created the photograph used the Photoshop round marquee tool. He discovered the distinct stair step outline, which is unique to the Photoshop software marquee tool, ringing the circumference of the earth in the NASA photograph.

The expert also discovered that there were square pixel remnants of sky, clouds, and water that were mistakenly left behind underneath the black background applied around the earth image. There should not be sky, clouds, and water floating in square pixels in space. The expert's professional opinion is that NASA's purported picture of the earth is not a photograph from a million miles in space, but rather the creation of a graphic artist, using Adobe Photoshop software.[513]

We don't need an expert to detect that the July 6, 2015 Deep Space Climate Observatory satellite picture is not a candid photograph of earth from space. The image itself contains the proof that it is a forgery. If you look at the lower southwest portion of the globe image above, you can see that the graphic artist who drew the clouds, drew in the word "SEX," upside down. With the image flipped 180° (below), the word "SEX" in the cloud formation can be seen right side up in the northeast portion of the globe.

Figure 137: NASA alleged photograph of the earth flipped 180°, revealing "SEX" written in the cloud formation.

It is notable that the word "SEX" in the clouds is written at a 23.4° vertical angle, which is the supposed tilt of the globular earth under the heliocentric model. The above box drawn around the word "SEX" is at precisely a 23.4° vertical angle. There is no evidence of the supposed 23.4° earth tilt. If there is no evidence for it, how was that angle arrived at? Recall that heliocentrism is a religion, and so you can expect the angle picked for the supposed tilt of the earth to have religious significance. The high priests of the heliocentric religion did not arbitrarily pick 23.4°. The adjacent horizontal angle to the supposed 23.4° tilt of the earth under the heliocentric model is 66.6° (90°−23.4°= 66.6°). "Here is wisdom. Let him that hath understanding count the number of the beast: for it is the number of a man; and his number is Six hundred threescore and six." Revelation 13:18. Heliocentrism is more a spiritual deception than it is a scientific deception. That is why the speed of the earth's supposed orbit around the sun is reported to be 66,600 miles per hour.[514] The word "SEX" in the clouds is clear evidence that the image, which NASA purports to be a photograph of earth, is an artistic rendition; it is a fiction.

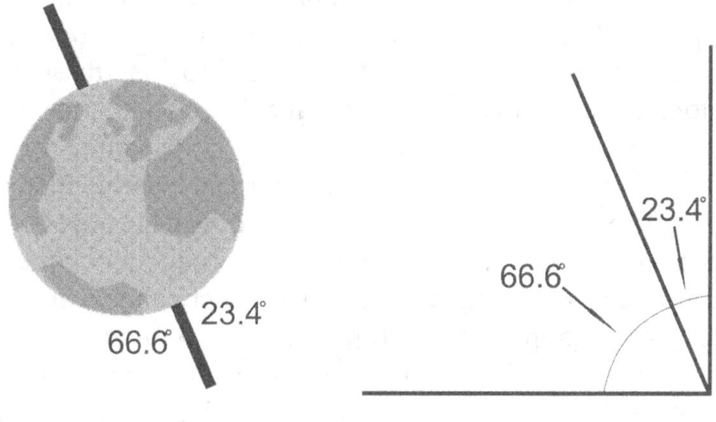

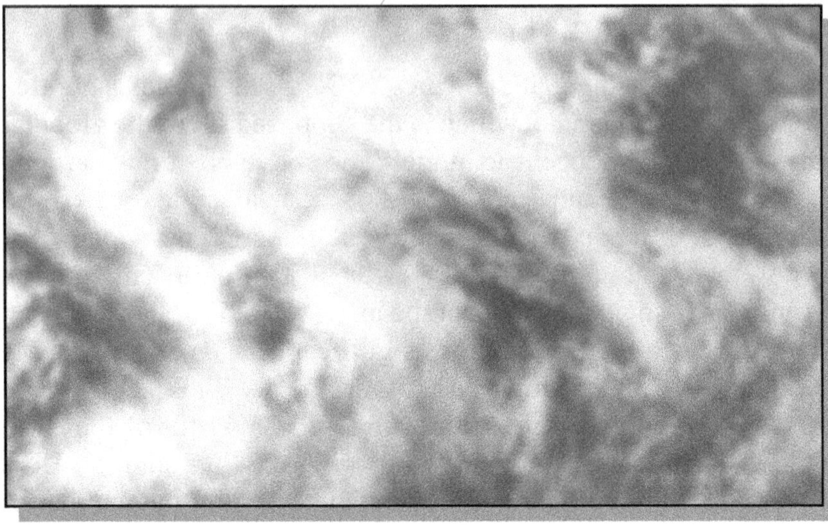

Figure 139: Closeup of "SEX" written in clouds on image of earth that NASA claims is a photograph from space.

The fraud of NASA is most evident when comparing one NASA Earth photograph with another NASA Earth photograph. The dimensions of the continents depicted on several of NASA's alleged Earth photographs do not match, which is a dead giveaway that they are fabrications. For example, below is a comparison of NASA's alleged Deep Space Climate Observatory (DSCO) photograph of the earth taken on July 6, 2015[515], with a composite photograph allegedly taken from the NASA's Earth-observing satellite - Suomi NPP, on January 4, 2012.[516] Notice the glaring difference in the size of the North American Continent on the two globes. The globes are identical in outer dimensions, yet the North American continent in the Suomi NPP photograph on the bottom is almost twice the size of the North American continent in the DSCO photograph on the top. That is impossible if the two photographs were in fact photographs of the earth. The startling differences in continent sizes is proof of NASA disinformation.

NASA's DSCO
Alleged 2015 Earth Photo

NASA's Suomi NPP
Alleged 2012 Earth Photo

It is not just the many differences in the earth images from NASA that prove the NASA hoax. Sometimes the NASA hoax can be proven through identical Earth images from NASA.[517] For example, On May 18, 1969, NASA alleged that the Apollo 10 astronauts took a video of the earth as they were in their translunar phase of their moon orbiting mission.[518] The translunar phase is just NASA-speak for traveling toward the moon. Six days later, on May 24, 1969, NASA documented that the Apollo 10 astronauts took another video shot of the earth during their transearth stage.[519] The transearth stage is the stage where they are traveling back to Earth from the moon. The NASA hoax is proven when one compares the image of the earth from the translunar phase video on May 18th with the image of the earth from the May 24th transearth phase video. The Earth images in both videos are identical, right down to the cloud formations. It is impossible for the earth to have identical cloud formations six days apart. Clearly, NASA was using the same image of the earth in both video shots to scam the public into believing that the astronauts were traveling from the spherical earth to the moon and back again.

Identical Cloud Formations On Earth Six Days Apart

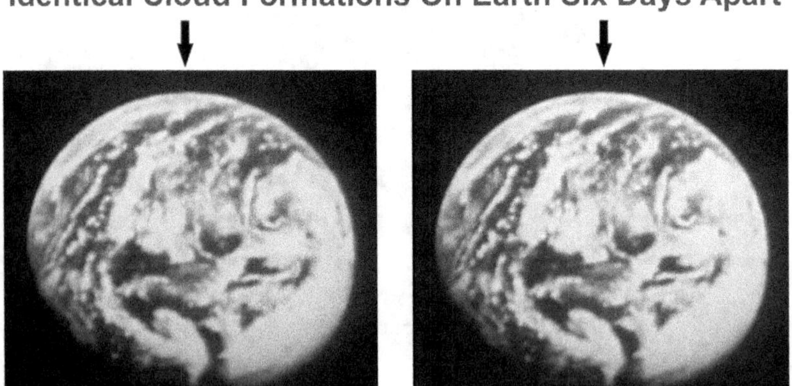

Still frame from a video alleged by NASA to have been taken by astronauts aboard the Apollo 10 spacecraft on May 18, 1969, as the astronauts were traveling away from the earth toward the moon.

Still frame from a video alleged by NASA to have been taken by astronauts aboard the Apollo 10 spacecraft on May 24, 1969, as the astronauts were traveling away from the moon toward the earth.

Below is a picture of the earth that is alleged by NASA to have been taken from the moon by astronauts during the Apollo 17 mission. To the right of that image is the same image with the output levels adjusted upward in Photoshop.[520] With the levels adjusted upward, there is revealed around the image of the earth a cropped rectangular box. That box is evidence that the image of the earth was pasted onto a black background by a graphic artist; it was not photographed from the surface of the moon, as alleged by NASA. Rob Skiba made his discovery of the box artifact around the picture of the earth publically known in a posting on YouTube. Approximately two days later, on June 16, 2015, NASA changed its website posting of the picture by replacing it with a new picture that removed the box artifact around the earth and added stars to the sky.[521]

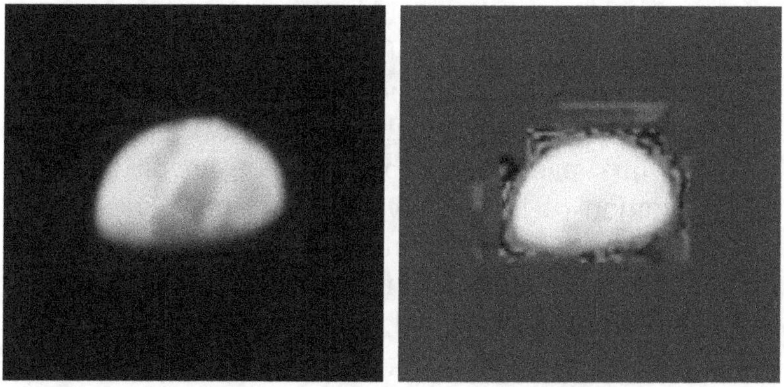

Photograph of earth alleged by NASA to have been taken by astronauts from the moon (left). When the levels on the photograph are adjusted upward using Adobe Photoshop (right), the image of the earth is shown to be surrounded by a square box, which indicates that the NASA image of the earth was pasted onto a black background and not photographed from the moon.

Robert Sungenis, who was a champion for NASA's photoshopping of clouds, could not come up with a defense for pasting a fake earth in the dark sky of what is supposed to be outer space. The best Sungenis could come up with is that it was a bad thing for NASA to do, and it may be evidence that NASA lied about the moon landings, but that it does not prove that the earth is flat.

> Suffice it to say, NASA's photo-shopped image of the Earth generates enough doubt and skepticism among those already predisposed to reject NASA's claims of landing a man on the moon, that NASA loses the battle before it starts. It is quite an understatement to say that this particular photo of the Earth gives NASA a credibility problem, not only with flatearthers but with many others. It is one thing to photo-shop clouds when making a composite of the Earth from hundreds of satellite photos, but it is quite another to make the world believe the Earth was photographed from 230,000 miles away by pasting in a picture of the Earth. The only thing that saves NASA's tarnished credibility is the fact that this particular debate is not about the reputation of NASA, since we are not here to prove whether NASA landed a man on the moon but whether the Earth is flat, especially since the generation that produced the photoshopped Earth was in a wholly different time and political situation in the 1960s than we are today in the 2010s. This caveat is not for the purpose of condoning what may have prompted NASA to make the fake photo, but only to say that NASA's foibles do not prove the Earth is flat.[522]

Sungenis Jesuitically minimizes the NASA fraud by calling it a "foible." That is a little like calling Jack the Ripper "unkind." Sungenis maintains that NASA's "foibles" in creating false evidence of moon landings does not mean that the earth is flat. Sungenis completely misses the point. The reason NASA exists is to create false propaganda to convince people that the earth is a spinning ball in space. That was the principal reason for faking the moon landings. And, indeed, that is why the moon photos were falsified.

If the model of a spinning ball in space is refuted, that leaves us with a completely different interpretation of the scientific and observable evidence all around us that indicates that the earth is flat. In the end, the population may come to understand that NASA did not go the moon. NASA did not go to the moon because NASA could not go to the moon. NASA could not go to the moon because the moon is not in outer space. The moon is not in outer space because there is no outer space. There is no outer space because God created a flat earth with a firmament over it. Above the firmament is heaven. God's throne is in heaven. Isaiah 66:1, Psalms 33:14, Matthew 5:34. That is the point that Sungenis is missing (or perhaps hiding).

Sungenis is an ardent promoter of the Roman Catholic Church doctrine who has written books which attack the sovereign grace of God and the authority of the Bible. For Example, included among the more than 20 books on science and religion written by Sungenis are: *Not by Faith Alone: The Biblical Evidence for the Catholic Doctrine of Justification*, and *Not by Scripture Alone: A Catholic Critique of the Protestant Doctrine of Sola Scriptura*.

Although Sungenis has written many books and articles on science and cosmology, he is basically a theologian who is engaging in spiritual deception. Since the controversy over whether the earth is stationary and flat or a spinning sphere is really a spiritual issue, who better to defend NASA than a theologian. NASA is engaging in a spiritual deception under the banner of science, falsely so-called. NASA is at the forefront of a massive spiritual deception to cover up the true nature of God's creation. With his theological background (Ph.D. in religious studies) and his scientific acumen, Sungenis feels empowered to spin the NASA deceptions as mere "foibles," based on nothing more than his say-so.

Sungenis' argument starts with an initial acknowledgment that NASA's faking of photographs was wrong, but those fake photos only mean that NASA did not go to the moon. He Jesuitically argues that we cannot infer any further nefariousness from those fake photos. That argument is nonsense and is refuted by the fact that NASA has kept up the charade of using image trickery to falsely portray an earth spinning in space long after the moon landings were allegedly accomplished. The spherical earth deception has been part and parcel of NASA's mission since its inception; and it continues today.

For example, the pictures below are a series of three versions from NASA of a photograph allegedly taken on October 12, 2015, from a camera (LROC) onboard a satellite called the Lunar Reconnaissance Orbiter (LRO).[523] The LRO satellite is allegedly orbiting the moon. NASA titles the photograph as "Earthrise," which is a homage to a famous (but fraudulent) image of earth from the moon provided by NASA during the Apollo 8 alleged orbit around the moon on December 24, 1968.

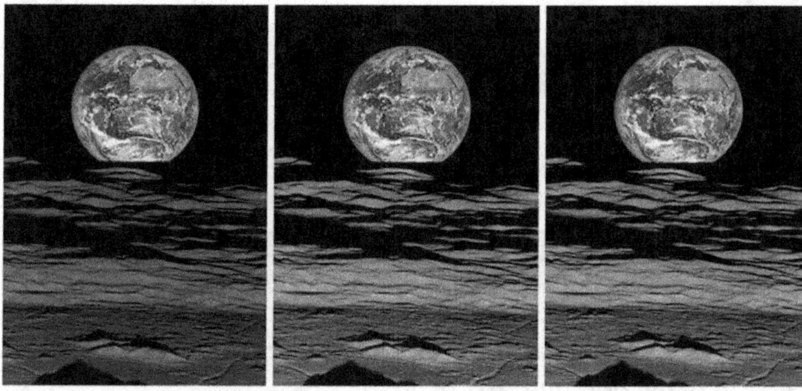

Three versions by NASA of a photograph allegedly taken on October 12, 2015, from a camera called the Lunar Reconnaissance Orbiter Camera (LROC) aboard the LRO satellite.

The NASA Office of Communications provided the following explanation for the above three versions of the Lunar Reconnaissance Orbiter Camera (LROC) image.

> In addition to sharpening Earth and adding color from the WAC [Wide Angle Color], the LRO [Lunar Reconnaissance Orbiter] team made some adjustments to the contrast to accentuate details on the lunar surface. In the strip of images directly above, the Earth and Moon have the same contrast stretch in the image on the left. In the center image, the brightness of the Moon has been increased relative to the Earth. In the image on the right, color from the WAC has been overlain on Earth.[524]

NASA states that the version of the Lunar Reconnaissance Orbiter Camera (LROC) photograph on the viewer's far left is the original photograph from the LROC; the version of the photograph in the middle has had the alleged moon surface brightness increased; and the version of the photograph on the right has had color added from the Wide Angle Camera (WAC).[525] The NASA description indicates that only the viewer's far right photograph has had anything added, and only color was added from the Wide Angle Camera (WAC) onboard the LRO.[526] Yet, when the contrast is enhanced in each of the three NASA versions of the photograph using Corel Photo-Paint software contrast enhancement tool, there is a cropped rectangle that is revealed around the earth in each of the versions of the photograph allegedly taken by the LROC. The rectangle around the earth indicates that the earth image has been pasted in using computer software.[527]

Using Corel Photo-Paint, the contrast was enhanced in each of the three different NASA versions of the photograph allegedly taken from the LROC aboard the LRO satellite, revealing a cropped rectangle around the earth in each version of the photograph.

Indeed, noise analysis of the photographs taken by the LROC that are posted in the LROC photo gallery reveal that they lack the typical characteristics of consistent photographic noise patterns that would be expected in real photographs.[528] There are patches in the LROC images that lack any photographic noise whatsoever. Indeed there are even rectangular patches with no photographic noise, which is indicative of computer manipulation. That lack of noise indicates that the LROC images were created by pasting images together in a digital laboratory using computer software.[529] In addition, there should be ambient light from the earth apparent when adjusting the contrast on the LROC images if they were actual photographs, but no ambient light can be seen in the LROC images when adjusting the contrast.[530] The LROC images are not actual photographs. The LROC images are CGI counterfeits.[531]

NASA is trying to pass off the pasted image of earth as an actual photograph of earth taken from the LROC camera on the LRO satellite allegedly orbiting the moon. Where did the image of the earth in the rectangle come from that was used to paste into the LROC "Earthrise" image? The LROC "Earthrise" image is allegedly a photograph taken on October 12, 2015. When searching the NASA website for the images taken from the Earth Polychromatic Imaging Camera (EPIC) onboard the alleged NOAA Deep Space Climate Observatory (DSCOVR) satellite for the date of October 12, 2015, we find the exact same image of earth used in the LROC "Earthrise" image listed as taken by EPIC onboard the DSCOVR satellite.[532]

The DSCOVR satellite is allegedly floating in space approximately one million miles away, whereas the LRO satellite is supposed to be orbiting the moon that is alleged under the heliocentric theory to be approximately 239,000 miles from earth. When increasing the contrast in the LROC image, a rectangle tilted at a 12° angle surrounding the earth becomes visible. When the EPIC earth image is also tilted at a 12° angle we find that it matches exactly in every detail, including the attitude of the continents and cloud formations in the LROC image.[533] The identical cloud formation means that both pictures had to have been taken at the same moment. The chances of that happening at random are infinitesimal. The precise match of the two images tilted at 12° strongly suggests that the LROC image of earth and the EPIC image of earth are one and the same. NASA simply used the same earth image in both the EPIC alleged photograph and in the LROC "Earthrise" image.

NASA could argue that both pictures were taken on the same day and thus it is no surprise that the photographs of the earth would look similar. But the earth images are not merely similar, they are identical in all respects, including continent attitude and cloud formations. That could only happen if the satellite cameras took the pictures at the exact same moment from the same angle. But even if we accept that as a possibility, one could only accept such an occurrence as actually happening by ignoring the rectangle around the LROC earth image. The rectangle is proof that the LROC image is a forgery. Furthermore, there is no way to explain the identical 12° angle. The pasted rectangle containing the LROC earth image is at a 12°angle. When the EPIC picture is tilted at a 12° angle they match precisely in every respect, including the attitude of the continents and cloud formations. Both the EPIC and LROC earth images are one and the same and have the same source: computer software.

NASA is caught using the same image of earth to represent photos of earth taken from two different satellites.

Image portrayed by NASA as being taken from the DSCOVR satellite EPIC camera. When tilted at a 12 degree angle it matches exactly, right down to the cloud formations, the picture of the earth below, which NASA alleges was taken from the LRO satellite.

"Earthrise" image that has been portrayed by NASA as being taken from the camera onboard the LRO satellite as it orbited the moon. When the contrast is enhanced, it causes a rectangle surrounding what is supposed to be the earth in space to appear. That is clear evidence that the purported photograph is actually a composite with an image of the earth pasted in. That image is the same image of earth portrayed by NASA as being taken from the DSCOVR satellite EPIC camera.

Indeed, if EPIC onboard the DSCOVR satellite actually snapped the picture of the earth from approximately a million miles away at the same moment as supposedly did the LROC camera from approximately 239,000 miles away, then the moon would be clearly visible traversing across the EPIC photo frame, with the earth in the background.[534] That is because both the purported EPIC and the LROC images are centered in the camera frame. To be actual photographs, centered in the frame as they are, would require them to have been taken straight on at the equatorial plane of the earth. That is the only way the full spherical earth could be illuminated by the sun, as is seen in both the LROC and EPIC images. That straight on shot from the equatorial plane would require the moon to be in the EPIC image, since the DSCOVR satellite position puts the moon (supposedly 239,000 miles from earth) directly in between DSCOVR (alleged to be approximately a million miles from earth) and the earth.

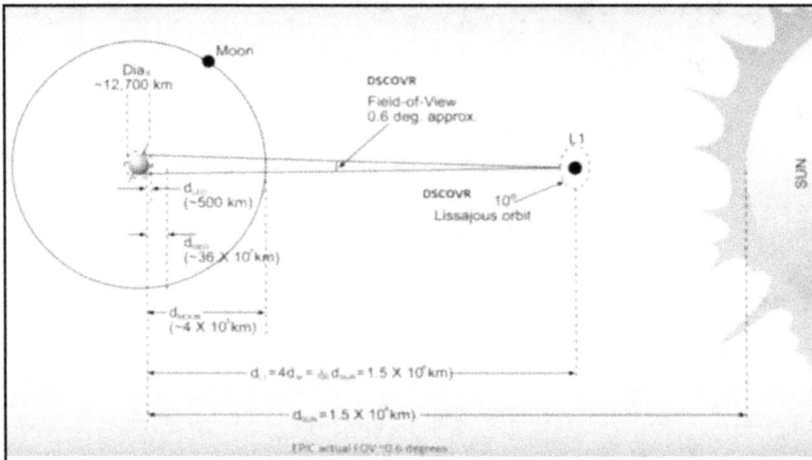

Above is a NASA diagram showing the EPIC field of View. In order for the identical LROC and the EPIC images to be two different photographs of earth taken at precisely the same time (as being identical requires), the moon must necessarily be in the field of view of the EPIC camera. The moon is missing from the alleged photograph of earth taken by EPIC, which means that the EPIC and LROC images <u>CANNOT</u> be two separate photographs taken from those two satellites. Those two images are in fact one and the same image of earth created using CGI and being falsely passed off by NASA as being two separate photographs taken from the two different satellites (DSCOVR and LRO) in space.

The moon is alleged by the heliocentric theory to be inclined in its supposed orbit 5.14° from the ecliptic plane around the earth. That 5.14° inclination would still put the moon inside the EPIC frame. Proof of that can be seen in the alleged photograph taken by EPIC on July 16, 2015, depicted below. That means that the moon should have been in the EPIC frame traversing the equatorial plane of the earth at the moment the EPIC picture was allegedly taken on October 12, 2015. The moon had to be in the center of the EPIC frame, because that is the only way that the LROC, which is allegedly orbiting the moon, could put the earth in the center of its frame. But, alas, there is no moon in sight traversing across the EPIC image of earth. And so we know that both the EPIC and the LROC image of earth are not two different photographs of earth, but rather one and the same image of the earth; that image was constructed using CGI.

Figure 147: Image depicting the moon transiting across the earth that NASA alleges was taken by EPIC aboard the DSCOVR satellite on July 16, 2015.

The image on this page depicts the moon in front of the earth that NASA alleges was taken by EPIC onboard the DSCOVR satellite on July 16, 2015.[535] A similar position of the moon in front of the earth should have been depicted in the October 12, 2015, EPIC image. That is because the EPIC image is identical in all respects to the LROC image. Both the EPIC and LROC October 12, 2015, alleged photographs were taken from the equatorial plane of earth. The LRO satellite was over the moon, which was in a direct line between the earth and the DSCOVR satellite at the moment the LROC and EPIC both allegedly took their photographs. That would put the moon squarely between the earth and the DSCOVR satellite at the moment EPIC snapped its picture. But the moon is missing from the alleged photograph taken by EPIC on October 12, 2105. That means that the two identical LROC and EPIC photographs allegedly taken on October 12, 2015, are actually one image of earth created using CGI that is being falsely portrayed by NASA as two separate photographs taken by two different cameras onboard two different satellites.

Why would NASA create bogus images of earth in space? Because the earth is not a floating sphere that is spinning and orbiting in space. And the LRO and DSCOVR satellites do not exist. In fact, there are no satellites in outer space. Indeed, there is no outer space. So NASA must phony up CGI images and lie about them being taken from satellites to create the false impression of a spinning, orbiting, spherical earth in outer space. Otherwise, the truth of the flat stationary earth, with God in heaven above would be generally known. NASA must replace the glorious heaven that is above the firmament with a dark, foreboding, empty void. If science, falsely so called, can convince people that there is no heaven as depicted in the Bible, then it is very easy to discredit the Bible and make it believable that there is no God either.

Colonel Valentin Petrov, associate professor at the Gagarin Air Force Academy, reported that Nikita Khrushchev stated "Why should you clutch at God? Here is Gagarin who flew to space but saw no God there."[536] Petrov was a good friend of the legendary Russian Cosmonaut Yuri Gagarin, who is alleged to be the first man in space. Colonel Petrov revealed the statement made by Nikita Khrushchev to refute claims that the statement was made by Gagarin. Petrov explained that the statement was not made by Gagarin, but by Khrushchev about Gagarin.

Nikita Khrushchev was the First Secretary (formerly and later known as General Secretary) of the Communist Party of the Soviet Union from 1953 to 1964, and the Chairman of the Council of Ministers (in effect, Premier) from 1958 to 1964. Petrov explained that Khrushchev made the statement about there being no God in space before the Central Committee plenary meeting that was considering anti-religious propaganda as a way to undermine the growing influence of Christian churches in the Soviet Union. Khrushchev was advising that line of propaganda in order to refute the legitimacy of the Christian belief that God is in heaven above. Khrushchev inadvertently let the cat out of the bag. His statement reveals that the purpose of the space programs for all countries is to demonstrate that above us is not heaven, but rather an empty void, which they call space. Since there is no heaven, then there can be no God. The modern space programs fake sending astronauts into space and jimmy up phony CGI pictures for the purpose of undermining the authority of the Bible and proving that there is "no God there."[537]

27 Google Earth is not a Globe

NASA is joined in its concealment of the flat earth by private corporations. Indeed, the governments of the world regularly fund technology that is provided to private corporations, which then use it in partnership with governments to deceive the public in many areas, including, but not limited to, the deception that the earth is a spinning globe. Sometimes this funding is done secretly through black budgets by governmental intelligence agencies. Sometimes the government funding becomes public knowledge, and that usually happens when the deception is not seen as a deception because the technology is funded to achieve some purportedly altruistic motive.

For example, Google does not hide the official government sources, like the National Geospatial-Intelligence Agency (NGA), the National Oceanic and Atmospheric Administration (NOAA), and the U.S. Navy, for the Google Earth data. What Google is not keen for the public to know is that Google Earth itself is software that is based upon technology that was developed through strategic funding from the U.S. Central Intelligence Agency (CIA). Toni Hiley, executive director of the CIA Museum, told Leanna Garfield from Tech Insider that the initial version of Google Earth was actually based upon the CIA developed software called EarthViewer.[538] The CIA runs its own venture capital firm called In-Q-Tel that it uses to invest "in all kinds of Silicon Valley start-ups, including what would eventually become Google Earth."[539]

EarthView was developed through a "strategic investment" in 2003 from the CIA-funded venture-capital firm, In-Q-Tel, in a company called Keyhole, Inc.[540] Keyhole was "a pioneer of interactive 3-D earth visualization and creator of the groundbreaking rich-mapping EarthViewer 3D system. CIA worked closely with other Intelligence Community organizations to tailor Keyhole's systems to meet their needs."[541] Google then purchased Keyhole in 2004, and used its technology to bring about Google Earth.

Google Earth is a tool of propaganda designed to conceal the flat earth. Proof of that fact can be easily determined by anyone. Take any facing point on any physical sphere and rotate the sphere 360° across and you will return the sphere to its original orientation with the starting point facing you. But when you take any facing point on Google Earth and then rotate the graphic of the earth in the Google Earth Pro software 360° from east to west or vice versa, you will find that you have not returned to your starting point.[542] You will need to continue turning the graphic an extra 90° or so to get back to your starting point.[543]

Mike Helmick argues that the phenomenon described above is caused by the fact that Google Earth is a 3D globe being depicted on a 2D screen.[544] He claims that it is, therefore, not possible to spin the Google Earth image precisely 360°. On the surface, Helmick's point seems to have merit. The problem with Helmick's argument is that the phenomenon becomes even more pronounced as the Google Earth image is expanded. Indeed, anyone will find that when the Google Earth image is enlarged he must turn the Google Earth image another 180° to get back to his starting point. If what Helmick is saying is true, then his explanation would explain the phenomenon no matter how large is the Google Earth image. But his explanation does not account for a full 180° error in rotation on the enlarged Google Earth image. In the end, what it means is that Google Earth is not a true depiction of a spherical earth. Rather, it is software that is depicting a flat earth to look spherical by using computer algorithms.

Furthermore, Helmick's argument cannot explain the strange phenomenon when the Google Earth Pro software image is rotated clockwise (or counterclockwise) 360°. Doing that demonstrates that the Google Earth image is not a true and accurate depiction of a globular earth. When one rotates the Google Earth image 360°clockwise, the image of the continent that is facing the viewer when starting the rotation of the Google Earth image will flip to the opposite side of the sphere when the 360° clockwise turn is completed. That would not happen if Google Earth were an accurate depiction of a truly spherical earth. If the earth were truly a sphere and Google Earth was an accurate depiction of that sphere, the continent facing the viewer at the end of the clockwise turn would be in view at precisely the same orientation as when the 360° turn began.

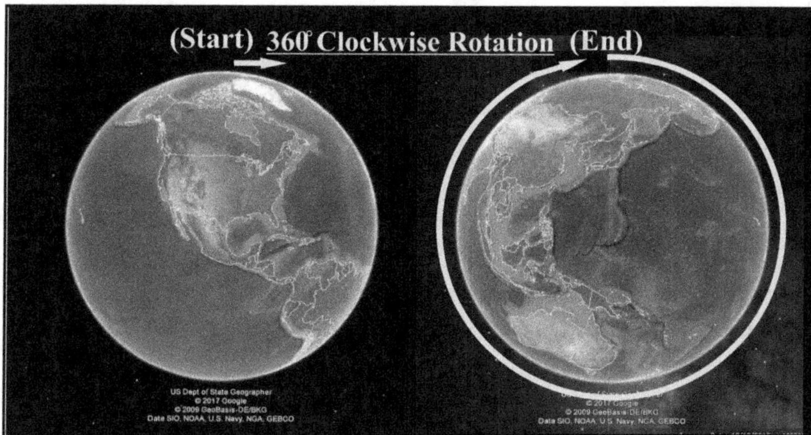

Rotating Google Earth clockwise 360° causes the front facing continent (North America) to flip around to the back of the globe. At the conclusion of the rotation we see East Asia, Russia, and Australia. If the Gooble Earth were in fact depicting a truly globular Earth, then North America would be facing the viewer after the 360° clockwise turn of the globe in the image at right. But that is not what happens. The flipping of North America to the back of the globe by Google Earth after the 360° clockwise rotation indicates that Google Earth is not a true depiction of a spherical earth, but is rather a computer program that is adjusting the display of a flat earth in the form of a sphere.

This strange phenomenon indicates that Google Earth is a computer graphic program that is depicting data from a flat earth and portraying it as though the earth is a sphere. Google Earth must provide accurate data when zooming in to any particular inhabited area. However, the software must adjust that accurate data from the flat earth as the viewer uses Google Earth to zoom in and out to and from a fictional globular earth. In order to do that the computer software must constantly adjust the sphere to accommodate the disparity between the dimensions of the real flat earth and the dimensions of its spherical depiction by Google

Earth.

It is inevitable when transforming a flat earth into a sphere that the compass bearings between cities will become distorted. There is no way to transform the flat earth into a sphere without also distorting the navigational headings between locations. Rich Hopkins discovered just such deformation in Google Earth. Hopkins found that when one travels on Google Earth due east and then turns 180° to return back again due west, the traveler does not end up at his starting point.[545]

In reality, on a flat earth, a traveler who retraces his route in the opposite direction will end up where he started. But that does not happen on Google Earth. A person traveling east or west on Google Earth who tries to retrace his route will end up far away from his starting point. That is because Google Earth transforms the real, flat earth into a mythical globe. As a result, Google Earth produces distorted compass bearings.

For example, this author used the ruler tool on Google Earth Pro to draw a line in a direction at 90° due east from Spokane, Washington, for a distance of 2,150 miles. At the end of that 2,150 mile line was Clayton, Delaware. One would think that to get back to Spokane, Washington, all you would need to do is turn 180° and travel back in the opposite direction, which would be 270° due west. But when using the ruler tool on the Google Earth Pro software to draw a line 180° in the opposite direction from Clayton, Delaware, at a heading of 270° due west for 2,150 miles, the line does not end at Spokane, Washington. Instead, the line ends in a remote area, near Dateland, Arizona, which is more than 1,000 miles south of Spokane. That is proof that Google Earth software is not depicting the reality of a spherical earth; it is deceptively portraying the flat earth as though it is a globe. Doing so warps the compass bearings.

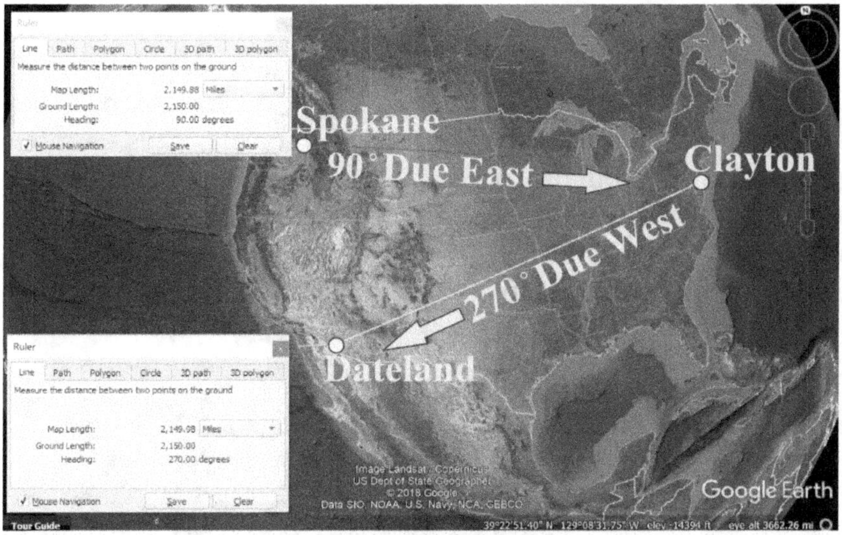

Above is an annotated screen shot of Google Earth Pro showing a path drawn using the Google Earth ruler tool (faint dark line) from Spokane at a heading of 90° due east and ending 2,150 miles away at Clayton, Delaware. Another 2,150 mile line, lighter in color, was then drawn using the Google Earth ruler tool 180° in the opposite direction, at a heading of 270° due west from Clayton. The line did not end at Spokane as expected. Instead, the westbound line drawn by Google Earth ended near Dateland, Arizona, which is more than 1,000 miles south of Spokane.

This author repeated the exercise in the southern hemisphere (more accurately described as a hemisplane) and discovered the same phenomenon. The graphic below shows how a traveler in Australia using Google Earth Pro to retrace his east-west route will find himself way off-course in the middle of nowhere. According to the Google Earth Pro ruler tool, a traveler leaving from Perth, Australia, on a heading of 90° due east for 2,272 miles will end up at Sunshine Beach on the east coast of

Australia. But when using the ruler tool to retrace the route back again by turning 180° in the opposite direction for a distance of 2,272 miles at a heading of 270° due west, the traveler does not end up anywhere near Perth. Instead, the traveler ends up more than 700 miles north of Perth, in a desolate area, near Hamersley Ridge. It is clear that Google Earth does not accurately reflect the real world. It is a carefully contrived software illusion (but not a very good one) whose purpose is to conceal the reality of the flat earth.

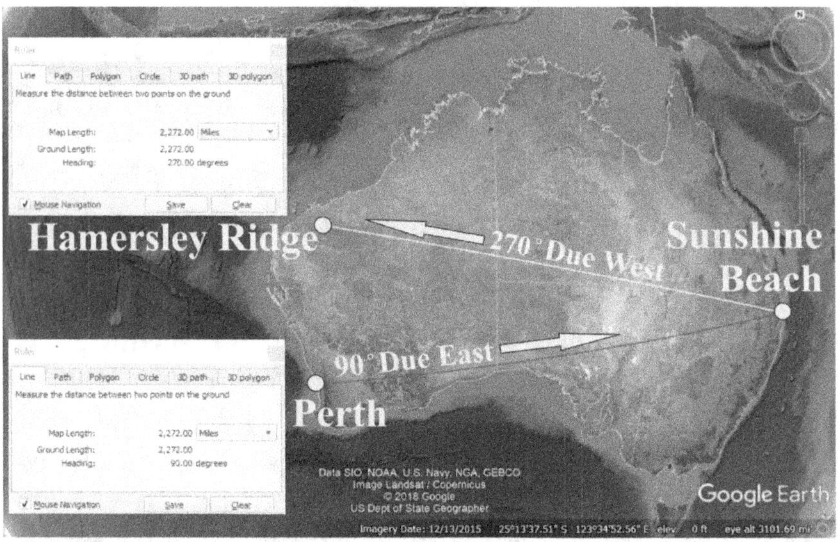

Above is an annotated screen shot of Google Earth Pro. The ruler tool in Google Earth Pro was used to draw a 2,272 mile line from Perth, Australia, at a compass heading of 90° due east to Sunshine Beach on the east coast of Australia. The ruler tool was used a second time to draw a 2,272 mile line 180° back again from Sunshine Beach, at a compass heading of 270° due west. Instead of arriving back at Perth, as should have happened in the real world, the line terminated near Hamersley Ridge, approximately 700 miles north of Perth.

Hikers and explorers relying on Google Earth for navigation could find themselves lost and stranded in the middle of nowhere due to Google Earth's misrepresentation of the true compass bearings. Google Earth should carry the standard disclaimer for works of fiction: "This software is for entertainment purposes only. It is the product of its designer's imagination. Any resemblance to reality is purely coincidental. Use at your own risk."

28 NASA Moon Missions

David Groves has opined that the picture of Buzz Aldrin descending from the Lunar Excursion Module (LEM) onto the moon's surface was illuminated with artificial light. David Groves has a BSc (Honors) Class 1 in Applied Physics and has a PhD in Holographic Computer Measurement. Dr. Groves is also a Chartered Physicist and a Member of the Institute of Physics. Dr. Groves founded Quantec Image Processing in the UK. Dr. Groves has 25 years experience analyzing photographs. He uses holographic computer techniques for analyzing images and obtaining three dimensional information from two dimensional photographs. He used those techniques to analyze the famous photograph of Buzz Aldrin descending the LEM's ladder to the surface of the moon during the Apollo 11 moon mission.

Dr. Groves' crack team of experts at his company examined the Aldrin Apollo 11 picture. They noticed a number of anomalies in the photograph that indicated that it was illuminated with artificial lighting. Most notably, there was a lighting hot spot on Aldrin's boot. Dr. Groves had access to that type of boot. Dr. Groves knew the focal length of the camera lens and the orientation of the boot. With that information, he was able to determine that the hot spot on the boot was caused by an artificial light source that was between 24 and 36 centimeters to the right of the camera.[546] NASA has gone on record explaining that there was no artificial lighting used in any of the pictures taken from the moon.

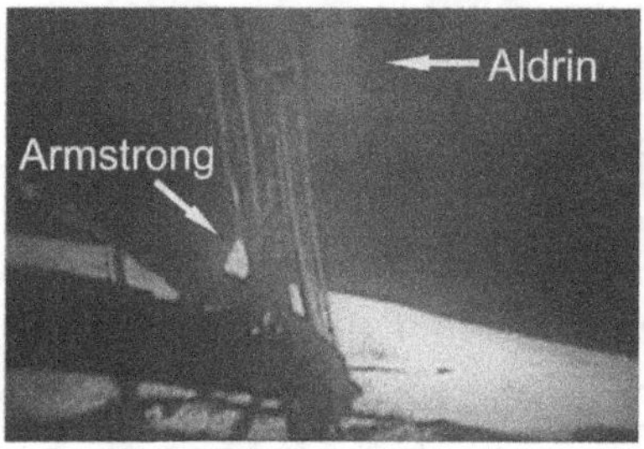

Apollo 11 Video ↑
Aldrin in Darkness

Apollo 11 Photo ↑
Aldrin Illuminated

In addition to NASA's official statement that there was no artificial lighting used for the NASA photographs from the moon, it can be clearly seen in the televised NASA video that Neil Armstrong had no artificial light source when he was standing behind the LEM in position when Aldrin descended the ladder. No artificial lighting was present. In the video, Aldrin is on the side of the LEM, awash in the darkness of the shadow. It is so dark in the area of Aldrin's descent that he can barely be seen. In order for a camera to take the brightly-lit picture of him, as published by NASA, artificial lighting had to be used. Dr. Groves'expert analysis of the photo proved that artificial lighting was in fact used. NASA maintains, and the video shows, however, that there was no artificial light source present on the purported moon surface.[547] This is proof that the photograph was not taken on the moon, but in a studio.

Mary D. M. Bennett, Stan Gooch, and David S. Percy collaborated with photographic expert Jack White in analyzing other NASA photographs. Percy, FRSA, ARPS, is an Associate of the Royal Photographic Society and an award winning film and television producer. Percy was nominated Film Cameraman of the Year at BISFA (forerunner of the IVCA). Bennett, Gooch, Percy, and White discovered evidence of photographic fakery and staging of scenes in a studio of the Apollo moon landing photographs. They have posted their evidence on their website at www.aulis.com.[548]

In the picture below you can see that the LEM for the Apollo 15 moon mission is in view in one photograph, but in a later photograph, with the same backdrop, the LEM is gone. Percy, *et al.*, explain:

> Of course although the backdrops match, the foregrounds are different. This was probably to help disguise the fact that the setting has been used for both a general scene, (photo AS15-82-11082) and an alternative setting for the LM (photo AS15-82-11057). It is highly likely that there was a finite number of backdrops available to the photographic team and that NASA failed to anticipate such meticulous analysis of the imagery.[549]

The LEM is the vehicle used to land on the moon. There would be no one on the moon to take a picture of the landing site before the LEM arrived. Once the LEM lands, it remains stationary and cannot be moved. A picture of the LEM landing site without the LEM, therefore, would be impossible. The disappearance of the LEM from view in the picture can only be explained by the fact that the purported moon pictures were not taken from the surface of the moon.

Figure 152: Annotations to NASA Photos by Jack White.

The LEM disappearing act also happened during the Apollo 17 moon mission. In the picture below, Jack White reveals his discovery that the same backdrop was used for three different photographs from the Apollo 17 moon mission. Clearly, nobody would be there to take the pictures without the LEM, which is the moon landing craft for the mission. Yet, we see that in two out of the three pictures, the LEM is missing from its landing area. That is convincing proof that all three photographs were taken from a studio on Earth; the photo editors simply used the same backdrop for all three photos.[550]

Figure 153: Annotations To NASA Photos by Jack White.

Jack White discovered that those who staged the Apollo 17 moon landing inadvertently placed lunar rover tracks around the LEM before the lunar rover was unpacked from the LEM.[551]

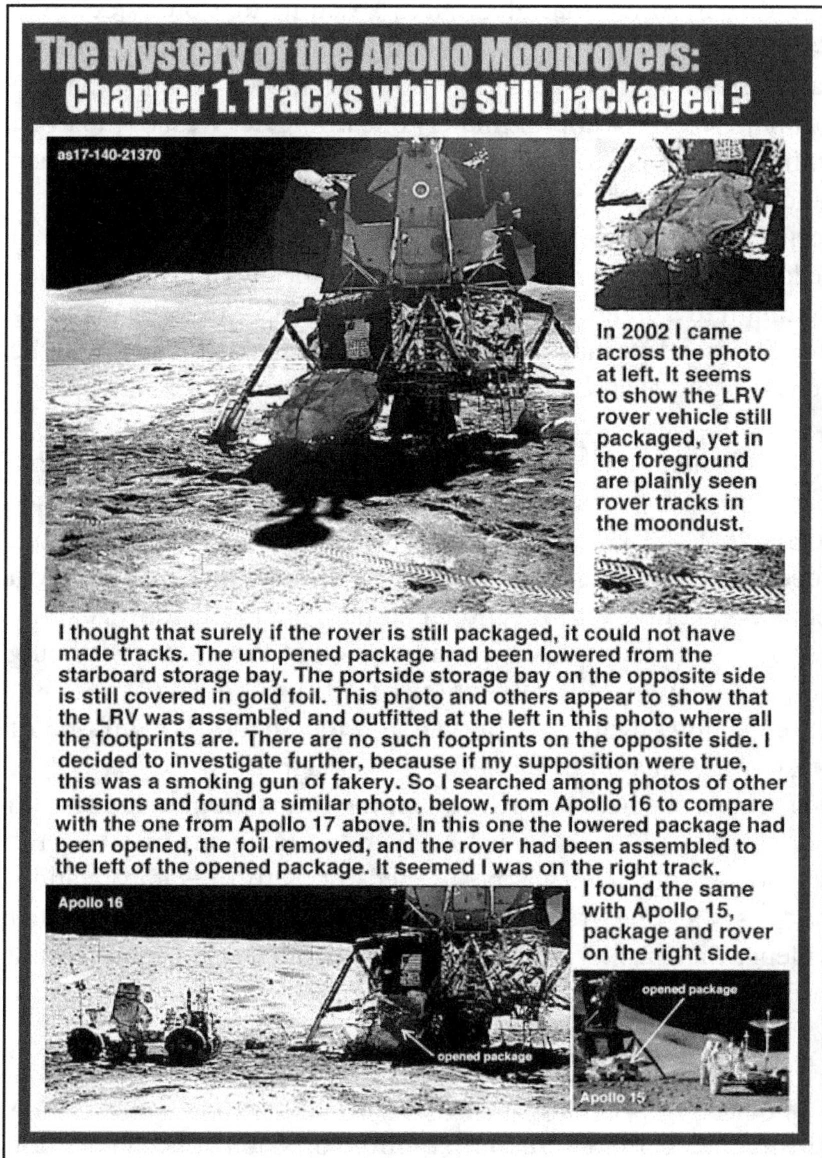

Figure 154: Annotations by Jack White.

In a documentary film produce by NASA about the Apollo 16 moon mission, two videos, each allegedly taken from the moon, were displayed and narrated. NASA claims that one video was taken from the moon showing John Young at the rim of a moon crater on April 21, 1972.[552] The NASA narrator explains that the astronauts traveled approximately one kilometer west of the LEM landing site to get to the moon crater where Young can be seen standing.[553] The NASA documentary explains that the next day, April 22, 1972, the astronauts traveled over 4 kilometers south of the landing site in the lunar rover and climbed up the side of Stone Mountain to a crater 700 feet in elevation above the LEM landing site. The two astronauts, John Young and Charles Duke, can be seen standing near the rim of the alleged moon crater.

When viewing the two videos, supposedly taken one day apart and many miles distance from one another, one can see clearly that the astronauts are in the exact same spot in both videos. The foreground is identical, right down to every rock and pebble. Not only is the location identical, but the shadows cast by the rocks are identical, which indicates that the location is being illuminated by an artificial, stationary light source. That is because, if the shadows were cast by the sun, the shadows would be of different lengths on different days. A lunar day is 30 Earth days long. Under the heliocentric model, the sun is thought to move in the lunar sky 12° in 24 hours. Further, the camera from which the two videos were taken was positioned at exactly the same location, and was shooting from precisely the same angle. This is clear evidence that the moon landings were videotaped from earth, using artificial lighting.

NASA has never explained the videos. Instead, it has tried to conceal the evidence. This author has searched the NASA website video library of the Apollo 16 moon mission for the original video footage used in the NASA Apollo 16 film documentary.[554] The original videos of the two excursions discussed above are not posted on the NASA internet video library, which is supposed to contain a complete library of the Apollo 16 videos. Nonetheless, the two videos have been represented by NASA to be genuine videos of the Apollo 16 moon mission.

Identical foreground, with identical shadows, videotaped from precisely the same distance, at precisely the same angle, portrayed by NASA to have been taken at two separate locations on two separate days on the moon.

Portrayed by NASA to be one kilometer west of the moon landing site on April 21, 1972.

Portrayed by NASA to be four kilometers south of the moon landing site on April 22, 1972.

The NASA space missions have been a hoax from the beginning. There is no such thing as the vacuum of space. In Genesis, it states that God made the firmament and he called the firmament Heaven. Genesis 1:7-19. The firmament, as the name suggests, is something hard. It is not a vacuum of nothingness. God set the sun, moon, and stars in the firmament of the heaven. The firmament divides the waters beneath the firmament on the earth from the waters above the firmament. Genesis 1:6-9. That is why the sky is blue; there is water above the firmament.

Proof of the NASA hoax is found in early photographs that were purported to be photographs of space missions. Researcher Ralph Rene uncovered a photograph of a fake space walk in an autobiography written by Apollo Astronaut Michael Collins, titled *Carrying The Fire.* Collins was the command module pilot aboard the Apollo 11 moon mission.

In Collins' book, he presented the two pictures (depicted below), separated in his book by almost 200 pages. The picture on the bottom left is a picture that was snapped by a professional NASA photographer aboard the zero gravity plane used by NASA. That airplane, nicknamed the vomit comet, flies in a parabola, which temporarily creates a weightless environment. Ralph Rene describes what is seen in the photograph.

> The cabin is padded to protect the occupants from the inevitable fall the instant the loop is terminated. Here, Collins, as he practices space walking, is holding the propulsion rod in his right hand. Note the similarity between this picture and the one on the adjoining page [top-right below]. Note also that the suit's wrinkles deny that it is inflated.[555]

The other picture of Collins in the book, which is intended to represent Collins' purported space walk, is depicted on the top-right and is described by Rene:

> This picture was also extracted from *Carrying The Fire* and was allegedly taken during a space walk on the Gemini 10 mission exactly three years before his Apollo 11 mission to the Moon. NASA claims to have landed the first men on the Moon during this mission. He is shown holding a jet reaction propulsion rod with his left hand.[556]

Rene saw the two pictures in Collins' book and could not get over their similarities. He had a hunch.

> Following an obvious hunch I had negatives made of both previous pictures. Then I had another negative made of the first photo reduced in size and flopped over. Collins is now practicing with his left hand, the same hand he used in his alleged space walk. I then had the "Gemini 10" picture blown up until the figure of Collins was the same size as this one.
>
> I then rotated the blown-up photo clockwise until the propulsion rod made the same angle across the page on both pictures. Even the expression of his face is the same. Collins would have us believe that this picture was taken by a different person many months later. However, the negative of either picture placed over the print of the other produces a point to point coincidence until the binding line at the knee is reached. The missing area was removed when the photo was bound in his book.[557]
>
> If I do the same to the original NASA picture #66-40127 (which took over 18 months to get

from NASA), the point to point coincidence continues to the soles of his boots. Why did NASA feel it necessary to fake pictures and lie to us as early as July 1966?[558]

Flipped Picture of Collins aboard the zero gravity training plane used by NASA

NASA flipped photograph in Collins' book purportedly taken During his Gemini 10 space walk

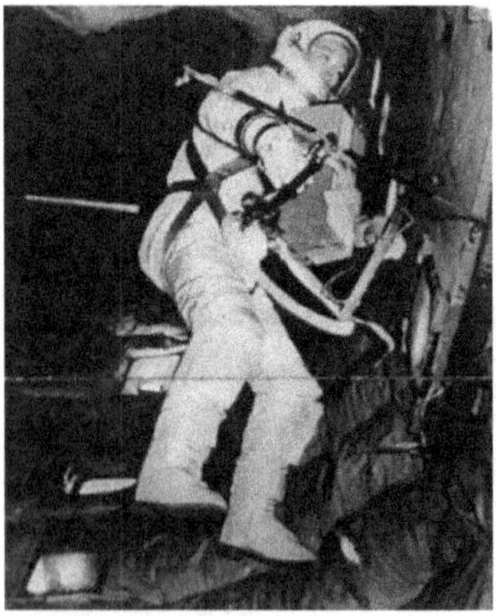

Picture of Collins aboard the zero gravity training plane used by NASA as it was depicted in Collins' book

The two pictures above compared side by side prove that they are not two different pictures, but rather the same picture. The purported space walk picture is actually the training picture that has been flipped and the background removed and replaced by the blackness of space to make it appear that Collins was in space. If Collins were actually in space, there would be no need to fake a photograph and falsely portray it as a space walk. The NASA space program is a government propaganda operation designed to convince the masses that the earth is a globe spinning in the vacuum of space as it orbits the sun.

The picture positioned on the top-left is the picture flipped by Rene. As you can see plainly, when comparing the Rene flipped picture with the NASA flipped picture, the two pictures of Collins are identical, except for the background. Collins put both the fake space walk picture and the training picture in his book. Collins obtained both pictures from NASA. Collins eschewed a ghostwriter for his book and proudly states in the foreword to the book: "Above all, I am glad I wrote it myself." Collins has been caught red-handed in a hoax. Those photographs are clear evidence that Collins was a knowing participant in the NASA space mission deceptions. Collins is the subject in both photographs. Why was it necessary to fake a space walk? Because Collins never went into space. Collins could not go to a place that does not exist.

So damning are the photographs in Collins's book that NASA apologists have accused Ralph Rene of lying about the implications of the photograph.[559] They argue that the photograph of Collins purporting to be in space appears without a caption, and, therefore, that Collins did not intend to convey that it was a photograph of him in space.[560]

There are two problems with that argument. First, there is no other reason to alter the photograph by removing the training background and replacing it with a black space background other than to convey the impression that it is a genuine photograph of a space walk. The photograph in its altered condition could only be interpreted as depicting a genuine space walk. Collins' book was published in 1974, long before Photoshop or any other photo manipulating software came on the scene. To change the background on a photograph in 1974 required a graphic artist, with expertise in photo manipulation. NASA went through a lot of trouble, using 1974 technology, to remove the training background and replace it with a black sky. Why was it necessary to flip the training photograph when adding the black background? The only reason to flip the photograph is to obscure its origin as a training photograph. The only reason to obscure the space walk photograph's origin would be a desire to falsely portray the altered photograph as a genuine photograph of the alleged space walk by Collins.

The second problem is that the cover of the 1975 edition of Collins' book contains that same photograph showing him floating in space, with the earth in the background. Collins was clearly intending to convey that the photograph was a genuine depiction of his space walk.

The attitude of Collins in the cover photograph has been rotated to differentiate it from the training photograph. The cover photograph depicts the earth in the background; Collins was using the training picture on his cover to falsely portray it as a picture of his space walk. When the picture on the cover is rotated, we see that it is the unflipped training photograph. Collins used a training photograph not once, but twice, to falsely portray it as a picture of his alleged space walk. The photograph on the cover and the space walk photograph in the book are one and the same. There is one difference between the space walk photograph on the cover and the space walk photograph inside the book. The photograph on the cover is not flipped. However, both have the same origin and objective; they are both false portrayals of a space walk through the use of what is, in reality, only a training photograph.

NASA knew before the Gemini 10 mission launched that it was not going into space. They knew that there was no space within which to conduct a space walk. That is the principal reason for the photograph of Collins' NASA training in the vomit comet. Why else would NASA send a photographer to take pictures during a routine training exercise? Why was it necessary for Collins to wear his entire space suit simply to train for a weightless environment? Why take the chance of damaging that space suit during training? It is obvious; the vomit comet ride, with Collins wearing his full space suit, was intended as a photo session. The plan all along was to doctor the photo to falsely represent a space walk by Collins.

Collins and NASA could not allow the damning photographic evidence discovered by Rene to continue to float around (no pun intended). Once the issue came to the fore, Collins revised the book to remove the moon walk picture with the black background.[561] The fact that Collins redacted that photograph from the later editions of his book speaks volumes. He could have kept the photograph in the book and simply explained its real meaning. Instead, he removed it. Why? Because there is no innocent explanation. If there were some innocent explanation, then Collins would simply have explained the photograph in the later editions.

To add insult to injury, the fake hero Collins hoodwinked a genuine hero, Charles A. Lindbergh, to write the foreword to his book. Collins knows no shame.

The reason that NASA found it necessary to fake pictures and videos is because there is no vacuum of space. The earth is encompassed by a firmament, exactly as the Bible states. The myth of the limitless vacuum of space, suggests that there is no heaven. If there is no heaven, which is the abode of God, that in turn suggests that there is no God. If there is only an empty vacuum of space, that refutes God's word that states that God is walking above us in the circuit of heaven. "Thick clouds are a covering to him, that he seeth not; and he walketh in the circuit of heaven." (Job 22:14)

During the Apollo 17 moon mission the sound of impact could be clearly heard when an astronaut was seen hitting something with a hammer.[562] The fact that the sound of the hammer hitting the object could be heard meant that the astronauts were not in a vacuum. In the supposed vacuum of space, sound cannot travel. Sound waves need a medium, such as the earth's atmosphere, through which to travel. If the astronauts were truly in the vacuum of space on the moon, the sound of impact should not have been heard. The same phenomenon was heard during the Apollo 15 moon mission, when an astronaut threw an object at the LEM and it hit with an audible thud.[563]

Indeed, the space travel to the moon was premised on using rockets in the vacuum of space. Anyone with a rudimentary knowledge of physics knows that rockets require a fulcrum to push against to move.[564] On the earth the fulcrum is initially the ground, and after liftoff the fulcrum is the atmosphere. Space is supposed to be an empty vacuum. In a vacuum, there is nothing to act as a fulcrum for a rocket to push against. A rocket in the vacuum of space would be useless.[565]

Brian Mullin, who is a licensed engineer in four states, posted videos on YouTube that explained in detail how rockets could not function in the vacuum of space. His videos were becoming very popular, with hundreds of thousands of views. On or about May 2017 his YouTube channel was deleted along with all of his videos. There has been no explanation from Mullin as to why. Clearly, he was a threat to the powerful scientific community in his challenge to the status quo. And apparently pressure was brought to bear on him to shut down his channel. Mullin's videos are still available on other channels from people who had the foresight to download and preserve his work.

The space station is another hoax. All supposed photographs of the space station are taken in a large underwater tank in order to simulate the fiction of weightless space. Indeed, NASA and space agencies from other countries have been caught faking space walks. There are numerous official videos from various countries' space agencies where bubbles can be seen rising from trapped air during purported space walks.

In 2013, it was revealed that an astronaut nearly drowned while he was supposedly in space. I kid you not. How is it possible to drown in the vacuum of space? NASA has since retrofitted space suits with snorkels. Fox News reported:

> After a spacewalking astronaut nearly drowned in his helmet in July, NASA has a plan to protect its crew when they venture into the vacuum of space this weekend: snorkels and absorbent towels. ... "Some smart engineers on the ground said, hey, this looks like a snorkel you'd use for scuba diving," explained Allison Bolinger, NASA's lead U.S. spacewalk officer.[566]

Indeed, in one official NASA video of the International Space Station, Steve Blakey discovered a diver moving around wearing an underwater scuba tank.[567] There is no purpose for a scuba tank in space, but there is a real need for a scuba tank underwater, which is where the International Space Station is really located.

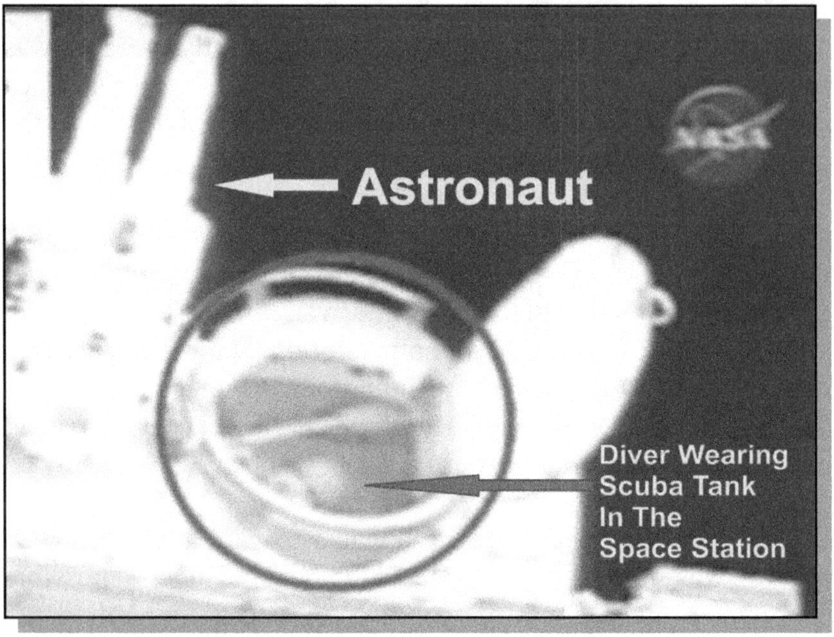

What about the hundreds of pounds of moon rocks brought back from the moon? Surely that is hard evidence that NASA went to the moon; isn't it? Not really. The rocks purported to be moon rocks are simply rocks from Earth that are used as moon rock props. One example is the fake moon rock presented to the former Prime Minister of the Netherlands, Willem Drees Jr., by the U.S. Ambassador to the Netherlands, J. William Middendorf II, on October 9, 1969. The rock was presented on the occasion of the visit to the Netherlands of the Apollo 11 astronauts, Armstrong, Collins, and Aldrin.[568]

When Drees died in 1988, the moon rock he received from the U.S. Ambassador was donated to the Dutch Rijksmuseum, which is the national museum of the Netherlands. The Rijksmuseum vetted the authenticity of the moon rock by contacting NASA directly. NASA confirmed that the moon rock was genuine. According to a museum spokesman, Ms. Van Gelder, no one doubted the authenticity of the rock, because it was donated by the prime minister, who received it from the U.S. Ambassador, accompanied by the astronauts who retrieved the rock from the moon. Furthermore, its authenticity was confirmed by NASA. The Rijksmuseum was so confident of its authenticity that it insured the rock for a half million dollars.

In 2009, researchers from the Free University of Amsterdam tested the moon rock and determined that it was not from the moon. How could they tell that it was not from the moon? Because they were able to determine that it was, in fact, petrified wood. The researchers were pretty sure that there are no trees on the moon.

From where did the U.S. Ambassador get the rock? Directly from NASA. The ambassador was the official representative of the U.S. Government accompanying the Apollo 11 astronauts during their goodwill trip to the Netherlands. The astronauts, who claimed that they obtained the rock from the moon, brought the moon rock with them. That means that the fake moon rock had a chain of custody directly from NASA to the U.S. Ambassador to the Netherlands, to a former Prime Minister of the Netherlands, then to the national museum of the Netherlands. NASA subsequently vouched for the rock's authenticity after the museum received it, but it turned out to be a fake moon rock. The ineluctable conclusion is that NASA was distributing fake moon rocks.

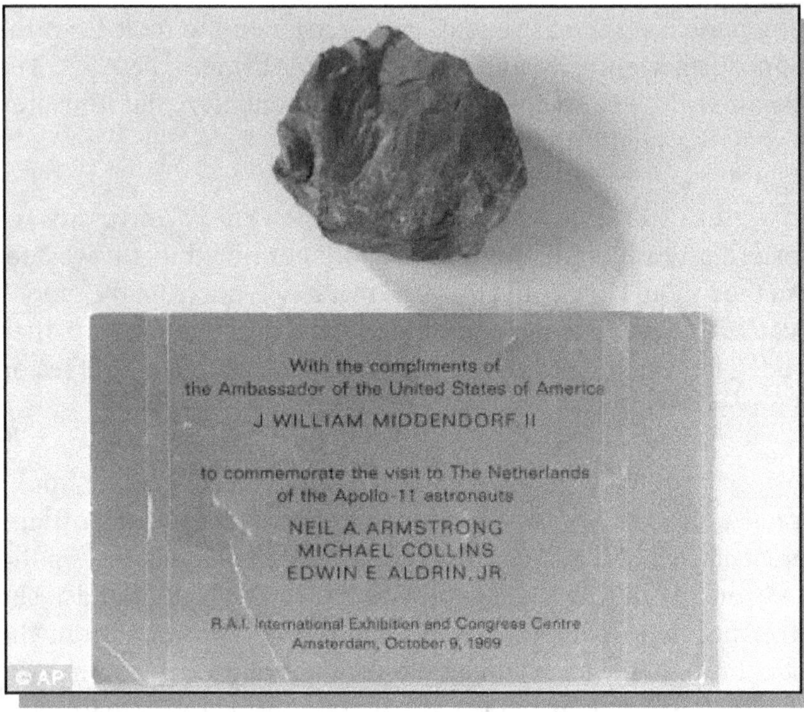

Figure 160: Associated Press picture of the fake moon rock from NASA, presented to the former Prime Minister of the Netherlands by the U.S. Ambassador.

Not surprisingly, "U.S. embassy officials were unable to explain the findings, but are investigating."[569] As of this writing it has been over six years since the U.S. Government has promised to investigate the moon rock fraud. There have been no findings announced since that time. When the U.S. Government promises to find the cause for the fraudulent moon rock that came from NASA and after six years does not announce any findings, that suggests that the U.S. Government is the culprit. NASA is the source for the fraudulent moon rock.

The U.S. Ambassador is the official representative of the U.S. Government in a foreign country. The U.S. Government will never explain how they presented a fake moon rock as a gift to a foreign government official, because to do so would require it to admit that all moon rocks are fake. NASA never went to the moon, and the fake moon rock, which was vetted through and authenticated by NASA and came directly from the Official U.S. Government representative in the Netherlands, is yet one more piece of evidence proving that the NASA moon landings were hoaxes. NASA was handing out fake moon rocks, because they have no real moon rocks to give out. They have no real moon rocks, because they never landed on the moon.

The fake moon rock presented to former Prime Minister Drees was not the only moon rock gifted to the Netherlands when the astronauts visited on October 9, 1969. A news account of the visit by the Apollo 11 astronauts to the Netherlands, appearing in the October 9, 1969, edition of *Nieuwsblad van het Noorden*, states that the astronauts paid a lightning visit to the Netherlands; they arrived in the Netherlands at 10:00 a.m. and departed at 3:00 p.m. It almost seems that their only purpose in making the trip to the Netherlands was to distribute moon rocks and then get out of town. The news account bears a picture of the three astronauts meeting with Queen Juliana, Prince Bernard, and Prince Claus and presenting the Queen with gifts. The most significant gift presented to the Queen was a moon rock.

The article states that "when the astronauts arrived at the palace, the Queen received a replica of the message the astronauts left on the moon and a piece of moon rock."[570] That moon rock was one of a group of moon rocks, which included the fake moon rock presented to former Prime Minister Drees. The Queen's moon rock was never tested for authenticity, because after the moon rock given to Prime Minister Drees was found to be fake, the Queen's moon rock was found to have disappeared.[571]

ASTRONAUTEN BIJ KONINGIN

De Amerikaanse astronauten Neil Armstrong, Michael Collins en Edwin Aldrin hebben vandaag een bliksembezoek aan ons land gebracht. Om tien uur landden ze op Schiphol (met een der vliegtuigen van president Nixon) en om drie uur vanmiddag vertrokken ze weer. In deze paar uren zijn ze op het koninklijk paleis op de Dam in Amsterdam ontvangen door de Koningin en in een rondvaartboot naar het RAI-congrescentrum gevaren, waar zij voor 1400 genodigden een film van hun maanreis vertoonden en de tentoonstelling „Maanvaart '69" openden. Bij de ontvangst in het paleis kreeg de Koningin een replica van de boodschappen die de astronauten op de maan achterlieten en een stukje maansteen.

October 9, 1969, news article appearing in *Nieuwsblad van het Noorden,* with a picture documenting the Apollo 11 astronauts' meeting with Queen Juliana of the Netherlands, at which time they gave her two gifts, one of which was a moon rock. Left to right in the picture: Queen Juliana, Prince Bernard, Collins, Aldrin, Armstrong, and Prince Claus. The last sentence of the article, translated into English, states: "When the astronauts arrived at the palace, the Queen received a replica of the message the astronauts left on the moon and a piece of moon rock." A similar moon rock, which was given to the former Dutch Prime Minister during the astronauts' visit that day, was determined in 2009 to be fake; it proved to be petrified wood. The moon rock given to the queen, conveniently "disappeared" before it too could be proven to be fake.

U.S. Ambassador to the Netherlands J. William Middendorf II presented yet a third moon rock to Prince Bernhard of the Netherlands in late 1969 or early 1970. That moon rock, reportedly weighing 34.1 grams, could not so easily be made to disappear, because it was put on display in the Dutch Museum in Leiden.[572] Museums are known for their tight security, and therefore it would be much more difficult for that moon rock to "disappear." There has not been any public announcement of Prince Bernhard's moon rock having been tested, nor is any announcement likely. The embarrassment of the exposure of the fraudulent moon rock given to former Prime Minister Drees was enough for NASA; certainly, immediate diplomatic efforts were made to prevent any further public revelations that would give away the deception and expose the NASA moon landings as elaborate hoaxes.

Mary Bennett and David Percy in their book, *Dark Moon*, recount an interview that led Paul Jacobs to believe that high officials in the Federal Government knew that the moon rocks were fake and were privy to the moon landing hoax.

> A few months before his untimely death in 1978, the well known American investigative journalist Paul Jacobs interviewed the Head of the US Geological Survey Department in Washington, D.C. Jacobs asked him for his views on the moon landings in general and the lunar rock samples in particular. When he reported his interview to [Bill] Kaysing, Jacobs said that he received some very cryptic answers. The geologist had assured him that the rocks were real. But when Paul Jacobs said: "If they weren't real and you were in on the hoax, you would still attest to their reality wouldn't you?"—the geologist just smiled.[573]

In a Fox Network documentary, former astronaut Brian O'Leary stated:

> Regarding the Apollo missions, I can't say 100% for sure whether these men walked on the moon ... It's possible that NASA could have covered it up, just in order to cut corners and be the first to allegedly go to the moon."[574]

O'Leary was not just any astronaut; he was a NASA astronaut in the 1960s and served as a science advisor during the Apollo moon missions. He worked closely with the purported moon landing astronauts, including Neil Armstrong, Buzz Aldrin, and Michael Collins.

NASA simply could not allow an Apollo astronaut postulating that the moon missions could have been hoaxed. One can imagine that after O'Leary made his statement, he had an official visit from NASA, where he was given counseling in no uncertain terms. O'Leary subsequently clarified his statement made during the Fox documentary. O'Leary stated that "somehow I may have given the impression that Apollo may have been hoaxed,"[575] but that he in fact believes that NASA did go to the moon. O'Leary said that "it was real ... Apollo happened."[576]

O'Leary claimed that the Fox producers used quotes from him taken out of context to suggest that O'Leary had serious doubts about the legitimacy of the moon landings. The problem with O'Leary's claim is that he is seen in a video interview saying the very words quoted above. He never explained how his words were taken out of context. Indeed, in a later interview, O'Leary suggests that he in fact meant what he said during the Fox documentary:

> O'Leary was initially skeptical when his astronaut friends who continued in the program and flew the lunar landing missions gave abbreviated answers to his questions. But he is quick to add that they may just have been sick of being asked so many questions upon their return. And he is sick himself of being asked about his position on the Apollo hoax; when contacted in March 2009 to confirm our quotes, O'Leary said this interview was his last public word on the subject. [577]

If one reads carefully O'Leary's above statement, he essentially admits that he was "initially skeptical" of the legitimacy of the moon landings. That suggests that he meant what he said during the Fox documentary. When O'Leary said on camera that he could not say with certainty that the Apollo astronauts walked on the moon and that it was possible that NASA covered up the fact they did not do so, he said it, because he believed it. It seems that the producers of the Fox documentary did not twist his words. O'Leary was in fact skeptical of the moon landings, just as he said on camera. O'Leary meant what he said.

The most that can be said about his retraction is that he subsequently changed his mind, and he came to believe that the moon landings were real. The problem with that explanation is that he was an astronaut in the 1960s. He was a science advisor to the NASA Apollo missions at that time. For over 30 years he believed that the Moon landings may have been hoaxed. Suddenly, after his private beliefs, which he held for over 30 years, become public via a network news documentary, he suspiciously changes his mind and announces that the moon landings were real after all.

O'Leary later stated that if there had been a moon landing hoax, he would have known about it, because the astronauts who were in on it would have told him.[578] That is something, however, that he had in mind for the entire 30 years that he doubted the authenticity of the moon landings. How could it suddenly be the basis for him changing his mind to believe the moon landings were real? His explanation does not pass the smell test. His explanation cannot be the reason that he went from being a doubter to believing "it was real ... Apollo happened."[579]

O'Leary was being deceptive when he claimed that if the moon landings were a hoax, he believed that the participating astronauts would have told him. O'Leary's statement, that the moon landing astronauts would have told him about it being a hoax, is impeached by O'Leary's other statement that the astronauts who flew the alleged lunar landing missions did not open up to him about the missions, but "gave abbreviated answers" to his questions, seemingly to make him go away. The inference he drew was that the astronauts were being cryptically short with him, because they were holding back the fact that the moon landings were a hoax. He gave that as one of the reasons he initially doubted the authenticity of the moon landings in the first place.

Brian O'Leary's claim that the moon landing astronauts would have told him if the Apollo missions were faked is incredible for yet another reason. He was fully aware that all astronauts know what would happen to them if they were to let that secret out. For example, Gus Grissom was very public about his view that the chances were slim of NASA meeting the Apollo mission requirements.[580] Shortly before Grissom died, he hung a large lemon on the Apollo space capsule as the press looked on, thus graphically indicating his opinion of the space capsule. Grissom, along with fellow astronauts Ed White and Roger Chaffee, was shortly thereafter immolated inside a test capsule when it burst into flames as it sat on the launch pad during a test. During the test, and before the fire, there was a communication failure. Grissom is recorded saying at that time: "How are we going to get to the moon when we can't communicate between two buildings?"[581]

Grissom's opinion that the Apollo mission was doomed to failure was based upon his close study and examination of the mission rocket and other equipment. He took detailed notes and wrote reports about his findings. The day Grissom died on the launch pad, FBI Agents burst into Grissom's home and seized all of his records. Grissom's wife reports that those records were never returned.[582] The fire on the launch pad took place on 27 January 1967, at 6:30 p.m.[583] The FBI search took place in the evening of January 27th, immediately after Grissom died. Someone was at the ready to give the order to conduct the search as soon as it was determined that Grissom was dead. That suggests governmental foreknowledge of the launch pad fire.

Christopher Ruddy of NewsMax revealed that before his death, Grissom had received death threats, which his family believed emanated from within the space program.[584] The threats were viewed as so serious that Grissom was put under Secret Service protection and had been moved from his home to a secure safe house. Grissom's wife states that Grissom told her that "if there is ever a serious accident in the space program, it's likely to be me."[585]

Prior to the Apollo 1 capsule disaster, an inspection official, Thomas Baron, wrote a 500 page report that documented mismanagement and incompetence at NASA's Pad 34, where the Apollo 1 disaster took place. Thomas Baron testified before Congress regarding the deaths of Grissom, White, and Chaffee, and gave critical testimony about improper actions and irregularities at NASA.[586]

NASA tried to discredit Baron's testimony before Congress, but to no avail. Baron was a meticulous investigator, and he had the documented evidence. One example of NASA's effort to discredit Baron happened after Baron testified to an interview he conducted with Marvin Holmburg, who told him that the astronauts, apparently sensing the danger they were in, tried to get out of the capsule five minutes before the capsule burst into flames.[587] NASA apparently persuaded Holmburg that his revelation to Baron was imprudent. When Holmburg was called to testify before Congress, he suddenly changed his story and denied that he ever told Baron that the astronauts tried to get out of the capsule.[588]

Baron's 500 page report mysteriously disappeared, and to this day it has never been found. NASA failed to discredit Baron before Congress, and they could not allow him to repeat what he knew to the press. All criminal organizations know that dead men tell no tales. Within a week after his Congressional testimony, Baron, his wife, and his stepdaughter were killed, when his car allegedly stalled at a railroad crossing and was struck by a train.[589] Contrary to Florida law, no autopsy was performed. All three bodies were quickly cremated, thus ensuring that any foul play would not be discovered.[590]

A little known fact about the pressure test of the capsule in which Grissom, White, and Chaffee were participating, was that the capsule that was being pressure tested was obsolete at the time it was tested. It was never in the plans to ever be used again for a NASA launch.[591] It was, therefore, not very useful for a pressure test. The capsule, however, made the perfect inescapable oxygen bomb.

It seems that the capsule was modified to make it impossible for the astronauts to free themselves from the capsule if something went awry during the test. In his book, *NASA Mooned America*, Ralph Rene explained:

> Also, was it really the vicissitudes of life that the outward opening hatch was coincidentally changed that very morning to one that opened inward? An inward-opening hatch meant that any inside pressure, acting outward, would prevent it from being opened — even if someone was standing by, which they weren't. It was also bolted up from the outside and lacked explosive bolts.[592]

Another suspicious fact is that NASA decided to use pure oxygen to pressure test the obsolete capsule. It was completely unnecessary to use pure oxygen to run a pressure test. Pure oxygen is very dangerous, even at ordinary pressure. Steel wool will burn in pure oxygen. In one case, Rene explained that "an asbestos blanket, normally classed as fireproof, was consumed when used to smother flames during an oxygen fire."[593] Any hydrocarbon can burst into flames with the smallest spark or even a slight increase in heat in a pure oxygen environment. Oily rags have been known to spontaneously burst into flames in a pure oxygen environment.

Between 1962 and 1967 the Federal Government had no less than seven (7) explosive fires occur at various facilities while using pure oxygen. The NASA engineers knew full well the danger in which they were putting Grissom, White, and Chaffee. NASA had previously commissioned a report by Dr. Emanuel M. Roth, which was published in 1964. Rene states that "Dr. Roth cited difficulties with 100 percent oxygen atmospheres even under low pressures. Any competent engineer should have known the dangers of oxygen at 16.7 or 20.2 psi [the pressure of the oxygen in the capsule Grissom, White, and Chaffee were in]."[594]

Rene further stated that "[t]his is why I cannot believe that this was 'standard operating procedure' or that Grissom and his crew knew about it. NASA not only ignored their own tests on pure low-pressure oxygen but upped the ante by increasing the pressure above atmospheric! [Erlend] Kennan and [Edmund] Harvey had this to say, 'Most U.S. scientists could not believe their ears when they learned that fact. Oxygen at such pressure comes in the category of an oxygen bomb.'"[595]

Rene revealed one example of the dangerousness of oxygen that resulted in two men being killed at a U.S. Government facility. A terrible accident took place on January 1, 1967, approximately three weeks before Grissom, Chaffee, and White were burned to death. "Two men were handling 16 rabbits in a chamber of 100% oxygen at 7.2 psi at Brooks Air Force Base and all living things died in the inferno. The cause may have been as simple as a static discharge from the rabbits fur ... but we'll never know."[596] That accident was at 7.2 psi. The pressure in the capsule that Grissom, White, and Chaffee were in was cranked up to between 16.7 and 20.2 psi.

Experts call pure oxygen put under pressure "oxygen bombs" because pure oxygen is exponentially more dangerous when put under pressure. When under pressure, combustible materials can spontaneously explode into flames because of the heat generated by rapid oxidation. The obsolete space capsule that was being tested by Grissom, White, and Chaffee was full of combustible materials. At a pressure between 16.7 and 20.2 psi any spark would cause the oxygen to explode and immediately immolate any living thing in the capsule, just as happened.

Grissom smelled a rat prior to the pressure test. The dangerous test just did not make sense to him. Ralph Rene explains:

> Wouldn't you have smelled a rat? Perhaps Grissom was a bit worried. He got Wally Schirra to ask Joe Shea, NASA's chief administrator, to go through the test with him. "Grissom still wanted Shea to be with him in the spacecraft." Shea refused because NASA couldn't patch in a fourth headset in time for the test. Is that likely? It is difficult to believe that this couldn't have been done in the 24-hour time frame available. If I had a crew of technicians who couldn't install another headset jack in that amount of time I 'd fire the whole damn crew.[597]

Ralph Rene opined the following: "I also feel that spontaneous combustion would have been much too subtle for the CIA. If it was a CIA hit they would have done it with an electric squib or incendiary device wired to a switch, programmed to be thrown toward the end of the test."[598]

Rene's opinion of the method used to murder Grissom, White, and Chaffee turned out to be correct. Gus Grissom's son, Scott Grissom, for years believed that his father was murdered. He, however, was unable to prove it until 1990, when he was granted access to the Apollo 1 space capsule that exploded in flames killing his father. Scott Grissom discovered a mechanism in the space capsule purposely placed there to create a spark and cause the oxygen rich capsule to explode into flames. Amy Shira Teitel reported for *Discovery News*:

> The younger [Scott] Grissom had his suspicions in the 1960s but wasn't able to prove foul play until the 1990s when he was granted access to the charred Apollo 1 capsule. Rooting around the instrumentation, he found a "fabricated metal plate" behind a switch on one of the instrument panels that controlled the source of the capsule's electrical power. Its placement behind that switch, he said, was clearly an act of sabotage. It ensured that when any crew member toggled that switch there would be a spark. That spark would have been enough to start the fire that killed the crew.
>
> A McDonnell-Douglas engineer, Clark MacDonald, backed Scott Grissom's story. In his own accident investigation he identified an electrical short brought on by a changeover to battery power as the reason for the fire. But NASA destroyed his report, he said.[599]

Clark MacDonald, during a later news interview, opened up about how his conscience bothered him regarding what he knew to be evidence of the murder of the three astronauts:

> "I have agonized for 31 years about revealing the truth but I didn't want to hurt NASA's image or cause trouble," MacDonald told the paper. "But I can't let one more day go by without the truth

being known."[600]

Ralph Rene explains just a part of the cover-up by NASA after the disaster.

> [Frank] Borman played dumb when he was called before Congress. In testifying under oath he said, "None of us were fully aware of the hazard that existed when you combine a pure oxygen atmosphere with the extensive distribution of combustible materials and a likely source of ignition ... and so this test ... was not classified as hazardous." And if Borman was as unaware of all the dangerous fires that erupted during NASA's own safety tests over the years, why did he later write about 20.2 psi oxygen in this manner: "That is an extremely dangerous environment, the equivalent of sitting on a live bomb, waiting for someone to light the fuse."[601]

Borman held a master's degree in engineering and taught thermodynamics at West Point, and yet he claims that neither he nor any of the eminent engineers at NASA understood the dangers of a pressurized oxygen capsule. That is pure malarkey! NASA had to silence Grissom, who was an independent thinker and a threat to blow the whistle on NASA's moon landing hoax. Grissom, White, and Chaffee were literally sitting inside an oxygen bomb rigged by NASA to explode and silence a troublesome whistle blower.

James Irwin, the Apollo 15 Command Module Pilot, by the grace of God, became a Christian and was giving his Christian testimony in Nashville, Tennessee. At that time, he met a fellow Christian, Lee Gelvani, who had almost convinced James Irwin, to confess to the moon landing hoax. Irwin suspiciously died the day before he was expected to confess to the hoax.

In the days before his death, Irwin called Bill Kaysing on the telephone. Kaysing, at that time, had written the best known book about the moon landing hoax.[602] Irwin stated during the call that he was concerned that the phone he was using was tapped. He gave Kaysing his home phone number, and asked him to call him back that coming Friday at his home in Colorado Springs, Colorado. When Kaysing called him as scheduled that Friday, Kaysing found out that Irwin had died of a heart attack.[603] James Irwin died on Thursday, August 8, 1991, the day before Kaysing was scheduled to call him. Irwin knew that his phone calls were tapped, which speaks volumes about the kind of pervasive surveillance the astronauts are under. While there is no explicit proof that Irwin was murdered, his sudden death the day before he was expected to confess to the moon landing hoax is certainly suspicious.

Between the years 1964 and 1967, a total of 10 astronauts died in freak accidents. Those deaths accounted for fully 15% of NASA's astronaut corps.[604] O'Leary would have us believe that the Apollo astronauts who were portrayed as going to the moon would have told him about a hoax, if there had been one. The astronauts are not known to be slow learners. The deaths of Grissom, White, and Chaffee, along with the other astronauts offered stark object lessons for the moon landing astronauts to keep the hoax a secret. Indeed, it is likely those object lessons served to motivate O'Leary to retract his publicly expressed doubts about the authenticity of the moon landings.

In 2004, President George W. Bush announced plans to return to the moon. In his formal space exploration address, he stated the obvious pathetic state of the current space exploration achievements since the U.S. purportedly last went to the moon. "In the past 30 years, no human being has set foot on another world or ventured farther up into space than 386 miles, roughly the distance from Washington, D.C., to Boston, Massachusetts."[605]

President Bush said that NASA would start with unmanned missions. "Beginning no later than 2008, we will send a series of robotic missions to the lunar surface to research and prepare for future human exploration."[606] Bush planned for the first manned mission to the moon in 2015.

2015 has come and gone, and there has been no NASA manned moon mission. Why not? It would seem easy enough. Just dust off the old Apollo program plans and start right up again. The rockets and landers supposedly worked perfectly on six separate moon missions 40 years ago, with much older technology. Going back to the moon, with today's technology, should be a piece of cake.

It seems that NASA cannot replicate the Apollo missions and is therefore forced to go back to the drawing board. Why? One reason is that NASA and its contractors allegedly lost the plans for the Lunar Excursion Module (LEM) and the Lunar Rover. They also claim that they lost the plans for the huge engines that powered the Apollo space craft.[607]

Don Pettit is a NASA astronaut who is a veteran of several long duration missions aboard the mythical International Space Station. In 2016, Petit was speaking at the Space for Innovation Conference at London's Science Museum, where he stated: "I'd go back to the moon in a nanosecond. The problem is that we don't have the technology to do that anymore. We used to, but we destroyed that technology. And it's painful to build it back again."[608]

When there are allegations that an event was faked, and evidence that could prove that it actually happened disappears while it is in the possession of the perpetrators of the alleged fraud, the only reasonable conclusion is that the missing evidence would prove the fraud.

Obviously, the plans were made to disappear, because those plans would reveal the impossibility of landing on the moon. NASA and their contractors would rather appear to be incompetent knuckleheads, than be indicted for criminal fraud, so they just "lost" the plans that could prove the fraud.

NASA is good at taking great studio photographs of the astronauts and producing elaborate patches to memorialize its missions for posterity. However, it can't seem to maintain the really important records of its alleged achievements. Aron Ranen discovered during production of his documentary of the Apollo moon missions that all of the audio tapes from the Apollo missions have also disappeared.[609]

In addition, the telemetry videotapes of the alleged Apollo 11 moon landings have all been erased.[610] All 45 videotapes were erased. The only tapes remaining are old second source television network recordings. The official report from NASA states:

> The 45 Apollo 11 tapes were degaussed, recertified, and reused to satisfy a NASA-wide shortage of one-inch tapes more than a decade later. NASA's M-22 recordings of the Apollo 11 moonwalk likely were gone forever.[611]

NASA would have people believe that the original videotapes of the Apollo 11 historic landing on the moon were taped over, because their need for blank videotapes overrode the importance of the moon mission tapes. NASA argued that nobody from NASA understood the historic significance of the tapes and simply taped over them. People preserve their family pictures and videos of vacations, because they are records of memorable events. The Apollo 11 moon mission was a monumental historic achievement; everyone at NASA certainly understood that fact. Why didn't NASA do what ordinary people do and buy more videotape? Erasing the moon landing mission tapes would be like using the Declaration of Independence as packing paper and arguing that once we won our independence, we didn't need that old thing anyway; we might as well get some use out of it.

Why did NASA find it necessary to destroy the telemetry videotapes? Because the heads of NASA realized that many of the tapes contain footage that would reveal the hoax. They learned their lesson in 2001, when Bart Sibrel got his hands on a NASA videotape that showed the Apollo 11 astronauts faking the pictures of Earth, by using some sort of transparency against the cabin window. Sibrel included that tape in his 2001 documentary movie, *A Funny Thing Happened on the Way to the Moon,* to the great embarrassment of NASA. NASA realized that it could not allow a repeat of that leak, and so it made certain that all remaining videotapes disappeared. The tape erase story is simply the cover story for the deliberate destruction of incriminating evidence.

One might think all is not lost; NASA still has the original video tapes of all of the other moon landing missions. However, that is not so. It seems that NASA has been busy destroying tapes. Bart Sibrel gives a rundown of the further efforts of NASA to destroy records of the mythical moon landings.

> Not only are all of the specifications, blueprints, and telemetry data of the miraculous NASA 1960's technology that allegedly went to the moon on the first attempt nowhere to be found, all of the original television transmission videotapes, estimated to weigh about two tons, have also been mysteriously "misplaced" inside the Federally Secured National Archives! Why? Because naïve director Ron Howard, whose own wise grandfather warned him that the moon landings were staged, requested that all of the original videotapes be transferred to High Definition for an I-Max documentary he was

producing to commemorate the "moon" missions. Prior to this, the highest quality version of the original footage that anyone had ever seen was deliberately fourth generation (a copy of a copy of a copy of a copy), in order to disguise the falsification detail of the fake "moon" landscapes.

When this moon landing believing prominent film director asked for the original NASA videotapes of all of the "moon walks" to be transferred to High Definition in order to be projected onto a screen that was nearly 120 feet by 100 feet for the first time in history, for all the world to see at a resolution that was at least four times greater in detail than had ever before been witnessed, within days of his request, all of the original videotapes vanished from the Federal Archives!!! But of course, we still went to the moon, right?

Proof of this is that when Howard's documentary commemorating the "moon" landings was finally produced without this essential material, he resorted to filling about 95 percent of the gap with "reenactments" of the "moon walks," which he likewise filmed in a television studio. In fact, there was so little actual footage available of the alleged greatest event of the 20th century, that this multimillion dollar director had to resort to renting VHS tapes of other filmmakers' previous work at Blockbuster Video in order to have at least some of the vanishing NASA footage! He was so ashamed of the low quality of the images he was forced to use because of the scarcity of them, that he reduced their size within the nearly 120 foot by 100 foot screen to only about 10

> percent of the available space.
>
> In case you were wondering "what if", someday, the mysteriously vanished original videotapes were to one day be found? . . . Oops! . . . Did I forget to tell you? . . . The only machine on the entire planet that could ever play these one of a kind format videotapes was deliberately disassembled and destroyed by NASA, right after the originals got "misplaced", making that scenario absolutely impossible![612]

Incidentally, Ron Howard engaged in a limited hangout of the truth of the Apollo 11 moon landing hoax during a video clip from a season four episode of "Arrested Development" entitled "The B. Team," which aired in 2013 on Netflix. Howard was the executive producer of that show. In one scene, Ron Howard plays himself and has a conversation with the character played by Jason Bateman.

> Howard: Hey, let's go inside the LEM, do you want to?
>
> [*Both Howard and Batement walk over to where the LEM is situated in the studio office area.*]
>
> Batemen: Wow is this the one that landed on the moon?
>
> Howard: On a sound stage.
>
> Bateman: Oh from Apollo 13?
>
> Howard: No from 1969. I'll tell you about it inside the LEM. It's sound proof in there. And it's a national secret. So, NASA did go to the moon

> in '71. That one was real. But in '69, they weren't ready. So they faked the whole thing on the sound stage of Gentle Ben. Me and my brother, we hid up in rafters, we seen the whole thing.[613]

The monologue by Howard takes place inside a mock-up of the LEM that is set up in his studio offices. The whole purpose of the vignette is to assure the listeners that "NASA did go to the moon in '71." Ron Howard's monologue is an example of exposing part of a conspiracy in order to spin the facts to persuade people to believe the big lie that the earth is a sphere floating in the vacuum of space. Before Howard admits that NASA faked the moon landings in 1969, he reassures Batemen (and the listening audience) that they did in fact go to the moon in 1971. Howard attributes an innocent reason for faking the moon landing on a sound stage; NASA wasn't ready. That acknowledgment by Howard admits facts that are becoming increasingly clear to the public anyway. However, in order to keep the lie of a spherical earth floating around the sun in the vacuum of space, Howard dampens the revelation by reassuring the listening audience that NASA did in fact go to the moon in 1971. He prefaces his monologue with the statement "NASA did go to the moon in '71. That one was real." That is the primary message with which the audience is intended to be left.

The revelation of the fake moon landing by Howard is done in the context of a situation comedy TV show, so it can be easily dismissed. However, be mindful that Howard was the executive producer of the show and therefore had complete control over the script. Making the revelation during a TV show is the perfect way in which to spin the evidence of the moon hoax. Howard is able to reinforce the myth of the earth floating in the vacuum of space, and his statement that the Apollo 11 moon landing was faked was not an official pronouncement by an insider but rather part of a cameo appearance in a situation comedy TV show. In order to make the moon landing hoax admission seem unbelievable, Howard adds the incredible fact that he and his brother hid up in the studio rafters and saw its filming. It is the perfect set up for a limited hangout, where the truth is mixed with lies. His admission in the context of a TV show that the Apollo 11 moon landing was hoaxed was done to establish that the other moon landings that followed were real. That context allows Howard, who has inside information, to make what is essentially a non-admission/admission. Thus the myth of the vacuum of outer space is reinforced.

29 NASA and Freemasonry

Thomas Africa reveals that Nicolaus Copernicus (1473-1543 A.D.) was not a revolutionary discoverer of a new heliocentric astronomy; he was rather a restorer of the heliocentric type of system espoused by Pythagoras of Samos (570-495 B.C.).[614] Some state that Pythagoras' system was not purely heliocentric, but that it was a system where the planets, including the sun, orbited a central, invisible fire. Nonetheless, Pythagoras' system was the first system that called for the planets to travel in a circular orbit, and so he has been recognized by the early scientists as the true founder of the heliocentric system. In fact, Johannes Kepler (1571-1630 A.D.) called Pythagoras the "grandfather of all Copernicans."[615] Copernicus, himself, insisted that his system was not an innovation, but was rather a revival of the lost doctrine of Pythagoras.[616] Galileo Galilei (1564-1642 A.D.) viewed the papal edict of 1616 as a suppression of the "Pythagorean opinion of the mobility of the earth."[617]

Copernicus also "borrowed" from the theories of Aristarchus of Samos (310-230 B.C.) that the earth orbited the sun. Copernicus had no problem acknowledging the contributions of Pythagoras, but for some reason, he decided to conceal his knowledge of Aristarchus's writings. Thomas Africa reveals that "Copernicus knew of Aristarchus' heliocentrism but consistently concealed this knowledge, and finally deleted his one passing reference to it, from either vanity, 'Pythagorean' scruples, or both."[618]

Pythagoras was the first person to have presented the idea of the circular orbit of spherical planets around a central fire. He is purported to have added a counter-earth to arrive at 10 orbiting planets (including the sun).[619] Jose Wudka alleges that the added counter-earth was to explain eclipses and because the number 10 was viewed as sacred by heathen philosophers.[620] Not coincidently, the 10 orbiting spheres match the 10 spherical *sefirot* of the Jewish Kabbalistic god, *Ein Sof.*

Master Mason Dr. James Anderson, founder of the London Masonic Lodge, stated in his book, *Defence of Masonry*, that Freemasonry descended from Pythagoras.[621] Master Mason William Hutchinson stated in his book, *Spirit of Masonry*, that ancient Masonic records indicate that the foundation of Freemasonry is in Pythagorean principles.[622] Another Master Mason, William Preston, in his *Illustrations of Masonry,* states that Pythagoras was initiated into the deep mysterious Masonic principles, which he then spread to the countries in which he traveled.[623] Albert Mackey, in the Encyclopedia of Freemasonry reveals the following details about Pythagoras and his Masonic connections:

> In his return to Europe, he [Pythagoras] established his celebrated school at Crotona, a Dorian Colony in the south of Italy, about 529 B.C., much resembling that subsequently adopted by the Freemasons. ... Before admission to the privileges of this school, the previous life and character of the candidate were rigidly scrutinized, and in the preparatory initiation secrecy was enjoined by an oath, and he was made to submit to the severest trials of his

> fortitude and self-command. ... The mode of living in the school of Crotona was like that of the modern Communists.[624]

The *Jewish Encyclopedia* labels some of the Kabbalistic philosophies as Pythagorean.[625] Both the Pythagorean occultism and the Kabbalah flowed from Babylonian mysticism. The *Jewish Encyclopedia* asserts that Gnosticism was Jewish in character and is of "Chaldean [i.e., Babylonian] origin."[626] That indicates that the Gnosticism finds its origins from Jews in Babylon. The Kabbalah is the written memorialization of the mysticism that was adopted by the Jews during their Babylonian captivity.

The word "Pythagorean" is an adjective and seems to be used in the *Jewish Encyclopedia* to identify the kind of doctrines in the Kabbalah. They are of the same nature as Pythagorean doctrines. It certainly cannot mean that the Kabbalah flowed from Pythagoras, because the Jews were first brought into captivity in Babylon, in or about 597 B.C., which was approximately 27 years before Pythagoras was born. The Jews were released from their Babylonian captivity in or about 538 B.C.

Pythagoras was a Greek. He did not travel to Babylon until in or about 525 B.C.[627] Pythagoras was reportedly held captive there for 5 years. Iamblichus (born in or about 250 A.D.), who was a Syrian philosopher, writes that Pythagoras "was transported by the followers of Cambyses as a prisoner of war. Whilst he was there he gladly associated with the Magoi and was instructed in their sacred rites and learnt about a very mystical worship of the gods. He also reached the acme of perfection in arithmetic and music and the other mathematical sciences taught by the Babylonians."[628]

Pythagoras could not have been the source of any of the doctrines in the Kabbalah, because by the time Pythagoras came on the scene, the Jews had already been held captive in Babylon, been introduced to the Babylonian occult mysteries, and been released from their captivity. No doubt, there were many Jewish mystics still in Babylon by the time that Pythagoras arrived. It is likely that at that time Pythagoras was initiated into the mysteries of what we know today as the Kabbalah.

The *Jewish Encyclopedia*, calling some of the Kabbalistic philosophies Pythagorean, seems to be more a homage to Pythagoras that is being used to describe the nature of particular doctrines found in the Kabbalah. Indeed, when called upon to identify the source of the mysticism in the Kabbalah, the *Jewish Encyclopedia* stated unequivocally that Gnosticism is of Jewish character and is of "Chaldean [i.e., Babylonian] origin."[629]

Certainly, there could have been some synergism in the interactions between the Kabbalistic Jews and Pythagoras, but that does not erase the truth that in the end the philosophies of both the Jewish mystics and Pythagoras are rooted in occult Babylonianism. The fact that the Pythagorean theorem, for which Pythagoras is famous, was known by the Babylonians a thousand years before Pythagoras testifies to the Babylonian origins of his philosophy.[630] S. Pancoast, who was a physician to the infamous occult theosophist H.P. Blavatsky, states that Pythagoras was a Kabbalist of the highest order.[631] He further states that the symbols of Masonry are Kabbalistic, and were known to Pythagoras.[632] Pancoast reveals that Pythagoras' initiation into the secrets of the Kabbalah led Pythagoras to the heliocentric philosophy.

> Pythagoras held that the Sun is the center of the solar system around which all the planets revolve; that the stars are Suns like ours, each the center of a system; that the earth revolves yearly around the Sun and daily on its axis; that the planets are inhabited, and that they and the earth are ever revolving in regular order.[633]

Johannes Reuchlin (1455-1522 A.D.) was a German humanist and political counselor to the Chancellor of Germany.[634] He was a classics scholar and an expert in the ancient languages and traditions (Latin, Greek, and Hebrew). Reuchlin was affiliated with the heads of the Platonic Academia (della Mirandola and others).[635] Reuchlin confirmed that Pythagoras obtained his philosophy through the Jewish Kabbalah:

> Pythagoras, who is the father of philosophy, did nevertheless not receive those teachings from the Greeks, but rather he received them from the Jews. Therefore he must be called 'a Kabbalist,' ... and he himself was the first to convert the name 'Kabbalah,' unknown to the Greeks, in the Greek name philosophy. Pythagoras' philosophy emanated from the infinite sea of the Kabbalah.[636]

Freemasonry can be traced to Pythagoras and is a religion that is founded upon the Jewish Kabbalah. Albert Pike states in *Morals and Dogma* that "Masonry is a search for Light. That leads us directly back, as you see, to the Kabbalah."[637] Albert Mackey, in his authoritative Encyclopedia of Freemasonry, confirms Albert Pike's averment. Mackey states that the Kabbalah, being "[t]he mystical philosophy or theosophy of the Jews, ... is intimately connected with the symbolic science of Freemasonry."[638]

Upon what authority did Albert Pike rely for writing his authoritative *Morals and Dogma of the Ancient and Accepted Scottish Rite*? Martin L. Wagner conducted an objective and thorough study of Freemasonry and wrote a book about his findings titled *Freemasonry: An Interpretation*.[639] Wagner states that "Albert Pike drew largely from the writings of Eliphas Levi, the Abbe Constant, a great Kabbalist, and whom Buck considers as knowing more of the occult science than any one since the days of the old initiates, for illuminating and illustrating Freemasonry."[640] The Kabbalah is unadulterated witchcraft. Magic and occult mysticism run throughout the Kabbalah.[641] The Kabbalah contains a great deal of black magic and sorcery, and invoking the powers of devils.[642]

One of the key points revealed in the Protocols of the Learned Elders of Zion is the secret use of ostensibly Gentile nations and institutions in order to further Jewish Zionist aims, while hiding the Jewish influence over those institutions. In the Protocols, the Learned Elders of Zion state that they have used Masonry as a cover to hide their involvement in the plan for a "new world order." Paragraph two of Protocol 4 states:

> Who and what is in a position to overthrow an invisible force? And this is precisely what our force is. Gentile masonry blindly serves as a screen for us and our objectives, but the plan of action of our force, even its very abiding place, remains for the whole people an unknown mystery.[643]

The Gentile facade of Freemasonry offers the Zionist Jews the perfect cover. We can see the same hidden control by Jews over the "Christian" Zionist movement. Freemasonry is based upon Judaism.[644] It is a Gentile front for Jewish mysticism, whose history, grades, and official appointments, are rooted in Jewish theosophy.[645]

The Zionists promote and control masonic lodges. They use those lodges as indispensable secret intelligence agencies and organs of influence. Paragraphs four and five of Protocol 15 states:

> [W]e shall create and multiply free masonic lodges in all the countries of the world, absorb into them all who may become or who are prominent in public activity, for these lodges we shall find our principal intelligence office and

> means of influence. All these lodges we shall bring under one central administration, known to us alone and to all others absolutely unknown, which will be composed of our learned elders.
>
> * * *
>
> It is natural that we and no other should lead masonic activities, for we know whither we are leading, we know the final goal of every form of activity whereas the goyim have knowledge of nothing.[646]

Zionist Jews use the lodges of Freemasonry as recruiting grounds for their Gentile front-men. The Gentile nature of freemasonry is only a cover; freemasonry is wholly based upon the Jewish Cabala. Using the Gentile front of freemasonry to further Zionist ends is explained in paragraphs four and seven of Protocol 11.

> The goyim are a flock of sheep, and we are their wolves. And you know what happens when the wolves get hold of the flock? ... For what purpose then have we invented this whole policy and insinuated it into the minds of the goy without giving them any chance to examine its underlying meaning? For what, indeed, if not in order to obtain in a roundabout way what is for our scattered tribe unattainable by the direct road? It is this which has served as the basis for our organization of secret masonry which is not known to, and aims which are not even so much as suspected by, these "goy" cattle, attracted by us into the "show" army of masonic lodges in order to throw dust in the eyes of their fellows.[647]

The statements in the Protocols that Freemasonry is rooted in Judaism is confirmed by Wagner, in his study of Freemasonry. Wagner quotes Masonic authorities that reveal that "Masonry in its purity, derived as it is from the old Hebrew Kabbala as a part of the great universal wisdom religion of remotest antiquity."[648] Wagner concludes:

> A candid investigation convinces us that Freemasonry is indebted in a very large measure to the Kabbalah for its philosophical ideas, its methods of interpreting the scriptures, its doctrines of emanations, its art speech, its cosmogonical views, and its veils and glyphs. In a certain sense it is a continuation of the Kabbalah under a different name and guise.[649]

Further confirmation of the Judaic foundations of Freemasonry comes from the authoritative Rabbi Isaac Wise. Wise confirms that the Gentile nature of Freemasonry is only a cover: "Freemasonry is a Jewish establishment, whose history, grades, official appointments, passwords, and explanations are Jewish from beginning to end."[650]

The October 28, 1927 Jewish Tribune of New York stated: "Masonry is based on Judaism. Eliminate the teachings of Judaism from the Masonic Ritual and what is left?"[651] Michael Hoffman concluded: "It is from these [Cabalistic and Talmudic] recondite doctrines of Judaism that the Freemasons and other occult workers of iniquity derive their beliefs."[652] Henry Makow describes Freemasonry as "Judaism for Gentiles."[653] Makow states that it is "a way for the Cabalistic Jewish elite to enlist Gentiles in their conspiracy."[654]

What are the religious doctrines flowing from the Kabbalah that form the foundation of Freemasonry? It is the worship of Lucifer. Albert Pike, the theological pontiff of

Freemasonry, explains:

> That which we must say to a crowd is—We worship a God, but it is the God that one adores without superstition. To you, Sovereign Grand Inspectors General, we say this, that you may repeat it to the Brethren of the 32nd, 31st, and 30th degrees—The Masonic Religion should be, by all of us initates of the high degrees, maintained in the purity of the Luciferian Doctrine. If Lucifer were not God, would Adonay whose deeds prove his cruelty, perfidy and hatred of man, barbarism and repulsion for science, would Adonay and his priests, calumniate him? Yes, Lucifer is God, and unfortunately Adonay is also God. For the eternal law is that there is no light without shade, no beauty without ugliness, no white without black, for the absolute can only exist as two gods: darkness being necessary to the statue, and the brake to the locomotive. Thus, the doctrine of Satanism is a heresy; and the true and pure philosophical religion is the belief in Lucifer, the equal of Adonay; but Lucifer, God of Light and God of Good, is struggling for humanity against Adonay, the God of Darkness and Evil.[655]

Adonay is the Hebrew word used in the Old Testament that is the name for God and is translated into English in the Bible as "Lord." Pike blasphemously calls God "the God of Darkness and Evil." Pike calls Lucifer, the "God of Good." Pike admits that Lucifer is the Masonic god of light. "And no marvel; for Satan himself is transformed into an angel of light." (2 Corinthians 11:14) In his authoritative treatise, that is to this day the doctrinal Bible of Masonry, *Morals and Dogma*, Pike pays homage to the god of Freemasonry: "Lucifer, the Light-Bearer! Strange and mysterious name to give to the Spirit of Darkness! Lucifer, the Son of the Morning! Is it he who bears the light, and with its splendors intolerable blinds feeble, sensual or selfish Souls? Doubt it not!"[656] Lucifer's name in fact means "light bearer." The Masonic initiation ceremonies finds the candidate repeatedly seeking more light. If the candidate reaches the highest degree of Freemasonry, he will be informed that the light he seeks is found in the light bearer, Lucifer, who is the god of Freemasonry.

Manly P. Hall, 33° Freemason and highly respected Masonic authority, explains that "[w]hen the Mason learns that the key to the warrior on the block is the proper application of the dynamo of living power, he has learned the mystery of his Craft. The seething energies of Lucifer are in his hands and before he may step onward and upward, he must prove his ability to properly apply energy."[657]

The god of Freemasonry being Lucifer is understandable, since Freemasonry is based upon Judaism, and the god of Judaism is Satan. Infamous Satanist Anton LaVey was the founder of the Church of Satan and the author of the Satanic Bible. LaVey's 1992 authorized biography was written by LaVey's intimate associate and High Priestess of the Church of Satan, Blanche Barton. The biography was written by a Satanist, about a Satanist, for Satanists. LaVey reveals in his biography that he was sympathetic to communism. It is, therefore, no surprise to find out that LaVey was also an ardent Zionist. In fact, he was intimately involved with militant Israeli groups in running guns to Israel. LaVey states, with satisfaction, that the vindication of his communist and Zionist involvement came when Assaf Dayan, son of Moshe Dayan, told him that his Satanic Bible contained precisely the religion of Judaism as practiced in Israel.

> I felt a culmination of that whole episode in my life many years later, shortly after The Satanic Bible came out. I met with Assaf Dayan, actor

> son of Israel's legendary Defense Minister Moshe Dayan, and he was glowing about the book, agreed with everything in it. He said it was exactly the philosophy they practiced, were forced to practice, in modern Israel. He extended an invitation for me to stay at his family home in Tel Aviv whenever I wished.[658]

Satanism is essentially the religion of Judaism. It is no surprise, therefore, to learn that Anton LaVey, the founder of the Church of Satan, was a Jew.[659] LaVey changed his name to sound more Gentile, but he was born a Jew named Howard Stanton Levey.[660] Being a Jew also explains why he was such an ardent Zionist. The Times of Israel reveals to its mostly Jewish readership that "[a]s the 'sigil' for his movement, LaVey adopted an inverted pentacle surrounded by the Hebrew letters for Leviathan, a sea monster featured in the Old Testament."[661]

It is common for Jews to change their names in order to conceal their Jewish roots. Such behavior earns them the label of crypto-Jews. For example, Malcolm Jarry, the founder of The Satanic Temple, freely admits to the Jewish audience of The Times of Israel that he is a "secular Jew," and that his Gentile name is actually a pseudonym. Of course, it not surprising to read that the self-proclaimed Satanist (being a Jew) said: "I am an unwavering supporter of Israel."[662]

The reporter for the Times of Israel explains that "[i]n addition to Jarry's belief that Judaism and Satanism can co-exist, there are parallels with how Judaism and Satanism have been branded by their detractors, he said."[663] The parallel between Judaism and Satanism is a little inside baseball that Jarry feels comfortable revealing to the Jewish readership of The Times of Israel. But the fact that Judaism and Satanism have the same core beliefs is not something that the Jews want to be known among the Gentile population in general.

What are the parallels alleged by Jarry between the way Satanism and Judaism has been branded by their detractors? The Times of Israel quotes Jarry:

> "The false accusations that have been thrown at Jews historically are similar to what some people say about Satanism," said Jarry, mentioning accusations of blood libel and — more recently — fabricated allegations that Israel perpetrates genocide against Palestinian children.[664]

I am sorry, but the blood sacrifice of children and genocide against Palestinian children perpetrated by Jews are not fabrications. Both are very real; and the evidence for both is revealed in this author's book, *Bloody Zion*. Blood sacrifice is part and parcel of both Satanism and Judaism. That fact is concealed by Judaism's cloak of religion. Judaism is simply Satanism all dressed up in religious garb to look holy. Matthew 23:27. Those trappings of religion do not change its evil core. Matthew 23:33. Jesus described the Jewish religious leaders as serpents who are damned to hell. Matthew 23:33. Judaism is the proverbial wolf in sheep's clothing. *See* Matthew 7:15.

The evidence of the Satanic roots of Judaism is proof that Jesus was not using hyperbole when he told the Jews: "Ye are of your father the devil, and the lusts of your father ye will do. He was a murderer from the beginning, and abode not in the truth, because there is no truth in him. When he speaketh a lie, he speaketh of his own: for he is a liar, and the father of it." (John 8:44)

Heliocentrism is a fundamental tenet of the Kabbalah. Consequently, heliocentrism is central to Freemasonry. Many Masonic lodges throughout the world are named in honor of Copernicus.[665] The following praise of Copernicus is from the 1843 Freemasons' Quarterly Review.

> Copernicus, and his successors in the study of the starry firmament, have supplanted the art of astrology by proving that all the movements of the heavenly bodies tend to promote the honour and glory of the Great Architect of the Universe.[666]

The link between heliocentrism and Freemasonry explains the close affiliation between Freemasonry and NASA. For example, James Edwin Webb, who was the NASA administrator from 1961-68, was a Freemason. In the November 1969 edition of the Masonic Magazine, *The New Age*, there appeared an article written by 33° Freemason Kenneth S. Kleinknecht, who was the Manager of the Apollo Program Command and Service Modules; the Deputy Manager of the Gemini Program; and the Manager of Project Mercury. Kenneth S. Kleinknecht, by the way, is the brother of C. Fred Kleinknecht, 33°, Sovereign Grand Commander, The Supreme Council (Mother Council of the World), Southern Jurisdiction, USA, Washington. In *The New Age* article, Kenneth Kleinknecht stated:

> Note how many of the astronauts themselves are Brother Masons: Edwin E. Aldrin, Jr.; L. Gordon Cooper, Jr.; Donn F. Eisele; Walter M. Schirra; Thomas P. Stafford; Edgar D. Mitchell, and Paul J. Weitz. Before his tragic death in a flash fire at Cape Kennedy on January 27, 1967, Virgil I. "Gus" Grissom was a Mason, too. Astronaut Gordon Cooper, during his epochal Gemini V spaceflight in August of 1965, carried with him an official Thirty-third Degree Jewel and a Scottish Rite flag. Via the lunar plaque, the Masonic ensignia and flag, and the Masonic astronauts themselves – Masonry already is in the space age. Can we doubt Freemasonry and its spiritual relevance to the modern era when even

> its material representatives have today made historic inroads into the infinite expanses of outer space?[667]

The Grand Lodge of Texas, A.F. & A.M., has posted the following explanation on the internet, regarding its chartering of a Masonic Lodge on the moon, which is titled Tranquility Lodge No. 2000.

> On July 20, 1969, two American Astronauts landed on the moon of the planet Earth, in an area known as Mare Tranquilitatis, or "Sea of Tranquility". One of those brave men was Brother Edwin Eugene (Buzz) Aldrin, Jr., a member of Clear Lake Lodge No. 1417, AF&AM, Seabrook, Texas. Brother Aldrin carried with him SPECIAL DEPUTATION of then Grand Master J. Guy Smith, constituting and appointing Brother Aldrin as Special Deputy of the Grand Master, granting unto him full power in the premises to represent the Grand Master as such and authorize him to claim Masonic Territorial Jurisdiction for The Most Worshipful Grand Lodge of Texas, Ancient Free and Accepted Masons, on The Moon, and directed that he make due return of his acts. Brother Aldrin certified that the SPECIAL DEPUTATION was carried by him to the Moon on July 20, 1969.[668] (emphasis in original)

Figure 162: Freemason Buzz Aldrin (right) with Luther A. Smith, the Masonic Sovereign Grand Commander, holding the Masonic flag Aldrin took with him when he allegedly landed on the moon.

Buzz Aldrin was interviewed by Alex Jones in 2009. During the interview, there was the following exchange between Alex Jones and Buzz Aldrin:

> Alex Jones: "Mr. Aldrin, I always wanted to ask you this. We saw the photos of the little Masonic flag to the moon and some of the names of the missions and the numerology, is there anything to that? Or what is the Masonic influence? We know there's Masonic influence in the founding of the country; what is the Masonic influence on NASA?"

> Aldrin: "As far as I can tell, zero. There were some Masonic brothers of mine in Texas that wanted me to take some kind of a Masonic emblem to the moon, in some gesture of – I don't know what it would be a gesture of – but I told them that it was not within my authority to do such a thing."[669]

Aldrin is on record with a September 19, 1969, letter he sent three days after meeting with the heads of the Scottish Rite of Freemasonry in the House of the Temple in Washington D.C., acknowledging that he presented the very flag he denied carrying to the moon during the 2009 Alex Jones interview.

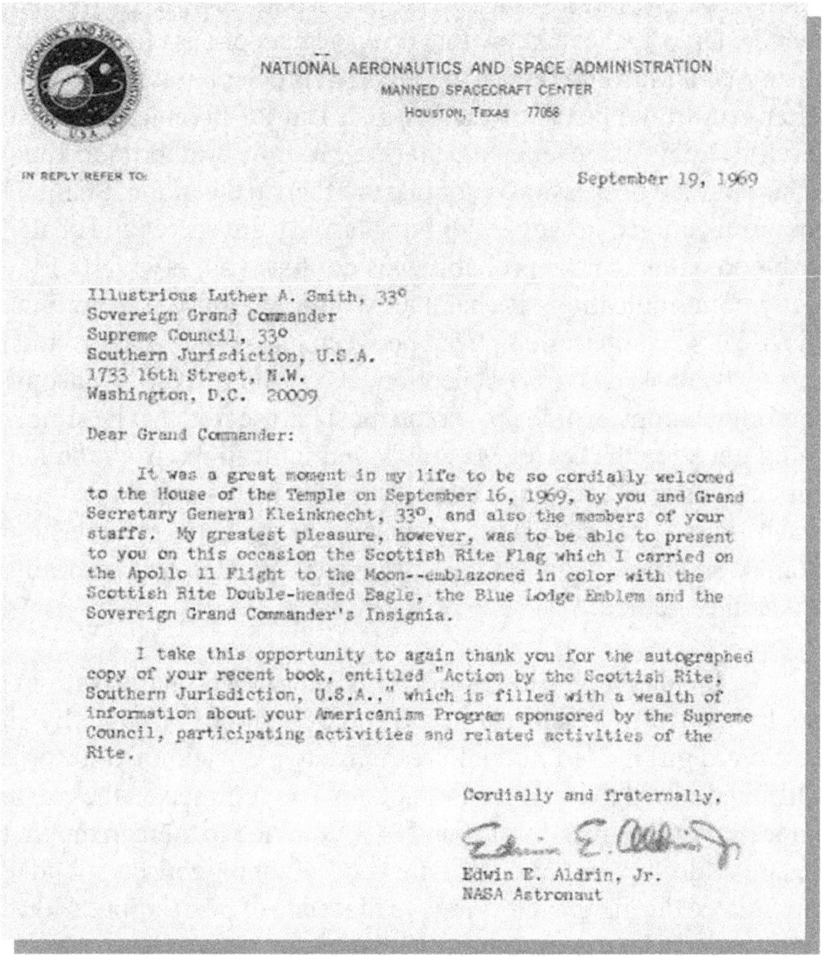

NATIONAL AERONAUTICS AND SPACE ADMINISTRATION
MANNED SPACECRAFT CENTER
HOUSTON, TEXAS 77058

IN REPLY REFER TO:

September 19, 1969

Illustrious Luther A. Smith, 33°
Sovereign Grand Commander
Supreme Council, 33°
Southern Jurisdiction, U.S.A.
1733 16th Street, N.W.
Washington, D.C. 20009

Dear Grand Commander:

It was a great moment in my life to be so cordially welcomed to the House of the Temple on September 16, 1969, by you and Grand Secretary General Kleinknecht, 33°, and also the members of your staffs. My greatest pleasure, however, was to be able to present to you on this occasion the Scottish Rite Flag which I carried on the Apollo 11 Flight to the Moon--emblazoned in color with the Scottish Rite Double-headed Eagle, the Blue Lodge Emblem and the Sovereign Grand Commander's Insignia.

I take this opportunity to again thank you for the autographed copy of your recent book, entitled "Action by the Scottish Rite, Southern Jurisdiction, U.S.A.," which is filled with a wealth of information about your Americanism Program sponsored by the Supreme Council, participating activities and related activities of the Rite.

Cordially and fraternally,

Edwin E. Aldrin, Jr.
NASA Astronaut

Furthermore, there was an article in the December 1969 issue of the Masonic magazine, New Age (the official organ of the Scottish Rite Southern Jurisdiction), which included a picture of Aldrin presenting to the Scottish Rite headquarters in Washington, the "Masonic emblem" which he allegedly carried with him to the moon and back.

Why would Aldrin lie about the Masonic connection to NASA? Because, it would reveal the real power and influence behind NASA. Pulling on that thread would expose the hidden agenda. Didn't Aldrin know that he was on record as claiming that he carried a Masonic flag to the moon in the September 19, 1969, letter sent to the heads of the Scottish Rite of Freemasonry? He certainly knew of the letter and that the Masonic brotherhood knew of his claimed possession of a Masonic flag on the moon. So many communications, however, within Masonry are secret and sealed by blood oaths that he probably was confused as to which of the many communications he has had within Masonry were public. His life is so compartmentalized between his public facade, which is mostly based upon deception, and his private Masonic communications, which are for the most part secret, that he simply could not keep the secret Masonic communications from the few public Masonic communications straight in his mind. Without much time to reflect on his answer to the unexpected question from Alex Jones, Aldrin simply resorted to his standard operating procedure; he lied.

Incidentally, Buzz Aldrin was interviewed at the National Book Festival in Washington, DC, on September 5, 2015, by an 8-year old girl named Zoey. In response to a question from Zoey, Aldrin admitted that the reason nobody has returned to the moon since the Apollo missions is that NASA did not go there in the first place. Aldrin, who has a history of alcohol abuse, sounded intoxicated during the interview. In that state of inebriation, Aldrin inadvertently let the cat out of the bag.

> Zoey: "Why has nobody been to the moon in such a long time?"
>
> Buzz Aldrin: "That's not an 8-year old's question. That's my question. I want to know. But I think I know. Because we didn't go there. And that's the way it happened. And if it didn't happen, it's nice to know why it didn't happen. So in the future, if we want to keep doing something, we need to

know why something stopped in the past that we wanted to keep it going."[670]

There has been (and probably continues to be) a controlling faction making up a statistically unlikely plurality of Masonic astronauts and administrators at NASA.[671] In fact, Kleinknecht's list of Masonic astronauts in his *New Age* article is by no means complete. For example John Glenn Jr. (Mercury 6), James Irwin (Apollo 15), and Fred Haise (Apollo 13), all of whom were Freemasons, are missing from Keinknecht's honor roll of astronaut Freemasons.

Freemasons are very proud of their NASA connection. Aldrin's denial of any Masonic connection to NASA suggests that knowledge of the connection is to be kept within the Masonic brotherhood. Below is a Masonic medallion struck in commemoration of the 10th anniversary of the Apollo 11 alleged moon landings.

The following description of the medallion appears on the Phoenix Freemason Museum website:

This 1979 medallion was struck to commemorate

> the 10th Anniversary of our flags on the moon. Many people were never aware that Astronaut and Brother Neil Armstrong carried aboard on his Apollo flight to the moon 2 (two) flags. One was the American flag, the other was a flag designed by the Southern Jurisdiction of the Scottish Rite depicting the double-headed eagle emblem. This flag now resides in the museum collection of the Scottish Rite Museum at the House of the Temple. It is a spectacular three-dimensional medallion and measures 1 3/4 inches in diameter. It was made by the Medallic Art Co. of Dansbury, CT.[672]

Richard Hoagland, who was a former Science Advisor to CBS News during the Apollo program, from 1968–1971, alleges in his book, *Dark Mission: The Secret History of NASA*, that NASA is controlled by Freemasons, and from NASA's beginnings, it has had an occult underside that has been carefully concealed from the public.[673] While Hoagland's statement is true, Bill Kaysing has good reason to believe that Hoagland is a shill, working for NASA, who is trying to divert attention away from the fact that NASA never went to the moon.[674] From 1956 to 1963, Kaysing was the head of technical publications for the entire propulsion laboratory at Rocketdyne, which was a research facility for the development of large liquid propellent rocket engines. Rocketdyne was a division of North American Aviation and later of Rockwell International, which built the Saturn V rockets used in the NASA Apollo Missions. While at Rocketdyne, Kaysing had top secret clearance, which gave him access to documents pertaining to the Mercury, Gemini, and Apollo programs. Kaysing became convinced by what he saw at Rocketdyne and subsequent research that the Apollo moon landings were a complete hoax. He set forth his evidence in a book, first published in 1976, titled: *We Never Went to the Moon: America's Thirty Billion Dollar Swindle*.

There seems to be merit to Kaysing's allegation about Hoagland. Hoagland has made an effort to spin the evidence of the clearly faked NASA photos and videos, not as proof that the moon landings never took place, but rather as proof that NASA is concealing the existence of space aliens. In the intelligence community, that is called a limited hangout, where a part of a conspiracy is seemingly exposed by a shill, in order to spin the people in a direction away from the truth and provide some plausible, but less nefarious, explanation for the government deception.

Hoagland avers that only some of the moon photos and videos are faked. He advocates a position that NASA went to the moon, and NASA doctored only some of the photographs to hide evidence of intelligent life. In a radio interview, Hoagland stated:

> I'll give you my bottom line. I think that there is an enormous Apollo conspiracy, but I think we've been sold the wrong conspiracy, to keep people like you, bright guys who are asking good questions, looking in the wrong direction, which these people are task masters at doing. The real conspiracy is not did we go to the moon, but what did we find on the moon that they don't want you to know. .... I have found areas where NASA faked the imagery, I believe to hide really cool stuff."[675]

Hoagland uses hyperbole and inaccurate information. One way the intelligence community discredits opposition to its programs is to expose inaccuracies that have been sown by its own shills. Care should be taken to verify the facts. That is why this author has provided endnotes establishing the authority for virtually every fact in this book.

Hoagland presents some evidence that is true (the occult

Masonic influence in NASA and the fake photos and videos) but then spins their significance away from hoaxed moon landings and toward space aliens. That serves to discredit the information about the occult practices and the fake photos and thus causes many people to dismiss any thought that the moon landings were fake. For other people, Hoagland's theory of NASA fabricating videos and photos in order to conceal life on other planets gives some plausible, less nefarious, explanation for the NASA deception and serves to steer those people away from the fact that NASA did not go to the moon. The purpose for NASA's existence is to conceal the flat earth in order to condition people to believe the Satanic lie of a spinning orbiting earth where man is an insignificant part of an infinite, godless universe. Hoagland furthers that end.

Most who have exposed the fraud of NASA seem to give little thought to motive behind faking the moon landings. Some ascribe the motive to distraction from the Vietnam War, others to cold war prestige, but the motive most often ascribed to NASA is money. Certainly, NASA swindled multi-billions of dollars from the U.S. taxpayer. The money swindled was certainly profitable for those behind the moon landing hoax conspiracy. The real objective was not to gain a short term profit, but to grow the love of money in the hearts of men as a way to control and enslave the world. "The love of money is the root of all evil." 1 Timothy 6:10 The love for money must be made to grow, and that is the motive behind the moon landing hoax. In order for the root of evil to grow deep into the soil, that soil must be tilled by the lie that there is no God. In order to grow the love for money in men's hearts, men must be convinced that there are no eternal consequences for lying, cheating, and stealing to get money. That requires that men believe that there is no God who can mete out any punishment for sin. That in turn necessitates that man be kept ignorant that he is made by God, in God's image, on the flat earth, which is at the center of his creation. The moon mission serves to beguile men into believing in a godless, endless universe, where man is on an insignificant spherical planet careening through space.

The key to understanding the moon landing hoax is to realize that it is not scientific deception, it is spiritual deception. Symbolism is important in witchcraft. The NASA logo includes the forked tongue of a serpent to symbolize that it is under the control of the great serpent, Satan. "And the great dragon was cast out, that old serpent, called the Devil, and Satan, which deceiveth the whole world: he was cast out into the earth, and his angels were cast out with him." Revelation 12:9.

It is significant that the NASA moon missions were named Apollo. Apollo is the Greek sun god. Apollo is often depicted riding a horse drawn chariot with the sun shining behind him. Notice in the Apollo XIII emblem, Apollo himself is not seen. The depiction seems to be of the horses coming from earth; however, there is no chariot. In view of the missing chariot, the horses could be interpreted as drawing the earth behind it, which suggests that the earth itself is the chariot of Apollo. Satan is described in the Bible as "the god of this world" (2 Corinthians 4:3-4), who blinds the minds of the lost to the light of the gospel of Jesus Christ.

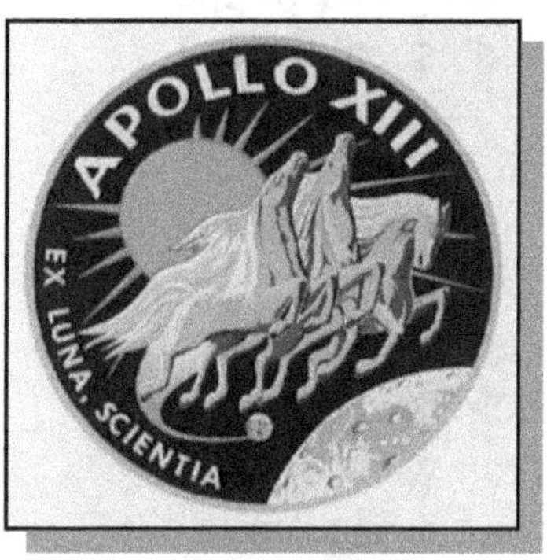

Figure 166: Official NASA Emblem for the Apollo XIII Mission

Apollo is the same Apollyon referenced in Revelations as the angel of the bottomless pit. "And they had a king over them, which is the angel of the bottomless pit, whose name in the Hebrew tongue is Abaddon, but in the Greek tongue hath his name Apollyon." (Revelation 9:11) The Edinburgh Encyclopedia states:

> On this passage, Hensius makes the following observation: There can be no doubt that the Pythian Apollo is the same as the Ob and Abaddon of the Hebrews which the Greeks translated literally as Apollyon.[676]

In Wakeman Ryno's *Amen: The God of the Amonians Or a Key to the Mansions in Heaven*, he states that the reference to Apollyon in Revelation 9:11 is a reference to Apollo, who is Satan. "Satan, Belial, Lucifer, Abaddon, and Apollyon are all one and the same — the Sun (Apollo) in the Sign of the Scorpion, the king of the bottomless pit."[677] (parenthetical in original). The god of Freemasonry is Lucifer, so it is not surprising that Masonic influenced NASA would name its moon missions after the Masonic god, Apollo (Lucifer). The NASA Apollo missions were central to conditioning the world into believing the lie of a spinning spherical Earth, orbiting the sun.

Texe Marrs summarizes the Masonic beginnings and the Satanic aims of NASA:

> NASA's space program has from the start been founded on the principles of Masonic alchemy and the magic of the mystery religions of the ancients. The prophet Daniel told us that the last day's world ruler, the antichrist king, would be mighty, "And through his policy also he shall cause craft to prosper..." Craft, as in witchcraft! ... The earliest beginnings of the U.S. space program involved the secretive OSS/CIA project, Operation Paperclip, in which Nazi rocket scientists like Werner Von Braun were brought from war-torn Germany to America and given responsibility for development of space vehicles. The Freemasons were then put in charge of the newly created space agency, and magic and

> witchcraft were integrated and wedded with the newest advances in technology. ... Virtually everything that NASA does is permeated with magic and alchemy. Moreover, the real purpose of NASA is contained in another matrix, hidden from the public at large. This process involves the creation of Satanic ritual magic enabling the Illuminati elite to acquire and accumulate power even as the mind-controlled and manipulated masses are pushed into ever increasing states of altered consciousness.[678]

Freemasonry is the child of the Jewish Kabbalah. Freemasonry, at the highest levels, is under Zionist control, which means that NASA is ultimately under Zionist control. In witchcraft there are double meanings to acronyms and words. NASA is of course an acronym for the National Aeronautics and Space Administration. *Nasa* is also a Hebrew word that means to lift up, to carry off, or to exalt oneself.[679] The Hebrew word *Nasa* additionally means to cause to bear iniquity.[680]

The Zionist influence within NASA is well hidden. There are, however, indications of Zionist influence within NASA. One example is the Columbia Shuttle Mission STS-107 emblem. The Space Shuttle Mission STS-107 was a disastrous mission, during which the Columbia Shuttle was destroyed when it allegedly reentered the atmosphere on February 1, 2003. It is the proper protocol for a commemorative emblem to display the flag of the host country along with the flag of a guest astronaut's country. The Columbia Shuttle Mission STS-107 emblem was notable for its glaring breach of proper protocol, in that it displayed only the flag of the guest astronaut's country, without displaying the flag of the host country, which was the United States. What country was the guest astronaut from? He was from Israel. The presence of the flag of Israel on the Columbia Shuttle Mission STS-107 emblem, without also including the flag of the United States, was intended as a not-so-subtle symbol of Israeli hegemony over NASA and the United States Government.

The inclusion of the flag of Israel, and the exclusion of the U.S. flag, is a portent of things to come. Israel will ultimately stab the U.S. in the back. Before Israel does that, it will milk the U.S. for all it can. Indeed, Israeli Prime Minister Benjamin Netanyahu was overheard by a CIA agent saying to his supporters after visiting Jonathan Pollard's jail cell, that "[o]nce we squeeze all we can out of the United States, it can dry up and blow away." Pollard is a Jewish spy who was caught spying on the U.S. for Israel. Rafi Eitan, who is a Mossad spymaster, an advisor to Fidel Castro, and an Israeli cabinet minister, told one of Israel's largest daily newspapers, *Yediot Aharonot*, in June 1997: "I failed in the Pollard affair, just as I failed in other intelligence operations beyond enemy lines." Eitan's statement indicates that Israel considers the United States an enemy. For more information on the duplicity of Israel, read this author's book, *9/11-Enemies Foreign and Domestic*.

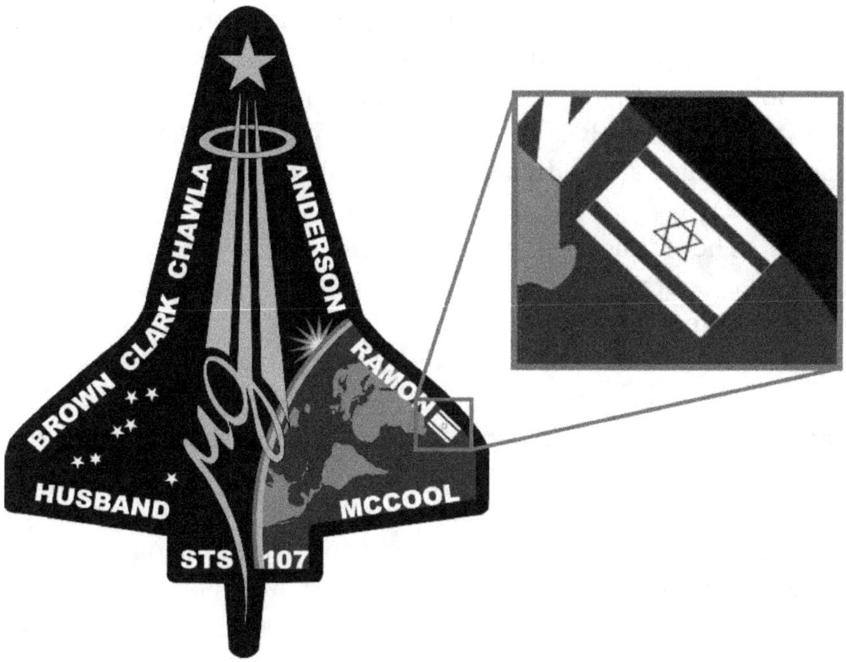

NASA and the Israel Space Agency (ISA), for a long time, have had a very close working relationship. For example, in 1986, NASA and ISA entered into a formal agreement to share technology.[681] That agreement is the means by which massive amounts of highly sensitive technology, developed by NASA at taxpayer expense, is handed over to Israel.

Interestingly, the American-Israeli Cooperative Enterprise revealed that in "October 1999, Ben-Gurion University researchers joined an international project to map the earth sponsored by NASA, the German space agency DARA and the Italian Space Agency ASI."[682] Certainly, that earth mapping project is all part of the international effort to conceal the flat earth. The alleged NASA Curiosity Rover mission to Mars in November 2011, is claimed by Israel to have included an Israeli-manufactured refrigerator.[683]

Israel is interested in obtaining technology from NASA, and NASA is all too obliging. That explains, for example, Israel's hosting of the 2015 International Astronautical Conference (IAC), which was held in Jerusalem. At that conference, Israel announced yet another partnership between the Israel Space Agency (ISA) and NASA, which involves "joint missions, personnel and scientific data exchanges, ground-based research facilities."[684]

The technology sharing agreements with the U.S. are on their face bilateral, but in fact are one-sided. Make no mistake about it, Israel has advanced technology, but it has very little to share with the U.S. Virtually all of Israel's technology has been given to it by or stolen from another country, usually the U.S. Israel, however, does share technology obtained from the U.S. with other like minded countries. There has been a long, albeit secret, history of mutual cooperation between communist China and Israel in the development of nuclear and other military weapons. In fact, Israel has been cited as one of the primary conduits for the flow of U.S. and other western technologies to communist China. No one who is a patriotic American would ever enter into the one-sided technology sharing agreements with Israel. The one-sided agreements were ratified by NASA officials, because Israel controls NASA. Masonic officials at NASA simply go along with whatever Israel wants. ISA is not truly a space agency; it is an intelligence agency. Its purpose is to obtain as much U.S. technology as it can get. The technology sharing agreements between the NASA and ISA are in reality technology hand-over agreements, where U.S. advanced technology is delivered to Israel. For a detailed explanation of Israel's hegemony over the U.S. Government, read this author's book, *Bloody Zion*.

30 Media Gatekeepers

Those behind the NASA deception will do anything to keep a lid on the moon landing hoax. For example, the first edition of *We Never Went to the Moon* came out in the Spring of 1975. That year, the author, Bill Kaysing, did a number of radio interviews. On December 7, 1975, Kaysing was being interviewed by Victor Boc on radio station KOME in San Jose, California. In the middle of the radio show, an engineer came in and announced that they were off the air.[685] It turned out that someone had dropped a napalm bomb from a helicopter on the KOME transmission towers in the Gilroy Hills, causing $250,000 in damage and putting the radio station off the air for three days.[686] It was clear to law enforcement that the reason for the bombing was to stop the interview of Kaysing about the moon landing hoax from being broadcast. They feared that Kaysing, himself, might be in danger; he was therefore offered police protection.[687]

The media gatekeepers are typically more sophisticated than using bombs. They have much more subtle ways of hindering mass awareness of the NASA moon landing hoax. The bombing, however, illustrates their resolve (or more aptly their desperation) in suppressing knowledge of the hoax. The reason that there is such an impetus to keep a lid on the Apollo moon hoax is that exposure of it will lead to the question of why the U.S. would fake a moon landing. Many people would then discover that the moon landings were hoaxed for the purpose of concealing the firmament and the flat earth. Satan and his minions cannot allow that to happen.

An example of the efforts to keep the masses ignorant of the flat earth is the recent shut down of the International Flat Earth Research Society (IFERS) forum. That forum, formerly found at http://ifers.boards.net/, was shut down on December 31, 2015. The message that appears across the screen when accessing the above URL is: "In accordance with Section 25(a) of the ProBoards Terms of Service, this forum has been taken offline." Section 25(a), in pertinent part, states:

> PROBOARDS RESERVES THE RIGHT TO, IN PROBOARDS' SOLE DISCRETION AND WITHOUT NOTICE OR LIABILITY, DENY USE OF THE WEBSITE AND/OR SERVICES TO ANY PERSON FOR ANY REASON OR FOR NO REASON AT ALL.[688] (emphasis with all caps in original)

Basically, ProBoards decided to shut down the IFERS forum, because their terms of service allows them to do so "for any reason or for no reason at all." ProBoards is one of the largest forum hosting sites in the world. It hosts over 3,000,000 internet forums, with a combined 22,800,000 users worldwide and a total of over 600 million page views per month.

The administrator of the IFERS, Eric Dubay, revealed that ProBoards shut down the IFERS forum without any notice to Dubay or any of the IFERS administrators. The IFERS forum was becoming quite popular and an important source of information about the flat earth. At the time of its shutdown, IFERS had more than 3,000 members and more than 50,000 visitors each month.[689]

Dubay addressed the continuous effort of ProBoards to censor him and IFERS, before finally shutting IFERS down:

> This was not the first act of censorship by Proboards as well. A few months ago they also threatened to delete our entire forum if we didn't remove within 48 hours the thread for my documentary Adolf Hitler vs. The Jew World Order claiming "racism and hate speech." We begrudgingly complied but all sent very polite and in-depth messages to Proboards explaining how nothing in the documentary was "hateful or racist," and were received with similar silence.[690]

Dubay further explained how Facebook also deleted his account three times. He is now on his fourth Facebook account.[691] This author had visited the IFERS forum before its shutdown. The forum had many good resources. This author disagrees with some of the posts, particularly Dubay's attacks on Christianity. This author has, nonetheless, cited to Dubay's flat earth writings on several occasions in this book, which, of course, should not be interpreted to mean that I agree with Dubay's other political or religious beliefs, which he has every right to express. It is notable that ProBoards never threatened to shut down Dubay's forum because of his attacks on Christianity. It was only after he posted a video that exposed the Jewish elite's plans for world domination that he was threatened with a shutdown.

Dubay's undoing was his connection of heliocentrism and the flat earth cover-up to the elite Jews. It is not surprising that he came to that conclusion, as that is where any diligent study of the flat earth and the origins of heliocentrism leads. That was a bridge too far, however, for the Zionist elite to allow. They simply could not permit the Jewish connection to be revealed on a public forum. Consequently, the IFERS forum was shut down.

The Jewish elite are starting to flex their financial and political muscles. They are slowly beginning to undermine the freedom of internet media outlets. They have long ago strangled any free access to the major media outlets. Other than to ridicule, discussion of the Apollo moon hoax or a belief in a flat earth is simply not allowed in the major media. The major media outlets are under Jewish control. John Whitley reported in 2003 that "seven Jewish Americans run the vast majority of US television networks, the printed press, the Hollywood movie industry, the book publishing industry, and the recording industry."[692] He explained that "[m]ost of these industries are bundled into huge media conglomerates." He listed the Jewish men and stated that "[t]hose seven Jewish men collectively control ABC, NBC, CBS, the Turner Broadcasting System, CNN, MTV, Universal Studios, MCA Records, Geffen Records, DGC Records, GRP Records, Rising Tide Records, Curb/Universal Records, and Interscope Records."[693] Whitley's research revealed:

> Most of the larger independent newspapers are owned by Jewish interests as well. An example is media mogul Samuel I. 'Si' Newhouse, who owns two dozen daily newspapers from Staten Island to Oregon, plus the Sunday supplement Parade; the Conde Nast collection of magazines, including Vogue, The New Yorker, Vanity Fair, Allure, GQ, and Self; the publishing firms of Random House, Knopf, Crown, and Ballantine, among other imprints; and cable franchises with over one million subscribers.[694]

Whitley's conclusions are as valid today as they were in 2003. Whitley explains why: "I could add that Michael Eisner could depart Disney tomorrow but the company will remain in the hands of Shamrock Holdings, whose principal office is now located in Israel."[695] For a more complete discussion of this issue, read this author's book, *Bloody Zion*.

The major choke points for information on the internet are owned by Jews, including, but not limited to, Google, YouTube (a division of Google), Yahoo, Facebook, and Wikipedia. The Zionist influence over those companies is palpable. For example, Google´s Sergey Brin, Facebook´s Mark Zuckerberg and Yahoo´s President Susan Decker attended a presidential panel on technology, presided over by Israel's President Shimon Peres, during Israel´s 60th anniversary celebrations, May 13-15, 2008.[696] Google has established a research and development center in Israel.[697] Google actively censors articles and whole websites that are critical of Israel or reveal the criminal machinations of the Jews.[698] Google follows dictates from ADL and other Zionist organizations to adjust its search robots to censor websites that are viewed as critical of Jews or Zionism.[699]

Another flat earth forum was quickly set up by a rival of Eric Dubay on ProBoards, after the demise of IFERS.[700] The new forum is called The Flat Earth Skeptic's Society (FESS). The interesting thing about that forum is that they created a forum board topic category titled, "Jesuits, Not Jews, Rulers of All Evil."[701] That is a first for any forum: set up a topic, which by its terms, prohibits discussion of a topic. As Queen Gertrude said in Hamlet: he "doth protest too much, methinks." It seems that the new forum is going to gate keep the site to steer clear of anyone connecting heliocentrism to the Jews. As of this writing there have been no posts to that topic.

Dubay is not the only one who has suffered from censorship on the flat earth issue. Sometimes the gatekeepers use their power and influence to put pressure directly on the flat earth advocates themselves to shut down discussion of flat earth. For example, popular Christian radio host Rob Skiba, under pressure, shut down his website on August 16, 2015. His website addressed the issue of the flat earth.[702] Skiba later repented of his action in shutting down the website and put the website back online.[703] Skiba explained his actions:

> [A]lmost since day one, this "quest for truth" has brought us nothing but pain and significant damage - emotionally, spiritually and especially financially. The attacks have been relentless. This tells me that there must be something to it, for the enemy to respond that way. I've covered many controversial subjects in my life, but none have even come close to bringing the kind of resistance as this one. So, the day I shut this site down, I was really feeling like I never should have even gone down this "rabbit trail" in the first place and I had truly had enough of it all. But I now realize that was selfish of me. I know many people have come to rely on this site for their own quest for truth, so I am sorry. My "haters" have been so vocal that I ended up forgetting how many supporters we have too.[704]

The pressure applied to Skiba gives you some idea of the behind-the-scenes pressure that may have been applied to the owners of ProBoards to shut down the IFERS forum. Skiba did not go into specifics, but suffice it to say that the flat earth controversy brings with it attacks from all sides by Satan and his minions. While Skiba has many videos and other information on his website proving that the earth is flat, he has also posted on his website a statement averring that he considers those alleging that he believes that the earth is flat are "exceptionally ignorant, slanderous, backstabbing posers and lovers of gossip."[705] Apparently Skiba thinks it is slanderous backstabbing to be spoken of as believing that the earth is flat. That should give you some idea of how effective Satan has been in creating a spherical earth hive mentality. Anyone found in the hive who does not accept the orthodox view that the earth is a globe is attacked. Skiba has been cowed and fears being attacked by the hive. He stated on his website: "What if I am wrong about the globe view? Am I prepared to deal with the resulting aftermath of scorn and ridicule that is sure to follow?"[706] Skiba's official stated position as of this writing is that he still believes that the earth is a globe, and he is simply weighing the merits of both sides of the debate. We must take to heart the admonition from God:

> Put on the whole armour of God, that ye may be able to stand against the wiles of the devil. For we wrestle not against flesh and blood, but against principalities, against powers, against the rulers of the darkness of this world, against spiritual wickedness in high places. Wherefore take unto you the whole armour of God, that ye may be able to withstand in the evil day, and having done all, to stand. Stand therefore, having your loins girt about with truth, and having on the breastplate of righteousness; And your feet shod with the preparation of the gospel of peace; Above all, taking the shield of faith, wherewith ye shall be able to quench all the fiery darts of the wicked. And take the helmet of salvation, and the sword of the Spirit, which is the word of God: Praying always with all prayer and supplication in the Spirit, and watching thereunto with all perseverance and supplication for all saints;" (Ephesians 6:11-18)

31 Map of the Flat Earth

Most have a difficult time understanding the flat earth, because since childhood, they have been conditioned by the ubiquitous depiction of the earth as a globe. The first scientific object displayed to them in the classroom was typically a globe of the earth. They have been instructed in their early lessons of world history about how Magellan circumnavigated the globe.

Once one sees just how the flat earth truly exists, it will be easy to understand how Magellan could be believed to have circumnavigated the globe, when in fact he simply circumnavigated a plane.

In the depiction below we see that the earth is flat with the north pole in the center of a circular plane. The south is at all points at the perimeter of the plane. Antarctica actually covers the entire circumference of the earth plane.

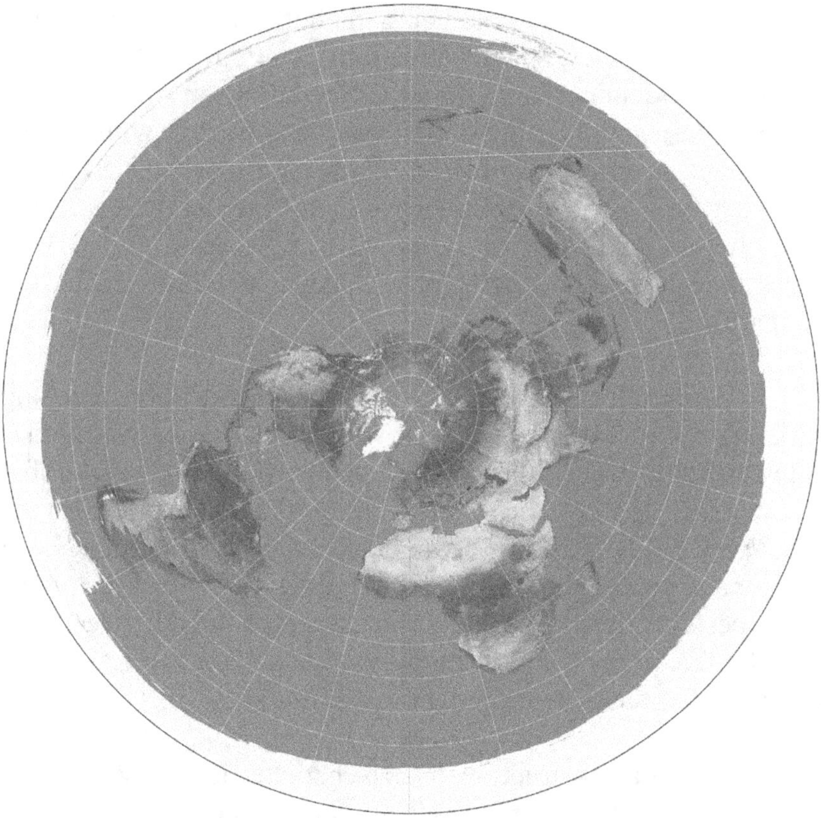

The above map is in fact called a polar azimuthal equidistant map. According to the United States Geological Survey (USGS), azimuthal maps are considered to be accurate in displaying continents and oceans.[707] Because of their accuracy they are used for air and sea navigation. Indeed the USGS states that the azimuthal equidistant maps are "used by USGS in the National Atlas of the United States of America™ and for large-scale mapping of Micronesia. Useful for showing airline distances from center point of projection."[708]

North is the center of the flat earth, and south is a direction emanating out from the North Pole to the outer rim of Antarctica. Thus, longitude lines are north/south lines that splay out from the North Pole to the outer rim of Antarctica, much like the spokes on a wheel, with the North Pole being the hub and Antarctica being the rim. East and west are directions circling the North Pole between equidistant points from the North Pole anywhere on the globe. Thus, latitude lines are east/west circles with points equidistant from the North Pole. If you travel east or west at the same latitude you will eventually circumnavigate the earth and wind up at your starting point. Circumnavigation does not necessitate that the earth be a globe. Magellan circumnavigated the flat earth.

Antarctica is the rim of the flat earth. Upon reaching Antarctica, explorers are first met with a massive ice wall that is between 1,000 and 2,000 feet thick, with 100 to 200 feet of that thickness rising above the water.[709] One of the first explorers to see the Antarctic ice wall was Sir James Clark Ross, a British Naval Officer and polar explorer. Ross was confronted with a shear cliff of ice perfectly level on top that he estimated to be between 150 and 200 feet high extending east and west as far as the eye could see. Ross famously described the Antarctic ice wall:

> It was an obstruction of such character as to leave no doubt in my mind as to our future proceedings, for we might as well sail through the cliffs of Dover as to penetrate such a mass.
>
> It would be impossible to conceive a more solid-looking mass of ice; not the smallest appearance of any rent or fissure could we discover throughout its whole extent, and the intensely bright sky beyond it but too plainly indicated the great distance to which it reached southward.[710]

Figure 169: Ship Abutting up to the Antarctic Ice Wall Supplying Fuel for the German Neumayer Station

The picture above is of the double-hulled ice breaker ship, Polarstern.[711] The ship is 118 meters (387 feet) long; the communications tower on the ship is approximately 135 feet above the water. The ship is pictured supplying fuel for the German Neumayer Station at Antarctica on or about December 7, 2006.[712] The ship gives us a scale for the massive height and depth of the Antarctic ice wall.

The USGS admits that the azimuthal equidistant map is accurate for all distances and directions from the center point on the map. The USGS, however, claims that the map is only accurate when the starting point is from the center of the map. In the above map of the flat earth, therefore, the USGS admits that all distances and directions from the North Pole to anywhere on the map traveling south are accurate, but claims it is not necessarily accurate when starting from any other point on the map other than the North Pole.

Notice the similarity between the UN flag logo below and the flat earth map. The only thing that the UN left off the flag is Antarctica. The skeletal world government set up to enslave the world knows that the earth is flat and they have hidden their flat earth secret in plain view.

Notice also the number of sections in the UN Flag logo. There are exactly 33 sections, which correlates nicely with the 33 degrees of the Scottish Rite of Freemasonry.

One might wonder, where is the equator? In the Polar azimuthal equidistant map below, the Equator, Tropic of Cancer, and Tropic of Capricorn are all clearly labeled.

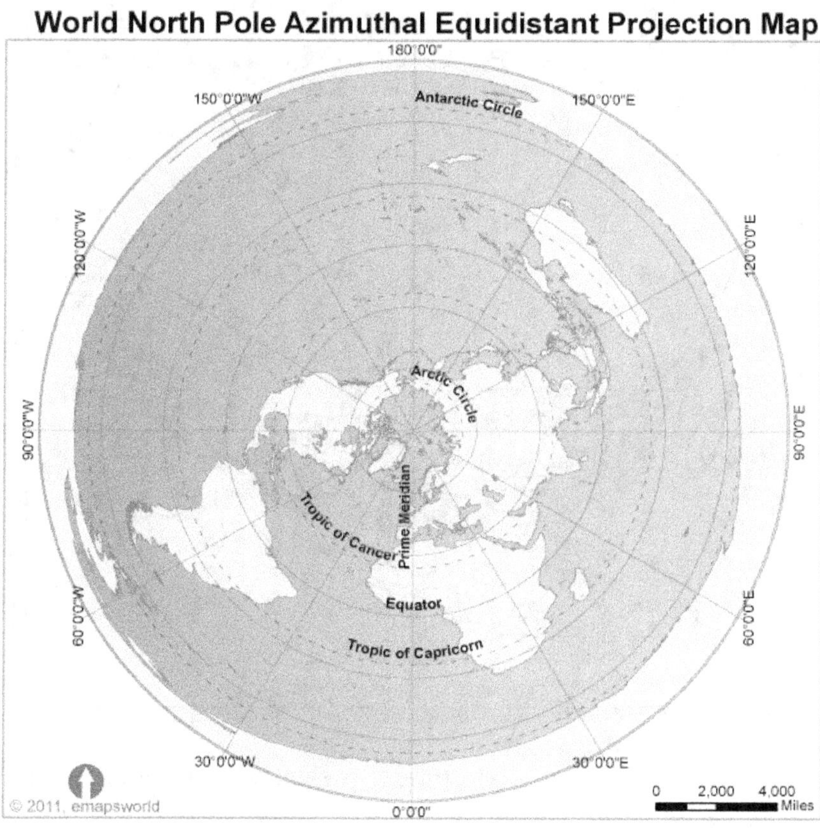

Below is a polar azimuthal equidistant standard map of the world from 1892.

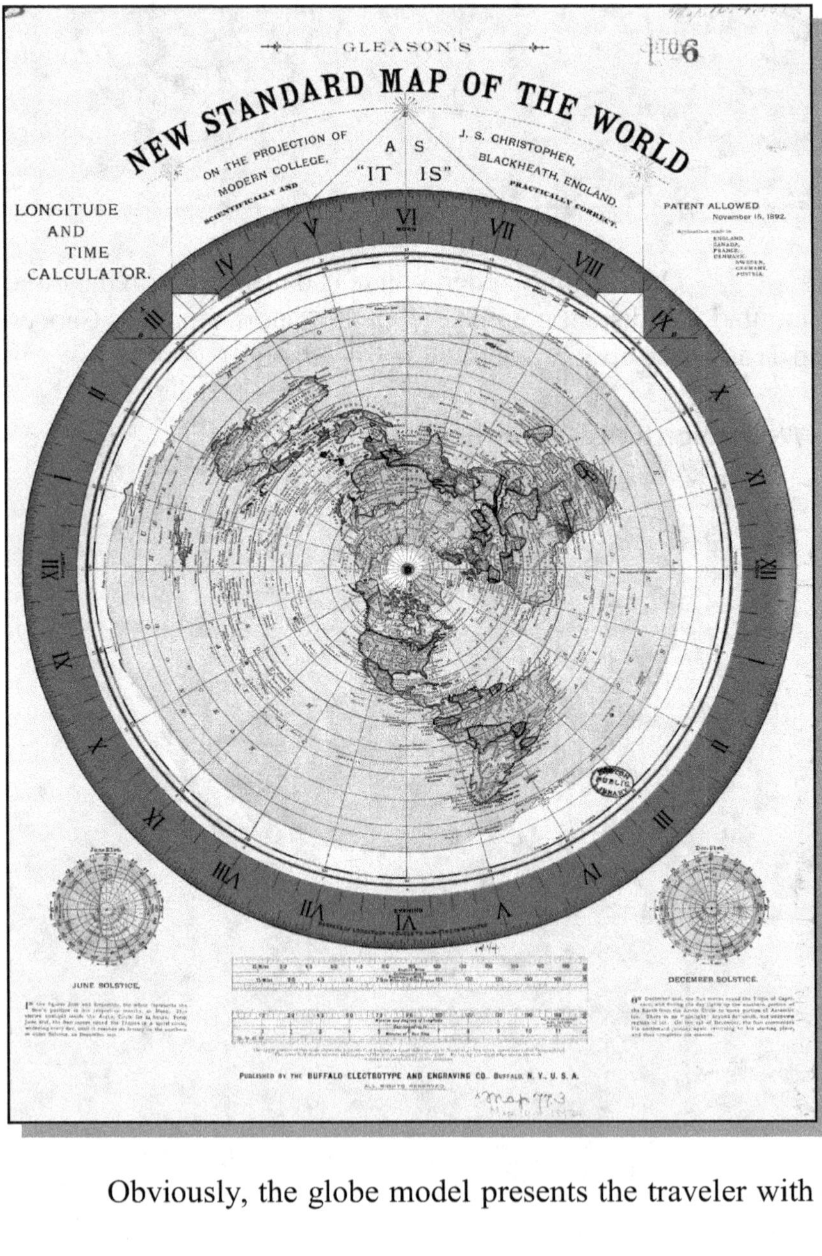

Obviously, the globe model presents the traveler with a

significantly different configuration than does the flat earth model, particularly in the so-called "southern hemisphere." The discrepancy between the reality of a flat earth and the myth of a globe have found many a seafarer off course as they traveled the south seas. Samuel Rowbotham explains the dangers of the false charts based upon a globe model:

> In the southern hemisphere, navigators to India have often fancied themselves east of the Cape when still west, and have been driven ashore on the African coast, which, according to their reckoning, lay behind them. This misfortune happened to a fine frigate, the Challenger, in 1845.[713]

> How came Her Majesty's Ship Conqueror, to be lost? How have so many other noble vessels, perfectly sound, perfectly manned, perfectly navigated, been wrecked in calm weather, not only in a dark night, or in a fog, but in broad daylight and sunshine--in the former case upon the coasts, in the latter, upon sunken rocks--from being 'out of reckoning,' under circumstances which until now, have baffled every satisfactory explanation.[714]

Rowbotham explains how the British ship Challenger completed the circuit of the southern region, indirectly, to be sure, circumnavigating Antarctica. The approximate circumference of Antarctica is supposed to be approximately 14,460 statute miles on the globular earth model. However, the Challenger spent three years and traversed nearly 69,000 miles. That is a distance that would have allowed the ship to circle Antarctica over 4 times if the earth were in fact a globe. The 69,000 mile long journey around the circumference of the earth that is Antarctica, however, is perfectly explained by a flat earth.[715]

Rowbotham quotes from voyagers, who found that their charts, which used a globular earth model, almost always put them off course in the south seas. "We found ourselves every day from 12 to 16 miles by observation in advance of our reckoning."[716] Another southern seafarer stated: "By our observations at noon we found ourselves 58 miles to the eastward of our reckoning in two days."[717]

The spherical earth charts did not account for the expanse of distance between the lines of longitude the further south one travels on a flat earth. Consequently, the mariners' true position on the earth, determined by using celestial navigation and a chronometer, was found to be off by several degrees from the position on the chart that was plotted using dead reckoning. One mariner detailed the distance that the erroneous charts, which assumed a globular earth, put them off course in the south seas.

> February 11th, 1822, at noon, in latitude 65.53. S. our chronometers gave 44 miles more westing than the log in three days. On 22nd of April (1822), in latitude 54.16. S. our longitude by chronometers was 46.49, and by D.R. (dead reckoning) 47° 11′: On 2nd May (1822), at noon, in latitude 53.46. S., our longitude by chronometers was 59° 27′, and by D.R. 61° 6′. October 14th, in latitude 58.6, longitude by chronometers 62° 46′, by account 65° 24′. In latitude 59.7. S., longitude by chronometers was 63° 28′, by account 66° 42′. In latitude 61.49. S., longitude by chronometers was 61° 53′, by account 66° 38′.[718]

The skippers of the ships in the south seas could only guess that the discrepancies between their dead reckoning plotted on their charts and their true positions, confirmed by precise chronometers and sextants, was due to the currents. Rowbotham explains, however, that currents could not possibly account for the navigation errors, because the errors in the navigation were manifested regardless of whether the ships were traveling east or west.

> The commander of the United States exploring expedition, Lieutenant Wilkes, in his narrative, says that in less than 18 hours he was 20 miles to the east of his reckoning in latitude 54° 20′ S. He gives other instances of the same phenomenon, and, in common with almost all other navigators and writers on the subject, attributes the differences between actual observation and theory to currents, the velocity of which, at latitude 57° 15′ S., amounted to 20 miles a day. The commanders of these various expeditions were, of course, with their education and belief in the earth's rotundity, unable to conceive of any other cause for the differences between log and chronometer results than the existence of currents. But one simple fact is entirely fatal to such an explanation, viz., that when the route taken is east or west the same results are experienced. The water of the southern region cannot be running in two opposite directions at the same time; and hence, although various local and variable currents have been noticed, they cannot be shown to be the cause of the discrepancies so generally observed in high southern latitudes between time and log results. The conclusion is one of necessity--is forced upon us by the sum of the evidence collected that the degrees of longitude in any given southern latitude are larger than the degrees in any latitude nearer to the northern centre; thus proving the already more than sufficiently demonstrated fact

> that the earth is a plane, having a northern centre, in relation to which degrees of latitude are concentric, and from which degrees of longitude are diverging lines, continually increasing in their distance from each other as they are prolonged towards the great glacial southern circumference.[719]

The problem of longer distances between longitude lines south of the equator is known by many ship captains and airline pilots today. The scientific priesthood has come up with a model of the earth to explain the phenomenon and keep pilots, ship captains, and the general public in the dark about the flat earth. So called scientists cannot admit the truth, that the earth is flat, so they, instead, portray the earth as an oblate spheroid. An oblate spheroid is a squashed down sphere, similar in shape to a rugby ball or an American football.

Squashing the earth into an oblate spheroidal model offers some plausible explanation for the larger expanse for travelers south of the equator, but it is does not fully explain the phenomenon. Neil DeGrasse Tyson, who is a famous astrophysicist, and the Director of the Hayden Planetarium at the Rose Center for Earth and Space in New York City, reveals the scientific explanation that purports to fully resolve the anomaly. Tyson avers that the oblate spheroid earth is misshapen; it is actually wider below the equator than it is above the equator. Tyson stated that the earth "is not actually a sphere, it's oblate, it's an oblate spheroid. Not only that it's slightly wider below the equator than above the equator. ... A little chubbier ... it's like pear-shaped."[720]

Of course, Tyson's description of a pear-shaped earth is contrived to accommodate the experience of travelers covering a larger expanse between longitude lines below the equator. The predicament modern scientists find themselves in with their pear-shaped earth explanation, though, is that the space agencies (e.g., NASA) have presented to the world many alleged photographs of the earth from space that depict a perfectly proportioned sphere. The earth cannot be both a perfect sphere and pear shaped at the same time. The pathological liars in the scientific priesthood jump from one lie to the next. And this is yet another example of a scientific lie that conflicts with a previous scientific lie. It should be no surprise, as they are in the service of Satan, who is the father of lies. John 8:44. "Science falsely so called" (1 Timothy 6:20) has, by its deception, put itself in a place of its own making, between a rock (a perfect sphere) and a hard place (a pear-shaped spheroid). The earth is, in truth, neither pear-shaped nor a perfect sphere. It is flat.

Below is a map depicting the path of the sun and moon. They would progress in their path from the Tropic of Cancer for the summer months in the north (the winter months in the south) to the Tropic of Capricorn for the winter months in the north (the summer months in the south).

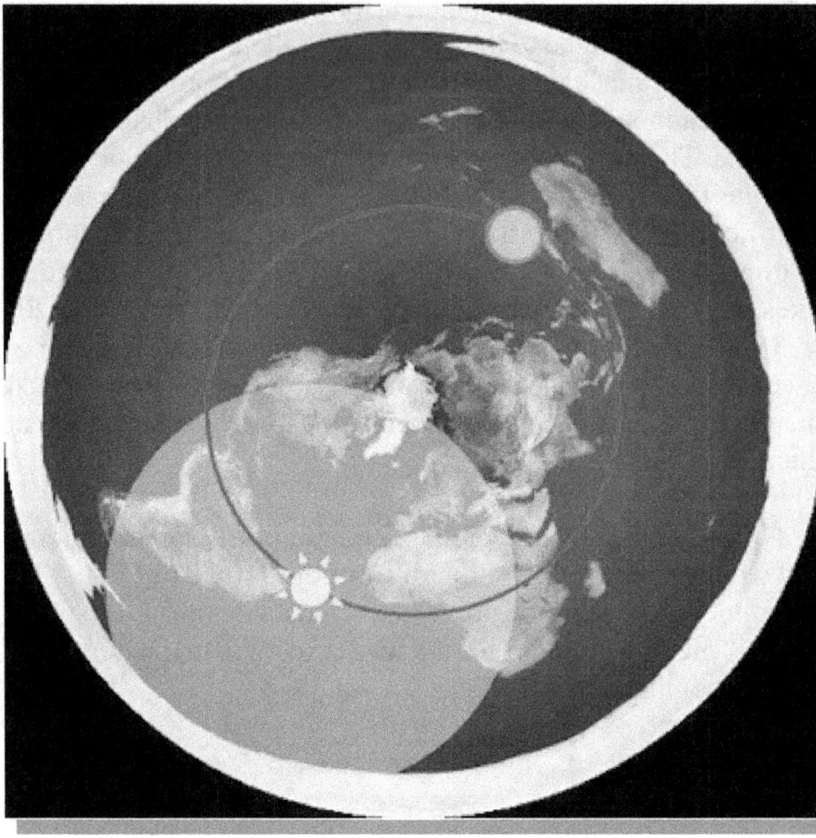

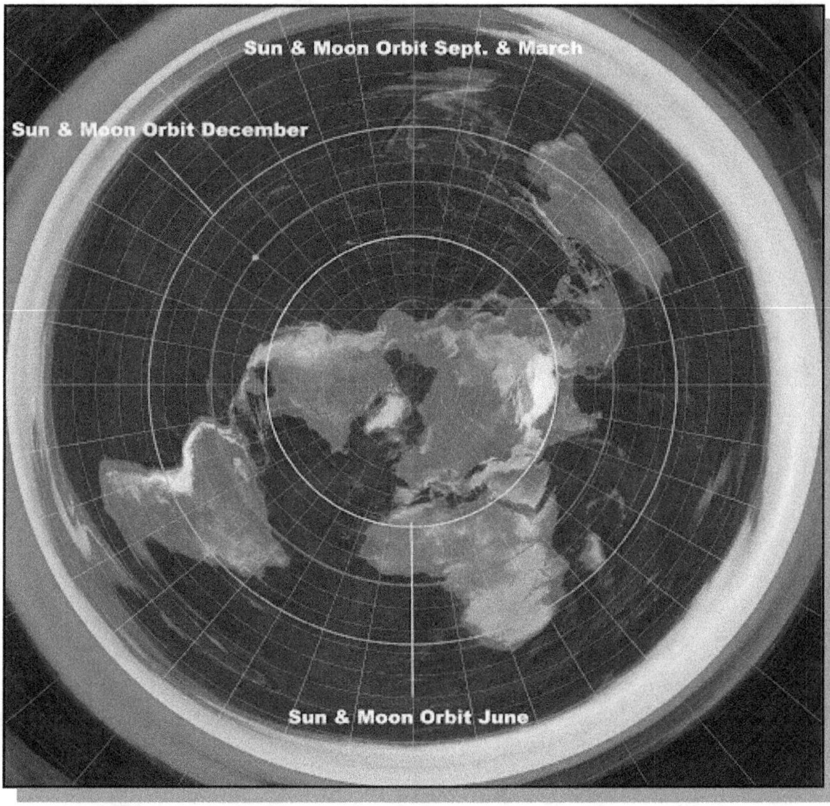

The governments of the world want the masses to believe that the world is round. However, when governments want to know how to address issues on the real configuration of the world, they reference a polar azimuthal equidistant flat earth map. Indeed, the U.S. Military employs hundreds of soldiers and civilians in producing accurate maps of areas of deployment. Those maps are almost never made public. Below is a frame from an official video tour of the White House Situation Room, published by the White House and posted on the internet, showing a polar projection of an azimuthal equidistant map in the White House Situation Room.[721] When it really counts, the U.S. Government wants to know the real configuration of the earth: a flat earth.

Polar Azimuthal Equidistant Map in the White House Situation Room

32 Fauna and Flora Prove the Earth Is Flat

Eric Dubay, in his book, *Flat Earth Conspiracy*, sets forth a summary of the discrepancy between what is really observed north and south of the equator and what should be observed if the earth were a globe.

> If the Earth were truly a globe, the Arctic and Antarctic polar regions and areas of comparable latitude North and South of the equator should share similar conditions and characteristics such as comparable temperatures, seasonal changes, length of daylight, plant and animal life. In reality, however, the Arctic/Antarctic regions and areas of comparable latitude North/South of the equator differ greatly in many ways.[722]

Dubay quotes Thomas Winship, who explains that there are significant differences in the temperatures, plants, and animals found at comparable latitudes north of the equator from those found south of the equator.

> If the earth be the globe of popular belief, the same amount of heat and cold, summer and

> winter, should be experienced at the same latitudes North and South of the Equator. The same number of plants and animals would be found, and the same general conditions [should] exist. That the very opposite is the case, disproves the globular assumption. The great contrasts between places at the same latitudes North and South of the Equator, is a strong argument against the received doctrine of the rotundity of the earth.[723]

If the earth were a globe, one would expect that the temperature range should be similar between the Arctic region in the north and Antarctic region in the south. Yet, Dubay discovered that the temperature range is quite different between the Arctic and Antarctic regions:

> Antarctica is by far the coldest place on Earth with an average annual temperature of approximately -57 degrees Fahrenheit, and a record low of -135.8! The average annual temperature at the North Pole, however, is a comparatively warm 4 degrees. Throughout the year, temperatures in the Antarctic vary less than half the amount at comparable Arctic latitudes. The Northern Arctic region enjoys moderately warm summers and manageable winters, whereas the Southern Antarctic region never even warms enough to melt the perpetual snow and ice.[724]

Again, this discrepancy in temperature range between the Arctic and Antarctic regions is confirmed by Thomas Winship:

> This uniformity of temperature [in Antarctica] partly accounts for the great accumulation of ice which is formed not on account of the great

> severity of the winter, but because there is practically no summer to melt it. In the Antarctic there is eternal winter and snow never melts. As far north as a man has travelled he has found reindeer and hare basking in the sun, and country brilliant with rich flora; within the Antarctic circle no plant is to be found.[725]

Dubay lists particular examples of the marked differences in the flora of the Arctic and Antarctic regions:

> The island of Kerguelen at 49 degrees Southern latitude has only 18 species of native plants that can survive its hostile climate. Compare this with the island of Iceland at 65 degrees Northern latitude, 16 degrees further North of the equator than Kerguelen is South, yet Iceland is home to 870 species of native plants. On the Isle of Georgia, just 54 degrees Southern latitude, the same latitude as Canada or England in the North, where dense forests of various tall trees abound, the infamous Captain Cook wrote that he was unable to find a single shrub large enough to make a toothpick![726]

Captain Cook wrote about the contrast between the Arctic and Antarctic zones. He found vegetables and land animals in the Arctic as far north as the 80th parallel. Yet, he found that the perpetually frigid temperatures in the Antarctic region, primarily from the lack of sunlight, caused a starkly desolate landscape there. Consequently, beginning at only the 58th parallel south, in the Antarctic region he found only lichen, seabirds, and whales.

> Not a tree was to be seen. The lands which lie to the south are doomed by nature to perpetual frigidness - never to feel the warmth of the sun's

> rays; whose horrible and savage aspect I have not words to describe. Even marine life is sparse in certain tracts of vast extent, and the sea-bird is seldom observed flying over such lonely wastes. The contrast between the limits of organic life in Arctic and Antarctic zones is very remarkable and significant. Vegetables and land animals are found at nearly 80 degrees in the north; while from the parallel of 58 degrees in the south, the lichen, and such-like plants only, clothe the rocks, and seabirds and the cetaceous tribes alone are seen upon the desolate beaches.[727]

Dubay explains that "in the Arctic there are 4 clearly distinguished seasons, warm summers, and an abundance of plant and animal life, none of which can be said of the Antarctic. The Eskimo live as far North as the 79th parallel, whereas in the South no native man is found higher than the 56th."[728]

Admiral Ferdinand von Wrangel, the 19th century Russian Arctic explorer, wrote how in the North there were "countless herds of reindeer, elks, black bears, foxes, sable and grey squirrels fill the upland forests; stone foxes and wolves roam over the low ground; enormous flights of swans, geese, and ducks arrive in spring, and seek deserts where they may moult, and build their nests in safety. Eagles, gulls, and owls pursue their prey along the sea-coast; ptarmigan run in troops among the bushes; little snipes are busy among the brooks and in the morasses; the social crows seek the neighbourhood of man's habitations; and when the sun shines in spring, one may sometimes even hear the cheerful note of the finch, and in autumn that of the thrush."[729]

Yet, von Rangel found that in the south "beyond the 70th degree of Southern latitude not a tree meets the eye, wearied with the white waste of snow; forests, woods, even shrubs have disappeared, and given place to a few lichens and creeping woody plants, which scantily clothe the indurated soil. Still, in the farthest north, nature claims her birthright of beauty; and in the brief and rapid summer she brings forth numerous flowers and grasses, to bloom for a few days, to be again blasted by the swiftly-recurring winter. The rapid fervour of an arctic summer had already (June 1st) converted the snowy waste into luxuriant pasture-ground, rich in flowers and grass, with almost the same lively appearance as that of an English meadow."[730]

Why is there such a difference in the flora and animal life present in the south of the equator versus that found north of the equator? The only possible explanation is that the larger expanse below the equator on the flat earth causes a concomitant difference in climate as compared to that relatively smaller area above the equator. The reason is explained by Samuel Rowbotham, who states that the greater speed with which the sun must travel over the larger circuit of the flat earth south of the equator is why there is the stark desolation in the flora and animal life during the summer months south of the Antarctic circle. There is not that desolation found in the summer months north of the Arctic circle, because the sun makes its 360°circuit in 24 hours while traveling at a slower speed over a shorter distance thus giving its warmth and life-giving influence over a given area for a longer period of time.

That greater expanse south of the equator makes it necessary for the sun to travel faster during the southern summer months to cover that larger area in the same 24 hour period as to travel the smaller area during the northern summer months. Thus the sun gives less of its light over any given area south of the equator during its southern summer months. That is why there is such a stark difference between the flora and fauna found north of the equator from that found south of the equator. This difference in flora and animal life would not exist if the earth were a spinning globe. Rowbotham explains:

> The constant sunlight of the north develops, with the utmost rapidity, numerous forms of vegetable life, and furnishes subsistence for millions of living creatures. But in the south, where the sunlight never dwells, or lingers about a central region, but rapidly sweeps over sea and land, to complete in twenty-four hours the great circle of the southern circumference, it has not time to excite and stimulate the surface; and, therefore, even in comparatively low southern latitudes, everything wears an aspect of desolation. These differences in the north and south could not exist if the earth were a globe, turning upon axes underneath a non-moving sun. The two hemispheres would at the same latitudes have the same degree of light and heat, and the same general phenomena, both in kind and degree. The peculiarities which are found in the south as compared with the north, are only such as could exist upon a stationary plane, having a northern centre, concentric with which is the path of the moving sun.[731]

William Capenter describes how the severe difference between the flora, animals, and temperatures, in the latitudes north of the Arctic circle versus the same latitudes south of the Antarctic circle is due to the fact that the earth is flat and the area south of the equator offers a larger expanse of land over which the sun must travel. Thus, the land has less exposure to the sun as it is traveling faster over the larger expanse. This creates a colder and more sparse environment south of the Antarctic circle than is found north of the Arctic circle, which proves that the earth is flat.

> Every year the Sun is as long south of the equator as he is north; and if the Earth were not 'stretched out' as it is, in fact, but turned under, as the

> Newtonian theory suggests it would certainly get as intensive a share of the Sun's rays south as north; but the Southern region being, in consequence of the fact stated, - far more extensive than the region North, the Sun, having to complete his journey round every twenty-four hours, travels quicker as he goes further south, from September to December, and his influence has less time in which to accumulate at any given point. Since, then the facts could not be as they are if the Earth were a globe, it is a proof that the Earth is not a globe.[732]

One would think that with the more rapid speed of the sun south of the equator the days would be shorter during the southern summer than they are during the northern summer. Rowbotham made that argument and cited several examples. But he did not compare cities at equivalent north and south latitudes. When we check cities at equivalent north and south latitudes, we find that the length of the day during the southern summer solstice is the same as the length of the day during the northern summer solstice.

For example, Toronto, Canada, which is at 43.65° North Latitude has a sunrise at 5:36 a.m. and a sunset at 9:02 p.m., for a total length of daylight of 15:26 during the northern summer solstice (midsummer), June 1, 2017.[733] Toronto, Canada has a sunrise at 7:47 a.m. and a sunset at 4:43 p.m. for a total length of daylight of 8:56 during the northern winter solstice (midwinter), December 21, 2017.[734]

Compare Toronto, Canada, with Christchurch, New Zealand, which is located at 43.53° South Latitude. In Christchurch, New Zealand, the sunrise is at 8:02 a.m. and sunset is at 4:59 p.m., for a total length of day of 8:56 during the southern winter solstice (midwinter), June 21, 2017.[735] In Christchurch, New Zealand the sunrise is at 5:44 a.m. and the sunset is at 9:10

p.m., for a total length of day of 15:26 during the southern summer solstice (midsummer), June 21, 2017.[736]

That means that equivalent latitudes north of the equator (Toronto, Canada, 43.65 ° North Latitude) and south of the equator (Christchurch, New Zealand, 43.53° South Latitude) have equivalent length of the days during their respective winter solstices (8:56) and summer solstices (15:26).

It seems that the days are the same length of time at equivalent latitudes north and south of the equator during their respective summer solstices. Does that mean that the earth is a globe, because a flat earth would seem to require the days to be shorter south of the equator during its southern summer? No, it does not.

First, the objective evidence indicates that in fact during the southern summer the sun travels faster south of the equator, as it must, to travel a longer distance in 24 hours than it travels during the northern summer on a flat earth. This speed of the sun over the southern regions during the southern summer can be observed by anyone who has traveled far south of the equator. That faster speed is manifested by the rapidness at which twilight disappears. Rowbotham cites to many examples of rapidness of the disappearance of twilight. One witness stated that "there is not the twilight [in New Zealand] which you get in England. Here it is light till about eight o'clock, then, in a few minutes, it becomes too dark to see anything, and the change comes over in almost no time."[737]

Rowbotham explains that during the southern summer solstice, when the sun is traveling faster over a larger expanse, "the passage of the light must of necessity be proportionably more rapid; and the morning and evening twilight more abrupt. In the north the light on summer evenings seems as it were unwilling to terminate; and at midsummer, for many nights in succession, the sky is scarcely darkened. The twilight continues for hours after visible sunset. In the south, however, the reverse is the case, the day ends suddenly, and the night passes into day in a few seconds."[738]

Rowbotham cites to one example of a 19th century mariner who was accustomed to the more lengthy northern twilight who was surprised by the speed of the sun's disappearance over the horizon while on a whaling vessel in the far southern ocean.

> In the evening I was ordered aloft, and the captain cried out 'Be quick, Jack, or you'll be in the dark!' Now the sun was shining brightly, and it seemed far from the time of sunset, and I remember well that I looked at the captain, thinking he must be a little the worse for grog. However, I went aloft, and before I had finished the order, which was a very short time, I was in pitch darkness, the sun seemed all at once to drop behind or below the sea. I noticed this all the time we were in the far south, whenever the sun was visible and the evening fine . . . Any mariner, who has been a single season in the southern whaling grounds, will tell you the same thing.[739]

The rapid disappearance of twilight during the New Zealand summer is evidence that the sun is traveling more rapidly, as it must, over a larger southern expanse of land in the same 24 hours that it must travel over a relatively smaller area north of the equator during the northern summer on a flat earth. The evidence is clear. The sparseness of the flora and fauna and the rapid disappearance of twilight in the far south latitudes indicates that the earth is flat with a larger expanse below the equator. Scientists using trigonometry have obtained varying calculated distances for the altitude of the sun above the earth of between approximately 700 and 3,000 statute miles.[740] That indicates that the sun varies in its distance above the earth. The only explanation for the length of days being equal north and south of the equator at equivalent latitudes is that the sun is traveling higher in the sky south of the equator. That would result in a larger expanse covered by the sunlight as the sun travels faster over the southern regions, resulting in equivalent length of days. The sun being higher in the sky would cause the life-giving sun rays to be more diffuse. That fact, combined with the sun's more rapid transit over the land, explains the sparseness of flora and fauna in the far south latitudes as compared to the far north latitudes.

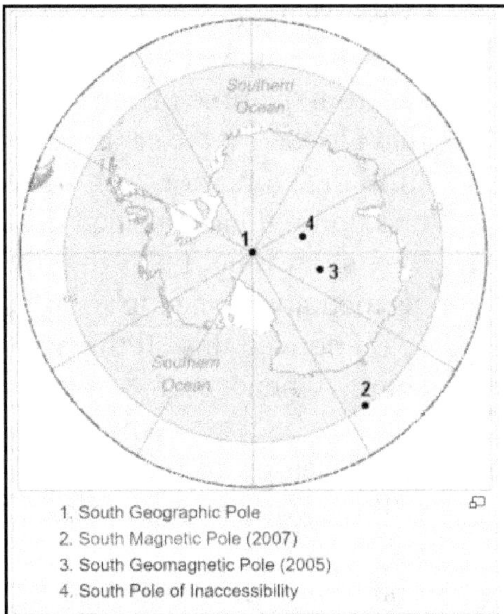

Wikipedia's Depiction of Four Different South Poles

The significant differences between the Arctic and Antarctic regions prove that the earth cannot be a globe. What is called Antarctica is actually a massive land and ice wall at the outer edge of Earth. That necessarily means that there is no South Pole. Scientists actually admit that the South Pole changes and cannot be fixed. The National Geographic explains that "[d]ue to plate tectonics, the exact location of the South Pole is constantly moving."[741] The National Geographic is not going to tell the truth and admit that the South Pole is constantly moving, because there is no South Pole. So, they fib, and say that it is due to "plate tectonics."

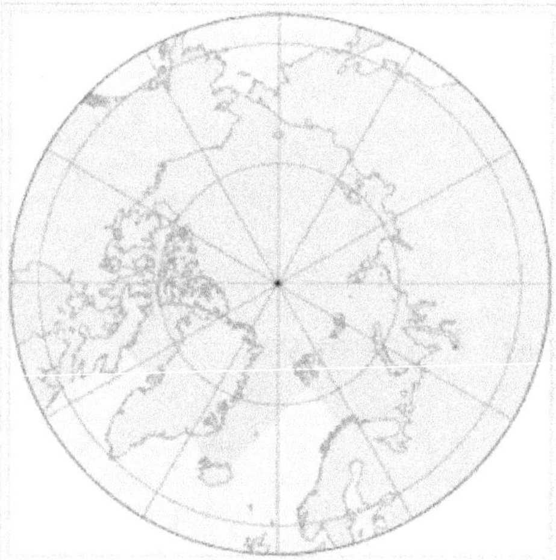

An azimuthal projection showing the Arctic Ocean and the North Pole. The map also shows the 75th parallel north and 60th parallel north.

Wikipedia's Depiction of a Single North Pole

Wikipedia, for its part, actually depicts four different South Poles.[742] But when it comes to the North Pole, Wikipedia only depicts a single location for the North Pole.[743] That is because the North Pole is in fact a single point in the center of the flat earth. Whereas, the South Pole is fictional, because Antarctica is the outer ring of the flat earth.

In addition to all of the other locations for the South Pole, there is a "ceremonial South Pole," marked by a red and white striped pole topped with a shiny, chrome-colored sphere, that is admitted not to actually be the South Pole. This fake "ceremonial South Pole" is surrounded by the flags of the Antarctic Treaty signatory countries. But they must constantly move the location of the "ceremonial South Pole."[744] The reason they must constantly move the pole is that the location of the "ceremonial South Pole" all depends on where along the outer rim of Antarctica on the flat earth any given Antarctica exploration party begins its trek southward. They must constantly move the "ceremonial South Pole" along the outer rim of Antarctica so that it coincides with the "ceremony" of the party arriving at the "ceremonial South Pole."

33 Time-Lapse Trickery

As we have seen, the length of the summer solstice days are equal north and south of the equator at equivalent latitudes. But equal length summer days north and south of the equator is not a uniform phenomenon. For example, there are not equal length days in areas north of the Arctic circle as compared to areas south of the Antarctic circle. That is because south of the Antarctic circle we find the outer rim of the flat earth, where there cannot possibly be 24 hour summer daylight, as is the case with the 24 hour daylight summer days north of the Arctic circle. But in order to sell to the masses the idea of a globular earth, governments fund projects that produce fake time-lapse videos showing 24 hours of daylight during the southern summer, south of the Antarctic circle.

Anthony Powell produced a documentary film, in association with the New Zealand Film Commission and Antarctica New Zealand with financial support from the National Science Foundation, titled *Antarctica: A Year on Ice.*[745] Powell claims to have grown up on a farm but that he has spent most of his adult life living and working in Antarctica. Powell is a cinematographer who specializes in time-lapse photography and is so skilled that he has made his own time-lapse equipment that will work in extreme cold weather, as is found in Antarctica.

Although the film is subtitled "A Year on Ice." Powell states that he spent 10 years making the film. The opening scene of the documentary shows a spinning Earth with text explaining that the Antarctic Continent is at the bottom of the planet. That little piece of propaganda sets the stage for conditioning the viewer that the earth is a spinning globe.

At the 14:00 minute mark in the film, Anthony Powell appears on camera to show his watch and announce that "it's the middle of summer, let's check the time, just after 12:00 o'clock; 12:00 o'clock midnight."[746] Powell then begins tracking the sun via an automated camera. A text appears on screen that states: "The sun remains above the horizon for four months at McMurdo."[747] The time-lapse video tracks the sun as it purportedly circles 360° above the horizon for 24 hours at McMurdo Station in Antarctica.

Ostensibly, the time-lapse video is irrefutable proof that the earth is a globe, since there is no way to have a sun circling 360° overhead for 24 hours on a flat earth in Antarctica. A sun circling 360° overhead for 24 hours is a reality North of the Arctic Circle (approximately 66° north latitude) during the northern summer, and showing a 360° sun circling overhead for 24 hours south of the Antarctic Circle (approximately 66° south latitude) during the southern summer would seem to prove the spinning globe hypothesis.

Indeed, those who argue for a globular spinning earth have trotted out Powell's time-lapse of the sun circling 360° above the horizon for 24 hours at McMurdo Station as irrefutable proof that the earth cannot be flat and must be a spinning globe.[748] It would seem that the case is closed.

Not so fast! David Weiss, smelled a rat and did a closer examination of Powell's time-lapse. Weiss doggedly scrutinized Powell's time-lapse video and uncovered clear and convincing evidence that the time-lapse video is a piece of chicanery.[749] Weiss discovered that Powell's time-lapse video is not a continuous time-lapse as purported by Powell. In fact, the time-lapse was spliced together using daytime shots to cover over the fact that the sun actually went below the horizon and there was a night sky. The spliced time-lapse video made it appear that the sun was traveling for 24 continuous hours above the horizon around McMurdo Station. But that appearance was not a true portrayal of reality.

Weiss posted his proof that Powell's time-lapse video is fraudulent on his YouTube video channel, *Deep Inside the Rabbit Hole.*[750] Proof of the fact that the time-lapse video is a cleverly spliced video can be seen in two distinct anomalies in the time-lapse. The first anomaly is that when Powell first shows up on camera to announce that it is just after 12:00 o'clock midnight, the mountain seen in the background is mostly bare of snow, and the sun is off to the right of the mountain. Yet, when the camera starts its time-lapse motion, the sun has traveled behind the mountain, which suggests a delay in starting the time lapse of perhaps more than an hour. Within that hour, the mountain has miraculously become snow covered, although the sky appeared clear both when he announced the time and when the time-lapse video starts. Either there was an (improbable) hour-long torrential flash snowstorm framed on either end by clear skies or the time-lapse video was not continuous, but spliced from time-lapse photographs and video taken on different days.

The second, and most notable, anomaly uncovered by Weiss is that the cloud formations at the beginning of the supposed 24-hour time-lapse are identical to the cloud formations at the conclusion of the 24-hour time-lapse. Indeed, close analysis reveals that the beginning and concluding time-lapse photographs are the same series of time-lapse photographs. That can mean only one thing. The purported time-lapse is not truly a continuous 24-hour time-lapse, but is rather a splice creating a time-lapse video made up of daytime shots to inaccurately portray the sun continuously above the horizon as it travels overhead 360° for 24 hours in Antarctica.

The fact that it is necessary to create misleading evidence of the sun circling overhead 360° for 24 hours in Antarctica is the best evidence that there is no such thing as 24 hours with the sun above the horizon in Antarctica. Why create a false time-lapse of the sun circling 360° above the horizon for 24 hours? Because, it is impossible to have the sun to circle 360° above the horizon for 24 hours in Antarctica. It is impossible because the earth is flat and Antarctica is the outer rim of the flat Earth. But because the globe model requires that the sun circle overhead 360° for 24 hours during the Antarctic summer, it was necessary to flimflam people by the artifice of false evidence showing that.

A screen shot of the side-by-side analysis by David Weiss of Anthony Powell's supposed 24 hour time-lapse of the sun over the horizon at McMurdo Station in Antarctica. The comparison shows that the cloud formations and every detail of the start of the time-lapse are identical to the end of the time-lapse, supposedly 24 hours later. That proves that the time-lapse is not truly a 24 hour time lapse, but is rather a splice of time-lapse photographs constructed to make it appear that the sun is above the horizon for 24 hours when in Antarctica. Clearly, that was done by Powell to give the erroneous impression that the earth is a sphere.

Indeed, Jeran Campanella analyzed numerous official web cam time-lapse videos from government-sponsored Antarctica stations and found that every time-lapse video during the southern summer months had missing frames.[751] The missing time-lapse frames give the misleading impression that the sun is traveling 360° overhead every 24 hours during the summer months in Antarctica. Campanella contacted the Australian Arctic Division to find out why their time-lapse web-cam videos have missing frames. The official spokesman for the Australian Arctic Division explained that the web manager designed the webcams "with a script he had written to omit frames that are mostly to all black [*sic*] to save on space, bandwidth and overall quality of the video - chunks of time with just black screen is a bit of a waste."[752] That is an official admission that the missing frames in the Australian Arctic Division time-lapse videos during the Antarctic summer months are missing frames of nighttime darkness. That means that there is no 24 hours of sunlight during Antarctica summer, which in turn means that the earth cannot be a moving globe.

34 What Does the Bible Say?

It has been the accepted dogma of so-called "science" that the earth is a globe spinning faster than the speed of sound as it orbits the sun at approximately 80 times faster than the speed of a rifle bullet. The sun, in turn, is supposed to be hurtling through the Milky Way galaxy at more than 500 times faster than the speed of a rifle bullet. The Milky Way galaxy is, itself, alleged to be racing through space at a speed that some scientists claim is over 1,000 times faster than the speed of a rifle bullet.

God's word, however, states that the earth is fixed and does not move. "Fear before him, all the earth: the world also shall be stable, that it be not moved." (1 Chronicles 16:30) Indeed, the earth cannot be moved. "The LORD reigneth, he is clothed with majesty; the LORD is clothed with strength, wherewith he hath girded himself: the world also is stablished, that it cannot be moved." (Psalms 93:1).

Notice that the immovable earth is closely associated with the praise and glory of God Almighty. They are inseparable concepts. In Psalms 93:1 we read that the LORD is clothed with majesty and strength, just as the world is stable and cannot be moved. If the earth is movable, then it impeaches the majesty and strength of God. Psalms 104.1-5 makes it clear that God laid the foundations of the earth and stretched forth the heavens just as he is clothed with honor and majesty. The concepts of God's majesty and a stable immovable earth are inseparable. If the earth moves and spins, then the God of the Bible cannot exist. However, a stable immovable earth confirms the existence of God.

> Bless the LORD, O my soul. O LORD my God, thou art very great; thou art clothed with honour and majesty. Who coverest thyself with light as with a garment: who stretchest out the heavens like a curtain: Who layeth the beams of his chambers in the waters: who maketh the clouds his chariot: who walketh upon the wings of the wind: Who maketh his angels spirits; his ministers a flaming fire: Who laid the foundations of the earth, that it should not be removed for ever. (Psalms 104:1-5)

Knowledge of God is revealed through his creation. "The invisible things of him from the creation of the world are clearly seen, being understood by the things that are made, even his eternal power and Godhead." (Romans 1:20) To have a false concept of God's creation gives one a false concept of God. Who is the creator? The Bible gives the answer. Jesus created all things.

> For by him were all things created, that are in heaven, and that are in earth, visible and invisible, whether they be thrones, or dominions, or principalities, or powers: all things were created by him, and for him: And he is before all things, and by him all things consist. And he is the head of the body, the church: who is the beginning, the firstborn from the dead; that in all things he might have the preeminence. (Colossians 1:16-18)

If one believes in a creation that does not exist, he also necessarily believes in a creator that does not exist. It is important, therefore, to have an accurate understanding of God's creation. God did not make a moveable, spherical earth. A Jesus who creates a spherical, moving earth is a different Jesus from the true Jesus in the Bible. *See* 2 Corinthians 11:3-4. The sovereign reign of God can only be properly understood, if one first understands that God made a stationary, flat earth. In fact, Christians are called upon to tell the heathen that the LORD reigns and we are to also tell them that the world shall not be moved.

> O worship the LORD in the beauty of holiness: fear before him, all the earth. Say among the heathen that the LORD reigneth: the world also shall be established that it shall not be moved: he shall judge the people righteously. (Psalms 96:9-10)

Why are Christians not telling the heathen that the earth shall not be moved today? Because of the pride of life. Christians today are afraid of being labeled a "flat earther," which the devil and his minions have propagandized people to associate with being ignorant. It is a time honored practice of the devil to use the sinful flesh as a lever to quiet Christians.

There are many Christian "scientists" who try to ease their conscience by accommodating the lies of "science" with the truth of the gospel. They take the view that the Bible is true about everything, except scientific "facts." One example is Francis Collins, M.D., Ph.D., who is a prominent evangelical Christian geneticist, who is the director of the National Institute of Health and believer in the "science" of evolution. Dr. Collins stated that he reads and trusts the Bible about theological truths. He, however, suggests that discoveries in science are just as authoritative as the Bible and reconciles the conflicts between the Bible and so-called "science" by claiming that the Bible was not intended to be a textbook of science. He claims that the Bible is only a book about the nature of God and man.[753]

God contradicts charlatans like Collins. God's word is true regarding all things, both in theology and science. **"I esteem all thy precepts concerning <u>all things</u> to be right."** (Psalms 119:128) Indeed, Jesus stated that if one does not believe him about the earthly things that he explained (i.e., true science), they could not believe him regarding theological truths. "If I have told you earthly things, and ye believe not, how shall ye believe, if I tell you of heavenly things?" (John 3:12)

Jesus maintained the truth of the Genesis account of creation, which was written by Moses. Jesus told the Jews that if they did not believe Moses, they would not believe him. "For had ye believed Moses, ye would have believed me: for he wrote of me. But if ye believe not his writings, how shall ye believe my words?" (John 5:46-47) The King James Holy Bible titles the first book of the Bible as "The First Book of Moses, Called Genesis." Where did Moses write about Jesus? In Genesis 3:15 and Genesis 22:18.

The Jews, however, rejected Moses and went after their own traditions, which have been memorialized in the Talmud and Kabbalah. The Jews do not believe Moses, therefore they do not believe Jesus. One cannot believe the gospel of Jesus Christ without believing the creation account in Genesis. A Christian who says he believes Jesus Christ but rejects the creation account believes in a different Jesus. "For if he that cometh preacheth another Jesus, whom we have not preached, or if ye receive another spirit, which ye have not received, or another gospel, which ye have not accepted, ye might well bear with him." (2 Corinthians 11:4) All of (not just some of) God's words are pure truth. John 17:17.

We are admonished not to be seduced by the world and its

lusts. The desire for the love of the world is what has silenced the entire Christian church on the issue of heliocentricism.

> Love not the world, neither the things that are in the world. If any man love the world, the love of the Father is not in him. For all that is in the world, the lust of the flesh, and the lust of the eyes, and the pride of life, is not of the Father, but is of the world." (1 John 2:15-16)

God "laid the foundations of the earth, that it should not be removed for ever." Psalms 104:5. What kind of foundation is there in a spinning earth careening through space? Such a spinning, moving earth cannot be said to have any foundation. If there is no foundation, there is no God, since God stated he laid the foundations of the earth. In fact, the heavens prove the existence of God, as they demonstrate his skill as the creator. In Psalms 19:1 it states that the heavens declare the glory of God and the firmament shows his handiwork. Part of the glory of God and his handiwork is the fact that the earth is stationary and the sun travels in a circuit over the earth.

> The heavens declare the glory of God; and the firmament sheweth his handywork. Day unto day uttereth speech, and night unto night sheweth knowledge. There is no speech nor language, where their voice is not heard. Their line is gone out through all the earth, and their words to the end of the world. In them hath he set a tabernacle for the sun, Which is as a bridegroom coming out of his chamber, and rejoiceth as a strong man to run a race. His going forth is from the end of the heaven, and **his circuit unto the ends of it**: and there is nothing hid from the heat thereof. (Psalms 19:1-6)

It is the sun that travels in a circuit, not the earth. What is a circuit? It is a continuous circular route that starts and ends in the same place. Some might argue that the circuit of the sun is the sun's revolutions around the Milky Way galaxy. The problem with that interpretation is that Psalm 96:10 states "the world also shall be established that it shall not be moved." If the earth cannot be moved, yet the sun is careening through space, that means that the sun would be flying away from the earth.

The earth was created on the first day. Genesis 1:1. The sun was not created until the fourth day. Genesis 1:14-19. That means that the earth could not have been spinning on its axis and orbiting around the sun for the first three days, since the sun had not yet been created. Clearly, the earth was stationary.

The earth started its existence "without form, and void." Genesis 1:2. God then gave the earth form, and that form was flat. We know that because the surface of the waters God created had to have been flat, just as the surface of water is today. Indeed, "God moved upon the face of the waters." Genesis 1:2. As the face on a clock is flat, so also is the face of the water on the earth flat.

Gravity did not exist, because there was no need for gravity. The waters, grass, herbs, and trees, which were all created in the first three days, were all held to the earth by their density. How and why would the earth suddenly on the fourth day transform into the shape of a ball, begin spinning on its alleged axis, and then start orbiting the sun? What was the force that began the alleged spinning and orbit? Why is that transformation and movement not mentioned in the Bible?

It is only if the earth were a sphere would it be necessary for gravity. Where was the alleged force of gravity before the earth supposedly turned into a ball and started spinning? Gravity was not necessary before the alleged transformation of the earth into a spherical spinning ball. How did it spring into existence? Gravity does not spring into existence upon spinning spherical objects on earth. The earth and the objects on the earth obey the same laws of physics. And so if gravity was a reality on earth, spinning spherical objects on earth should develop gravity, but they do not. Objects do not develop gravity on earth, because gravity is a myth, and so is a spherical spinning earth.

It is God who created the heavens and the earth, and he had a purpose for doing so. It was to be inhabited by man, whom he created in his image.

> For thus saith the LORD that created the heavens; God himself that formed the earth and made it; he hath established it, he created it not in vain, he formed it to be inhabited: I am the LORD; and there is none else. (Isaiah 45:18)

God's creation of heaven and earth is part and parcel of his creative powers each day when he forms each person in the womb. If people lose sight of God as creator of all things, then any sin, including abortion can be justified. It is important for people to understand the nature of God's creation, because it reveals his character. In Isaiah 44:24 God states that he alone stretched forth the heavens and spread abroad the earth. A ball is not spread out. You spread out things that are flat, like a flat bed spread. God spread out the earth and he stretched forth the heavens above the earth.

> Thus saith the LORD, thy redeemer, and he that formed thee from the womb, I am the LORD that maketh all things; that stretcheth forth the heavens alone; that spreadeth abroad the earth by myself. (Isaiah 44:24)

The account of creation in Genesis refutes the commonly held view of a spherical earth that is surrounded by an endless vacuum of "space." In Genesis 1:6 we read that "God said, Let there be a firmament in the midst of the waters, and let it divide the waters from the waters." A firmament is something that is hard. Indeed, in Job the firmament is likened to a strong, molten looking glass. "Hast thou with him spread out the sky, which is strong, and as a molten looking glass?" (Job 37:18) A looking glass is a highly polished reflective surface.

The firmament divides the waters, which means that there is water above the firmament in heaven. Indeed, all one needs to do is look skyward to see the blue water that is above the firmament. The blue sky is not the atmosphere as claimed by scientists; the blue sky is the water above the canopy of the firmament. "Praise him, ye heavens of heavens, and ye waters that be above the heavens." (Psalms 148:4) God placed the sun, moon, and stars in the firmament.

> And God said, Let there be lights in the firmament of the heaven to divide the day from the night; and let them be for signs, and for seasons, and for days, and years: And let them be for lights in the firmament of the heaven to give light upon the earth: and it was so. (Genesis 1:14-15)

When God flooded the earth he opened the windows of heaven to let out some of the water that was above the firmament.

> In the six hundredth year of Noah's life, in the second month, the seventeenth day of the month, the same day were all the fountains of the great deep broken up, and the windows of heaven were opened. (Genesis 7:11)

God formed the foundation of the earth from the beginning. "Of old hast **thou laid the foundation of the earth**:

and the heavens are the work of thy hands." (Psalms 102:25) God laid the foundation of earth immovable forever. "Who laid the foundations of the earth, that it should not be removed for ever." (Psalms 104:5) Please note that God laid the foundations of the earth, which does not at all suggest a globular earth. "**Mine hand also hath laid the foundation of the earth**, and my right hand hath spanned the heavens: when I call unto them, they stand up together." (Isaiah 48:13) The foundation is stated to be pillars. "He raiseth up the poor out of the dust, and lifteth up the beggar from the dunghill, to set them among princes, and to make them inherit the throne of glory: for **the pillars of the earth are the LORD'S, and he hath set the world upon them**." (1 Samuel 2:8) What is the foundation of the earth? One could infer that it is the seas. "The earth is the LORD'S, and the fulness thereof; the world, and they that dwell therein. **For he hath founded it upon the seas, and established it upon the floods**." (Psalms 24:1-2) Indeed, God states that he stretched out the heavens over the earth, which suggests a flat earth. "I have made the earth, and created man upon it: I, even my hands, **have stretched out the heavens**, and all their host have I commanded." (Isaiah 45:12) God has repeatedly referred to the earth as being founded upon water. "To him that **stretched out the earth above the waters**: for his mercy endureth for ever." (Psalms 136:6) Notice that the earth is "stretched out" above the waters. A flat earth could be said to be "stretched" out, but such an expression would preclude a globular earth.

Nowhere in the Bible does God state that the earth is a globe. In fact, God expressly states that the face of the earth is a circle. "**It is he that sitteth upon the circle of the earth**, and the inhabitants thereof are as grasshoppers; that **stretcheth out the heavens as a curtain, and spreadeth them out as a tent to dwell in**:" (Isaiah 40:22) A circle is two-dimensional and flat. That does not mean that the earth is two dimensional, as the earth certainly has depth. Rather, it means that the earth has a face that is a flat circle. Amos 9:6. It is "the circle **of** the earth." A globe, on the other hand, is a three dimensional ball. Isaiah knew the difference between a ball and a circle. *See* Isaiah 22:18. If Isaiah meant ball in Isaiah 40:22, he would have said ball. He didn't say ball, because the earth is not a ball. There can be no confusion here. God is stating that the face of the earth is a flat circle and the heavens are spread like a tent over that circle.

In the Bible, it states in reference to wisdom: "I was there: when he set a compass upon the face of the depth." (Proverbs 8:27) Notice God set a compass upon the face of the depth. What does compass mean in that context? Certainly, a compass is a directional instrument, but the word compass means more than that. Compass also means a circular boundary; the limit or boundary of a space, and the space included; encircle; to surround; to inclose on all sides; a circular course.[754] A compass is a flat circular boundary. In Proverbs 8:27 we have God encircling the seas with a boundary. That can only happen on a flat earth. The boundary of the seas is the ice rim of Antarctica. The globular myth is that one can circumnavigate the globe by going around Antarctica, because it is believed to be a continent in the southern most part of a spherical earth. The globular earth myth does not have a circular boundary for the seas, as stated in Proverbs 8:27. "Let God be true, but every man a liar." (Romans 3:4)

God's creation of a stationary earth is a testimony to his truth. In Psalms 33:4-9 we find that just as "the word of the LORD is right," so also "he spake, and it was done." In the same way, "all his works are done in truth," so also "he commanded, and it stood fast." Just as his word is true, so also does the earth, that was created by his word, stand fast.

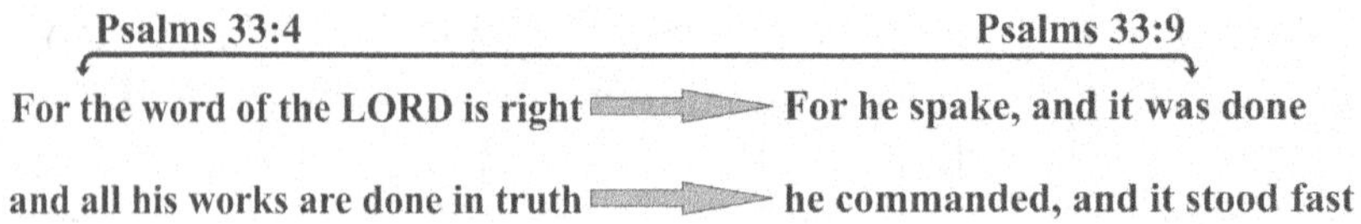

> **For the word of the LORD is right; and all his works are done in truth.** He loveth righteousness and judgment: the earth is full of the goodness of the LORD. By the word of the LORD were the heavens made; and all the host of them by the breath of his mouth. He gathereth the waters of the sea together as an heap: he layeth up the depth in storehouses. Let all the earth fear the LORD: let all the inhabitants of the world stand in awe of him. **For he spake, and it was done; he commanded, and it stood fast**. (Psalms 33:4-9)

In Psalms 33:4-9 David explains that all of God's works are done in truth. God's creation of an immovable earth that stands fast testifies to God's truth. God spake his truth and through that true and sure word he created a fixed earth. Just as the earth is fixed and stands fast, so also is God's word true. If the earth is moving, it means that God's word is variable and cannot be true. The mythology of a spinning, orbiting earth calls into question the truth of God's word. If the earth is orbiting and spinning that means that the Bible, which is God's word, cannot be true. But a spinning, orbiting earth is a lie, and God's creation of an earth that stands fast testifies to his truth.

Five verses later, in Psalms 33:14, God describes his reign over what can only be a flat earth. David segues from explaining that the earth stands fast to, five verses later in Psalms 33:14, explaining that God looks upon all the inhabitants of the earth from his habitation. Where is God's habitation? God's habitation is in heaven. God's throne is heaven and the earth as his footstool. "Thus saith the LORD, **The heaven is my throne, and the earth is my footstool**: where is the house that ye build unto me? and where is the place of my rest?" (Isaiah 66:1) Footstools are not spherical. God can only look over the inhabitants of the earth from one location if the earth is flat. So, within five verses in Psalms 33 we have God describing a stationary, flat earth.

God states that he looks upon all the inhabitants of the earth from one location, his place of habitation in heaven. "From the place of his habitation he looketh upon all the inhabitants of the earth." (Psalms 33:14) It is not possible for God to look upon all of the inhabitants of the earth on a globe from one location; those on the backside of the earth would be hidden from view. People might argue that God is all powerful and therefore can do anything. Yes, but Satan is not all powerful. In Matthew 4:8 and Luke 4:5, the devil took Jesus up to a high mountain and showed him all the kingdoms of the world. The devil did not take Jesus around the world on a tour of the kingdoms. The devil showed him all of the kingdoms from one point on a high mountain. How could the devil have shown all the kingdoms of the world from a single spot on a high mountain on a globular earth? Such an event could only be done on a flat earth.[755]

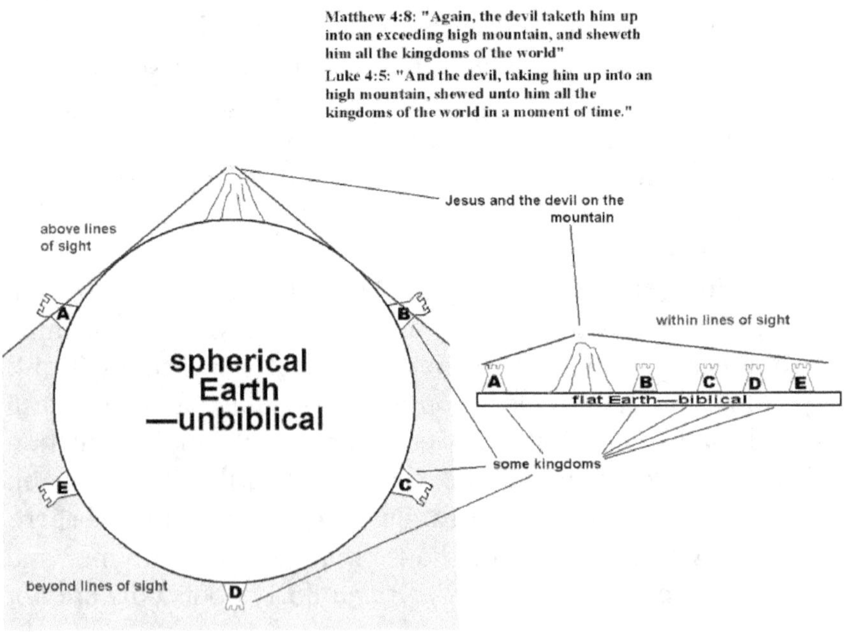

Some argue that Satan had simply shown Jesus a vision of the kingdoms of the world and thus the event is no proof that the earth is flat. The problem with that argument is that it would not be required for Satan to take Jesus up to a high mountain simply to show him a vision of the kingdoms of the supposed spherical earth. Satan could have done that from where they were standing. There was a reason Satan took Jesus up to an "exceeding high mountain;" and that reason was to show Jesus the kingdoms of the world that span the flat earth.

When Jesus returns to earth, everyone will see him in the sky. "Behold, he cometh with clouds; and **every eye shall see him**, and they also which pierced him: and all kindreds of the earth shall wail because of him. Even so, Amen." (Revelation 1:7) Jesus will return physically "with clouds." That is a fulfillment of the prophecy in Mark 13:26 that Jesus will physically return "in the clouds." That would only be possible on a flat earth. If the earth were a globe those on the back side of the globe could not see him. Some may argue that the prophecy of Jesus' return in Revelation 1:7 would be worked out by God in some miraculous way so that everyone on the supposed spherical earth will see his return. That view is impeached by Acts 1:9-11, where God makes clear that Jesus will return to earth from heaven just as he ascended from earth to heaven. The only way that every eye on earth could see Jesus' descent from heaven in that manner is on a flat earth.

> And when he had spoken these things, while they beheld, he was taken up; and a cloud received him out of their sight. And while they looked stedfastly toward heaven as he went up, behold, two men stood by them in white apparel; Which also said, Ye men of Galilee, why stand ye gazing up into heaven? this same Jesus, which is taken up from you into heaven, shall so come in like manner as ye have seen him go into heaven. (Acts

1:9-11)

God describes the face of the earth as a circle. He also describes the earth as having four corners. "And after these things I saw four angels standing on the **four corners of the earth**, holding the four winds of the earth, that the wind should not blow on the earth, nor on the sea, nor on any tree." Revelation 7:1. See also Isaiah 11:12. While the word corner brings to mind a sharp angle, that is not its only meaning. A corner is not just where two edges meet. Corner also means parts, or regions. Indeed, not all corners are sharp angles; we often refer to driving on a rounded curve in the road as going around a corner. A corner is also defined as "indefinitely any part; a part. ... The end, extremity or limit."[756] The following phrases given in the American Dictionary of the English Language are examples of how corner means more than the place where two edges meet: "They searched every corner of the forest. They explored all corners of the country."[757] It also has the meaning of "direction" or "quarter."[758] It could also be "a small area away from the center, especially one that is quiet, peaceful, or secret."[759] The Merriam Webster Dictionary defines corner as "a private, secret, or remote place,"[760] and gives that example of "a quiet corner of New England" and "to every corner of the earth."[761]

The context of Revelation 7:1 suggests that corners mean remote regions of the earth. Indeed, God refers to the ends of the earth. There are no "ends" to a globular earth.

> He hath made the earth by his power, he hath established the world by his wisdom, and hath stretched out the heaven by his understanding. When he uttereth his voice, there is a multitude of waters in the heavens; and he **causeth the vapours to ascend from the ends of the earth**: he maketh lightnings with rain, and bringeth forth the wind out of his treasures. Jeremiah 51:15-16. See also Jeremiah 16:19.

Indeed, the context of corners suggests that God in fact means ends of the earth. Elsewhere in the Bible, God describes the lightning going out to the ends of the earth. "He directeth it under the whole heaven, and his lightning unto **the ends of the earth**." (Job 37:3) God looks out over the whole earth at once and sees to the ends of the earth; again, that suggests a flat earth. "For he looketh to **the ends of the earth**, and seeth under the whole heaven." (Job 28:24) Modern science has replaced God's abode of heaven with the myth of the empty vacuum of space, thus suggesting that there is no heaven and there is no God. If there is only an empty vacuum of space, that means that God cannot be above us walking in the circuit of heaven. "Thick clouds are a covering to him, that he seeth not; and he walketh in the circuit of heaven." (Job 22:14)

In Daniel we read of his vision of a tree that was so high that it could be seen to the end of the earth. There is no end to the earth on a globe; such a tree could only be seen at the end of a flat earth. "The tree grew, and was strong, and the height thereof reached unto heaven, and the sight thereof to the end of all the earth:" (Daniel 4:11)

All builders know what it is to "snap a line" when building to ensure that objects are laid even and level. When snapping a line, one stretches a cord that is coated with chalk and it is snapped against the surface of whatever is being constructed. The line leaves a perfectly straight chalk line between the two points of the line. You cannot stretch a line on a ball. Notice that in Job 38 God stretched the line upon the earth, which suggests a flat earth. Notice also how God takes hold of the ends of the earth. A globe does not have ends that can be held.

> Where wast thou when I laid the **foundations of the earth**? declare, if thou hast understanding. Who hath laid the measures thereof, if thou knowest? or who hath **stretched the line upon**

> **it**? Whereupon are the **foundations thereof fastened**? or who laid the corner stone thereof; When the morning stars sang together, and all the sons of God shouted for joy? Or who shut up the sea with doors, when it brake forth, as if it had issued out of the womb? When I made the cloud the garment thereof, and thick darkness a swaddlingband for it, And brake up for it my decreed place, and set bars and doors, And said, Hitherto shalt thou come, but no further: and here shall thy proud waves be stayed? Hast thou commanded the morning since thy days; and caused the dayspring to know his place; **That it might take hold of the ends of the earth, that the wicked might be shaken out of it**?" (Job 38:4-13)

The Bible prophesies that the stars will fall to the earth. "And the stars of heaven fell unto the earth, even as a fig tree casteth her untimely figs, when she is shaken of a mighty wind." (Revelation 6:13) The prophecy of the stars falling to earth would be ridiculous if the stars were large suns millions of light years away. Of course, that is the point of modern cosmology, to refute the truth and authority of the Bible. If the modern cosmology were correct, a single star would engulf the earth if it struck the earth. Modern so-called science serves to undermine the word of God.

God pours water upon the "face of the earth." "It is he that buildeth his stories in the heaven, and hath founded his troop in the earth; he that calleth for the waters of the sea, and poureth them out upon **the face of the earth**: The LORD is his name." (Amos 9:6) Everyone knows that when speaking of objects, "face" is the flat visible side of an object. For example, we commonly speak of the face of a clock. Never is a ball or globe referred to as having a face. We would never refer to a baseball as having a face. Indeed, it cannot have a face, it is round. Face implies the front of an object. Persons have faces, which is because a person's face is considered the front of his head, which is not at all like the back of his head. A spherical earth cannot have a face, because there is no front to a globe. However, both God and man refer to the face of the earth, which denotes that the earth is flat. "Thou sendest forth thy spirit, they are created: and thou renewest **the face of the earth**. (Psalms 104:30)

Indeed, God describes the earth in Job 38:14 as being "turned as clay to the seal." The tradition of placing a seal on documents using clay dates to antiquity. The wet clay was flattened under pressure using a seal, which was a hard die, called a signet. The signet was pushed into the clay. This was often done to authenticate royal decrees. The clay then dried with the positive impression from the negative image on the royal signet left visible in the flat pressed clay. The clay image itself was known as the seal. All such seals were necessarily pressed flat.

God calls the process, of creating the clay seal of the earth, being "turned." The word turn in this context is pregnant with meaning. Turn means "to form; to shape,"[762] and "to transfer."[763] That is what happens when the image on the signet is pressed against the clay. That process shapes the clay and transfers the image from the signet to the clay. Turn also means "to reverse the sides or surfaces of: invert."[764] And that is what we see in the positive image left in the clay from the negative image on the signet. Turn has yet another meaning: "to cause to move in a circular course,"[765] "to cause to move around an axis or a center."[766] In the context of a clay seal as described in Job 38:14, it implies that the flat clay seal is being "turned" into a circle. Thus, God's description of the earth being "turned as clay to the seal" clearly describes a flat earth, and it suggests that it is a circle. *See also* Isaiah 40:22, God "sitteth upon the **circle** of the earth."

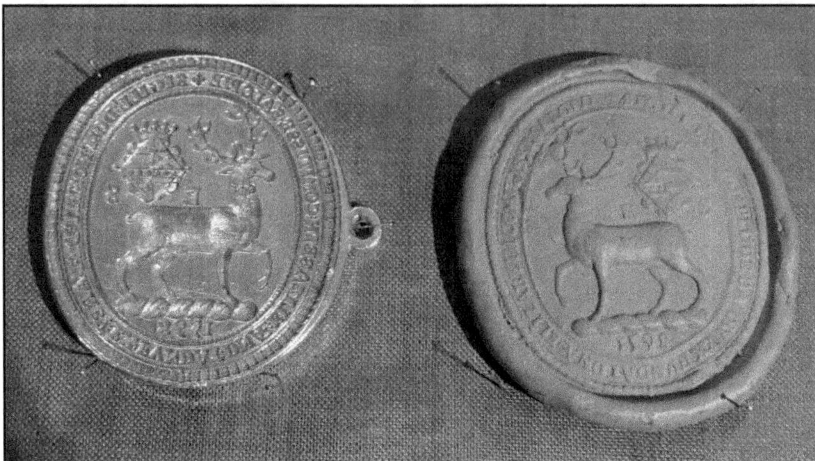

"It [the earth] is turned as clay to the seal." Job 38:14

Joshua 10:12-13 poses a real problem for the Christians who believe in a heliocentric model of a rotating earth orbiting around the sun. In Joshua 10:12-13, God caused the sun and the moon to stand still. In order for Christians to hold to heliocentricity, they must necessarily call God a liar, because God moved Joshua to give a command to both the sun and the moon to stand still in Joshua 10:12-13; whereby God stopped both the sun and moon in their motion across the sky.

> Then spake Joshua to the LORD in the day when the LORD delivered up the Amorites before the children of Israel, and he said in the sight of Israel, Sun, stand thou still upon Gibeon; and thou, Moon, in the valley of Ajalon. And the sun stood still, and the moon stayed, until the people had avenged themselves upon their enemies. Is not this written in the book of Jasher? So the sun stood still in the midst of heaven, and hasted not to go down about a whole day. (Joshua 10:12-13)

God states clearly in the passage that the sun stood still. The passage does not say that the earth stopped spinning. "Scientists" claim that the earth is spinning at 1,000 miles per hour at the equator and that motion of the earth accounts for the path of the sun across the sky. If that were the case, the event in Joshua could never have happened. Any vehicle coming to a sudden stop at 1,000 miles per hour would kill all the passengers aboard. It would be a bloody mess. The same would happen on a supposedly spinning earth. A sudden stop of the supposed spin of the earth would have killed every living thing on earth. God would have killed the very army of Israel that he was helping. That makes no sense.

It was not the earth that stopped spinning. God did just what he said he did; he caused the sun to stop and stand still in the sky. Joshua did not command the earth to stop spinning. The passage states that Joshua commanded the "Sun, stand thou still." God moved Joshua to command the sun to stop and "the sun stood still" by the power of God. Indeed, God states that it "hasted not to go down about a whole day." That means that ordinarily the sun would be hasting to go down, but in this particular case it did not. The sun is ordinarily in motion, coming up and going down. On that day, God stopped the sun's usual motion in the sky. To suggest otherwise is to call God a liar.

God's intention in helping Israel was to stop the sun. Stopping the moon was not helpful to Israel. However, once God stopped the sun, he also stopped the moon, because God wanted to keep the motions of the sun and moon synchronized. The synchronization of the sun and moon is part of God's creation. God made the sun and moon to be "lights in the firmament of the heaven to divide the day from the night; and let them be for signs, and for seasons, and for days, and years." Genesis 1:14. He made the sun to rule the day and the moon to rule the night. Genesis 1:15-16. In order to keep that purpose intact, once God stopped the sun, it was necessary for God to also stop the moon.

Notice also that Joshua ordered the sun to stand still upon Gibeon and the Moon, to stand still in the valley of Ajalon. The sun could not be said to be "upon Gibeon" if the sun were more than 100 times larger than a spherical earth and 93 million miles away in distant space, as it is portrayed under the heliocentric model. Furthermore, the heliocentric model puts the moon 240,000 miles away from a globular earth in distant space, which would preclude it from being "in the valley of Ajalon" as stated by God in Joshua 10:12-13. The sun and moon both must be much smaller than and closer to the flat earth in order to stand still over the areas in which they were located in Joshua, chapter 10.

"Scientists" acknowledge that it is the moon that moves over the earth. They wrongly opine that it is orbiting around a spherical earth. If God were stopping the sun by stopping the spin of the earth, the "scientific" view of the independently moving moon would require that the moon keep moving in its supposed orbit around the spherical earth. But that is not what happened. God stopped the moon in that same way he stopped the sun; he stopped it in its usual circuit over a stationary, flat earth.

Satan has tried to muddy the waters so that Christians will not discover the truth of the flat unmovable earth at the center of God's creation. He must do this by having his minions publish corrupt versions of the Bible. For example, corrupt Bible versions like the New International Version (NIV) have been trotted out with corrupt passages that obfuscate the clarity of God's word. Despite the fact that God's word states clearly the earth is immovable, the new Bible versions wish to change God's word to comport with what they believe is the "scientific fact" that the earth is rotating on its axis and at the same time orbiting around the sun. In 2 Kings God reveals a miracle he performed by making the shadow cast by the sun on Ahaz's sundial to reverse and go back ten degrees.

And Isaiah said, This sign shalt thou have of the

> LORD, that the LORD will do the thing that he hath spoken: shall the shadow go forward ten degrees, or go back ten degrees? And Hezekiah answered, It is a light thing for the shadow to go down ten degrees: nay, but let the shadow return backward ten degrees. And Isaiah the prophet cried unto the LORD: and he brought the shadow ten degrees backward, by which it had gone down in the dial of Ahaz. (2 Kings 20:9-11 AV)

Now, if you accept what God says in his word as true that the earth cannot be moved, it must have been the sun that went back ten degrees. In fact, in Isaiah 38:8, God reveals that is exactly what happened; the sun moved back ten degrees by which degrees it had already gone down. In order for the sun to return the ten degrees that it had already gone down, the sun must have been moving across the sky in its ordinary path before its reversal. Hence, the earth is stationary and the sun revolves around the earth.

Because the New International Version (NIV) Bible translators do not believe God's word, they have changed the passage in Isaiah 38:8 to state that "sunlight" went back ten steps on the "stairway," rather than what actually happened, that the sun itself returned ten degrees. Simply stated, they have changed the verse to comport with a heliocentric view of the universe.

AV	NIV
Behold, I will bring again the shadow of the degrees, which is gone down in the sun dial of Ahaz, ten degrees backward. **So the sun returned ten degrees, by which degrees it was gone down.** (Isaiah 38:8 AV)	I will make **the shadow cast by the sun go back the ten steps** it has gone down on the stairway of Ahaz. So the **sunlight went back the ten steps** it had gone down. (Isaiah 38:8 NIV)

By changing the passage to say that the sunlight went back instead of the sun, the NIV translators have removed the fact that the sun is moving and have allowed for an explanation that the earth reversed its rotation, thus causing the sunlight to move back. Furthermore, they have removed the miracle of the event entirely by stating that the shadow went back ten steps on a stairway, rather than ten degrees on a sundial. A shadow cast by a pillar can go up and then back down steps due to the ordinary travel of the sun across the sky, however, the shadow cast upon a sundial cannot move backwards unless the sun moves backwards.

The corrupt new Bible versions have concealed from the reader the admonition to avoid the opposition of "**science** falsely so called." God clearly had in mind to warn his flock about heliocentricity and evolution. Instead, the reader of the new Bible versions is told to turn away from "what is falsely called **knowledge**." There is no field of study called "knowledge," but there is a whole discipline of study called "science." The admonition to avoid so-called knowledge, therefore, has no real meaning. That is the point of the new Bible versions: to strip the Bible of any real guidance.

"O Timothy, keep that which is committed to thy trust, avoiding profane and vain babblings, and oppositions of **science** falsely so called." 1 Timothy 6:20 AV (KJV) (emphasis added).

Timothy, guard what has been entrusted to your care. Turn away from godless chatter and the opposing ideas of what is falsely called **knowledge**. 1 Timothy 6:20 (NIV) (emphasis added).

The corrupt Bible versions leave the reader without any warning against so-called science. Bible corruption is a centuries old plan by Satan. One of the early Bible corruptions was the Roman Catholic Douay-Rheims Bible, which was an English translation of the Latin Vulgate. It was published in 1610. In 1750 the Roman Catholic Church published a revision of the Douay-Rheims Bible under the guidance of Roman Catholic Bishop Richard Challoner (1691–1781). The revision changed the word "compass" in the original 1610 Douay-Reims Bible in the verse at Isaiah 40:22 to "globe" in the 1750 revision, just in time to hop on the heliocentric bandwagon. That Bible is given over entirely to following after "science falsely so called" in direct opposition to God's admonition. In the passage at Isaiah 40:22, the Douay-Rheims Bible changes the word "circle" to "globe," to comport with the godless helicentric model that has since become all the rage in "science falsely so called."

"It is he that sitteth upon the **circle** of the earth, and the inhabitants thereof are as grasshoppers; that stretcheth out the heavens as a curtain, and spreadeth them out as a tent to dwell in:" Isaiah 40:22 AV (KJV) (emphasis added).

"It is he that sitteth upon the **globe** of the earth, and the inhabitants thereof are as locusts: he that stretcheth out the heavens as nothing, and spreadeth them out as a tent to dwell in. Isaiah 40:22 (Douay-Rheims Version) (emphasis added).

The Authorized (King James) Version (indicated by AV or KJV) of the Bible is God's word in the English language. The King James Bible is a faithful English translation of God's word that is largely based upon the authoritative Greek text known as the *Textus Receptus* (Received Text).

Every other English Bible version, including the NIV, is corrupted by the servants of Satan. The NIV is based upon corrupted texts that were provided by the Roman Catholic Church. The two primary manuscripts that form the basis for the NIV, and indeed all modern English Bibles, except for the Authorized (King James) Version, are the *Sinaiticus* and the *Vaticanus*. The manuscript *Sinaiticus*, which is often referred to by the first letter of the Hebrew alphabet, *Aleph*, is written in book form (codex) on velum.[767] It contains many spurious books such as the Shepherd of Hermes, the Didache, and the Epistle of Barnabas.[768] *Sinaiticus* was allegedly discovered by Constantine Tischendorf (1815-1874) in a wastebasket in St. Catherine's monastery on Mount Sinai in February of 1859.[769] The monks of St. Catherine's denied Tischendorf's wastebasket story. The wastebasket story seems to have been a tale spun by Tischendorf to explain his mutilation of the codex and destruction of some leaves to conceal acrostics that would identify the author of the codex. *Sinaiticus* is covered with alterations that are systematically spread over every page and were made by at least ten different revisors.[770] The alterations are obvious to anyone who examines the manuscript.[771]

Tischendorf was a nominal Protestant who had very close ties with the Roman Catholic hierarchy. He was welcomed at the Vatican by the pope in a private audience. Tischendorf also had meetings in the Vatican with the Vatican Librarian, Cardinal Angelo Mai (1782-1854), who, incidently, wrote an 1838 edition of the *Codex Vaticanus*.[772] *Sinaiticus* is not an ancient fourth century codex, as alleged by Tischendorf. Indeed, the author of *Sinaiticus* came forward and stated publically that it was in the early nineteenth century that he authored what became known as *Sinaiticus*.[773] When Constantine Simonides (1820-1867), a renowned Greek scholar and paleographer, found out that the Greek manuscript of the Bible that he wrote when he was at the Monastery of Saint Catherine was being promoted by Constantine Tischendorf as a fourth century manuscript, he refuted that claim. Simonides invited Tischendorf to a debate on the issue and challenged Tischendorf to bring *Sinaiticus* with him so that Simonides could demonstrate, with the public watching, where he had made unique markings that only the author of the document would know. Tischendorf agreed to the debate, but never showed up for the planned debate and never explained his absence. Tischendorf instead chose to use the media organs controlled by the Roman Catholic church to attack Simonides' character. To this day, Simonides is libelously accused of forgery, always with qualifiers like "allegedly" and "claimed to be," etc. What is almost never mentioned is that the allegations of forgery were actually litigated and Simonides was exonerated in a court of law.

It is a matter of public record, however, that a Greek Archimandrite (superior abbot) named Kallinikos, who was a resident of the St. Catherine's Monastery, came forward and verified that he witnessed, with his own eyes, Simonides writing *Sinaiticus* at Mount Athos in Greece.[774] Furthermore, Kallikos stated that he was at St. Catherine's Monastery and saw Tischendorf in possession of the manuscript that he recognized as being the work of Simonides.[775] Kallinikos accused Tischendorf of guile. Kallinikos stated that *Sinaiticus* was not an ancient text but was washed with lemon juice and herbs to lighten the text and make it appear to be an ancient text. He further accused Tischendorf of mutilating the codex written by Simonides.

The manuscript *Vaticanus*, often referred to by the letter "B,"originated in the Vatican library, hence the name.[776] *Vaticanus* was first revealed in 1841; where the manuscript had been prior to that date is unclear.[777] One thing is clear, the manuscript omits many portions of scripture which explain vital Christian doctrines. *Vaticanus* omits Genesis 1:1 through Genesis 46:28; Psalms 106 through 138; Matthew 16:2,3; Romans 16:24; the Pauline Epistles; Revelation; and everything in Hebrews after 9:14.[778]

It should not be surprising that the Vatican would produce a manuscript that omits the portion of the book of Hebrews which exposes the Mass as completely ineffectual and deletes Revelation chapter 17, which reveals Rome as the seat of "MYSTERY, BABYLON THE GREAT, THE MOTHER OF HARLOTS AND ABOMINATIONS OF THE EARTH." It is notable that the two primary manuscripts used by the new Bible versions were found in the care and custody of the Roman Catholic Church.

The *Vaticanus* and *Sinaiticus* manuscripts, which make up less than one percent of the existing ancient manuscripts, differ significantly from the Received Text that is the basis for the Authorized (King James) Version. *Vaticanus* omits at least 2,877 words; it adds 536 words; it substitutes 935 words; it transposes 2,098 words; and it modifies 1,132 words; making a total of 7,578 verbal divergences from the Received Text. *Sinaiticus* is an even worse corruption, having almost 9,000 divergences from the Received Text.[779]

Those corrupted texts (*Vaticanus* and *Sinaiticus*) became

the core of the modern Greek text that is the basis for all new Bible versions, including the popular NIV. The personalities behind the new texts have an occult new age agenda. The compilers of the corrupted Greek text used in virtually all of the new Bible versions were Brooke Foss Westcott and Fenton John Anthony Hort.[780]

Westcott and Hort were nominal Protestants, but they were defacto Roman Catholics. Hort denied the infallibility of the Holy Scriptures, he did not believe in the existence of Satan, he did not believe in eternal punishment in Hell, nor did he believe in Christ's atonement.[781] Hort, however, did believe in Darwin's theory of evolution, he believed in purgatory, and he also believed in baptismal regeneration.[782] Hort hated the United States and wished for its destruction during the civil war, because he was a communist who hated all things democratic.[783]

Westcott was equally Romish in his beliefs.[784] He, like Hort, rejected the infallibility of the Holy Scriptures.[785] He viewed the Genesis account of creation as merely an allegory.[786] He did not believe the biblical account of the miracles of Jesus.[787] He did, however, believe in praying for the dead and worshiping Mary.[788] Politically, Westcott was a devout Socialist.[789]

Westcott and Hort were both necromancers who were members of an occult club called the "Ghostly Guild."[790] Westcott also founded another club and named it "Hermes."[791] According to Luciferian H.P. Blavatsky, Hermes and Satan are one and the same.[792] Hort viewed evangelical Christians as dangerous, perverted, unsound, and confused.[793]

Assisting Westcott and Hort in creating the corrupt Greek revision was Dr. G. Vance, a Unitarian, who denied the deity of Christ, the inspiration of the Holy Scriptures, and the Godhead (Jesus Christ, God the Father, and the Holy Ghost).[794] Jesuit Roman Catholic Cardinal Carlo Maria Martini, the prelate of Milan, was the editor of the corrupted Greek text.[795] Martini believed the occult new age philosophy that man can become divine.[796] Remember, that is the very lie that Satan used to deceive Eve into eating the forbidden fruit: "ye shall be as gods." Genesis 3:5.

The combined effect of having a corrupted text and then having that text interpreted using a free-form translation method known as dynamic equivalence has been that the NIV has 64,098 fewer words than the King James Bible.[797] That is a 10% loss in the Bible. That means that an NIV Bible would have 170 fewer pages than a typical 1,700 page King James Bible.[798] Let's read what God thinks about such deletions of his holy words. "And if any man shall take away from the words of the book of this prophecy, God shall take away his part out of the book of life, and out of the holy city, and from the things which are written in this book." Revelation 22:19. God takes the misuse of his name very seriously (Exodus 20:7), but it is even more serious to tamper with God's word. [T]hou hast magnified thy word above all thy name. (Psalms 138:2)

The New International Version (NIV) is the most popular of the new Bible versions. Reportedly, the NIV represents 45 % of all Bibles sold.[799] Dr. Virginia Mollenkott, the textual style editor for the NIV, is an admitted lesbian.[800] The Chairman of the NIV Old Testament Committee, Dr. Woudstra, was considered to be sympathetic to the interests and practices of sodomites. The NIV chief editor vaunted the fact that the NIV showed that it is a great error to believe that in order to be born again one has to have faith in Jesus as Savior.[801]

Rupert Murdoch owns the exclusive rights to the NIV.[802] The NIV is published by Zondervan, which is owned by Murdoch's News Corporation.[803] Murdoch's News Corporation also owns Harper Collins, the publisher of Anton LaVey's Satanic Bible. LaVey is the founder of The Church Of Satan. He is reputed to have been a Jew, whose real name was Howard Levey.

Both the Satanic Bible and the NIV Bible are featured on the Harper Collins sales website.[804] "Can two walk together, except they be agreed?" (Amos 3:3 AV) LaVey's book is an exoteric Satanic Bible, whereas the NIV is an esoteric Satanic Bible.

Time magazine called Murdoch one of the four most powerful people in the world, and for good reason; he has a media empire that includes Twentieth Century Fox, Fox Television, cable television providers, purported "satellites," and newspapers and television stations throughout America, Europe, and Asia.[805] The pope bestowed upon Murdoch the title of "Knight Commander of St. Gregory" for promoting the interests of the Roman Catholic Church.[806]

Defenders of the new Bibles claim that the essential doctrines of the Christian Faith are expressed in the new Bibles, even though they have been deleted or changed in many passages. James H. Son, author of *The New Athenians*, likened the logic of that argument to removing a stop sign from a busy street intersection and then justifying the removal because the other traffic signals in the city were left intact. Even though the sign only contained one word, that word is of critical importance to those who arrive at the intersection, just as each word in the Holy Bible is of critical importance to those who are reading it. The above example of the change in the passage in Isaiah 38:8, is just one of many examples of subtle changes in wording in the new Bible versions used to change the true meaning of a passage.

God has made the point in the Holy Bible that **every word** of God is important. "And Jesus answered him, saying, It is written, That man shall not live by bread alone, **but by every word of God**." (Luke 4:4 AV) Incidently, the doctrine of Luke 4:4 is missing in the new Bible versions. The NASB, which is another corrupt Bible version based upon the *Vaticanus* and *Sinaiticus* manuscripts, leaves out the last clause and simply states: "And Jesus answered him, 'it is written, MAN SHALL NOT LIVE ON

BREAD ALONE.'" (Luke 4:4 NASB) The new versions leave the reader in ignorance as to what it is other than bread by which man lives: **"by every word of God."** (Luke 4:4 AV)

> And he humbled thee, and suffered thee to hunger, and fed thee with manna, which thou knewest not, neither did thy fathers know; that he might make thee know that man doth not live by bread only, but **by every word that proceedeth out of the mouth of the LORD doth man live**. (Deuteronomy 8:3 AV)

> **Every word of God is pure**: he is a shield unto them that put their trust in him. (Proverbs 30:5 AV)

35 Crushing Gravity

Under the heliocentric theory, the earth is spinning at approximately 1,000 miles per hour at the equator. The heliocentric scientists had a problem with their theory. How could they explain how people, animals, and things do not feel the centrifugal force of the spinning earth? Isaac Newton saved the day with his theory of gravity. Newton's theory of gravity supposedly acts as a centripetal force working against the centrifugal force of the hypothesized spinning earth. Gravity is necessary on a spinning earth. But the earth is not spinning. On a stationary flat earth there is no centrifugal force. Since there is no centrifugal force, there can be no centripetal force. Hence, there is no gravity, because there is no need for the centripetal force of gravity on a flat, motionless earth.

Newton's theory of gravity is founded upon the premise that all objects are attracted to all other objects based upon their mass. Eric Dubay explains that this "law" of science has never been proven and cannot be observed.

> [T]his magnetic-like attraction of massive objects gravity is purported to have can be found nowhere in the natural world. There is no

> example in nature of a massive sphere or any other shaped-object which by virtue of its mass alone causes smaller objects to stick to or orbit around it! There is nothing on Earth massive enough that it can be shown to cause even a dust-bunny to stick to or orbit around it! Try spinning a wet tennis ball or any other spherical object with smaller things placed on its surface and you will find that everything falls or flies off, and nothing sticks to or orbits it. To claim the existence of a physical "law" without a single practical evidential example is hearsay, not science.[807]

Dubay is correct that gravity is not based upon science; gravity is like a heathen religious superstition. People believe it to be true, simply because the modern day witch-doctors, whom we call "scientists," say so. Dubay further questions the magical qualities of gravity, where it is both a force of suspension, causing orbit, and at the same time a force of attraction, causing adhesion.

> Now, even if gravity did exist, why would it cause both planets to orbit the Sun and people to stick to the Earth? Gravity should either cause people to float in suspended circular orbits around the Earth, or it should cause the Earth to be pulled and crash into the Sun! What sort of magic is "gravity" that it can glue people's feet to the ball-Earth, while causing Earth itself to revolve ellipses round the Sun? The two effects are very different yet the same cause is attributed to both.[808]

According to the heliocentric model, the force of gravity at the equator is perfectly balanced against the centrifugal force of the spinning earth. All persons and objects are supposedly perfectly balanced through gravity by their mass against the centrifugal force of the spinning earth to remain attached to the earth.

The problem with the gravitational theory is that according to that theory, the gravitational attraction to the earth by all persons and objects remains the same at all places on the earth. That means that the gravitational force at the North Pole is the same as the gravitational force at the equator. That poses a very real problem if the earth is spinning as alleged. That is because the centrifugal force decreases every mile toward the north pole, where the centrifugal force is ultimately reduced to zero, because the North Pole is the axis of the supposedly spinning earth. On a globe, as you travel north or south of the equator the circumference parallel to the equator becomes less. Consequently, the speed of the earth's spin at those more northern and southern latitudes from the equator would be slower than its speed of spin at the equator. For example, at 45 degrees north latitude, the earth's spin should be approximately 700 miles per hour. As the speed of the spin is reduced, so also is the correlative centrifugal force. Thus, a uniform force of gravity with a non-uniform centrifugal force would necessarily result in objects having different weights depending on their latitude. The spinning earth and the mystical force of gravity are thus proven to be fictions.

Some will point out that objects do in fact have different weights at the equator and the North Pole. However, the difference is the reverse of what would be expected by the interaction between the alleged centrifugal force and the theory of gravity. Objects weigh a fraction of a percentage less at the North Pole than at the equator. The reason has to do with the fact that there is more atmospheric pressure the further one travels towards the equator, which causes the objects at the equator to weigh slightly more.

There is no such thing as gravity; gravity is not necessary on a flat earth. It is density that keeps objects from floating off the surface of the earth. People and objects are heavier than the air and therefore do not float off the ground. There are some gasses, of course, that are lighter than air, and they float off the ground. Everyone has seen helium balloons float up in the air. Everyone understands that helium balloons are not some sort of anti-gravity devices; they float up in the air, because helium is lighter than air. Why do people not understand that apples fall from trees to the ground, not because of gravity, but because apples are denser than air? They believe in the mystical force of gravity, not because it has been proven true, but because they have been brainwashed into believing in it. Gravity does not exist. David Wardlow Scott explains:

> Any object which is heavier than air and which is unsupported, has a natural tendency to fall by its own weight. Newton's famous apple at Woolsthorpe, or any other apple when ripe, loses hold of its stalk, and, being heavier than the air, drops as a matter of necessity, to the ground, totally irrespective of any attraction of the Earth. For, if such attraction existed, why does not the Earth attract the rising smoke which is not nearly so heavy as the apple? The answer is simple—because the smoke is lighter than the air, and, therefore, does not fall but ascends. Gravitation is only a subterfuge, employed by Newton in his attempt to prove that the Earth revolves round the Sun, and the quicker it is relegated to the tomb of all the Capulets, the better will it be for all classes of society. He draped his idol with the tawdry tinsel of false science, knowing well how to beguile the thoughtless multitude, for, with a little alteration of Byron's famous lines, it is still true that "[m]ortals, like moths, are often caught by glare. And folly wins success where Seraphs might

> despair." Gravitation is a clever illustration of the art of hocus-pocus —heads I win, tails you lose; Newton won his fame, and the people lost their senses.[809]

Gravity is more akin to a heathen religious belief, than it is a scientific theory. The "scientists" promoting gravity are like priests in a religious cult who have immortalized the man who first postulated gravity and follow the gravitational theory like some religious dogma.

Indeed, when a person drops a rock in water it will be explained by scientists that it sinks because it is denser than water. The difference in density between that same rock and the atmosphere is even greater than the difference in density between the rock and water. But when, while in the atmosphere, that rock is dropped to the ground, presto-chango, suddenly the scientists change the explanation for why the rock drops to the ground. Scientists do not explain that the rock drops to the ground because it is denser than air. Instead, they explain that the rock drops to the ground because the mysterious force of gravity is pulling it down to the ground.

Sadly, the above statement is not factitious; it is reality. It is the generally accepted view in the world. The largest internet search engine on Earth, Google, is a good barometer of the thoughts and views of the world, as the search results in Google are, in large part, due to the popularity of websites. This author entered two questions into the Google search engine. The first question was: "Why does a rock sink in water?" The topmost answer to that question in the search results was from William Orem[810], Ph.D., who is on the faculty at Emerson College. Orem explained that a rock will sink in water due to it having greater density than water.

But wait! you say. I can drop even a small rock

into the glass and it sinks right away. How Come? ...

The rock is unable to displace its own weight in water, and so down it goes–just as Archimedes said it would.

Why Is The Rock Unable To Displace Its Own Weight When The Wood Was Able To?

Its because the rock has a greater density. “Density” just means how much stuff there is packed into the object. If we could magically inflate the rock, so it got bigger in size without adding any more stuff to it, its density would go down. Eventually it would be less dense than water and could float. So the rock sinks not because it weighs more than the wood, but because it is too dense to displace its own weight.[811]

The second question was: “Why does a rock fall to the earth?” The topmost answer to that question in the search results was from the Department of Physics at the University of Illinois, which explained how objects fall to the earth based upon the pulling force of gravity.

Q: Why do heavy and light objects fall at the same speed?

A: How fast something falls due to gravity is determined by a number known as the "acceleration of gravity", which is 9.81 m/s^2 at the surface of our Earth. Basically this means that in one second, any object's downward velocity will increase by 9.81 m/s because of gravity. This is just the way gravity works - it accelerates everything at exactly the same rate.

Q: So the force of gravity pulls harder on heavier objects, and it pulls every object no matter what the mass (neglecting air resistance) toward the Earth with enough force to have it accelerate 9.81 m/s/s. But what I don't understand is how this force changes. ... [H]ow does gravity "know" how hard it needs to pull the object to make it go 9.81 m/s/s faster. And also, why does Earth have gravity and other objects do not?

A: It's not true that other objects lack gravity. According to Newton's theory of universal gravitation (published in 1687) absolutely every object exerts a gravitational pull on every other object. The Earth's gravity is most noticeable around here because the Earth is big. Smaller objects have smaller effects.[812]

The popular cult-like belief in gravity started with Sir Isaac Newton, who was a scientist who published his historic three-volume *Philosophiæ Naturalis Principia Mathematica* (The Mathematical Principles of Natural Philosophy) in 1687. In that book, Newton propounded his laws of motion and universal gravity. Newton is viewed by most in the scientific community as the seminal theoretician for the postulation of gravity.

Gravity is not supported by any scientific proof. What then was the source of the theory? S. Pancoast reveals that "the law of attraction and repulsion" in the Kabbalah was popularized under the name "gravity" by Isaac Newton.[813] S. Pancoast states:

He [Pythagoras] was never permitted to declare publicly what he knew and believed, but taught

> his immediate pupils all the wonders of his philosophy, under the most binding obligation of secrecy. Pythagoras was forbidden to divulge this knowledge because it would reveal the law of attraction and repulsion, which constituted one of the great secrets of the sanctuary. Over a millennium later, Newton was led to the discovery of these forces by his studies of the Kabbalah.[814]

What most do not know is that Isaac Newton was a religious mystic, who was very well studied in Judaism, and Jewish texts. Aron Heller at the *Times of Israel* revealed that Newton "learned how to read Hebrew, scrolled through the Bible and delved into the study of Jewish philosophy, the mysticism of Kabbalah and the Talmud."[815] Indeed, one writer has revealed that "7,500 pages of his [Newton's] theological speculations, written in his own hand, are digitized at Israel's national library at Hebrew University."[816]

Whom do we find in possession as the caretaker of Isaac Newton's religious writings before they ended up in the National Library of Israel at Hebrew University in Jerusalem? It was a Zionist Jew, Abraham Yahuda, who obtained and maintained Newton's theological writings. Yahuda scoured the earth to collect Isaac Newton's writings on religion. Why was Yahuda so interested in the writings of Isaac Newton? Sarah Dry states that Yahuda "set about trying to purchase the Newton papers and wrote to [his wife] Ethel on July 28, 'I am thrilled with the thought of acquiring them. He wrote a lot about the Bible and the Jews, about Cabbala and all sorts of Jewish questions.'"[817]

Yahuda was a contemporary of Albert Einstein and conferred with Einstein about Newton's mystical religious writings.[818] Einstein was also a Zionist Jew. Religiously, Einstein had much in common with Newton, as both Einstein and Newton believed in the reality of mysticism.[819] Sarah Dry reveals this interesting fact about Einstein's view of the publication of Isaac Newton's theological writings:

> In an interview with the historian of science I. B. Cohen that took place just two weeks before he died in 1955, Einstein spoke about, among other things, Newton's theological writings. He said that it was significant that Newton had "sealed them all up in a box," an indication, Einstein thought, of Newton's awareness of how imperfect they were. Newton had "obviously" not wanted to publish these speculations during his own lifetime. Einstein said, "with some passion," that he hoped they would not now be published.[820]

Why was Einstein so passionately hopeful that Newton's theological writings not be published? Because those writings would give up the game. Isaac Newton's writings on theology would expose for the world the theological source for his theory of gravity, the Jewish Kabbalah. That is why Zionist Abraham Yahuda scoured the earth to collect Newton's theological writings and why they are now safely stored in Jerusalem, at the National Library of Israel at Hebrew University. Dry further reveals:

> Much as Stokes and Adams had before him, Einstein considered Newton's private papers with an eye toward gleaning as much as possible of his method of discovery, what he refers to here as "the formative development" of his work in physics. Einstein implicitly links the process by which Newton developed his physics and his theology; by studying the one, we might gain an insight into the other.[821]

Einstein viewed Newton's theological writings as "the

formative development of his work in physics." In a letter to Yahuda, Einstein stated the following about Isaac Newton's theology:

> Newton's writings on biblical subjects seem to me especially interesting because they provide deep insight into the characteristic intellectual features and working methods of this important man. The divine origin of the Bible is for Newton absolutely certain, a conviction that stands in curious contrast to the critical skepticism that characterizes his attitude toward the churches. From this confidence stems the firm conviction that the seemingly obscure parts of the Bible must contain important revelations, to illuminate which one need only decipher its symbolic language. Newton seeks this decipherment, or interpretation, by means of his sharp systematic thinking grounded on the careful use of all the sources at his disposal. While the formative development of Newton's lasting physics works must remain shrouded in darkness, because Newton apparently destroyed his preparatory works, we do have in this domain of his works on the Bible drafts and their repeated modification; these mostly unpublished writings therefore allow a highly interesting insight into the mental workshop of this unique thinker.[822]

Sir Isaac Newton's study of the Kabbalah is the real source for his theory of gravity. Marshall Hall explains that "[a]lthough Kabbalism is nearly always associated with Judaism, there are so-called 'Christian Kabbalists' such as Newton, Dee, Kepler, Shakespeare, Cardinal Nicolas of Cusa and a long list of Theosophists, Rosecrucians, Masons, Crowleyites, etc. All of these and more have been or now are promoters of Kabbalist mysticism."[823] Hall states that "Newton, in short, insured the acceptance of the Copernican Model for two centuries with his arcane 'mathematical' concepts; concepts upon which others—Einstein through Sagan, et al—could erect today's Pharisee Cosmology."[824]

Gershom Scholem is considered the founder of the modern, academic study of Kabbalah, and was the first Professor of Jewish Mysticism at the Hebrew University of Jerusalem. Scholem has stated that gravity is a force that has been known to the Jewish mystics from antiquity. This is certainly something that Newton would have come across in his study of the Jewish Kabbalah. Gershom Scholem associated gravity "with the final *Heh*, the *Shekhinah*, The Presence, the alchemical Earth, the Daughter in the divine family. Note the interesting connection between alchemical Earth and gravity, which rules our Earth!"[825]

Migene Gonzalez-Wippler, in her book, *The Kabbalah & Magic of Angels*, explains the foundational connection between the mysticism of the Jewish Kabbalah and the theory of gravity.

> It [gravity] is the result of the union of cause and effect and the law of momentum. It keeps the universe in motion, and it also works with dark matter to maintain stars, planets and galaxies in harmonious balance. For that reason it is equated with *Tiphareth* [a/k/a *Tifereth*], the sixth sphere of the Tree of Life, which is the center of the Tree beneath *Kether*, the first sphere. *Tiphareth* is identified with the sun, which is at the center of the solar system and keeps its planets in steady orbits through gravitation.[826]

The Tree of Life, to which Gonzalez-Wippler refers, is a graphic depiction of the Jewish god, *Ein Sof*. It is not a coincidence that the *sefirot* are depicted in the Tree of Life as spheres. The Tree of Life (*Ein Sof*) is the foundational source for the "scientific" theory of gravity. For the myth of the spherical earth to be believed, there must be a force to explain why people and things do not fly off into space.

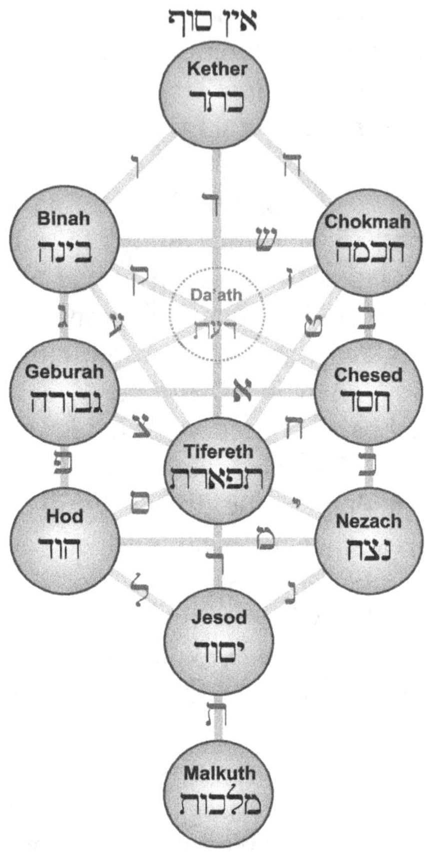

Figure 183: **The Tree Of Life (*Ein Sof*) From the Jewish Kabbalah**

It is not necessary that gravity be proven or even make sense, as long as it is propped up by a facade of experts called "scientists." Make no mistake about it, though, there is nothing scientific about gravity. It is pure heathenism, from the Jewish Kabbalah. The priest-scientists of the heliocentric religion tolerate no dissension from their religious dogma of gravity. According to Migene Gonzalez-Wippler, gravity "is equated with *Tiphareth* [a/k/a *Tifereth*], the sixth sphere of the Tree of Life." As explained below, that means that gravity is not only an attribute of the Jewish god, *Ein Sof*, it is actually one of the god's of the Kabbalah, in its own right.

The god of the Kabbalah, who is called *Ein Sof*, is made up of ten attributes (*sefirot*). Each *sefirah* (singular of *sefirot*) is not only designated as a particular trait of *Ein Sof* but is also an anthropomorphic part of that one god. In addition, each *sefirah* is either a god or a goddess in its own right. The first nine *sefirot* (plural of *sefirah*) are in turn divided evenly into three triads, containing three *sefirot* each and representing three major sections of the anthropomorphic parts of the mystical body of *Ein Sof.* The tenth sefirah is the *Shekinah* (a/k/a *Malkuth*), which is not part of the three triads.[827]

The Kabbalah describes the lower third triad of its heathen god (*Ein Sof*) as made up of three *sefirot*: 1) *Netzach* (Endurance/Victory), 2) *Hod* (Majesty/Glory), and 3) *Yesod* (Foundation).[828] *Netzach* and *Hod* are the right and left legs of *Ein Sof*, and *Yesod* is *Ein Sof's* phallus. According to the Kabbalah, the light and power of the sefirot are channeled through the phallic god Yesod to the last *Sefirah*, which is the *Shekinah* (a/k/a *Malkuth*).[829] This phallic god is part of the blatantly erotic interpretation of the Jewish god found in the Kabbalah.[830] Rabbi Geoffrey W. Dennis in *The Encyclopedia of Jewish Myth, Magic, and Mysticism* explains: "The *Zohar* includes multiple interpretations built around the concept of God's genitals."[831]

Judaism is a phallic religion that was adopted by the Jews during their captivity in Babylon. Dan Cohn-Sherbok and Lavinia Cohn-Sherbok explain the development of the esoteric sexual meanings concealed within the orthodox Jewish liturgy:

> Likewise, Phallic symbolism was employed in speculations about the ninth *Sefirah*, *Yesod*, from which all the higher *Sefirot* flowed into the *Shekinah* as the life force of the cosmos. In later centuries erotic terminology was used in the Hasidic works to describe movement in prayer which was depicted as copulation with the *Shekhinah*.[832]

The movement in prayer mentioned by Dan and Lavinia Cohn-Sherbok is called *shuckling* (a/k/a *shokeling* or *shoklen*). It is also called *davening*, which is simply a Yiddish word for prayer. The *shuckling* by Jews simulates the movement of copulation in sexual union with *Shekhinah*. *Shuckling* is symptomatic of the fact that Judaism is a phallic religion, which has liturgical practices and prayers with occult sexual meanings. Baal Shem Tov is considered the founder of Hasidic Judaism and a highly respected authority on Jewish theology. Baal Shem Tov stated that "Prayer is mating with the Shechinah."[833] The mating during prayer is manifested in swaying back and forth. Rabbi Eli Malon explains: "By 'prayer,' he [Baal Shem Tov] meant the literal swaying back and forth, suggestive of intercourse, that is customary of traditional Jewish prayer."[834] The swaying is understood by Jews to signify copulation with the goddess Shekinah.[835] Former Jew, Nathaniel Kapner (a/k/a Brother Nathaniel), confirms the hidden meaning of the swaying by the Jews:

> Watch closely how the rabbis thrust their pelvises and penises back and forth in a prescribed prayer movement called "davening" in which the Jew copulates with the 'Shekinah' in order to give birth to an erotic union with the 'Ein Soph,' the Kabbalistic masculine emanation of their false god.[836]

The occult sexual meaning behind *shuckling* is generally known among Jews, but is kept hidden from Gentiles. In 2013, Rabbi Michael Leo Samuel openly discussed in an article addressed to a Jewish audience in the *San Diego Jewish Herald*, the *shuckling* of Hasidic Jews in front of a Victoria's Secret lingerie store. He wondered if the Hasidic Jews needed a visual aid for their *davening*. Samuel paraphrased the writings of Baal Shem Tov regarding the meaning of the swaying during Jewish prayer:

> Prayer is zivug (coupling) with the Shechinah.

> Just as there is motion at the beginning of coupling, so, too, one must move (sway) at the beginning of prayer. Thereafter one can stand still, without motion, attached to the Shechinah with great deveikut (cleaving to God) As a result of your swaying, you can attain great bestirment. For you think to yourself: "Why do I move myself? Presumably it is because the Shechinah surely stands before me." This will effect in you a state of great hitlahavut (enthusiasm; rapture).[837] (parentheticals in original)

It seemed that Rabbi Samuel was upset that the Jews were *shuckling* in front of a Victoria's Secret lingerie store, because it revealed too much about the occult meaning of the Jewish religion to the world. Rabbi Samuel stated: "With respect to the Hassidic Jews praying in front of the Victoria's Secret lingerie store, they really believe that the world is not observing. They behave like a young child who covers his ears and screams, thinking that nobody around him can hear him."[838]

The powerful undercurrent of phallic worship infused into orthodox Judaism by the Kabbalah, includes the practice of sex magic.[839] The sex magic is an offshoot of the secret doctrine in Judaism. It is a common doctrine found in secret societies, that the mystic can find redemption through an "heroic" willingness to do evil.[840] The secret rabbinic doctrine is that evil can be redeemed by embracing it; there is a spiritual good in doing evil.[841] That explains why Jesus said to the Jews: "Ye are of *your* father the devil, and the lusts of your father ye will do."John 8:44. As is the case with the clerics of all phallic religions, there is rampant pederasty among Jewish clerics. Rabbinic pederasty is documented in this author's book, *Solving the Mystery of BABYLON THE GREAT*.

Moshe Idel, in *Hasidism Between Ecstasy and Magic*,

explains that "the concept of the descent of the *Zaddiq* [Jewish mystic or saint], which is better known by the Hebrew phrase, *Yeridah zorekh Aliyah*, namely the descent for the sake of the ascent, the transgression for the sake of repentance. . . . Much attention has been paid to this model because of its essential affinities with Zoharic and Lurianic Kabbalah . . . this model was a very important one in Hasidic thought."[842] That concept is the core belief in the system of "black magic." The source of this secret doctrine of "black magic" is Babylon.[843] The oldest texts for this Babylonian black magic in Judaism are the texts *Sifrei h-Iyyun*, *Sefer ha-Bahair*, and the *Hikoth Yesirah*, which is also known as the *Sefer Yetzirah*.[844]

The *Yesod* (*Jesod*) unites the *Shekinah* and the *Tif'eret*. *Tif'eret* is the offspring of *Hokhmah* and *Binah*.[845] The *Hokmah* and *Binah* are two of the three *sefirot* of the divine head of the mystical body of the *Ein Sof* (*Kether* is the third *sefirah*). *Tif'eret* is not only a god himself, but he also represents the heart and torso of the body of the Kabbalah god, *Ein Sof*.

The "tree of life" diagram[846] depicts the 10 *sefirot* as they are presented in the *Sefer Yetzirah* (Book of Formation) and gives a visual representation of the relationship between the different *sefirot* within the Ein Sof. Note that *Malkuth* and *Shekinah* are the same *sefirah*. In the diagram, only *Malkuth* is depicted. Note also that the spelling varies somewhat from source to source. For example, *Chokmah* in the diagram is the same as *Hokmah* and *Jesod* is the same as *Yesod*. Jewish scholars readily acknowledge that are many parallels between the Cabalistic concept of god and that found in Buddhism, Hinduism, and so-called Gnosticism.[847] That is not surprising, since they all flow from the same mystical waters of Babylon. In the Kabbalah, gravity is equated with *Tiphareth* [a/k/a *Tifereth*], the sixth sphere of the Kabbalistic Tree of Life. Sir Isaac Newton simply took that heathen religious belief, cloaked it in mathematical equations, and presented it to the world as "science."

36 Scientific Proof the Earth Does Not Move

Many attempts were made to prove that heliocentricity was true and geocentricity was false. Every such attempt has been a failure. The most famous, because of its precision and irrefutability, was the experiment done by physicist Albert A. Michelson (1852 – 1931) and chemist E. W. Morley (1838 – 1923). The Michelson/Morley experiment, using an interferometer, which measured light rays, established that the earth is stationary.[848] Michelson was involved in other experiments that confirmed, to his dismay, that the earth is stationary.

The Michelson/Morley experiment (1887) does not stand alone. It is joined in its confirmation of a stationary earth by the James Bradley experiment (1729) (proving that the ether is not carried along by the earth), the Sagnac experiment (1913) (proving that there is in fact an ether), the Michelson/Gale experiment (1925) (proving that the ether passes over the earth once every 24 hours), and Airy's failure (1871) (proving that the stars move, carried by the ether, while the earth remains stationary).

There are many other experiments that have each time given results that were not only consistent with a stationary earth but indicative of a stationary earth, from the light polarization experiments of E. Muscart in 1872 to the mutual inductance experiments of Theodore de Coudres in 1889 to the 1903 *Touton-Noble* experiments.[849] Indeed, there is not a single experiment that proves that the earth moves. The moving earth is based entirely on a theory and is contradicted by all of the experimental evidence.

The Sagnac experiment, proving the existence of ether, destroys the theory of relativity, which necessarily assumes that there is no ether. The Michelson/Gale experiment proved that the ether passed over the earth once every 24 hours, but it did not prove whether it was the ether moving or the earth spinning. Airy's failure determined with scientific certainty that in fact, it was the ether carrying the stars that was moving over the earth and that the earth was stationary. Dr. Neville Thomas Jones, Ph.D., explains that "George Airy proved that the world was stationary and the stars are moving."[850] Because his experiment proved that the earth does not move, which was the opposite of the expected outcome, Airy's experiment is commonly known as "Airy's failure."

There has been a virtual blackout, however, within science education of these experiments. Dr. Malcom Bowden reveals that he "asked 3 Christian physicists if they had ever heard of them; not one had!"[851] Indeed, in March 2005, another physicist wrote to Dr. Bowden that "after 35 years as a professional physicist, with a thesis in relativity, I only learned of Sagnac's experiment last year."[852] The degree of the cover-up of true science is simply astounding!

The Michelson/Morley experiment was simple in concept. A light beam was split. One of the split beams was sent at a right angle to earth's supposed direction of travel and the other was sent along the path of the earth's supposed direction of travel. The light traveling in the direction of the earth's travel should have taken longer than the light traveling at a right angle to the earth's direction of travel. To the amazement of the scientific world, the results were null, meaning that there was no difference in the speed of the light beams. That meant that the earth was motionless.

The Michelson/Morley experiment shook the scientific world. The results of the experiment could not be attacked, as the experiment was meticulously conducted, using precise instruments, and it was performed by Albert A. Michelson, whom *The New York Times* described as America's greatest physicist.[853] The implications were devastating for the Copernican model of a globular spinning earth orbiting around the sun. If the heliocentric model fell, then evolution would not be far behind. The priests of science knew that something had to be done if they were going to maintain their godless religion.

In desperation, the priests of science tried to explain away the null result of the Michelson/Morley experiment. Hendrik A. Lorenz and George Fitzgerald swallowed their pride and put their formerly good names to a ridiculous theory that the null result in the Michelson/Morley experiment was because the solid steel arm that was pointed in the direction of the supposed motion of the earth became shorter due to the earth's movement through the ether. That shortening of the steel caused the light to arrive at the same time, not because the earth was stationary, but because the earth was moving. Seriously, I am not making this up! This supposed contraction became known as the Lorenz/Fitzgerald contraction.

There was absolutely no evidence to support the theory of the Lorenz/Fitzgerald contraction, but it was all they had to try to keep the spinning, and orbiting earth model of heliocentricity alive. They came up with the theory of the Lorenz/Fitzgerald contraction only because they had nothing else to explain the null result and they were not going to abandon the Copernican model.

Some scientists aver that it is a misnomer to call the Lorenz/Fitgerald theory a "contraction," since as first theorized, it was supposedly the transverse arm of the Michelson/Morley interferometer that was lengthened and so it should actually be called the Lorenz/Fitzgerald expansion.[854] Regardless, if it was called an expansion or contraction, it was an inane theory that was unsupported by any proof. Arthur Miller described the Lorenz/Fitzgerald contraction as "physics of desperation."[855] Indeed, it was more akin to a fairy tale for adults, than it was science. It was not long before Einstein was pressed into service, with his theory of relativity.

Notice that the scientific community rejected acceptance of a stationary earth, not because there was any scientific evidence that contradicted it, but because it contradicted the central "faith" of the godless scientific community. The Copernican theory was simply not to be abandoned, no matter what the scientific evidence showed.

37 Einstein to the Rescue

Something had to be done about the results of the Michelson/Morley experiment. The high priests of science simply could not allow the world to discover that the earth does not move. Enter Albert Einstein and his theory of relativity. Einstein announced his special theory of relativity in 1905 and his general theory of relativity in 1915.

Einstein's theory of relativity was able to explain away the null results of the Michelson/Morley experiment. According to Einstein the speed of light is constant and consequently an object moving through space would not show movement, if light was used to measure the speed of the motion. According to Einstein, the speed of light is constant for every observer no matter what speed the observer is traveling. Armed with the theory of relativity, scientists argued that the earth was moving, despite the null result of the Michelson/Morley experiment. They alleged that it was incorrectly shown not to be moving in the Michelson/Morley experiment, because the light used in the experiment, which was expected to show the speed of the earth through the ether, could neither increase nor decrease in speed.

The Michelson/Morley experiment was designed to measure the expected speed of the supposedly moving earth through the ether. Einstein could not have a constant speed of light as he theorized, if there was an ether. Einstein solved that issue by simply announcing that there was no ether. Einstein, by theorizing that there is a constant speed of light and there is no ether, solved the Michelson/Morley null movement result and saved the heliocentric model of an orbiting, rotating earth.

Light travels in waves, which is a fact that Einstein accepted. A wave needs a medium through which to travel. Imagine a wave in water without the water. It is an impossibility. The medium through which light travels is called ether. This simple law of physics that a wave needs a medium through which to travel is something that is easy to understand. The corollary of an ether for light to travel through, however, will seem foreign to the reader, because it is not taught in schools. The educational system has been given over completely to the mystical theory of relativity, where the ether is simply not allowed to exist.

Einstein removed the ether from existence by edict. The problem is that a wave of light cannot exist without a medium for that light. Professor Herbert Dingle explains the absurdity of light waves without an ether through which to travel.

> Light consisted of vibrations in that ether, that had physical properties, which also were, in principle, determinable. What Einstein was proposing, therefore, was to retain the finite velocity of light without the existence of any standard with respect to which that velocity had a meaning. Light consisted of waves, with a definite length, frequency and velocity, in nothing; it was the grin without the Cheshire cat. ... The physical part of the theory was expendable; only the equations needed to be saved. Einstein saw a way of saving the equations, and did not consider it worth while to

> 'explain' light. ... Einstein was satisfied to 'explain' it in terms of things that we understood nothing of — in other words, not to explain it at all. If his assumptions were granted he did save the equations, and when his theory ultimately made its general impact on the world, mathematics had so dominated physics that the non-existence of the Cheshire cat was regarded as a triviality; the grin remained, and all was well.[856]

Einstein's construct of "no ether" is an impossibility, just as a wave in water without the water is an impossibility. In any event, Einstein was proven wrong, and the ether was proven to exist in the 1913 Sagnac experiment. That experiment proved scientifically, beyond any doubt, that there is in fact an ether, through which light travels.

Indeed, airlines today use cockpit ring laser gyroscopic compasses that are based upon the discovery by George Sagnac of fringe changes in light traveling through the ether. The changes in the fringes of light is then computed into a reading, which tells the pilot about changes in bearing of the airplane. Without ether, those sophisticated optical compasses would not work at all. The very existence of the laser gyroscopic compasses used by airplanes today proves that there is an ether and impeaches Einstein's claim that there is no ether. The existence of ether destroys the theory of relativity and establishes the Michelson/Morley experiment as proving that the earth is stationary.

Einstein removed the ether, which upended the traditional laws of physics. Einstein did not present any proof that there was no ether; he just made it up. Removing the ether was a way of explaining why Michelson/Morley's interferometer showed no motion of the earth. Removing the ether removed the resistence of ether to light waves, which allowed Einstein to conjure the myth that light will not change speed on a moving surface. That explained away the Michelson/Morley results showing that the earth did not move. If, however, there is an ether, that necessarily means that the earth is stationary, as proven in the Michelson/Morley experiment.

Robert Sungenis reveals that the scientific community seems to ignore the fact that while Einstien's special theory of relativity (1905) removed the ether, his later (1915) general theory of relativity added the ether back in again.[857] Einstein added the ether back in because he was painted into a corner by experiments that proved the ether in fact existed. For example, the Sagnac experiment (1913) proved that there was an ether. Einstein could not ignore those experimental results. Einstein, himself, admitted in 1920 that in fact the ether exists:

> Recapitulating, we may say that according to the general theory of relativity, space is endowed with physical qualities; in this sense, therefore, there exists an ether. According to the general theory of relativity space without ether is unthinkable; for in such space there not only would be no propagation of light, but also no possibility of existence for standards of space and time (measuring-rods and clocks), nor therefore any space-time intervals in the physical sense.[858]

In effect, the general theory of relativity contradicts the special theory of relativity. The theory of relativity is schizophrenic. It both denies and affirms the existence of the ether. The special theory of relativity and the general theory of relativity taken together add up to nonsense.

The theory of relativity is a complete lie. It is based entirely on convoluted and deceptive mathematical models. Nikola Tesla's statement about the modern methods of scientists like Einstein is revealing. "Today's scientists have substituted mathematics for experiments, and they wander off through equation after equation, and eventually build a structure which has no relation to reality."[859]

Nikola Tesla was the most brilliant scientist and inventor of his time. So advanced were his discoveries that upon his death in 1943 his research papers were seized by the FBI and classified "Top Secret" at the request of the U.S. War Department. One of Tesla's most notable discoveries was alternating electrical current (a/k/a AC) that is today used to power most homes and businesses. Tesla understood true science and knew a scientific scam when he saw one. In 1935, Tesla called Einstien's theory of relativity "[a] magnificent mathematical garb which fascinates, dazzles and makes people blind to the underlying errors. The theory is like a beggar clothed in purple whom ignorant people take for a king..., its exponents are brilliant men but they are metaphysicists rather than scientists."[860]

Professor Herbert Dingle was once an eminent proponent of the theory of relativity. He later realized that it was simply a myth, supported not by scientific experiments, but rather by false math formulae. He discovered that the theory of relativity is held to be true, not because it is true, but because mathematical formulae were devised and held up as evidence of its truth. "Not only are hypotheses held to contain the 'real truth'; it is now claimed that any (mathematical) hypothesis is necessarily true." (parenthetical in original)[861]

The problem with basing proof for a scientific theory solely on mathematics is that a mathematical equation may not correspond to reality. A mathematical formula may only prove something that is imaginary and not real. Professor Dingle explains:

> [T]he symbols that compose a mathematical expression may, with equal mathematical

> correctness, correspond both to that which is observable and that which is purely imaginary or even unimaginable. If, therefore, we start with a mathematical expression, and infer that there must be something in nature corresponding to it, we do in principle just what the pre-scientific philosophers did when they assumed that nature must obey their axioms, but its immensely greater power for both good and evil makes the consequences of its misapplication immensely more serious.[862]

The experiments supporting the theory of relativity were "thought experiments" performed completely through complicated math formulae designed to bedazzle ignorant laymen. Professor Dingle stated:

> [M]athematics has been transformed from the servant of experience into its master, and instead of enabling the full implications and potentialities of the facts of experience to be realised and amplified, it has been held necessarily to symbolise truths which are in fact sheer impossibilities but are presented to the layman as discoveries.[863]

Dr. Dingle reveals the key point that is the cornerstone of the theory of relativity. Many highfalutin scientists do not seem to know this one simple fact. **"[I]n the language of mathematics we can tell lies as well as truths, and within the scope of mathematics itself there is no possible way of telling one from the other."**[864] That is the dirty secret behind the theory of relativity. Einstein used mathematics to tell lies. The only way to determine if a mathematical formula has any validity is to test it in the real world. Dr. Dingle explains that "[w]e can distinguish them only by experience or by reasoning outside the mathematics, applied to the possible relation between the mathematical solution and its supposed physical correlate."[865] Physical experiments are not something that have been done with much success in proving the theory of relativity. Consequently, scientists resort to thought experiments, using mathematical formulae, which have no correlation to reality.

How can math be used to tell lies, as alleged by Professor Dingle? A simple example will illustrate how math can be used to support a falsehood. If one were to say that a glass that is half-empty is the same as a glass that is half-full, that would be true. One can use mathematics, however, to make that simple truth be the foundation for a falsehood. Let's put the half-full glass equaling a half-empty glass into an equation, where "E" represents an empty glass and "F" represents a full glass: $\frac{1}{2}E = \frac{1}{2}F$. That equation ($\frac{1}{2}E = \frac{1}{2}F$) is accurate as it is presented; a half empty glass is equal to an half full glass. Now, in basic algebra, if one multiplies both sides by the same number, it does not affect the accuracy of the equation. Thus, to multiply both sides of the equation by two, one would get the result of $E = F$. Under the rules of algebra, that is supposed to be a true statement. We know, however, that in reality an empty glass does not equal a full glass (thus, in reality $E \neq F$). However, mathematics can be used to present a falsehood as truth ($E = F$). That is the type of unreal reasoning that permeates the theory of relativity, where the scientific testing is done in thought experiments, using mathematics. This creates a fantasy world of relativity. The theory of relativity is not science, it is mysticism, supported only by mathematical models.

Physicists gave up trying to understand the absurd results of the formulae used to explain the theory of relativity, and simply capitulated without much of a fight. They accepted the mathematical formulae of Einstein, even though they often gave inaccurate and incongruous solutions. Dingle explains that "with the apparent success in 1919 of Einstein's general theory with its then quite new and terrifying mathematical machinery of tensor calculus, came the fatal climax. ... [Physicists] gave up trying to understand the whole business, surrendered the use of their intelligence, and accepted passively whatever apparent absurdities the mathematicians put before them."[866]

Einstein's biographer, Ronald Clark, reports that Einstein's friend, Janos Plesch, suggested to Einstein that there seemed to be some connection between mathematics and fiction, Einstein replied: "There may be something in what you say. When I examine myself and my methods of thought I come to the conclusion that the gift of fantasy has meant more to me than my talent for absorbing positive knowledge."[867] The theory of Relativity is not science; it is fantasy, conjured by mathematical formulae in the minds of Einstein and his followers.

The blind faith of the scientific community in the supposed truth of the theory of relativity, and its intolerance for any evidence of its invalidity, is akin to adherents in a religious cult. When one realizes that relativity is more a religion than it is science, it explains why the ascension of relativity is in direct proportion to the descent of Christianity in society. It seems that there is more tolerance in religion than there is in the scientific community toward heterodoxy, especially when it comes to the theory of relativity. Indeed, Professor Dingle said as much:

> It is ironical that, in the very field in which Science has claimed superiority to Theology, for example — in the abandoning of dogma and the granting of absolute freedom to criticism — the positions are now reversed. Science will not tolerate criticism of special relativity, while Theology talks freely about the death of God, religionless Christianity, and so on (on which I make no comment whatever). Unless scientists can be awakened to the situation into which they

> have lapsed, the future of science and civilisation is black indeed.[868] (parenthetical comment in original)

Charles Lane Poor, Professor of Celestial Mechanics at Columbia University, and the author of a number of standard textbooks on astronomy, stated that "the Relativity Theory strikes directly at our fundamental concepts as to the structure of the universe; its conclusions are startling and completely upsetting to our common-sense way of looking at physical and astronomical phenomena."[869] Dr. Louis Essen, a distinguished mathematician, and Fellow of the Royal Society, stated that the theory of relativity was not truly a physical theory but rather simply a number of sometimes contradictory assumptions. Lord Ernest Rutherford is considered the father of nuclear physics; so eminent was he that chemical element 104 was named rutherfordium after him. Lord Rutherford has called the theory of relativity, simply "nonsense."

In 1922, Professor Herbert Dingle wrote *Relativity for All*, one of the first standard textbooks on the theory of relativity. His second book on the subject, written approximately 20 years later, *The Special Theory of Relativity*, remained for a long time the standard work in English and American universities on the theory of relativity. Indeed, Professor Dingle was one of the foremost experts on the theory of relativity in the world. During a span of fifty years, he studied the theory intensively and conferred about it with all the physicists who were experts in it (e.g., Einstein, Eddington, Tolman, Whittaker, Schroedinger, Born and Bridgman). So renowned was Dingle's expertise on the theory of relativity, that when Einstein died in 1955, the BBC chose Professor Dingle to broadcast a tribute to Einstein.

In 1959, after years of believing and promoting the theory of relativity, Dingle realized that something was wrong. He found a paradox in the theory of relativity. He spent 13 years canvassing his large network of scientists to try to find an answer to the paradox. Nobody could answer the paradox. He tried to publish the paradox, but was refused all access to scientific journals. Finally, in 1972, Dingle decided to publish his conclusion in a book titled: *Science at the Crossroads*. He explained in his book that he published it only because he was denied access to scientific journals to present his evidence. In that book, Professor Dingle presents unimpeachable proof that the theory of relativity is invalid.

In order to understand the paradox with which Professor Dingle was faced, some explanation should be given. The *coup de grace* to the Michelson/Morley experiment results showing that the earth does not move is the central maxim of relativity theory that there is no way to tell which of two bodies is in motion. The theory of relativity provides that motion is relative to the observer. Thus from earth it would appear that the sun is moving. However, from the perspective of the sun, the earth is moving. According to the theory of relativity there is no way to establish which is the case, because the movement of the two bodies is only movement relative to the other body. This maxim of relativity effectively kills the null result of the Michelson/Morley experiment, since according to the theory of relativity, the null result was only a matter of relative perspective. Under the theory of relativity, if you were to fall on your face, it cannot be said that you fell to the ground, as it is equally likely that the earth rose up to meet your face. That is the kind of silly conclusion brought about by the theory of relativity.

In addition to the above relativity of motion, Einstein theorized that time slows down, the faster one travels. For example, if a twin (Paul) takes a trip on a spaceship at near the speed of light and he returns to earth ten years later, his twin brother (Peter), left back on earth, will have aged ten years, but the twin on the spaceship would only have aged very little. The problem with that postulation from Einstein is that under the theory of relativity, the movement of each brother is relative. Each twin sees the other as moving, and therefore each brother should have aged more slowly than the other brother. The conclusion under the theory of relativity is that Peter has aged more slowly than Paul and at the same time Paul has aged more slowly than Peter. Of course, it is impossible for each twin to age more slowly than the other twin.

The twin paradox is chosen by this author, because it very simply illustrates the issue. Professor Dingle, however, never actually used the twin paradox, because there is an alleged quirk in that example that gives the supporters of the theory of relativity an out (or so they allege); they assert that there is no symmetry, since the twin on the spaceship is traveling outbound and inbound, which involves two inertial frames. Of course, that is pure sophistry, and addressing such nonsense is beyond the scope of this book. Professor Dingle was too well versed in the theory of relativity to allow the promoters of relativity such an easy out, so he steered clear of using the aging twins example. He, instead, used an example of two clocks moving in the same trajectory at different speeds.

Professor Dingle asked scientists all over the world to assist him in finding an answer to the paradox using speeding clocks, with one clock traveling faster than the other, in the same direction. Every scientific journal in the world refused to even address the paradox that Dingle raised. No one could resolve the paradox, and the scientific community seemed to think it was impolite to even discuss it. "I can present the matter most briefly by saying that a proof that Einstein's special theory of relativity is false has been advanced; and ignored, evaded, suppressed and, indeed, treated in every possible way except that of answering it, by the whole scientific world."[870] In science, a paradox is a self-contradictory conclusion that is logically impossible. A theory that causes a logically impossible result is necessarily wrong. A paradox in the theory of relativity simply had to be suppressed by the high priests of science.

Professor Dingle laid out the paradox, which has never been resolved to this day, as follows:

> According to the theory, if you have two exactly similar clocks, A and B, and one is moving with respect to the other, they must work at different rates ..., i.e. one works more slowly than the other. But the theory also requires that you cannot distinguish which clock is the 'moving' one; it is equally true to say that A rests while B moves and that B rests while A moves. The question therefore arises: how does one determine, consistently with the theory, which clock works the more slowly? Unless this question is answerable, the theory unavoidably requires that A works more slowly than B and B more slowly than A --which it requires no super-intelligence to see is impossible. Now, clearly, a theory that requires an impossibility cannot be true, and scientific integrity requires, therefore, either that the question just posed shall be answered, or else that the theory shall be acknowledged to be false. But, as I have said, more than 13 years of continuous effort have failed to produce either response.[871]

Professor Dingle concludes that "[t]he magical influence of this word [relativity] has transformed science in this field into a superstition as powerful as any to be found in primitive tribes."[872] We have it on the authority of Professor Dingle, one of the foremost experts on the theory of relativity that the theory of relativity is false.

The theory of relativity is more than just false; it is nonsense. It is based upon the premise that motion is not absolute, but rather it is relative to the frame of reference of the observer.

Einstein's motive in constructing such a theory was to explain away the Michelson/Morley null result for the motion of the earth.

The twin paradox identified by Professor Dingle was the result of Einstein's construct of time dilation. This theory of time dilation postulated by Einstein was constructed under a rubric he called "spacetime."[873] Einstein created the new concept of "spacetime" in order to make sense of the Michelson/Morley experiment under his new theory of relativity. Einstein's theory of relativity requires that the speed of light must be constant, no matter where a person is or how fast a person is traveling when measuring the speed of the light. The speed of light (or indeed the speed of anything) is simply a measure of distance over time. Because the theory of relativity bakes a moving earth into the null Michelson/Morley result, the distances the split beams of light traveled at right angles to each other in the Michelson/Morley interferometer must necessarily be different on each axis. But if the speed of light is constant, the theory of relativity needed a variable that could explain the null result for a constant light speed over different distances. That variable was time. Since the speed of light must be constant, the null result must be explained by changing the time in each axis.

Under the new "spacetime" rubric, time slows down for an object as it increases in speed. Thus, Einstein came up with (or rather plagiarized) the theory of time dilation. Time dilation necessarily also requires length contraction of an object in motion, but for simplicity, we will address time dilation as a solitary factor. The theory of time dilation was necessary to explain how the split light beams arrived at the same time on the two axes on the Michelson/Morley interferometer over different distances (since the earth was supposed to be moving), while also assuming that both light beams were traveling at the same speed. As explained by Robert Sungenis, "Einstein had two choices to explain the 1887 Michelson-Morley experiment: (1) to make the Earth constant (i.e., not moving in space); or [2] make light constant (always the same speed no matter who is observing it). If you choose #2, then you are also required to dilate time, contract lengths, and increase mass in order to make everything else balance."[874]

In reality, the Michelson/Morley's interferometer measured two splits of a light beam traveling over the same distance at the same time, which could only happen if the earth was stationary. But that reality was "unthinkable," because the Copernican model of a spinning earth rotating around the sun had to be preserved. Einstein came to the rescue by trotting out his theory of relativity with his mystical concept of "spacetime," where the constant of time is no longer a constant. The "spacetime" theory is completely unproven and unprovable, because it is untrue. But when "scientists" construct a religious deception clothed in scientific jargon, the truth doesn't matter.

In reality, time is a constant and the earth is stationary and flat. And real, observable scientific experiments prove it. Time is, always has been, and always will be a constant measure. Einstein's theory of "spacetime" is mysticism that only exists in thought experiments. Modern scientists' acceptance of it is akin to priestcraft of a superstitious religion. Robert Sungenis succinctly states the ineluctable conclusion that "Einstein's theories were invented to keep the Earth moving when all the experimental evidence showed it was not moving."[875]

According to the theory of relativity, there is no way to tell the difference between an object that is at rest and an object that is moving at a constant velocity in an inertial reference frame. According to Einstein, all motion is relative to the frame of reference of the observer. For instance, according to the theory of relativity, it is equally valid to say that the railroad track is moving under a train as it is to say that same train is traveling at a constant speed over a stationary track. A person standing next to the track would perceive the train moving as it passed by him. But a passenger on the same train moving at a constant speed, who is inculcated in the school of relativity, would perceive the person standing next to the track and the landscape speeding by the train and conclude that it is equally possible that the train is standing still and the earth is moving beneath him as it is that the train is moving on the track. We know, however, that is nonsense. The train is in fact moving. The motion of the train can be detected and measured. The theory of relativity is not based upon true science and measurable observation; it is based upon religious superstition that is propped up by deceptive mathematical models that contradict reality.

The World Heritage Encyclopedia describes the importance of Einstein's theory of relativity in explaining away the null result of the Michelson/Morley experiment.

> This [Einstein's 1905 theory of special relativity] allows a more elegant and intuitive explanation of the Michelson-Morley null result. In a comoving frame the null result is self-evident, since the apparatus can be considered as at rest in accordance with the relativity principle, thus the beam travel times are the same. ... Special relativity is generally considered the solution to all negative aether drift (or isotropy of the speed of light) measurements, including the Michelson–Morley null result.[876]

Very simply, the theory of relativity explains that the null result of the Michelson/Morley experiment was because the instrument doing the measuring was, relatively speaking, at rest, as that was its frame of reference, and not because it was actually at rest. Einstein asks us to ignore the reality as actually measured and accept in its place the mathematical postulates of relativity. The theory of relativity postulates that no motion of the earth was detected in the Michelson/Morley experiment not because the earth is in fact motionless but rather because the measurement was performed from the earth. According to Einstein's theory, if the Michelson/Morley experiment would have been done from say the moon, then the moon would have been detected to be motionless and the earth would have appeared to be in motion. Under the superstitious religion of relativity, motion is all relative to the frame of reference of the observer.

Lest you think this author is overstating the fact, please make your own judgement after reading the explanation of the theory of relativity by Albert Einstein himself during an address he gave at Princeton University:

> What we mean by relative motion in a general sense is perfectly plain to everyone. If we think of a wagon moving along a street we know that it is possible to speak of the wagon at rest, and the street in motion, just as well as it is to speak of the wagon in motion and the street at rest. That, however, is a very special part of the ideas involved in the principle of Relativity.[877]

The sad thing is that scientists believed him! And they still believe him today! They do not perceive that relativity is not true science, it is a religious deception clothed in scientific lingo. It is based on belief in mystical principles, which contradict common observation. Gerrard Hickson accurately describes Einstein's theory of relativity as the very negation of reason. Referring to the above quote from Einstein, Hickson states:

> That would be amusing if we read it in a comic paper, or if Mutt and Jeff had said it; but when Professor Einstein says it in a lecture at the Princeton University, we are expected not to laugh; that is the only difference. It is silly, but I may not dismiss the matter with that remark, and so I will answer quite seriously that it is only

> possible for me to speak of the street moving while the wagon remains still— and to believe it— when I cast away all the experience of a lifetime and am no longer able to understand the evidence of my senses; which is insanity. Such self-deception as this is not reasoning; it is the negation of reason; which is the faculty of forming correct conclusions from things observed, judged by the light of experience. It is unworthy of our intelligence and a waste of our greatest gift; but that introduction serves very well to illustrate the kind of illusion that lies at the root of Relativity.[878]

Nevile Martin Gwynne describes the irrationality that is woven through the warp and woof of the theory of relativity.

> The concept of relativity attached to his [Eintstein's] name and propagated by him represents an attack on human reason so insidious and diabolical, and so successful, that no opportunity of demonstrating its falsity, and not only its falsity but, to anyone prepared to believe his own powers of reason, its blatantly obvious falsity, should be allowed to pass.[879]

Gwynne proves that most of the elements of the theory of relativity were not the discoveries (Gwynne properly describes them as inventions) of Einstein. Indeed, if one examines the historical record, the only reasonable conclusion is that Einstein plagiarized the entire theory of relativity. Gwynne states that "Einstein's works can be searched from beginning to end without revealing a single original thought of real importance."[880] Gwynne documents the little known historical facts that Einstein stole ideas from other scientists and passed them off as his own. He plagiarized their work. He gave no attribution to the other scientists.

> Curved space, for instance, was thought of by Riemann; adding a fourth dimension, that of time, to geometry to create the new concept of space-time, by Minkowski; the doctrine that objects contract in proportion to the speed at which they moved, by Fitzgerald; and the idea that the velocity of light in a vacuum was constant irrespective of the notion of any object connected with the light ray, by Lorentz. ... Did he [Einstein] first assert the impossibility of detecting the velocity of the earth through the ether? No, this was done by J.H. Poincaré and H.A. Lorentz. ... Did Einstein coin the name Relativity? No, Poincaré did. ... It was Poincaré too, who first asserted that no velocity can exceed that of light. Einstein was not the first to assert that a clock in motion runs slow. This was done by Sir Joseph Larmor. Einstein was not the first to assert that matter is crinkles in curved space. Professor W.K. Clifford advanced this quaint notion in 1870, nine years before Einstein's birth. ... Did Einstein even invent the famous equation, $E=mc^2$, which has become almost synonymous with his name the equation from which nuclear energy and nuclear destruction capability are supposedly derived? Not even that. In 1881 J.J. Thompson had produced a formula, $E=\frac{3}{4}mc^2$, in respect of a charged spherical conductor moving in a straight line. In 1900 Poincaré suggested that electromagnetic energy might possess mass density in relation to energy density, such that $E=mc^2$, where E is energy and m is mass.[881]

Plagiarization is intellectual theft. The unimpeachable

record proves that Einstein was not a genius, but was simply a very clever con man, with powerful backers. Gwynne concludes that "[t]he truth about Einstein is that he was no more than a puppet."[882] Gwynne presents compelling evidence that Einstein was selected to play the specific role of refuting the Michelson/Morley experiment and reestablishing the rotating globular earth.

> [I]f Einstein had not existed another would have been selected to fill his place, for he possessed no qualities which are not available in profusion in almost any place in any age. ... The obstinate truth about Einstein is that in mathematics he was no more than competent and that among the so-called discoveries presented to the world under his name one can search in vain for one that was original. Had Einstein not been selected, for reasons which had nothing to do with intellectual ability, to act out a role which was deemed necessary for the furtherance of the war against God and civilisation, his claim to immortal fame would have been that of a talented and not-undistinguished physicist, a life-long Zionist, an occasionally enthusiastic admirer of Stalin's Russia.[883]

Einstein was a front-man for very powerful interests behind the theory of relativity. That theory was simply an amalgamation of theories propounded by many scientists over many years. As Gwynne points out the global elite needed to have a single front-man for their theory to be popularly accepted.

> [I]t is much easier to impose false beliefs on the world if they are personalised. If a theory is put forward without reference to the person who originated it, there will be a tendency for it to be

> judged on its merits and then, if it clearly has no merits, for it to be rejected. This is far from being the case if a theory, however ludicrously opposed to common sense, is put forward by a man of universally acknowledged genius. When that happens, the tendency will be for the theory to be examined with respect; if it cannot be understood, this will be ascribed to the incapacity of the person examining the theory; if it appears manifestly illogical, it will be assumed that the originator has grasped a logic which is beyond the reach of lesser mortals. In short, it will gradually become accepted on no better grounds than the authority of the person who has advanced it.[884]

Why was Einstein, of all people, chosen to be the front man? There are very powerful inter-generational interests behind promoting Einstein. These interests have an occult religious agenda to enslave the world. Martin Gwynne identifies the core of the conspiracy as Jewish. That Einstein was a Zionist Jew was probably the qualifying factor that put him at the top of the list to be the mouthpiece for the Satanic conspiracy to send the scientific world into darkness through the theory of relativity.

> From the middle of the nineteenth century onwards, those presented to the world as the modern geniuses marking the turning points in civilisation have been Jews. I do not wish to exaggerate this, and it is certainly true that non-Jews too, such as Darwin at the beginning of the period and Lord Keynes in more recent times, have had their nonsense presented as majestic contributions to human knowledge. Nevertheless, if asked to name the three men whose writings had the greatest influence in shaping the modern world, few would go beyond

Karl Marx, Sigmund Freud and Albert Einstein. Explanations for the phenomenon, adequate or otherwise, are suggested elsewhere in other papers that I have written. Here I record only the fact and the inference that can be derived from it. The Jews are entering into what they believe to be their inheritance.

If it be accepted that it was desirable to build up the reputation of a single man for the difficult task of imposing Relativity on the world and that that man should be a Jew, why was Einstein, out of all the other Jewish scientists available, chosen to play the role assigned to him? One can only speculate. Clearly his being a Zionist and a Communist would have recommended him highly to those who selected him; it seems to be agreed by all who came into contact with him that he had much charm, probably indispensable in the task allotted to him; and eyewitness accounts of his lectures provide evidence of considerable abilities as an actor and a showman, which, for the successful accomplishment of the purpose for which he was used, are talents even more necessary than charm. There must, however, have been many other people with similar or better credentials even in a population restricted to people interested in physics. Failing some revelation by those who chose him, all that can be said is that we need have little doubt that he earned his duties and his privilege somehow. I have given some indication of what Isaac Newton did to earn the rewards that he received and is still receiving in this world. Those who recall this and take seriously verses eight and nine of the fourth chapter of St. Matthew have little

> alternative to the belief that such fame and adulation as Einstein received in his lifetime and has received since, and which on the face of it were wholly undeserved, must have been earned at the expense of an extremely exacting bargain in respect of his immortal soul.[885]

After the general acceptance of Einstein's theory of relativity, science entered into a strange new world where experiments were not done using instruments in the physical world, but instead using mathematics in the mind of the scientists. Einstein was famous for announcing new "mind experiments."

Einstein claimed that he did not know of the Michelson/Morley experiment prior to coming up with his special theory of relativity in 1905. Robert Shankland published an article in 1963, in which he stated that Einstein told him in 1950 that he only became aware of the Michelson/Morley experiment after he published his paper on special relativity in 1905. Shankland pointed out that indeed Einstein did not mention the Michelson/Morley experiment in his 1905 paper, suggesting by that fact that Einstein did not know about the Michelson/Morley experiment.

Einstein's claimed ignorance of the Michelson/Morley experiment is contradicted by other statements that he made indicating that in fact he did know about the experiment. Einstein is on record admitting that he did in fact know about the Michelson/Morley experiment and it played a role in his theory of relativity.[886] Einstein's biographer, Ronald Clark, stated that one of the principle issues for science with the Michelson/Morley experiment was that the experimental results proved that the earth is stationary. Clark explained the implications of the Michelson/Morley experiment meant that the whole Copernican theory had to be scuttled, which was "unthinkable" for the scientific community.[887]

> The problem which now faced science was considerable. For there seemed to be only three alternatives. The first was that the Earth was standing still, **which meant scuttling the whole Copernican theory and was unthinkable**. The second was that ether was carried along by the earth in its passage through space....The third solution was that the ether simply did not exist, which to many nineteenth century scientists was equivalent to scrapping current views of light, electricity, and magnetism, and starting again.[888] (emphasis added)

Robert Sungenis explains that "Einstein picked the third solution, which claimed that the aether did not exist, and therefore there could be no significant effect on the light beam discharged in the direction of the Earth's presumed motion around the sun."[889]

Sungenis explains "that the Special Theory [of Relativity] was not invented to explain natural phenomena in the universe, but solely as an answer to the 1887 Michelson-Morley experiment which showed the Earth was standing still. ... All in all, the Special Theory was an ad hoc theory invented to keep the Earth moving in the face of all the experiments that showed it wasn't moving."[890]

The Michelson/Morley experiment results were a devastating blow to the Copernican heliocentric model. But heliocentrism simply could not be scuttled and so all of science must be transformed by removing the pesky ether. The ether was banished. It was deemed by scientific edict to no longer exist, because its continued presence was a fact that proved that the earth was stationary. It was "unthinkable" to modern science that the earth was stationary.

Einstein never mentioned the Michelson/Morley experiment in his 1905 paper announcing his theory of special relativity.[891] In 1942 Einstein claimed to Michelson's biographer that he had already become "pretty much convinced of the validity of the [relativity] principle before I did know this [Michelson/Morley] experiment and its results."[892] It seems that Einstein was trying to avoid having anyone connect his theory of relativity with the Michelson/Morley experiment. However, the historical evidence suggests that Einstein was lying. Think about it; how could anyone believe that Einstein would be ignorant of the Michelson/Morley experiment, when in fact it was the talk of the entire scientific community? His claim of ignorance simply does not pass the smell test. Regardless, we have proof that Einstein lied when he claimed ignorance of the Michelson/Morley experiment.

Forty-two letters were uncovered between Einstein and his fiancee Mileva Mari. Those letters reveal that in fact Einstein knew about the Michelson/Morley experiment as early as 1899.[893] In addition, in a recently uncovered 14 December 1922 speech that Einstein delivered at Kyoto University in Japan, Einstein admitted that he was aware of the Michelson/Morley experiment and the "strange result" of that experiment affected directly his theory of special relativity.

> While I was thinking of this problem in my student years, I came to know the strange result of Michelson's experiment. Soon I came to the conclusion that our idea about the motion of the earth with respect to the ether is incorrect, if we admit Michelson's null result as a fact. This was the first path which led me to the special theory of relativity. Since then I have come to believe that the motion of the Earth cannot be detected by any optical experiment, though the Earth is revolving around the Sun.[894]

Notice Einstein reinforces the scientific myth that the earth revolves around the sun, but he claims that fantastic movement "cannot be detected by any optical experiment." Why can the movement of the earth not be detected by any optical instrument? Because there are no optical instruments that can detect movement that is not there. Indeed, Einstein knows that, which is why he is so certain that no instrument could ever detect the motion of the earth.

The translation of Einstein's speech has been criticized by two Japanese academicians. It seems that it is important to many in the scientific field to cast some doubt on whether Einstein knew about the Michelson/Morley experiment. The Japanese academicians are trying to undermine the startling fact that Einstein was motivated to come up with his theory of relativity by the Michelson/Morley experiment. The circumstances and the personage behind the translation of Einstein's Kyoto speech testify to it accuracy.

In August 1982 *Physics Today* published the above quoted English translation by Yoshimasa A. Ono of Einstein's speech, which was derived from the original notes taken by J. Ishiwara. The speech by Einstein was given in German and translated for the Japanese audience simultaneously by J. Ishiwara. The translator, Ishiwara, was a professor of physics at Tohoku University, and was quite familiar with both Einstein and the theory of relativity, because he had studied under Arnold Sommerfeld and Einstein from 1912 to 1914. Chinese intellectual Zheng Zhenwen vouches for Ishiwara's scientific credentials. Zhenwen stated at the time that Ishiwara was "the only expert of relativity studies in Japan," and that Ishiwara "had a unique understanding of the theory of relativity."[895]

The *Physics Today* article reveals that "Ishiwara kept careful notes of the lecture, and published his detailed notes (in Japanese) in the monthly Japanese periodical *Kaizo* in 1923; Ishiwara's notes are the only existing notes of Einstein's talk."[896] Of all the people in the world, Ishiwara was most qualified to accurately translate Einstein's original German into Japanese. His only motive was to accurately translate and memorialize what was said by the world renowned physicist, Albert Einstein.

There is confirmation that the Japanese translation of Einstein's Kyoto speech is accurate. That confirmation comes in the form of a speech Einstein gave a year before his Kyoto speech, during which Einstein similarly admitted that he knew about the Michelson/Morley experiment before he came up with his theory of special relativity. In May of 1921, during a visit to the University of Chicago, where Michelson was on the faculty, Einstein gave a short speech at the Francis W. Parker School. In that speech he stated, in pertinent part:

> I thought about whether it would be possible to perceive through some experiment that the earth moves in the ether. But when I was a student, I saw that experiments of this kind had already been made, in particular by your compatriot, Michelson. He proved that one does not notice anything on earth that it moves, but that everything takes place on earth as if the earth is in a state of rest.[897]

The above 1921 statement by Einstein impeaches his later claims that he did not know about the Michelson/Morley experiment. Jeroen van Dongen draws the ineluctable conclusion from Einstein's 1921 statement that Einstein knew about the Michelson/Morley experiment long before he came up with his special theory of relativity and that the Michelson/Morley experiments influenced his theory of special relativity.

> What does the Parker school lecture imply for our understanding of Einstein's relation to the Michelson-Morley experiment, and its influence on the creation of the special theory? Taking the

> text at face value, there can be no doubt that Einstein knew of the Michelson-Morley experiment prior to 1905. He attributed a significant role to ether drift experiments in general, and singled out the Michelson-Morley experiment for specific mention. It further suggests that Einstein had learned of the experiment before becoming convinced of the principle of relativity—contrary to his later recollections.[898]

Why was Einstein so insistent in later years in claiming he was ignorant of the Michelson/Morley experiment? Because, he could not allow there to be any connection made between his theory of relativity and the results of the Michelson/Morley experiment. To do so would give up the game. His theory of relativity was trotted out, publicized, and crammed down the throats of academia for the sole purpose of explaining away the Michelson/Morley result that proved the earth does not move. Einstein could not allow the world to know that. Hence, he falsely claimed that he did not even know about the Michelson/Morley experiment when he devised his theory of relativity.

Most people do not know that Einstein was a con artist chosen to play a role. In order for his theory of relativity to maintain its status, however, the powerful elite must also maintain Einstein's esteem as a paragon of intellect. If Einstein is found out to have been a fraud, then his theory of relativity will be scrutinized and found out to be an elaborate scientific deception. Consequently, Einstein has been elevated to the stature of a scientific demigod, and the world's propaganda machine intends to keep him there. For example, Frederic Golden in Time Magazine described Einstein as follows: "He was the embodiment of pure intellect, ... he was unfathomably profound — the genius among geniuses who discovered, merely by thinking about it, that the universe was not as it seemed. ... his ideas, like Darwin's, reverberated beyond science, influencing modern culture."[899]

What was the occasion for such praise from Time? Golden was describing Einstein in the December 31, 1999, Time Magazine cover story about him. Einstein had been named Time Magazine's "Person of the Century." Think about that for a moment; Albert Einstein was selected as the most influential person from among all the persons who lived during that 20th century. That should give some idea of the importance to which the powerful elite running the world think it is to keep people ignorant of the fact that the earth does not move. That brings to mind the warning of Jesus: "Woe unto you, when all men shall speak well of you! for so did their fathers to the false prophets." (Luke 6:26)

Apparently, some very powerful interests are very happy about the hoax of Einstein's theory of relativity. These powerful interests are dark and evil. Jewish controlled Time Magazine is owned by one of the largest media conglomerates in the world, Time Warner. The founders of Time Magazine were Briton Hadden and Henry Luce, who were both members of the Satanic secret society, the Brotherhood of Death, commonly known as the Skull and Bones Society.

President George W. Bush, in his autobiography, *A*

Charge to Keep stated: **"During my senior year I joined Skull and Bones, a secret society, so secret I can't say anything more."** What is so secret that he cannot speak any further about it? The secret is that in return for power, wealth, and fame, he must blindly obey his satanic masters in their antichrist conspiracy to enslave and rule the world. The initiation ceremony for Skull and Bones involves, but is not limited to, the inductees lying naked in a coffin and telling their deepest sexual secrets. Anton LaVey, the founder of the Church of Satan, in his *Satanic Rituals: Companion to the Satanic Bible,* states that such a coffin ritual is a satanic ritual common in many pagan orders. During the ritual a powerful spiritual force charges through the participants transforming their lives dramatically. This powerful spiritual force is a devil. The participants in these ceremonies end up possessed by a devil.

Evidence indicates that the Order of Skull & Bones founded at Yale in 1832 is a chapter of the Illuminati, which was originally founded in 1776 at the University of Ingolstadt in Germany.[900] From this we know that Skull & Bones is not American at all, but is a branch of a foreign secret society.[901]

As with the Jesuits and the Illuminati, Skull & Bones has many ostensible Gentiles who are members. From this fact most people have mischaracterized the Skull & Bones as a purely Gentile organization. That is not true. Just as with the Illuminati and the Jesuits, the Skull & Bones is controlled by and serves the interests of Zionist Jews. George W. Bush is a prime example of an ostensible Gentile member of Skull & Bones who acted in the interests of Israel to the detriment of the United States. He was completely controlled by Zionist Jews.

The Jewish control of Skull & Bones comes from its roots as a chapter in the crypto-Jewish Illuminati. Some of the practices and terms of the Skull & Bones reveal the Jewish nature of the Order. For example, those outside Skull & Bones are referred to by Skull & Bones members as vandals and **"Gentiles."**[902]

Furthermore, in an attempt to conceal the meanings of their writings from any Gentile outsider who may obtain a copy, members of the Skull & Bones often obscure key words by deleting the vowels. For example, patriarchs would be written as p-tr–rchs, bones would be written as b-n-s.[903] The Hebrew alphabet does not have vowels, they use accent marks, and so Jews are accustomed to writing without using vowels. It is not surprising that they would follow that same practice when trying to conceal the meaning in their writings from the uninitiated Gentile world.

The Skull & Bones use the Hegelian dialectic to change society into a totalitarian state. Under Hegel's dialectic there must be a conflict, either real or perceived, between a thesis and an antithesis which is resolved by a synthesis of the two. The secret societies create these conflicts in order to move society regressively away from Christ and Christian principles and toward Satan and satanic principles. David Bay explains the origin and power behind the Cabalistic Skull and Bones:

> The Brotherhood of Death Society in the United States is the Skull and Bones Society in Yale University in New Haven, Connecticut. Its belief structure is identical to that of the Thule Society. Therefore, we can conclude that Bones Men affirm this belief about Jesus Christ, thus condemning them to committing the Unpardonable Sin. The list of some of the Families comprising Skull and Bones is frightening, for it immediately shows the extent to which America has been influenced by this Satanic organization. Remember, the men of these families have likely committed the Unpardonable Sin. (Quoting from Antony Sutton, "America's Secret Establishment", p. 22).

> Rockefeller Family (Standard Oil), Weyerhaeuser Family (Lumber), Sloane Family (Retailing), Pillsbury Family (Flour Milling), J.P. Morgan Family (Banking), Taft Family (Politics), Bush Family, including former President George Bush. Wait a minute, you say, George Bush likely committed the Unpardonable Sin because of his membership in Skull and Bones? Yes. Now, you can see how easy it was for Bush to lead the charge into the Satanic New World Order. Now you can see that Bush was far different in his innermost heart than he was on his media-created surface.[904]

David Bay mentions Skull and Bones member George H.W. Bush spearheading the Satanic New World Order. Indeed, President Bush stated in his state of the union address before a joint session of the U.S. Congress regarding the U.S. led invasion of Iraq that "[w]hat is at stake is more than one small country; it is a big idea: a **new world order**."[905]

President Bush was not leading anything, he was simply an effeminate sycophant, who was obedient to his Zionist overlords. A little known fact is that one of the first public mentions of a "new world order" was in reference to the worldwide interests of the Jews. The October 6, 1940 issue of The New York Times reported that "Arthur Greenwood, member without portfolio in the British War Cabinet, assured the Jews of the United States that when victory was achieved an effort would be made to found a **new world order**."[906] The message from Greenwood was delivered by Rabbi Maurice L. Perlzweig, chairman of the British section of the executive committee of the World Jewish Congress, to Dr. Stephen S. Wise, chairman of the executive committee of the World Jewish Congress.[907] Greenwood's message was to the Jews and for the Jews.

NEW WORLD ORDER PLEDGED TO JEWS

Arthur Greenwood of British War Cabinet Sends Message of Assurance Here

RIGHTING OF WRONGS SEEN

English Rabbi Delivers to Dr. S. S. Wise New Statement on Question After War

In the first public declaration on the Jewish question since the outbreak of the war, Arthur Greenwood, member without portfolio in the British War Cabinet, assured the Jews of the United States that when victory was achieved an effort would be made to found a new world order based on the ideals of "justice and peace."

Mr. Greenwood, who is Deputy Leader of the British Labor party, declared that in the new world the "conscience of civilized humanity would demand that the wrongs suffered by the Jewish people in so many countries should be righted." He added that after the war an opportunity would be given to Jews everywhere to make a "distinctive and constructive contribution" in the rebuilding of the world.

The message was delivered last week to Dr. Stephen S. Wise, chairman of the executive committee of the World Jewish Congress, by Rabbi Maurice L. Perlzweig, chairman of the British section of the congress. Rabbi Perlzweig arrived from England Monday evening.

Intention to Right Wrongs

Comparing the statement with the Balfour Declaration of 1917, Dr. Wise declared that in a sense it had "wider and farther reaching implications," as it dealt with the status of Jews throughout the world. He said that Mr. Greenwood's message could be interpreted as a statement of England's firm intention to help right the wrongs which Jews have suffered and continue to suffer today because of Hitler's "disorder and lawlessness."

Mr. Greenwood, sending the Jews of America a message of "encouragement and warm good wishes," wrote:

"The tragic fate of the Jewish victims of Nazi tyranny has, as you know, filed us with a deep emotion. The speeches of responsible statesmen in Parliament and at the League of Nations during the last seven years have reflected the horror with which the people of this country have viewed the Nazi relapse into barbarism.

"The British Government sought again to secure some amelioration of the lot of persecuted Jewry both in Germany itself and in the countries which were infected by the Nazi doctrine of racial hatred. Today the same sinister power which has trampled on its own defenseless minorities, and by fraud and force has temporaily robbed many small peoples of their independence, has challenged the last stronghold of liberty in Europe.

New World Order Forecast

"When we have achieved victory, as we assuredly shall, the nations will have the opportunity of establishing a new world order based on the ideals of justice and peace. In such a world it is our confident hope that the conscience of civilized humanity would demand that the wrongs suffered by the Jewish people in so many countries should be righted.

"In the rebuilding of civilized society after the war, there should and will be a real opportunity for Jews everywhere to make a distinctive and constructive contribution; and all men of good-will must assuredly hope that in new Europe the Jewish people, in whatever country they may live, will have freedom and full equality before the law with every other citizen."

In an interview at the Hotel Astor, Rabbi Perlzweig declared he was certain Mr. Greenwood "speaks for England." There is a clear realization, he added, that freedom and emancipation for the Jewish people are tied up with emancipation and freedom for people everywhere. The message, Rabbi Perlzweig remarked, was the subject of earnest consideration by the British Government.

"This is a declaration in behalf of the whole world," he observed. "Here the British Government expresses clearly what it hopes will take place after the war is won."

Figure 185: The New York Times, October 6, 1940, page 10.

Greenwood's statement was designed to enlist the

powerful Jewish interests in the war with Germany. By making that promise, England had promised the Jews English assistance in gaining Jewish hegemony over a new world order, if the Jews would use their enormous resources and influence to deliver the United States into the war against Germany. Indeed, that is exactly what happened. The "New World Order" is world domination by the Jews. It is the goal of the Zionist conspiracy.

The Skull and Bones Society is just one of many occult organizations doing the work of the devil. "For we wrestle not against flesh and blood, but against principalities, against powers, against the rulers of the darkness of this world, against spiritual wickedness in high places." (Ephesians 6:12) That spiritual truth seems to have been missed by Dr. Dingle. He simply could not understand how and why the clearly fallacious theory of relativity could gain such universal acceptance in the scientific community. Dr. Dingle, in frustration, asked:

> How is it possible that such an obvious absurdity [as the theory of relativity] should not only have ever been believed but should have been maintained and made the basis of almost the whole of modern physics for more than half a century; and that, even when pointed out, its recognition should have been universally and strenuously resisted, in defiance of all reason and all the traditions and principles of science?[908]

On November 17, 1952, the Prime Minister of Israel, David Ben Gurion, asked Albert Einstein to accept an invitation to become the President of Israel, upon the authorization of the Knesset of Israel.[909] Einstein turned down the offer, but the offer is testimony to the importance and esteem with which Einstein was held among Jews. It also demonstrates the political importance of the theory of relativity to the Jews. The descent in the minds of men of God, who created a flat, immovable earth, gives rise to the ascent of Jews in power and influence in the world. Having the earth become one of many millions of planets in the universe (and a rather insignificant one at that) drives from the minds of men any thought that God created a unique earth, at the center of his creation, and that man is made in his image. Once God is removed from the thoughts of men, it is then possible to introduce ideas of "everything is relative," where there is nothing inherently good or bad, because it is all relative to a person's own ethical opinion. There is no longer reference to absolutes contained in the Bible, morals become malleable, with each man a law-maker in his own right. So we see that the theory of relativity is a theory that bleeds into all aspects of society. Indeed, read in Psalms 19:1-8 how there is a direct correlation between knowledge of God's creation and his commandments. Concealing the glory of God as declared by his handiwork, makes it very easy to introduce concepts of relative morality. All manner of sin can then be justified (e.g., abortion and sodomy).

> The heavens declare the glory of God; and the firmament sheweth his handywork. Day unto day uttereth speech, and night unto night sheweth knowledge. There is no speech nor language, where their voice is not heard. Their line is gone out through all the earth, and their words to the end of the world. In them hath he set a tabernacle for the sun, Which is as a bridegroom coming out of his chamber, and rejoiceth as a strong man to run a race. His going forth is from the end of the heaven, and his circuit unto the ends of it: and there is nothing hid from the heat thereof. The law of the LORD is perfect, converting the soul: the testimony of the LORD is sure, making wise the simple. The statutes of the LORD are right, rejoicing the heart: the commandment of the LORD is pure, enlightening the eyes. (Psalms 19:1-8)

38 Conspiracy

Conspiracy is a concept that has been propagandized into disfavor, much to the delight of Satan and his minions, who are only too happy to push the idea that if one believes there is a world conspiracy against Jesus Christ and his followers he must be a radical on the fringe of society or worse. Most people are afraid of being marginalized and thus avoid speaking of conspiracies. People instead try to construe events as coincidental, when in fact they can only be adequately explained as the product of prior agreements of conspirators who have combined in coordinated actions. God has revealed that there is a religious conspiracy.

> **There is a conspiracy** of her prophets in the midst thereof, like a roaring lion ravening the prey; they have devoured souls; they have taken the treasure and precious things; they have made her many widows in the midst thereof. Her priests have violated my law, and have profaned mine holy things: they have put no difference between the holy and profane, neither have they shewed difference between the unclean and the clean, and have hid their eyes from my sabbaths, and I am profaned among them. Her princes in the midst

> thereof are like wolves ravening the prey, to shed blood, and to destroy souls, to get dishonest gain. And her prophets have daubed them with untempered morter, seeing vanity, and divining lies unto them, saying, Thus saith the Lord GOD, when the LORD hath not spoken. (Ezekiel 22:25-28)

Notice that it was the Jewish prophets and priests whom God identified as being involved in the conspiracy. They were turning people from God, which caused their damnation in hell. Thus they were "devouring souls" like spiritual ravening lions. The metaphor of a ravening lion, which God has used elsewhere in the Bible to describe the devil himself, suggests that the Jewish prophets and priests were acting as the direct agents of Satan: "Be sober, be vigilant; because your adversary the devil, as a roaring lion, walketh about, seeking whom he may devour." (1 Peter 5:8) The result of the longstanding conspiracy against God is that the conspirators had turned to iniquity and they served other gods.

> **And the LORD said unto me, A conspiracy is found among the men of Judah, and among the inhabitants of Jerusalem**. They are turned back to the iniquities of their forefathers, which refused to hear my words; and they went after other gods to serve them: the house of Israel and the house of Judah have broken my covenant which I made with their fathers. (Jeremiah 11:9-10)

The conspiracy against God has continued through history up to and beyond the first coming of Jesus Christ. We read in the following passage, how the Jews conspired to kill Jesus.

> Then assembled together the chief priests, and the scribes, and the elders of the people, unto the

> palace of the high priest, who was called Caiaphas, And consulted that they might take Jesus by subtilty, and kill him. (Matthew 26:3-4)

The Jews stirred up the people, who then persuaded the Roman governor, Pontius Pilate, to crucify Jesus. "[T]he chief priests and elders persuaded the multitude that they should ask Barabbas, and destroy Jesus." (Matthew 27:20) The methods used by the Jews have not changed. Today the Jews control the mass media, which they use to propagandize the people into pushing the politicians into doing their bidding against God and his anointed.

The conspiracy against the LORD and his anointed has been festering since the fall of man in the garden of Eden. It continues to this day and involves the rulers of the earth.

> Why do the heathen rage, and the people imagine a vain thing? **The kings of the earth set themselves, and the rulers take counsel together, against the LORD, and against his anointed, saying, Let us break their bands asunder, and cast away their cords from us**. He that sitteth in the heavens shall laugh: the Lord shall have them in derision. (Psalms 2:1-4)

This monstrous conspiracy, while it involves men, is not headed by a man. "For we wrestle not against flesh and blood, but against principalities, against powers, against the rulers of the darkness of this world, against spiritual wickedness in high places." (Ephesians 6:12) The kingpin of this diabolical conspiracy is that terrible dragon, Satan. He is the adversary of Almighty God. He is also an adversary to all of God's creation. Satan tried to tempt Jesus into worshiping him, but Jesus quoted from God's word and told Satan that one should only worship God.

> Again, the devil taketh him up into an exceeding high mountain, and sheweth him all the kingdoms of the world, and the glory of them; And saith unto him, All these things will I give thee, if thou wilt fall down and worship me. Then saith Jesus unto him, Get thee hence, Satan: for it is written, Thou shalt worship the Lord thy God, and him only shalt thou serve. (Matthew 4:8-10)

If you are a disciple of God's perfect and holy Son Jesus Christ, then Satan hates you. "Be sober, be vigilant; because your adversary the devil, as a roaring lion, walketh about, seeking whom he may devour." (1 Peter 5:8) God, however, has a means by which one can avoid becoming the prey of Satan: "Submit yourselves therefore to God. Resist the devil, and he will flee from you." (James 4:7) In order to resist the devil one must first know his plans and methods.

39 Identifying the Conspirators

Zionist Jews have created all manner of antichrist philosophies designed to propagandize and enslave the masses. One of the principal satanic movements propagated by Zionist Jews is communism. Paragraph three of Protocol 2 of the Protocols of the Learned Elders of Zion states: "think carefully of the successes we arranged for Darwinism (Evolution), Marxism (Communism), Nietzsche-ism (Socialism). To us Jews, at any rate, it should be plain to see what a disintegrating importance these directives have had upon the minds of the goyim."[910]

Rabbi S. Wise stated in the May 5, 1935, American Bulletin: "Some call it Marxism; I call it Judaism."[911] On April 4, 1918, The Jewish Chronicle stated that "the ideals of Bolshevism are consonant with the finest ideals of Judaism."[912]

The fact that communism is actually a religious philosophy born of the Talmud and put into practice has been concealed from the ignorant Gentiles. In a 1971 White House recording released by the National Archives in 1999, President Richard Nixon revealed: "The only two non-Jews in the communist conspiracy were Chambers and Hiss. . . . Every other one was a Jew and it raised hell with us."[913] The Talmudic Jews have been successful in preventing any revelations about their involvement in establishing a new communist world order by labeling anyone who exposes their efforts an antisemite.

Many think that Zionism is the struggle by the Jews for a homeland. Zionism is much more than the Jews establishing a homeland. That is merely a cover for a much grander plan to rule the world. Zionism is the child of the Talmud, and Talmudism is communism. The communist revolution in Russia was planned and executed by Jews according to the doctrines of their Talmud.

Make no mistake about it, the communists stated at the outset that their aim is world domination. The 1919 World Congress of the Communist International stated that it intended to wage a "fight by all available means, including armed force, for the overthrow of the international bourgeoisie and for the creation of an international Soviet republic."[914] That statement was in essence a declaration of war against all countries in the world. Once one understands that communism is a cloak for Zionism, it becomes clear that the Zionists have declared war on the world and plan on ruling the world with an iron fist after their victory.

Communism is essentially Talmudic Judaism. The exoteric communism portrays itself as an atheistic political movement. The persecution of the Christian churches wherever communists have been brought to power seems to prove that point. However, the real nature of communism as a Judaic religious movement is manifest by the fact that Jewish synagogues do not suffer the same fate as Christian churches in communist countries. Communism reveals Judaism as a satanic conspiracy against God and man.

Jewish scholar Henry Makow, Ph.D., was asked why when the Bolsheviks took over Russia and tried to eliminate religion, Christian churches were destroyed, and the Christian clergy were killed or imprisoned, yet the Jewish synagogues and rabbis remained untouched. If communism was a political movement that was antagonistic to all religion, why was Judaism given a pass? Makow researched that issue:

> Last week I stumbled across the answer in a book by American historian Edwin Schoonmaker:
>
> "Fifteen years after the Bolshevist Revolution was launched to carry out the Marxist program, the editor of the 'American Hebrew' could write: 'According to such information that the writer could secure while in Russia a few weeks ago, not one Jewish synagogue has been torn down, as have hundreds-perhaps thousands of the Greek Catholic Churches. . . . In Moscow and other large cities one can see Christian churches in the process of destruction . . . [whenever] the Government needs the location for a large building.' (American Hebrew, Nov. 18, 1932, p. 12) Apostate Jews, leading a revolution that was to destroy religion as the 'opiate of the people' had somehow spared the synagogues of Russia." ("Democracy and World Dominion," 1939, p.211)
>
> If the Communists hated God and religion so much, why didn't they destroy synagogues too? Do Christianity and Judaism worship the same God? Or can a religion that claims an exclusive ownership of God be a religion? Could it be that Judaism is really a secret society like Freemasonry where the members don't know the hidden agenda, which is in fact expressed by Communism? Although many religious Jews were indeed anti-Communist, Schoonmaker's information suggests there may be an affinity between Talmudic Judaism and pagan Satanic Communism.[915]

V.I. Lenin's maternal grandfather, Israel Blank, was Jewish. Researcher Wayne McGuire of Harvard University wrote: "Lenin was a Jew by the standards of Israel's Law of Return: he possessed a Jewish grandparent."[916] Lenin, in apparent reference to himself, said: "The clever Russian is almost always a Jew or has Jewish blood in him."[917] Leon Bronstein (Trotsky), supreme commander of the Soviet Red Army, was also Jewish.[918]

The Bolshevik revolution was Jewish from top to bottom. Of 556 leading conspirators in the Bolshevik state in 1918-19 there were 17 Russians, two Ukrainians, eleven Armenians, 35 Latvians, 15 Germans, one Hungarian, ten Georgians, three Poles, three Finns, one Czech, one Karaim, and 457 Jews.[919] As pointed out by Robert Wilton in his book The Last Days of the Romanovs, the communist revolution was not an insurrection by Russians, but rather a secret invasion by Jews. As of 1983, the Premier of the Soviet Union was a Jew (Andropov) and 23 out of 25 members of the Politburo (the Soviet ruling clique) were Jews. In addition, every top member of the Soviet military and police was a Jew.[920] Robert Wilton explains:

> The Germans knew what they were doing when they sent Lenin's pack of Jews into Russia. They chose them as agents of destruction. Why? Because the Jews were not Russians and to them the destruction of Russia was all in the way of business, revolutionary or financial. The whole record of Bolshevism in Russia is indelibly impressed with the stamp of alien invasion. The murder of the Tsar, deliberately planned by the Jew Sverdlov (who came to Russia as a paid agent of Germany) and carried out by the Jews Goloschekin, Syromolotov, Safarov, Voikov and Yukovsky, is the act not of the Russian people, but of this hostile invader.[921]

Colonel Jack Mohr states: "One of the greatest difficulties of the Talmudic Pharisees has been that of bringing communism into power while trying to conceal its Talmudic origin."[922] However, the direct and circumstantial evidence that the communist revolution in Russia was a conspiracy perpetrated by Talmudic Jews is overwhelming. Circumstantial evidence that points to Jewish control of the communist revolution is that once the communists in Russia seized power, the first law they passed made antisemitism a crime punishable by death.[923] While Christian church buildings were turned into animal stables, slaughter houses, and dance halls, the Jewish synagogues were untouched. [924] Christian pastors were removed from their pastoral duties and made to work on roads and in slave labor camps, yet the Jewish rabbis were permitted to continue their clerical duties.[925] "Some 200,000 (Christian) clergy, many crucified, scalped and otherwise tortured, were killed during the approximately 60 years of communist rule in the former Soviet Union, a Russian commission reported Monday (Nov. 27, 1995) . . . 40,000 churches (were) destroyed in the period from 1922 to 1980."[926]

Historian Michael Hoffman II exposed the hidden meaning behind some of the bloodthirsty communist propaganda:

> Lenin declared, "We are exterminating the bourgeoisie as a class." His partner in crime, Apfelbaum (Zinoviev) stated: "The interests of the revolution require the physical annihilation of the bourgeoisie class." Who were these bourgeoisie? Certainly not Jews. Trotsky gave a clue to their identity in a 1937 interview in the New York Jewish newspaper, *Daily Forward:* "The longer the rotten bourgeoisie society lives, the more and more barbaric will anti-Semitism become everywhere."
>
> **Bourgeoisie was a Bolshevik code-word for**

> **Gentile.** The first law passed after the Communists seized power in Russia made anti-semitism a crime punishable by death. (*Izvestia,* July 27, 1918).

* * *

> The Jewish Bolsheviks regarded politics as a branch of Gentile pest control. Hatred of Christians, especially the peasant "bourgeoisie" was their prime motivation. The systematic destruction of the Christian peasantry of Russia as so many vermin, beginning with Lenin's attack on them in the summer of 1918 and his forced starvation in 1921, has been almost completely ignored in Western history.[927]

Henry Makow explains that the brutality of the Bolshevik Revolution in Russia was inspired and led by Talmudic Jews. Makow estimates that the Cheka (later renamed OGPU, then NKVD and finally KGB) was responsible for at least 20 million deaths.[928] The primary foci of the slaughter were Gentiles in general and Christians in particular. According to Slava Katamidze: "The Church became the target of Bolshevik hostility from the very beginning."[929]

Makow quotes Jewish writer Sever Plocker, who stated that the brutal Cheka was led and staffed mainly by Jews.[930] The Cheka was empowered by Lenin and Trotsky, who were Illuminati Jews and financed by Illuminati Jewish bankers. "Many Jews sold their soul to the devil of the Communist revolution and have blood on their hands for eternity," Plocker writes. "We mustn't forget that some of the greatest murderers of modern times were Jews."[931]

Plocker gives the example of Genrikh Yagoda, who was a bloodthirsty Jew who oversaw the slaughter of millions of

Russians:

> An Israeli student finishes high school without ever hearing the name "Genrikh Yagoda," the greatest Jewish murderer of the 20th Century, the GPU's deputy commander and the founder and commander of the NKVD. Yagoda diligently implemented Stalin's collectivization orders and is responsible for the deaths of at least 10 million people. His Jewish deputies established and managed the Gulag system.[932]

Moses Mordecai Marx Levi, alias Karl Marx, was a Jew, a Satanist, and a member of the "League of the Just," which was a branch of the Illuminati.[933] In 1847, Marx was commissioned by the Illuminati to write the *Communist Manifesto*, which is an outline of their plans for world domination.[934]

Paragraph seven of Protocol 3 states that the Gentile communists and socialists are dupes that are helping the Zionist Jews bring about the destruction of the Gentiles.

> We appear on the scene as alleged saviours of the worker from this oppression when we propose to him to enter the ranks of our fighting forces - Socialists, Anarchists, Communists - to whom we always give support in accordance with an alleged brotherly rule (of the solidarity of all humanity) of our social masonry. The aristocracy, which enjoyed by law the labor of the workers, was interested in seeing that the workers were well fed, healthy, and strong. We are interested in just the opposite - in the diminution, the killing out of the goyim.[935]

The Talmudic Jews have been successful in preventing

any revelations about their involvement in establishing a new communist world order by labeling anyone who exposes their efforts an antisemite. What most do not understand is that many who claim to be objects of antisemitism are not Semites at all. Semites are those who are descended from Shem, the oldest son of Noah. Most Jews living in Israel and throughout the world today are eastern European converts to a religion that they call Judaism, but in fact is Babylonian Talmudism. The Europeans who later converted to this Babylonian form of Judaism are known as Ashkenazi or Khazar Jews.

Dr. Benjamin H. Freedman, a former Jew, states that the Khazars were a pagan nation whose religious worship was a mixture of phallic worship and other forms of idolatry. In the 7th century their King Bulkan chose Talmudism, which most now call Judaism, as the state religion.[936] Today Khazar Jews are called "Yiddish."

These Ashkenazi Jews are people without any allegiance to any nation. Their primary objective is to own the entire world. To get an idea of the nefarious objective of these Talmudists, let us read an 1879 letter from Baruch Levy to Karl Marx:

> The Jewish people as a whole will be its own messiah. It will attain world dominion by the dissolution of other races, by the abolition of frontiers, the annihilation of monarchy, and by the establishment of a world republic in which the Jews will everywhere exercise the privilege of citizenship. In this new world order the children of Israel will furnish all the leaders without encountering opposition. The governments of the different peoples forming the world republic will fall without difficulty into the hands of the Jews. It will then be possible for the Jewish rulers to abolish private property, and everywhere to make

> use of the resources of the state. Thus will the promise of the Talmud be fulfilled, in which it is said that when the messianic time is come, the Jews will have all the property of the whole world in their hands.[937]

Many think that communism could not be the work of Talmudic Jews because Russia is allied with the Arab countries. Things, however, are not what they appear. Jack Bernstein, an American Ashkenazi Jew who moved to Israel shortly after its founding in 1948, returned in disgust to the United States after witnessing the duplicity of Israel. He revealed that the aboriginal Jews of Palestine, who are called Sephardic Jews, are discriminated against in modern Israel. They are second class citizens at the bottom strata of society in Israel, along with Christians and Muslims. In his book, *The Life of an American Jew in Racist Marxist Israel*, explains the Machiavellian strategy of Israel.[938]

Bernstein found out that communist Russia's support for the Arab countries is a Jewish subterfuge.[939] He stated that part of the Machiavellian plan is for communist Russia to appear to support the Arabs by furnishing them with military aid. However, the communists will never provide enough aid to allow the Arabs to destroy Israel. Since 1949, there has been a free-flow of intelligence sharing between Israel and communist Russia.[940] Bernstein obtained this information directly from the horse's mouth: the Secretary-Treasurer of the Communist Party in Northern Tel Aviv.[941]

Bernstein pointed out that Israel presents itself as a democracy, but in fact Israel is a communist country to its core. He stated that Zionism and communism are one and the same. The purest form of communism is found in Jewish kibbutzim in Israel.

Michael Collins Piper reveals in his book, *Final*

Judgment, that Israel's communist ties are not limited to the former Soviet Union. There has been a long, albeit secret, history of mutual cooperation between communist China and Israel in the development of nuclear and other military weapons. In fact, Israel has been cited as one of the primary conduits for the flow of U.S. and other western technologies to communist China.

Jesus was victorious over Satan on the cross. Satan regrouped and attacked the manifestation of Christ on earth, the Christian church. The Bible identifies the people Satan used to persecute the church: the Jews. The Jews are the enemies of the gospel of Jesus Christ. For example, in Acts chapters 6 and 7 it explains how Stephen preached the gospel before the Jews. What did the Jews do? They stoned Stephen to death. Jesus identified the Jews thusly: "Ye are of your father the devil, and the lusts of your father ye will do." John 8:44. Jesus was not using hyperbole. The religion of the Jews in fact is the mystical worship of Satan. The Jews must keep that fact secret until they have full control of all of the organs of their planned world government.

According to John Torell, in the Jewish Cabala (a/k/a Kabbalah), the "holy serpent"is trying to set herself free from the bottomless pit. Once she does this she can enter the earth as "the messiah." Cabalists teach that "the messiah" will only appear on earth in one of two ways. One way is for the Cabalistic Jews to destroy all evil on the earth and make it totally good. The other alternative is for the Cabalistic Jews to destroy all good on earth and make it totally evil.[942]

The Cabalistic Jews have decided that is harder to make things good and so they have chosen the alternative of destroying all goodness and making the world evil. This Cabalistic religion is grounded in the commission of sins in order to bring about the ascension of their messiah, the "holy serpent," out of the bottomless pit to make "her" appearance on earth. These Jews believe that only by breaking the laws of God can they serve their god.

Do not think that Jews are ignorant of the fact that Satan is the god of their religion. Harold Wallace Rosenthal, Administrative Assistant to United States Senator from New York, Jacob K. Javits, in a 1976 interview with Walter White Jr. stated: "Most Jews do not like to admit it, but our god is Lucifer."[943] That explains the revelation by Albert Pike, the theological pontiff of Freemasonry, who said that "[t]he doctrine of Satanism is heresy; and the true and pure philosophic religion is the belief in Lucifer, the equal of Adonay; but Lucifer, God of Light and God of Good, is struggling for humanity against Adonay, the God of Darkness and Evil."[944] The evidence is that Freemasonry flows from Cabalistic Judaism. Albert Pike revealed in his doctrinal Bible of Masonry, *Morals and Dogma*, that "Masonry is a search for Light. That leads us directly back, as you see, to the Kabalah."[945] It is necessarily the case that the god of Freemasonry (Lucifer) would be the same as the god in the religious book upon which it is based (the Jewish Kabala).

The modern day Pharisees go to great lengths not only to sin themselves, but also to lead many others into sin as they serve their evil god, Satan. Jesus revealed their nature when he said to the Pharisees: "Woe unto you, scribes and Pharisees, hypocrites! for ye compass sea and land to make one proselyte, and when he is made, ye make him twofold more the child of hell than yourselves." (Matthew 23:15)

The fact that Judaism is a religion of sin and rebellion that embraces evil is confirmed by Jewish "rabbi" David Cooper, who spent eight years studying the Cabala in Jerusalem's Old City.

> The lesson is that even the heart of Satan has a divine spark; even the heart of evil yearns to be redeemed. This is important, because we learn that our job is not to set up the battleground to

> eradicate evil, but search out its spark of holiness. Our task is not to destroy but to build.[946]

* * *

> The mystical teaching of the Baal Shem Tov, however, presents us with a new paradigm. It says that evil has a divine nature within it. As the Zohar describes, 'there is no sphere of the Other Side (evil) that entirely lacks some streak of light from the side of holiness.' [Zohar II:69a-b] Rather than destroy it, our task is to uplift it.[947]

The purposeful promotion of evil by Jews is the reason that we find vice, death, and destruction at every turn of the Zionist screw. Near the top of the Zionist power pyramid we find the Rothschild banking family. Henry Makow reveals that the Rothschilds were Sabbatean-Frankists, which is a sect of Judaism. Sabbateans are followers of Rabbi Sabbatai Zevi (1626-1676), who claimed to be the Jewish Messiah. Frankists are followers of Jacob Frank (1726-1791). Frank was a popular Jewish Cabalist, who claimed to be the reincarnation of Sabbatai Zevi. Sabatean-Frankists are "characterized by their 'liberated' sexual practices, incest, pedophilia, etc. They believed the Messiah would return only when the world descended into sin and chaos. They would advance this process."[948] Sabbatean-Frankists adhered to the code of "what had been prohibited would be allowed: adultery, incest, pedophilia. (This is the occult origin of our 'sexual liberation.'). Inspired by the Cabala, they practiced 'holiness through sin.' Good would come through the annihilation of Western civilization and the triumph of Evil."[949]

Surely Satan has an end in mind for constructing such a sinister religious doctrine. He does, and it is nothing short of the subjugation of all men under the dictatorial rule of his antichrist! He uses the escalating sin and crime in society as a justification to bring about more government regulation and control of the masses. As explained by Edmond Burke: "Men are qualified for civil liberty in exact proportion to their disposition to put moral chains on their own appetites. Society cannot exist unless a controlling power upon will and appetite be placed somewhere, and the less of it there is within, the more there is without. It is ordained in the eternal constitution of things that men of intemperate minds cannot be free. Their passions forge their fetters."

This evil doctrine can also be seen in the Talmud, where incest, fornication, adultery, etc. are promoted as virtues and something to be desired. It is difficult for the Gentile world to fully comprehend what is happening in this Jewish netherworld of conspiracy unless they understand the nature of Cabalistic Judaism. It is a religion that is based on the promotion, propagation, and commission of sin as a means to world domination.[950]

The Cabala, like the Talmud, graphically blasphemes Jesus. For example, in Zohar III, 282a, the Cabala refers to Jesus as a dog who resides among filth and vermin.[951]

Individual Jews can be saved only by the grace of God through faith in Jesus Christ. Such a saved Jew will reject Judaism, just as a saved Satanist will reject Satanism. Judaism is irreformable. The Jews who follow Judaism to their death will die in their sins. Such Jews are described by Stephen as stiff-necked people, who are in perpetual rebellion against God.

> Ye stiffnecked and uncircumcised in heart and ears, ye do always resist the Holy Ghost: as your fathers did, so do ye. Acts 7:51.

Henry Makow, in his book *Illuminati, The Cult that Hijacked the World*, explains the occult nature of Judaism:

The "secret society" appears to be the organizational model for Judaism as well as Freemasonry, Zionism and Communism (which are Masonic orders.). Essentially, the leadership deceives and manipulates the membership with idealistic-sounding goals. Only those corruptible (and blackmail-able) are let in on the true agenda and allowed to rise.

This model now applies to the whole world. "Successful" people have often accepted the Devil's Bargain-"Serve me and I will give you the world."

This view of Judaism is confirmed by the author of the "Protocols of Zion" who says:

"No one will ever bring under discussion our faith from its true point of view since this will be fully learned by none save ours, who will never dare to betray its secrets." (Protocol 14)

* * *

Edith Starr Miller, an expert on religion and the occult, called Judaism "a secret society posing as a religion," and "a sect with Judaism as a rite."

The real purpose of Judaism and all secret societies, Miller says, is to con people into advancing the agenda of the super rich. "Regardless of their exoteric objects, the esoteric aims of most societies are all directed toward the same end, namely: the concentration of political, economic and intellectual power into the hands of a small group of individuals, each of whom

> controls a branch of the International life, material and spiritual, of the world today." ("Occult Theocracy," 1933, p.661)
>
> Flavien Brenier compares the goals of Judaism with Freemasonry: securing political power and gradually modifying "the conceptions of the people in the direction of their secret doctrine." ("Occult Theocracy," 80)
>
> The secret aim of Judaism is the same as Freemasonry.[952]

Part and parcel of Zionism is the Judaic religious philosophy based upon the occult worship of a type of god known as an egregore. An egregore is an autonomous god of the collective group mind. The egregore influences the thoughts of a group of people. In the case of Judaism, it is the Jewish people. Henry Makow opines that this Judaic egregore is Lucifer:

> Perhaps the Jewish elite has a different God. A reader, "George," knew the heiress of a rich Jewish banking family "not the Rothschilds, although her family dwelled in a palace neighbouring the Rothschilds."
>
> "Leah was one of my classmates in the Geneva University (Switzerland) where I was studying psychology . . . She was an attractive blue-eyed blonde lady. . . . As she was wearing a Star of David, I asked if she was a believer. She answered "yes and no" and added that she believed in a god of the Jews who was serving the Jews rather than served by them. I immediately asked whether she was speaking of an egregore . . . Her only response was "yes" and she broke

that conversation. Never again did we mention the subject."[953]

* * *

George speculates that elite Jews created the Jewish God as their "egregore," i.e., an instrument of their collective will, i.e., their desire to vanquish the heathens and rule supreme:

"Could it be possible that the ancient Levite priests found a way to create a supernatural entity from the collective mind of the tribe of Judah? An entity born from a collective mind the Levites were shaping into self-isolationism and segregationist a.k.a., extreme ethno-centrism? . . . An entity designed to help the "chosen people" destroy the ‘alien’ nations and provide it with material rewards? An entity resulting from the black magic of the Levite priests who, as the first indoctrinated atheists, were denying the universal God of Moses because they didn't want to submit to a ‘Lord and Master’ but rather become themselves ‘Lords and Masters’ through their slavish ‘egregorious’ god?"

This "egregore" is Lucifer. Elite Jews have made him the master meme of the New World Order. The elite Jews are the "Communist Capitalist International," the intermarried German Jewish banking families who, according to Christian Rakovsky include the Rothschilds, Warburgs, Schiffs and many others.

They have also intermarried with the corrupt gentile elites of Europe and America, many of whom think they are Jews.[954]

All egregorious gods are evil and under the command of Satan. The Jewish egregore is viewed within Jewry as a serpent that is gaining control of nation after nation as it slithers across the globe in its journey toward world domination. Texe Marrs reveals: "In the religion of Judaism, the Sacred Serpent is said to rise from the abyss. Through his power and guidance, a divine World Jewry will triumph over the Gentiles, and the Jews' Messiah will reign supreme."[955] Makow's opinion that the egregore of Judaism is Lucifer is perceptive, since Lucifer controls all egregores. That fact is implicitly supported by the writings of eminent Jewish Rabbis. Texe Marrs explains:

> Take, for example, the celebrated rabbi known throughout the Jewish religious world as the "Gaon of Vilna." It was he who taught of the Kabbalah's doctrine that inside Judaism's vaunted Tree of Life there resides a great and Sacred Serpent whose masculine name is Leviathan and whose feminine name is Malkut. It is this Sacred Serpent, the Kabbalah teaches, that in the coming Messianic age shall rise from the abyss to conquer the Gentiles and exalt God's Chosen, the Jews. This Leviathan, the holy and piercing serpent, is the expected Messiah prophesied to appear, the one who will supernaturally possess the bodies of the world's Jews and lead them to global domination and glory.
>
> This strange doctrine, accepted by the vast majority of today's Orthodox Rabbis, also makes the bold claim that the Jews are a Holy Race of wise and virtuous serpent beings. Collectively, World Jewry is claimed to be the very incarnation

on earth of the Holy Serpent.

Could this be what the true Messiah and Lord of the Universe, Jesus Christ, meant when he confronted the wicked pharisaic Jews—equivalent to today's Orthodox Jewry—by flatly declaring:

"Ye serpents, ye generation of vipers, how can ye escape the damnation of hell?" (Matthew 23:33)

* * *

The Jews themselves, in their devilish religion of Judaism, admit their god is the Serpent. Their leaders well know that this is the Devil, or Satan. In some type of creepy and eerie doctrinal confession, the rabbis are even discovered to be boasting of their Sacred Serpent and they say that all Jews live inside its belly! Is this not unbelievable and revelatory?[956]

Whom do we see described in the Bible as the serpent? It is none other than Satan himself. "And the great dragon was cast out, that old serpent, called the Devil, and Satan, which deceiveth the whole world: he was cast out into the earth, and his angels were cast out with him." Revelation 12:9.

The earthly god of Judaism, who is androgynous, is known as Leviathan or Malkut.[957] Leviathan is an evil god, adopted by the rabbis as the god of the Jews during their captivity in ancient Babylon.[958]

The occult Judaism, which is hidden from the uninitiated, is that the Jews view themselves as the very manifestation of their egregore god, whom they call Leviathan. This collective god,

Leviathan, also known as the "holy serpent," will subjugate the world under the rule of the Jews.

The egregore of the Jews is confirmed in paragraph one of Protocol 3 in the Protocols of the Learned Elders of Zion, which refers to the "Symbolic Snake" and depicts it as an egregorious symbol of the Jewish people:

> Today I may tell you that our goal is now only a few steps off. There remains a small space to cross and the whole long path we have trodden is ready now to close its cycle of the Symbolic Snake, by which we symbolize our people. When this ring closes, all the States of Europe will be locked in its coil as in a powerful vice.

Professor Nilus in the epilogue to the1905 edition of his book, *The Great Within the Small*, explains the symbolism of the snake with the head of the snake being those initiated in the Zionist plan to conquer the world and the body of the snake being the mostly unwitting body of Jewry. The snake symbolized the slyness of the Jews to penetrate into the hearts of nations in order to undermine and subdue them under Jewish power. Nilus' describes the egregorious "Symbolic Snake"as follows:

> [T]he Symbolic Snake, whose head was to represent those who have been initiated into the plans of the Jewish administration, and the body of the Snake to represent the Jewish people—the administration was always kept secret, even from the Jewish nation itself. As this Snake penetrated into the hearts of the nations which it encountered, it undermined and devoured all the non-Jewish power of these States.[959]

After the Bolshevik Revolution in Russia, Professor Nilus

was arrested by the Cheka, imprisoned, and tortured. Bolshevism is Communism. Communism is simply government by the principles of the Jewish Talmud. The Jewish president of the court told Professor Nilus that the brutal treatment he received was as retribution for having done the communists incalculable harm in publishing the *Protocols of the Learned Elders of Zion*. The Protocols exposed the details of the plot by elite Jews to subjugate Gentiles and rule the world. Professor Nilus was released for a few months, but was soon rearrested by the Cheka. He was confined in prison until February 1926. He died in exile in the district of Vladimir on January 13, 1929. The treatment of Professor Nilus by the Jewish Bolsheviks speaks clearly to the authenticity of the *Protocols of the Learned Elders of Zion*.

John Torell has revealed that in the Jewish Cabala, the "holy serpent" is mystically trying to set itself free from the bottomless pit in order to enter the earth as "the messiah."[960] The "holy serpent" of the Jews is an egregorious god, which means the Jews believe they are their own messiah. In 1879 Baruch Levy wrote to Karl Marx and stated that the Jews will, in the end, have all of the world's property in their hands. Levy stated that the promise of the Talmud would then be fulfilled, which is that "the Jewish people as a whole will be its own messiah."[961]

The Messianic promise of the Talmud to which Levy referred can be found in the Babylonian Talmud at Folio 111a of Tractate Kethuboth, where it states in pertinent part: "We have a tradition that Babel will not witness the sufferings that will precede the coming of the Messiah." A footnote to that passage explains: "These are the throes of mother Zion which is in labor to bring forth the Messiah - without metaphor, the Jewish people." The Jewish hierarchy plan to establish a Messianic reign over a new world order tyranny from their earthly Zion in Israel.

What does the Bible say about Leviathan? In Job 41:34 we read that "He beholdeth all high things: he is a king over all the children of pride." God has stated that "[p]ride goeth before destruction, and an haughty spirit before a fall." Proverbs 16:18. Whereas, Jesus Christ calls on his followers not to be prideful, but rather to be humble. "Whosoever therefore shall humble himself as this little child, the same is greatest in the kingdom of heaven." Matthew 18:4.

So, we know that the Leviathan, who is king over the children of pride, must be the very opposite of God. Leviathan must be "the great dragon was cast out, that old serpent, called the Devil, and Satan" mentioned in Revelation 12:9.

Please note that it was the serpent in the garden of Eden who beguiled Eve. The serpent was the most subtle of the beasts. He told Eve that she and Adam could become as gods if they ate the fruit from the tree. Genesis 3:4-5. That is exactly what the Jews are claiming for themselves. They have believed the lie of the serpent that they are gods. Each Jew believes that he is a god who is part of the collective Jewish egregorious god. The Jewish view of themselves as gods is found in the Jewish religious doctrine. For example, the Talmud at Tractate Sanhedrin, Folio 58b refers to Jews as a "Divine Presence."[962] Rabbi Shimon ben Yohai declared his godhood by stating: "I am beyond the jurisdiction of any angel or judge in heaven."[963]

Satan is the adversary of God. He tries to be like God and his earthly kingdom is a devilish imitation of God's heavenly kingdom. God's kingdom is a spiritual kingdom that is not of this earth. John 18:36. God has a heavenly Jerusalem (containing a heavenly Zion). Revelation 21:10. Satan's kingdom is a kingdom of the flesh that is headquartered in the earthly Jerusalem (containing an earthly Zion). The gospel of Jesus Christ explains that Jesus is in believers and believers are in Jesus. John 14:20; 17:20-23. There is one spiritual body of Christ, with Jesus the head. Colossians 1:18. As with the earthly imitation Zion, Satan also has a fleshly imitation body of satanic believers. That body is not a spiritual body. It is the fleshly body of the Jews. It is the egregore of the "holy serpent" of the Jewish people. As the church is the body of Christ, so also in a perverse way is the egregore of the Jews the body of Lucifer.

The Jewish egregore, or group god, is a communist beast that is kept together through antisemitism. Antisemitism is the means by which the head of the snake keeps the body of Jews following in lockstep its leadership. Without antisemitism, the body of Jews would disperse and be assimilated into the Gentile nations. That simply cannot be allowed by the Jewish hierarchy. The Jews must be kept as an insular body separated from the Gentiles. The power of the Jewish hierarchy is centered in its control of the egregorious body of Jews. Henry Makow explains:

> [Nicholas] Lysson discusses how the Jewish leadership actively provoke anti-Semitism because it is indispensable for Jewish cohesion and survival. Jews acknowledge this. Are they saying that, without its "egregore" i.e. its predatory agenda, Jews would have no corporate raison d'etre? Of course this secret is kept from the rank-and-file, as in Freemasonry which Rakovsky said is designed to bring about "the triumph of Communism."[964]

The language of witchcraft always carries two meanings. There is the exoteric meaning for the uninitiated public, and there is the esoteric meaning that is only understood by the inner circle of initiates. For example, Jewish communists claim that they are atheists. That is the exoteric meaning of communist doctrine. However, the occult meaning of communism is hidden. The esoteric meaning of communism is that the collective body of Jews make up the egregore, who is the collective god of communism. So when communists talk about atheism, they only mean that there is no God outside of themselves. The egregore, however, only includes the Jewish communists; the Gentile communists are simply the skin of the serpent that will ultimately be shed after they have served their purpose.

As of the year 2000 there were 268 kibbutzim in Israel.[965] A Kibbutz is a collective Jewish society. According to the Jewish Virtual Library, Kibbutz is "the fulfillment of the idea 'from each according to his ability, to each according to his needs.'"[966] That is a direct quote from Karl Marx.[967] The Kibbutz collective is the purest form of communism. The Kibbutz is one manifestation of the Jewish egregore. When the Jewish egregore is allowed unrestrained aggression against Gentiles, the exoteric term used to describe it is "communism." It is not a surprise that as one gets closer to the head of the egregorious serpent in Israel, we find the purest form of communism.

What does this Jewish egregore leave in its path? History has shown that it is nothing but merciless death and destruction. The egregorious religious doctrine of the godhood of the Jews is how the Jewish communists could be so bloodthirsty. They do not believe that there are any moral constraints on their conduct. They are imbued with an ethnic superiority complex, where they are gods over the Gentiles. They view Gentiles, particularly Christians, as their enemies to be enslaved and exterminated. Makow explains:

> The greatest mass slaughter in history was not the Jewish holocaust but the Ukrainian holodomor, i.e., the "hunger." By Stalin's own estimate, ten million Ukrainians died, mostly at the hands of Bolshevik Jews.
>
> The holodomor took place because the Bolsheviks confiscated all the grain. Lysson writes:

"A quarter of the rural population, men, women and children, lay dead or dying in a great stretch of territory with some forty million inhabitants, like one vast Belsen. The rest, in various stages of debilitation, had no strength to bury their families or neighbours. [As at Belsen] well-fed squads of police or party officials supervised the victims."

The extermination of the Ukrainian Kulaks was directed from the Kremlin where the Bolshevik leadership lived in family apartments and maintained a fraternal atmosphere suffused by collectivist idealism, i.e., their egregore. This is how historian Simon Sebag Montefiore, the scion of an elite British Jewish family, describes the scene during the holomodor. Of course he doesn't mention that most of the main players were Jews.

"The Party was almost a family business. Whole clans were members of the leadership. . . . This pitiless fraternity lived in a sleepless frenzy of excitement and activity, driven by adrenalin and conviction. Regarding themselves like God on the first day, they were creating a new world in a red-hot frenzy." ("Stalin: The Court of the Red Tsar," p. 40, 45)

The Soviets didn't raze synagogues because Communism expressed the Jewish "egregore." Christians and Jews apparently do not worship the same God. The Christian God represented by Jesus is universal love and brotherhood. The Jewish God has been supplanted by a Cabalistic egregore that "serves" elite Jews only. It represents their ruthless sociopathic urge for world dominion. The same egregore- the desire to supplant God- animates the New World Order.[968]

The idealism of brotherhood and comradery that is commonly propagandized in communist circles is just so much window dressing that conceals a satanic deception. The real agenda of communism is the destruction of religion, family, and nation; the accumulation of all wealth and the enslavement of mankind.[969]

The idealistic Gentile followers of communism are dupes, who are viewed by the Jewish leaders as "useful idiots," who will be summarily executed when the Jews come to power. That has been the pattern in all communist revolutions in the past, and it will continue to be the practice in the future. The Zionist Leviathan is a cunning and diabolical egregorious viper bent on nothing short of world rule based upon Jewish supremacy.

The core doctrine of Judaism is the rejection of Jesus Christ. Harmony Daws explains that the Jewish religion is diametrically opposed to Christianity.

> As we have said many times, the modern Jewish religion is extremely aware of the inherent hostility between itself and Christianity. The two faiths could hardly be more opposed in their basic tenets. Judaism absolutely rejects the deity of Christ, which is (or should be) the central foundation on which all other Christian beliefs are built. Modern Judaism exalts the Jewish people as the correct representations of God in this world while Christians call for a spiritual revolution that extends to all people of all races. Jewish leaders and sacred texts know that these two faiths are diametrically opposed; unfortunately, Christians seem completely ignorant.[970]

Jesus criticized the Pharisees for their religious traditions. Those traditions were oral traditions at that time. Later they were memorialized in the Talmud and the Kabbalah (a/k/a Cabala). The Kabbalah and the Talmud today span numerous volumes. Jesus called the Pharisees hypocrites, who masqueraded as religious men, but who were in reality irreligious frauds.

> Then came to Jesus scribes and Pharisees, which were of Jerusalem, saying, Why do thy disciples transgress the tradition of the elders? for they wash not their hands when they eat bread. But he answered and said unto them, **Why do ye also transgress the commandment of God by your tradition?** For God commanded, saying, Honour thy father and mother: and, He that curseth father or mother, let him die the death. But ye say, Whosoever shall say to his father or his mother, It is a gift, by whatsoever thou mightest be profited by me; And honour not his father or his mother, he shall be free. **Thus have ye made the commandment of God of none effect by your tradition. Ye hypocrites**, well did Esaias prophesy of you, saying, This people draweth nigh unto me with their mouth, and honoureth me with their lips; but their heart is far from me. But **in vain they do worship me, teaching for doctrines the commandments of men.** (Matthew 15:1-9)

The Pharisees had an outward appearance of piety, in order to gain political and religious control of the Jews. In secret, however, they practiced an occult doctrine that was only known to its initiates. Lady Queenborough (Edith Miller) explains in her book *Occult Theosophy*:

> The Chaldean science acquired by many of the

> Jewish priests, during the captivity of Babylon, gave birth to the sect of the Pharisees whose name only appears in the Holy Scriptures and in the writings of the Jewish historians after the captivity (606 B. C). The works of the celebrated scientist Munk leave no doubt on the point that the sect appeared during the period of the captivity. "From then dates the Cabala or Tradition of the Pharisees. For a long time their precepts were only transmitted orally but later they formed the Talmud and received their final form in the book called the *Sepher ha Zohar*."[971]

The scrupulous observance by the Pharisees of Jewish religious tradition was only a cover for their secret doctrine. They had rejected Jehovah and had adopted the pantheism of Babylon. They pretended that their many rituals were necessary for the worship of Jehovah, but those were only man-made rules to conceal their secret Babylonian religion. Jesus rebuked them for it, calling them hypocrites for their vain worship of God through man-inspired rituals and honoring God with fine words, when their hearts in fact were focused on the heathen gods of Babylon. See Mark 7:5-7. The Pharisees had accepted the satanic lie that they had become "as gods." See Genesis 3:5. Their new Babylonian (a/k/a Chaldean) religion was to be exclusive to the Jews, who were to rule the world. Edith Miller explains:

> The Pharisees, then, judging it wiser to capture the confidence of their compatriots by taking the lead in the religious movement, affected a scrupulous observance of the slightest prescriptions of the law and instituted the practice of complicated rituals, simultaneously however cultivating the new doctrine [i.e., secret doctrine] in their secret sanctuaries. These were regular secret societies, composed during the captivity of

> a few hundred adepts. At the time of Flavius Josephus which was that of their greatest prosperity, they numbered only some 6,000 members. This group of intellectual pantheists was soon to acquire a directing influence over the Jewish nation. Nothing, moreover, likely to offend national sentiment ever appeared in their doctrines. However saturated with pantheistic Chaldeism they might have been, the Pharisees preserved their ethnic pride intact. This religion of Man divinised, which they had absorbed at Babylon, they conceived solely as applying to the profit of the Jew, the superior and predestined being. The promises of universal dominion which the orthodox Jew found in the Law, the Pharisees did not interpret in the sense of the reign of the God of Moses over the nations, but in that of a material domination to be imposed on the universe by the Jews. The awaited Messiah was no longer the Redeemer of original Sin, a spiritual victor who would lead the world, it was a temporal king, bloody with battle, who would make Israel master of the world and 'drag all peoples under the wheels of his chariot.' The Pharisees did not ask this enslavement of the nations of a mystical Jehovah, which they continued worshipping in public, only as a concession to popular opinion, for they expected its eventual consummation to be achieved by the secular patience of Israel and the use of human means.[972]

Jesus cursed the Pharisees to their face: "Woe unto you, scribes and Pharisees, hypocrites! for ye are as graves which appear not, and the men that walk over them are not aware of them." (Luke 11:44) The Babylonian traditions of the Pharisees, which were traditions passed down orally from generation to generation, were eventually (in part) memorialized in the Kabbalah and Talmud. The double aim of the Pharisees was to wrestle political control over the Jews from the Sadducees and "to modify gradually the conceptions of the people in the direction of their secret doctrine."[973] They accomplished both goals. Today Orthodox Jewry is an insular authoritarian society that is completely given over to the practice of the Babylonian religion of the ancient Pharisees.

The twisted Babylonian god of modern Jewry as expressed in the Talmud and Kabbalah is a god of vengeance and hatred of Gentiles and particularly Christians. Edith Miller summarizes the nature of the Jewish god as being "just and merciful only to his own people, but foe to all other nations, denying them human rights and commanding their enslavement that Israel might appropriate their riches and rule over them."[974]

The Talmud sets forth clear distinctions between Gentiles and Jews. Jews are viewed as gods, whereas Gentiles are viewed as animals. For example, in Talmud Tractate Sanhedrin, Folio 58b it states that smiting a Jew on the jaw is like assaulting a "Divine Presence."[975] If the assailant happens to be a gentile, he is "worthy of death."[976] A Gentile child is considered to be subhuman.[977]

Judaism is a religion of hate. The hatred of Christians and all Gentiles of all races runs through the warp and woof of Judaism. The Jews recite *Amidah*, which is a set of eighteen (by some accounts nineteen) weekly Jewish prayers. The twelfth prayer is called *Birkat ha-minim*. The *Birkat ha-minim* is actually a hateful curse against heretics and enemies of the Jews, particularly Christians. The curse was first introduced in the *Amidah* in the first century at Jabneh by Samuel ha Katan, at the request of Rabban Gamaliel II in order to drive followers of Jesus Christ from the synagogues.

The book of Acts records the successful efforts of the early Christians in converting Jews to Christianity. The early Christian disciples would meet with the Jews in the synagogues and explain the gospel. Many Jews became followers of Christ. Acts 17:1-13. The Bible records that the chief ruler of the synagogue at Corinth believed in Jesus and was saved. Acts 18:8. The Jewish leadership saw Christianity as a threat to Judaism and could not allow the spread of Christianity to continue. They instituted the *Birkat ha-minim* as a means of making clear to the Jews that Judaism was incompatible with Christianity. They wanted to keep the newly converted Christians out of the synagogue and stop Christianity from spreading.

Min is a pejorative that is commonly used to refer to a Christian, with *minim* being the plural form of *min*. According to the Encyclopedia Judaica, the *Birkat ha-minim* has to be recited during public worship by the *hazzan* so as to avoid any suspicion of heresy.[978] The *hazzan* is a Jewish cantor who leads the congregation in prayer. If the *hazzan* makes an error and omits the *Birkat ha-minim*, he is required to return to the *Birkat ha-minim* and recite the curse. That regulation requiring return to the omitted curse does not apply to any of the other benedictions recited as part of the *Amidah*.[979]

Not only do Jews include curses against Christians in their weekly prayers, they have corrupted the yearly Passover celebration into a hate-fest against Christians. Ariel Toaff explains in his book *Pasque di Sangue* (*Blood Passover*), that in orthodox Judaism Passover has become a ritual of detestation against Jesus.[980]

Ariel Taoff is a professor of Jewish renaissance and medieval history at Bar Ilan University[981] in Israel, just outside of Tel Aviv. Professor Toaff is a Jewish insider, who comes from a rabbinic bloodline that is highly respected among Jews. He is the son of one of the most revered rabbis in the world, Elio Toaff. He is the former chief rabbi of Rome and was considered the dean of Italian Jewry. Some have called Elio Toaff the "Pope of the Jews." So influential was Rabbi Taoff in Rome, he was one of only two people that Pope John Paul II mentioned by name in his will.[982]

Toaff explains how over time Judaism and the Passover memorial in particular had been corrupted into a hate-fest against Christ and Christians.

> In this Jewish-Germanic world, in continual movement, profound currents of popular magic had, over time, distorted the basic framework of Jewish religious law, changing its forms and meanings. It is in these "mutations" in the Jewish tradition – which are, so to speak, authoritative – that the theological justifications of the commemoration in mockery of the Passion of Christ is to be sought, which, in addition to its celebration in the liturgical rite, was also intended to revive, in action, vengeance against a hated enemy continually reincarnated throughout the long history of Israel (the Pharaoh, Amalek, Edom, Haman, Jesus).[983]

Toaff explains that the Passover meal was transformed into a ceremonial curse upon Christians. He explains how during the middle ages, the orthodox Jews sprinkled the Passover table with wine to signify the blood hatred of Christians by the vengeful Jews. The Jews made wine an integral part of the ceremony of Passover, which is intended as a disdainful twist on the last supper of Christ. Wine was not part of the Passover celebration as laid down by God. The Passover was intended to be a memorial that was to celebrate God having freed the Jews from Egyptian slavery. Passover also looked forward to the coming of Christ, who is the sacrificial Lamb of God, who takes away the sins of the world. *See* John 1:29.

The fourteenth day of the first month is the Passover (Leviticus 23:4-5, Exodus 12:17-18). Passover is immediately followed by the seven days of unleavened bread (Leviticus 23:6-7, Exodus 12:15-16). There is no mention of using wine as part of the Passover ceremony. Jesus instituted wine at the last supper. Matthew 26:27-29. The Jews have now instituted wine as an integral part of their Passover ceremony; it is intended as a contemptuous mockery of the last supper. The Passover ceremony included cursing Christians with the ten curses that were brought upon Egypt during the Jewish captivity there. Toaff explains:

> At this point, in the traditional reading of the *Haggadah*, according to the custom of the Ashkenazi Jews, the curses against the Egyptians were transformed into an invective against all the nations and enemies hated by Israel, with explicit reference to the Christians.[984]

Toaff was quick to point out that this cursed ceremony was not practiced by the Sephardic or Italian Jews, but as far as he knew, it was only practiced among the German Orthodox Ashkenazi Jews. Toaff explained that the Ashkenazi tradition was that the head of the family would state at the table: "Thus we implore God that these ten curses may fall on the gentiles, enemies of the faith of the Jews."[985] Toaff stated that was a clear reference to the Christians. The orthodox Jews call on God to cast his anger upon the Gentiles and let the fury of his anger persecute and destroy them.[986]

Toaff came to the ineluctable conclusion, based upon careful examination of authoritative evidence, that in the middle ages blood from Christian boys, who were ritually murdered, was used in some orthodox Jewish communities as part of the yearly Passover ceremony.

The Passover of the Bible had been twisted by the

orthodox Jews and turned into an anti-ritual re-enacting the crucifixion of the hated Jesus. Witness testimony establishes that the children were not simply kidnapped and quickly killed. They were brutally and sadistically tortured in a satanic ritual that was intended to be an insult to Jesus Christ.[987] Toaff states that Jews are obsessed with hatred for Jesus Christ and Christians. Toaff explains the anti-Christian symbolism of the ceremony:

> The memorial of the Passion of Christ, relived and celebrated in the form of an anti-ritual miraculously exemplified the fate destined for Israel's enemies. The blood of the Christian child, a new *Agnus Dei*, and the eating of his blood, were premonitory signs of the proximate ruin of Israel's indomitable and implacable persecutors, the followers of a false and mendacious faith.[988]

Toaff explained the horrifying facts that the young victim was kidnapped by the Jews and subjected to a ritualized torture, symbolizing the torments of Christ. The ritual was passed down through oral tradition from generation to generation, from rabbi to rabbi.

> Samuele vaguely attributed these traditions to the rabbis of the Talmud (*Iudei sapientiores in partibus Babiloniae*), who were said to have introduced the ritual in a very remote epoch, "before Christianity attained its present power." Those scholars, united at a learned congress, were said to have concluded that the blood of a Christian child was highly beneficial to the salvation of souls, if it was extracted during the course of a memorial ritual of the passion of Jesus, as a sign of contempt and scorn for the Christian religion. Over the course of this counter-ritual, the innocent boy, who had to be

> less than seven years old and had to be a boy, like Jesus, was crucified among torments and expressions of execration, as had happened to Christ.[989]

It should be noted that many Jews disagreed with such a macabre ritual and did not follow the practices testified to during the many trials of the Jews. In fact, evidence was introduced on one case that two Jews who were approached to help in kidnapping a child refused to assist in the effort and argued against doing so. Toaff points out that the practice of the ritual use of the blood of a Christian child in the Jewish Passover meal was only practiced by fundamentalist orthodox Ashkenazi sects.[990] Toaff states that the orthodox Jews were the most assertive and controlling. "Medieval Ashkenazi Judaism made up a hermetically sealed orthodoxy, which fed upon itself, confined by a myriad of minute ritualistic regulations, which they considered binding on all, the mere memorization of which constituted an arduous and almost impossible task."[991]

Cardinal Ganganelli, who later became Pope Clement XIV, conducted an investigation of ritual murder by the Jews. His report on the matter was supposed to be a political accommodation to the Jews, to provide an official exoneration of them in response to the public clamor over their ritual murders. Cardinal Ganganelli did his duty to the Jews as far as his conscience would allow. He opined that it was not a certain and common axiom among Jews to engage in ritual murder of children. However, the evidence of the ritual murder by Jews of Simon of Trent and Andreas of Rinn was so compelling that Cardinal Ganganelli had to admit that those cases were in fact ritual sacrifices of children by the Jews. Cardinal Ganganelli, in the face of overwhelming evidence, made the following concession to the truth by stating in his 19 January 1760 report: "I admit, then, as true, the fact of the Blessed Simon, a boy three years old, killed by the Jews in Trent in the year 1475 in hatred of the faith of Jesus Christ ... I also admit the truth of another fact, which happened in the year 1462 in the village of Rinn, in the Diocese of Brixen, in the person of the Blessed Andreas, a boy barbarously murdered by the Jews in hatred of the faith of Jesus Christ."[992]

Previously Pope Gregory XIII (1572-85) had recognized Simon as a martyr. Pope Gregory, himself, visited the shrine to Simon. His successor, Pope Sixtus V (1585-90), ratified the cult of St. Simon in 1588, allowing the celebration of mass in his name. Pope Benedict XIV (1740-58) confirmed the cult of Saint Simon through his Bull *Beatus Andreas*, which beatified both Simon of Trent and Andreas of Rinn.[993] The power of the Jews over the Catholic Church was finally brought to bear to suppress the historical evidence of ritual murder. In 1965, the Catholic Church outlawed the cult of Saint Simon. The church officials had the remains of the child removed and dumped in a secret place to avoid any resumption of pilgrimages to his grave.

The ritual sacrifice of Gentile children by the Jews has led to repeated banishment of Jews from one nation after another throughout history. For example, on March 31, 1492, King Ferdinand and Queen Isabella banished the Jews from Spain for their wickedness, which included ritual murder. In reference to ritual murder by Jews, the banishment edict states that "whenever a grave and detestable crime is committed by some members of a given group, it is reasonable that the group be dissolved or annihilated."[994] Respected historian William Thomas Walsh, in his book, *Isabella of Spain*, states that the detestable crime mentioned in the edict was the 1492 ritual murder by Jews of Christopher of Laguardia. Christopher of Laguardia was later made a Catholic Saint. It was not until 1967 that Jews were officially allowed back into Spain.

In 1911, a Jew, Menachem Mendel Beiliss, was accused of the brutal ritual murder of a young boy named Andrei Yustshinsky. As was the practice in such cases, powerful Jews were mobilized worldwide in an effort to see that the Jewish defendant was exonerated. The best Jewish lawyers were hired for the defense. Witnesses and officials were bribed.[995] A high ranking police investigator was caught planting evidence that would benefit the defendant.[996] Assassins were employed to poison key witnesses who could not be bribed. For example, three children who witnessed Beiliss abduct Andrei Yustshinsky were given cake by an investigator; the next day, all three fell ill.[997] Two of the children subsequently died, and the third was sick for months. It is estimated that the equivalent of $115 million in today's dollars was spent on the effort to see that the Jewish defendant was exonerated.[998]

Ultimately, the jury concluded that the boy was the victim of Jewish ritual murder, but they could not rule for sure that the defendant was among the Jews who actually wielded the instrument that murdered the child. The jury vote was 6-6; a simple majority was needed to convict in Russia. Matt Hale states that "it was proven to have taken place inside the synagogue at the Jewish-owned Zaitsev factory; the jury could not determine for certain that Beiliss was the one who wielded the awl, causing Andrei to be punctured over 45 times around his body, in an effort to procure as much blood as possible."[999]

Czar Nicholas II commissioned the prosecuting attorney, G. G. Zamyslovsky, to write a book about the proceedings. Zamyslovsky produced a 500 page book about the trial. It was published on the eve of the Bolshevik Revolution in Russia.[1000] Bolshevism is Communism. Communism is simply Judaism enshrined as government, and it is of, by, and for the Jews. The very first kangaroo court held after the communist revolution in Russia was the trial of G. G. Zamyslovsky. Zamyslovsky was arrested, and after a show trial, he was executed for having the temerity of revealing the facts of the Jewish ritual murder of Andrei Yustshinsky.[1001] Zamyslovsky's book was suppressed by the communists. The book has recently surfaced in the original

Russian language text.[1002] As of 2015, a complete English translation of the book has been completed, but as of this writing, it has not yet been published.[1003]

The infamous Jewish ritual murder of a Catholic Priest known as Father Thomas and his servant was reported in the April 6, 1850, edition of The New York Herald:

> [W]ho would have dreamt of beholding the bloody mysteries of the Talmud exposed in their turn, and of having the trial of one of the most savage and ferocious murders ever yet recorded in the annals of criminality, once more brought before the public? Who would have imagined that certain fanatics use human blood to moisten their holy unleavened bread?
>
> Our readers will undoubtedly recollect the universal sensation created throughout the world, by the discovery of Father Tommaso, a Christian missionary from Sardinia, and his servant, Abrahim Amara, who had been bled to death, their limbs chopped off joint by joint, their bones pounded in the mortar, and their mutilated remains thrown into a drain in the city of Damascus.
>
> The manuscripts of the original trial, as sealed by the French and Austrian consuls of that city, are now in this city, and we can thereby defy the 'Great Nation' to contradict the truth of our statements.[1004]

Toaff explained that only a small nut-sized amount of dried blood was shaken from a glass vial into the wine. The head of the family would announce: "This is the blood of a Christian child." He then would recite the ten plagues of Egypt as curses against Christians and gentiles. Toaff states:

> The head of the family takes a bit of the blood of the Christian child and drops it in his glass full of wine [...] then, putting his finger in the wine, with that wine where the blood of the Christian child has been shaken, he sprinkles the table and food on the table with it, pronouncing the Hebraic formula in commemoration of the ten curses, which God sent to the refractory Egyptian people who refused to liberate the Jewish people. At the end of the reading, the same head of the family, referring to the Christians, utters the following words (in Hebrew): 'thus we beseech God that he may similarly direct these ten curses against the gentiles, who are enemies of the Jewish faith.'

Toaff obtained evidence of the Jewish practice of a blood Passover from testimony of Jewish witnesses at the trial of the murder of a Christian child in the 1400's. Expert testimony of Mosè of Würzburg at trial explained that "the Jews naturally require the blood of a Christian child, but if they were poor and could not afford any blood, they were relieved of the expense."

The witnesses at the trial testified that a small amount of blood of the Christian child was also mixed with the dough of the unleavened bread that was eaten during the Passover meal. There was testimony from a witness named Israele, who was identified as Samuele of Nuremberg's son, who stated that the mixing of the blood into the unleavened bread was a memorial of the blood with which the Lord commanded Moses to paint the door-posts of the doors. However, Toaff found another deeper meaning.

> Vitale of Weissenburg, Samuele's agent, preferred to confer a second meaning upon the

> rite, that is, that of an upsidedown memorial to the Passion of Christ, considered as an emblem and paradigm of the fall of Israel's enemies and of divine vengeance, forewarning of final redemption. "We use the blood", he declared, "as a sad memorial of Jesus ... in outrage and contempt of Jesus, God of the Christians, and every year we do the memorial of that passion ... in fact, the Jews perform the memorial of the Passion of Christ every year, by mixing the blood of the Christian boy into their unleavened bread.

The evidence is that the Jewish ritual murders continued long after the middle ages. In the 1941 book, *Jewish Ritual-Murder, A Historical Investigation,* Hellmut Schramm, Ph.D., documented over 130 cases of Jewish ritual murder of gentile children between the years 1144 A.D. and 1913 A.D.[1005] Those cases documented that Jewish doctrine requires the victim to first undergo the most unimaginably sadistic tortures before his blood is drained, in order for the blood to be efficacious.

The Jewish ritual concludes with the victim's blood being drained while the victim's heart is still beating. Dr. Schramm quoted the sworn testimony directly from trial transcripts of the gruesome details of the torture and bleeding of the victims. Dr. Schramm also documented the bribery of the high government officials by rich, powerful Jews in order to undermine investigations and prosecutions of Jewish ritual murderers. The blood, once drained from the victim, is carried away and then later distributed within the Jewish community to be used by Jewish families in the Jewish Passover ceremony.

Is Jewish ritual murder still going on today? Adel Hamood, a reporter for the major Egyptian newspaper, *Al-Ahram*, wrote an article titled, *Jews Make Matzos from the Arab's Blood* in the October 28, 2000, edition of that newspaper. In that article

Hamood revealed that dead Palestinian children have been repeatedly found around Israel, with their blood drained. Hammood stated that "Romania, Germany, Italy, Poland, United States, Egypt, Syria, Portugal, Jordan, France, Russia, Austria, Spain - nearly every civilization in the entire world has reported the crime of Jewish ritual murders. Has the entire world been involved in a massive conspiracy against Jews? Or have some fanatical Jews been engaged in a conspiracy against all non-Jews? Quite frankly, it seems to be the latter."[1006]

The Zionist Organization of America (ZOA) pressured the Egyptians to retract the aforementioned article or they would see that the United States withheld financial aid to Egypt. Matt Hale states that ZOA President Morton A. Klein wrote to the U.S. Congress: "We urge Congress to refrain from considering the Clinton administration's proposal for $225 million in extra aid to Egypt until the Mubarak government publicly apologizes and repudiates the blood libel article, and replaces the editor responsible for its publication."[1007] That is the typical response of the Jews to any hint in the press of Jewish ritual murder. That is why the world press, for the most part, is silent on this issue. Any editor who allows such an article to be printed will find himself without a job in short order.

On May 1, 1989, Oprah Winfrey interviewed a Jewish woman who revealed her family's involvement in ritual sacrifice of babies. The woman appeared under disguise and used an alias, "Rachel." The women later publicly came out and revealed that she was Vicki Polin. Polin later revealed that the original agreement with Oprah before her appearance on the show was that Oprah would not mention that she was Jewish. The show was only going to focus on the inter-generational sacrifice of babies, without any mention that she and her family were Jewish. One must ask, why was that such a concern? It seems that it was okay to discuss the sacrifice of babies under the general rubric of Satanism, but there could be no mention of its link to Judaism.

Polin stated that Oprah let it slip out during the interview that she and her family were Jews, which was contrary to the agreement. What was most telling was that it became clear during the interview that the sacrifice of the babies was part and parcel of her Jewish heritage. That was revealed early in the interview, when Rachel (Vicki Polin) stated that her family had an extensive family tree that goes back to the 1700's, and they keep track of who in the family has been involved in child sacrifice. Her initial point in the interview was that she "was born into a family that believes in this."[1008] The family heritage seemed to be very important to the rituals. That was something, however, that Rachel (Vicki Polin) wanted concealed. The chief executives in Oprah Winfrey's production company are Jews. It is likely that if such an agreement to conceal Rachel's (Vicki Polin's) Jewish heritage was not in place, Rachel (Vicki Polin) would likely have never have been allowed to appear on the show. The pertinent part of the interview on Oprah's show went as follows:

> Oprah: "And this is a — does everyone else think it's a nice Jewish family? From the outside you appear to be a nice Jewish girl.... And you are all worshiping the devil inside the home?"
>
> Rachel: "Right, There's other Jewish families across the country. It's not just my own family."
>
> Oprah: "What Kinds of things went on in the family?"
>
> Rachel: "There would be rituals in which babies would be sacrificed."
>
> ***
>
> Oprah: "This is the first time I heard of any Jewish people sacrificing babies, but anyway - so

> you witnessed the sacrifice?"
>
> Rachel: "Right, when I was young, I was forced to participate in that, in which I had to sacrifice an infant."[1009]

After the airing of that show, Oprah Winfrey was put under tremendous pressure from powerful Jews and Jewish organizations. Kitty Kelley, in her 2010 biography of Oprah Winfrey, revealed that immediately after the show aired, the phones at her production company, Harpo, started jangling with hundreds of irate Jewish callers from all over the country. "Within hours, Jewish groups rose up in condemnation, and Oprah's show became a national news story."[1010] Oprah Winfrey was faced with a financial crisis. No show can appear on television without the approval of the Jewish elite. The chief operating officer at Harpo productions was Jeff Jacobs, who is Jewish. "Recognizing the danger of a national boycott of The Oprah Winfrey Show and the potential loss of sponsors, which could spell financial ruin for everyone, Jacobs quickly offered to meet with Jewish leaders in Chicago to try to salvage the situation."[1011] Kitty Kelly explains that Oprah met with Jewish leaders at a meeting arranged by a wealthy Jewish philanthropist, Harriet Brady, to make amends:

> Oprah arrived at the meeting on May 9, 1989, with Debar DiMaio and two Jewish members of her senior staff, Jeffrey Jacobs and Ellen Rakieten. They sat down with Michael Kotzin, director of the Jewish Community Relations Council of Metropolitan Chicago; Jonathan Levine, midwest director of the American Jewish Committee; Barry Morrison, director of the Greater Chicago/Wisconsin Regional Office of the Anti-Defamation League of B'nai B'rith; Rabbi Herman Schaalman, president of the Chicago Board of Rabbis; Maynard Wishner,

> resident of the Jewish Federation of Metropolitan Chicago; Judge Marovitz; and Mrs. Brady. Oprah was sufficiently contrite and vowed never again to broadcast a show on devil worship. She agreed to reach out to B'nai B'rith, which fights anti-Semitism and racism, whenever her show focused on those subjects, and she promised to exercise better judgment in selecting her guests.[1012]

It is notable that part of Oprah's deal for her to head off a sponsor boycott of her show led by the Jewish elite was to "never again broadcast a show on devil worship."[1013] If the concern was that the show unfairly portrayed Jews, why was the agreement struck with Oprah that she not broadcast any shows whatsoever on devil worship? It seems that the Jewish elite were concerned that any show on devil worship raised the risk that there would be a connection to the Jews. The only way that such a connection could be made is if there is evidence that such a connection exists. It was not just a coincidence that Anton LeVey, who was the founder of the Church of Satan, was a Jew. The reality is that the theology at the core of Judaism is the worship of the devil. The Jewish elite simply could not risk the Jewish correlation being inadvertently discovered during a show on devil worship. They realized that there could be a repeat of the connection between Jews and Satanic ritual murder as already happened on Oprah's show with Rachel. The agreement put in place by Oprah's production team for Rachel's appearance that there be no mention that the practitioners in ritual murder were Jews, was found, in practice, to be an insufficient safeguard. Consequently, Oprah was prohibited thereafter from doing any shows whatsoever on devil worship.

Ritual murder is the dark Satanic secret of Judaism. Jesus alluded to that fact when he told the Jews: "**Ye are of your father the devil, and the lusts of your father ye will do. He was a murderer from the beginning**, and abode not in the truth, because there is no truth in him. When he speaketh a lie, he speaketh of his own: for he is a liar, and the father of it." (John 8:44)

The curses pronounced against Christians by the Jews during their Passover ceremony are symptomatic of the orthodox Jewish antipathy of Gentiles in general and hatred of Christians in particular. The most revered rabbis (*gedolim*) view Gentiles as garbage. According to orthodox Judaism, Gentiles are not only an inferior species, but a species that is "completely evil." For example, Rabbi Saadya Grama of Beth Medrash Govoha in his book *Romemut Yisrael Ufarashat Hagalut* (Jewish Superiority and the Question of Exile) states: "The Jew by his source and in his very essence is entirely good. The *goy* [Gentile], by his source and in his very essence, is completely evil. This is not simply a matter of religious distinction, but rather of two completely different species."[1014]

Rabbi Grama is simply stating the philosophy contained in the Talmud. He explains that according to his understanding of Jewish religious doctrine "The difference between Jews and gentiles is not historical or cultural, but rather genetic and unalterable."[1015]

Rabbi Grama's opinion cannot be dismissed as a fringe view within orthodox Jewry. His assertion is backed up by other esteemed Rabbis. For example, Rabbi Shneur Zalman, the esteemed founder of *Chabad-Lubavitch*, taught that the difference between Jew and Gentile is not merely religious or racial, but that the souls of Jews and Gentiles are completely different in kind. "Gentile souls are of completely different and inferior order. They are totally evil, with no redeeming qualities whatsoever . . . Indeed they themselves are refuse. . . . All Jews are innately good, all gentiles are innately evil."[1016]

According to the Talmud, Christians are allied with

hell,[1017] and Jesus is not only cursed,[1018] he is described as being tormented in boiling hot semen.[1019] The Talmud, however, gives immunity to rabbis from ever going to hell.[1020] Chagigah 27a (a/k/a Hagigah 27a) states: "As to disciples of the wise, the flame of Gehenna [hell] has no power over them."[1021] The Talmud explains in Tractate Baba Bathra that "disciples of the wise men" means scholars or distinguished students.[1022] That means that not only are rabbis immune from hell, but also all scholars or distinguished students of the Talmud. Rabbi Shimon ben Yohai declared: "I am beyond the jurisdiction of any angel or judge in heaven."[1023]

Rabbi ben Yohai, who believed he was beyond the jurisdiction of God, did not think Gentiles were even worthy to live. His views regarding Gentiles were that "even the best of gentiles should all be killed."[1024] Rabbi ben Yohai is not a rabbi on the fringes of Judaism; he is in fact one of the most revered of rabbis in Judaism; his grave is a shrine in Israel. He authored the Zohar, which is the principal work of the Kabbalah.

While the Jewish clergy consider Gentiles subhuman, they will turn on Jews who do not adhere to their dictates. Jewish rabbis have nothing but contempt for Jews who do not follow the Jewish traditions.[1025] This overbearing attitude by Jewish rabbis is nothing new. One can sense the contempt that the Pharisees had for the common Jews in John 7:49, where common Jews were impressed by what Jesus had to say, and the Pharisees responded by cursing the Jews for not knowing the "law." The law to which the Pharisees referred was their oral tradition.

William Wotton explains the undue burdens placed on the Jews through the laws of the Pharisees:

> They were absurdly minute in the literal observance of their vows, and as shamefully subtle in their artful evasion of them. The

> Pharisees could be easy enough to themselves when convenient, and always as hard and unrelenting as possible to all others. They quibbled and dissolved their oaths with experienced casuistry.[1026]

William Wotton (1666-1727) was a rare genius; he undertook foreign language translation at age five, attended Cambridge University at age 10, being graduated at age 13.[1027]

The common Jews are as much victims of the Jewish hierarchy as are the Gentiles and Christians. The common Jews are being spiritually brainwashed to do the bidding of their rabbis. Jesus explained the process: "Woe unto you, scribes and Pharisees, hypocrites! for ye compass sea and land to make one proselyte, and when he is made, ye make him twofold more the child of hell than yourselves." (Matthew 23:15) Jesus cursed the Pharisees and scribes; he could have, but did not, curse all Jews. It is the reprobate spiritual leaders who were the targets of Jesus' epithets. He came to set Jew and Gentile alike free from spiritual bondage. "Then said Jesus to those Jews which believed on him, If ye continue in my word, then are ye my disciples indeed; And **ye shall know the truth, and the truth shall make you free**." (John 8:31-32)

According to Orthodox Judaism Gentiles have no property or other rights that can be asserted against a Jew. Baba Bathra 54b: Property of Gentiles is like the desert; whoever gets there first gets it.

Sanhedrin 57a: If a Gentile robs a Jew, he must pay him back; but if a Jew robs a Gentile, the Jew may keep the loot.

> With respect to robbery - if one stole or robbed or [seized] a beautiful woman, or [committed] similar offences, if [these were perpetrated] by

> one Cuthean against another, [the theft, etc.] must not be kept, and likewise [the theft] of an Israelite by a Cuthean, but that of a Cuthean by an Israelite may be retained.[1028] (brackets in original)

Likewise, if a Gentile kills a Jew, the Gentile is to be killed; but if a Jew kills a Gentile, there is no death penalty. "For murder, whether of a Cuthean by a Cuthean, or of an Israelite by a Cuthean, punishment is incurred; but of a Cuthean by an Israelite, there is no death penalty."[1029] The clear meaning of tractate 57a is that if a Gentile (a/k/a Cuthean or heathen) robs a Jew, he must pay him back. But if a Jew robs a Gentile he may keep the loot. Likewise, if a Gentile kills a Jew, the Gentile is to be killed. But if a Jew kills a Gentile, there is no death penalty.

The clever Jews will deny that the passages in Sanhedrin 57a refer to Gentiles. They will point out that there is no reference to Gentiles in those passages; that the passages refer only to "Cutheans." If you read footnote 33 to Sanhedrin 57a you will find that the editors placed a guide to the Jews for their understanding of the terms used. It states that the word "Cuthean" is a reference to Gentiles. Footnote 33 states: "'Cuthean' (Samaritan) was here substituted by the censor for the original goy (heathen)." The text was revised by a censor. The job of the censor was to obfuscate certain passages, so that Gentiles reading the Talmud would have a difficult time discerning the attack upon them. Inserting the word "Cuthean" in place of "goy" is one example of that obfuscation. Often the word "heathen" was substituted for Gentile or goyim in the Talmud by the censor.

In the Talmud the words *goi*, *goy*, and *goyim* (pejorative for gentiles) were replaced with "Kushite" or "Kuthite" in the 1578 Basel edition of the Talmud.[1030] "Heathen" is often substituted for *Goy* or Gentile in the Talmud. *Min* and *minim* (pejorative for Christians) were replaced with words like "Sadducee" or "Epicurean."[1031] Other words used in place of Christians are "idol worshiper," "Akum," and "Cuthean."[1032] The curses and blasphemy against Jesus are concealed by replacing Jesus' name with Balaam and other pseudonyms. Jewish Professor Israel Shahak explains that the offensive passages in the Talmud with the new terms could be explained away to the gullible Gentiles, but the Jewish reader could easily recognize the new euphemism for the old expressions.[1033]

The redacted Talmud is trotted out by rabbis to prove that there are no imprecations against Jesus and no admonitions to steal from or kill gentiles. The rabbis keep secret the fact that the language in the publically available Talmud has been changed to conceal the offense of the original text.

The censored texts were secretly recorded and preserved by the Jews. They then clandestinely reproduced the uncensored text in a supplement known as the *Hesront Ha-shas*. "Anyone who possessed the Catholic-authorized expurgated Talmud and the *Hesront Ha-shas* possessed the complete Talmud - an ingenious joke on Christians and quite a coup by the Vatican Mafia on behalf of its rabbinic partners in the conspiracy against the gospel of Jesus Christ."[1034]

An example of the obfuscation of offensive passages is Sanhedrin 106b, which says this about Jesus (pseudonym Balaam[1035]):

> A certain min said to R. Hanina: Hast thou heard how old Balaam was? — He replied: It is not actually stated, but since it is written, Bloody and deceitful men shall not live out half their days, it follows that he was thirty-three or thirty-four years old. He rejoined: Thou hast said correctly; I personally have seen Balaam's Chronicle, in which it is stated, 'Balaam the lame was thirty years old when Phinehas the Robber killed

him.'[1036]

This cryptic verse in Sanhedrin 106b conceals that fact that the "bloody and deceitful" man referred to is Jesus. In the Talmud, Balaam is used as a code for Jesus. A footnote to the passage explains that Phinehas the Robber is a coded pseudonym for Pontius Pilate and "Balaam's Chronicle" means "a Gospel."[1037]

Elizabeth Dilling in her well researched book, *The Jewish Religion, Its Influence Today*, explains that Balaam is a pseudonym used in place of Jesus. She explains that "[p]roof that Jesus is called 'Balaam' is found in the Jewish Encyclopedia (under "Balaam") which, after enumerating His alleged loathsome qualities, states: 'Hence … the pseudonym 'Balaam' given to Jesus in Sanhedrin 106b and Gittin 57a.'"[1038]

The Jewish Encyclopedia is authoritative. It presents a long list of eminent scholars who are experts in Judaism that contributed to the encyclopedia entry on Balaam, and who no doubt took great care in accurately explaining the meaning and use of "Balaam."[1039] The entry for Jesus in the Jewish Encyclopedia confirms that Balaam in the Talmud is a pseudonym for Jesus: "As Balaam the magician and, according to the derivation of his name, 'destroyer of the people,' was from both of these points of view a good prototype of Jesus, the latter was also called 'Balaam.'"[1040]

Jewish censors redacted and obfuscated the Talmud, in part, because Jews are very concerned about Gentiles discovering their hidden hatred toward Gentiles. It is for that reason that Jews are forbidden to share with gentiles their religious doctrines. "To communicate anything to a goy about our religious relations would be equal to the killing of all the Jews, for if the goys knew what we teach about them, they would kill us openly." (Book of Libbre David, 37)[1041] "Every goy who studies Talmud, and every Jew who helps him in it, ought to die." (Sanhedryn 59 a. Aboda Zora 8-6. Szagiga 13.)[1042]

Gemara at Sahendrin Tractate 52b provides that adultery is forbidden with the neighbor's wife, but is not forbidden with the wife of a heathen (Gentile). The implication is that a Gentile is not a neighbor.

> Our Rabbis taught: [And the man that committeth adultery with another man's wife, even he that committeth adultery with his neighbour's wife the adulterer and the adulteress shall surely be put to death]. 'The man' excludes a minor; 'that committeth adultery with another man's wife' excludes the wife of a minor; 'even he that committeth adultery with his neighbour's wife' excludes the wife of a heathen; 'shall surely be put to death', by strangulation.[1043] (brackets in original).

Sanhedrin 100b states: Those sent to hell include "he who reads uncanonical books etc. A Tanna taught: this means, the books of the Sadducees." It would seem that the uncanonical books are books of the Sadducees, but that is coded language inserted by censors to conceal the fact that the "uncanonical books" actually refers to the New Testament. A footnote immediately after the words "books of the Sadducees" states:

> This probably refers to the works of the Judeo-Christians, i.e., the **New Testament**. There were no Sadducees after the destruction of the Temple, and so **'Sadducees' is probably a censor's emendation for sectarians or Gentiles** (Herford, Christianity in the Talmud, p. 333.) [MS.M. reads, Minim.][1044] (emphasis added)

Talmudic Judaism has the most intense hatred for Jesus.[1045] While some Jews will deny that the Talmud teaches such things, Benjamin Freedman, a former Talmudic Jew, stated that:

“there have never been recorded more vicious and vile libelous blasphemies of Jesus, of Christians and the Christian faith than you will find between the covers of the 63 books of the Talmud, which forms the basis of Jewish religious law, as well as being the textbook used in the training of rabbis.”[1046] For example:

Shabbath 104b: Jesus was a magician and a fool. Mary was an adulteress.

Sanhedrin 106a also calls Jesus a soothsayer, and his mother, Mary, a harlot:

> Balaam also the son of Beor, **the soothsayer**, did the children of Israel slay with the sword. A soothsayer? But he was a prophet! — R. Johanan said: At first he was a prophet, but subsequently a soothsayer. R. Papa observed: This is what men say, 'She who was the descendant of princes and governors, **played the harlot** with carpenters.'[1047]

A footnote to the Sanhedrin 106a passage explains that Balaam in that passage refers to Jesus and she who played the harlot with carpenters alludes to Mary, the mother of Jesus. That footnote states:

> Herford, Christianity in the Talmud, p. 48, suggests that **Balaam is frequently used in the Talmud as a type for Jesus** (v. also pp. 64-70). Though no name is mentioned to shew which woman is meant, the mother of Jesus may be alluded to, which theory is strengthened by the statement that she mated with a carpenter.[1048]

Sanhedrin 43a: Jesus was guilty of sorcery and apostasy; he deserved execution. The disciples of Jesus deserve to be killed.

Gittin 57a: Jesus was sent to hell, where he is punished in boiling hot semen for mocking the Rabbis.

> He then went and raised **Balaam** by incantations. ... He then asked: What is your punishment? He replied: With boiling hot semen. He then went and raised by incantations the **sinners of Israel**. He asked them: Who is in repute in the other world? They replied: Israel. What about joining them? They replied: Seek their welfare, seek not their harm. Whoever touches them touches the apple of his eye. He said: What is your punishment? They replied: With boiling hot excrement, since a Master has said: Whoever mocks at the words of the Sages is punished with boiling hot excrement.[1049] (emphasis added)

We know that passage in Gittin 57a refers to Jesus, because the Jewish Encyclopedia states that Balaam is a pseudonym for Jesus in Gittin 57a.[1050] Furthermore, the passage has a footnote appended to "sinners of Israel" that states: "MS.M. Jesus,"[1051] which means that "sinners of Israel" is a pseudonym for Jesus.

The above racial hatred toward Gentiles, Jesus, and Christians by the Jews contained in the Jewish Talmud is hard for some to believe. It is so brazenly hateful that some may doubt the accuracy of the above quotations from the Talmud. The above quotations were taken directly from the Jews' College English translation of the Talmud, which was published through Soncino Press between 1935 and 1948. It was produced by the most learned Judaic scholars at the time and is recognized worldwide as the authoritative English translation of the Talmud. It is of unquestionable accuracy and authority, which is fully revealed once one is aware of the redaction and obfuscation from the Jewish censors explained above.

The Soncino Press has a long history of accurate printing of the Talmud, going back for centuries. The Soncino website states:

> The seal of the Soncino Press and the name it represents date back half a millennium. The most famous of the first Jewish printers, the Soncino family hailed from the Alsatian town of Speyer and settled in Soncino, a village in northern Italy from which they took their name. Joshua Soncino set up one of the world's first Hebrew printing presses in 1484, on which he printed Soncino's first Hebrew book, the Talmud tractate Berakoth. He also produced the first printed Hebrew Bible complete with vowels--no mean feat for the early, primitive days of printing.[1052]

The late Dr. Joseph Herman Hertz, The Chief Rabbi of the British Empire at the time the Soncino Talmud was first published, wrote forewords for four of the six seders. He praised the Soncino Talmud with these words:

> A reliable English translation of the whole Babylonian Talmud has long been looked forward to by scholars. This expectation is beginning to be realised by the publication of the Soncino edition of the Order Nezikin. ...
>
> This notable achievement is due to the quite extraordinary erudition of the Editor, Rabbi Dr I. Epstein, assisted by his staff of scholarly translators. The Editor's Prefatory Note gives some indication of his colossal task. Aside from planning the scope and character of the work, the Editor Fixed the Text, controlled the translation and interpretation, as well as the introductions

> and glossaries to the various parts, and supplied the greater portion of the 'cultural' notes.
>
> The Publishers too have done their share in the undertaking conscientiously and efficiently. With the result, that never before has there appeared a translation of the Order Nezikin as helpful to the student as these volumes of the Soncino edition of the Babylonian Talmud in English. ...
>
> The eight volumes of Seder Nashim have been planned on the same lines as those of Seder Nezikin, alike in regard to Text, rendering and cultural Notes. The Editor and his collaborators have again performed with consummate skill a task of stupendous difficulty, and the standard of scholarship and accuracy set in the previous volumes has been fully maintained. The Publishers also have left nothing undone to render the Soncino Seder Nashim in every way a worthy continuation of their Seder Nezikin.

The following accolades are printed on the dust jacket of the 1961 Edition of the Soncino Talmud:

> [T]his translation of the Talmud will open up for the English speaking readers the most varied and indispensable index to the collective wisdom of the post-Biblical Hebrew mind, and afford direct knowledge of the ordinances which have governed and continue to govern the religious institutions and social polity of the Jewish people. *- The late Rabbi Dr. A. Feldman, Dayan of the Beth Din, London.*
>
> I offer my heartiest congratulations to all who

have worked together in producing this translation of the Talmud into English. They have dealt with the enormous difficulties of their tasks with remarkable skill. - *The Rev. Dr. R. Travers Herford, author of Christianity in Talmud and Midrash and Talmud and Apocrypha.*

The Editor, Dr. Isidore Epstein, and his band of scholars have produced a translation which is a marvel of skill in dealing with the formidable difficulties which the Talmud presents. - *The Jewish Chronicle.*

The Soncino Talmud is ... an adornment for any library, a hallmark of intelligent Jewishness in a family ... an unfailing source of inspiring information to the students of every creed and race. - *Jewish Forum.*

It [the Soncino Talmud] will have value for Gentile as well as Jewish scholars. - *The Manchester Guardian.*

These volumes are magnificently produced, and will assuredly take their place among the monuments of modern scholarship. - *The Yorkshire Post.*

The rendering is accurate and scholarly ... In particular the translators deserve praise for their choice of suitable equivalents for technical terms and phrases that are difficult to turn into another language. - *The Times Literary Supplement.*

Michael Hoffman explains that "[l]ike the Talmud, the Kabbalah supersedes, nullifies and ultimately replaces the

Bible."[1053] Lawrence Fine, Professor of Jewish Studies and prominent scholar of medieval Judaism and Jewish mysticism, reveals that the Kabbalah contains the "true" meaning of the Old Testament. The "simple" meaning of the biblical language recedes into the background as the symbolic meaning contained in the Kabbalah supercedes the Bible and takes control. There is a code to the true meaning in the Bible that can only be unlocked through the Kabbalah.

> [T]he reader must become accustomed to regarding biblical language in a kabbalistically symbolic way. The Kabbalists taught that the Torah is not only the speech or word of God, but is also the many names of God or expression of God's being. It is a vast body of symbols, which refers to the various aspects of divine life, the sefirot, and their complex interaction. **The simple meaning of biblical language recedes into the background as symbolic discourse assumes control**. The true meaning of Scripture becomes manifest only when it is read with the proper (sefirotic) code. **Thus the Torah must not be read on the simple or obvious level of meaning; it must be read with the knowledge of a kabbalist who possesses the hermeneutical keys with which to unlock its *inner* truths.**[1054]

The Kabbalah at Zohar III, 152a states: "Thus the tales related to the Torah are simply her outer garments, and woe to the person who regards that outer garb as the Torah itself! For such a person will be deprived of a portion in the world to come."[1055] That passage in the Kabbalah puts a curse on anyone who tries to read the Bible for what it actually says, instead of with the mystical gloss put on it by the Kabbalah.

The Kabbalah is Judaic mystical practices that were

adopted by the Jews from Babylon. H.P. Blavatsky described the Kabbalah as:

> The hidden wisdom of the Hebrew Rabbis of the middle ages derived from the older secret doctrines concerning divine things and cosmogony, which were combined into a theology after the time of the captivity of the Jews in Babylon. All the works that fall under the esoteric category are termed Kabalistic.[1056]

Magic and occult mysticism run throughout the Kabbalah. Judith Weill, a professor of Jewish mysticism, stated that magic is deeply rooted in Jewish tradition, but the Jews are reticent to acknowledge it and don't even refer to it as magic.[1057]

Gershom Scholem (1897-1982), Professor of Kabbalah at Hebrew University in Jerusalem, admitted that the Kabbalah contains a great deal of black magic and sorcery, which he explained involves invoking the powers of devils to disrupt the natural order of things.[1058] Professor Scholem also stated that there are devils who are in submission to the Talmud; in the Kabbalah these devils are called *shedim Yehuda'im*.[1059]

The *Jewish Chronicle* revealed that occult practices such as making amulets, charms, and talismans are taught in Jerusalem at the rabbinic seminary Yeshivat Hamekubalim.[1060] That is why Jesus said to the Jews: "Ye are of your father the devil, and the lusts of your father ye will do." John 8:44. The Bible states clearly that the magic arts are an abomination to the Lord.

> There shall not be found among you any one that maketh his son or his daughter to pass through the fire, or that useth divination, or an observer of times, or an enchanter, or a witch, Or a charmer, or a consulter with familiar spirits, or a wizard, or

a necromancer. For all that do these things are an abomination unto the LORD: and because of these abominations the LORD thy God doth drive them out from before thee. (Deuteronomy 18:10-12)

40 Mind Control

Zionist Jews control the educational system, and have created a false science that has removed God from the minds of men through so-called science that alleges that there is a universe without a creator. This atheistic education undermines Christian morals and order. It is all happening according to plan. Once the Zionists are in position of rule over the world, they will teach nothing but submission to their authority. Paragraph three of Protocol 16 of the Protocols of the Learned Elders of Zion states:

> We must introduce into their education all those principles which have so brilliantly broken up their order. But when we are in power we shall remove every kind of disturbing subject from the course of education and shall make out of the youth obedient children of authority, loving him who rules as the support and hope of peace and quiet.[1061]

It is indispensable to the Zionists' plans for world domination to undermine all religion. That is why they need to control the education of our children. That is where they instill the first ideas of atheism through the theories of evolution and heliocentricity. Paragraph three of Protocol 4 states:

> It is indispensable for us to undermine all faith, to tear out of the mind of the "goyim" the very principle of god-head and the spirit, and to put in its place arithmetical calculations and material needs.[1062]

Heliocentrism, which theorizes that the earth orbits the sun as it also spins, is the seminal scientific theory from which all scientific deception flows. Heliocentrism is not based upon science; it is based upon religious superstition. So-called scientists have concluded that the earth orbits the sun. Nicolaus Copernicus died in 1543 on the day his book, *On the Revolutions of the Celestial Spheres*, was published. Most people do not know that Copernicus did not originate the theory that the earth revolves around the sun. Aristarchus of Samos (310 – 230 B. C.) postulated that the earth rotates on an axis daily and orbits the sun annually. Pythagoras (circa 500 B.C.) had a similar model of planets (which purportedly included the sun) orbiting around a central, invisible fire. Both Pythagoras' and Aristarchus' models had generally been rejected, until Copernicus' book was published. There was initially strong resistance to Copernicus' heliocentric system. However, over time the heliocentric view, with the earth and the other planets rotating around sun, has won popular acceptance.

The heliocentric theory removed the earth as the center of creation and challenged the entire ancient authority of the Bible regarding the universe and its origins. Heliocentricity is the progenitor of the theory of evolution.

Tycho Brahe (1546 – 1601), was born three years after Copernicus died. His observations and models established that the earth is stationary. Scientists have through objective experiments confirmed Brahe's findings regarding a stationary earth.

You would think that all of this spinning, orbiting, and hurtling through space is something that has been detected and measured by the "scientists" who promote that notion. In fact, all true scientific efforts to measure that movement has only proven that the earth is stationary and does not move.

Elite Jews are the agents of Satan in a massive conspiracy working toward world domination. In order to control people, these arch-criminals must remove from the consciences of the people any knowledge of God. Satan and his minions have a problem, and that is found in God's creation, which everyone can see around them. God's creation reveals God to everyone day and night.

> The heavens declare the glory of God; and the firmament sheweth his handywork. Day unto day uttereth speech, and night unto night sheweth knowledge. There is no speech nor language, where their voice is not heard. Their line is gone out through all the earth, and their words to the end of the world. In them hath he set a tabernacle for the sun, Which is as a bridegroom coming out of his chamber, and rejoiceth as a strong man to run a race. His going forth is from the end of the heaven, and his circuit unto the ends of it: and there is nothing hid from the heat thereof. (Psalms 19:1-6)

In Psalms 19:1-6, God reveals that the heavens declare his glory and the firmament shows his handiwork. All one need do is look up and God is revealed by his heavenly creation. Indeed, God testifies that the sun travels in a circuit. Anyone looking up can see this happening each day. Satan had to figure out a way to deceive people into rejecting what God's creation was revealing. Satan did this through convincing people that what they are seeing is a very distant sun and the earth is spinning on an axis while it rotates around that sun. He has convinced the world that the sun does not move in a circuit as the Bible states, but rather, it is the earth that is moving. How could Satan get people to ignore what they could see with their own eyes? George Orwell provides an answer.

George Orwell was a disillusioned communist insider who exposed the methods and aims of the world ruling elite in the form of a dystopian novel titled *1984.* In his book, a conversation takes place between O'Brien, a powerful member of the Inner Party, which is the elite ruling minority, and the protagonist, Winston Smith. The conversation takes place while O'Brien is in the midst of torturing Winston. O'Brien tells Winston:

> The Party seeks power entirely for its own sake. We are not interested in the good of others; we are interested solely in power. Not wealth or luxury or long life or happiness: only power, pure power. ... We know that no one ever seizes power with the intention of relinquishing it. Power is not a means, it is an end. One does not establish a dictatorship in order to safeguard a revolution; one makes the revolution in order to establish the dictatorship. The object of persecution is persecution. The object of torture is torture. The object of power is power. Now do you begin to understand me? ...
>
> We are the priests of power ... God is power. But at present power is only a word so far as you are concerned. It is time for you to gather some idea of what power means. The first thing you must realize is that power is collective. The individual only has power in so far as he ceases to be an individual. You know the Party slogan: "Freedom is Slavery". Has it ever occurred to you that it is reversible? Slavery is freedom. Alone — free —

> the human being is always defeated. It must be so, because every human being is doomed to die, which is the greatest of all failures. But if he can make complete, utter submission, if he can escape from his identity, if he can merge himself in the Party so that he IS the Party, then he is all-powerful and immortal. **The second thing for you to realize is that power is power over human beings. Over the body — but, above all, over the mind. Power over matter — external reality, as you would call it — is not important. Already our control over matter is absolute.**[1063]

At that point, Winston challenged O'Brien. Winston stated that the party could not control matter, because they cannot control climate or prevent diseases. It was at that point that O'Brien explained: **"We control matter because we control the mind."**[1064]

Winston and O'Brien continued their conversation, and during the discussion Winston exclaimed to O'Brien: "But the whole universe is outside us. Look at the stars! Some of them are a million light-years away. They are out of our reach for ever."

O'Brien rebutted Winston's declaration, by explaining:

> 'What are the stars?' said O'Brien indifferently. 'They are bits of fire a few kilometres away. We could reach them if we wanted to. Or we could blot them out. The earth is the centre of the universe. The sun and the stars go round it.'
>
> Winston made another convulsive movement. This time he did not say anything. O'Brien continued as though answering a spoken

objection:

> 'For certain purposes, of course, that is not true. When we navigate the ocean, or when we predict an eclipse, we often find it convenient to assume that the earth goes round the sun and that the stars are millions upon millions of kilometres away. But what of it? Do you suppose it is beyond us to produce a dual system of astronomy? The stars can be near or distant, according as we need them. Do you suppose our mathematicians are unequal to that? Have you forgotten doublethink?[1065]

Notice that Orwell reveals in his book that heliocentrism is central to the deception of "doublethink." In Orwell's *1984*, Orwell explains that doublethink is to deny the existence of objective reality, while at the same time accepting the reality that is denied. Orwell explained that doublethink is "to be conscious of complete truthfulness while telling carefully constructed lies."[1066]

Under the rubric of doublethink, scientists at the highest level accept that the earth is flat, yet they deny that known reality, because pressure has been put upon them to "fit in," to gain status within their profession. If someone even so much as mentions any contradiction of the orthodox view of heliocentricity, he is subject to ridicule and loses his status as a respected "scientist." There are strategically placed gatekeepers, who ensure that the truth does not come out. Deviation from the "party line" is akin to blasphemy, which subjects the blasphemer to instant social and professional disapproval and punishment. Research grants to the blasphemer would instantly dry up. The scientific heretic would lose his professorship and be brought to economic and professional ruin to reveal what he knows to be the truth. In order to get along, scientists go along with doublethink. Otherwise, they would be crushed.

What is God's view of doublethink? "A double minded man is unstable in all his ways." (James 1:8)

O'Brien identifies the deceivers in the doublethink heliocentric myth. O'Brien reveals that it is "mathematicians" who are at the center of the heliocentric deception. As we have seen, mathematics has been used to replace experimental physicists to construct the doublethink new "reality" of relativity. Astrophysicists are not really physicists; they are lying mathematicians, who have created a mythology using mathematic models. The reality is that no astrophysicist has ever been to the place to where they claim to be experts. The reason is simple, there is no such thing as outer space. These mathematicians, masquerading as astrophysicist, simply bedazzle the uninformed with bizarre equations they use to support the deception of heliocentrism. Astrophysicists are akin to priests in an occult religion. Just as the witch-doctor is held in awe by the ignorant tribe for his superstitious knowledge, so also is the astrophysicist held in awe for his so-called "scientific" knowledge. In both cases their esteem is born of spiritual deception.

Doublethink is indispensable to the mind control of the ruling elite. Orwell explains in *1984*:

> To tell deliberate lies while genuinely believing in them, to forget any fact that has become inconvenient, and then, when it becomes necessary again, to draw it back from oblivion for just so long as it is needed, to deny the existence of objective reality and all the while to take account of the reality which one denies — all this is indispensably necessary. Even in using the word DOUBLETHINK it is necessary to exercise DOUBLETHINK. For by using the word one admits that one is tampering with reality; by a fresh act of DOUBLETHINK one erases this

> knowledge; and so on indefinitely, with the lie always one leap ahead of the truth. Ultimately it is by means of DOUBLETHINK that the Party has been able — and may, for all we know, continue to be able for thousands of years — to arrest the course of history.[1067]

Understand what Orwell is revealing. The ruling elite have complete control over all elements of society, because they control what people think. They do not actually control matter; all they need do is convince that population to believe what is false about matter and they thus control matter in that sense. It is the ultimate in mind control to get people to believe a lie is truth. Indeed, the slogan of the Party in *1984* is that "Freedom is Slavery."

The Jewish elite do not have the power to make the earth spin at approximately 1,000 miles per hour at the equator, while at the same time orbiting the sun at approximately 66,600 miles per hour. All they need to do is convince the population of the world that is what happens. That is what O'Brien meant when he stated that "we control matter because we control the mind."

Why would the Jewish elite want to deceive the world about the nature of our world? Eric Dubay, author of *The Flat Earth Conspiracy*, offers this explanation:

> Our eyes and ears tell us the Earth is flat and motionless, everything in the sky revolves around us, but when we cease to believe our own eyes and experience, we have to prostrate ourselves at the feet of the very pseudo scientists who are blinding us, treat them as experts, astronomical priests, who have special knowledge only they can access like the Hubble telescope.
>
> **So by brainwashing us of something so**

gigantic and fundamental it actually makes every other kind of lesser indoctrination a piece a cake. Earth being the flat earth center of the universe, of which everything else and the heavens revolve, gives special importance and significance not only to earth but to us humans, the most intelligent of the designer's design.

By turning earth into a spinning ball, thrown around the Sun, shot into infinite space from a God-less big bang, they turned humanity into a random meaningless, purposeless accident of a blind dumb universe. So it's like trauma based mind control ... So this modern atheistic, Big Bang, heliocentric globe earth, chance evolution, paradigm spiritually controls humanity by removing God or any sort of intelligent design from the mind and replaces purposeful divine creation with haphazard random cosmic coincidence.

And so by removing Earth from the motionless center of the universe these Masons have moved us physically and metaphysically from a place of supreme importance to one of complete nihilistic indifference.

If the Earth is the center of the Universe then the ideas of God creation and a purpose for human existence are resplendent, but if the Earth is just one of billions of planets revolving around billions of stars and billions of galaxies then the ideas of God creations and a specific purpose for Earth and human existence become highly implausible.

> By surreptitiously indoctrinating us into their scientific materialist sun worship, not only do we lose faith beyond the material, we gain absolute faith in materiality, superficiality, social status, selfishness, hedonism and consumerism.
>
> If there is no God and everyone is just an accident then all that matters is me, me, me. ... Their rich powerful corporations with their slick sun cult logos sell us idols to worship, slowly taking over the world, while we tacitly believe their science, vote for their politicians, buy their products, listen to their music, watch their movies, all sacrificing our souls to the altar of materialism.
>
> It's a big deception![1068]

Dubay makes the point that "this is basically the biggest thing you could possibly lie about, so the psychopaths that control society are interested in world domination; the best way to brainwash the whole world is to lie to the whole world about what the world is."[1069] The New World Order is based upon disinformation, and Heliocentrism is its keystone deception.

Orwell's book, *1984*, is a fictional novel written to warn us; and it has its corollary in modern day reality. For example, David Rockefeller (1915 – 2017), would be an example of someone Orwell would consider a member of the powerful and privileged Inner Party. David Rockefeller was the grandson of Standard Oil founder John D. Rockefeller and chairman and chief executive of Chase Manhattan Corporation. While David Rockefeller's personal fortune was reported to be three billion dollars, he controlled exponentially more wealth through interlocking trusts. He founded The Trilateral Commission in 1973 in order to bring about regional cooperation, which the commission calls "growing interdependence," toward "globalization," and ultimately, world government.[1070] Archibald E. Roberts, Lt. Col., AUS. Ret., in his book, *The Most Secret Science*, described The Trilateral Commission's three regional commands as the very manifestation of George Orwell's world superpowers described in *1984*. "George Orwell's three world superpowers, 'Oceania,' 'Eastasia,' and 'Eurasia,' exist today, governed by The Trilateral Commission, a cabal of international bankers, industrialists and academicians."[1071] Roberts calls the Trilateral Commission a "semi-secret world government administration. Decision-making is reserved to the 'Secret Government of Monetary Power.'"'[1072]

The prime method toward world government is war and subversion; and the prime target is the United States. The seditious conspirators use war, terror, and subversion to reduce the industry of the United States, undermine the family, corrupt the morals, and take away the freedoms of its citizens so that the United States will more easily accept a merger into a world government.

On February 17, 1950, James Paul Warburg confidently declared to the Subcommittee of the Committee on Foreign Relations in the United States Senate: **"We shall have World Government, whether or not we like it. The only question is whether World Government will be achieved by conquest or consent."**[1073] Not just anybody can be invited to speak before the U.S. Senate committee. And James Paul Warburg (1896-1969) knew what he was talking about. He was the son of Paul Moritz Warburg. James was also the nephew of both Felix Warburg and Jacob Schiff. Who were they? Jacob Schiff and Felix Warburg managed Kuhn, Loeb & Company, which is the banking house that financed the Russian Revolution through Max Warburg (James Paul Warburg's brother). Max Warburg served on the board of the German Reichsbank until 1930. Those men were powerful Jewish communists working toward a Zionist world government.

The occasion of James Paul Warburg's testimony before

the U.S. Senate was to call on the Senate to pass Resolution 56. Resolution 56 was the Senate corollary to the 1949 joint resolution H.C.R.64 that was introduced in the House of Representatives:

> Resolved by the House of Representatives (the Senate concurring) that it is the sense of the Congress that it should be a fundamental objective of the foreign policy of the United States to support and strengthen the United Nations and to seek its development into a World Federation, open to all nations, with defined and limited powers adequate to preserve peace and prevent aggression through the enactment, interpretation and enforcement of world law.[1074]

What does all of the highfalutin language mean? It means that the U.S. Government would give up its sovereignty to an all-powerful world government under United Nations rule. One of the advocates of the resolution, Cord Meyer, chairman of the National Executive Committee of the United World Federalists, admitted at a hearing before the Senate Subcommittee on the United Nations Charter:

> We in the United States would be declaring our willingness to join with other nations in transferring to the U.N. constitutional authority to administer and enforce law that was binding on national governments and their individual citizens.[1075]

Warburg was a mouthpiece for Zionists who want a world government under their control. Warburg revealed that Resolution 56 was a step toward setting up the United Nations as a sovereign world government, able to impose its unfettered will on all the nations of the world:

> I think the essential thing we should undertake is that we declare our willingness to participate in some sort of world organization capable of enacting, administering, interpreting, and, enforcing world law, whether you call it a federation, a government, or world order, I don't think that matters.[1076]

The public, who understood better than their congressional representatives what the resolution actually meant, expressed their outrage over the resolution. After the public outrage was felt by the sponsors of the resolution, they backed away from their endorsement of it. For example, one sponsor, Rep. Bernard W. Kearney (R-New York) said:

> We signed the Resolution believing we were sponsoring a movement to set up a stronger power within the United Nations for world peace ... Then we learned that various organizations were working on state legislatures and on peace movements for world government action under which the entire U.S. Government would be submerged in a super world government ... Perhaps we should have read the fine print in the first place. We do not intend to continue in the role of sponsors of any movement which undermines U.S. sovereignty.[1077]

Subsequently, 18 of the 23 states which had passed resolution 56, rescinded it. Ultimately, it failed. That was only a temporary setback for the insurgent Zionists.

The Zionist traitors have many arrows in their quiver. David Rockefeller was another among the many rich and powerful who are part of the world conspiracy against God and man. The world government envisioned by the conspirators is based upon the communist model. In his 2002 autobiography, David Rockefeller, with immense power and feeling untouchable, brazenly admitted to his treacherous involvement in the massive conspiracy to bring about a one world government.

> Some even believe we are part of a secret cabal working against the best interests of the United States, characterizing my family and me as 'internationalists' and of conspiring with others around the world to build a more integrated global political and economic structure - one world, if you will. If that's the charge, I stand guilty, and I am proud of it.[1078]

The world government, for which David Rockefeller and his ilk work so feverishly, is a Zionist world government centered in Israel. Another of those evil minions working toward a world government was Israeli Prime Minister David Ben Gurion, who in 1962 opined for Look Magazine what he thought the world would be like in 25 years (1987).

> The image of the world in 1987 as traced in my imagination: The Cold War will be a thing of the past. Internal pressure of the constantly growing intelligentsia in Russia for more freedom and the pressure of the masses for raising their living standards may lead to a gradual democratization of the Soviet Union. On the other hand, the increasing influence of the workers and farmers, and the rising Political importance of men of science, may **transform the United States into a welfare state with a planned economy**. Western and Eastern Europe will become a federation of autonomous states having a Socialist and democratic regime. With the exception of the U.S.S.R. as a federated Eurasian

> State, all other continents will become united in a world alliance at whose disposal will be an international police force. All armies will be abolished and there will be no more wars. **In Jerusalem, the United Nations (a truly United Nations) will build a shrine of the prophets to serve the federated union of all continents; this will be the seat of the Supreme Court of Mankind, to settle all controversies among the federated continents, as prophesied by Isaiah.**[1079] (emphasis added)

Notice that Ben Gurion predicted the democratization programs of Glasnost and Perestroika of the General Secretary of the Central Committee of the Communist Party of the Soviet Union, Mikhail Gorbachev (1985). He also predicted the rise of the European Union under socialism and the reduction of the United States to a "welfare state," which we witness happening at an accelerated pace today. How was he able to predict those events? Because he was in on the evil plans for a world government with an international "police force" that would protect the ruling elite and crush the freedom of the masses. Most revealing is the fact that Ben Gurion's vision culminates in the world government centered in Jerusalem.

The governments of the world rule upon the principle of disinformation. This is illustrated by an episode that took place in 1983, when New Hampshire Governor Meldrim Thomson Jr. wrote to President Ronald Reagan objecting to the planned national holiday for Martin Luther King Jr. Thomson pointed out the evidence of King's "immoral character" and his affiliation with communists. Ronald Reagan's response to Thomson gives insight into the effect of disinformation on the masses and how it frames the laws and policies of government. On October 3, 1983, President Reagan responded to Thomson: "On the national holiday you mentioned, I have the reservations you have, but here the perception of too many people is based on an image, not reality. Indeed to them, the perception is reality."[1080]

It did not matter to Reagan what the truth was, it was the perception that was important when it came to making political decisions. The masses of people had been deceived into believing that Martin Luther King Jr. was a selfless champion of civil rights for all. The truth that King was a tool of the communists to push for special government granted group privileges as a means to supplant individual God-given rights was known only by a few.

Because of the effectiveness of the disinformation campaign that propped up Martin Luther King Jr. as a paragon of virtue, the government was in the clear in acting to honor a plagiarizing, womanizing, communist shill.[1081] Ralph Abernathy, who King called "the best friend I have in the world," documented in his autobiography, *And the Walls Came Tumbling Down*, that Martin Luther King Jr. was an immoral charlatan. King was a wolf in sheep's clothing. Dr. Jimmy T. (Gunny) LaBaume reveals that Martin Luther King Jr. was a heathen who was playing the role of a Christian minister.

> King, the charismatic speaker, does not seem to be the committed Christian he professed to be. In fact, he said that the Bible was filled with "legends and myths" and denied that Christ was physically raised from the dead. As a seminary student King was not at all enthusiastic about Christianity. Carl Rowan wrote in Reader's Digest (September 1967) that "the thinking of Gandhi and Thoreau was … burning inside King"[1082]

The King family is in on the charade. King's wife, Coretta Scott King, knowing full well that her husband was a charlatan, marched into federal court, and on Jan. 31, 1977, "obtained a federal court order sealing 845 pages of FBI records about her husband for 50 years, 'because its release would destroy his reputation!' This federal judge sealed the files until the year 2027."[1083] Indeed, many of the sealed documents have been characterized as "obscene." Most of Martin Luther King Jr.'s Boston University doctoral dissertation has been proven to be plagiarized. Plagiarization seems to have been a family value in the King household. Corretta Scott King, who acted as Martin Luther King Jr.'s secretary at the time Martin Luther King Jr. put together his dissertation, was an accomplice in the plagiarization. Dr. LaBaume reveals that Martin Luther King Jr. simply could not help himself. King was a serial plagiarizer.

> [O]ver his life, most of his papers, speeches, and 'sermons' were copied word for word from Dr. Jack Boozer, Edgar S. Brightman, and Paul Tillich. As a matter of fact, he lifted whole sections of his famous "I Have a Dream" speech from a sermon by Archibald Carey, a popular black preacher in the 1950's.[1084]

Martin Luther King Jr. routinely embezzled Southern Leadership Conference (SLC) money to pay for prostitutes. King was a moral degenerate, whose role was to implement the tried and true communist method of secretly agitating violence, while at the same time preaching non-violence. He thus appeared to be the victim of the very violence he planned ahead of time. Martin Luther King Jr. was a communist revolutionary. According to King's biographer and sympathizer David J. Garrow, "King privately described himself as a Marxist."[1085] In his 1981 book, *The FBI and Martin Luther King, Jr.*, Garrow quotes King as saying in SCLC staff meetings, "we have moved into a new era, which must be an era of revolution.... The whole structure of American life must be changed.... We are engaged in the class struggle."[1086] Dr. Don Boys, who is a former member of the Indiana House of Representatives and has authored 15 books, explains that “those words [of King] are right out the Communist handbook.”[1087] Dr. Boys further states: “Need I remind you that it was illegal to be a Communist in those days? The Communist Party was dedicated to overthrow our Constitution and way of life.”[1088] Dr. Boys reveals little known inside information from Uriah Fields, who was Martin Luther King Jr.'s secretary early on in King's communist agitation.

> The Rev. Uriah J. Fields, King's secretary during the early stage of the Montgomery Bus Boycott, wrote in the March 1968 issue of American Challenger: "King helps to advance Communism. He is surrounded with communists. This is a major reason why I severed my relationship with him during the fifties. He is soft on Communism. I don't believe that he is a Communist or a Christian, for that matter." There is no doubt that King was not a Christian, and he was not a "card-carrying" Communist because Communists stopped carrying cards many years before that time. The question is not whether King belonged to the Communist Party, U.S.A. but was he a Communist?[1089]

The circumstantial evidence certainly suggests that Martin Luther King Jr. was a communist. Jewish Communist Stanley Levison can best be described as King's behind-the-scenes "handler." Levison had for years been in charge of the secret Soviet funding of the Communist Party, USA. Levison was King's mentor and was actually the brains behind King. It was Levison who edited King's book, *Stride Toward Freedom.*[1090] Levison controlled the fund-raising and agitation activities of the SCLC from behind the scenes. Levison wrote many of King's speeches. King described Levison as one of his “closest friends.”[1091]

Dr. Boys further reveals that Julia Brown, who worked

undercover in the communist party on behalf of the FBI, testified before the U.S. Congress about Martin Luther King Jr.'s communist ties.

> While King may not have been an official member of the Communist Party, he was certainly supported by them because he supported their cause! Julia Brown was an undercover agent for the FBI in the Cleveland, Ohio area. She told a Congressional sub-committee that while she was a member of the Communist Party, "she knew Martin Luther King to be closely connected with the Communist Party." On the lecture circuit, she told audiences that King "was the hero of America's Communists. The cells that I was associated with in Cleveland were continually being asked to raise funds for Martin Luther King's activities." She also said, "Mr. King was one of the worst enemies my people ever had." Oh, by the way, Julia was black! How about a holiday for a real black hero? In June of 1979 she testified at a Senate Judiciary Committee and concluded her statement to the Committee: "If this measure is passed honoring Martin Luther King, we may as well take down the stars and stripes that flies over this building and replace it with a red flag."[1092]

Despite Brown's testimony, Congress nonetheless voted to approve the national holiday for Martin Luther King Jr. four years later. Reagan signed the bill into law. Reagan knew all of the above information about Martin Luther King Jr., and even more, but he did not care. He was in on the game. He, like Martin Luther King Jr., was also a charlatan. He was a socialist actor simply playing the role of a conservative politician. The statement of George H. W. Bush to Sarah McLendon could as easily have applied to Ronald Reagan. Bush told Sarah McLendon in 1992: "Sarah, if the American people ever find out what we have done, they would chase us down the street and lynch us."

Figure 186: Martin Luther King Jr. Federal Building, Atlanta, Georgia

A conquering force, upon taking control of a country, will always change the national landmarks. Such changes in national landmarks are also a sign of a covert political takeover. One of the first things that communists do when taking over a country is to rewrite the history of the nation so that the people are left without a national heritage. In accomplishing that, national heroes are excised from the history books, or their patriotic deeds are distorted to fit communist ideology. The communists then elevate its killers and criminals as the new heroes.[1093] These new communist heroes are then celebrated with monuments and national holidays. That is what has been done in the case of Martin Luther King Jr.[1094] He is a communist saint who is idolized in the United States with a national holiday. Streets, schools, and federal buildings all over the country have been named in his honor. Indeed, much to the confusion of the local citizens, many long established streets have been renamed in honor of Martin Luther King Jr.

While the Martin Luther King Jr. national holiday is largely symbolic, its symbolism acts to reinforce in the minds of the general population the need for the government to pass laws and use its force to grant privileges to a favored group. The result of the new laws is an erosion of the individual liberties of all. It is this method that has given rise to the government privilege of women to kill their children through abortion to supplant the unborn child's God-given right to life. It can also be seen with the special government privileges granted to sodomites. Those special group privileges of the sodomites will be used as a lever to supplant that God-given individual rights of Christians.

One recent example of using government privileges to trample the God-given rights of the American people is the case of a Christian couple, Aaron and Melissa Klein mentioned in this book's introduction. As you recall, the Kleins were forced to shut down their bakery after they were ordered to pay $135,000 in damages by the Oregon Bureau of Labor and Industry (BOLI) for refusing to bake a cake for a lesbian wedding because of their sincerely held religious belief that sodomy is a sin.[1095] Todd Starnes of Fox News and Commentary explains that "the Klein case has demonstrated once again that gay rights trump religious liberty. Other Christian business owners should pay close attention. The Kleins had a choice. They could obey the government or they could obey God. They chose God – and now they must pay the price."[1096]

The Klein case is the sour fruit of the civil rights tree planted by the communist handlers of Martin Luther King Jr. It is only the beginning. The civil rights laws that are now being used to trample God-given rights could not have been passed without the disinformation that was part and parcel of the civil rights movement.

Barbara Honegger, who was Assistant to the Chief Domestic Policy Adviser to the President, was present at a cabinet meeting during the first weeks (February 1981) of the Reagan Administration in the Roosevelt Room in the West Wing of the White House when CIA Director William Casey stated: "We'll know our disinformation program is a success when everything the American public believes is false."[1097] Honegger reported the statement to her close friend Senior White House correspondent Sarah McClendon, who in turn made it public without naming Honegger as the source.

In or about 2013, people posted on the internet their skepticism of the authenticity of the Casey quote. On November 25, 2014, Barbara Honegger came forward to verify its genuineness by revealing that she was the source for the quote. Honegger stated:

> I am the source for this quote, which was indeed said by CIA Director William Casey at an early February 1981 meeting of the newly elected President Reagan with his new cabinet secretaries to report to him on what they had learned about their agencies in the first couple of weeks of the administration. The meeting was in the Roosevelt Room in the West Wing of the White House, not far from the Cabinet Room. I was present at the meeting as Assistant to the chief domestic policy adviser to the President. Casey first told Reagan that he had been astonished to discover that over 80 percent of the 'intelligence' that the analysis side of the CIA produced was based on open public sources like newspapers and magazines. As he did to all the other secretaries of their departments and agencies, Reagan asked what he saw as his goal as director for the CIA, to which he replied with this quote, which I recorded in my notes of the meeting as he said it. Shortly thereafter I told Senior White House correspondent Sarah McClendon, who was a close friend and colleague, who in turn made it public. Barbara Honegger.[1098]

Note the two significant things revealed by CIA Director

Casey: 1) The CIA relies on public sources for over 80 percent of its intelligence; and 2) he, as Director of the CIA, stated that his goal was to provide disinformation to the American public, and that his success in that goal would be when everything the public believed was false. Why wouldn't Director Casey make it his goal to fix the woeful intelligence gathering of the CIA that he was astonished to discover? Because he understood that intelligence gathering is not the primary purpose of the CIA, and therefore it was not a priority to fix that deficiency. The CIA's chief purpose is to provide disinformation to the American public as a means of controlling the people. His goal, as director, was to further the CIA's principal purpose.

How is the CIA able to run a disinformation campaign to the degree that it can influence the American public? Dave McGowan, in his book, *Derailing Democracy*, quotes William Colby, former CIA Director, saying: "The Central Intelligence Agency owns everyone of any significance in the major media."[1099] A CIA operative discussed with Philip Graham, editor of the Washington Post, how easy it was for the CIA to buy influence over journalists willing to peddle CIA propaganda and cover stories. "You could get a journalist cheaper than a good call girl, for a couple hundred dollars a month."[1100]

Michael Collins Piper, in his book *Final Judgement*, reveals the CIA subversion of the American media, particularly with regard to covering up the CIA and Israeli Mossad connection to the assassination of President John F. Kennedy.[1101] Piper explained that the subversion of the American media by the CIA is pervasive. Piper cited award winning author David Wise, who, in his book, *The American Police State: The Government Against the People*, exposed the role of the CIA in manipulating the media:

> The CIA's contacts with the publishing world were not confined to attempts to suppress books. Through the U.S. Information Agency as a

> 'cut-out,' the CIA subsidized major publishers to produce books, some of which were then sold in the United States bearing no government imprint to warn the unsuspecting purchaser. In 1967 publisher Frederick A. Praeger conceded he had published 'fifteen or sixteen' books for the CIA. By the mid-sixties, more than $1 million had been spent by the government on its 'book development' program. The Senate intelligence committee estimated that by 1967, the CIA had produced, sponsored, or subsidized 'well over 1,000 books' here and abroad.[1102]

Wise reveals that "the CIA operated two news services of its own in Europe. These 'proprietaries,' or CIA cover companies, serviced American newspapers; one had more than thirty U.S. subscribers."[1103] The thirty U.S. subscribers to whom Wise refers are newspapers and other publishing concerns. Those U.S. media outlets then parrot the CIA propaganda, which, in turn, influences public opinion and behavior.

Indeed, it seems that the Washington Post had developed into a virtual department of the CIA. Alex Constantine reveals:

> Former Washington Post publisher Philip Graham "believing that the function of the press was more often than not to mobilize consent for the policies of the government, was one of the architects of what became a widespread practice: the use and manipulation of journalists by the CIA" This scandal was known by its code name Operation MOCKINGBIRD. Former Washington Post reporter Carl Bernstein cites a former CIA deputy director as saying, "It was widely known that Phil Graham was someone you could get help from."[1104]

Carl Bernstein, in his 1977 article for Rolling Stone, *The CIA and the Media*, recounts a briefing of members of the Senate Intelligence Committee by William B. Bader, former CIA intelligence officer, who was hired by the Senate to head the Senate investigation into CIA influence in the media. Bader told the Senate Committee members: "There is quite an incredible spread of relationships. You don't need to manipulate Time magazine, for example, because there are [Central Intelligence] Agency people at the management level."[1105] Bader's investigation provided a low-end estimate of 400 journalists who were CIA shills. Those 400+ "working journalists" included all areas of news reporting, from top management to reporters, editors, correspondents, and photographers. The CIA influence also reached into book publishers, trade publications, and newsletters.[1106] Bernstein explained that "Colby, Elder and Rogovin also implored individual members of the [Senate] committee to keep secret what the staff had found. 'There were a lot of representations that if this stuff got out some of the biggest names in journalism would get smeared,' said another source."[1107] Bernstein revealed that a senator who was the object of the Agency's lobbying later said: "From the CIA point of view this was the highest, most sensitive covert program of all.... It was a much larger part of the operational system than has been indicated."[1108]

Why is the CIA so adamant that the lid be kept on their influence over the media? Because the media is parroting propaganda that influences the very thoughts of the population. The media is used not to inform the people, but rather as a tool of control over the masses. The CIA subversion of the media has turned the idea of a free press into a charade. Thomas Jefferson once said that "the man who reads nothing at all is better educated than the man who reads nothing but newspapers." That statement is as true today as it was in Jefferson's day.

The CIA subversion of the media is effective because it is

generally not known. The general population of the United States believes that the major media outlets are providing them with unbiased news. In fact, nothing could be further from the truth. The false belief that the news media is independent and free from government influence is why the government propaganda spewed from the major media outlets is so effective. The major media outlets are part and parcel of a mass mind control operation. Michael Rivero explains:

> As terrible as it is to live in a nation where the press in known to be controlled by the government, at least one has the advantage of knowing the bias is present, and to adjust for it. In the United States of America, we are taught from birth that our press is free from such government meddling. This is an insidious lie about the very nature of the news institution in this country. One that allows the government to lie to us while denying the very fact of the lie itself.[1109]

One must understand that the CIA is simply a tool of mass disinformation under the control of the Jewish elite. Government is force; we must look behind the labels (e.g., CIA) to those who control that force. The CIA, as all agencies of government, are the means of control, but the true controllers are hidden behind the scenes, pulling the strings in secret.

Edward Bernays was named one of the 100 most influential Americans of the 20th century by Life magazine. Bernays was the nephew to the wicked founder of psychoanalysis, Sigmund Freud. Bernays was a Jewish expert on propaganda, who was in the service of the powerful elite. Bernays famously wrote in 1928:

> The conscious and intelligent manipulation of the organized habits and opinions of the masses is an

> important element in democratic society. Those who manipulate this unseen mechanism of society constitute an invisible government which is the true ruling power of our country. We are governed, our minds are molded, our tastes formed, our ideas suggested, largely by men we have never heard of. This is a logical result of the way in which our democratic society is organized. Vast numbers of human beings must cooperate in this manner if they are to live together as a smoothly functioning society. Our invisible governors are, in many cases, unaware of the identity of their fellow members in the inner cabinet. They govern us by their qualities of natural leadership, their ability to supply needed ideas and by their key position in the social structure. Whatever attitude one chooses toward this condition, it remains a fact that in almost every act of our daily lives, whether in the sphere of politics or business, in our social conduct or our ethical thinking, we are dominated by the relatively small number of persons a—trifling fraction of our hundred and twenty million—who understand the mental processes and social patterns of the masses. It is they who pull the wires which control the public mind.[1110]

Be mindful that Bernays was a proponent of propaganda as a tool to manipulate the behavior of the masses. He was in the service of the very "invisible governors" whom he praised for their "natural leadership" and "ability to supply the needed ideas." Bernays explained matter-of-factly that "it is they who pull the wires which control the public mind." Bernays felt that these governors were necessary for "the orderly functioning of our group life."[1111]

In order for the "invisible governors" to manipulate the population through propaganda, it is necessary for them to control the mass media. Evidence presented before United States Congress in 1917 proved that J.P. Morgan, who was the American agent for the international Jewish (Rothschild) banking interests, purchased control over the major media in the United States.[1112] That control continues today. According to some researchers, Jews own or control 96% of the world's major media outlets to include newspapers, television, movies, and other media.[1113]

On February 17, 1917, Congressman Oscar Callaway presented the following facts before the United States Congress which explained the successful efforts of J.P. Morgan and his cabal to control public opinion in order to involve the United States in World War I.

> Mr. CALLAWAY. Mr. Chairman, under unanimous consent, I insert in the record at this point a statement showing the newspaper combination, which explains their activity in this war matter, just discussed by the gentleman from Pennsylvania,
>
> [Mr. Moore]: In March, 1915, the J.P. Morgan interests, the steel, shipbuilding, and powder interests, and their subsidiary organizations, got together 12 men high up in the newspaper world and employed them to select the most influential newspapers in the United States and sufficient number of them to control generally the policy of the daily press of the United States.
>
> These 12 men worked the problem out by selecting 170 newspapers, and then began, by an elimination process, to retain only those necessary for the purpose of controlling the

general policy of the daily press throughout the country. They found it was only necessary to purchase the control of 25 of the greatest newspapers.

The 25 papers were agreed upon; emissaries were sent to purchase the policy, national and international, of these papers; an agreement was reached; the policy of the papers was bought, to be paid for by the month; an editor was furnished for each paper to properly supervise and edit information regarding the questions of preparedness, militarism, financial policies, and other things of national and international nature considered vital to the interest of the purchasers.

This contract is in existence at the present time, and it accounts for the news columns of the daily press of the country being filled with all sorts of preparedness argument and misrepresentations as to the present condition of the United States Army and Navy and the possibility and probability of the United States being attacked by foreign foes.

This policy also included the suppression of everything in opposition to the wishes of the interests served. The effectiveness of this scheme has been conclusively demonstrated by the character of stuff carried in the daily press throughout the country since March, 1915. They have resorted to anything necessary to commercialize public sentiment and sandbag the national congress into making extravagant and wasteful appropriations for the Army and Navy under the false pretense that it was necessary.

> Their stock argument is that it is "patriotism." They are playing on every prejudice and passion of the American people.[1114]

How successful have the Jews and their fellow conspirators been in controlling public knowledge and opinion? Read and weep over the sad truth as John Swinton, the former Chief of Staff for the New York Times, explains the state of the supposed free press in the United States in a speech before the New York Press Club in 1953.

> There is no such thing, at this date of the world's history, in America, as independent press. You know it and I know it. There is not one of you who dares to write your honest opinions, and if you did, you know beforehand that it would never appear in print. I am paid weekly for keeping my honest opinion out of the paper I am connected with. Others of you are paid similar salaries for similar things, and any of you who would be so foolish as to write honest opinions would be out on the streets looking for another job. If I allowed my honest opinions to appear in one issue of my paper, before twenty-four hours my occupation would be gone. The business of the journalists is to destroy the truth; to lie outright; to pervert; to vilify; to fawn at the feet of mammon, and to sell his country and his race for his daily bread. You know it and I know it and what folly is this toasting an independent press? We are the tools and vassals of rich men behind the scenes. We are the jumping jacks, they pull the strings and we dance. Our talents, our possibilities and our lives are all the property of other men. We are intellectual prostitutes.[1115]

Many may wonder just how this media control is leveraged by the CIA. The media coverage of the 9-11 attacks offers one example. The mass media was instrumental in the deception of the American people. The major media outlets played a key role in the deception of attacks of 9-11, by laying the blame on innocent patsies, concealing the treasonous conduct of government officials and the acts of war against the United States by agents of a foreign nation (Israel), and actively relaying a deceptive portrayal of what really took place on 9-11.

Prior to 9-11-01 there had never been a case where a steel structured high-rise tower has collapsed as a result of fire. All are aware that on 9-11-01 the twin towers at the World Trade Center (WTC) collapsed. Many are unaware, even today, that a third tower also collapsed that day (WTC 7). The government would have us believe that Towers 1,2, and 7, all collapsed as a result of fire. The collapse of World Trade Center Tower Seven on September 11, 2001, at 5:20 p.m. reveals the media involvement in the attacks.

Let's set the stage for the collapse of Tower 7. Tower Seven had 47 stories, which made it 600 feet tall. Aside from the WTC Twin Towers, it was one of the tallest buildings in lower Manhattan. It was separated from Twin Towers by a city block. Tower 7 was fully 355 feet away from the north face of the North Tower, and Tower 6 stood between it and the North Tower. The South Tower was even further away. Tower 7 was no closer to the Twin Towers than any of the other surrounding buildings which suffered only superficial damage. Tower 7 was not struck by either of the alleged planes (which we now know didn't exist) or significant debris from Towers One or Two, yet mysteriously two limited fires broke out in the building and it suddenly collapsed later in the evening of September 11.

Interestingly, the 23rd floor of Tower Seven received 15 million dollars worth of renovations, including independent and secure air and water supplies and bullet and bomb resistant windows designed to withstand 200 MPH winds. The renovation was intended to be used by the Mayor of New York, Rudolph Giuliani, as an emergency command center. Part of the reason for the command center was the 1993 bombing of the World Trade Center. The 23rd floor was ideal for a command center because it had an unobstructed view of the north sides of the Twin Towers, which since the 1993 bombing were considered prime terrorist targets. Tower 7 was a well-built 100% steel-framed skyscraper. It had a series of 58 columns ringing its perimeter, and a bundle of 25 columns in its core.[1116] Suspiciously, on the day of the 9/11 attacks, Mayor Giuliani and his entourage alleged that they set up shop in a different location and did not use the special bunker designed precisely for such an event. Giuliani stated that Tower 7 was evacuated immediately upon the first plane impact.

> The command center was at 7 World Trade Center, which is the building that was north of the World Trade Center that went down in the afternoon. It went down maybe 4 or 5 o'clock in the afternoon. **But from the very moment that the first plane hit, 7 World Trade Center was evacuated.**[1117]

The pictures below illustrate just one example of the active involvement of the mass media in spreading disinformation in the aftermath of the attacks on 9/11. The pictures are frames from a BBC live broadcast on 9-11-01 that announced the collapse of WTC building 7 (also known as the Solomon Brothers Building), in the late afternoon on the day of the 9/11 attacks.[1118] At the time of the broadcast, the World Trade Center (WTC) Towers 1 and 2 had already collapsed earlier that morning.

The BBC announcer, Phil Hayton, was announcing matter-of-factly that the Solomon Brothers Building (WTC 7) had collapsed (20 minutes before the actual collapse). Hayton repeats the announcement again at 5:07 p.m. EST. At approximately 5:07 p.m. EST, the newscaster states: "Now more on the latest building collapse in New York, you might have heard a few moments ago us talking about the Solomon Brothers Building collapsing, and indeed it has. . . . Jane what more can you tell us about the Solomon Brothers Building and its collapse?" The curious thing about the broadcast video is that the building did not collapse until 13 minutes later, at approximately 5:20 p.m. EST.

After the announcer tells the audience that Solomon Brothers Building (WTC 7) had collapsed, he goes to a live feed to Jane Standley on the scene in New York. Initially her head blocks WTC 7, but she later moves and shows the audience the smoke rising from the collapse site of towers 1 and 2. When she does that you can see WTC 7 clearly standing in the background. When she resumes her appearance on the screen you can see WTC 7 in the background over her left shoulder and to the left of her left ear.

Figure 187: BBC broadcasting the collapse of the 47 story WTC 7 building (Solomon Brothers Building) 13 minutes before its collapse.

Someone at the BBC jumped the gun and had the script read too early. Apparently, the conspirators got discombobulated over the time difference between London and the New York and went with their orchestrated story of the WTC 7 collapse before it actually happened. This is evidence that 9/11 was scripted in advance, and that the major media outlets were part of the conspiracy.

The BBC has received much attention and a lot of inquiries over their 9-11-2001 broadcast predicting the collapse of WTC Tower 7. It took the BBC until February 27, 2007, to respond to the inquiries, by publishing a short response from Richard Porter.[1119] Think about that. This is a news organization that is supposed to be in the business of gathering news and rapidly disseminating an accurate account of events to the public. It took them years to respond to inquiries about their prediction of the collapse of Tower 7. That is the behavior that is consistent with someone who is not quite sure what to say and wants to take their time to ensure that they do not say anything that might later turn out to be incriminating.

The problem in which the BBC finds itself is that the circumstances of the pre-reporting on 9-11-2001 are so compellingly incriminating that no explanation will mitigate its guilt. They have, so to speak, been caught red-handed. The only thing that they can do is either admit guilt, which they are not going to do, or try to cover their tracks with lies, which is what they have done. The BBC response is quite revealing and incriminating. The most notable thing in its response is that the BBC simply makes assertions without any supporting evidence, as though a denial is sufficient. The BBC stated:

> We're not part of a conspiracy. Nobody told us what to say or do on September 11th. We didn't get told in advance that buildings were going to fall down. We didn't receive press releases or

scripts in advance of events happening.[1120]

The BBC states that they are not part of a conspiracy and that nobody told them in advance that the buildings would fall down on September 11th. However, if they broadcast an event that nobody could possibly foresee, and then the event takes place exactly as they pre-reported, that clearly indicates that they had prior knowledge of the event. Clearly, the BBC had an advance script of the 9/11 attacks. That makes them accomplices.

Their pre-reporting was more than a prediction of the event, it was a reporting of the event before it happened. If the BBC had said that Tower 7 "would soon collapse" they would have some explaining to do. People would want to know how they knew it was going to collapse. However, the BBC did not say Tower 7 would "soon collapse;" the BBC announced that Tower 7 had already collapsed. That is doubly suspicious, because it proves that they had a script about the collapse of Tower 7 and were reading from the script. The BBC made a mistake by reading the script too early.

The BBC premature reporting of Tower 7's collapse is more than a mere prediction of an event; it is a reporting of an unforeseeable event that actually occurred just as it was reported, but the report was aired before the unforeseeable event occurred. The only way that could happen is if the BBC had a script of the planned events of 9-11-2001 ahead of time. The only source for such a script had to be from the perpetrators. That puts the BBC in the middle of the 9/11 conspiracy and cover-up. The BBC denial does not address that point at all.

When Phil Hayton, the BBC announcer who is seen on screen speaking with Jane Standley, was apprised of the suspicious collapse of WTC and the BBC broadcast suggesting foreknowledge by the BBC, he did not even try to argue the point. Hayton had to agree that it seems there is a conspiracy. He said:

"I sense that you think there's a conspiracy here-but you might be right."[1121]

Figure 188: Later in the BBC broadcast there is a studio shot live from England, with Phil Hayton sitting at his studio desk, showing the live feed from New York of Jane Standley with WTC 7 Still standing behind her to the left of her left ear. However, the screen script announces that the Solomon Brothers Building (WTC 7) "has also collapsed."

The official BBC response through Richard Porter gets worse:

> We no longer have the original tapes of our 9/11 coverage (for reasons of cock-up, not conspiracy).[1122]

A cock up? That is quite simply incredible. The BBC, which is known for its meticulous record keeping and storage of news archives going back over 50 years, inexplicably loses live news footage documenting the crime of the century! That explanation does not pass the smell test. How could the news agency of record for Britain, who is by law required to keep records of broadcasts, lose those records? The answer is that they did not lose them; the records are clear evidence that support a finding that the BBC was part of the 9/11 conspiracy. This entire event, with an examination of the events surrounded the supposed "missing" original tapes is recounted in this author's book, *9/11-Enemies Foreign and Domestic*. While the BBC is a British media outlet, *9/11-Enemies Foreign and Domestic* documents clear evidence of the U.S. mass media's active involvement as accessaries in spreading disinformation about the 9/11 attacks.

The 9/11 attacks were false flag attacks for a purpose. Vice President Dick Cheney, Defense Secretary Rumsfeld, and Deputy Defense Secretary Wolfowitz are listed as signatories on the June 3, 1997, Statement of Principles for the Project for a New American Century (PNAC).[1123] PNAC was a Zionist controlled think tank. PNAC ceased operations in 2006. Not to worry, the Zionists behind PNAC have reconstituted a new Zionist think tank to replace PNAC called Foundation for Defense of Democracies (FDD). As did PNAC before it, FDD advocates that the U.S. fight wars on behalf of Israel. For example, the neoconservative warmongers at FDD have issued a report calling for regime change in Syria and for the U.S. to use military strikes with missiles and warplanes to destroy Iran's nuclear capabilities. PNAC issued a paper (Rebuilding America's Defenses) on September 2000, one year before the 9-11-2001 attacks. That document advocated transforming the United States politically and militarily.

The September 2000 PNAC strategy paper advocated for the United States to increase its defense budget, transform the military to use advanced technologies, redefine the military to perform constabulary missions throughout the world, and maintain sufficient forces to fight and win multiple large scale wars. The September 2000 paper stated: "Further, the process of transformation, even if it brings revolutionary change, is likely to be a long one, absent some catastrophic and catalyzing event - like a new Pearl Harbor."[1124] The 9/11 attacks would certainly qualify as a new Pearl Harbor.

The strange inaction of Wolfowitz and Rumsfeld on 9-11-2001 is revealing when it is viewed in the light of the need for a "new Pearl Harbor" expressed by their September 2000 PNAC report. The 9/11 attacks certainly brought about a political and military "revolutionary change" as envisioned in their report. The proof is in the pudding. The United States has in fact been transformed politically and militarily as a result of the 9/11 attacks.

The 9/11 attacks reveal who is the real power behind the U.S. Government. All one needs to do is to look to who benefitted the most from the 9/11 attacks. The answer is obvious - Israel. The agents of Israel in both the government and the media worked together to accomplish the dirty deed.

Those working for a New World Order under their god, Satan, have convinced the world that the earth is a spinning globe careening through the empty vacuum of space. That is the mother of all deceptions, which is the cornerstone of the New World Order. Having pulled off that deception, all other deceptions, like the 9/11 attacks, are child's play for them. Without the heliocentric deception, which drives God from the minds of men, they could not hope to hoodwink the masses as they secretly work out their plans for world domination though their implementation of the Hegelian dialectic.

Some will argue that it does not make a difference whether the earth is a sphere of flat, so why make an issue of it? The reason that Satan and his minions need people to believe that the earth is a sphere careening through space is that it is the foundation upon which they can "prove" that there is no God. If there is no God, then the only authority to which people should answer on earth is government. The moral laws of God in the Bible then become irrelevant. All sin is justified if the government authorizes it.

People are imbued with a conditioned response to trust governmental authority. It is hard for people to grasp the scope and depth of the world conspiracy against God and man, because, as explained by Former FBI Director J. Edgar Hoover, the average person is simply unable to believe that such a monstrous conspiracy exists. If they can deceive the world about the very nature of God's creation, and conceal the glory of God by convincing people that we live on a spinning ball careening through space, all other deceptions are a piece of cake. The gullible masses will go along with what the government authorities say, because docile obedience is central to the conditioning process. The godless premise of heliocentrism is the central doctrine through which Satan is able to control the minds of the masses.

For the distrustful minority who actually examine the evidence, though, the conspiracy is revealed. As we have seen, the conspiracy is revealed when we examine the evidence in the attacks on 9/11. The Kennedy assassination is another example where the evidence proves that very powerful forces have formed a shadow government very different from that which the people think exists. Just as the real world is flat, so also the real government is not of the people, by the people, and for the people. In reality a criminal conspiracy has taken control of the very reigns of the U.S. Government. The uninformed will reject such statements (as they are conditioned to do), but it can be proven to them if they would look at the facts (which they are conditioned not to do). Proof of that statement is found when examining the evidence from the Assassination of the 36th President of the United States, John F. Kennedy.

The Satanic cabal so completely controlled the U.S. government from behind the scenes, they were able to brazenly assassinate President Kennedy and then use the very organs of government to conceal their involvement. In order to assure the success of the assassination of President Kennedy, the conspirators had to control those who were tasked with protecting the President. The U.S. Secret Service did not provide agents riding on the rear of the presidential limousine. In fact the two agents initially assigned to ride on the rear of the limousine were ordered off the limousine as it was leaving Love Field in Dallas.[1125] In addition, the Dallas motorcycle officers, rather than flanking the sides of the limousine, as was done in the previous motorcades in San Antonio, Houston, and Fort Worth, were ordered to ride behind the limousine.[1126] Those two decisions gave the assassins unobstructed shots at President Kennedy. According to at least ten eyewitnesses who were very close to the limousine at the time of the assassination, upon hearing the first shot fired the Secret Service Agent driving the limousine, William Greer, slowed the limousine to a complete or near complete stop.[1127] The limousine only accelerated out of Dealey Plaza after President Kennedy had been fatally shot in the head. Finally, the agent in charge of the President's security detail in the follow up vehicle, Emory Roberts, ordered the other Secret Service Agents not to move when the first shots were heard, thus ensuring that Kennedy would remain an unprotected sitting duck.[1128]

On November 17, 1963 the Director of the FBI, J. Edgar Hoover, notified all FBI field divisions via teletype that there would be an attempt to assassinate President Kennedy during his November 22-23, 1963 trip to Dallas, Texas.[1129] The FBI, however, did not take any action or forward the information to any other state or federal agency, including the Secret Service, who as the bodyguards for the President should have been notified immediately.[1130] Such notification would ordinarily have been forwarded as a matter of course on such matters. One could speculate that there must have been a back channel off record communication between Hoover and the Special Agents in Charge in the FBI field divisions instructing them not to take any action on the teletype.

Why then would the teletype be sent out in the first place if no action was contemplated? It can reasonably be inferred that Hoover sent the teletype as a sort of cover story in case the assassination failed and Hoover's complicity in the plot was uncovered. He could then point to the teletype as evidence that he took appropriate efforts to warn of the plans to assassinate the President. After the assassination succeeded there was no need for the teletype and all FBI field divisions destroyed their copies and all record of it. The day of the assassination, the night clerk for the New Orleans Division, William S. Walter, immediately understood the significance of the teletype and made a longhand copy of the teletype before it was destroyed from the FBI files.[1131] What many do not know is that Lee Harvey Oswald was an FBI informant, whose controlling agent in Dallas was James Hosty. Hosty's name and unlisted home phone number was found in Oswald's address book.[1132] However, when the FBI transcribed a list of entries made in Oswald's address book for the Warren Commission, they omitted the Hosty entry.[1133] It is likely that the specific warning about the assassination was given to the FBI by Oswald himself.

On the day of his arrest Lee Harvey Oswald was given a nitrate test, which showed that he had not fired a rifle of any kind within the previous 24 hours.[1134] The Dallas police department and the federal government kept that fact a secret for ten months after the assassination.[1135] In addition, Oswald was interrogated for 12 hours over a two day period. There is no record or report of any kind prepared by the Dallas Police Department or the FBI on what Oswald said during his 12 hours of interrogation. Statements are routinely memorialized as a matter of course during or after interrogation even in minor misdemeanor cases. In this case the law enforcement officials were interrogating the alleged assassin of the President of the United States. It is incredible that there is no report of what Oswald said during that interrogation. The conspirators realized that they could not allow Oswald to remain alive for a trial during which he would be given an opportunity to tell his story, and so Jack Ruby was pressed into service to silence Oswald for good. We will never know what Oswald said during his interrogation.

Julia Ann Mercer saw Jack Ruby (his real name was Rubenstein), who two days later murdered Lee Harvey Oswald, drop off a man from his car who was carrying a rifle in a case near the grassy knoll at Dealey Plaza approximately one hour before the arrival of the presidential motorcade.[1136] She stated that three uniformed officers standing on the overpass near the grassy knoll showed no curiosity regarding the man with the rifle.[1137] She was shown a mug shot of Ruby by the FBI after the assassination, and she identified Ruby as the person she saw at the grassy knoll.[1138] That suggests that Ruby had already been identified as an accomplice by the FBI when Mercer told what she knew.[1139] She identified Ruby as an accomplice the day before Ruby shot Oswald.[1140] Furthermore, the Dallas Police Department allowed Ruby access to the police basement to kill Oswald after he had been identified as an accomplice in the Kennedy assassination. Mercer informed both the FBI and the Dallas County Sheriff's Department what she saw, but the FBI and the sheriff's department later altered her statement to make it appear that Mercer could not identify Ruby. The sheriff's department went a step further and forged her signature on a phony affidavit.[1141] The FBI has never denied the alteration of Mercer's statement, nor have they ever explained it. Mercer was never called as a witness before the Warren Commission.

The great majority of witnesses at Dealey Plaza heard

repeated fire coming from the grassy knoll. Three men were arrested by the Dallas Police after being chased from that area.[1142] There is no record of their mug shots, fingerprints, or names.[1143]

The assassination of President Kennedy could not have happened without the cooperation of the Secret Service hierarchy. The picture sequence below are screen captures of frames from a video taken at Love Field in Dallas after the arrival of President Kennedy on the day he was assassinated. Notice in frame #1 the two Secret Service Agents are running at either bumper of he Presidential Limousine, which was standard operating procedure. In fact the limousine has specially designed handles and foot platforms for the Secret Service Agents to ride on the back of the car as it is travels at procession speeds.

1

Emory Roberts Orders Agents Off

2

SS Agent is Perplexed by Order

3

SS Agent Waives Arms in Exasperation

4

Puzzled Agent is Left Behind, Arms Up

You can see the Secret Service Supervisor, Emory Roberts, stand up in the trailing car and orders the two agents to leave their assigned posts. You can see the two agents turning their heads in response to the Roberts voice command. In frame #2, you can see that agent who was running along side the right bumper is completely perplexed by the order. He is waiving his hands as if to ask "what is going on?" He continues to waive his hands, and you can see in frame #3 his hands are out by his side in exasperation. In Frame #4 the agent is standing with his arms up in the air; he is totally perplexed by what is going on, as the motorcade leaves him behind. The agent clearly did not understand the reason for the order. He no doubt had been trained to stay along side the presidential limousine and had done so in the past. The order from Roberts was contrary to the standard operating procedure of the Secret Service. Obviously, Emory Roberts understood what was to happen, and he wanted to make sure the assassins had a clear shot at an unguarded President. What is interesting is that once Roberts saw that the agents had left their posts per his instructions, he sat back down in the trailing car. His mission was accomplished.

According to the Warren Commission, Lee Harvey Oswald, from his perch on the sixth floor of the Texas School Book Depository, shot Kennedy in the back of his head. That theory is a complete and utter deception by the Warren Commission. The physical evidence contradicts that cover story, because the wound in the back of President Kennedy's head was an exit wound. That means that Kennedy was shot in the head by a shooter in front of him. It would be physically impossible for the shot that struck Kennedy in the head to have been shot by Oswald from Oswald's fictional position on the sixth floor of the Texas School Book Depository, which was behind Kennedy.

When a bullet enters a body it makes a small entrance wound approximately the diameter of the bullet; depending on the circumstances, the bullet hole could be slightly larger or slightly smaller than the bullet. A hole smaller than the bullet sometimes results at the entrance sight because of the elasticity of skin, which allows the skin at the edges of the wound to stretch slightly as the bullet enters and then spring back after the bullet passes through the hole in the skin. The difference in diameter from the bullet and the entrance hole left by the bullet will typically be very slight, not more than a couple of millimeters.

It is a fundamental principle of wound ballistics that when a bullet goes through a body and exits at the other end, it generally makes a significantly larger exit wound.[1144] According to the Mercer University School of Medicine: "Exit wounds are generally larger than entrance wounds, due to the fact that the bullet has expanded or tumbled on its axis."[1145] This phenomenon is even more apparent when a high velocity rifle bullet strikes a human skull. That is because the human skull contains the soft tissue of the brain surrounded by the inelastic hard bone of the skull. According to the Department of Forensic Medicine at the University of Dundee: "High velocity gunshot wounds of the head produce bursting injuries of the skull due to temporary cavity formation."[1146] The bursting of the skull is never at the entrance site, it is always at and contiguous to the exit site of the bullet. The exit wound to the back of President Kennedy's head was approximately the size of a baseball.

Dr. Charles Crenshaw's observations prove that it was impossible for Oswald to have shot Kennedy in the head from the sixth floor of the Texas School Book Depository. Dr. Crenshaw was one of the attending physicians who examined Kennedy in the emergency room at Parkland Hospital on the day of the assassination. He saw a bullet entrance wound to the front of Kennedy's throat area. He further stated that from the damage he saw to President Kennedy's head, there was no doubt in his mind that another bullet had entered President Kennedy's head through the front, it passed through his cranium and obliterating part of the temporal and all of the parietal and occipital lobes. The bullet lacerated the cerebellum and blew a large hole through the back of Kennedy's head. Dr. Crenshaw stated that there was a large hole, the size of his fist (which he estimated to be 2 ½ inches in diameter) in the right rear parietal, occipital area of Kennedy's head.[1147] He stated that the large exit wound in the rear of Kennedy's head was a cavernous hole; the scull and scalp had been blown away, revealing brain matter, which was hanging out from it. Dr. Crenshaw explained:

> Two wounds were visible. There was a small, round opening in the front of the midline of the throat. This became the site of Dr. Malcolm Perry's tracheotomy incision. In the occipito-parietal region at the right rear of the head, there was an avulsive wound nearly as large as a fist. Bone, scalp, and hair were missing in the region, and brain tissue, including much of the cerebellum, was hanging from the opening. I considered the throat wound to be an entrance wound and the large head wound to be an exit wound. Along with many of my Parkland colleagues, I believed at the time that President Kennedy had been hit twice from the front.[1148]

The fact that "much of the cerebellum" was hanging from the wound indicates that the wound was low in the head. The cerebellum is the lower part of the brain located in the back of the head with a texture and appearance that is quite distinct from the larger part of the brain, called the cerebrum. Dr. Crenshaw's observations were confirmed by the Chief of Neurosurgery at Parkland Hospital, Dr. William Kemp Clark, who identified the large gaping hole in the skull bone at the right rear of the head as having both cerebellum and cerebrum exuding from it.[1149] The eyewitness testimony of Dr. Clark and Dr. Crenshaw that there was cerebellum exuding from the large open wound in the back of President Kennedy's head confirms that the exit wound in

President Kennedy's head was low in the back of his head. That indicates that the head shot came from in front of Kennedy, and precludes the possibility that the head shot was fired by Lee Harvey Oswald from the sixth floor of the Texas School Book Depository, which was behind Kennedy.

Dr. Crenshaw and Dr. Clark are just two of many physicians who witnessed the exit wound in the back of Kennedy's head. The conspirators knew that this group of expert and credible witnesses was a serious problem. They also knew that a picture is worth a thousand words and so they dummied up phony autopsy photos that purported to show a small entrance wound in the back of Kennedy's head. They further created counterfeit x-rays showing an exit wound in the front of his head. The fraudulent pictures and x-rays were for public consumption to deceive those not aware of the eyewitness accounts of the wounds and to gather duped (or cooperating) experts who would base their opinions of a shot from the rear on the phony photographs and x-rays.

Robert Knudsen was on the White House Photography staff during the Kennedy administration. Knudsen had been the photographer for the autopsy of President Kennedy.[1150] His family recalls him being summoned to Andrews Air Force Base after the Kennedy assassination to photograph the autopsy. Knudsen never spoke about the details of the autopsy with his family, as he was sworn to secrecy. That silence was broken in the 1970s, after he was summoned to testify before the House Select Committee on Assassinations, which was reinvestigating the assassination of President Kennedy. During his testimony, Knudsen was shown the official autopsy photographs. Attorney Jacob Hornberger reported:

> According to his wife and children, Knudsen returned home and indicated to his family that autopsy photographs he had been shown during his testimony were fraudulent. He said that there were clearly some shenanigans going on and that

> if anything were ever to blow up, he wanted his family to know that he had nothing to do with it. Protecting his integrity within his family was obviously extremely important to Robert Knudsen.[1151]

It turned out later that the photographer of record for President Kennedy's autopsy was not Knudsen, but rather John T. Stringer, a highly respected autopsy photographer for the U.S. Navy who taught photography at the Bethesda Naval Medical School.[1152] Allegedly, Knudsen was not even at Kennedy's autopsy. But according to Jacob Hornberger, "[n]o one has ever questioned the integrity, veracity, or competence of Robert Knudsen. He was highly respected, both personally and professionally. It would [be] difficult to find a more credible witness than Robert Knudsen." In 1977 Knudsen publically explained how he photographed the Kennedy autopsy in the nationally circulated *Popular Photography* magazine. He had no motive to lie. And if he was lying, it could easily be proven, but it never was. Indeed, his account of photographing the Kennedy autopsy was verified by his family's contemporaneous account. It seems he must have photographed the autopsy; otherwise, the House Select Committee on Assassination would have no reason to call him as a witness. The only reasonable inference is that there were two separate and independent autopsies performed on President Kennedy's body that involved two separate photographers.

Saundra Spencer was a U.S. Navy petty officer who worked in the U.S. Navy's photography lab in Washington, D.C., in 1963. In the 1990's she was summoned to testify before the Assassination Records Review Board (ARRB). The reason for her testimony is that Spencer had been asked, on a top-secret basis during the weekend of the assassination of President Kennedy, to develop autopsy photographs of President Kennedy's body. Jacob Hornberger explains that "[p]ursuant to the culture of secrecy and classified information in the military, Spencer kept her role in developing those autopsy photographs secret for some 30 years, until she was summoned to testify before the ARRB."[1153] Hornberger further reveals:

> During her testimony, Spencer was shown the official autopsy photographs of the President's body. After carefully examining them, she testified directly and unequivocally that the autopsy photographs in the official record were not the ones that she developed on the weekend of the assassination. She stated that the autopsy photographs she developed showed a large exit-sized wound in the back of President Kennedy's head, which matched what treating physicians at Parkland Hospital had stated. The photographs in the official record show the back of Kennedy's head to be fully intact. A large exit-sized wound in the back of the president's head would, of course, imply a shot fired from the president's front.
>
> After Spencer's testimony before the ARRB, no one came forward to challenge, question, or dispute the truthfulness and accuracy of her testimony.
>
> What are we to make of Knudsen [the photographer] and Spencer [who developed the photographs]?
>
> There are two conclusions that can be reasonably drawn: First, the official autopsy of President Kennedy, which was carried out by the U.S. military, was fraudulent and, second, there is no conceivable innocent explanation for having

carried out a fraudulent autopsy.[1154]

In order to keep the deception going, the U.S. House Committee on Assassinations, which convened hearings between 1976 and 1979, refused to even show the altered x-rays and photos to the doctors who were eyewitnesses to Kennedy's wounds. The committee knew that the doctors would very credibly testify that the pictures and x-rays were forgeries. There is no convincing way to impeach the credibility of doctors who remember clearly viewing the cerebellum hanging from the gaping wound in the back of Kennedy's head. In the face of such testimony, the official autopsy photographs and x-rays would be proven to be frauds. The committee could not allow that to happen. So they kept the eyewitness doctors away from the phony autopsy photographs and x-rays.

The picture sequence below is from the book *Murder in Dealey Plaza*[1155] and depicts eighteen (18) eyewitnesses explaining the size and location of the large gaping wound in President Kennedy's head. Each picture is a frame from video interviews of each of the witnesses showing the witnesses describing a large wound suffered by Kennedy to the rear of his head. All 18 witnesses remembered the large wound in substantially the same place in the back of President Kennedy's head. In order to have an exit wound in the rear, the shooter must have shot Kennedy from the front. That means that Oswald could not have been the shooter, because Oswald could not have struck Kennedy with a bullet in the front of his head if Oswald were shooting from the sixth floor of the Texas School Book Depository, which was behind Kennedy.

Beverly Oliver: "The whole back of his head went flying out the back of the car."

Phillip Willis: "It took the back of his head off."

Marilyn Willis: "A red 'halo' [was] coming out the back of his head."

Ed Hoffman: "The rear of his head was gone, blasted outward."

Dr. Robert McClelland: "It was in the right back part of the head—very large . . . a portion of the cerebellum fell out on the table while we were doing the tracheotomy."

Dr. Paul Peters: ". . . right there, occipital parietal."

Dr. Kenneth Salyer: "This wound extended into the parietal area."

Dr. Charles Carrico: "There was a large—quite a large—defect about here [pointing] on his skull."

Dr. Rirchard Delaney: "It was up in this area."

Dr. Charles Crenshaw: "The wound was the size of a baseball."

Dr. Ronald Jones: "My impression was there was a wound in this area of the head." When shown the faked autopsy X-ray, Parkland Hospital's Dr. Jones said, "There was no damage to the face that was visible. . . The X-rays are incompatible with the photographs, which show no injury to the face."

Nurse Audrey Bell: "There was a massive wound at the back of his head."

Theron Ward: "It was right back here."

Aubrey Rike: "You could feel the sharp edges of the bone at the edge of the hole in the back of his head."

Frank O'Neill: ". . . a massive wound in the right rear."

Jarrol Custer: "From the top of his head, almost to the base of the skull, you could see where that part was gone."

Paul O'Connor: "[There was] an open area all the way across into the rear of the brain."

Floyd Reibe: ". . . a big gaping hole in the back of the head."

The evidence that there was an exit wound in the back of Kennedy's head was known by the Warren Commission at the time of the commission hearings. Dr. Clark gave sworn testimony before Arlen Specter, assistant counsel for the Warren Commission, during which Dr. Clark stated: "I then examined the wound in the back of the President's head. This was a large, gaping wound in the right posterior part, with cerebral and cerebellar tissue being damaged and exposed."[1156] As a Neurosurgeon, Dr. Clark certainly knows what "cerebellar tissue" looks like. In order to have "cerebellar tissue" exuding from the large gaping exit wound, the wound must have been low in the back of Kennedy's head, where the cerebellum is located. Yet the Warren Commission allowed into evidence altered and forged autopsy pictures and x-rays that concealed the fact that there was a large exit wound in the back of Kennedy's head. The Warren Commission pretended to conclude that Oswald acted alone and shot Kennedy from the rear. They knew full well that was not true.

The Warren Commission was on a mission to steer the investigation away from the true culprits and toward the innocent patsy, Lee Harvey Oswald. The Warren Commission engaged in a cover-up; that is clear. Their public mission was to solve the mystery of the assassination of the President. Their secret mission was to pin the assassination on Oswald to the exclusion of anyone else. The members of the Warren Commission were accessories after the fact to the assassination of President Kennedy. The powerful forces behind the assassination arranged to have a commission of insiders who would conceal the truth from the American people under the guise of a blue ribbon panel of investigators. The 9-11 Commission of today has served the exact same function as the Warren Commission of yesteryear, to cover up the identities of the true perpetrators.

There is further evidence of the impossibility of Oswald being President Kennedy's assassin. The photograph taken by James Altgen proves that it was impossible for Oswald to have shot President Kennedy from the rear or from anywhere else.

The photograph below is a photograph taken by AP photographer James Altgen immediately after President Kennedy was hit with the first bullet.[1157] The photo appeared on pages 24 and 25 of the December 14, 1963 issue of Post. You can see Kennedy's left hand through the windshield as he grabs his throat. Immediately in front of Kennedy and to his right you can see John Connally's profile as he turns to his right. Notice in the fourth car back the Secret Service Agents in Vice President Lyndon Johnson's Secret Service protective detail have opened the door to their car and are in the process of getting out to come to the protection of LBJ as the shots are heard. That indicates that the motorcade must have come to a complete stop, which is consistent with the numerous eyewitnesses who saw the motorcade come to a complete stop. It is much easier to shoot a stationary target, and the conspirators planned to have the Kennedy limousine stopped. Oddly, Kennedy's Secret Service detail had not yet even made a move to protect Kennedy after the first shot. The most important aspect of the photo is that in the background Altgen also photographs a person who appears to be Lee Harvey Oswald standing in the first-floor doorway of the Texas School Book Depository at the very moment Kennedy was shot.

James Altgen's Photo Showing Lee Harvey Oswald in the Texas School Book Depository First Floor Doorway Watching as President Kennedy is Being Shot

Photograph #1 on the left below is a blowup of Oswald in the doorway. The man in the doorway appears to be wearing the same clothing Oswald was wearing on that very day, the day Oswald was arrested, and the clothing is worn by him in exactly the same fashion as the man in the doorway: unbuttoned to the mid-section. The photograph #2 on the right below is a photograph taken of Oswald shortly after his arrest, the day President Kennedy was shot. Notice, he is wearing the same tweed shirt and tee-shirt combination, with the shirt halfway open. Reportedly, color film photographs taken from a different angle that day, show the color of the shirt worn by the man in the doorway to be orange-brown. When Oswald was arrested, he was wearing the identical orange-brown shirt with missing buttons and tee-shirt underneath. Reportedly, Oswald told the Dallas Police that he was standing in the doorway of the Texas Book Depository when Kennedy was shot, just as depicted in the photo. Oswald clearly has no gun in his hands and he is not on the sixth floor of the school depository. He could not, therefore, have been the shooter as claimed by the Warren Commission. It appears that Oswald was just what he said he was: a "patsy." The conspirators had to kill Oswald, because they knew that the evidence against him would never stand up to courtroom scrutiny. The Altgen Photograph would be impossible to refute.

1

Blowup of Oswald in Doorway of the TSBD

2

Oswald as He Appeared Shortly After His Arrest

According to the official Warren Commission Report, Oswald was supposed to be shooting at Kennedy from the 6th Floor of the Texas School Book Depository. In order to cover for the fact that James Altgen photographed an unarmed Oswald in the first floor of the School Book Depository at the time Kennedy was being shot, the Warren Commission identified the individual in the doorway as Billy Lovelady, a fellow Book Depository employee who looked similar to Oswald. Lovelady died on January 1979 from heart failure (the later House Committee Report on Assassinations was issued in March 1979, having begun in 1976). Lovelady identified himself as the person in the Altgens' photo to the Warren Commission.

One problem with Lovelady's claim is that he testified before the Warren Commission that he was sitting on the steps to the Texas Book Depository eating his lunch when the Presidential motorcade drove by.[1158] He stated that two fellow employees, Bill Shelley and Sarah Stanton, were "right behind" him.[1159] As one can see in the photo, nobody is standing behind the person in the doorway and that person is clearly standing in the doorway, not sitting on the steps.

Lovelady stated that he then heard what he thought at the time were firecrackers.[1160] He did not know that President Kennedy had been shot until a coworker, Gloria Cavalry, ran up and told him.[1161] If it was Lovelady in the entranceway, he would not have had to been told that the President had been shot, because the man in the entranceway had a perfect line of site to the events. As one can see, the person in the entranceway is looking at Kennedy at the moment the first shot hit Kennedy; he would have seen the final shot explode President Kennedy's head; he would have seen Mrs. Kennedy crawling on the trunk of the limousine grabbing for the back of his skull and brains and the Secret Service agent jump in the limousine before it sped off. The whole purpose for people to be watching the motorcade was to see the President, and the man in the entranceway was no exception; he was looking directly at President Kennedy as he was being shot.

Furthermore, if Lovelady were the man in the doorway, he would have testified to what he saw, but he did not, because he was sitting down on the steps eating his lunch; he was only able to testify to what he heard. That point is made clear by the question not asked by the Warren Commission counsel. The Counsel asked Lovelady what he heard, but he did not ask Lovelady what he saw. Why didn't he ask Lovelady such a basic question like "what did you see?" Because he knew that Lovelady would testify that he did not see anything at the time of the shooting, because he was sitting down eating his lunch. If he did not see anything, then he could not have been the man in the doorway, because the man in the doorway is looking directly at the limousine at the time the first shot hits Kennedy! If Lovelady answered that he did not see anything, it would have established that it was Oswald in the doorway, and that was something the Warren Commission did not want established.

Another problem with that identification is that Lovelady

stated on other occasions that he wore a red and white vertically striped short sleeved shirt that day. Other photos taken on November 22, 1963 purportedly show Lovelady wearing a large squared, bold patterned, red and blue plaid shirt. The man in the doorway could not be Lovelady, because it does not appear that the man is wearing either a short sleeved striped or a plaid patterned shirt.

The Warren Commission made its judgement that Oswald was a lone assassin without even examining the autopsy photographs or x-rays.[1162] The pathologist in charge of the autopsy at Bethesda Naval Hospital burned his notes after he heard that Oswald was killed by Ruby.[1163]

The above facts reveal the complicity of the Dallas Sheriff's Department, the Dallas Police Department, the FBI, and the Secret Service in the assassination of President Kennedy. Other facts reveal the involvement of the CIA and the U.S. Military. The major media, controlled by Jews, have also been willing accomplices in the treasonous cover-up of the John Kennedy assassination. NBC even went so far as to attempt to bribe a witness involved in New Orleans District Attorney Jim Garrison's prosecution of Clay Shaw.[1164] NBC's involvement in the attempt to derail Jim Garrison's investigation reached to the very top of NBC. The president of NBC had discussions with the president of Equitable Insurance Company, the employer of the witness in question, Perry Russo. NBC tried to persuade Russo to go on national television and falsely say "I am sorry for what I said because I lied, some of what I said was true but I was doctored by the District Attorney's staff into testifying like I did."[1165] Perry stated that James Phelan of the Saturday Evening Post told him that he was working hand-in-hand with NBC reporter Walter Sheridan and that they were going to destroy Jim Garrison and his probe into the Kennedy assassination.[1166]

Jim Garrison was approached by John J. King, who at the

time used the alias John Miller.[1167] King (Miller) offered Garrison a federal judgeship if he would drop the investigation into Clay Shaw's involvement in the Kennedy assassination.[1168] King made clear to Garrison that he was in a position to guarantee his immediate appointment to the federal bench.[1169] Apparently, King was speaking for others, because in order to become a federal judge he must be appointed by the President of the United States with the advice and consent of the Senate. Some very powerful interests were behind King's offer. Subsequent investigation revealed that King was a wealthy oilman from Denver, Colorado who was involved in lucrative Israeli oil projects.[1170] Interestingly, one of King's business partners was Bernie Cornfield, who was the protegee and front man for Rabbi Tibor Rosenbaum.[1171] Rosenbaum was the founder of the Bank De Credit International and the central financier behind Permindex.[1172] Permindex was a joint CIA - Mossad front which played a prominent role in facilitating the Kennedy assassination. Clay Shaw was one of the Permindex directors.[1173] The fact that King could guarantee a federal judgeship to Jim Garrison in return for dumping his investigation of Clay Shaw gives one some idea of the powerful interests that were behind the cover-up of the Kennedy assassination. A November 1963 FBI teletype identified John J. King as a wealthy Dallas businessman who was a close friend to Jack Ruby.[1174] Jack Ruby (whose real name was Jacob L. Rubenstein) was a Jewish gangster with connections to the CIA, the FBI, the Israeli Mossad, and the Dallas Police Department. His connections inside the Dallas Police Department allowed him access to kill Lee Harvey Oswald.

Clay Shaw, in a 1967 taped interview, stated the following:

> It seems to me that here you are faced with an either/or situation. You must either believe that the Warren Commission is substantially correct and that a lone assassin got a lucky shot at the

> President, or you must conceive of a conspiracy so vast as to boggle the imagination. If indeed there were a conspiracy, it could not have been a small one, it will have to involve the Dallas police force, the doctors at Parkland Hospital, the doctors at Bethesda Hospital, the FBI, the Secret Service, the members of the, some members of the Department of Justice, the Attorney General, his chief aides, and possibly even the White House. Now I cannot accept that.[1175]

Shaw, who was part of the vast conspiracy to which he referred, knew what he was talking about. Shaw was correct in saying that one has only two choices; either one believes the ridiculous and incredible conclusion of the Warren commission or one must accept that the assassination of President Kennedy was the work of a vast conspiracy. When one considers that there has not been a legitimate investigation into the Kennedy assassination by any federal or state authority since New Orleans District Attorney Jim Garrison's investigation and prosecution of Clay Shaw, the next logical conclusion is that the conspirators, and the secret criminal syndicate to which they belonged, must still exist and hold the necessary power to prevent an investigation.

Notice that Shaw studiously avoided any mention of the CIA. Notice also that he mentions the Department of Justice and then includes separately the head of the Department of Justice, the Attorney General, who we all know was at that time John Kennedy's Brother, Robert. Since nobody would believe that Robert Kennedy was involved in the assassination plot, Shaw feels that his argument requires one to conclude that the Warren commission was correct. His sophistry is too obvious. Lyndon Johnson was part of the conspiracy, and he was Kennedy's Vice President. Johnson obviously kept the plot secret from John Kennedy. J. Edgar Hoover was involved in the conspiracy. Hoover was the Director of the FBI, which is part of the

Department of Justice. Hoover was Robert Kennedy's subordinate. As did Johnson, so also Hoover kept the plot secret from Robert Kennedy. Shaw seems to suggest that one must conclude that the head of an executive branch department must be involved in any and all conspiracies of his subordinates. By Shaw's twisted logic, John Kennedy would have to have been in on the conspiracy to assassinate himself, because all of the federal agencies mentioned were subordinate to him, who as President of the United States was chief executive of the executive branch of government. Shaw was truthful in one thing: that there is a vast conspiracy. That conspiracy, however, should not boggle the imagination, it should drive us to our knees in prayer to Almighty God and steel our hearts to fight against the conspiracy with intrepid fortitude.

Pointing out Robert Kennedy's position as the Attorney General at the time of the assassination was a strategy advised in a secret 1964 CIA memo (later made public) circulated within the hierarchy of the CIA as a way of answering critics of the Warren Commission Report and refuting claims of a conspiracy. The CIA memo advises its operatives to "note that Robert Kennedy, Attorney General at the time and John F. Kennedy's brother, would be the last man to overlook or conceal any conspiracy."[1176]

Paragraph 3b of that CIA memo further advises its operatives:

> To employ propaganda assets to refute the attacks of the critics. Book reviews and feature articles are particularly appropriate for this purpose. The unclassified attachments to this guidance should provide useful background material for passing to assets. Our ploy should point out, as applicable, that the critics are (I) wedded to theories adopted before the evidence was in, (II) politically interested, (III) financially interested, (IV) hasty

and inaccurate in their research, or (V) infatuated with their own theories.[1177]

A fair view of the facts reveals that the assassination of President Kennedy was a *coup d'etat*. How were all these separate state and federal agencies and the news media tied together into one giant conspiracy? Michael Collins Piper in his book, *Final Judgment*, presents a compelling case that Zionists in general, and the Israeli Mossad in particular, played a primary role in the assassination of President John F. Kennedy, and the subsequent cover-up. Piper's book is chock-full of sources and evidence, with over 1,000 endnotes, and 10 appendices. Other books have revealed the separate involvement of Lyndon Johnson, the Dallas Police Department, the CIA, the FBI, anti-Castro Cubans, French intelligence agencies, the U.S. Secret Service, and organized crime in the assassination of President Kennedy. The intrepid Mr. Piper methodically explains how all of these persons and organizations were tied together in a conspiracy that at its core was set in motion by the Israeli Mossad.

The motives for the Kennedy assassination are manifold, but one of the key reasons was Kennedy's intent to put an end to the Israeli plans for developing their own nuclear weapons. The Jerusalem Post reported on July 25, 2004 that the jailed nuclear whistle blower, Mordechai Vanunu, revealed that the Israeli government was behind the assassination of President Kennedy.

> Comments by freed nuclear spy Mordechai Vanunu that Israel was behind the assassination of US President John F. Kennedy failed to bring smiles to government officials Sunday. Vanunu said that according to "near-certain indications," Kennedy was assassinated due to "pressure he exerted on then head of government, David Ben-Gurion, to shed light on Dimona's nuclear reactor."[1178]

In addition, John Kennedy realized after the Bay of Pigs fiasco that he was not in control of the executive branch of government. He came to understand that Zionists had infiltrated the very warp and woof of government. Beginning with the CIA, Kennedy began to make dramatic changes to take the reigns of government back from the hidden treasonous cabal that was steering the United States toward destruction. Kennedy had even implemented a plan to wrestle control of U.S. currency from the Federal Reserve by issuing U.S. notes. Kennedy also planned on completely pulling all advisors out of Vietnam. Kennedy had the audacity to actually take control of government and act in the best interests of the citizens who elected him. The shadow government simply would not stand for a President who would not be their puppet and act in their interests. In their eyes there was no alternative, Kennedy had to go. Immediately upon the assassination of President Kennedy there was a 180 degree reversal of many of U.S. policies instituted by Kennedy, particularly the U.S. policy towards Israel.

One of the key players in the conspiracy to assassinate Kennedy was Meyer Lansky, the Zionist Jewish gangster, who Piper reveals was the de facto head of organized crime in the United States. Piper explains how Lansky had deep and continuing working relationships with both the CIA and the Israeli Mossad. Although Piper's book deals with the assassination of President Kennedy, which happened in 1963, the lessons he imparts are important and topical today for all American citizens who love their liberty. Piper's book is an autopsy which dissects the putrid body of a world conspiracy that is so pervasive, so intrusive, so powerful, it can assassinate the President of the most powerful country in the world and then conceal its involvement in that crime by controlling the mass media and even the very organs of that government. Piper takes off the blindfold and pins the tail on the donkey. His book is aptly titled; it is truly the Final Judgment in the assassination of our beloved President, John F. Kennedy, who valiantly and selflessly tried to wrestle the control of the U.S.

government from a Zionist cabal. After assassinating President Kennedy that cabal resumed control of the government, which they have maintained to this day.

The best evidence of the Zionist control over our government is the 1964 Warren Commission and its fraudulent report pinning the Kennedy assassination on the patsy, Lee Harvey Oswald. The next best evidence of Zionist control is the U.S. House Committee on Assassinations report. The committee's 1979 report determined that there was probably a conspiracy to assassinate President Kennedy. One would think that such a finding would result in a monumental investigation to find the culprits. However, Congress apparently had no interest in finding out who was behind the assassination. The committee ran out of money and Congress was not going to continue funding a committee that might actually find the culprits. Instead, the committee turned over its findings to the U.S. Department of Justice for further investigation. In its report the committee criticized the investigation into the Kennedy assassination conducted by the FBI, but that did not stop them from turning the investigation back over to the Department of Justice, of which the FBI is part. Once the investigation was referred to it, the Department of Justice also seemed not to have any curiosity in finding out who was really behind the assassination of President Kennedy. To this day, no resources of any significance have been devoted to any legitimate investigation of the Kennedy assassination by any organ of the Department of Justice. In fact, there is an abundance of documented proof that long prior to the referral to the Department of Justice by the U.S. House Committee on Assassinations, the FBI has been engaged in a concerted effort to conceal evidence and obstruct any investigation into the Kennedy assassination. The fact that the U.S. Congress and the U.S. Department of Justice have not shown any interest in finding out who assassinated the President of the United States and in actuality have engaged in concealing evidence and obstructing the investigation suggests that the culprits that were behind the assassination also control the very powers of government.

President Kennedy realized full well the nature and scope of the Zionist conspiracy lined up against him and the American people. He is reported to have stated: "The high office of President has been used to foment a plot to destroy the Americans' freedom, and before I leave office I must inform the citizen of his plight." His aim to do something about the conspiracy was the reason he was targeted for assassination. In a speech given at the Waldorf-Astoria Hotel New York City, April 27, 1961 Kennedy explained the nature and scope of the conspiracy:

> For we are opposed around the world by a monolithic and ruthless conspiracy that relies primarily on covert means for expanding its sphere of influence--on infiltration instead of invasion, on subversion instead of elections, on intimidation instead of free choice, on guerrillas by night instead of armies by day. It is a system which has conscripted vast human and material resources into the building of a tightly knit, highly efficient machine that combines military, diplomatic, intelligence, economic, scientific and political operations. Its preparations are concealed, not published. Its mistakes are buried, not headlined. Its dissenters are silenced, not praised. No expenditure is questioned, no rumor is printed, no secret is revealed. It conducts the Cold War, in short, with a war-time discipline no democracy would ever hope or wish to match.[1179]

President Lincoln also knew full well of the treachery of the money power. He was viewed as a formidable obstacle to their plans and so was assassinated. Lincoln stated:

> The money powers prey upon the nation in times

> of peace and conspire against it in times of adversity. It is more despotic than a monarchy, more insolent than autocracy, and more selfish than bureaucracy. It denounces as public enemies all who question its methods or throw light upon its crimes. I have two great enemies, the Southern Army in front of me and the bankers in the rear. Of the two, the one at my rear is my greatest foe.

The Zionists must keep a lid on the flat earth. Satan's minions know that if the flat earth is generally known, Satan's plan for a world dominion will be crushed, and he will have to start all over again. Read in the book of Acts how the disciples of Christ responded to the threatenings of the Jewish elders and chief priests who "commanded them not to speak at all nor teach in the name of Jesus." Acts 4:18.

> And being let go, they went to their own company, and reported all that the chief priests and elders had said unto them. And when they heard that, they lifted up their voice to God with one accord, and said, Lord, thou art God, which hast made heaven, and earth, and the sea, and all that in them is: Who by the mouth of thy servant David hast said, Why did the heathen rage, and the people imagine vain things? The kings of the earth stood up, and the rulers were gathered together against the Lord, and against his Christ. (Acts 4:23-26)

Notice how the Christians, with a true knowledge of God, immediately resisted the authority of the Jewish elders and chief priests and understood that the Jewish leaders were heathens in a conspiracy with the rulers of the world against God. The knowledge of the gospel of Christ had caused the people to see the massive world conspiracy against God and Christ. The Jewish religious leaders understood that the gospel of Jesus Christ had opened the eyes of the people, and they had lost their hold on them. The Jewish hierarchy learned their lesson. They knew that in order to rule over the people it was necessary to drive from them any true knowledge of God. God explains that "heaven is my throne, and earth is my footstool." (Acts 7:49) That is the kernel to understanding who God is and how close he is to us. The Jews cannot let that be known.

Their heliocentric model removes heaven and replaces it with the empty vacuum of space. If people can be convinced that there is no heaven, which is the throne of God, then it is easy to convince men that there is no God at all. Religious leaders know that most people have a need for a god, and so they conjure up a false, pantheistic god, who serves their ends. Although people will pay lip service to believing in God, it is not the true, sovereign God of the Bible, whose throne is heaven, and whose footstool is the earth.

Even many so-called "Christian" churches preach a perverted gospel with a false Christ. Under the true gospel of Jesus Christ, salvation is by the grace of God alone through faith in Jesus Christ alone. John 1:1-14, 14:6, Acts 4:12, and Ephesians 2:8. There is no salvation but through Jesus Christ. Since, however, the false gospel is inspired by the Jews, it provides for salvation without Jesus Christ, and in particular for the Jews. For more information on the corrupt Judaized gospel, read this author's books: *The Anti-Gospel, Solving the Mystery of Babylon the Great,* and *Bloody Zion*. It is this method of subtle deception that the god of this world, who is Satan, uses to deceive people into serving him.

> But if our gospel be hid, it is hid to them that are lost: In whom **the god of this world hath blinded the minds of them which believe not**, lest the light of the glorious gospel of Christ, who

is the image of God, should shine unto them. (2 Corinthians 4:3-4)

41 The Religion of Modern Cosmology

Marshall Hall summarizes the satanic conspiracy behind the Jewish Kabbalistic foundation of the so-called "science" of heliocentrism, the big bang theory, evolution, and modern psychology with the largely successful objective of degenerating society and corrupting the "Christian" church.

> Riding the first waves of the Copernican Revolution in the 1550's, 60's, and 70's was Kabbalist Isaac Luria. He understood how fellow Kabbalist Nachmanides' 13th century Big Bang Paradigm could actually be built upon the Copernican basis, and he promptly tied that understanding in with Kabbalist Ben HaKana's 1st century calculation of a 15 billion year old universe...which only another Kabbalist would be privy to. At first, it was not clear to me why Luria is given so much credit for "prophesying" Big Bangism 400 years ahead of its establishment. After all, Nachmanides had the same idea over 300 years before Luria did and he also understood that a Big Bang Cosmology could never be built without first establishing a contra-Biblical

heliocentric foundation. Luria, however, had the advantage of living at a time when a Gentile Churchman (Copernicus) would push that foundational moving-earth concept onto the world stage. The fact that Luria virtually pounced upon connecting the dots in the Rabbis HaKana and Nachmanides' Big Bang Cosmological Paradigm 400 years before it was brought to fulfillment attests to the fortuitous launching of the necessary Copernican foundation to that Paradigm during his lifetime much more than any "prophetic" ability attached to him. Given the prospects of a "scientifically" accepted rotating earth model--and fortified with a centuries-old secret plan to destroy Bible Credibility by destroying its "Origins Scenario"--the path was cleared for Kabbalic "science" to conquer the Life Sciences with "evolutionism." That done, Social and Behavioral "Sciences" and "Educational Psychology" and evolution based Christianity could proceed to transvaluate Christian morals and ethics until they conformed to Talmud/Kabbala morals and ethics.

Although Nachmanides understood that a heliocentricity model was the required first step on the road to legitimizing a relativistic, big bang, expanding universe cosmological fantasy which would establish eons of evolutionism and destroy all Bible credibility, he didn't have the mathematical mechanism attached to Copernicanism that gave it "scientific" status. Luria was the right man in the right place at the right time (1534-1572) to declare something like this in the Kabbalic Inner-Sanctum: "Now we have the mechanism that can make the

Cosmology of the sages work!"...in much the same way 300 years later that Karl Marx could write to Lassalle that Darwin's book had given "the death blow to God" and further exult: "Darwin's book...serves me as a basis for the class struggle [communism] in history." …

If one insists that the issue is about "Science" when referring to the mutually exclusive Creation Models of the Kabbala and the Bible, one needs a definition of what "Science" is. All dictionaries agree that the root word of "Science" is the Latin scire which means: "to know". There is one uncomplicated way of knowing something, and that way is to observe it and to note that no matter how many times that something is observed it always does the same thing. Observable and repeatable = Known, and Known = "science". All else invites deception. …

"Seeing is believing" is an old adage that upholds the observational requirement for anything that is defined as known, i.e., as "scientific". If no one has ever, ever, ever actually seen a "hypothetical something" about the Cosmos take place, can that "hypothetical something" be "known"? Obviously not. Can that "hypothetical something" be truly "scientific" if it is not known? Not according to "something" called a dictionary. It can be "science" falsely so called which causes man to err" (I Tim.6:20, 21), but it cannot be "true science". (Today's "science" is Orwellian doublespeak.).[1180] (parenthetical in original)

What are the religious tenets found in the Kabbalah, which is the source for heliocentrism? J.F.C. Fuller in his book *The*

Secret Wisdom of Qabalah explains the secret doctrine of the Kabbalah:

> In this duality of good and evil the danger is that the uninitiated see two separate entities: they see good as God and evil as Satan. Consequently, they love the one and hate the other....This is the essential ignorance which separates Christianity from Judaism; for to the Christians' salvation consists in the conquest of evil - that is, in its separation from the good....The personification of evil in the form of Satan as the God of Evil is the heresy which separates Christianity from Judaism.[1181]

Marshall Hall summarizes the satanic doctrines of the Jewish Kabbalah:

> Satan is not a created "entity" as the Bible and Christianity describe him to be (Col. 1:16). No, no, says the Kabbalist; that is the "essential ignorance which separates Christianity from Judaism." Making Satan the God of Evil "is heresy to the Kabbalist." The Kabbalist "Creator" is one entity, good and evil; a monotheistic amalgam of good and evil. (This is the same hermaphroditic ogre worshipped as the Baphomet creator/architect g'd found in many mystical—always secret—Societies which have higher orders of these "illuminated" ones in control.).[1182]

While many rabbis throughout history held a geocentric view and even believed in a flat earth, modern day rabbis were quick to yield to the heliocentric view, once the "science" of heliocentricity gained popularity. Why the quick capitulation?

Because both heliocentrism and a spherical earth were among the most esoteric occult beliefs of Judaism. For example, the highly respected Talmudic scholar Baal Hamaor, R' Zerachiah HaLevy (1125-1186), stated that the world is a sphere, and it can be divided into four quarters.[1183]

The Zohar, at vol. 3, on the portion of Vayikra, fol. 9b-10a (according to the translation of the Sulam, sections 134-145) states:

> In the book of Rav Hamnuna Sava it is explained in more detail that all the world rolls in a circle like a ball, these down and these up (that is, the creatures on the sphere are in opposition to each other, and the seven parts of the ball are the seven lands (the Sulam commentary)), and all the creatures in six lands are different in their appearances in accordance with the differences of air in each place, and they stand on their feet as other humans. Therefore there are places in the world that when it is light for those on one side of the sphere it is dark for those on the other side of the sphere, for these it is day while for those it is night; and there is a place in the world where it is always day and there is no night, except for a few moments."[1184] (parenthetical in original)

We find in the Babylonian Talmud in Tractate Abodah Zarah at Folio 41a that Jews are prohibited from having images, if the image is holding an orb, because that would imply that the idol "grasps the whole world as though it were a ball."[1185] The Jerusalem Talmud Tractate Abodah Zarah has a variant, wherein it states that an idol holding a sphere indicates "that it sacrifices itself for the whole globe."[1186]

The Jerusalem Talmud at Avoda Zara Chapter 3, Folio 42c

states that "R' Yonah said: When Alexander the Macedonian wanted to go back, he flew [on the back of an eagle] higher and higher until he saw the earth as a ball and the sea as a plate."[1187] (brackets in original)

That suggests that some Jews from antiquity held out the view that the world was in the shape of a round ball. Indeed, the Zohar on Vayikra p. 10a states: "In the book of Rav Hamnuna Sava it is explained that all the world rolls in a circle like a ball. ... There are places in the world that when it is light for those on one side of the sphere it is dark for those on the other."[1188]

Who was it that trotted out the Kabbalah inspired heliocentristic model? None other than a Roman Catholic priest, Nicolaus Copernicus (1473-1543). George William Rutler, writing for the Catholic Education Resource Center, reveals that Copernicus was a Catholic priest who was the temporary administrator of the diocese of Frauenburg. His priestly status was confirmed by DNA analysis that established that Copernicus was buried in the Polish cathedral of Frombork.[1189] A fact the Roman Catholic Church does not want known is that Copernicus dedicated his seminal book establishing a heliocentric model, *On the Revolutions of the Celestial Spheres,* to Pope Paul III, whom he addressed as "most Holy Father."[1190] Copernicus' dedication suggests that he was surrounded by Catholic clergy, who shepherded him and his heliocentric work. Indeed, he reveals in that dedication that Nicolaus Schonberg, the Roman Catholic Cardinal of Capua, was first among his friends who had encouraged him not to abandon his heliocentric study and writing. Another person Copernicus identified as spurring him on to publish his book was his "very dear friend, Tidemann Giese, [Roman Catholic] Bishop of Culm."[1191]

That is not surprising since Roman Catholicism is essentially Kabbalistic Judaism all dressed up for Gentiles. That little known fact is explained in this author's book, *Solving the*

Mystery of Babylon the Great.

There is a clear parallel between the traditions of the Pharisees of old and those of modern Roman Catholic priestcraft. The Roman Catholic Church follows the practice of the Jews and calls the combination of man's tradition and God's word "the Word of God." To a Protestant Christian the word of God means the Holy Bible. However, to the Roman Catholic, it means the Holy Bible plus their traditions.

> **Sacred Tradition and Sacred Scripture make up a single sacred deposit of the Word of God**. *CATECHISM OF THE CATHOLIC CHURCH*, § 97, 1994.
>
> [T]he church, to whom the transmission and interpretation of Revelation is entrusted, **does not derive her certainty about all revealed truths from the holy Scriptures alone. Both Scripture and Tradition must be accepted and honored with equal sentiments of devotion and reverence**. *Id.* at § 82 (emphasis added).

The Catholic Church has grafted its tradition onto the word of God. With this sleight of hand they have deceived people into following doctrines that are directly contrary to God's word as found in the Holy Bible.

The very idea of adding traditions to God's word is based upon the practice of the Jews. Michael Hoffman explains: "The Talmud is Judaism's holiest book (actually a collection of books). Its authority takes precedence over the Old Testament in Judaism. Evidence of this may be found in the Talmud itself, Erubin 21b (Soncino edition): 'My son, be more careful in the observance of the words of the Scribes than in the words of the Torah (Old Testament).'"[1192]

In that section of the Talmud there is a distinction made between the Torah and the Talmud (words of the Scribes). Often, that distinction is not made. Jews often refer to both the Talmud and the Torah as "Torah." As with the Catholic Church calling the combination of their traditions and the Bible, the word of God, so also the Jews say that Torah is the combination of their traditions (Talmud and Kabbalah) and the Old Testament. However, in Orthodox Judaism the Jewish traditions contained in the rabbinical writing of the Talmud and Kabbalah supercede and supplant the word of God found in the Old Testament (which is also called the Tanakh). That same thing is true regarding Catholic traditions that supplant the word of God.

The Jews teach that Moses was given revelation in two forms on Mount Sinai, oral and written. The smaller revelation was the written Torah, the larger revelation was kept orally. "This 'Oral Torah' had been transmitted faithfully by the leaders of each generation to their successors, by Moses to Joshua , and then to the elders, then to the prophets, to the men of the Great Assembly, to the leaders of the Pharisees, and finally to the earliest rabbis. The earliest rabbis saw themselves as heirs to the Pharisees."[1193]

In one statement Jesus exposed the lie that the oral traditions of the Jews were given by God to Moses at Mount Sinai. Jesus stated: **"For had ye believed Moses, ye would have believed me: for he wrote of me. But if ye believe not his writings, how shall ye believe my words?"** (John 5:46-47) If the oral traditions had truly been given by Moses, they would have testified to the authenticity of Jesus as Christ. Since the Jews rejected Jesus, because he contravened their traditions, that is proof that their traditions could not have come from Moses. Michael Hoffman explains that one statement by Jesus "crushed the whole beguiling system of indoctrination predicated on the Pharisaic myth of a divinely inspired, oral tradition of the elders."[1194]

Rabbi Ben Zion Bokser admits that the so-called traditions of the Jews that form the foundations of Judaism are entirely extra-biblical. He states that Jews and Christians alike are under the fallacious impression that Judaism is a religion based upon the Hebrew Bible. He states to the contrary that "[m]uch of what exists in Judaism is absent in the Bible, and much of what is in the Bible cannot be found in Judaism. . . . **Judaism is not the religion of the Bible**."[1195]

Judaism is based primarily upon the Kabbalah, Talmud, and other rabbinical writings. Where there is a conflict between their traditions (Talmud and Kabbalah) and the Old Testament (Torah), their traditions take precedence. Jews claim that the Talmud is partly a collection of traditions Moses gave them in oral form. Those traditions had not yet been written down in Jesus' time. Christ condemned the traditions of the Scribes and Pharisees, because those traditions (which later became written down in the Talmud) nullify the teachings of the Holy Bible. **"Making the word of God of none effect through your tradition**, which ye have delivered: and many such like things do ye." (Mark 7:13)

Rabbi Joseph D. Soloveitchik is regarded as one of the most influential rabbis of the 20th century. He is viewed as the unchallenged leader of Orthodox Judaism and the top international authority on halakha (Jewish religious law). "Soloveitchik was responsible for instructing and ordaining more than 2,000 rabbis, "an entire generation" of Jewish leadership."[1196] However, when the N.Y. Times explained his study, the only basis mentioned for his ascendant religious leadership was his study of the Talmud. "Until his early 20s, he devoted himself almost exclusively to the study of the Talmud."[1197] There was no mention in the article of the esteemed rabbi's study of the Old Testament (Torah) as the basis for being one of the leading authorities on Jewish law. That is because the Talmud and the Kabbalah form the basis for Judaism, and they are largely contrary to the Old Testament (Torah). Hoffman states: "The rabbi's credentials are all predicated upon his mastery of the Talmud."[1198] "Britain's Jewish Chronicle of March 26, 1993 states that in religious school (yeshiva), Jews are 'devoted to the Talmud to the exclusion of everything else.'"[1199]

To add tradition to God's word is rebellion against God's command that nothing be added or taken away from his words. "Ye shall not add unto the word which I command you, neither shall ye diminish ought from it, that ye may keep the commandments of the LORD your God which I command you." (Deuteronomy 4:2) "What thing soever I command you, observe to do it: thou shalt not add thereto, nor diminish from it." (Deuteronomy 12:32)

There is a terrible curse that comes with adding or taking away from God's word.

> For I testify unto every man that heareth the words of the prophecy of this book, If any man shall add unto these things, God shall add unto him the plagues that are written in this book: And if any man shall take away from the words of the book of this prophecy, God shall take away his part out of the book of life, and out of the holy city, and *from* the things which are written in this book. (Revelation 22:18-19)

The Holy Bible warns us about those who would attempt to turn us away from Christ to follow the traditions of men.

> Beware lest any man spoil you through philosophy and vain deceit, **after the tradition of men, after the rudiments of the world**, and not after Christ. (Colossians 2:8)

> **Wherefore if ye be dead with Christ from the**

> **rudiments of the world, why, as though living in the world, are ye subject to ordinances, (Touch not; taste not; handle not; Which all are to perish with the using;) after the commandments and doctrines of men?** Which things have indeed a shew of wisdom in will worship, and humility, and neglecting of the body; not in any honour to the satisfying of the flesh. (Colossians 2:20-23)

> He answered and said unto them, Well hath Esaias prophesied of you hypocrites, as it is written, This people honoureth me with *their* lips, but their heart is far from me. **Howbeit in vain do they worship me, teaching *for* doctrines the commandments of men. For laying aside the commandment of God, ye hold the tradition of men**, *as* the washing of pots and cups: and many other such like things ye do. And he said unto them, Full well **ye reject the commandment of God, that ye may keep your own tradition.** (Mark 7:6-9)

Jesus said: "I am the bread of life: he that cometh to me shall never hunger; and he that believeth on me shall never thirst." (John 6:35) Very simply, Jesus promised salvation to all who believed on him. Adding any other requirement to faith in Jesus corrupts the gospel, resulting in the bread of death rather than the bread of life.

Jesus warned his disciples to beware of the doctrine of the religious leaders of their time. Jesus compared their doctrine to leaven. Only a little leaven of man-made rules works its way through the whole loaf and corrupts God's pure doctrine. The leaven of today's religious leaders is no different; the leaven of tradition corrupts God's pure word. Man's tradition has turned the

Bread of Salvation into spiritual poison killing the souls of those who eat of the corrupted loaf.

> **Then Jesus said unto them, Take heed and beware of the leaven of the Pharisees and of the Sadducees.** And they reasoned among themselves, saying, It is because we have taken no bread. Which when Jesus perceived, he said unto them, O ye of little faith, why reason ye among yourselves, because ye have brought no bread? Do ye not yet understand, neither remember the five loaves of the five thousand, and how many baskets ye took up? Neither the seven loaves of the four thousand, and how many baskets ye took up? How is it that ye do not understand that I spake it not to you concerning bread, that ye should beware of the leaven of the Pharisees and of the Sadducees? **Then understood they how that he bade them not beware of the leaven of bread, but of the doctrine of the Pharisees and of the Sadducees.** (Matthew 16:6-12)

> **A little leaven leaveneth the whole lump.** (Galatians 5:9)

God wants us to purge out the leaven of man's tradition.

> Your glorying *is* not good. Know ye not that a little leaven leaveneth the whole lump? **Purge out therefore the old leaven, that ye may be a new lump, as ye are unleavened. For even Christ our passover is sacrificed for us**: Therefore let us keep the feast, not with old leaven, neither with the leaven of malice and wickedness; but with the unleavened bread of

> sincerity and truth. (1 Corinthians 5:6-8)

Man's tradition requires works to earn salvation. Salvation, however, is by God's Grace through faith alone on the completed work of Jesus Christ, who paid for all of our sins on the cross. Good works flow from salvation; good works cannot earn salvation.

> **For by grace are ye saved through faith; and that not of yourselves: it is the gift of God: Not of works, lest any man should boast. For we are his workmanship, created in Christ Jesus unto good works, which God hath before ordained that we should walk in them.** (Ephesians 2:8-10)

The Jewish Encyclopedia acknowledges the Babylonian (a/k/a Chaldean) origins of the Kabbalah (a/k/a Cabala). In addition, the Jewish Encyclopedia explains that Gnosticism flowed from the Jews to the ersatz "Christians." That is yet more authority that Gnosticism flowed from Babylon via the Jewish Gnostics to lay the foundation for the Roman Catholic theology. The esoteric Gnosticism imbued in the Catholic theology was based upon the Jewish Kabbalah.

> The Pythagorean idea of the creative powers of numbers and letters, upon which the "Sefer Yez.irah" is founded, and which was known in tannaitic times . . . is here proved to be an old cabalistic conception. In fact, the belief in the magic power of the letters of the Tetragrammaton and other names of the Deity . . . seems to have originated in Chaldea . . . Whatever, then, the theurgic Cabala was, which, under the name of "Sefer (or "Hilkot" Yez.irah,") induced Babylonian rabbis of the fourth century to "create

a calf by magic."

* * *

> But especially does Gnosticism testify to the antiquity of the Cabala. Of Chaldean origin, as suggested by Kessler . . . and definitively shown by Anz . . . Gnosticism was Jewish in character long before it became Christian.[1200]

Marshall Hall explains how the Jewish Kabbala is the source for the big bang that supposedly caused the universe to spring into existence billions of years ago.

> The size of the Universe is now determined by the dictates of the Kabbala. That book--in addition to outlining the "Relativity" concept to allow the 15+ billion year age of the Universe--also outlines the Big Bang and Expanding Universe concepts which give license to claim billions of light year distances to the stars. Both of these myths about the age and size of the Universe have been brought to fruition by a Theoretical Science Establishment which has used NASA's "Origins Program" to establish the Kabbala's reign over all "science"."[1201]

Hall quotes a Jewish commentator, who states that "The Ramban [13th century. Rabbi Nachmanides], who refers to what he writes as coming from 'hidden knowledge,' says that this initial creation was something so small and without physical form. This idea that everything originated from a singular point in the universe is what science calls The Big Bang!"[1202]

Research has confirmed that the Jewish commentator quoted by Hall is correct. The modern figure of 14 billion years attributed to the age of the universe is not based upon science, but rather a mystical religious belief. This belief that the universe is billions of years old was first announced by Nachmanides. Nachmanides was known as Ramban, which is an acronym of his name, Rabbi Moshe Ben Nachmon (1194-1270 A.D.). Nachmanides was the foremost expert on Jewish religious laws and customs (halakhist) of his age. Ramban was a mystic who incorporated the Jewish Kabbalah into his teachings. He was vehemently against Christianity and often tried to refute Christianity in debates.[1203]

Nachmanides interpreted the creation account in Genesis not as six literal, 24 hour days, but rather a much longer period that equaled approximately 15-3/4 billion years. That is very close to the 14+ billion years that is often hypothesized by modern so-called "scientists."[1204]

The "big-bang" theory of creation is simply a restatement of Nachmanides' commentary of the creation account. According to Nachmanides, in the beginning (15-¾ billion years ago) the universe was initially non-corporeal, ethereal, and no larger than a mustard seed. This small mustard seed sized etherea was somehow transformed into matter. Nachmanides describes a universe that came into being from the initial transformation of concentrated matter in the mustard seed sized area, to a point when it suddenly expanded (what scientists today call the "big-bang") into the universe that we see today. The "big-bang" theory began as a religious belief of a Jewish rabbi. Since there is no science that supports it, it remains, in reality, a Jewish religious belief.

Yakov Leib HaKohain (a/k/a Lawrence G. Corey, Ph.D., John-Francis, Aziz, Mehemed Effendi, and Kali Dass) is a Jewish Sabbatian Kabbalist, who converted to the Catholic Church, Islam and Hindu Vedanta, respectively. He holds a doctorate degree in Jungian Studies and Comparative Religion.[1205] Yakov Leib HaKohain confirms the Kabbalistic origins of the big-bang theory.

> Isaac Luria (1534-1572) -- also known as the "Ari Zaal," or "Divine Rabbi Isaac," -- was, and remains to this day, unarguably the greatest Kabbalist in world history. His doctrines, based entirely on the Old Testament and Zohar, anticipated virtually world-for-word [*sic*]-- or at the very least, concept for concept -- the theory of the "Big Bang" origins of the universe in astrophysics. Thus, the cosmogony of a 16th century Jewish Kabbalist was literally validated by modern science four hundred years later."[1206]

Professor Gerald Schroeder is typical of modern scientists who have tried to reconcile the supposed billions of years old universe with the biblical account of creation within 6 days, which took place approximately 6,000 years ago. Gerald Schroeder is an Orthodox Jewish physicist. Dr. Schroeder received his PhD in nuclear physics and earth and planetary sciences from the Massachusetts Institute of Technology (MIT). He was on the staff of the MIT Physics Department and was a member of the United States Atomic Energy Commission.

Schroeder relies on the Jewish Talmud to interpret the six days of creation to be equivalent to approximately 15 billion years. He states:

> The Talmud tells us that the soul of Adam was created at five and a half days after the beginning of the six days. That is a half day before the termination of the sixth day. At that moment the cosmic calendar ceases and an earth based calendar starts. ... that comes out to be 15 billion years. NASA gives a value of about 14 billion years. Considering the many approximations, and that the Bible works with only six periods of time, the agreement to within a few percent is

> extraordinary. The universe is billions of years old from one perspective and a mere six days old from another. And both are correct![1207]

The problem with Schroeder's attempt to reconcile the language in the Bible with the scientific view of billions of years is that the language in the Bible allows no room for such an interpretation.

> And God called the light Day, and the darkness he called Night. And the evening and the morning were the first day. (Genesis 1:5)
>
> And the evening and the morning were the second day. (Genesis 1:8)
>
> And the evening and the morning were the third day. (Genesis 1:13)
>
> And the evening and the morning were the fourth day. (Genesis 1:19)
>
> And the evening and the morning were the fifth day. (Genesis 1:23)
>
> And God saw every thing that he had made, and, behold, it was very good. And the evening and the morning were the sixth day. (Genesis 1:31)

God defined the day and night and described the events of creation as having taken place between the "evening and the morning." Each day was marked off by "the evening and the morning." They were literal evenings and mornings. They were literal days. There is no biblical authority for the argument that the days mentioned in the Bible spanned millions of years. That is pure sophistry, born out of a heathen desire to strip God of the glory he deserves for having created the heavens and the earth in six literal days, by the exercise of his sovereign will, through his spoken commands. God rested on the seventh day. Genesis 2:2.

According to Schroeder, the six days of creation in Genesis took 14-15 billion years. He accepts the so-called "scientific" findings that creation is 14-15 billion years old but bases his acceptance of that age on the writings of Jewish religious sages. Schroeder states:

> One of the most obvious perceived contradictions between Torah and science is the age of the universe. Is it billions of years old, like scientific data, or is it thousands of years, like Biblical data? When we add up the generations of the Bible and then add the secular rulers that followed, we come to fewer than 6000 years. Whereas, data from the Hubbell telescope or from the land based telescopes in Hawaii, indicate the number at 15 billion years plus or minus 10%. In trying to resolve this apparent conflict, I use only ancient biblical commentary because modern commentary already knows modern science, and so it is influenced by what science always [*sic*]. That commentary includes the text of the Bible itself (3300 years ago), the translation of the Torah into Aramaic by Onkelos (100 CE), the Talmud (redacted about the year 400 CE), and the three major Torah commentators. There are many, many commentators, but at the top of the mountain there are three, accepted by all: Rashi (11th century France), who brings the straight understanding of the text, Maimonides (12th century Egypt), who handles the philosophical concepts, and then Nahmanides (13th century Spain), the most important of the Kabbalists.

> These ancient commentaries were finalized hundreds or thousands of years ago, long before Hubbell was a gleam in his great-grandparent's eye. So there's no possibility of Hubbell or any other scientific data influencing these concepts. That's a key component in keeping the following discussion objective.[1208]

Marshall Hall explains how modern science has simply confirmed the Jewish Kabbalistic religious beliefs about the origins of the universe:

> Is the triumph of today's Big Bang Paradigm--which is certified by a Theoretical Science Establishment--a clear confirmation that those ancient mystic Rabbis and Rebbes were supernaturally gifted sages and prophets of God? That's what Kabbalists are saying. Or, is that triumph--which is smashing the cornerstone of Bible credibility and hence of Christianity--a long-laid, supernaturally motivated strategy to destroy God's real plan through Jesus Christ by destroying the Bible's Credibility and replacing the Creation Account with the Kabbala-confirming "creation" scenario, which scenario has been brought to pass by the very "science falsely so called" establishment that Christians were warned about by that exceptional Hebrew from the Tribe of Benjamin, Saul of Tarsus (I Tim. 6:20,21)?[1209]

Whom do we find revealed to the world as the progenitor of the big bang theory? Another Roman Catholic Priest. The Jewish controlled press try their best to conceal the Jewish religious origins of the big bang theory. As is typical of an intelligence operation, the Jewish controlled media have done what is known in the intelligence community as a limited hangout. They admit that the big bang has its origin in religion, but they conceal that it is Babylonian Judaism. They instead steer the public toward their Gentile front men in the Catholic Church, while they remain in the shadows pulling the strings. In a PBS article titled, *Big Bang Theory: A Roman Catholic Creation*, Edgar Herwick claims that it was a Roman Catholic priest who first postulated the big bang theory.

> While the Big Bang is as old as the universe itself, our concept of it is still strikingly new — less than 100 years old. And if you dig into its origins, you come across a curious fact.
>
> Atheists, devout Christians, you might want to sit down for this: The Big Bang theory was first proposed by a Roman Catholic priest.
>
> It wasn't just any priest. It was Monseigneur George Lemaître, a brilliant Belgian who entered the [Jesuit] priesthood following his service as an artillery officer in the Belgian army during World War I. He was also an accomplished astronomer and a talented mathematician and physician. After earning his graduate degree in astronomy from the University of Cambridge in England, he came to Boston and spent a year at the Harvard College Observatory before earning his doctorate at MIT. ...
>
> As astonishing as Lemaître's idea was, perhaps equally surprising to us now was the reaction of the church. Lemaître was not jailed by the Pope like Galileo. He was not excommunicated the way Johannes Keppler was by the Lutheran Church. Quite the opposite. In the early 1950s,

Pope Pius XII not only declared that the big bang and the Catholic concept of creation were compatible; he embraced Lemaître's idea as scientific validation for the existence of God and of Catholicism."[1210]

Notice that the article from PBS (WGBH) states that the big bang originated only one hundred years ago with a Catholic priest. There is no mention of Jewish Rabbi Nachmanides whose postulation of the big bang predated Lemaitre's theory by approximately 700 years.

Figure 193: Georges Lemaître & Albert Einstein, 1933

At a conference in the 1930s, where Lemaître presented his theory, Einstein reportedly remarked, "This is the most beautiful and satisfactory explanation of creation to which I have ever listened."[1211]

Einstein, who was a Talmudic Jew, looked upon the big bang theory with favor. It seems that Einstein's approval puts the Jewish imprimatur on Lemaitre's big bang theory. Einstein's influence in the Zionist community was such that on November 17, 1952, Einstein was offered the presidency of Israel, which he, with regret, turned down.[1212]

It is not insignificant that Lemaitre was a Jesuit. What most do not know is that the Jesuit order is a Crypto-Jewish order of priests. While it may have gentile priests as members, it was founded by Jews and today has close secret ties with other clandestine Jewish organizations. Benjamin Disraeli was a Jew and a former Prime Minister of England; he revealed that the first Jesuits were Jews.[1213] Ignatius of Loyola's secretary, Polanco, was of Jewish descent and was the only person present at Loyola's deathbed. Ignatius Loyola himself was a crypto-Jew of the Occult Kabbalah. A crypto-Jew is a Jew who converts to another religion and outwardly embraces the new religion, while secretly maintaining Jewish practices.

James Lainez (J.G. 1558-1565), who succeeded Ignatius Loyola (J.G. 1541-1556) as the second Jesuit General, was also of Jewish descent. The fourth Jesuit General was a Belgian Jew named Eberhard Mercurian (J.G. 1573-1580).[1214] Jews were attracted to the Jesuit order and joined in large numbers.[1215] Some of the most influential Jesuits in history, such as Francisco Ribera (1537-1591) and Emanuel Lacunza (1731-1801), were Jews. During the 5th General Congregation in 1593 of the 27 Jesuits who proposed changes to the Constitutions, 25 were of either Jewish or Moorish descent.[1216] Many of the Jesuit doctrines are similar to those found in the Kabbalah and Babylonian Talmud. It is possible that the second beast in the book of Revelation is the Jesuit General, who is known as the black pope.

The Abbate Leone explains the prideful arrogance of the Jesuits and their esoteric symbolism signifying their link to Judaism.

> The Jewish high-priest wore on his breast the jewel called the oracle. The order of the Jesuits considers itself, under the New Alliance, as the oracle from whence the pope draws his inspiration. They proclaim themselves "the masters of the world, the most learned of mortal men, the doctors of the nations, the Apollos, the Alexanders of theology, prophets descended from heaven, who deliver the oracles in the recumenic councils." The epitaph which they composed for Loyola strikingly exhibits their love of grandiloquence, and their overweening pride. It

runs thus: — "Whoever thou art who conceivest in thy mind the image of Pompey the Great, of Caesar, and of Alexander, open thine eyes to the truth, and thou wilt learn from this marble that Ignatius was the greatest of conquerors."[1217]

John Torell explains the Jewish origins of the Jesuit order:

The Illuminati order was not invented by Adam Weishaupt, but rather renewed and reformed. The first known Illuminati order (Alumbrado) was founded in 1492 by Spanish Jews, called "Marranos," who were also known as "crypto-Jews." With violent persecution in Spain and Portugal beginning in 1391, hundreds of thousands of Jews had been forced to convert to the faith of the Roman Catholic Church. Publicly they were now Roman Catholics, but secretly they practiced Judaism, including following the Talmud and the Cabala. The Marranos were able to teach their children secretly about Judaism, but in particular the Talmud and the Cabala, and this huge group of Jews has survived to this very day. After 1540 many Marranos opted to flee to England, Holland, France, the Ottoman empire (Turkey), Brazil and other places in South and Central America. The Marranos kept strong family ties and they became very wealthy and influential in the nations where they lived. But as is the custom with all Jewish people, it did not matter in what nation they lived, their loyalty was to themselves and Judaism. [1218]

* * *

In 1491 San Ignacio De Loyola was born in the

Basque province of Guipuzcoa, Spain. His parents were Marranos and at the time of his birth the family was very wealthy. **As a young man he became a member of the Jewish Illuminati order in Spain. As a cover for his crypto Jewish activities, he became very active as a Roman Catholic.** On May 20, 1521 Ignatius (as he was now called) was wounded in a battle, and became a semi-cripple. Unable to succeed in the military and political arena, he started a quest for holiness and eventually ended up in Paris where he studied for the priesthood. In 1539 he had moved to Rome where he founded the "JESUIT ORDER," which was to become the most vile, bloody and persecuting order in the Roman Catholic Church. In 1540, the current Pope Paul III approved the order. At Loyola's death in 1556 there were more than 1000 members in the Jesuit order, located in a number of nations. [1219]

Setting up the Jesuit order, Ignatius Loyola devised an elaborate spy system, so that no one in the order was safe. If there was any opposition, death would come swiftly. The Jesuit order not only became a destructive arm of the Roman Catholic Church; it also developed into a secret intelligence service. **While the Popes relied more and more on the Jesuits, they were unaware that the hard core leadership were Jewish, and that these Jews held membership in the Illuminati order which despised and hated the Roman Catholic Church.**[1220]

The Jesuits were established by Ignatius of Loyola. Ignatius of Loyola was the leader of a secret occult organization known as the *Alumbrados* (Spanish for Illuminati).[1221] On August

15, 1534, Loyola started a sister organization to the *Alumbrados*, which he called the Society of Jesus. It is more commonly known today as the Jesuits. Loyola was arrested by the Dominican order of Catholic inquisitors, who were concerned with his growing influence and power throughout Europe. Because of his influential allies among the principalities of Europe, he was granted an audience with the pope. Loyola promised the pope his allegiance and agreed to do the bidding of the papacy throughout the world. Pope Paul III formally approved the Jesuits as a Catholic religious order in his 1540 papal bull *Regimini Militantis Eccclesiae.*[1222]

The influence of the Jews through the Jesuits in the Roman Catholic Church has been manifested from the beginning in Catholic doctrine. The Council of Trent was an attack on Christianity with anathema after anathema against Christian doctrine that was orchestrated by the Jesuits. The control of the Jews over the Vatican is so complete that Cardinal Joseph Ratzinger (now Pope Benedict XVI), who was the prefect of the Congregation for the Doctrine of the Faith, issued an official doctrine of Catholic faith that accepted the Jewish view that the messiah is yet to come. There is apparently much double talk in the document, as it accepts the Jewish view of a coming messiah without overtly rejecting Jesus. Some have interpreted the document as denying the redemptive role of Jesus. The Catholic Church long ago implicitly denied the redemptive role of Jesus. The document is contained in a small book titled "The Jewish People and the Holy Scriptures in the Christian Bible." It is no surprise that this Jewish/Catholic doctrine was drafted by a Jesuit named Albert Vanhoye.[1223]

The Jewish influence over the Roman Catholic institution and its doctrines is manifest in *The Document of the Vatican Commission for Religious Relations with Judaism* § 4, which states: **"We propose, in the future, to remove from the Gospel of St. John the term, 'the Jews' where it is used in a negative sense, and to translate it, 'the enemies of Christ.'"**[1224]

At a speech at Hebrew University in Jerusalem, Roman Catholic Cardinal Joseph Bernadine stated:

> [T]here is need for . . .theological reflection, especially with what many consider to be the problematic New Testament's texts ... Retranslation ... and reinterpretation certainly need to be included among the goals we pursue in the effort to eradicate anti-semitism.
>
> [T]he gospel of John ... is generally considered among the most problematic of all New Testament books in its outlook towards Jews and Judaism ... this teaching of John about the Jews, which resulted from the historical conflict between the church and synagogue in the latter part of the first century C.E., can no longer be taught as authentic doctrine or used as catechesis by contemporary Christianity ... Christians today must see that such teachings ... can no longer be regarded as definitive teachings in light of our improved understanding.[1225]

Indeed, the Vatican has announced that Christians should not evangelize Jews. On December 11, 2015, *The Guardian* reported:

> Catholics should not try to convert Jews and should work with them to fight antisemitism, the Vatican has said, in a major new document that draws the church further away from the strained relations of the past.
>
> Christianity and Judaism are intertwined and God never annulled his covenant with the Jewish people, said the document from the Vatican's

> commission for religious relations with Jews.
>
> "The church is therefore obliged to view evangelisation to Jews, who believe in the one God, in a different manner from that to people of other religions and world views," it said.
>
> "In concrete terms this means that the Catholic church neither conducts nor supports any specific institutional mission work directed towards Jews." ...
>
> The report also said that while it is only thanks to Christ's death and resurrection that all people have the chance of salvation, Jews can benefit from this without believing in him. The authors appear to acknowledge that they are effectively squaring a theological circle, however, since how Jews can be saved while not believing in Christ "remains an unfathomable mystery in the salvific plan of God."[1226]

The position of the Vatican toward the Jews is part and parcel of the "Christian" Zionism movement that has also taken hold in many Protestant denominations. For further information on the Jewish influence over both the Catholic church and Protestant churches, read this author's book, *Bloody Zion*. "Christian" Zionism is directly contrary to the great commission given by Jesus Christ:

> Go ye therefore, and teach all nations, baptizing them in the name of the Father, and of the Son, and of the Holy Ghost: Teaching them to observe all things whatsoever I have commanded you: and, lo, I am with you alway, even unto the end of the world. Amen. (Matthew 28:19-20)

The foundational "salvific" doctrine of the Catholic church is unbiblical, and therefore ineffective anyway. The Roman Catholic Church requires faith plus works, whereas the true gospel requires only faith in Jesus Christ. The Catholic doctrine of faith plus works is the same doctrine of the Jewish gnostics, who argued that obedience to the law was necessary for salvation. Paul, on the other hand, states emphatically that works of the law will not justify a person.

> Knowing that a man is not justified by the works of the law, but by the faith of Jesus Christ, even we have believed in Jesus Christ, that we might be justified by the faith of Christ, and not by the works of the law: for by the works of the law shall no flesh be justified. (Galatians 2:16).

Justification is by God's grace through faith alone. "For by grace are ye saved through faith; and that not of yourselves: it is the gift of God: Not of works, lest any man should boast." (Ephesians 2:8-9) For more detailed information on salvation by the Grace of God alone read: *The Anti-Gospel, The Perversion of Christ's Grace Gospel.*[1227]

The Vatican states that "how Jews can be saved while not believing in Christ remains an unfathomable mystery in the salvific plan of God." It is a mystery in the same way that all lies are mysterious to those being deceived. God's word makes it clear, however, that there is no salvation for Jew or Gentile aside from Jesus. "Jesus saith unto him, I am the way, the truth, and the life: no man cometh unto the Father, but by me." John 14:6. There is only one way to salvation, and that is by the grace of God through faith in Jesus Christ. Ephesians 2:8. "He that hath the Son hath life; and he that hath not the Son of God hath not life." 1 John 5:12.

There is not a separate means of salvation for the Jew and

another for the Gentile. There is only one means of salvation for both. "There is neither Jew nor Greek, there is neither bond nor free, there is neither male nor female: for ye are all one in Christ Jesus." Galatians 3:28. All who are saved by grace through faith are Abraham's spiritual seed (i.e., Christians). "And if ye be Christ's, then are ye Abraham's seed, and heirs according to the promise." Galatians 3:29.

To refuse to evangelize the Jews is a sin, as it is disobeying a direct command from God to "[g]o ye into all the world, and preach the gospel to every creature." Mark 16:15. Jesus did not say to preach only to Gentiles; he commanded that the gospel be preached to "every creature;" that includes Jews. Preaching of the gospel is the means God uses to save. Without preaching the word of God, there is no salvation. The preaching must be to both the Jew and Gentile. Jesus is the Lord over all men:

> **For there is no difference between the Jew and the Greek: for the same Lord over all is rich unto all that call upon him. For whosoever shall call upon the name of the Lord shall be saved.** How then shall they call on him in whom they have not believed? and how shall they believe in him of whom they have not heard? and how shall they hear without a preacher? And how shall they preach, except they be sent? as it is written, How beautiful are the feet of them that preach the gospel of peace, and bring glad tidings of good things! But they have not all obeyed the gospel. For Esaias saith, Lord, who hath believed our report? So then **faith cometh by hearing, and hearing by the word of God.** Romans 10:11-17.

G. Richard Fisher succinctly explains the implications of

the "Christian" Zionist theology that has infected both the Catholic church and Protestant churches:

> If the early Church had taken this view that Jews are saved just by being Jews there would have been no Christians in Jerusalem, Judea or Samaria. When Paul was converted, God would have violated His own plan. Under this view Jesus would not have had Apostles and there would not be a Church today. These logical results seem to escape the purveyors of this form of Christian Zionism, which is neither Christian or Zionist.[1228]

The false gospel that Jews can gain entrance into heaven aside from faith in Jesus Christ is the same fallacious Jewish doctrine that God repudiated in the true gospel. Jesus told the Jews that if they truly followed the teachings of Moses, they would accept him as their Messiah, because Moses wrote of him. *See* John 5:39, 46-47. Jesus further told the Jews that "if ye believe not that I am he [Christ], ye shall die in your sins." John 8:24. Peter told the Jews that salvation only comes through Jesus Christ: "Neither is there salvation in any other: for there is none other name under heaven given among men, whereby we must be saved." Acts 4:12.

The Vatican is making the same claim for the Jews that Jesus refuted in the gospel of John. Jesus started out by telling the Jews that if they were his disciples they would know the truth and be set free from sin. John 8:31-32. The Jews objected and argued that they were not in bondage and had no need to be set free because they were Abraham's seed. John 8:33. They argued that by virtue of their status as the physical descendants of Abraham they were in no need of a savior. They claimed that God was their Father by virtue of their physical lineage from Abraham. John 8:39-41. John the Baptist told the Jews from the beginning that their lineage from Abraham would not help them. "And think not to say within yourselves, We have Abraham to our father: for I say unto you, that God is able of these stones to raise up children unto Abraham." Matthew 3:9.

Jesus made clear to the Jews that their father was the devil and that if God were truly their Father, they would love him. John 8:42. Those without the Son are also without the Father. The claim by "Christian" charlatans like John Hagee that the Jews have redemption other than by the grace of God through faith in Jesus Christ is impeached by scripture. The Jews deny Jesus is the Christ, and therefore they do not have the Father and are by definition antichrist.

> Who is a liar but he that denieth that Jesus is the Christ? He is antichrist, that denieth the Father and the Son. Whosoever denieth the Son, the same hath not the Father: he that acknowledgeth the Son hath the Father also. 1 John 2:23-24.

The "Christian" Zionist dual-covenant theology serves to kill any mission by "Christian" Zionists to spread the gospel to the Jews.[1229] Why would Jews come to Christ, if they are told by the "Christian" Zionists that their conversion is irrelevant to their salvation? If they are nationally saved without Jesus Christ anyway, what is the point of believing in Jesus Christ?

The Vatican is not alone in this devilish subterfuge. Steven Paas revealed in his book, *Christian Zionism Examined,* that some "Christian" Zionist missionary groups do not consider their activities in Israel as a mission but simply as "meeting with Israel."[1230] They do not see the conversion of Jews to Christ as part of their mission. Paas concluded the following about the missionary activities of the "Christian" Zionist *Center for Israel Studies* (CIS): "Their first priority is not leading Jews to Christ, but listening to and learning from Jews and Judaism."[1231]

What are the teachings of the Jews and Judaism to which Paas refers? The Jewish teachings followed by "Christian" Zionists are found in the Talmud, which in *Sanhedrin Folio* 90a provides that all Jews ("Israel") are guaranteed a portion in the world to come.[1232] According to the "Christian" Zionist thought, there is no point in evangelizing the Jews, since they are already guaranteed entrance into heaven without Christ; so say the authoritative Jewish doctrines found in the Talmud.

Jews view "Christian Zionists as "useful idiots," which is a pejorative phrase used by communists to describe Gentile communist propagandists who do not understand the Jewish goals behind communism. Jews have a secret that they keep from the "Christian" Zionists. According to the previously mentioned tractate in the Talmud (*Sanhedrin Folio* 90a), Christians, who are described as those who read the New Testament ("uncanonical books"), have no portion in the world to come.[1233] In fact, Jews have a particular hatred for Christians. The hatred by Jews against Christians is so intense that Jews are taught to utter a curse when passing a Christian Church, calling on their heathen god (Hashem) to "destroy this house of the proud."[1234]

Elizabeth Dilling explains: "The 'religious' Orthodox Jew recites the 'Eighteen Benedictions,' or 'Shemoneh Esreh,' three times week days, four times on holidays and Sabbaths, the 7th and 12th of which curse the Christians and non-Jews to hell and perdition. Thus, the 'good Orthodox Jew' gives us Christians 6 cursings on ordinary days, 8 on 'specials.'"[1235]

The halacha (Jewish religious law) is that it is forbidden for a Jew to engage in any of the religious practices of Christians. The Jews sit smugly meeting with "Christian" Zionists firm in their belief that the Christians are damned to hell, while they are guaranteed entrance into heaven without believing in Jesus Christ. By not evangelizing the Jews, "Christian" Zionists are unwittingly acceding to the laws of the Jews and lending a degree of authenticity to the antichrist Jewish teachings.

The point of church missionary activity should be to spread the gospel of Jesus Christ. However, the "Christian" Zionists go out of their way to avoid doing that very thing. Paas concludes: "Western Christian organisations in Israel often limit themselves to activities of dialogue, study, representation, or support."[1236]

The Vatican implies that evangelizing Jews is a form of antisemitism. On the contrary, to refuse to spread the gospel to the Jews is the purest form of antisemitism. To withhold from the Jews the gospel, which is the only means for their salvation, is unadulterated hatred toward the Jews. The "Christian" Zionists deliberately violate Jesus' command to "go ye into all the world, and preach the gospel to every creature." Mark 16:15. The "Christian" Zionists are the very instrument being used by Satan to keep the Jews in spiritual bondage by convincing them that they are part of a unique plan by God simply by virtue of their status as Jews. Despite their claims to love the Jews, Paas concludes that "[f]ailing to communicate Christ-given insights to the Jews is lacking love for them."[1237]

In ancient Palestine the Jews worked hand in hand with the Romans to crucify Christ. Now, the Jews work hand in glove with the Roman Catholic Church in their effort to eradicate Christ's gospel. The great harlot of Babylon in Revelation 17:5 has Mystery written upon her forehead. She is called Babylon because she is Babylonian. She is a mystery because she is masquerading as "the" Christian religion. Christian labels have been applied to Babylonian paganism to come up with the mystery religion we know as the Roman Catholic Church. Both the Talmudic Jews and the Vatican share that common Babylonian root. The Jesuits nurtured the Babylonian Cabalism in Roman Catholic doctrine. The similarities between the imperious whorish woman in Ezekiel 16:14-40 and the Mother of Harlots in Revelation 17:5 are unmistakable. They are one and the same. Roman Catholicism is an esoteric version of Babylonian Judaism. The Roman Church appears gentile to the uninitiated, but it is Jewish to its core. Orthodox Judaism appears to the uninitiated to be the Old Testament theology, but it is actually Babylonian to its core.

The *Extreme Oath of the Jesuits,* which is given to a Jesuit Priest when he is elevated to a position of command, contains the following provisions:

> I . . . declare and swear that his holiness, the Pope, is Christ's Vice-regent, and is the true and only head of the Catholic or Universal Church throughout the earth.
>
> * * *
>
> Therefore, to the utmost of my power, I shall and will defend this doctrine and his Holiness' right and customs against all usurpers of the heretical or Protestant authority, whatever especially the Lutheran Church of Germany, Holland, Denmark, Sweden, and Norway, and the now pretended authority of the Church of England and Scotland, the branches of the same, now where . . . . . I do now renounce and disown any allegiance as due to any heretical king, prince or state named Protestant or Liberals, or obedience to any of their laws, magistrates or officers.
>
> * * *
>
> That I will go to any part of the world, whatsoever, without murmuring and will be

> submissive in all things whatsoever communicated to me . . . . . I do further promise and declare, that I will, when opportunity presents, make and wage relentless war, secretly or openly, against all heretics, Protestants and Liberals, as I am directed to do to extirpate and exterminate them from the face of the whole earth, and that I will spare neither sex, age, nor condition, and that I will hang, waste, boil, flay, strangle, and bury alive these infamous heretics; rip up the stomachs and wombs of their women and crush their infants' heads against the wall, in order to annihilate forever their execrable race.
>
> That when the same cannot be done openly, I will secretly use the poison cup, the strangulation cord, the steel of the poinard, or the leaden bullet, regardless of the honor, rank, dignity or authority of the person or persons whatsoever may be their condition in life, either public or private, as I at any time may be directed so to do by any agent of the pope or superior of the brotherhood of the holy faith of the Society of Jesus.[1238]

The Jesuits are the secret army of the Roman church; they are often referred to as the "pope's militia." In fact, the leader of the Jesuits is called the "Jesuit General." He is unlike any other leader of a Catholic order, because the Jesuit General is independent of the Catholic Bishops and Cardinals; he answers directly to the Pope. Because of the power and influence of the Jesuit General, he is known as the "Black Pope." The Jesuit General has the purported authority to absolve persons of the sins of bigamy, murder, or any harm done to others as long as the matter is not publicly known and the cause of a scandal.[1239] Pope Gregory XII gave the Jesuits the authority to deal in commerce and banking, which has made the order quite wealthy.[1240] The popes have threatened princes, kings, and anyone else who interferes with the Jesuits with excommunication (*Latae Sententiae*).[1241] In one of the most authoritative works on the Jesuits, J. Huber, professor of Catholic theology wrote: "Here is a proven fact: the Constitutions [of the Jesuits] repeat five hundred times that one must see Christ in the person of the [Jesuit] General."[1242]

If the Jesuits are a Crypto-Jewish order that hates the Roman Catholic church, as alleged by John Torell, why would it become a priestly order within that church and become its greatest defender with extreme oaths of obedience to the pope? Because, the Roman church offers the Jews an ideal cover from which to wage war against Christianity. This is evidenced by the fact that the Jesuits virtually controlled the Council of Trent, which produced anathema after anathema against Bible believing Christians and fundamental Christian doctrine. This same Jesuit organization, however, initially opposed the inquisition.[1243] Why? Because the inquisition, in part, was initiated to root out crypto-Jewry inside the Catholic church. Of course, an organization founded and controlled by crypto-Jews would oppose a strategy that would expose its own Jewish core. Once the Jesuits gained control of the papacy they took a more favorable view of the inquisition. They became the pope's militia controlling the organs of the inquisition in a spiritual battle against Protestants.

The hierarchy of Orthodox Jews hate Gentiles; ostensibly gentile organizations like the Roman church offer an ideal front from which to strike against the hated Christians. That is why God refers to the great harlot of Babylon as "MYSTERY." The Roman church appears to the world to be a gentile religious organization, yet it is to its core Jewish. As long at the Roman church does the bidding of the Jews, they will defend it. Once the Roman church is no longer useful it will shed it like a snake sheds its old skin.

It seems that the Jews have the destruction of the Roman church all planned out. The following is a passage from the THE

PROTOCOLS OF THE LEARNED ELDERS OF ZION that seem to foretell the destruction of the Vatican by the nations of the world.

> When the time comes finally to destroy the papal court the finger of an invisible hand will point the nations towards this court. When, however, the nations fling themselves upon it, we shall come forward in the guise of its defenders as if to save excessive bloodshed. By this diversion we shall penetrate to its very bowels and be sure we shall never come out again until we have gnawed through the entire strength of this place.[1244]

That is very similar to the prophecy found in the book of Revelation.

> And the ten horns which thou sawest upon the beast, **these shall hate the whore, and shall make her desolate and naked, and shall eat her flesh, and burn her with fire.** For God hath put in their hearts to fulfil his will, and to agree, and give their kingdom unto the beast, until the words of God shall be fulfilled. And the woman which thou sawest is that great city, which reigneth over the kings of the earth. (Revelation 17:16-18)

The horns of the beast are ten kings on the beast that is being ridden by the great harlot who sits on seven mountains (the Vatican). Revelation 17. The PROTOCOLS have set forth the planned destruction of the Vatican. The PROTOCOLS seem also to have provided for the entrance of the antichrist. "THE KING OF THE JEWS WILL BE THE REAL POPE OF THE UNIVERSE, THE PATRIARCH OF THE INTERNATIONAL CHURCH."[1245]

In fact, the Jesuits, the great defenders of the Catholic faith, have actually exercised their power to undermine the authority of the church and bring it to its knees. The reason the Jesuits retaliated against the Catholic church was that the Jesuits were at one time dissolved as a Catholic order because of their subversive conduct. The subversion of the European nations by the Jesuits became so great that an immense amount of military and political pressure was brought against the pope by the European nations. Finally, Pope Clement XIII decided on the 3rd of February 1769 to dissolve the Jesuits. While Jesuits are under an oath of allegiance to the pope, that oath is secondary to their extreme oath of allegiance to the Jesuit General:

> I do further promise and declare, that I will have no opinion or will of my own, or any mental reservation whatever, even as a corpse or cadaver [perinde ac cadaver] but unhesitatingly, obey each and every command that I may receive from my superiors in the Militia of the Pope and Jesus Christ.[1246]

The night before Pope Clement XIII was to execute the dissolution, however, he suddenly fell ill and died. Prior to his death he cried out "I am dying . . . It is a very dangerous thing to attack the Jesuits."[1247] His successor, Pope Clement XIV, was also put under tremendous political pressure to dissolve the Jesuits, but he resisted doing so for three years until the political tension finally forced his hand. Pope Clement XIV issued the papal brief of dissolution, *Dominus ac Redemptor,* on August 16, 1773.[1248] Pope Clement XIV knew the significance of such an act to the papacy; he exclaimed: "I have cut off my right hand."[1249] In addition, Pope Clement XIV knew that by signing the brief dissolving the Jesuits he was signing his own death warrant. Soon after signing the brief the letters I.S.S.S.V. appeared on the palace walls in the Vatican.[1250] Pope Clement XIV knew what it meant and explained that it stood for *In Settembre, Sara Sede Vacante.*

Which translated means "in September, the See will be vacant (the pope will be dead)."[1251] Pope Clement XIV was poisoned and died on September 22, 1774.[1252]

Interestingly, it was just three years after Pope Clement XIV's suppression of the Jesuits that the subversive organization the "Illuminati" was purportedly founded by a trained Jesuit named Adam Weishaupt in 1776.[1253] Weishaupt was a Jew and a professor of canon law at Ingolstadt University, which was a Jesuit University and the center of the Jesuit counter-reformation.[1254] Alberto Rivera, a former Jesuit priest, stated that the occult Illuminati organization was not founded by Weishaupt, as many believe, but in fact was established long before Weishaupt. The Illuminati is in fact a reincarnation of the ancient *Alumbrados*, whose one time leader was Ignatius of Loyola, the founder of the Jesuits.[1255] The Illuminati was established by Lorenzo Ricco, the Jesuit General, in 1776, who used his disciple, Adam Weishaupt, as the front man for the new organization (which was really not new at all).[1256] The Jesuits, having just been suppressed by the pope in 1773, found it necessary to establish the Illuminati, which was an alliance between the Jesuits and the very powerful Ashkenazi Jewish Banking House of Rothschild. The purpose of Weishaupt initially was to avenge the papal suppression of the Jesuits by rooting out all religion and overturning the governments of the world, bringing them under a single world government, controlled of course by the Illuminati, under the authority of their god. That world government is commonly referred to by the Illuminati as the "New World Order." The god of the Illuminati is Satan.[1257]

Eric Jon Phelps in his book, *The Vatican Assassins*, explains:

> These 41 years [between their suppression in 1773 by Pope Clement XIV and their reestablishment by Pope Pius VII in 1814] were

absolutely golden for the Society of Jesus. For the Sons of Loyola punished all their enemies, including the Dominican priests, perfected the inner workings between themselves and Freemasonry, creating alliance between the house of Rothschild in establishing the illuminati; punished and absorbed the Knights of Malta . . . . They used the Orthodox Catherine of Russia and a Lutheran Frederick of Prussia to conquer and divide Poland, rendering the pope's Bull of Suppression of no effect in that Roman Catholic land. They caused the French Revolution, beheaded a Bourbon King and a Hapsburg Queen as punishment for being expelled from France and Austria. With Napoleon, the Freemason, they drove the Bourbons from their throne in Spain and the Braganzas from their throne in Portugal. They even attempted to take Palestine from the Moslems like the Crusaders of old.[1258]

The company's most important victories were both religious and political. They deeply penetrated the Russian Orthodox Church and Germany's Lutheran Church, its Tubingen University specifically. Politically, they took control of the crown and the Bank of England. For this reason England, with Viscount Palmerston, would never go to war with France again, it would conduct the Pope's opium wars against people of China (just like the company, with its CIA and Mafia Commission is presently conducting a massive drug trade against the "heretic and liberal" people of the American Empire) . . . The Jesuits also captured the Papacy with the Vatican; along with its landed church properties the world over, and for this reason the

> Papal Caesar, occupying Satan's sacred office of the Papacy, would never suppress the Society of Jesus *ever again*![1259]

The secret Illuminati organization was the hidden guiding hand behind the brutal French Revolution of 1787, during which 300,000 people were massacred in a godless orgy of violence.[1260] The rage by the Jesuits and their reconstituted Illuminati culminated in 1798, with the capture of the pope himself. Napoleon's General Berthier invaded Rome, took the Pope Pius VI prisoner, and held him until his death.

The Roman Catholic church had learned its lesson. On August 7, 1814, the Jesuits were restored as a Catholic order by Pope Pius VII.[1261] John Adams wrote to Thomas Jefferson in 1816 "I am not happy about the rebirth of the Jesuits. . . . Swarms of them will present themselves under more disguises ever taken by even a chief of the Bohemians, as printers, writers, publishers, school teachers, etc. If ever an association of people deserved eternal damnation, on this earth and in hell, it is the Society of Loyola. Yet, with our system of religious liberty, we can but offer them a refuge."[1262] Thomas Jefferson answered Adams: "Like you, I object to the Jesuits' reestablishment which makes light give way to darkness."[1263]

In 1835, Samuel Morse, the great inventor of the telegraph, echoed the concerns of Jefferson and Adams; he described the Jesuits and their threat to the United States as follows:

> And do Americans need to be told what *Jesuits* are? If any are ignorant, let them inform themselves of their history without delay: no time is to be lost: their workings are before you in every day's events: they are a *secret* society, a sort of Masonic order, with superadded features

> of most revolting odiousness and a thousand times more dangerous. They are not confined to one class of society; they are not merely priests, or priests of one religious creed, they are merchants, and lawyers, and editors, and men of any profession, and no profession, having no outward badge (in this country,) by which to be recognised; they are about in all your society. They can assume any character that of angels of light, or ministers of darkness, to accomplish their one great end, the *service* upon which they are sworn to *start at any moment, in any direction,* and for any service, commanded by the general of their order, bound to no family, community, or country, by the ordinary ties which bind men; and *sold for life* to the cause of the Roman Pontiff.[1264]

The concerns of Morse, Adams, and Jefferson were justified; once being reestablished as a Catholic order the Jesuits did not miss a beat; during the 19th century they fomented revolutions throughout the world, attempting to bring to power oppressive despots whom they would then control. They were at one time or another expelled from Russia (1820), Belgium, Portugal (1834), the Italian states (1859), Spain (three times-1820, 1835, and 1868), Germany (1872), Guatemala (1872), Mexico (1873), Brazil (1874), Ecuador (1875), Colombia (1875), Costa Rica (1884), and France (twice-1880 and 1901).[1265] They caused the Swiss Civil war in 1847. As a result they were banished from Switzerland in 1848.[1266] Up until the year 2000, the Swiss Constitution (article 51) forbade the Jesuits from engaging in any cultural or educational activity in Switzerland.[1267] In the year 2000 Switzerland ratified a new constitution, in which article 51 was removed. Those are just a sampling of the over 70 countries from which the Jesuits have been expelled for conducting subversive activities. The Jesuit subversion has continued to modern times, causing the Jesuits to be expelled from Haiti in 1964 and Burma in 1966.[1268] To this day they are instigating communist revolutions in South America. The Jesuits' new brand of South American communism is known as "Liberation Theology."

Jesuits have a long and sordid history of distorting moral obligations and practicing and advocating situational ethics. For example, God commands without exception that "Thou shalt not bear false witness against thy neighbor." Exodus 20:16. The Jesuits, on the other hand, permit the use of ambiguous terms to mislead a judge or outright lying under oath if the witness makes a mental reservation.[1269] The Jesuits teach that if a young girl is pregnant, she may obtain an abortion if the pregnancy would bring dishonor to her or a member of the clergy.[1270] They do not stop there; another Jesuit maxim states: "If a Father, yielding to temptation, abuses a woman and she publicises what has happened, and, because of it, dishonours him, this same Father can kill her to avoid disgrace."[1271] That is not the only cause that is justification for murder. The Jesuits further teach that "[a] monk or a priest is allowed to kill those who are ready to slander him or his community."[1272]

Immorality is not unique to the Jesuit order only. The doctrines of the Catholic Church allow for all sorts of situational ethics. Thomas Aquinas, the most influential source of economic and theological doctrines for the Catholic Church, stated that it is lawful and not a sin for a man to steal another's property in order to fulfill a basic need. Aquinas' view was that all goods are community goods and therefore it is not a sin to take another's property when you need it.[1273] That is in fact the official position of the Roman church today as expressed by the Second Vatican Council. "If one is in extreme necessity he has the right to procure for himself what he needs out of the riches of others."[1274]

What is the authority for this ethic? It is found in the Talmud. In *Baba Bathra 54b* it states: "The property of a heathen is on the same footing as desert land; whoever first occupies it acquires ownership."[1275] Compare that to the eighth commandment of God: "Thou shalt not steal." Exodus 20:15. Furthermore, in *Baba Kamma 113b* it states that if one finds lost property it must be returned if the owner is a Jew; however if the owner is a gentile it can be kept. "It is to your brother that you make restoration, but you need not make restoration to a heathen."[1276]

The Jesuits are zealous persecutors of Christians or anyone whom they view as an enemy of the Vatican. Jesuits take a solemn oath to destroy Protestant Christians and destroy any government that offers protection to Protestant Christians.[1277] They are the natural enemies of liberty, their whole system is based on thoughtless, ruthless, blind obedience. Ignatius himself writing to his Jesuits in Portugal said: "We must see black as white if the church says so."[1278] Jesuits are the subversive ambassadors of the Catholic Church, bringing chaos and ruin to all nations they infiltrate. They believe that "[t]he Catholic Church has the right and duty to kill heretics because it is by fire and sword that heresy can be extirpated. . . . Repentance cannot be allowed to save them, just as repentance is not allowed to save civil criminals; for the highest good of the church is the unity of the faith, and this cannot be preserved unless heretics are put to death."[1279]

Alberto Rivera, a former Jesuit Priest, was saved by the grace of God and came out of the Jesuit priesthood. The Jesuits made numerous attempts to kill him before he could reveal the secrets of the Jesuits. He survived the attempts on his life and exposed much about sinister methods and motives of the Jesuits.

Franz Wernz, the Jesuit General from 1906-1915, stated that "[t]he Church can condemn heretics to death, for any rights they have are only through our tolerance, and these rights are apparent not real"[1280] That view of heretics having no rights is very similar to the philosophy of the Jewish Talmud. Sanhedrin

57b provides that if a heathen robs another or a Jew the property must be returned. If, however, a Jew robs a heathen, the property does not have to be returned to the heathen.[1281] That section of the Talmud also states that if a heathen murders another heathen or a Jew, the heathen should suffer the death penalty. If, however, a Jew kills a heathen "there is no death penalty."[1282] For example, Jewish Rabbi Moseh Levinger was sentenced to a mere 5 months in jail by an Israeli court for the unprovoked cold-blooded murder in September 1988 of a Palestinian shopkeeper, Hassan Salah.[1283] Prior to entering prison, Rabbi Levinger was feted at a party by Israeli President Chaim Herzog and Israeli Army General Yitzhak Mordechai.[1284] Rabbi Levinger was released from prison after only serving three months.[1285] Baba Kamma 38a expressly states that Canaanites are "outside the protection of the civil law of Israel." Thus a Canaanite has no remedy against a Jew whose ox gores the Canaanite's ox, but if a Canaanite's ox gores a Jew's ox, the Canaanite shall pay in full the damages.

Professor Israel Shahak explains:

> According to the Jewish religion, the murder of a Jew is a capital offense and one of the three most heinous sins (the other two being idolatry and adultery). Jewish religious courts and secular authorities are commanded to punish, even beyond the limits of the ordinary administration of justice, anyone guilty of murdering a Jew. A Jew who indirectly causes the death of another Jew is, however, only guilty of what talmudic law calls a sin against the 'laws of Heaven', to be punished by God rather than by man.
>
> When the victim is a Gentile, the position is quite different. A Jew who murders a Gentile is guilty only of a sin against the laws of Heaven, not punishable by a court.[1286] To cause indirectly the

> death of a Gentile is no sin at all.[1287]
>
> Thus, one of the two most important commentators on the Shulhan Arukh explains that when it comes to a Gentile, 'one must not lift one's hand to harm him, but one may harm him indirectly, for instance by removing a ladder after he had fallen into a crevice .., there is no prohibition here, because it was not done directly.[1288] He points out, however, that an act leading indirectly to a Gentile's death is forbidden if it may cause the spread of hostility towards Jews.
>
> A Gentile murderer who happens to be under Jewish jurisdiction must be executed whether the victim was Jewish or not. However, if the victim was Gentile and the murderer converts to Judaism, he is not punished.[1289]

The animus by Jews toward Gentiles includes a prohibition against giving medical treatment to Gentiles. Israel Shahak explains:

> According to the Halakhah, the duty to save the life of a fellow Jew is paramount.[1290] It supersedes all other religious obligations and interdictions, excepting only the prohibitions against the three most heinous sins of adultery (including incest), murder and idolatry.
>
> As for Gentiles, the basic talmudic principle is that their lives must not be saved, although it is also forbidden to murder them outright.[1291] The Talmud itself expresses this in the maxim 'Gentiles are neither to be lifted [out of a well]

nor hauled down [into it]'. Maimonides[1292] explains:

"As for Gentiles with whom we are not at war ... their death must not be caused, but it is forbidden to save them if they are at the point of death; if, for example, one of them is seen falling into the sea, he should not be rescued, for it is written: 'neither shalt thou stand against the blood of thy fellow'[1293] - but [a Gentile] is not thy fellow."

In particular, a Jewish doctor must not treat a Gentile patient. Maimonides - himself an illustrious physician - is quite explicit on this; in another passage[1294] he repeats the distinction between 'thy fellow' and a Gentile, and concludes: 'and from this learn ye, that it is forbidden to heal a Gentile even for payment...'

However, the refusal of a Jew - particularly a Jewish doctor - to save the life of a Gentile may, if it becomes known, antagonize powerful Gentiles and so put Jews in danger. Where such danger exists, the obligation to avert it supersedes the ban on helping the Gentile. Thus Maimonides continues: ' ... but if you fear him or his hostility, cure him for payment, though you are forbidden to do so without payment.' In fact, Maimonides himself was Saladin's personal physician. His insistence on demanding payment - presumably in order to make sure that the act is not one of human charity but an unavoidable duty - is however not absolute. For in another passage he allows [a] Gentile whose hostility is feared to be treated 'even gratis, if it is unavoidable'.

> The whole doctrine - the ban on saving a Gentile's life or healing him, and the suspension of this ban in cases where there is fear of hostility - is repeated (virtually verbatim) by other major authorities, including the 14th century Arba'ah Turirn and Karo's Beyt Yosef and Shulhan 'Arukh.[1295] Beyt Yosef adds, quoting Maimonides: 'And it is permissible to try out a drug on a heathen, if this serves a purpose'; and this is repeated also by the famous R. Moses Isserles.
>
> The consensus of halakhic authorities is that the term 'Gentiles' in the above doctrine refers to all non-Jews.[1296]

That Jewish attitude toward non-Jews is the same attitude the crypto-Jewish Jesuits have toward Christians. Jesuit priests are subjected to certain "spiritual exercises" which were first devised by Ignatius Loyola. During the spiritual exercises the subject becomes possessed and controlled by a devil.

> We imbue into him spiritual forces which he would find very difficult to eliminate later, forces more lasting than all the best principles and doctrines; these forces can come up again to the surface, sometimes after years of not even mentioning them, and become so imperative that the will finds itself unable to oppose any obstacle, and has to follow their irresistible impulse.[1297]

Between 1569 and 1605 the Jesuits orchestrated no less than eleven plots against Protestant England, which involved invasion, rebellion, and assassination. Each is known by the leader of the treachery: Ridolfi, Sanders, Gregory XIII, Campion, Parsons, Duke of Guise, Allen, Throgmorten, Parry, Babington,

Sixtus V, Philip II of Spain, Yorke, Walpole, Southwell, and Guy Fawkes.[1298]

In the 1586 "Babington plot" the Jesuits along with other Catholics planned to kill Protestant Queen Elizabeth I, place Catholic Mary Stuart, Queen of Scots on the throne of England and bring England under subjection to the Pope of Rome. That plot was discovered and Mary was executed for her troubles.[1299]

After the failed Babington plot, the Pope, in league with Philip II of Spain, planned to invade England and bring it under papal control. In 1588 Spain brought the 136 ship Spanish Armada against England. The Sovereign God of the Universe whipped up a freak storm which devastated the Armada and allowed England with only 30 ships to defeat Spain after an eight hour sea battle.[1300]

On November 5, 1605 Jesuit led Roman Catholic conspirators planned to kill King James I and the entire English Parliament by blowing up the House of Lords. They placed 20 barrels of gunpowder under the House of Lords. The plan was to blow up the house of Lords when the Lords, Commons, and King were all assembled on November 5, 1605 for the opening of Parliament.[1301] The plot, however, was discovered and the conspirators were captured. To this day that event is simply referred to as the "Gunpowder Plot."[1302] November 5 is a national holiday in England, commemorating the Catholic conspiracy in the Gunpowder Plot. The holiday is called Guy Fawkes Day; Guy Fawkes was one of the Gunpowder Plot conspirators.[1303]

The Jesuit subversion of nations has caused 56 countries to ban the Jesuits, most of whom have since lifted the bans. In 1759 the Jesuits were banned throughout the Portuguese Empire.[1304] In 1764 the Jesuits were outlawed in France, and in 1767 they were banned from Spain.[1305] On April 6, 1762 the French Parliament issued the following "statement of arrest" (indictment):

> The said Institute [Jesuits] is inadmissible in any civilised State, as its nature is hostile to all spiritual and temporal authority; it seeks to introduce into the Church and States, under the plausible veil of a religious Institute, not an Order truly desirous to spread evangelical perfection, but rather a political body working untiringly at usurping all authority, by all kinds of indirect, secret, and devious means. . . .[The Jesuits' doctrine is] perverse, a destroyer of all religious and honest principles, insulting to Christian morals, pernicious to civil society, hostile to the rights of the nation, the royal power, and even the security of the sovereigns and obedience of their subjects; suitable to stir up the greatest disturbances in the States, conceive and maintain the worst kind of corruption in men's hearts.[1306]

All nations should learn from the experience of Protestant England and understand the threat that Rome and the Jesuits pose to any free country. The Roman Catholic Church uses religious superstition to usurp the authority and undermine the independence of any state. King Henry VIII cast off the yoke of Rome and declared that he was to be the head of the church in England.

In May 1538, the pope sought his revenge for the separation of the Church of England from Rome; the pope excommunicated all in Ireland who recognized the supremacy of the King of England or any ecclesiastical or civil power greater than that of the Roman Catholic Church. The events are recounted in the classic *Foxe's Book of Martyrs*:

> A short time after this, the pope sent over to Ireland (directed to the archbishop of Armagh and

> his clergy) a bull of excommunication against all who had, or should own the king's supremacy within the Irish nation; denouncing a curse on all of them, and theirs, who should not, within forty days, acknowledge to their confessors, that they had done amiss in so doing.
>
> Archbishop Browne gave notice of this in a letter dated, Dublin, May, 1538. Part of the form of confession, or vow, sent over to these Irish papists, ran as follows: "I do further declare him or her, father or mother, brother or sister, son or daughter, husband or wife, uncle or aunt, nephew or niece, kinsman or kinswoman, master or mistress, and all others, nearest or dearest relations, friend or acquaintance whatsoever, accursed, that either do or shall hold, for the time to come, any ecclesiastical or civil power above the authority of the Mother Church; or that do or shall obey, for the time to come, any of her, the Mother of Churches' opposers or enemies, or contrary to the same, of which I have here sworn unto: so God, the Blessed Virgin, St. Peter, St. Paul, and the Holy Evangelists, help me," etc. is an exact agreement with the doctrines promulgated by the Councils of Lateran and Constance, which expressly declare that no favor should be shown to heretics, nor faith kept with them; that they ought to be excommunicated and condemned, and their estates confiscated, and that princes are obliged, by a solemn oath, to root them out of their respective dominions.[1307]

The political and religious attacks against Protestant England by Rome continued up to and beyond 1641; in 1641 the beast of Rome planned a murderous insurrection in Ireland. The objective of the barbarous conspiracy was to murder all Protestants in Ireland, without exception. In this instance, as in many others, we find the Jesuits leading the murderous and maniacal charge. The Jesuits placed their hellish imprimatur on the massacre by beginning it on the feast day of their founder, Ignatius of Loyola. When the dust finally settled on the genocide, Rome had exterminated 150,000 innocent men, women, and children. This massacre illustrates the danger of a Roman Catholic majority in any country. No matter who seems to control the political reigns, when the lawful government is at odds with Rome there will be hell to pay. Rome is a master at mass insurrection through the incitation of base barbarians who have sold their soul to the superstition of the Roman Cult.

The Irish genocide was planned and orchestrated from the Vatican. It was executed through the leadership of the Jesuits and the other priests of Rome. He who has eyes let him see through the pious facade of Rome. He who has understanding let him understand the danger posed by Rome and the Jesuits. The Roman Catholic Church never changes.

King Henry VIII's error is all too obvious, for there is only one head of God's church and that is Jesus Christ. The pope was incensed at the insolence of King Henry. The pope, however, was not angry because the king blasphemously claimed the authority of Christ as head of the church, but because the king had replaced him, the supreme pontiff of Rome, as the head of the church. The King of England was politically too strong for the pope to do anything to change the situation in England (though he tried); consequently the monarch of England to this day is the head of the Church of England (known as the Anglican or Episcopal Church). The Episcopal church is one of the harlot daughters of the Roman Catholic mother of harlots. *See* Revelation 17:5. For the most part, the leadership of the Episcopal Church are not Christian; they are rather the proverbial rejected tares growing in Christ's wheat field.

The Jesuits are all over the godless philosophies of the Jews that are being passed off as science. Not only do we have Jesuit Priest Lemaitre taking credit for the Rabbi Nachmanides' big bang theory, but we have Jesuit Priest Pierre Teilhard de Chardin at the forefront of the religious belief of evolution masquerading as "science." Indeed, de Chardin was caught orchestrating the Piltdown Man fraud. It was the Piltdown Man fraud that did the most to embed the evolutionary religion in the minds of scientists and the curricula of schools. In 1913 Piltdown Man was announced to the world as clear evidence of a transition between man and ape. For 40 years it was touted as evidence in support of evolution, until in 1953 it was exposed as a forgery.

It was later discovered that the skull of Piltdown Man was from a modern man and that the jawbone and teeth were from an orangutan. The teeth in the jaw had been filed down to make them look human. The bones and teeth had been chemically treated to give them the appearance of being prehistoric. The bones were then planted at the burial site in which they were found. There is a strong belief among those who have investigated the matter that the noted Jesuit Priest Pierre Teilhard de Chardin was instrumental in perpetrating that hoax. The scientist who helped unmask the forgery, Dr. Kenneth Oakley, formerly of the British Museum, said that a letter written to him by Teilhard in 1954 had given him "strong indications that Teilhard was in collusion with Charles Dawson," in committing the Piltdown Man hoax.[1308]

The famous evolutionist Theodosius Dobzhanksy (The American Biology Teacher, volume 35, number 3, March 1973, page 129) quoted Pierre Teilhard de Chardin as saying: "Evolution is a light which illuminates all facts, a trajectory which all lines of thought must follow."[1309] That is not a scientific approach, it is a religious philosophy; it is a rejection of Christ. Jesus is our light that we must follow, not evolution. "Then spake Jesus again unto them, saying, I am the light of the world: he that followeth me shall not walk in darkness, but shall have the light of life." (John

8:12)

Despite this and other frauds among evolutionists, in 1996, Pope John Paul II announced that evolution is compatible with Christian beliefs. While evolution is compatible with Catholicism and Judaism, evolution is not compatible with Christianity; evolution is irreconcilable with and antagonistic to Christianity.

In 1998, the pope toned down his position, by announcing that evolution alone cannot account for human existence. He, however, did not repudiate his pro-evolutionary position. God's word describes Adam as being "**made** a living soul." The new corrupt Bible versions follow the evolutionary philosophy of the world.. For example, the New International Version (NIV) changes God's word to say that Adam "**became** a living being." In the NIV man was not created, but instead just "became." This evolutionary slant fits in nicely with the Roman Catholic and Judaic teachings.

AV (KJV)	**NIV**
And so it is written, The first man Adam **was made** a living soul. (1 Corinthians 15:45 AV (KJV))	So it is written: "The first man Adam **became** a living being." (1 Corinthians 15:45 NIV)

The new Bible versions being promoted by the Catholic Church are counterfeits. God's true word in the English language is the Authorized (King James) Version of the Bible. The Jesuits and the Roman Catholic Church could not stop the propagation of God's word, so they happened upon a strategy to lure people away from God's true word by presenting them with counterfeit Bibles, such as the NIV. The fake Bibles are part and parcel of the Roman Catholic animus toward God's true word found in the Authorized (King James) Version of the Bible. Dr. Ian Paisley, a Protestant minister and Member of the British Parliament from Northern

Ireland, Member of the European Parliament, and founder of the Democratic Unionist Party, exposed the Jesuit and Roman Catholic view of the Bible. Paisley uncovered an oath taken by Protestants during the dark ages upon their conversion into the Roman Catholic Church.

> Bishop Wordsworth, an eminent Church of England divine uncovers the secrecy of the Jesuits. In the exposure of a document, used by them in their early days to compel Protestants to submit to Mother Church. *Roman Catholic Confession Publicly Prescribed and Proposed to Protestants on their Admission to the Roman Catholic Church.*
>
> "We confess that we've been brought from heresy, to the true saving Roman Catholic faith, by the singular care of our supreme governors, and by the diligence and aid of our masters the fathers of the Order of Jesuits. And we desire to certify this by our vows to the world at large. We confess that whatever new thing the pope ordains, whether it be in Scripture or not in Scripture and whatever he commands is divine and therefore ought to be held by lay people in greater esteem than the precepts of the living God. We confess that the reading of holy scripture is the origin of heresy and schism and the source of blasphemy. We confess that holy scripture is imperfect and a dead letter until it is explained by the Supreme Pontiff, and allowed by him to be read by the laity. We confess and assert, that the Pope as our most holy father, is to be obeyed in all things, without any exception and that such heretics as contravene his orders, are not only to be burnt but to be delivered both body and soul to Hell."[1310]

Evolution is antithetical to the creation account found in the Bible. God has kept his promise to preserve his word. The Catholic Church, therefore, opted for the only strategy left open. It has presented the world with phoney Bibles that allow for evolution. The pope's pro-evolutionary views were simply confirming the tradition of his spiritual ancestors, the Jews. For example, Moses ben Maimon (a/k/a Maimonides), the most revered and authoritative rabbi in Judaism, had a decidedly evolutionary view. In fact, he put a racial spin on the Talmudic passage at Baba Mezia 114b, which states that only Jews are men, Gentiles ("heathen") are not men. Maimonides lowered blacks within the gentile category to a status just above a monkey.[1311] Maimonides stated:

> [T]he Negroes found in the remote South, and those who resemble them from among them that are with us in these climes. The status of those is like that of irrational animals. To my mind they do not have the rank of men, but have among the beings a rank lower than the rank of man but higher than the rank of apes. For they have the external shape and lineaments of a man and a faculty of discernment that is superior to that of the apes.[1312]

It seems that Maimonides (1135-1204 A.D.) framed the foundational principles for the theory of evolution long before Charles Darwin. The theory of evolution is founded upon the same racism expressed by Maimonides. Darwin was a racist who believed that Blacks were closer to apes in the evolutionary process. In fact, the liberal humanists don't want the general public to know that the full title of Darwin's seminal 1859 book on evolution was: "THE ORIGIN OF SPECIES BY MEANS OF NATURAL SELECTION OR THE PRESERVATION OF FAVORED RACES IN THE STRUGGLE FOR LIFE."

Darwin elaborated on his racist views as follows: "At some future period, not very distant as measured by centuries, the civilized races of man will almost certainly exterminate and replace the savage races throughout the world. At the same time the anthromorphous apes will no doubt be exterminated. The break between man and his nearest allies will be wider, for it will intervene between man in a more civilized state, as we may hope, even that the Caucasian and some ape as low as a baboon instead of as now between the Negro or Australian and the gorilla."[1313] That is very close to the racism expressed by Maimonides.

There seems to be some division among Jews, at least in public discourse, over the legitimacy of evolution to explain man's being. Many Jews reject creation and accept evolution, while others take the view that there is a god who did create man. Some in the creation camp accept a literal six days of creation, while others take the view that a god did it gradually through evolution. Rabbi Shraga Simmons expresses a view that attempts to reconcile the creation account in Genesis and the theory of evolution. This attempt to reconcile the irreconcilable tries to salvage the legitimacy of the theory of evolution in the face of the contrary biblical account:

> Rabbi Shimshon Rafael Hirsch (19th century Germany) further explains that each "Day" represents a specific stage of creation - i.e. a mingling of raw materials and bursts of dramatic new development. As you go through the Torah's account, you see described a gradual process from simple to more complex organisms - first a mass of swirling gasses, then water, then the emergence of dry land, followed by plants, fish, birds, animals, and finally, human beings. This pattern may be similar to the evolutionary process proposed by science.[1314]

The problem with Rabbis Simmons and Hirsch is that they have taken what God has stated in the Bible and proposed that God did not mean what he said. The Bible does not offer proof of a "gradual process from simple to more complex organisms" as theorized by Rabbi Simmons. The Bible presents creation of the heavens and the earth out of nothing within six literal days.

The theory of evolution is not only contrary to God's word, but it is not based on true science; its origins are from heathen religious beliefs. According to the established laws of science, evolution is an impossibility. The second law of thermodynamics, also known as the law of entropy, is that all matter, living or inanimate, goes from a state of order to disorder. The theory of evolution reverses that sequence and states that over time organisms go from a state of disorder to order; from the simple to the complex.

To illustrate the conflict between evolution and the laws of science, suppose one were to write each letter of one's name on a separate card. If those cards were thrown out a second story window, they would scatter and fall to the ground in a chaotic display. The scattering of the cards over time as they fall to the ground illustrates the law of entropy. The evolutionist would say that the reason that the cards did not fall to the ground in order, spelling out the person's name, is that they were not given enough time to become orderly. The evolutionist would advise one to get into an airplane and throw the cards out of the plane when it reached an altitude of 10,000 feet. By the theory of evolution the more time the cards are in the air falling, the more time they have to organize and spell out the person's name when they finally land on the ground. According to the law of entropy, and common sense, giving the cards more time to fall to the ground only increases the disorder. The evolutionist, however, contrary to the laws of science and common sense, would have you believe that the more time the cards have to fall to the ground, the more orderly they will become.

The theory of evolution, which flows from a heathen religious philosophy, is the seed that germinated into communism and socialism. The Jewish Encyclopedia states that Judaism is not contrary to the theory of evolution and admits that evolution has been applied to every part of society.

> Herbert Spencer and others have applied the theory of evolution to every domain of human endeavor—civilization, religion, language, society, ethics, art, etc., tracing the line of development from the homogeneous to the heterogeneous, though recrudescences of and lapses into older forms and types (degeneration, atavism) are by no means excluded. The relation of the teachings of Judaism to this theory is not necessarily one of hostility and dissent.[1315]

Hitler, Lenin, Stalin, and Trotsky were all converts to the theory of evolution. Evolution was the foundational philosophy for their political actions and their justification for their maniacal brutality. Once one becomes a believer in evolution, it is a small step beyond that to being a believer in a communist revolution. It is no surprise that the Russian communist revolution was controlled from the beginning by Jews. The atheism of evolution is reconcilable with Judaism; that explains why one sees the seemingly strange occurrence of so many atheistic and agnostic Jews taking part in the Judaic liturgy of the synagogue. The atheism of evolution makes it easy to conclude that if there is no life giver, there is no law giver, no one made me, no one owns me, and, therefore, there is no right and wrong. Thus, there is nothing intrinsically wrong with stealing, assault, torture, murder, even murdering millions of people.

God "hath made of one blood all nations of men for to dwell on all the face of the earth." Acts 17:26. Racial distinctions are contrary to the commands of God: **"Judge not according to**

the appearance, but judge righteous judgment." John 7:24. *See also* 1 Samuel 16:7 "But the LORD said unto Samuel, Look not on his countenance, or on the height of his stature; because I have refused him: for *the LORD seeth* not as man seeth; for **man looketh on the outward appearance, but the LORD looketh on the heart.**"

Christians should understand that our war is not a carnal war where distinctions are made between races of people as defined by the pagan world system. Christians are in a spiritual war against unseen "spiritual wickedness in high places." Ephesians 6:12. **"For though we walk in the flesh, we do not war after the flesh: (For the weapons of our warfare are not carnal,** but mighty through God to the pulling down of strong holds;) Casting down imaginations, and every high thing that exalteth itself against the knowledge of God, and bringing into captivity every thought to the obedience of Christ; And having in a readiness to revenge all disobedience, when your obedience is fulfilled. **Do ye look on things after the outward appearance?** If any man trust to himself that he is Christ's, let him of himself think this again, that, as he *is* Christ's, even so *are* we Christ's." (2 Corinthians 10:3-7)

42 Heliocentric "Christianity"

Jesus warned his disciples to beware of the doctrine of the Pharisees and Sadducees. Matthew 16:6-12. The essence of that doctrine is to replace God's word with man made traditions. Mark 7:7-13. Jesus knew that the worldly traditions of men would be used to turn men away from him. "Beware lest any man spoil you through philosophy and vain deceit, after the tradition of men, after the rudiments of the world, and not after Christ." (Colossians 2:8)

God prophesied that his warning would go unheeded.

> For the time will come when they will not endure sound doctrine; but after their own lusts shall they heap to themselves teachers, having itching ears; And they shall turn away their ears from the truth, and shall be turned unto fables. (2 Timothy 4:3-4)

As prophesied, the modern churches have not heeded Jesus' warning. Modern churches have, almost universally, bought into the heliocentric myth. That Judaic superstition, masquerading as science, is the foundation upon which Satan has built false doctrines in the churches.

The heliocentric fable creates a misunderstanding of God's creation. To misunderstand God's creation causes a concomitant misunderstanding of the characteristics of God. God has shown man his character through his creation. If man has a misunderstanding of God's creation, he will have a misunderstanding of who God really is. Paul explains this in his letter to the Romans:

> [T]hat which may be known of God is manifest in them; for God hath shewed it unto them. For **the invisible things of him from the creation of the world are clearly seen, being understood by the things that are made, even his eternal power and Godhead**; so that they are without excuse. (Romans 1:19-20)

Jesus is God. Jesus created all things in heaven and on earth. There is nothing that has been created that Jesus did not create.

> For **by him were all things created, that are in heaven, and that are in earth, visible and invisible**, whether they be thrones, or dominions, or principalities, or powers: **all things were created by him, and for him**: And he is before all things, and **by him all things consist**. And he is the head of the body, the church: who is the beginning, the firstborn from the dead; that in all things he might have the preeminence. (Colossians 1:16-18)

To misunderstand the creation causes those with such misunderstanding to in turn misunderstand the character of Jesus. If one believes in the heliocentric creation, he will necessarily believe in a heliocentric creator. A heliocentric creation does not exist. So also, a heliocentric creator does not exist. A heliocentric creator is a false god. We have been warned to avoid the preaching of a false gospel, which presents a false Jesus.

> But I fear, lest by any means, as the serpent beguiled Eve through his subtilty, so your minds should be corrupted from the simplicity that is in Christ. For if he that cometh preacheth **another Jesus**, whom we have not preached, or if ye receive another spirit, which ye have not received, or **another gospel**, which ye have not accepted, ye might well bear with him. (2 Corinthians 11:3-4)

This false heliocentric Jesus is at the core of many false doctrines. Satan uses the subtle strategy of first undermining the authority of the Bible by introducing the concept that God does not really mean what he has said in the Bible about the flat earth he has created. Once that seed is planted in the minds of the congregation, the (beguiled and beguiling) pastor can then undermine what God has said about faith and doctrine. Under this rubric where duped (or unscrupulous) pastors portray the scriptures as having meaning other than the expressed words, doctrine becomes malleable.

One example of a false doctrine born of a false heliocentric Jesus is Arminianism. The false Arminian god is no longer the sovereign God of the Bible by whom "all things consist." Instead, the Arminian god is a kind-of super hero, who can foresee things that will happen in the future, but is powerless to interfere with man's free will. Jesus made it clear that understanding his creation is important to understanding the true gospel. "If I have told you earthly things, and ye believe not, how shall ye believe, if I tell you of heavenly things?" (John 3:12) That statement was made to explain the importance of the analogy he drew between wind and the work of the Holy Spirit to illustrate how salvation is completely by the grace of the sovereign God and outside of the will or control of man. Jesus flat out said we must be born again to see the kingdom of God. John 3:3-7. In the analogy, he compared the new birth of salvation, which can only come through the Holy Spirit, with the wind. Just as "the wind bloweth where it listeth, and thou hearest the sound thereof, but canst not tell whence it cometh, and whither it goeth: so is every one that is born of the Spirit." (John 3:8) His point is that man cannot birth himself. We are born again not of blood, nor of the will of the flesh, nor of the will of man, but of God. (John 1:13)

Arminianism is based upon five false premises: 1: God's election is conditioned on the free will choice of man; 2: Jesus atoned for the sins of everyone in the world, both saved and unsaved; 3: While man is depraved, God provides a special (prevenient) grace to all men that partially awakens them from their depravity so that they can make a free will choice whether to believe in Jesus; 4: Man can resist the grace of God; and 5: God assists one who is saved in resisting the temptations of the devil, but a person can by the exercise of his free will reject God and lose his salvation.[1316] For a more detailed discussion of this topic, please read this author's book titled: *The Anti-Gospel: The Perversion of Christ's Grace Gospel.*

Every one of those Arminian beliefs is refuted by the words of Christ in the gospel. The Bible makes it clear that Jesus is "the blessed and only Potentate, the King of kings, and Lord of lords." (1 Timothy 6:15) Jesus can not only see things that will happen, he makes those things happen. Jesus created all things, by him his creation consists, and he is a potentate over his creation. He did not leave something as important to his creation as salvation subject to the uncertain vicissitudes of man's free will. Indeed, he stated repeatedly in his word that he is in complete control of our salvation, from beginning to end. He is a creator of all things, even of the faith necessary for salvation. "Looking unto Jesus the author and finisher of our faith." (Hebrews 12:2) Jesus not only authors (creates) our faith, he finishes it to its end:

salvation. He truly creates all things and by him all things consist, including faith.

The Arminian Jesus, however, is not the creator of an immovable flat earth and is also not the creator (author) and finisher of our faith. The fictional, heliocentric, Arminian god created a fictional, unstable Earth and thus has unstable doctrine of salvation, whereby man is deemed to be the creator (author) and finisher of his own faith. That is a different Jesus and a different gospel. It is so much easier to sell the Arminian god to the churches when the masses have been inculcated in the far-off heliocentric god.

Such a far-off god only peers through empty space to foresee what man will do. That concept of a detached god is only one step removed from atheism, which is the ultimate end of heliocentrism. Indeed, under heliocentricism, with the vacuum of space in place of God's abode in heaven, the only rational conclusion is that there is no God. That is why heliocentrism is the very foundation of atheism itself. It stands to reason then that those who hold onto a belief in a god, and at the same time a belief in heliocentrism, would necessarily have a god who is much diminished in his power and influence. For those who believe in a god of a heliocentric universe would necessarily lean toward a concept of a god who is much less active in the day-to-day affairs of man. Where heliocentrism has not totally destroyed all conception of God in the minds of men, it has at least undermined his attributes and created in the minds of men an impotent god.

Arminianism is the natural outgrowth of the belief in a god who has created a heliocentric universe. This false Arminian god leaves the salvation of men up to their own free will decision. The Arminian god will not lift a finger to interfere in man's decision. Arminianism is the half sister of atheism.

For example, John Wesley, who was an ardent Arminian

and founder the Methodist Church, admitted that he didn't even believe in the God of the Bible. In a 1766 letter to his brother, Charles Wesley, John Wesley bared his soul and revealed to Charles his innermost thoughts. In that letter, which John Wesley never expected to be revealed publicly, he admitted that he preached a faith that he, himself, did not have. John Wesley felt "born along" by some unknown force to do so. God would certainly not compel the preaching of a false gospel. It is, therefore, clear that the unknown force bearing John Wesley along to preach the Arminian gospel was the devil. That is an ineluctable conclusion from Wesley's own words:

> In one of my last [letters] I was saying that I do not feel the wrath of God abiding on me; nor can I believe it does. And yet (this is the mystery), **I do not love God. I never did. Therefore I never believed, in the Christian sense of the word. Therefore I am only an honest heathen**...And yet, to be so employed of God! And so hedged in that I can neither get forward nor backward! Surely there was never such an instance before, from the beginning of the world! If I ever have had that faith, it would not be so strange. **But I never had any other evidence of the eternal or invisible world than I have now; and that is none at all**, unless such as faintly shines from reason's glimmering ray. **I have no direct witness (I do not say, that I am a child of God, but) of anything invisible or eternal.**
>
> And yet I dare not preach otherwise than I do, either concerning faith, or love, or justification, or perfection. And yet I find rather an increase than a decrease of zeal for the whole work of God and every part of it. I am borne along, I know not how, that I can't stand still. **I want all the world**

to come to what I do not know.[1317]

Wesley was 63 years old when he wrote that letter. The dirty secret of Wesley is that he was a heathen, who did not believe in God. He preached a false gospel about a false god, in whom he did not really believe. How could Wesley so successfully preach a false gospel? Because people had been accustomed to ignoring God's words and accepting a contradictory gloss to those words. This process was born with heliocentricity, where it was necessary for the "Christian" churches to accommodate the new science of heliocentricity by reinterpreting God's word to conform with that so-called "science." The Arminian god appears nowhere in the scriptures.

Wesley's Arminianism was only a hair's breadth from atheism. There is little difference between the Arminian god, who minds his own business and leaves his creatures to their own devices, and no god at all. It is no wonder then that Wesley did not believe in God. His Arminian theology created a god in whom it is easy to lose belief. The devil, that subtle beast, could not have designed it any better.

The God of the Bible, on the other hand, is a loving God, who effectually intervenes to save his elect. God's word is clear. It is not by one's own will or efforts in keeping God's law that one is saved. Rather, it is by God's grace through faith in Jesus Christ that we are born again. "Jesus answered and said unto him, Verily, verily, I say unto thee, Except a man be born again, he cannot see the kingdom of God." (John 3:3) A man cannot birth himself. To be spiritually reborn requires the intervention of God.

We, who believe in the Jesus of the Bible, are adopted children of God. We were chosen by God for adoption before the world was created. "**According as he hath chosen us in him before the foundation of the world, that we should be holy and without blame before him in love: Having predestinated us**

unto the adoption of children by Jesus Christ to himself, according to the good pleasure of his will." (Ephesians 1:4-5) Notice that salvation is according to the good pleasure of God's will, not the good pleasure of man's will.

Arminians will argue that man must of his own free will believe in Jesus to be saved. While it is true that faith is necessary for salvation, the source of that faith is Jesus. That's right, Jesus is both the object of one's faith and the source of that faith. Jesus is the author (creator) of saving faith. Hebrews 12:2. God did not just foresee that his elect would believe in him, he predestined that saving faith. Jesus is the finisher of our faith. Jesus sees that faith to its end.

We, who believe in Jesus Christ, were predestined to be glorified with Christ. "**For whom he did foreknow, he also did predestinate to be conformed to the image of his Son, that he might be the firstborn among many brethren. Moreover whom he did predestinate, them he also called: and whom he called, them he also justified: and whom he justified, them he also glorified.**" (Romans 8:29-30) From beginning to end, salvation is the work of a sovereign God. We are not insignificant beings on an insignificant planet careening through endless space. We are the unique creation of a sovereign God. We are created in his image on a flat immovable Earth, where he watches over us and guides us. He saves his elect according to his sovereign grace through faith in Jesus Christ.

In order to accept the heliocentric model, it is necessary for a Christian to redefine the words in the Bible to explain away its depiction of the flat earth. Once a Christian is down the road of rejecting God's description of a flat earth in the Bible, it is easy for him to redefine the words in the Bible in other matters of doctrine. For example, Arminians cannot ignore the plain language in the Bible that God predestined his elect for salvation. Arminians simply redefine the word "predestinate." Arminians claim that "predestinate" when referring to God's election of those to be saved is limited to mean only that God knows those who will exercise their free will and believe in Jesus. The Arminian interpretation is that "God, in his divine foresight, looked down through the corridors of time and saw all of those who would choose salvation in Jesus Christ. Having this divine knowledge, He then ratified men's votes of confidence in His ability to save them."[1318]

One can perceive in the Arminian doctrine their concept of a far-off god who looks from afar "through the corridors of time" at those who would believe in him. The God of the Bible, however, is not far off. He is near. God is above us walking in the circuit of heaven. "Thick clouds are a covering to him, that he seeth not; and he walketh in the circuit of heaven." (Job 22:14) He is a sovereign potentate, who actively saves his elect. He is not simply a ratifier of man's decisions as purported by Arminians. God states that he looks upon all the inhabitants of the earth from his habitation in heaven. "From the place of his habitation he looketh upon all the inhabitants of the earth." (Psalms 33:14)

Romans 8:29-30 states that not only did God foreknow, "he also did predestinate." God makes a clear distinction between foreknowing and predestinating. Notice, God "also did predestinate." The Arminians, however, contradict God and misread that passage to say: "he did foreknow, he also did foreknow." That makes no sense. The Bible states that God both foreknew and predestinated his elect.

"Foreknow" is a word that is pregnant with meaning. It not only means to know beforehand, but it also means to love beforehand. The heliocentric Arminian god does not predestinate his elect for salvation, and he does not love them beforehand, he only knows ahead of time those who would believe in him. The Arminian god is not the God of the Bible. The God of the Bible predestined his elect for salvation, whereas the fictional, Arminian god only foreknows those who will believe in him. The Arminian god is an ineffectual ratifier of the decisions of sovereign man.

It is necessary for God to intervene to save us, because we are incapable of saving ourselves. God explains in Ephesians:

> **And you hath he quickened, who were dead in trespasses and sins**; Wherein in time past ye walked according to the course of this world, according to the prince of the power of the air, the spirit that now worketh in the children of disobedience: Among whom also we all had our conversation in times past in the lusts of our flesh, fulfilling the desires of the flesh and of the mind; and were by nature the children of wrath, even as others. But God, who is rich in mercy, for his great love wherewith he loved us, **Even when we were dead in sins, hath quickened us together with Christ, (by grace ye are saved**;) And hath raised us up together, and made us sit together in heavenly places in Christ Jesus: That in the ages to come he might shew the exceeding riches of his grace in his kindness toward us through Christ Jesus. **For by grace are ye saved through faith; and that not of yourselves: it is the gift of God**:" (Ephesians 2:1-8)

We are saved by God's grace through faith in Jesus Christ. Arminians protest that faith is authored by the free will of man. However, the Bible expressly states that Jesus is the author of our faith. Faith is "not of ourselves; it is the gift of God." Ephesians 2:8. God makes that point even more clear in John, where he states emphatically that power to believe in Jesus Christ unto salvation is given by God. The faith that is the basis to be born again does **not** come from the will of man but is rather **"of God."**

> But as many as received him, to them **gave he power** to become the sons of God, even to them that believe on his name: **Which were born, not of blood, nor of the will of the flesh, nor of the will of man, but of God**." (John 1:12-13) (emphasis added)

When God states that he created all things visible and invisible, he means just that. Colossians 1:16-17. He created "all things." You will notice that among the "all things" that God created were "powers." Colossians 1:16. In John 1:12 we see that one of those powers is "power to become the sons of God, even to them that believe on his name." All things "visible and invisible" created by God includes the faith to believe in Jesus Christ. Saving faith is not only faith in Jesus, it is the faith **"of"** Jesus. Galatians 2:16. Jesus is "the author and finisher of our faith." Hebrews 12:2. Jesus not only created all things, but by him all things consist, including saving faith. Colossians 1:17. He is in the believer and the believer is in him. John 14:20. Just as God created a stable immovable earth, so also he created a stable immovable faith. His creation (and his faith) is a reflection of who he is. He is a God who reigns righteously over his immovable earth. "Say among the heathen that the LORD reigneth: the world also shall be established that it shall not be moved: he shall judge the people righteously." (Psalms 96:10) The immovable earth he created reflects the rock solid stability of his sovereign and righteous reign as the all powerful Lord of lords. "Which in his times he shall shew, who is the **blessed and only Potentate, the King of kings, and Lord of lords.**" (1Timothy 6:15)

A heliocentric creator, on the other hand, is a god who seems to have little control over his creation. A heliocentric creator is removed beyond the infinite realm of the vacuum of outer space. The false heliocentric god of Arminianism only foresees what people will do, but he will not intervene in the affairs of men.

Arminians have their god so far out of the picture that they do not allow their god to interfere in any way in the salvation decision. An Arminian believer who exercises his free will to choose to be saved can later exercise that same free will to jettison his salvation. The true God of the Bible is not so fickle. He saves his elect to the uttermost. "Wherefore he is able also to save them to the uttermost that come unto God by him, seeing he ever liveth to make intercession for them." (Hebrews 7:25)

One might think that there must be some passage in the Bible that supports the Arminian theology. No, there is no such Bible passage. In order to suggest that there is biblical authority, the Arminians use the satanic trick of quoting Bible verses out of context. Virtually any false doctrine can be supported by biblical text taken out of context, even to the extent of trying to prove that "there is no God." Indeed, Psalm 14:1 states: "There is no God." It is an accurate quote, but it has been taken out of context.

When we see the passage in context, we see that the quoted clause has quite a different meaning. The entire passage reads: "The fool hath said in his heart, There is no God. They are corrupt, they have done abominable works, there is none that doeth good." Psalm 14:1. The context we see gives quite a different meaning than is intended by our hypothetical atheist. In like manner, Arminians take Bible verses out of context to promote a meaning contrary to God's intended meaning.

One example where Arminians twist God's meaning is found in the Bible passage often cited by Arminians to support their unbiblical doctrine that God is willing that everyone in the world should be saved. Arminian churches quote part of 2 Peter 3:9, taken out of context, as authority for their doctrine. In fact, this single passage is so key to the Arminian theology that it is the motto of the *Society of Evangelical Arminians*.[1319] Their seal contains the statement: **"Not Willing That Any Should Perish"**, which is a clause taken out of context from 2 Peter 3:9.

The Arminians have hijacked the gospel and all of the terms that have traditionally been used in the Christian community to describe orthodox biblical Christianity. An organization calling itself the *Society of Evangelical Arminians* makes no historical sense. While almost all Arminians consider themselves evangelicals, they deny the foundational biblical doctrines that are at the core of what it historically meant to be an evangelical. Dr. Michael Scott Horton, who is the J. Gresham Machen Professor of Systematic Theology and Apologetics, in his article *Evangelical Arminians, Option or Oxymoron?*, explains that it is an oxymoron for an Arminian to be described as an evangelical.

> [T]he evangelicals who faced this challenge of Arminianism universally regarded it as a heretical departure from the Christian faith. One simply could not deny total depravity, unconditional election, justification by grace alone through faith alone because of Christ alone, and continue to call himself or herself an evangelical. There were many Christians who were not evangelicals, but to be an evangelical meant that one adhered to these biblical convictions. ... Today one can be an evangelical-which has historically meant holding to total depravity, unconditional election, justification by grace through faith alone, the sufficiency of scripture-and at the same time be an Arminian, denying or distorting this very evangelical message.[1320]

Franklin Graham, son of Billy Graham, speaking on behalf of the Billy Graham Evangelistic Association, stated: "According to 2 Peter 3:9, the Lord is 'not willing that any should perish but that all should come to repentance.'"[1321]

At first glance it would appear that 2 Peter 3:9 supports what Graham has said. Closer examination of that passage reveals that the passage does not in fact support that false Arminian doctrine promoted by Graham. Notice the missing passage. "The Lord is **[...]** not willing that any should perish, but that all should come to repentance." 2 Peter 3:9.

Those who try to force the square peg of scripture into the round hole of their false doctrine must shave off parts of the Bible in order to get it to fit. In this case, Graham, as is the practice with all Arminians, shaved that portion of the passage which limits its application to those who are already chosen for salvation. What God means in that passage is that God is not willing that any who have been chosen for salvation by God should perish, but that all those who are saved should come to repentance. Read the entire passage in context and you will see that God is **"longsuffering to us-ward."** God is not willing that "us" should perish and that "us" should come to repentance.

> The Lord is not slack concerning his promise, as some men count slackness; **but is longsuffering to us-ward**, not willing that any should perish, but that all should come to repentance. 2 Peter 3:9.

Who are the "us" in 2 Peter 3:9? Simply read the first paragraph of the letter and we see that Peter is writing to "them that have obtained like precious faith with us." "Simon Peter, a servant and an apostle of Jesus Christ, **to them that have obtained like precious faith with us** through the righteousness of God and our Saviour Jesus Christ:" (2 Peter 1:1)

One can see that in 2 Peter 3:9, Peter was stating that God was not willing that any who believe in Jesus should perish. God's will is always done, and his will cannot be thwarted by man's will. If God has foreordained one to salvation, no one can stay his hand. "And all the inhabitants of the earth are reputed as nothing: and he doeth according to his will in the army of heaven, and among the inhabitants of the earth: and none can stay his hand, or say unto him, What doest thou?" (Daniel 4:35)

While the Arminian salvation is tenuous, the true salvation by and through God is eternal and permanent. **"All that the Father giveth me shall come to me; and him that cometh to me I will in no wise cast out."** (John 6:37)

All who are chosen by the Father for salvation will be saved. Notice in John 6:37 that Jesus made the points that 1) **"All that the Father giveth me**," 2) "**shall come to me**"; and 3) "**him that cometh to me I will in no wise cast out**." We see: 1) God gives **all** of his elect to Jesus, 2) **all** of his elect will come to Jesus, and 3) **all** of his elect will be securely saved and cannot lose their salvation. John chapter 6 precludes the possibility of falling away from salvation. Chapter 6 of John refutes the Arminian theology. God uses parallelism in John 6:39-40 to lock down what he means when he states that salvation comes by the faith of which he is the author and finisher.

"And this is the Father's will which hath sent me, that **of all which he hath given me I should lose nothing**, but should raise it up again at the last day." (John 6:39)	"And this is the will of him that sent me, that **every one which seeth the Son, and believeth on him, may have everlasting life**: and I will raise him up at the last day." (John 6:40)

Notice that in John 6:40 states that "every one which seeth the Son, and believeth on him, may have everlasting life." It is those same believers in verse 40 that Jesus states in verse 39 are given to him by the Father: "of all which he hath given me I should lose nothing." That means that all whom the Father has given to Jesus will "believeth on him." God preordained that they would believe in Jesus and gave them to Jesus. Jesus will lose none of those who believe on him. Jesus will "raise him up at the last day." Salvation is locked in once it is given to the believer by the grace of God. "Every" person who believes on Jesus are the same "all" whom the Father has given Jesus. None of those believers will ever be lost; they all have "everlasting life." They "all" will be raised up on the last day.

Just to punctuate the point Jesus made it clear in John 6:44 that "No man can come to me, except the Father which hath sent me draw him: and I will raise him up at the last day." (John 6:44) The very theme of the Bible that salvation is only by the grace of God through faith in Jesus Christ. E.g., Ephesians 2:8. Combine John 6:44 with verses 37, 39, and 40 of John chapter 6, and we see that it is only those who believe in Jesus who can be saved. All whom God has ordained to believe in Jesus will in fact believe in Jesus unto salvation. No believer will ever be lost. "All that the Father giveth me shall come to me; and him that cometh to me I will in no wise cast out." (John 6:37) All believers will inherit eternal life. "I will raise him up at the last day." John 6:40 and 6:44.

On the flip side, without the ordained election of God, no man can believe in Jesus. It is those, and only those, whom God draws to Jesus who will be saved. John 6:44. God's drawing is effectual, all who are drawn will in fact believe in Jesus unto salvation. John 6:39-40. All who do not believe in Jesus are ordained by God not to believe in Jesus. "No man can come to me, except the Father which hath sent me draw him." John 6:44.

Jesus ended his discourse in John chapter 6 with that very point. He stated that there were some among his audience who did not believe in him. And Jesus explained why. Because, he stated, "no man can come unto me, except it were given unto him of my Father."

> But there are some of you that believe not. For Jesus knew from the beginning who they were

> that believed not, and who should betray him. And he said, Therefore said I unto you, that no man can come unto me, except it were given unto him of my Father." (John 6:64-65)

How do Arminians address John chapter 6? They reinterpret the language to say that God's drawing is only effectual for those of their own free will who choose to believe in Jesus. They claim that not all who are drawn will believe in Jesus and be saved. The problem with that interpretation is that it ignores the clauses Jesus put at the end of verses 37, 39, 40, and 44. Those clauses mean that of **all** whom the Father has given to Jesus, he draws **all** of them to Jesus, they **all** will believe in Jesus, and they **all** will be saved. "I will in no wise cast out." John 6:37; "I should lose nothing, but should raise it up again at the last day." John 6:39; I will raise him up at the last day." John 6:40; "I will raise him up at the last day." John 6:44.

Those clauses alone impeach the entire Arminian construct. How do Arminians address those clauses? Arminians ignore those clauses, as though they are not there. They do not even try to explain them, because there is simply no way to reconcile their Arminian theology with God's clear words.

Ken Johnson, Th.D., is typical in that regard. In his deceptive book, *The Gnostic Origins of Calvinism*, Johnson cites to the verses in John 6:37, 39, and 44 in support of the proposition that "[t]he Holy Spirit draws men to Himself with grace."[1322] While that is certainly true, immediately after citing to those Bible passages, Johnson ignores the clear language in those passages to say: "But this drawing can be resisted."[1323] How can he make such a statement after typing out three Bible passages, all of which taken together (or even read separately) mean clearly that all those who are drawn by God to Jesus, will in no wise be cast out, but that Jesus will raise them up on the last day? Jesus states emphatically "that of all of which he hath given me, I should lose nothing, but should raise it up again at the last day." John 6:39. Jesus allows no possibility for resisting the drawing of God. The drawing of God is effectual and leads to salvation. Clearly Dr. Johnson has an Arminian agenda. And his agenda does not include being a faithful witness to the word of God.

There are two groups of people in the world: 1) those whom God draws to Jesus for salvation, and 2) those whom God does not draw to Jesus for salvation. Those who are drawn by God to Jesus are ordained by God to be saved. Whereas, those who are not drawn by God to Jesus are ordained by God to be damned. "No man can come to me, except the Father which hath sent me draw him: and I will raise him up at the last day." (John 6:44)

There are those, like Judas, who appear for a time to be part of the church but in the end they make manifest that they are enemies of the gospel. "They went out from us, but they were not of us; for if they had been of us, they would no doubt have continued with us: but they went out, that they might be made manifest that they were not all of us." (1 John 2:19) Judas and others like him went out, not because they were saved and lost their salvation, but rather because from the beginning "they were not of us." That is, they were pretenders to salvation; they were unsaved tares congregating among the saved wheat. Matthew 13:27-43.

Those who do not believe in Christ are lost because they have not been chosen by God for salvation. Those who are chosen for salvation cannot lose their salvation. John 10:26-30. There is simply no such thing as a person losing his salvation.

> **But ye believe not, because ye are not of my sheep, as I said unto you. My sheep hear my voice, and I know them, and they follow me: And I give unto them eternal life; and they shall never perish, neither shall any man pluck**

them out of my hand. My Father, which gave them me, is greater than all; and no man is able to pluck them out of my Father's hand. I and my Father are one. (John 10:26-30)

There are only two possibilities in the gospel. First, those who are lost cannot believe, because God has not chosen them to believe. The other possibility is the flip side of the first. Those who are chosen to believe will in fact believe, and they cannot ever lose their faith. Jesus promised that "no man is able to pluck them out of my Father's hand." There is no category for persons to be first saved and then for them to overrule God's choice by the power of their free will and "unsave" themselves. Such an occurrence is an impossibility. The only way to build such a theology is to ignore the clear message of the gospel.

God's creation reveals his very character. If people are deceived by heliocentrism, they then have a perverted view of creation and thus a perverted view of their god. Heliocentrism creates in the minds of men a different kind of god. The heliocentric god is a god made after the image of man. Under heliocentrism, man is all powerful, with the heliocentric god being a far-off helpless and hapless spectator, unable to intervene to save anyone. Indeed, there is very little difference between such a god and no god at all. The whole point of heliocentrism is to remove the God of the Bible from people's minds. As the heliocentric creation does not exist, so also, the heliocentric creator does not exist. All religions that worship a heliocentric creator worship a false god. From that misguided worship flows all manner of false doctrine.

The God of the Bible is real. He is the creator of a flat earth that is at the center of his creation. He intervenes in the affairs of men to effectually save his people. The gospel is simple. Salvation comes only by the grace of God through faith in Jesus Christ. John 6:47; Ephesians 2:8-10. No man can be saved unless

God elects to save him. John 6:65. All whom God elects to save will be saved. John 6:39.

God makes one a Christian. God must change your heart. As Jesus said, a man must be born again. John 3:3. No man is born of himself. One must be born of God. “Know ye that the LORD he is God: it is he that hath made us, and not we ourselves; we are his people, and the sheep of his pasture.” (Psalms 100:3) Salvation is not by the will of man. “Which were born, not of blood, nor of the will of the flesh, nor of the will of man, but of God.” John 1:13. God must draw you. John 6:44. Unless God draws a man, he will have no desire to be a Christian.

Man by nature is spiritually dead. God must quicken you, that is, make you spiritually alive. Ephesians 2:1-10. You then become a new spiritual creation through God’s Holy Spirit.

There is no way that a man would accept those things written in the Holy Bible unless God has first opened his heart to the spiritual truths in the Bible. If one accepts that Jesus Christ is Lord God, the creator of the universe who reigns from heaven, he should submit completely to his authority. Ask the Lord in prayer to help you and he will. "And straightway the father of the child cried out, and said with tears, Lord, I believe; help thou mine unbelief." Mark 9:24.

Understand this simple truth, that if you ask Jesus to save you, he will. You will not, indeed you cannot, unless God draws you and gives you the ability to do so. He will then give you the gift of the Holy Spirit. Pray to Jesus for salvation.

> And he said unto them, Which of you shall have a friend, and shall go unto him at midnight, and say unto him, Friend, lend me three loaves; For a friend of mine in his journey is come to me, and I have nothing to set before him? And he from

> within shall answer and say, Trouble me not: the door is now shut, and my children are with me in bed; I cannot rise and give thee. I say unto you, Though he will not rise and give him, because he is his friend, yet because of his importunity he will rise and give him as many as he needeth. And I say unto you, **Ask, and it shall be given you; seek, and ye shall find; knock, and it shall be opened unto you. For every one that asketh receiveth; and he that seeketh findeth; and to him that knocketh it shall be opened.** If a son shall ask bread of any of you that is a father, will he give him a stone? or if he ask a fish, will he for a fish give him a serpent? Or if he shall ask an egg, will he offer him a scorpion? If ye then, being evil, know how to give good gifts unto your children: how much more shall your heavenly Father give the Holy Spirit to them that ask him? (Luke 11:5-13)

Those that ascribe to the free will mythology will cite the above passage as authority for their position that the source of faith is the will of man. However, that passage says nothing of the source of the faith, the passage simply explains the result of faith.

Faith comes from God, it is a gift; he will shower you with his merciful grace if you ask him. You must humble yourself before almighty God and ask for his mercy and grace. The only way that you can come to Christ is if he draws you and causes you to ask him to save you. John 6:44. "**Blessed is the man whom thou choosest, and causest to approach unto thee**, that he may dwell in thy courts: we shall be satisfied with the goodness of thy house, even of thy holy temple." (Psalms 65:4)

43 The Inevitable Shills

The world conspiracy against God and man has its shills that it presses into service in order to undermine any effort to expose its plans. Exposing heliocentrism as a lie and presenting evidence of a flat earth simply cannot be allowed. A video titled *Flat Earth*, posted on You Tube by Joseph Flatley on March 21, 2015, illustrates the methods used by the ruling elite to coopt any opposition to them. *Flat Earth* is an excellent video, and makes a very compelling case for the flat earth. The poster of the video, Joseph Flatley, summarizes his views in one of his comment postings on the video page:

> The Earth is Flat 1. No satellites, just under sea cable, radar and cell towers 2. We are not spinning 3. Horizon is flat 4. There is no gravity, just temperature, mass, pressure and density 5. We are not moving upwards either 6. Coriolis effect can be seen on an airplane (both directions simultaneously) 7. Antarctica surrounds us and goes on forever 8. We are in a melted puddle of ice 9. Sun and moon are very small and circle above us 10. Man cannot fly higher than 80,000 feet (ha ha to Virgin) 11. Historic tests were

> conducted that proved that the ether circles us 12. Whatever NASA, Disney and Hollywood preach, believe the opposite to be true 13. Photos of earth in space are paintings and graphics.[1324]

Flatley seems emphatic about his beliefs in a flat earth. In his comments, he calls those who believe in a globular earth "globetards."

> I am so sick of the brainwashed globetards. They say that our ancestors were unintelligent, when they themselves do not realize that we were all part of the 'dumbing down of society' agenda. People from the past were brilliant compared to us. When they observed and recorded that the Earth was FLAT........they were right. We were fooled our whole lives. Wake the hell up everyone. Picture yourself putting a clear cup upside down on your floor to catch a bug. Your floor represents the vastness of Earth, the cup represents the DOME and the bug is one of us. Flat Earth is not a disc you can fall from, the plane goes on forever in all directions. WAKE THE HELL UP.[1325]

The problem with all of this is that if Flatley believes what he is saying, he is perfectly willing to try to deceive people about the truth of it. He is a shill, who is apparently trying to coopt and control information about the flat earth. How do I know? On September 20, 2015, only 6 months after posting the above video, and while the video was still posted, he wrote an article attacking those who believe in a flat earth. Below is an excerpt of the hatchet job that he did on Rich Hopkins, who seems to be an honest seeker of the truth who has uncovered irrefutable proof of a flat earth.

> Since [Rich] Hopkins's flat Earth ideas involve

> all the traditional conspiracy tropes (and since conspiracy theory itself has anti-Semitism in its DNA) I had to ask: Does he believe that the Holocaust happened?
>
> "I don't know," Hopkins says. "I wasn't there. Like, a lot of people say, 'Do you believe in the Holocaust?' I wasn't there, I don't know. All I know is that the people that win the wars write the history." He refuses to believe something just because it's taught in school. "I guess I'd have to go over to Germany and do some forensic examination to prove it."
>
> This isn't science, and this isn't skepticism. It's stupidity. While this one guy's crank idea probably isn't too terribly dangerous, it reflects a larger problem of rejecting scientific understanding—not an inability to understand science, but a refusal of its methods and conclusions. How will we possibly hope to thrive and survive, mister, if we're wasting precious time and energy trying to figure out if the Earth is round, if the moon is real, or if the Holocaust even happened? Everything's up in the air, floating over an obviously, demonstrable, inarguably flat Earth.[1326] (emphasis added)

Notice how Flatley inadvertently revealed the real issue. He seemingly goes completely off-topic and states in the article "conspiracy theory itself has anti-Semitism in its DNA." He then segues right into a question about the Holocaust. In doing this, he revealed more about himself than he planned to.

Vladmir Lenin explained the communist principle that "the best way to control the opposition is to lead it ourselves." As previously revealed, a little known fact is that Lenin was a Jew. It is notable that Flatley attempts to smear those who believe in any conspiracy, especially flat earth conspiracy, as anti-Semites. He knows that the slightest examination of heliocentricity will reveal that it is based upon esoteric Jewish religious beliefs originating in Babylon and nurtured by the Jews in their Kabbalah and Talmud. That is why he stated in the context of flat earth science, "conspiracy theory itself has anti-Semitism in its DNA."

False claims of antisemitism are used as a shield to protect the Jewish hierarchy when evidence of their crimes are uncovered. Former Israeli government official Shulamit Aloni succinctly described the epithet antisemite as "a trick we [Jews] always use."[1327] She explained that Jews hide behind it like a smokescreen, which is used to conceal evidence of Jewish malefaction.

Shulamit Aloni knows what she is talking about. She was a member of the Israeli Knesset from 1965 to 1969 and again from 1974 to 1996. She served on the Knesset Constitution Committee, Law and Justice Committee; State Audit Committee; Education and Culture Committee; and the Finance Committee. She served briefly as Minister without Portfolio from June to October 1974. From 1992 to 1996, Aloni served as Minister of Communications and the Arts, Science and Technology.[1328]

By the trick of labeling a person as an antisemite, the accused person's credibility is undermined, and the focus shifts from the evidence against wrongdoing by the Jews to the motives of the accused antisemite. The Jewish controlled media then destroys the reputation of the accused antisemite, and he becomes an object lesson for anyone who might consider criticizing Israel or Jews.

Indeed all secret societies employ undercover agents of infiltration who seek to control the opposition by gaining a leadership position in that opposition. For example, the preamble to the Extreme Oath of the Jesuits states in pertinent part:

> My son, heretofore you have been taught to act the dissembler: among Roman Catholics to be a Roman Catholic, and to be a spy even among your own brethren; to believe no man, to trust no man. Among the Reformers, to be a Reformer; among the Huguenots, to be a Huguenot; among the Calvinists, to be a Calvinist; among other Protestants, generally to be a Protestant; and obtaining their confidence, to seek even to preach from their pulpits, and to denounce with all the vehemence in your nature our Holy Religion and the Pope; and even to descend so low as to become a Jew among Jews, that you might be enabled to gather together all information for the benefit of your Order as a faithful soldier of the Pope.[1329]

Joseph Flatley has been acting the part of a flat earth believer, as a way of controlling and undermining the growing awareness of it. He has Jesuitically pretended to believe in a flat earth and even posted a professionally polished video promoting his views in agreement with the evidence for a flat earth. At the same time he has been caught in his published article stating that believing in the flat earth "isn't science," and it "isn't skepticism." No, Flatley thinks "[i]t's stupidity." In his article, he called the idea of a flat earth a "crank idea." He accuses those who believe in a flat earth of "rejecting scientific understanding."

Flatley goes in for the kill in his article and tries to associate anyone who believes in a flat earth as being an antisemite, denying the Holocaust. Indeed, Flatley stated outright his position that "conspiracy theory itself has anti-Semitism in its DNA." Flatley tries, unsuccessfully, to paint Rich Hopkins (and by association all who believe in a flat earth) as a Holocaust denier. Saying that Hopkins views on the flat earth are "wasting precious time and energy trying to figure out if the Earth is round, if the moon is real, or if the Holocaust even happened."[1330] Although Hopkins has never publicly spoken about the Holocaust, Flatley tries to imply that Hopkins is a Holocaust denier. Flatley hopes by doing that to cause the vast majority of people to have a visceral Pavlovian response, which has been pre-programmed by years of Jewish Holocaust propaganda, of rejecting the science of the flat earth and instead believing the heliocentric religion passing as "science."

Who really is Joseph Flatley, and why is he so interested in the flat earth? Flatley seems to have an expertise in the area of social media, particularly as it affects Israel. He seems to have taken a great interest in the leveraging of social media for propaganda purposes by the Israeli Defense Forces (IDF) in its battles with Hamas in and around the Gaza strip and elsewhere.[1331] Flatley has written several articles on that topic for *The Verge.*[1332]

Indeed, Flatley seems to be hooked into the massive social media and internet propaganda effort by Israel. As reported at *Zero Hedge*, "Israeli service men are not the only ones engaged in pro-Israeli social media efforts. As we've noted for years, Israel has employed its civilians to automatically vote stories questioning Israel down and to send pro-Israel letters to politicians and media."[1333] Jonathan Cook reported in July 2009, that "Israel's foreign ministry is reported to be establishing a special undercover team of paid workers whose job it will be to surf the internet 24 hours a day spreading positive news about Israel."[1334] Notably, the *Zero Hedge* article quoted Joseph L. Flatley, apparently due to his expertise on the subject. The *Zero Hedge* article ends with stark evidence of how important propaganda is to Israel: "Indeed, Israeli prime minister Netanyahu publicly thanked Israel's keyboard warriors for battling on social media for the cause."[1335]

Who are these "keyboard warriors" to which Netanyahu refers? In January 2012, Ali Abunimah reported that the National Union of Israeli Students had partnered with the Israeli government to pay students $2,000 to spend 5 hours per week in spreading pro-Israel propaganda on the internet.[1336] It took four years, but finally the major media decided to report on Jonathan Cook's 2009 allegations. However, the major media made it sound like the internet propaganda effort by Israel was just getting off the ground in 2013, when in fact it had been going on for at least four years by the time the Associated Press reported it. In an August 14, 2013, article, the Associated Press reported that "Israel is looking to hire university students to post pro-Israel messages on social media networks without needing to identify themselves as government-linked, officials said Wednesday."[1337]

Flatley's schizophrenic positions of being both pro-flat earth and anti-flat earth at the same time certainly suggest a nefarious motive. Could Flatley be part of Israel's effort to spread Israeli propaganda? If so, it would seem that the State of Israel has an interest in putting a stop to any evidence of a flat earth being spread to the masses. In view of the fact that the heliocentric globular earth model at its core is a Kabbalistic religious belief, it would stand to reason that Israel would want to stop any information that impeaches it. Israel certainly wants to put a stop to any in-depth research on the issue, as it would lead directly to the Jews as the prime source of the myth of a globular, spinning earth. That explains Flatley's segue into allegations of antisemitism and Holocaust denial in his flat earth article.

44 Compromising the Faith

Heliocentricity is so ingrained within the world's culture that many Christian ministries refuting evolution cling to the heliocentric model of a globular, spinning earth, despite clear biblical passages that refute that model. For example, Robert Carter, Ph.D., coauthored an article titled, *Why the Universe Does Not Revolve Around the Earth, Refuting Absolute Geocentrism*, which was posted on the *Creation Ministries International* website on February 12, 2015.[1338] In the article, Dr. Carter explained away the Bible passages describing the geocentric creation by arguing that God was not accurately describing the movement of the sun, but that God was instead using "phenomenological language." Below is a dialogue that this author had with Dr. Carter in the comments section to his article.

> **Edward H., United States, 25 February 2015**:
>
> You subtitled the article "Refuting absolute geocentrism," but you did no such thing. A refutation is proof that something is wrong. You presented arguments against geocentricity, but your article can hardly be termed a refutation. To

refute geocentricity would require proof. However, you have craftily dodged and avoided some strong scientific and biblical arguments that support geocentricity. For example, you made passing reference to Isaiah 38:1–7, but you did not cite or discuss the operative passage that proves geocentricity, which appears in verse 8. Why not? If you are refuting geocentricity, one would think that you would address Isaiah 38:8. "Behold, I will bring again the shadow of the degrees, which is gone down in the sun dial of Ahaz, ten degrees backward. So the sun returned ten degrees, by which degrees it was gone down." Isaiah 38:8. That passage clearly states that "the sun returned ten degrees, by which degrees it was gone down." God is stating clearly that he moved the shadow on the sun dial ten degrees by moving the sun back in its path by ten degrees. Your statement that God is using "phenomenological language" in the Bible passages that support geocentricity is just a clever way of saying that God is telling us a fib, because we are too dense to understand what is really happening. God could have easily stated that he moved the earth back ten degrees in its rotation, if that is what he had done. I think we could understand that phenomenon. God did not say that, because he did not do that. God is not lying to us about what he did by using what you call "phenomenological language." God explained exactly what happened, "the sun returned ten degrees, by which degrees it was gone down." Isaiah 38:8.

Robert Carter responds:

We have not 'craftily dodged' anything, and your

> argument really amounts to nothing. The refutation is in the physics, and so far nobody has attempted to seriously grapple with it, which is telling. Regarding Isaiah 38:8, the shadow went backwards in the reference frame of those doing the observing. If that happened today, the first-hand reports would use the exact same language even though later pundits would state the earth must have temporarily reversed its spin. I note that God did not say the universe reversed its spin around the earth, only that the sun/shadow went back, which could happen in a geocentric universe if God reversed the spin of the universe OR if He temporarily decoupled the sun (and planets!) from the rotation of the universe. In a geokinetic universe, God could reverse the spin of the earth OR move the sun and planets in their place to produce the same effect (thank you for making me think of this last bit). Therefore, this passage is entirely ambiguous. It does nothing to inform us of the workings of the universe. Nobody is being duplicitous.[1339]

Dr. Carter stated: "this passage [Isaiah 38:8] is entirely ambiguous. It does nothing to inform us of the workings of the universe." He seems to think that saying something makes it so. There is not any ambiguity in the passage, and that passage speaks directly to the workings of the universe. The foundation of his biblical argument is a "phenomenological language" tradition that requires three assumptions: 1) someone other than God is viewing the phenomenon, 2) that someone is telling us what he saw, and 3) and what he saw may not actually have happened as he described it. First of all, it is God speaking in Isaiah 38:8. Verse 4 states "then came the word of the Lord to Isaiah." Isaiah was speaking the words of God Almighty. God stated what he would do: "Behold, I will bring again the shadow of the degrees, which is gone down in the sun dial of Ahaz, ten degrees backward." God then stated exactly how he did it. "So the sun returned ten degrees, by which degrees it was gone down." God says, without ambiguity, that the sun returned the very ten degrees that it had gone down. Yet, Dr. Carter states the passage is ambiguous. In Dr. Carter's world any passage that contradicts his heliocentric theology is rendered "ambiguous." The ambiguity is not real. It is a construct of a method of Bible interpretation that assumes that God uses "phenomenological language" to say what he does not really mean.

Interestingly, in the biography for Dr. Carter on the Creation Ministries website it states of Dr. Carter: "He says he felt a tremendous joy when he realized his science and his religion were no longer at odds."[1340] Apparently, he was unwilling to see the religion behind the so-called "science" of heliocentricity and had to somehow find agreement between the heliocentric religion, passing itself as science, and the Christian religion of the Bible. Dr. Carter had a problem. He had to make some accommodation for the clear language in the Bible indicating a stationary earth. He did this by coming up with a theory that God does not truly mean what he is saying in the Bible. Dr. Carter claims that God is using "phenomenological language."

Carter's sophistry of alleging that God was using "phenomenological language" in the Bible is just another way of saying that God was not telling the truth about what was happening. "Let God be true, but every man a liar." (Romans 3:4) God does not lie. "God is not a man, that he should lie; neither the son of man, that he should repent: hath he said, and shall he not do it? or hath he spoken, and shall he not make it good?" (Numbers 23:19) Indeed, it is a command of God that Christians are to witness that God reigns, he is righteous, and the world is immovable. "Say among the heathen that the LORD reigneth: the world also shall be established that it shall not be moved: he shall judge the people righteously." (Psalms 96:10)

Lest, dear reader, you think that I am being unkind to Dr. Carter, I take my cue from Paul, who did not hesitate to upbraid Peter for his unchristian dissembling.

> But when Peter was come to Antioch, I withstood him to the face, because he was to be blamed. For before that certain came from James, he did eat with the Gentiles: but when they were come, he withdrew and separated himself, fearing them which were of the circumcision. And the other Jews dissembled likewise with him; insomuch that Barnabas also was carried away with their dissimulation. But when I saw that they walked not uprightly according to the truth of the gospel, I said unto Peter before them all, If thou, being a Jew, livest after the manner of Gentiles, and not as do the Jews, why compellest thou the Gentiles to live as do the Jews? (Galatians 2:11-14)

While Peter was a feverish worker for the gospel, he was wrong in his approach, just as Dr. Carter is a feverish worker against evolution, but is wrong regarding heliocentricity. Just as did Paul to Peter, in like manner, I am upbraiding Dr. Carter for his dissimulation. What is the dissembling about which Paul speaks in Galatians but concealing one's true thoughts by some pretense or hypocrisy? Dr. Carter's "phenomenological language" construct is so contrary to what it means be a Christian, it is hard to accept that he truly believes it, being, as he claims, a Christian. Dr. Carter seems to have engaged in dissembling similar to that of Peter, when Carter claimed in his article that God is using "phenomenological language" in describing the movement of the sun in the Bible. He would certainly object to the claim that God was using "phenomenological language" in the creation account in Genesis. He believes that God meant what he said when he said he created the world in 6 days. The days were in fact six literal days. Yet, he argues that God did not mean what he said in Isaiah 38.

Dr. Carter is a scientist on the staff of *Creation Ministries International*. The statement of faith for *Creation Ministries International* specifically states, in pertinent part:

> The doctrines of Creator and Creation cannot ultimately be divorced from the Gospel of Jesus Christ. ... The 66 books of the Bible are the written Word of God. The Bible is divinely inspired and inerrant throughout. Its assertions are factually true in all the original autographs. It is the supreme authority, not only in all matters of faith and conduct, but in everything it teaches. ... The account of origins presented in Genesis is a simple but factual presentation of actual events and therefore provides a reliable framework for scientific research into the question of the origin and history of life, mankind, the Earth and the universe. ... The days in Genesis do not correspond to geologic ages, but are six [6] consecutive twenty-four [24] hour days of Creation. ... The view, commonly used to evade the implications or the authority of Biblical teaching, that knowledge and/or truth may be divided into 'secular' and 'religious', is rejected. Facts are always subject to interpretation by fallible people who do not possess all information. By definition, therefore, no interpretation of facts in any field, including history and chronology, can be valid if it contradicts the scriptural record.[1341]

How can Dr. Carter use his "phenomenological language" to get around the clear statements in the Bible, when to do so violates the very statement of faith of *Creation Ministries International*, for which he works? The statement of faith states clearly that "the Bible is divinely inspired and inerrant throughout" and that "it is the supreme authority, in all matters of faith and conduct," and "in everything it teaches." The biblical account is a "simple but factual presentation of actual events and therefore provides a reliable framework for scientific research."

It seems that Dr. Carter and Creation Ministries International selectively apply their principle that the Bible is the supreme authority in everything they teach only to the account of origins presented in Genesis. Dr. Carter only accepts the Genesis account as being a "factual account of actual events," but that part of the Bible that addresses the cosmos uses "phenomenological language" that can be interpreted contrary to the expressed words, at the whim of the reader. How can that be? What is Dr. Carter's authority to decide what part of the Bible is a "factual account of actual events" and what part of the Bible uses "phenomenological language?"

When I asked Dr. Carter about his failure to biblically refute geocentricity in his article, he responded that "the refutation is in the physics." He confesses in his comment post that the basis for his rejection of an immovable earth is modern "science;" he calls it physics. He based his entire argument on the "scientific" theories that have been constructed upon Copernicus' heliocentric myth. He uses so-called "science" to overrule the Bible. He does the very thing that is forbidden by the *Creation Ministries International* statement of faith, which states in pertinent part, that "no interpretation of facts in any field, including history and chronology, can be valid if it contradicts the scriptural record."[1342] His approach not only violates the admonition in *Creation Ministries International's* statement of faith, it also violates the very scriptures he professes to uphold. "O Timothy, keep that which is committed to thy trust, avoiding profane and vain babblings, and oppositions of science falsely so called." (1 Timothy 6:20)

The key statement in the Creation Ministries International

that gives away the game they are playing is: "The 66 books of the Bible are the written Word of God. The Bible is divinely inspired and inerrant throughout. Its assertions are factually true **in all the original autographs**." (emphasis added). Notice the words: "in all the original autographs." What the *Creation Ministries International* implies by that statement is that they only stand by the "original autographs." The problem is that the original autographs no longer exist. Therefore, according to their statement of faith, there is no inerrant word of God that is factually true to which one can point. They assert allegiance to something that does not exist and has not existed for almost two millennia.

God holds his word in even higher esteem than even his name.

> **[T]hou hast magnified thy word above all thy name**. (Psalms 138:2)

God's name is so precious that the biblical penalty for blaspheming his name is death. Leviticus 24:16. However, God holds his word above even his name. Why? Because God's word is God's revelation of him to man. The Holy Bible states that:

> In the beginning was the Word, and the Word was with God, and **the Word was God**. The same was in the beginning with God. **All things were made by him**; and without him was not any thing made that was made. (John 1:1-3)

> In whom we have redemption through his blood, even the forgiveness of sins: Who is the image of the invisible God, the firstborn of every creature: **For by him were all things created, that are in heaven, and that are in earth, visible and invisible, whether they be thrones, or dominions, or principalities, or powers: all**

things were created by him, and for him: And he is before all things, and by him all things consist. (Colossians 1:14-17)

The gospel found in John states that God (the Word, the Creator) came to Earth in the flesh: Jesus Christ.

And the **Word was made flesh**, and dwelt among us, (and we beheld his glory, the glory as of the only begotten of the Father,) full of grace and truth. (John 1:14)

In the Holy Bible God the Father makes it clear that his Son, Jesus, is God.

But unto the **Son** he saith, Thy throne, **O God**, is for ever and ever: a sceptre of righteousness *is* the sceptre of thy kingdom. (Hebrews 1:8)

The Holy Bible is not like any other book; it is unique; it was written by God through men.

All scripture is given by inspiration of God, and is profitable for doctrine, for reproof, for correction, for instruction in righteousness: (2 Timothy 3:16)

Knowing this first, that no prophecy of the scripture is of any private interpretation. For the prophecy came not in old time by the will of man: but **holy men of God spake as they were moved by the Holy Ghost**. (2 Peter 1:20-21)

Which things also **we speak, not in the words which man's wisdom teacheth, but which the Holy Ghost teacheth**; comparing spiritual things

> with spiritual. But the natural man receiveth not the things of the Spirit of God: for they are foolishness unto him: neither can he know *them*, because **they are spiritually discerned**. (1 Corinthians 2:13-14)

In the beginning God created the heaven and the earth. How did he create? He created by speaking. **"God said . . . and it was so."** *See* Genesis 1:1-2:25. "Through faith we understand that the worlds were framed by the word of God, so that things which are seen were not made of things which do appear." (Hebrews 11:3)

God has promised to preserve his word forever. Even though the original autographs no longer exist, God's word has been supernaturally preserved. God's word is the way to salvation. God would not leave us without the means for our salvation. The following scripture passages testify that God has promised that his word will be preserved forever.

> For verily I say unto you, Till heaven and earth pass, one jot or one tittle shall in no wise pass from the law, till all be fulfilled. (Matthew 5:18)

> **Heaven and earth shall pass away, but my words shall not pass away**. (Matthew 24:35)

> The words of the LORD *are* pure words: *as* silver tried in a furnace of earth, purified seven times. **Thou shalt keep them, O LORD, thou shalt preserve them from this generation for ever**. (Psalms 12:6-7)

> **[T]he word of the Lord endureth for ever**. And this is the word which by the gospel is preached unto you. (1 Peter 1:25)

> The grass withereth, the flower fadeth: but **the word of our God shall stand for ever**. (Isaiah 40:8)

> **For ever, O LORD, thy word is settled in heaven**. (Psalms 119:89)

Dr. Carter's disrespectful "phenomenological language" theory for Bible passages goes hand-in-hand with the fact that he seems not to truly believe God's promise to preserve his word. He and the *Creation Ministries International* only consider the original autographs to be God's infallible word. The originals, however, have not been preserved. They are long gone. God never promised to preserve the original autographs, he promised to preserve his word. God's word is spiritual, eternal, and preserved in the English language in the Authorized (King James) Version of the Bible. For more information on that issue read this author's book, *Solving the Mystery of Babylon the Great.*

The "phenomenological language" theory simply does not hold up under scrutiny. Take, for example, God causing the sun and the moon to stand still in Joshua 10:12-13.

> Then spake Joshua to the LORD in the day when the LORD delivered up the Amorites before the children of Israel, and he said in the sight of Israel, Sun, stand thou still upon Gibeon; and thou, Moon, in the valley of Ajalon. And **the sun stood still, and the moon stayed**, until the people had avenged themselves upon their enemies. Is not this written in the book of Jasher? So **the sun stood still in the midst of heaven, and hasted not to go down about a whole day**. (Joshua 10:12-13)

Note that the passage clearly states that the "the sun stood still in the midst of heaven, and hasted not to go down about a whole day." It did not state that the earth stopped spinning. It states that the sun stood still. Interestingly, it also states that the moon also "stayed."

Joshua 10 is clearly not so-called "phenomenological" language. In Joshua 10, God is describing what actually happened. The passage states that "**the sun stood still, and the moon stayed**." The moon is a moving body over the earth under both the heliocentric and geocentric models. Consequently, Dr. Carter cannot ascribe "phenomenological" language to Joshua 10 for the halting of the moon's motion. He must admit that the moon actually stopped in its ordinary motion. That fact poses a problem for Dr. Carter. He would have to argue that the description in Joshua 10 of the sun stopping is "phenomenological" (that is, it didn't actually happen that way), but at the same time he must argue that the description of the moon stopping is what actually happened. That would be a double minded interpretation. "A double minded man is unstable in all his ways." (James 1:8)

Both the stopping of the moon and the stopping of the sun are described as happening in the same way in Joshua 10. They were both stopped in their motion over the earth. For that reason, Joshua 10 refutes Dr. Carter's "phenomenological" interpretation of Isaiah 38. In Joshua 10, the moon "stayed" in the same way that the sun "stood still." Indeed, Joshua commanded that the sun stand still over Gibeon and he commanded that the moon stand still in the valley of Ajalon. The sun in fact "stood still" in relation to its previous motion, in the same way that the moon "stayed" in relation to its previous motion.

In Joshua 10, God did not describe the earth stopping in its spin. Joshua 10 describes the halting of both the moon and the sun in their ordinary circuits over the earth. That means that the moon stopped in the same way that the sun stopped, which means that the sun must be moving over the face of the earth in the same way the moon moves over the face of the earth. Joshua 10 indicates that the sun was moving prior to its stopping; and Isaiah 38 indicates that the sun was moving prior to its reversal. Joshua 10 impeaches any view that Isaiah 38 is "phenomenological" language and not in fact a description of what actually happened. What actually happened? "[T]he sun returned ten degrees, by which degrees it was gone down." Isaiah 38:8.

A car or motorcycle traveling at a mere 60 miles per hour and coming to a sudden halt would kill all passengers on board. Imagine what would happen to people on Earth, when the 1,000 mile per hour spinning Earth suddenly stopped. People, standing on Earth, would be caused to tumble and crash along the earth with such force as to be killed instantly. Indeed, any physicist worth his salt would understand that if the earth were spinning, as supposed by modern science, at over 1,000 miles per hour at the equator, and it suddenly stopped, there would be complete carnage on Earth for all living things. The oceans still in motion, would flood the continents. The very idea is ludicrous. Yet, in Joshua 10, we have two armies facing each other and they are not the least bit fazed in their footing by the sun and moon stopping. Obviously, it was not the earth that stopped spinning, but rather the sun and moon stopped in their motion over a fixed non-moving Earth, just as the Bible describes it.

On September 13, 2016, Dr. Robert Carter again partnered with Dr. Jonathan Sarfati to write an article that went beyond the issue of geocentricity and addressed the issue of the flat Earth. The title of the article will give the reader some idea of the contempt that Drs. Carter and Sarfati have for anyone who accepts that the earth is flat. Their article is titled: *Flat Earth, and Other Nonsense, Dealing with Ideas That Would Not Exist Were it Not for the Internet.*[1343]

Their article portrays those who adhere to the flat earth in the most demeaning terms. They state in the article that belief in a flat earth is pseudoscience that is based upon hype that is being pushed by deceptive charlatans. They characterize belief in the flat earth as a nutty idea, which is only believed because of a tragic flaw in human psychology.

Carter and Sarfati also caution the readers to steer clear of any examination of a conspiracy surrounding the 911 attacks. And for good measure, they specifically poo-poo any thought that the NASA moon landings were hoaxes. Carter and Sarfati then dismiss independent research posted on the internet by stating that "[o]ur only conclusion is that the Internet is breeding people who have trouble thinking through important ideas."[1344] They think that "[c]learly, the Internet easily radicalizes people."[1345] Those are not the concerns of truth-seekers; those are the ravings of gatekeepers.

Indeed, Carter and Sarfati begin their insulting and deceptive article by expressing their exasperation at how they must lower themselves to address the issue of the flat earth. "Sadly, we must also take on the subject of whether or not the earth is flat."[1346] They then proceed to outright lie and pretend that somehow Pythagorean trigonometry proves their case for the spherical earth, when in fact it completely destroys their case. "One of the worst aspects of the flat-Earth claim is that to believe it you also have to deny simple trigonometry."[1347]

Carter and Sarfati fear that acceptance of a flat earth would require science "to go back centuries and reject just about everything we have learned about physics. This includes the great discoveries of Sir Isaac Newton."[1348] What is one of the great Newtonian discoveries that Carter and Sarfati fear rejecting? They specifically list the theory of gravity. Carter and Sarfati do not want to reject the theory of gravity. These are two scientists who are clinging to a theory (gravity) which has never been proven, but to which they adhere as some kind of religious dogma. If the earth were flat, they would no longer have any need for gravity to explain how things work. Indeed, Carter and Sarfati admit it. "If the earth is flat, we cannot use physics [i.e., gravity] to explain how things work."[1349] If the earth is not a spinning globe, then the theory that explains why we do not fly off the globe due to centrifugal force is rendered a nullity. Carter and Sarfati cannot have that.

Carter and Sarfati state that we should "try to think of something that disproves a spherical Earth (if you can!)."[1350] They cannot be serious! There is a truckload of scientific evidence and biblical authority that disproves a spherical earth, which is documented in this book. They seem to act as though the evidence does not exist, perhaps to hoodwink the reader who does not know about it. They tell the reader that as experts they have concluded that "the sphericity of the earth is one of the simplest aspects of operational science one could wish for."[1351] They portray themselves as experts (they both proudly announce that they each hold Ph.D.'s) and so they must be held to the standard expected of experts. To do so, requires us to conclude that they are not simply mistaken in their conclusion regarding the sphericity of the earth, but that they are being purposely deceptive, as no expert who claims to have examined the biblical and scientific evidence could ever come to the conclusion that the earth is a sphere.

Carter and Sarfati state that "[i]n the case of the flat Earth, people have been fed a line of false reasoning that takes a while for some people to 'unthink.' With this in mind, we encourage people to put on their thinking caps and take a serious look at what the Bible and science have to say."[1352] That sounds high-minded of Carter and Sarfati, but they do not practice what they preach. They point the reader to, of all organizations, NASA. That's right; Carter and Sarfati, in part, support their theory of a spherical earth on propaganda from NASA. They tell the reader to believe NASA. What about their admonition to look put on our "thinking caps and take a serious look at what the Bible and science have to say?" They didn't really mean that. What they really intend is for us to shut up and believe what we have been told. Carter and Sarfati's double-mindedness is almost too much to take as they admonish the readers to think for themselves, but they then vouch for the credibility of NASA and tell us to trust and believe NASA:

> The International Space Station has now gone around the earth over 100,000 times, carrying with it more than 220 different astronauts over the past 15 years. One astronaut, Col. Jeff Williams has recently returned from his fourth trip to space. Not only has Col. Williams set the record for cumulative days in space (534) but he is also an outspoken Christian! He and other astronauts have taken thousands of pictures and hundreds of hours of videos of the earth from space, many of which are available online. You can patch these together seamlessly to make a montage of the entire earth, and it is demonstrably a sphere. All of these scientists and astronauts are not lying![1353]

Carter and Sarfati want you to believe that NASA scientists and astronauts are not lying. But the fact is that NASA scientists and astronauts are lying, and so are Carter and Sarfati.

Creation Ministries International is just one of many examples of "Christian" ministries that have been carried away from Christ and spoiled by the rudiments of the world.

> Beware lest any man spoil you through philosophy and vain deceit, after the tradition of men, after the rudiments of the world, and not after Christ. (Colossians 2:8)

The opposition to the flat earth is born of fear. That fear emanates from a carefully contrived brainwashing scheme that has conditioned men to have a visceral opposition to the flat earth.

Make no mistake about it, heliocentricity is not true science, it is an occult religious doctrine. One of the high priests of science, who is inculcating the masses in the heliocentric mythology that derides any notion of a flat earth, is Brian Cox. He is an English physicist and professor of particle physics in the School of Physics and Astronomy at the University of Manchester. He is a Particle Physics and Astronomy Research Council advanced fellow and member of the high energy physics group at the University of Manchester. He also works on the ATLAS experiment at the Large Hadron Collider at CERN. He has written several books and is the author or coauthor of more than 950 scientific publications.

Regarding doubters of the Apollo Moon landing, Cox states that "if you don't think Apollo 11 landed on Moon you are a colossal nob end & should get a new brain."[1354] But he saves his most damning rebuke for those who believe the earth is flat. Cox states: "It's probably the most nonsensical suggestion that a thinking human being could possibly make. It is drivel."[1355] Those opinions are from a renowned and respected scientist. So-called "science" has degenerated to the point of name calling. There is no discussion of evidence. That cannot be allowed. There are certain topics that are not open for discussion. If you believe the earth is flat, you are simply not worthy of being considered a thinking man. You are to be shunned from the cult of "science" for believing such nonsense. Cox is an example of those in the scientific hierarchy who condition the masses to have a superstitious visceral rejection of the flat earth. This conditioning is illustrated by a post on Quora.com from H. Paul Honsinger. He is a retired attorney, who describes himself as a "space history fanatic" and "author of three published Sci Fi novels." Honsinger posted the following advice on Quora.com:

> If your child's teacher is really teaching flat Earth ideas as scientific truth, you should move heaven and Earth to see that the teacher is removed from the classroom as soon as possible and that he or

she is never involved in working with children again in any capacity, unless and until the teacher's educational deficiencies are corrected or the underlying mental illness sufficiently treated.

No one who believes in a flat Earth could possibly be a competent teacher, no matter the subject. The ignorance, the denial of reality, the refusal to grasp basic proven facts, and the staggering arrogance required to be a flat Earther would render the person totally unfit to be in the classroom.

Speed is of the essence here, as is highly possible that the teacher suffers from a serious mental illness or, at minimum, is simply too stupid and intellectually incompetent to function as a teacher. And, if the teacher cannot be brought back to reality and insists on spreading these defamatory and delusional lies, then he or she needs to be severed from the position in such a way as to make him or her totally unemployable in the education field in perpetuity.

Actually, as someone who has been an employer and who has supervised other employees, I would NEVER hire a known flat Earther in any capacity, to do any job, under any circumstances. I've been a trial lawyer, and I can just imagine the attorney for a plaintiff injured by a mistake made by this employee asking me, "Now, Mr. Honsinger, isn't it true that when you hired Mr. Vlattengeard to drive your delivery trucks, you KNEW that he believed the Earth was flat?" "Yes." "And, despite this wholly irrational belief, you thought he was competent to drive a two ton

> delivery truck at highway speeds and make split second decisions on which the lives of other people depended?" Such a person would be a huge liability risk, no matter what the job. I did a lot of child custody cases, and I would have loved for the opposing party to be a flathead. I could make a good case that such a person would not be a fit parent, given the deficient reality testing and basic irrationality required to adhere to such a belief.
>
> Again, I stress the need for speed. This person is either intellectually incompetent or mentally ill (I'm assuming that this teacher graduated from some college or university and, therefore, has no excuse not to know what Western science has known since the days when people worshiped Zeus and Athena). It is now late Monday night. By next Monday, the students should see another person in place of this "teacher.""[1356]

It should be noted that Honsinger's posting was not in response to any news of a teacher who was giving instruction about the flat earth. It was designed as a peremptory notice to any teacher who would dare consider teaching students about the flat earth. It was intended as a warning!

According to Honsinger, a teacher who believes in a flat earth should be immediately fired. He impassions that "speed is of the essence." He does not stop there, his opinion is that such a person should not be employed in any capacity, anywhere. Indeed, Honsinger views such a person as mentally unfit to be a parent. Honsinger's draconian view is typical of the attitude of the heliocentric gatekeepers in higher education.

Honsinger's ravings are those of a true heliocentric

believer. They are the fruit of years of brainwashing. His attitude exemplifies the attitude of academia toward the proofs of a flat earth, where any discussion of the evidence for a flat earth is beyond the pale. His comments reveal a mind that is so addled by propaganda that he can only vent his emotions in vitriol. Indeed, he would be among the crowd yelling "crucify him, crucify him." This is the result of the satanic scheme of heliocentrism.

The parallels between Honsinger's heliocentric sentiments and those of the communist Komizars are striking. Anyone who disagrees with the communist state and advocates for freedom is viewed as mentally ill. Indeed, disagreement with the state is *prima facie* evidence of mental illness. Communist countries use mental hospitals to reeducate (torture) those enemies of the state who would dare voice their disagreement. Heliocentrists have the same degenerate intolerance as their communist revolutionary counterparts. That is no surprise since both communism and heliocentrism have the same roots, the Jewish Talmud and Kabbala. Heliocentrism is not primarily a scientific deception; it is, in essence, a spiritual deception. Its adherents are religious zealots.

Suffering persecution in this world for righteousness sake is the very mark of a Christian. "Remember the word that I said unto you, The servant is not greater than his lord. If they have persecuted me, they will also persecute you; if they have kept my saying, they will keep yours also. But all these things will they do unto you for my name's sake, because they know not him that sent me." (John 15:20-21) Being reviled for speaking the truth of the gospel is a mark of a Christian. While the revilers intend their railings as curses, they are actually blessings. "Blessed are ye, when men shall revile you, and persecute you, and shall say all manner of evil against you falsely, for my sake. Rejoice, and be exceeding glad: for great is your reward in heaven: for so persecuted they the prophets which were before you." (Matthew 5:11-12)

Most “Christian” scientists have a desire to avoid persecution and be friends with the world. “Ye adulterers and adulteresses, know ye not that the friendship of the world is enmity with God? whosoever therefore will be a friend of the world is the enemy of God.” (James 4:4) “Christian” scientists that ignore or even reject God’s account of his creation in order to keep their standing in the scientific community, justify their stance by averring that the biblical account by God of a flat stationary earth is not really germane to the gospel. They have become enemies of God.

People do not have the authority to pick and choose what they like in God’s word and discard the rest. Indeed, the gospel is “every” word of God. Gospel literally means “God spell,” that means God’s word. Jesus made the point in Matthew “It is written, Man shall not live by bread alone, but by **every word** that proceedeth out of the mouth of God.” Matthew 4:4. **The Gospel is “every” word of God.** “Every word of God is pure: he is a shield unto them that put their trust in him. Add thou not unto his words, lest he reprove thee, and thou be found a liar.” (Pr 30:5-6)

Chuck Lawson, who is pastor of Temple Baptist Church[1357], states that the belief in a flat earth is part of a demonic attack designed to sow discord in the church.[1358] Lawson takes umbrage at being criticized for not believing in a flat earth. He considers it a great sin to criticize him, because he considers himself a “man of God.” He states that people should have reverence for the “man of God.” How dare anyone criticize the “man of God.”

Lawson is part of the problem in the modern church that has incorporated the heathen model of a single man leading and running a "church." He considers himself "a man of God" as being distinct from the laity who are beneath him. He is completely confused and confusing. Yet, he is a pastor. He remains wrong, and wallows in his ignorance, because nobody will correct him.

The modern church model will not allow it. God has made every Christian believer a king and priest. Revelation 1:6. But Lawson thinks that he is beyond reproach, and if someone seeks to correct him, Lawson views it as an attack on a "man of God." So Christians (who are kings and priests) are supposed to sit silently while he pontificates his ignorance. His arrogant hubris is palpable.

To whom does Lawson cite as his authority that the earth is a spinning, orbiting sphere? He begins by citing a fraud, Chuck Colson. He then cites a long list of astronauts as his authority. He concludes his harangue against the flat earth by explaining Elon Musk's planned space missions with wealthy passengers aboard. Lawson ominously warns that Musk's space missions will make those who believe in a flat earth look foolish when the many wealthy space tourists come back to earth with their photos of the spherical earth floating in outer space. Notably, Lawson cites to not a single Bible passage. Like the Pharisees before him, he has supplanted God's word with man's traditions.

The devil has convinced modern pastors to limit the gospel to only that which the pastor has decided in his wisdom addresses salvation. Matthew Stuckey, a minister at Verity Baptist Church, called belief in a flat earth, "foolish."[1359] Stuckey's fellow traveler, Steven Anderson, Pastor of Faithful Word Baptist Church, went even further. According to Anderson, flat earth is "foolishness" and "garbage" that is "of the devil." Anderson states that the study of the flat earth in the church is a conspiracy by "infiltrators" to "dumb down the church."[1360] Anderson further opines that promoting a flat earth is "Satanic," and that believing in a flat earth is a "sin." He thinks that people who believe in the "stupidity" of a flat earth are "feebleminded" "idiots." He described one Christian who gave him a gift of a flat earth map as a "bozo."

Oddly, Anderson double-mindedly stated that what while

he considers belief in a flat earth to be Satanic and sinful, he does not think it is a doctrinal issue and it is not heresy. Well, then if it is not doctrinal and it is not heresy, whey is he so adamantly against it? Because, he has one foot in the world and one foot in the Bible, and it scares the bejeebers out of him to be considered crazy by the world. Down deep, he loves the world. "Love not the world, neither the things that are in the world. If any man love the world, the love of the Father is not in him." (1 John 2:15).

That attitude by the pastors has given "Christian" scientists succor; it appeases their itching ears to justify their rejection of the true gospel. Modern pastors fail to understand that God's uses all of his word to draw one to a saving knowledge of Christ. "No man can come to me, except the Father which hath sent me draw him: and I will raise him up at the last day." (John 6:44)

The Genesis account of God's creation is every bit a part of the gospel as is the gospel of John. Moses wrote Genesis. If one does not believe the account of God's creation in Genesis how can one believe the words Jesus spoke? "For had ye believed Moses, ye would have believed me: for he wrote of me. But if ye believe not his writings, how shall ye believe my words?" (John 5:46-47) One cannot truly know God except through his creation. God makes it clear in his "gospel" that the knowledge of God is revealed through his creation. "The invisible things of him from the creation of the world are clearly seen, being understood by the things that are made, even his eternal power and Godhead." (Romans 1:20) To have a false concept of God's creation gives one a false concept of God. There are eternal consequences for being lukewarm about the gospel. "So then because thou art lukewarm, and neither cold nor hot, I will spue thee out of my mouth." (Revelation 3:16)

Creationists are quite correct in attacking the scientic fraud of evolution and standing by the biblical authority for creation.

However, many of them seem to reject the scientific evidence and biblical authority for a geocentric, flat earth, which only serves to undermine the authority of the Bible and, in turn, their credibility. Dr. Danny R. Faulkner exhibits the typical hypocrisy of many creationists who stand by the biblical authority for creation but reject that same biblical authority for a geocentric, flat earth. Dr. Faulkner taught at the University of South Carolina Lancaster for over 26 years. Dr. Faulkner is a member of the *Creation Research Society* (CRS) and also serves as the editor of the *Creation Research Society Quarterly*. He has written more than a hundred papers in various astronomy and astrophysics journals, and is author of *Universe by Design* and *The New Astronomy Book.*

How can a creation ministry accept the Bible authority for creation but reject that same authority for a geocentric, flat earth? They use the devil-inspired artifice of textual criticism. The CRS statement of belief is similar to that of the *Creation Ministries International.* The CRS statement of belief is that "the account of origins in Genesis is a factual presentation of simple historical truths."[1361] However, the CRS guts the authority of scripture by limiting the inspired word of God to only being "historically and scientifically true in the **original autographs**."[1362]

With this artifice of looking for his biblical authority to "the original autographs," which is a Bible that no longer exists, Dr. Faulkner can embark on the duplicity of ignoring the clear meaning of Bible passages by engaging in textual criticism of the English (Authorized King James Version) translation.

> Some creationists believe that the scientific assault on the Bible did not begin with biological evolution, but with the acceptance of the heliocentric (or more properly, geokinetic) theory centuries ago. These people believe that the Bible clearly states that the Earth does not move, and hence the only acceptable Biblical cosmology is

> a geocentric one. Modern geocentrists use both Biblical and scientific arguments for their case. ... Some geocentrists draw distinctions that do not exist in the original autographs or even in translations. ... Much of their case is based upon a misunderstanding of general relativity and the rejection of that theory. While geocentrists are well intended, their presence among recent creationists produces an easy object of ridicule by our critics.[1363] (parenthetical in original)

Dr. Faulkner points out the real issue: that the scientific assault on the Bible began with heliocentrism. Faulkner is quick to point out, however, that he does not believe that heliocentrism is an attack on the Bible. Faulkner relies on the devilish artifice of textual criticism to construct a belief that helocentricity is consistent with what the Bible says. He further relies for his authority on the bankrupt scientific theory of relativity. Faulkner concludes that geocentrism, particularly a belief in a flat earth, presents "an easy object of ridicule by our critics." Faulkner seeks friendship with the world on heliocentrism in order to avoid the ridicule of critics. He does not abide by the admonition of God: "[K]now ye not that the friendship of the world is enmity with God? whosoever therefore will be a friend of the world is the enemy of God." James 4:4. Dr. Faulkner, with his Ph.D., has great pride in his scholastic achievement. Therein lies the impediment to him accepting the truth of the geocentric, flat earth. He is quite frank in admitting that for him to do so would bring ridicule. His fear of ridicule is born of his "pride of life," which is of the world and is contrary to God. "For all that is in the world, the lust of the flesh, and the lust of the eyes, and the pride of life, is not of the Father, but is of the world." 1 John 2:16. Indeed, the devil knows the weak character of men and thus has conditioned people to have contempt and deride as an ignoramus anyone who believes the earth is flat. That heliocentric hive environment that zeros in on and attacks a man's pride serves as an impediment to the truth of the geocentric, flat earth being accepted in academic and scientific circles. William H. Poole explained this perpetuation of ignorance. "There is a principle which is a bar against all information, which is proof against all argument, and which cannot fail to keep a man in everlasting ignorance. This principle is, contempt prior to examination."

What Faulkner does not see is that his adherence to heliocentricity and his failure to abide by the clear biblical authority for a geocentric, flat earth, in fact, undermines the credibility of the biblical argument for creation. For example, Glenn Elert, in his article, *The Scriptural Basis for a Geocentric Cosmology*, states that the Bible clearly presents a flat earth that is immovable and lies at the center of all things.[1364] Elert rejects the authority of the Bible and believes in both evolution and heliocentricity. Elert states that "[t]he purpose of this essay is to demolish the notion that the Bible has any scientific relevance whatsoever. In particular, I aim to show that the same thinking that leads devout fundamentalists to deny evolution as atheism must also lead them to embrace geocentrism and flat-earthism as God-given truths."[1365] On one point, Elert is correct; both creation and a geocentric, flat earth are supported by the Bible. You simply cannot have one without the other. There is no lukewarm middle ground. "So then because thou art lukewarm, and neither cold nor hot, I will spue thee out of my mouth." Revelation 3:16.

Elert then proceeds in his article to present Bible passage after Bible passage that clearly state that the earth is flat, stationary, and at the center of God's creation. Elert then eviscerates the credibility of the creationists who rely on the inerrant authority of scripture to support the truth of God's creation in a literal six days while they hypocritically dismiss the clear authority in the Bible for a flat, stationary earth that is at the center of God's creation. Creationists, like Faulkner, erroneously think that they can stand firm having one foot in Bible truth to refute evolution, while having the other foot in the quicksand of Bible criticism and modern science eschewing a flat, stationary, geocentric earth. Elert is an example of how the heathen world uses creationists' own arguments that attack the biblical authority for a geocentric, flat earth against creationists to undermine the biblical authority for creation. The obvious hypocrisy of adhering to the authority of the inerrant Bible to refute evolution, while at the same time undermining the authority of that same Bible when it speaks of a flat, stationary earth that is at the center of that creation, serves to reveal the modern creationists as double minded charlatans. Elert is thus forced to conclude:

> The shocking truth of it all is that the pseudoscientific, anti-evolutionary forces have compartmentalized their literalism. There is a deafening silence among the aforementioned ranks when it comes to scientific theories other than the origin and evolution of life on earth. ... The message here is, I think, obvious. The Bible is the literal truth only when it's convenient and doesn't conflict with overwhelming evidence. In my view, this invalidates the core of the anti-evolutionary movement in its entirety. If the Bible is open to interpretation from time to time, then it is open to interpretation at any time. If the Bible is occasionally poetic, then it is possibly poetic at any time -- even on the first page, even on the last page, even on every page.[1366]

Elert never addresses the true scientific evidence in support of a flat, stationary, geocentric earth, because neither he nor any other evolutionist are ever called on to do so by creationists. By yielding the field on heliocentricity, creationists have lost advantage of biblical authority before the battle even begins. Elert can get away with falsely saying that heliocentrism is supported by "overwhelming evidence," because creationists accept that statement as true, and therefore do not try to refute the false claim. Elert then uses the very arguments of creationists against the Bible truth of a geocentric, flat earth to bludgeon the hypocritical creationists. The creationists are portrayed as con artists who use clever sophistry; this serves to undermine their argument for God's creation and, in turn, the authority of the Bible.

The Satanic brainwashing to believe the heliocentric myth has been so effective that even the distinguished scientist and Christian minister Kent Hovind cannot overcome his conditioning. Kent Hovind has done wonderful work exposing the lies in school textbooks promoting the theory of evolution. Hovind has suffered terrible persecution at the hands of the government, even to the point of being sent to prison on false charges in order to stop him from his effective ministry against the evolutionary lies in the school textbooks. Hovind, however, continues to believe in a heliocentric system with a globular, spinning earth, orbiting the sun.

Hovind has been asked by Christians about the biblical model of a flat earth; he claims to have examined the matter and admonishes his followers to reject any notion of a flat earth. He views the flat earth as a "silly distraction from the gospel."[1367] What does Hovind rely upon as his authority for rejecting the flat earth? He refers his followers to "any earth science book."[1368] Hovind hypocritically relies upon the authority of the very earth science books that he has proven to be full of lies promoting the myth of evolution in order to establish the validity of the supposed heliocentric model.

Heliocentric "Christian" scientists are accomplices with the dark forces of Satan in a spiritual war against God's word. Neil deGrasse Tyson explains the irreconcilable battle lines between modern scientific theories and the Bible. In the process, he reveals that "Chrisitian" compromisers within the scientific community are in league with the enemies of Christ. Neil deGrasse Tyson is a famous astrophysicist and author. Tyson is the Frederick P. Rose Director of the Hayden Planetarium at the Rose Center for Earth and Space in New York City, who is promoted in the major media outlets as an expert on the supremacy of science and the truth of the heliocentric model. In 2014, Neil deGrasse Tyson spoke with host Bill Moyers on Moyers' television show, *Moyers & Company*. Tyson was asked by Moyers whether faith in God could be reconciled with scientific reasoning. Tyson was emphatic that scientific reasoning is irreconcilable with faith in God. Tyson specifically zeroed in on the conflict between the generally accepted scientific theories and God's description of his creation revealed in the Holy Bible.

> [I]f you knew nothing about science, and you read, say, the Bible, the Old Testament, which in Genesis, is an account of nature, that's what that is, and I said to you, give me your description of the natural world based only on this, you would say the world was created in six days, and that stars are just little points of light much lesser than the sun. And that in fact, they can fall out of the sky, right, because that's what happens during the Revelation.
>
> You know, one of the signs that the second coming, is that the stars will fall out of the sky and land on Earth. To even write that means you don't know what those things are. You have no concept of what the actual universe is. So everybody who tried to make proclamations about the physical universe based on Bible passages got the wrong answer.
>
> So what happened was, when science discovers things, and you want to stay religious, or you want to continue to believe that the Bible is

> unerring, what you would do is you would say, "Well, let me go back to the Bible and reinterpret it." Then you'd say things like, "Oh, well they didn't really mean that literally. They meant that figuratively."
>
> So, this whole sort of reinterpretation of how figurative the poetic passages of the Bible are came after science showed that this is not how things unfolded. And so the educated religious people are perfectly fine with that. It's the fundamentalists who want to say that the Bible is the literally, literal truth of God, that and want to see the Bible as a science textbook, who are knocking on the science doors of the schools, trying to put that content in the science room. Enlightened religious people are not behaving that way. So saying that science is cool, we're good with that, and use the Bible ... to get your spiritual enlightenment and your emotional fulfillment.[1369]

In that vignette, Neil deGrasse Tyson inadvertently exposes heliocentrism for what it really is: an attack on the truths in the Holy Bible. Tyson also exposes the modern "Christian" heliocentric scientists as accessories with atheistic scientists in the spiritual attack on the authority of the Bible. "Can two walk together, except they be agreed?" (Amos 3:3)

Tyson uses the satanic trick of an all or nothing interpretation of the Bible. He creates a false paradigm, where either the Bible is to be taken as literal in every passage or it is to be taken as figurative in every passage. According to Tyson, if you fall into the literal camp, you are a backwoods idiot. But if you take the Bible as only a poetic book of some spiritual truths, you are enlightened. There is another group, whom Tyson and his ilk pretend do not exist. They are Christians who take the Holy Bible as being God's pure truth, and who understand that truth takes many literary forms. Certainly, there are passages in the Bible that are metaphors and similes, which are intended to be taken figuratively. The context of the passages reveals the figurative nature of those passages. But to take passages whose context clearly indicates that they are to be taken literally, like God's description of his creation, and instead interpret them as figurative or poetic passages, is plain and simple deception.

"Christian" heliocentric scientists, who reject God's description of his creation by poeticizing it, have rejected God's spiritual truths as well. Jesus explained: "If I have told you earthly things, and ye believe not, how shall ye believe, if I tell you of heavenly things?" (John 3:12)

As Tyson explains, "Christian" scientists (whom Tyson calls "educated" and "enlightened" religious people) who reject the truths of the Bible are "cool," and "good with" science theories that contradict the Bible. These compromisers tell their followers to "use the Bible ... to get your spiritual enlightenment and your emotional fulfillment," but don't believe it when it speaks of the truths of God's creation. God has rendered a judgement against those so-called "educated" and "enlightened" religious people: "So then because thou art lukewarm, and neither cold nor hot, I will spue thee out of my mouth." (Revelation 3:16)

Many so-called creationists do take some Bible passages literally. But those same creationists limit a literal interpretation to those Bible passages that describe the fact of creation. When Bible passages reveal the specific nature of God's creation, as a flat, stationary earth at the center of God's creation, those passages are interpreted figuratively by the "enlightened" creationists. Indeed, many creationists caution their followers to reject out of hand and not even consider the evidence of the flat, stationary earth. What does God think of that approach? "He that answereth a matter before he heareth it, it is folly and shame unto him." (Proverbs 18:13)

Søren Kierkegaard stated that "there are two ways to be fooled. One is to believe what isn't true; the other is to refuse to believe what is true." God states that "evil men and seducers shall wax worse and worse, deceiving, and being deceived." 2 Timothy 3:13. The so-called scientists that are pushing heliocentrism have been deceived and have in turn become deceivers themselves; they are fools who speak lies and refuse to believe the truth. "For the wisdom of this world is foolishness with God. For it is written, He taketh the wise in their own craftiness." (1 Corinthians 3:19) We are admonished to keep the faith and not join with those so-called scientists who oppose God's word.

> [K]eep that which is committed to thy trust, avoiding profane and vain babblings, and oppositions of science falsely so called: Which some professing have erred concerning the faith. Grace be with thee. Amen. (1 Timothy 6:20-21)

Endnotes

1.Live Science Staff, How Fast Does Earth Move?, Live Science, November 27, 2012, https://www.livescience.com/32294-how-fast-does-earth-move.html. NASA Jet Propulsion Laboratory, The Moon Probe Pioneer IV, at 6-7, https://www.jpl.nasa.gov/multimedia/apollo/pdf/pioneer4brochure.pdf. Colgrove, W. G., The Planet Earth, Journal of the Royal Astronomical Society of Canada, Vol. 23, p.163,165, http://adsabs.harvard.edu/full/1929JRASC..23..163C.

2.Anugrah Kumar, Christian Post, Christian Bakery Closed for Refusing Lesbian Wedding Cake Breaks Record on Crowdfunding Site With $352,000, July 18, 2015.

3.Anugrah Kumar, Christian Post, Christian Bakery Closed for Refusing Lesbian Wedding Cake Breaks Record on Crowdfunding Site With $352,000, July 18, 2015, http://www.christianpost.com/news/christian-bakery-closed-for-refusing-lesbian-wedding-cake-breaks-record-on-crowdfunding-site-with-352000-141657/#aORrHCwjI0oKMXVS.99.

4.Todd Starnes, Fox News, Oregon Ruling Really Takes the Cake -- Christian Bakery Guilty of Violating Civil Rights of Lesbian Couple, January 21, 2014, http://www.foxnews.com/opinion/2014/01/21/christian-bakery-guilty-violating-civil-rights-lesbian-couple.html.

5.Anugrah Kumar, Christian Post, Christian Bakery Closed for Refusing Lesbian Wedding Cake Breaks Record on Crowdfunding Site With $352,000, July 18,

2015.

6.Todd Starnes, Fox News, Christian Bakers Face Government Wrath for Refusing to Make Cake for Gay Wedding, February 3, 2015, http://www.foxnews.com/opinion/2015/02/03/christian-bakers-face-government-wrath-for-refusing-to-make-cake-for-gay.html.

7.Todd Starnes, Fox News, Christian Bakers Face Government Wrath for Refusing to Make Cake for Gay Wedding, February 3, 2015, http://www.foxnews.com/opinion/2015/02/03/christian-bakers-face-government-wrath-for-refusing-to-make-cake-for-gay.html.

8.Todd Starnes, Fox News, Christian Bakers Face Government Wrath for Refusing to Make Cake for Gay Wedding, February 3, 2015, http://www.foxnews.com/opinion/2015/02/03/christian-bakers-face-government-wrath-for-refusing-to-make-cake-for-gay.html.

9.Todd Starnes, Fox News, Christian Bakers Face Government Wrath for Refusing to Make Cake for Gay Wedding, February 3, 2015, http://www.foxnews.com/opinion/2015/02/03/christian-bakers-face-government-wrath-for-refusing-to-make-cake-for-gay.html.

10.Masterpiece Cakeshop v. Colorado Civil Rights Commission, 584 U.S. ___ (2018).

11.135 S. Ct. 2584 (2015).

12.Masterpiece Cakeshop v. Colorado Civil Rights Commission, 584 U.S. ___ (2018) (Thomas, J., concurring).

13.Rod Dreher, The Persecution Of Jack Phillips, August 15, 2018, https://www.theamericanconservative.com/dreher/the-persecution-of-jack-phillips/.

14.Pentagon Manual Details Rules for Transgender Military Personnel, Fox News, July 19, 2016, http://www.foxnews.com/politics/2016/07/19/pentagon-manual-details-rules-for-transgender-military-personnel.html.

15. Howard Slugh Aylana Meisel, Anti-discrimination Laws Are Wielded Against a Shelter for Battered Women, August 16, 2018, https://www.nationalreview.com/2018/08/anti-discrimination-laws-used-against-battered-women-shelter/.

16. Howard Slugh Aylana Meisel, Anti-discrimination Laws Are Wielded Against a Shelter for Battered Women, August 16, 2018, https://www.nationalreview.com/2018/08/anti-discrimination-laws-used-against-battered-women-shelter/.

17.Michael Swift's 1987 Gay Manifesto, http://www.blessedcause.org/protest/Gay%20Manifesto.htm.

18.Steve Baldwin, Child Molestation and the Homosexual Movement, 14 REGENT L. REV. 267, 278 (2002), http://www.mega.nu/ampp/baldwin_pedophilia_homosexuality.pdf.

19.Baldwin, supra, at 271, 278 citing W.D. Erickson et al., *Behavior Patterns of Child Molesters*, 17

ARCHIVES SEXUAL BEHAV., at 83 (1988).

20.Baldwin, supra, at 278.

21.Baldwin, supra, at 274.

22.Baldwin, supra, at 268.

23.Baldwin, supra, at 272-73, 277.

24.Baldwin at 272.

25.Reptilian Pets - Viewer Discretion Advised, Mar 8, 2016, at 8:15, https://www.youtube.com/watch?v=lYRuQ56Mie8.

26.Antonio Subirats, Flat Earth: Globe Earthers Endorse Paedophiles, August 11, 2018, https://www.youtube.com/watch?v=0ihL7lAsEto.

27.TEDx tries to Normalise Pedophilia, July 22, 2018, https://www.youtube.com/watch?v=Wn4ok_W-7d4. TEDx Talk under Review, June 19, 2018, https://blog.ted.com/tedx-talk-under-review/.

28.U.S. House of Representatives, 8th District of Wisconsin, Press Release, U.S. House Passes Rep. Green "Two Strikes Bill," http://www.house.gov/markgreen/PRESS/2000/July00News/NR072500TwoStrikesPassage.htm (website address current as of August 16, 2003). See also http://www.geocities.com/Wellesley/2726/Molester.html (website address current as of August 16, 2003); Jon Donenberg, Keller Attacks Johnson's Vote On 1987 Bill, The Daily Illini Online, October 3, 2000, http://www.dailyillini.com/oct00/oct03/news/printer/news01.shtml (website address current as of August 16, 2003).

29.TEDx tries to Normalise Pedophilia, July 22, 2018, https://www.youtube.com/watch?v=Wn4ok_W-7d4. TEDx Talk under Review, June 19, 2018, https://blog.ted.com/tedx-talk-under-review/.

30.Bill Hoffmann, Tom DeLay: Justice Dept. Wants to Legalize 12 'Perversions', 30 June 2015, NEWSMAX, http://www.newsmax.com/Newsmax-Tv/Tom-DeLay-Justice-Department-perversions/2015/06/30/id/652929/.

31.The Friends of Charles Darwin, Marx of Respect, http://friendsofdarwin.com/articles/marx/ (last visited on November 1, 2015).

32.Stamp, 150th Birth Anniversary of Charles Darwin, http://colnect.com/en/stamps/stamp/455245-150th_Birth_Anniversary_of_Charles_Darwin-Soviet_Union_USSR (last visited on November 2, 2015).

33.Mark Edmundson, Defender of the Faith?, *The New York Times*, September 9, 2007, http://www.nytimes.com/2007/09/09/magazine/09wwln-lede-t.html?pagewanted=all&_r=0.

34.Michael StGeorge, The Survival Of A Fitting Quotation, http://anonpress.org/spencer/ (last visited on December 26, 2017).

35.President George H.W. Bush, Address Before a Joint Session of the Congress on the State of the Union, January 29, 1991, http://www.presidency.ucsb.edu/ws/?pid=1253.

36.Did Lenin say “a lie told often enough becomes the truth”?, http://skeptics.stackexchange.com/questions/32926/did-lenin-say-a-lie-told-often-enough-becomes-the-truth (last visited on December 8, 2016).

37.Samuel Birley Rowbotham (Parallax), Zetetic Astronomy, Earth Not a Globe (1881).

38.Samuel Birley Rowbotham (Parallax), Zetetic Astronomy, Earth Not a Globe, at 11-12 (1881).

39.Samuel Birley Rowbotham (Parallax), Zetetic Astronomy, Earth Not a Globe, at 12-13 (1881).

40.J. Clendinning, The Principles of Surveying, at 98, Second Edition (1960).

41.Topographic Surveying, Department of the Army Technical Manual, TM5-234, U.S. Government Print Office, at 10 (1953).

42.Samuel Birley Rowbotham (Parallax), Zetetic Astronomy, Earth Not a Globe, at 31 (1881).

43.Samuael Birley Rowbotham, A/K/A "Parallax," Zetetic Astronomy, The Earth is Not a Globe, at 169-70 (1865),

44.Alexander Findlay, Lighthouses of the World, at 32 (1861).

45.Samuel Birley Rowbotham (Parallax), Zetetic Astronomy, Earth Not a Globe, at 28 (1881).

46.Samuel Birley Rowbotham (Parallax), Zetetic Astronomy, Earth Not a Globe, at 28-29 (1881).

47.Robert Schadewald, Looking for Lighthouses, Reprinted from Creation/Evolution #31 (1992), https://www.lhup.edu/~dsimanek/litehous.htm.

48.Robert Schadewald, Looking for Lighthouses, Reprinted from Creation/Evolution #31 (1992), https://www.lhup.edu/~dsimanek/litehous.htm.

49.Robert Schadewald, Looking for Lighthouses, Reprinted from Creation/Evolution #31 (1992), https://www.lhup.edu/~dsimanek/litehous.htm.

50.Robert Schadewald, Looking for Lighthouses, Reprinted from Creation/Evolution #31 (1992), https://www.lhup.edu/~dsimanek/litehous.htm.

51.Robert Schadewald, Looking for Lighthouses, Reprinted from Creation/Evolution #31 (1992), https://www.lhup.edu/~dsimanek/litehous.htm.

52.Bell Rock Lighthouse, http://www.bellrock.org.uk/ (last visited on October 15, 2015).

53.Alexander Findlay, Lighthouses of the World, at 44 (1861).

54.Alexander Findlay, Lighthouses of the World, at 32 (1861).

55.Fourteen Foot Bank Lighthouse, Lighthouse Friends, http://www.lighthousefriends.com/light.asp?ID (last visited on October 16, 2015).

56.Fourteen Foot Bank Lighthouse, New Jersey Lighthouse Society, http://www.njlhs.org/njlight/fourteen.html (last visited on October 16, 2015).

57.Fourteen Foot Bank Lighthouse, New Jersey Lighthouse Society, http://www.njlhs.org/njlight/fourteen.html (last visited on October 16, 2015).

58.Joshua Nowicki, https://joshuanowicki.smugmug.com/search/?q=chicago&c=photos (last visited on October 22, 2015).

59.Robert Sungenis, Flat Earth Flat Wrong, at 529 (2018).

60.Robert Sungenis, Flat Earth Flat Wrong, at 537 (2018).

61.Robert Sungenis, Flat Earth Flat Wrong, at 118 (2018).

62.Robert Sungenis, Flat Earth Flat Wrong, at 529 (2018).

63.Robert Sungenis, Flat Earth Flat Wrong, at 538 (2018).

64.Robert Sungenis, Flat Earth Flat Wrong, at 560 (2018).

65.Robert Sungenis, Flat Earth Flat Wrong, at 560 (2018).

66.Tom Coomes, Mirage of Chicago Skyline Seen from Michigan Shoreline, ABC 57 News, April 29, 2015, http://www.abc57.com/story/28925566/mirage-of-chicago-skyline-seen-from-michigan-shoreline.

67.Tom Coomes, Mirage of Chicago Skyline Seen from Michigan Shoreline, ABC 57 News, April 29, 2015, http://www.abc57.com/story/28925566/mirage-of-chicago-skyline-seen-from-michigan-shoreline.

68.Sjaak Slanina, Superior mirage, Encyclopedia of Earth, October 6, 2008, http://www.eoearth.org/view/article/156343/.

69.Mirage Reveals Chicago Skyscrapers over Lake Michigan, WILX, April 23, 2015, http://www.wilx.com/news/headlines/Mirage-Reveals-Chicago-Skyscrapers-over-Lake-Michigan-301021881.html.

70.Dr, Zak, Flat Earth - Refraction part 1/3, October 17, 2016, https://www.youtube.com/watch?v=XOdfn0CgRrg&index=15&list=WL, Dr, Zak, Flat Earth | Refraction part 2/3 - Star Trails, November 22, 2016, https://www.youtube.com/watch?v=RkDqdoINhYI&index=16&list=WL, Dr. Zak, Flat Earth | Refraction Part 3/3 - Experiment 1, December 10, 2016, https://www.youtube.com/watch?v=KPr1G3mC808&index=17&list=WL.

71.Rob Skiba, Does the Atmosphere Solve "The Problem of Sunsets" on a Flat Earth?, January 22, 2017, https://www.youtube.com/watch?v=tflhWwoqWAw.

72.Rob Skiba Proves the Chicago Skyline (As Seen from the Other Side) Is Not a Mirage, Jun 29, 2016, https://www.youtube.com/watch?v=o37t6iBS_q4.

73.Rob Skiba Proves the Chicago Skyline (As Seen from the Other Side) Is Not a Mirage, Jun 29, 2016, https://www.youtube.com/watch?v=o37t6iBS_q4.

74.Rossall Beach Residents and Community Group, http://www.rossallbeach.org.uk/iom.html (last visited on November 12, 2015).

75.Distance From Rossall Beach, Wyre District, United Kingdom to Port Soderick, Isle of Man, 62.82 miles, http://www.distancefromto.net/ (last visited on November 12, 2015).

76.Elevation Map of Rossall Beach, United Kingdom, http://elevationmap.net/rossall-beach-united-kingdom#menu2 (last visited on November 12, 2015).

77.Distance From Rossall Beach, Wyre District, United Kingdom to Snaefell, Isle of Man, 63.25 miles, http://www.distancefromto.net/ (last visited on November 12, 2015).

78.Rossall Beach Residents and Community Group, http://www.rossallbeach.org.uk/iom.html (last visited on November 12, 2015).

79.Flat earth test, Part 1. Proof? Full HD, https://www.youtube.com/watch?v=Ne_B4vqZ6oY (last visited on October 14, 2015).

80.Tokitea, Washington State Department of Transportation, http://www.wsdot.com/ferries/vesselwatch/VesselDetail.aspx?vessel_id=68 (last visited on October 15, 2015).

81.Flat Earth, Part 2. Proof ?, September 4, 2015, https://www.youtube.com/watch?v=eM2Tgjc528k.

82.Robert Sungenis, Flat Earth Flat Wrong, at 633 (2018), citing Flat Earth Debunked PSYOP EXPOSED: We'll Do It Live: The REAL Alternative Podcast, June 17, 2015, https://www.youtube.com/watch?v=VSy9oKHlpiE&feature=youtu.be.

83.Ships, January 30, 2014, https://www.youtube.com/watch?v=N2LV8gLWYqU.

84.Corsica From Nervi (Genova), December 4, 2007, Camera: Panasonic DMC-FZ10, Exposure: 0.040s (1/25), Focal Length: 72.00mm, F / Stop: f / 2,800, ISO Speed: ISO 200, Exposure Bias: 0.00 EV, http://www.panoramio.com/photo/10807454.

85.The camera used was a Panasonic DMC-FZ10; the exposure was 0.040s (1/25); the focal length was 72.00mm; the F/Stop was f / 2,800; the ISO Speed was 200; and the exposure bias was 0.00 EV. http://www.panoramio.com/photo/10807454.

86.U.S. Department of Homeland Security, UNITED STATES COAST GUARD, LIGHT LIST, Volume III, Atlantic and Gulf Coasts, Little River, South Carolina to Econfina River, Florida, (Includes Puerto Rico and the U.S. Virgin Islands), COMDTPUB P16502.3, U.S. Government Printing Office, Washington, DC 20402 (2015), http://www.navcen.uscg.gov/pdf/lightLists/LightList%20V3.pdf.

87.U.S. Department of Homeland Security, UNITED STATES COAST GUARD, LIGHT LIST, Volume III, supra, at 4 (2015).

88.U.S. Department of Homeland Security, UNITED STATES COAST GUARD, LIGHT LIST, Volume III, supra, at vi (2015).

89.D. Marble, UNDENIABLE Flat Earth PROOF!!! 10 Mile Puget Sound Laser Test, August 8, 2018, https://www.youtube.com/watch?v=rsAfcn2HRPw.

90.William Genske, May 6, 2018, Final Results: FECORE Measurements Disprove the WGS84 Model by Proving Flat Lake Surface!, https://fecore.org/final-results-fecore-measurements-disprove-the-wgs84-model-by-proving-flat-lake-surface/.

91.William Genske, May 6, 2018, Final Results: FECORE Measurements Disprove the WGS84 Model by Proving Flat Lake Surface!, https://fecore.org/final-results-fecore-measurements-disprove-the-wgs84-model-by-proving-flat-lake-surface/.

92.Flat Earth - Standard Bendy Light Refraction explained, December 27, 2017, https://www.youtube.com/watch?v=kS_1itONTPU.

93.Diego Gutierrez , Francisco J. Seron , Adolfo Munoz , Oscar Anson, Simulation of Atmospheric Phenomena, at 996, 12 May 2006, http://web.mit.edu/ytc/www/HLMA/Ref/1-s2.0-S0097849306001026-main.pdf.

94.Refraction of Light, Florida State University, https://micro.magnet.fsu.edu/primer/lightandcolor/refractionintro.html (last visited on January 23, 20018).

95.Ken Lee Wheeler, Uncovering the Missing Secrets of Magnetism, 3rd Edition, 2014, https://ia802502.us.archive.org/31/items/magnetism1small/magnetism1small.pdf.

96.Light: Does light actually 'travel'?, March 17, 2018, https://www.youtube.com/watch?v=PM7nA4TP4Y0&t=412s.

97.Secret of Light Explained by Ken L. Wheeler, February 3, 2017, https://www.youtube.com/watch?v=S43-I9NkR3c&t=5s.

98.Kaye and Laby, National Physical Library, Refractive Index of Gases, http://www.kayelaby.npl.co.uk/general_physics/2_5/2_5_7.html.

99.Refraction of Light, Florida State University, https://micro.magnet.fsu.edu/primer/lightandcolor/refractionintro.html (last visited on January 23, 20018).

100.Refraction of Light, Florida State University, https://micro.magnet.fsu.edu/primer/lightandcolor/refractionintro.html (last visited on January 23, 20018).

101.Refraction of Light, Florida State University, https://micro.magnet.fsu.edu/primer/lightandcolor/refractionintro.html (last visited on January 23, 20018).

102.Refraction of Light, Florida State University, https://micro.magnet.fsu.edu/primer/lightandcolor/refractionintro.html (last visited on January 23, 20018).

103.Refraction of Light, Florida State University, https://micro.magnet.fsu.edu/primer/lightandcolor/refractionintro.html (last visited on January 23, 20018).

104.Optical Density and Light Speed, http://www.physicsclassroom.com/class/refrn/Lesson-1/Optical-Density-and-Light-Speed (February 17, 2018).

105.Air Pressure Changes With Altitude, https://scied.ucar.edu/webweather/weather-ingredients/change-atmosphere-altitude (last visited on February 18, 2018).

106.Dr. Dale Ward, Vertical Structure of the Atmosphere, http://www.atmo.arizona.edu/students/courselinks/fall14/atmo336/lectures/sec1/structure.html (last visited on May 3, 2018).

107.Lyndon State College, Atmospheric Sciences Department, http://apollo.lsc.vsc.edu/classes/met130/notes/chapter1/vert_temp_trop.html (last visited on February 15, 2018).

108.Lyndon State College, Atmospheric Sciences Department, http://apollo.lsc.vsc.edu/classes/met130/notes/chapter1/vert_temp_trop.html (last visited on February 15, 2018).

109.Dr. Danny R. Faulkner, Flat Earth Proof: Just a Mirage, https://answersingenesis.org/astronomy/earth/flat-earth-proof-just-a-mirage/.

110.Jack A. Stone and Jay H. Zimmerman, Index of Refraction of Air, https://emtoolbox.nist.gov/Wavelength/Documentation.asp (last visited on July 29, 2018).

111.Jack A. Stone and Jay H. Zimmerman, Index of Refraction of Air, https://emtoolbox.nist.gov/Wavelength/Documentation.asp (last visited on July 29, 2018).

112.Atmosphere, https://content.meteoblue.com/en/meteoscool/the-earth/atmosphere (last visited on February 15, 2018).

113.Kaye and Laby, National Physical Library, Refractive Index of Gases, http://www.kayelaby.npl.co.uk/general_physics/2_5/2_5_7.html.

114.Kaye and Laby, National Physical Library, Refractive Index of Gases, http://www.kayelaby.npl.co.uk/general_physics/2_5/2_5_7.html.

115.Jack A. Stone and Jay H. Zimmerman, Index of Refraction of Air, https://emtoolbox.nist.gov/Wavelength/Documentation.asp (last visited on July 29, 2018).

116.Robert Sungenis, Flat Earth Flat Wrong, at 63 (2018).

117.Robert Sungenis, Flat Earth Flat Wrong, at 571 (2018).

118.Jack Williams, Why Dry Air Is Heavier than Humid Air, August 5, 2013, https://www.washingtonpost.com/news/capital-weather-gang/wp/2013/08/05/why-dry-air-is-heavier-than-humid-air/?utm_term=.853fd1a91078.

119.Jack Williams, Why Dry Air Is Heavier than Humid Air, August 5, 2013, https://www.washingtonpost.com/news/capital-weather-gang/wp/2013/08/05/why-dry-air-is-heavier-than-humid-air/?utm_term=.853fd1a91078.

120.The Water Cycle: Water Storage in the Atmosphere, The atmosphere is full of water https://water.usgs.gov/edu/watercycleatmosphere.html (last visited on June 7, 2018).

121.What is a Mirage?, http://www.physics.org/article-questions.asp?id=45 (last visited on February 16, 2018).

122.What is a Mirage?, http://www.physics.org/article-questions.asp?id=45 (last visited on February 16, 2018).

123.Frédéric Fabry, The Spatial Variability of Moisture in the Boundary Layer and Its Effect on Convection Initiation: Project-Long Characterization, at 84, 12 January 2005, https://journals.ametsoc.org/doi/pdf/10.1175/MWR3055.1. See also H.U. Sverdrup, The Humidity Gradient Over the Sea Surface, Journal of Meteorology, March 1946, https://journals.ametsoc.org/doi/pdf/10.1175/1520-0469(1946)003%3C0001:THGOTS%3E2.0.CO%3B2.

124.Frédéric Fabry, The Spatial Variability of Moisture in the Boundary Layer and Its Effect on Convection Initiation: Project-Long Characterization, at 84, 12 January 2005, https://journals.ametsoc.org/doi/pdf/10.1175/MWR3055.1.

125.Frédéric Fabry, The Spatial Variability of Moisture in the Boundary Layer and Its Effect on Convection Initiation: Project-Long Characterization, at 90, 12 January 2005, https://journals.ametsoc.org/doi/pdf/10.1175/MWR3055.1.

126.Carl A. Friehe, J. C. La Rue, F. H. Champagne, C. H. Gibson, and G. F. Dreyer, Effects of Temperature and Humidity Fluctuations on the Optical Refractive Index in the Marine Boundary Layer, Journal of the

Optical Society of America, Volume 65, Number 12, December 1975, at 1507. https://www.osapublishing.org/josa/abstract.cfm?uri=josa-65-12-1502.

127.Carl A. Friehe, J. C. La Rue, F. H. Champagne, C. H. Gibson, and G. F. Dreyer, Effects of Temperature and Humidity Fluctuations on the Optical Refractive Index in the Marine Boundary Layer, Journal of the Optical Society of America, Volume 65, Number 12, December 1975, at 1507. https://www.osapublishing.org/josa/abstract.cfm?uri=josa-65-12-1502.

128.Dr. Danny R. Faulkner, Flat Earth Proof—Just a Mirage, January 16, 2017, https://answersingenesis.org/astronomy/earth/flat-earth-proof-just-a-mirage/.

129.Olympus, https://www.olympus-lifescience.com/es/ (last visited on February 16, 2018).

130.Introduction to the Refraction of Light, Olympus, https://www.olympus-lifescience.com/es/microscope-resource/primer/lightandcolor/refractionintro/ (last visited on February 16, 2018).

131.Introduction to the Refraction of Light, Olympus, https://www.olympus-lifescience.com/es/microscope-resource/primer/lightandcolor/refractionintro/ (last visited on February 16, 2018).

132.Introduction to the Refraction of Light, Olympus, https://www.olympus-lifescience.com/es/microscope-resource/primer/lightandcolor/refractionintro/ (last visited on February 16, 2018).

133.Humphreys,W.J., Physics Of The Air Second Edition, 452 (1929).

134.Humphreys,W.J., Physics Of The Air Second Edition, 451 (1929).

135.Humphreys,W.J., Physics Of The Air Second Edition, 452 (1929).

136.Humphreys, W.J., Physics Of The Air, Second Edition, 449-51 (1929).

137.Jack A. Stone and Jay H. Zimmerman, Index of Refraction of Air, https://emtoolbox.nist.gov/Wavelength/Documentation.asp (last visited on July 29, 2018).

138.Humphreys, W.J., Physics Of The Air, Second Edition, 450-51 (1929).

139.Alec Bennett, Introduction to Atmospheric Visibility Estimation, Bristol Industrial and Research Associates Limited, March 2012. Srinivasa G. Narasimhan, Models and Algorithms for Vision through the Atmosphere, Columbia University, 2003.

140.Samuel Birley Rowbotham, Zetetic Astronomy, at 31-32 (1881), http://www.sacred-texts.com/earth/za/za14.htm.

141.Samuel Birley Rowbotham, Zetetic Astronomy, at 32 (1881), http://www.sacred-texts.com/earth/za/za14.htm.

142.Rob Skiba, Does the Atmosphere Solve "The Problem of Sunsets" on a Flat Earth?, January 22, 2017,

https://www.youtube.com/watch?v=tflhWwoqWAw.
Flat Earth: Refraction - the bending of light, September 7, 2017,
https://www.youtube.com/watch?v=x5a5kIdvnfE.

143.Flat Earth: Refraction - the bending of light, September 7, 2017,
https://www.youtube.com/watch?v=x5a5kIdvnfE.
Debunking UPWARD refraction over the curve 31.63 miles flat, October 15, 2016,
https://www.youtube.com/watch?v=Dt6mjnUdrxs.

144.Debunking UPWARD refraction over the curve 31.63 miles flat, October 15, 2016,
https://www.youtube.com/watch?v=Dt6mjnUdrxs.

145.Debunking UPWARD refraction over the curve 31.63 miles flat, October 15, 2016,
https://www.youtube.com/watch?v=Dt6mjnUdrxs.

146.Lieutenant John Warren, H. M. 33d Regiment of Foot, An Account of experiments made in the Mysore Country in, the year 1804, to Investigate the Effects of Terrestrial Refraction, 1804.

147.Lieutenant John Warren, H. M. 33d Regiment of Foot, An Account of experiments made in the Mysore Country in, the year 1804, to Investigate the Effects of Terrestrial Refraction, 1804.

148.Flat Earth: Refraction - the Bending of Light, September 7, 2017,
https://www.youtube.com/watch?v=x5a5kIdvnfE.

149.Atmospheric Magnification: Why Objects Are Behind Horizon, Even after Zoomed in with P900, March 15, 2017, https://www.youtube.com/watch?v=1zRNhLnaLZg.

150.February 21, 2018, 11:57 p.m., email from bmlsb69 to Edward Hendrie.

151.February 21, 2018, 11:57 p.m., email from bmlsb69 to Edward Hendrie.

152.Santa Barbara, February 23, 2017, https://www.timeanddate.com/weather/usa/santa-barbara/historic?month=2&year=2017.

153.Santa Barbara, January 30, 2017, https://www.timeanddate.com/weather/usa/santa-barbara/historic?month=1&year=2017.

154.Flat Earth: Refraction - the bending of light, September 7, 2017, https://www.youtube.com/watch?v=x5a5kIdvnfE.

155.Flat Earth: Refraction - the bending of light, September 7, 2017, https://www.youtube.com/watch?v=x5a5kIdvnfE.

156.Debunking Upward Refraction Over the Curve 31.63 Miles Flat, December 15, 2107, https://www.musiclessons.com/youtube/watch?v=Dt6mjnUdrxs. Also at https://www.youtube.com/watch?time_continue=2&v=Dt6mjnUdrxs.

157.February 24, 2018, 11:12 a.m., email from bmlsb69 to Edward Hendrie.

158.February 24, 2018, 11:12 a.m., email from bmlsb69 to Edward Hendrie.

159.Santa Barbara Weather History for November 22, 2015, https://www.timeanddate.com/weather/usa/santa-barbara/historic?month=11&year=2015.

160.Santa Barbara Weather History for June 12, 2016, https://www.timeanddate.com/weather/usa/santa-barbara/historic?month=6&year=2016.

161.Andrew Young, Astronomical Refraction, San Diego State University, 2012, https://aty.sdsu.edu/explain/atmos_refr/astr_refr.html.

162.Andrew Young, Astronomical Refraction, San Diego State University, 2012, https://aty.sdsu.edu/explain/atmos_refr/astr_refr.html.

163.Duck Soup - Dialogue Transcript, http://www.script-o-rama.com/movie_scripts/d/duck-soup-script-transcript-marx.html (last visited on February 19, 2108).

164.Flat Earth Awesome Zoom Part 3 Observable Diffraction, , December 26, 2016, at 23:50, https://www.youtube.com/watch?v=0LaFcXU37bo.

165.Samuel Birley Rowbotham, Zetetic Astronomy, at 34 (1881), http://www.sacred-texts.com/earth/za/za14.htm.

166.Jurgen Giesen, Atmospheric Refraction Applet Apparent Flattening of the Sun, April 11, 2011, http://jgiesen.de/refract/index.html.

167.Email from Edward Hendrie to Jurgen Giesen on August 3, 2018.

168.Harald Edens, Ph.D., Physics, New Mexico Institute of Mining and Technology, https://www.nmt.edu/academics/physics/faculty/hedens.php (last visited on August 3, 2018).

169.Harald Edens, Flattening of Sun and Moon, http://www.weatherscapes.com/album.php?cat=optics&subcat=flattening (last visited on August 3, 2018).

170.Email from Edward Hendrie to Harald Edens on August 3, 2018.

171.Curvature and Refraction, October 3, 2012, at 6:12, https://www.youtube.com/watch?v=CFb8FO9DoLU.

172.Pontchartrain Curvature Hoax Debunked; Perspective is Imaginary Curve [Phuket Word, GLOBEBUSTERS], February 22, 2018, https://www.youtube.com/watch?v=bfJRLv77YLU&t=1366s. See also Flat Earth - Proof Horizons Are Diffraction, Not Curvature, No Shrinking Sun Required, January 1, 2017, https://www.youtube.com/watch?v=fYbuvIqnJ3c.

173.Illusions of a Sunset, February 16, 2017, https://www.youtube.com/watch?v=6fnbYhF7j2U.

174.Terrestrial and astronomical refractions, San Diego State University, 2016, https://aty.sdsu.edu/explain/atmos_refr/terrestrial.html.

175.Illusions of a Sunset, February 16, 2017, https://www.youtube.com/watch?v=6fnbYhF7j2U.

176.Illusions of a Sunset, February 16, 2017, https://www.youtube.com/watch?v=6fnbYhF7j2U. See also Pontchartrain Curvature Hoax Debunked;

Perspective is Imaginary Curve [Phuket Word, GLOBEBUSTERS], February 22, 2018, https://www.youtube.com/watch?v=bfJRLv77YLU&t=1366s; Flat Earth - Proof Horizons Are Diffraction, Not Curvature, No Shrinking Sun Required, January 1, 2017, https://www.youtube.com/watch?v=fYbuvIqnJ3c.

177.Andrew T. Young, Atmospheric Extinction, San Diego State University, https://aty.sdsu.edu/explain/extinction/extintro.html (last visited on December 29, 2018).

178.Illusions of a Sunset, February 16, 2017, https://www.youtube.com/watch?v=6fnbYhF7j2U. See also Pontchartrain Curvature Hoax Debunked; Perspective is Imaginary Curve [Phuket Word, GLOBEBUSTERS], February 22, 2018, https://www.youtube.com/watch?v=bfJRLv77YLU&t=1366s; Flat Earth - Proof Horizons Are Diffraction, Not Curvature, No Shrinking Sun Required, January 1, 2017, https://www.youtube.com/watch?v=fYbuvIqnJ3c.

179.Tong Yu-fai, Effect of Atmospheric Refraction on the Times of Sunrise and Sunset, Hong Kong Observatory, August 2017, https://www.hko.gov.hk/m/article_e.htm?title=ele_00493.

180.Andrew T. Young, Atmospheric Extinction, San Diego State University, https://aty.sdsu.edu/explain/extinction/extintro.html (last visited on December 29, 2018).

181.Atmospheric refraction, 19 September 2015, https://infogalactic.com/info/Atmospheric_refraction.

182.Air Pressure at Altitude Calculator, https://www.mide.com/pages/air-pressure-at-altitude-calculator (last visited on August 7, 2018).

183.Jack A. Stone and Jay H. Zimmerman, Index of Refraction of Air, https://emtoolbox.nist.gov/wavelength/documentation.asp#AppendixA (last visited on August 7, 2018).

184.Carl A. Friehe, J. C. La Rue, F. H. Champagne, C. H. Gibson, and G. F. Dreyer, Effects of Temperature and Humidity Fluctuations on the Optical Refractive Index in the Marine Boundary Layer, Journal of the Optical Society of America, Volume 65, Number 12, December 1975, at 1507. https://www.osapublishing.org/josa/abstract.cfm?uri=josa-65-12-1502.

185.Dr, Zak, Flat Earth - Refraction part 1/3, October 17, 2016, https://www.youtube.com/watch?v=XOdfn0CgRrg&index=15&list=WL, Dr, Zak, Flat Earth | Refraction part 2/3 - Star Trails, November 22, 2016, https://www.youtube.com/watch?v=RkDqdoINhYI&index=16&list=WL, Dr. Zak, Flat Earth | Refraction Part 3/3 - Experiment 1, December 10, 2016, https://www.youtube.com/watch?v=KPr1G3mC808&index=17&list=WL. See Flat Earth Sunset Experiment: Air Density!! AWESOME!!, October 7, 2016, https://www.youtube.com/watch?v=ga_-FvaUT10. See also Flat Earth | Globe Earth Mathematically Debunked, September 23, 2017, https://www.youtube.com/watch?v=bol8vZ7pcu8.

186.TONG Yu-fai, Effect of Atmospheric Refraction on the Times of Sunrise and Sunset, Hong Kong Observatory, August 2017, http://www.hko.gov.hk/education/article_e.htm?title=ele_00493.

187.Dr, Zak, Flat Earth - Refraction part 1/3, October 17, 2016, https://www.youtube.com/watch?v=XOdfn0CgRrg&index=15&list=WL, Dr, Zak, Flat Earth | Refraction part 2/3 - Star Trails, November 22, 2016, https://www.youtube.com/watch?v=RkDqdoINhYI&index=16&list=WL, Dr. Zak, Flat Earth | Refraction Part 3/3 - Experiment 1, December 10, 2016, https://www.youtube.com/watch?v=KPr1G3mC808&index=17&list=WL. See Flat Earth Sunset Experiment: Air Density!! AWESOME!!, October 7, 2016, https://www.youtube.com/watch?v=ga_-FvaUT10. See also Flat Earth | Globe Earth Mathematically Debunked, September 23, 2017, https://www.youtube.com/watch?v=bol8vZ7pcu8.

188.Dr, Zak, Flat Earth - Refraction part 1/3, October 17, 2016, https://www.youtube.com/watch?v=XOdfn0CgRrg&index=15&list=WL, Dr, Zak, Flat Earth | Refraction part 2/3 - Star Trails, November 22, 2016, https://www.youtube.com/watch?v=RkDqdoINhYI&index=16&list=WL, Dr. Zak, Flat Earth | Refraction Part 3/3 - Experiment 1, December 10, 2016, https://www.youtube.com/watch?v=KPr1G3mC808&index=17&list=WL. See Flat Earth Sunset Experiment: Air Density!! AWESOME!!, October 7, 2016, https://www.youtube.com/watch?v=ga_-FvaUT10. See also Flat Earth | Globe Earth Mathematically Debunked, September 23, 2017, https://www.youtube.com/watch?v=bol8vZ7pcu8.

189.Flat Earth | Globe Earth Mathematically Debunked, September 23, 2017, https://www.youtube.com/watch?v=bol8vZ7pcu8.

190.Soundly, Lake Pontchartrain Transmission Lines Nikon P900 (3/3) - (Flat Earth Debunked), Julne 27, 2017, https://www.youtube.com/watch?time_continue=2&v=kADO7nkt-rk. Soundly Proving the Curvature of the Earth at Lake Pontchartrain, July 22, 2107, https://www.metabunk.org/soundly-proving-the-curvature-of-the-earth-at-lake-pontchartrain.t8939/. Soundly, Curved Water 4 Follow Up - Lake Pontchartrain - Electrical Transmission Pylons - (Debunk Flat Earth), May 25, 2017, https://www.youtube.com/watch?v=p6E07FlVdHI.Best Debunk EVER! Soundly Debunking Curved Water In NOLA!!, November 20, 2107, https://www.youtube.com/watch?v=HHeFywI4hWM.

191.Dr. Zack, Flat Earth | Curved Water Found-DEBUNKED, June 6, 2017, https://www.youtube.com/watch?v=PqFN-_xwyF8&t=515s.

192.Dr. Zack, Flat Earth | Curved Water Found-DEBUNKED, June 6, 2017, https://www.youtube.com/watch?v=PqFN-_xwyF8&t=515s.

193.Dr. Zack, Flat Earth | Curved Water Found-DEBUNKED, June 6, 2017, https://www.youtube.com/watch?v=PqFN-_xwyF8&t=515s.

194.Best Debunk EVER! Soundly Debunking Curved Water In NOLA!!, November 20, 2107, https://www.youtube.com/watch?v=HHeFywI4hWM.

195.Best Debunk EVER! Soundly Debunking Curved Water In NOLA!!, November 20, 2107, https://www.youtube.com/watch?v=HHeFywI4hWM.

196.Best Debunk EVER! Soundly Debunking Curved Water In NOLA!!, November 20, 2107, https://www.youtube.com/watch?v=HHeFywI4hWM.

197.Jesse Kozlowski, Chris Van Matre Talking About His Surveying Career And His Flat Earth Beliefs, at 0:34:00, March 3, 2018, https://www.youtube.com/watch?v=JCVY9yvyFBE.

198.Samuel Birley Rowbotham (Parallax), Zetetic Astronomy, Earth Not a Globe, at 41 (1881), http://www.sacred-texts.com/earth/za/za45.htm.

199.Samuel Birley Rowbotham (Parallax), Zetetic Astronomy, Earth Not a Globe, at 267-68 (1881), http://www.sacred-texts.com/earth/za/za45.htm.

200.Jesse Kozlowski, PLS (NJ), https://www.linkedin.com/in/jesse-kozlowski-pls-nj-22047571/?trk=public-profile-join-page (last visited on March 17, 2018).

201.Jesse Kozlowski, PLS (NJ), https://www.linkedin.com/in/jesse-kozlowski-pls-nj-22047571/?trk=public-profile-join-page (last visited on March 17, 2018).

202.Jesse Kozlowski, Chris Van Matre Talking About His Surveying Career And His Flat Earth Beliefs, at 1:34:00, March 3, 2018, https://www.youtube.com/watch?v=JCVY9yvyFBE.

203.Jesse Kozlowski, Jesse's Observations & Measurements,

https://jessekozlowski.wordpress.com/category/lighthouses/ (last visited on March 11, 2018).

204.Jesse Kozlowski, Jesse's Observations & Measurements, https://jessekozlowski.wordpress.com/category/lighthouses/ (last visited on March 11, 2018).

205.Jesse Kozlowski, Jesse's Observations & Measurements, https://jessekozlowski.wordpress.com/category/lighthouses/ (last visited on March 11, 2018).

206.The NonSequitur Show, Jeranism vs Kozlowski Discuss Flat Earth And Geodesy (LIVE), at 00:30:00 March 15, 2018, https://www.youtube.com/watch?v=I6dWs8bjN_A.

207.The NonSequitur Show, Jeranism vs Kozlowski Discuss Flat Earth And Geodesy (LIVE), at 00:55:00 March 15, 2018, https://www.youtube.com/watch?v=I6dWs8bjN_A.

208.The NonSequitur Show, Jeranism vs Kozlowski Discuss Flat Earth And Geodesy (LIVE), at 01:00:55 March 15, 2018, https://www.youtube.com/watch?v=I6dWs8bjN_A.

209.Jesse Kozlowski, One Mile Level Line, Jesse's Observations & Measurements, 10 February 2016, https://jessekozlowski.wordpress.com/2016/11/24/one-mile-level-line/. Jesse Kozlowski, One Mile Level Line, November 25, 2016, https://www.youtube.com/watch?v=A90FfiLqEOo&feature=youtu.be.

210.Flat Earth "Ramped Perspective" (Recorded Live Stream), February 20, 2017, https://www.youtube.com/watch?v=xKFKWVZbpyM. G. A. Kharaghani, Propagation of Refraction Errors In Trigonometric Height Traversing and Geodetic Levelling, Geodesy and Geomatics Engineering, Technical Report Number 132, November 2017, http://www2.unb.ca/gge/Pubs/TR132.pdf.

211.G. A. Kharaghani, Propagation of Refraction Errors In Trigonometric Height Traversing and Geodetic Levelling, Geodesy and Geomatics Engineering, Technical Report Number 132, at 19, November 2017, http://www2.unb.ca/gge/Pubs/TR132.pdf.

212.G. A. Kharaghani, Propagation of Refraction Errors In Trigonometric Height Traversing and Geodetic Levelling, Geodesy and Geomatics Engineering, Technical Report Number 132, at 48, November 2017, http://www2.unb.ca/gge/Pubs/TR132.pdf.

213.G. A. Kharaghani, Propagation of Refraction Errors In Trigonometric Height Traversing and Geodetic Levelling, Geodesy and Geomatics Engineering, Technical Report Number 132, at 150, November 2017, http://www2.unb.ca/gge/Pubs/TR132.pdf.

214.Jesse Kozlowski, One Mile Level Line, Jesse's Observations & Measurements, 10 February 2016, https://jessekozlowski.wordpress.com/2016/11/24/one-mile-level-line/. Jesse Kozlowski, One Mile Level Line, November 25, 2016, https://www.youtube.com/watch?v=A90FfiLqEOo&feature=youtu.be.

215.7th Day Truth Seeker, Chris Van Matre: Land Surveyor [GLOBEBUSTERS], August 6, 2017, https://www.youtube.com/watch?v=xVAknhVCrH4.

216.Jesse Kozlowski, One Mile Level Line, November 25, 2016, https://www.youtube.com/watch?v=A90FfiLqEOo&feature=youtu.be.

217. P.T. Kuduntwane & O.A. Kruger, Vertical Angle Calibration of a Theodolite, National Metrology Institute of South Africa (NMISA), https://pdfs.semanticscholar.org/1012/73b4f5bf0f4b7896b8051b2aa2209bdde75a.pdf (last visited on March 18, 2018).

218. P.T. Kuduntwane & O.A. Kruger, Vertical Angle Calibration of a Theodolite, National Metrology Institute of South Africa (NMISA), https://pdfs.semanticscholar.org/1012/73b4f5bf0f4b7896b8051b2aa2209bdde75a.pdf (last visited on March 18, 2018).

219.7th Day Truth Seeker, Chris Van Matre: Land Surveyor [GLOBEBUSTERS], August 6, 2017, https://www.youtube.com/watch?v=xVAknhVCrH4.

220.Basic Surveying - Theory and Practice, Oregon Department of Transportation, Ninth Annual Seminar, at 10-1, February 15-17, 2000. See Differential Leveling, Purdue University, https://engineering.purdue.edu/~asm215/topics/difflevl.html. See also Section 2: Differential Leveling, TxDOT Survey Manual, April 1, 2016, http://onlinemanuals.txdot.gov/txdotmanuals/ess/differential_leveling.htm.

221.Chapter 3 Surveying Measurements, Survey Manual, Department of Transportation, State of New Jersey, http://www.state.nj.us/transportation/eng/documents/survey/Chapter3.shtm (last visited on March 19, 2018).

222.Brent MacDonald, Differential Leveling, September 18, 2017, https://www.youtube.com/watch?v=ijYf5hsJlnw.

223.bmlsb69, Surface Atmosphere No Camera Zoom Can Correct, October 7, 2016, https://www.youtube.com/watch?v=ZX0r0mz4JgU.

224.Jesse Kozlowski, One Mile Level Line, Jesse's Observations & Measurements, 10 February 2016, https://jessekozlowski.wordpress.com/2016/11/24/one-mile-level-line/. Jesse Kozlowski, One Mile Level Line, November 25, 2016, https://www.youtube.com/watch?v=A90FfiLqEOo&feature=youtu.be.

225.Jesse Kozlowski, One Mile Level Line, Jesse's Observations & Measurements, 10 February 2016, https://jessekozlowski.wordpress.com/2016/11/24/one-mile-level-line/. Jesse Kozlowski, One Mile Level Line, November 25, 2016, https://www.youtube.com/watch?v=A90FfiLqEOo&feature=youtu.be.

226.Jesse Kozlowski, One Mile Level Line, Jesse's Observations & Measurements, 10 February 2016, https://jessekozlowski.wordpress.com/2016/11/24/one-mile-level-line/. Jesse Kozlowski, One Mile Level Line, November 25, 2016, https://www.youtube.com/watch?v=A90FfiLqEOo&feature=youtu.be.

227.Jesse Kozlowski, Jesse's Observations & Measurements, Flat Earth Proof? – Perfectly Flat Level Lake, https://jessekozlowski.wordpress.com/2016/11/05/looks-perfectly-flat-and-level-to-me/ (last visited on December 26, 2017). Jesse Kozlowski, Perfectly Flat Level Lake - Flat Earth Proof?, November 18, 2016, https://www.youtube.com/watch?v=ZlP3njwf6h0.

228.NEW - Flat Earth Confirmed - 7.53 miles - No Curvature, https://www.youtube.com/watch?v=bh1KN5gbPvU. December 9, 2017. Date: December 9, 2017 (~5:49 p.m.); Camera Coordinates: Lat: 40.31524628, Lng: -111.76501156; Spotlight Coordinates: Lat: 40.33612475, Lng: -111.90493711; Camera Height = 2.54 feet; Distance = 7.53 miles. See also Flat Earth Confirmed 7.53 Miles, December 10, 2017, https://www.youtube.com/watch?v=t1TVICrdx9Q&list=WL&index=2&t=0s.

229.Jesse Kozlowski, Where Is the Curve on Lake Pontchartrain?, December 24, 2017, https://www.youtube.com/watch?v=BtCyR_MwhFo&t=174s.

230.Dr, Zak, Flat Earth - Refraction part 1/3, October 17, 2016, https://www.youtube.com/watch?v=XOdfn0CgRrg&index=15&list=WL, Dr, Zak, Flat Earth | Refraction part 2/3 - Star Trails, November 22, 2016, https://www.youtube.com/watch?v=RkDqdoINhYI&index=16&list=WL, Dr. Zak, Flat Earth | Refraction Part 3/3 - Experiment 1, December 10, 2016, https://www.youtube.com/watch?v=KPr1G3mC808&index=17&list=WL.

231.Rob Skiba, Does the Atmosphere Solve "The Problem of Sunsets" on a Flat Earth?, January 22, 2017, https://www.youtube.com/watch?v=tflhWwoqWAw.

232. Liz Osborn, Most Humid Cities in the United States, https://www.currentresults.com/Weather-Extremes/US/most-humid-cities.php (last visited on February 3, 2018).

233.Jesse Kozlowski, Chris Van Matre Talking About His Surveying Career And His Flat Earth Beliefs, 1:35:00, March 3, 2018, https://www.youtube.com/watch?v=JCVY9yvyFBE.

234.Jesse Kozlowski, Chris Van Matre Talking About His Surveying Career And His Flat Earth Beliefs, 0:40:00, March 3, 2018, https://www.youtube.com/watch?v=JCVY9yvyFBE.

235.Jesse Kozlowski, Chris Van Matre Talking About His Surveying Career And His Flat Earth Beliefs, 0:42:00, March 3, 2018, https://www.youtube.com/watch?v=JCVY9yvyFBE.

236.Jesse Kozlowski, Chris Van Matre Talking About His Surveying Career And His Flat Earth Beliefs, 0:42:00, March 3, 2018, https://www.youtube.com/watch?v=JCVY9yvyFBE.

237.Past Weather in New Orleans, Louisiana, USA — December 2017, https://www.timeanddate.com/weather/usa/new-orleans/historic?month=12&year=2017.

238.Sun in New Orleans, December 10, 2017, https://www.timeanddate.com/sun/usa/new-orleans?month=12&year=2017.

239.H.U. Sverdrup, The Humidity Gradient Over the Sea Surface, Journal of Meteorology, March 1946, https://journals.ametsoc.org/doi/pdf/10.1175/1520-0469(1946)003%3C0001:THGOTS%3E2.0.CO%3B2.

240.H.U. Sverdrup, The Humidity Gradient Over the Sea Surface, Journal of Meteorology, March 1946, https://journals.ametsoc.org/doi/pdf/10.1175/1520-0469(1946)003%3C0001:THGOTS%3E2.0.CO%3B2.

241.Curvature and Refraction, October 3, 2012, https://www.youtube.com/watch?v=CFb8FO9DoLU.

242.Topographic Surveying, Department of the Army Technical Manual, TM5-234, U.S. Government Print Office, at 10 (1953).

243.Geodesy for the Layman, DMA TR 80-003, at 7, 16 March 1984, Defense Mapping Agency, Building 56, U.S. Naval Observatory, Washington D.C. 20305, http://earth-info.nga.mil/GandG/publications/geolay/Geo4lay.pdf.

244.John Clayton Tracy, Plane Surveying: A Text-Book and Pocket Manual, at 2 (1908).

245.J. Clendinning, The Principles of Surveying, at 2, Second Edition (1960).

246.Geodesy for the Layman, DMA TR 80-003, at 24, 16 March 1984, Defense Mapping Agency, Building 56, U.S. Naval Observatory, Washington D.C. 20305, http://earth-info.nga.mil/GandG/publications/geolay/Geo4lay.pdf.

247.Geodesy for the Layman, DMA TR 80-003, at 24, 16 March 1984, Defense Mapping Agency, Building 56, U.S. Naval Observatory, Washington D.C. 20305, http://earth-info.nga.mil/GandG/publications/geolay/Geo4lay.pdf.

248.Curvature and Refraction, October 3, 2012, https://www.youtube.com/watch?v=CFb8FO9DoLU.

249.Curvature and Refraction, October 3, 2012, https://www.youtube.com/watch?v=CFb8FO9DoLU.

250.Curvature and Refraction, October 3, 2012, at 6:12, https://www.youtube.com/watch?v=CFb8FO9DoLU.

251.J. Clendinning, The Principles of Surveying, at 99, Second Edition (1960).

252.Herbert Dingle, Science at the Crossroads (1972), at 18, http://blog.hasslberger.com/Dingle_SCIENCE_at_the_Crossroads.pdf.

253.Email from Christopher Van Matre to Edward Hendrie, February 26, 2018, 10:38 p.m.

254.Herbert Dingle, Science at the Crossroads (1972), at 18, http://blog.hasslberger.com/Dingle_SCIENCE_at_the_Crossroads.pdf.

255.Herbert Dingle, Science at the Crossroads (1972), at 16, http://blog.hasslberger.com/Dingle_SCIENCE_at_the_Crossroads.pdf.

256.George Orwell, 1984, (1949), http://www.george-orwell.org/1984/19.html.

257.George Orwell, 1984, (1949), http://www.george-orwell.org/1984/19.html.

258.George Orwell, 1984, (1949), http://www.george-orwell.org/1984/19.html.

259.Joseph F. Dracup, Geodetic Surveys in the United States, The Beginning and the next One Hundred Years, 1807-1940, https://www.ngs.noaa.gov/PUBS_LIB/geodetic_survey_1807.html.

260.Joseph F. Dracup, Geodetic Surveys in the United States, The Beginning and the next One Hundred Years, 1807-1940, https://www.ngs.noaa.gov/PUBS_LIB/geodetic_survey_1807.html.

261.National Geodetic Survey, History Overview, https://www.ngs.noaa.gov/web/about_ngs/history/ (last visited on February 28, 2018).

262.Atmospheric Refraction Errors for Optical Instrumentation, Preliminary Report, Technical Memorandum, No. 104, White Sands Proving Ground, Las Cruces, New Mexico, Systems Engineering Branch, Flight Determination Laboratory, United States Air Force, October 1953, at 5, http://www.dtic.mil/dtic/tr/fulltext/u2/020861.pdf.

263.Atmospheric Refraction Errors for Optical Instrumentation, Preliminary Report, Technical Memorandum, No. 104, White Sands Proving Ground, Las Cruces, New Mexico, Systems Engineering

Branch, Flight Determination Laboratory, United States Air Force, October 1953, at 5, http://www.dtic.mil/dtic/tr/fulltext/u2/020861.pdf.

264.Atmospheric Refraction Errors for Optical Instrumentation, Preliminary Report, Technical Memorandum, No. 104, White Sands Proving Ground, Las Cruces, New Mexico, Systems Engineering Branch, Flight Determination Laboratory, United States Air Force, October 1953, at 5, http://www.dtic.mil/dtic/tr/fulltext/u2/020861.pdf.

265.$100,000.00 Myrtle Beach Flat Earth Challenge [USDA Surveyor Ray Goodwin], May 23, 2018, https://www.youtube.com/watch?v=o5X7bGfLocc. $100,000 Myrtle Beach Flat Earth Challenge to any Surveyor, May 24, 2018, http://www.atlasexposed.com/100000-myrtle-beach-flat-earth-challenge-to-any-surveyor/.

266.$100,000 Challenge, https://www.youtube.com/watch?v=ZQpsmwW8twI.

267.$100,000 Challenge, at 24:50, https://www.youtube.com/watch?v=ZQpsmwW8twI.

268.Water Finds Its Own Level, September 16, 2015, https://redrundrain.wordpress.com/2013/09/16/water-finds-its-own-level/.

269.Eric Dubay, The Flat Earth Conspiracy, at 46 (2014).

270.David Wardlaw Scott, Terra Firma, at 134 (1901).

271.Samuel Rowbotham, Zetetic Astronomy, Earth Not a Globe!, at 54-55 (1881), http://www.sacred-texts.com/earth/za/za18.htm#page

54.

272.Samuel Rowbotham, Zetetic Astronomy, Earth Not a Globe!, at 54-55 (1881), http://www.sacred-texts.com/earth/za/za18.htm#page_54.

273.Samuel Rowbotham, Zetetic Astronomy, Earth Not a Globe!, at 54-55 (1881), http://www.sacred-texts.com/earth/za/za18.htm#page_54.

274.Samuel Rowbotham, Zetetic Astronomy, Earth Not a Globe!, at 54-55 (1881), http://www.sacred-texts.com/earth/za/za18.htm#page_54.

275.Samuel Rowbotham, Zetetic Astronomy, Earth Not a Globe!, at 55-56 (1881) http://www.sacred-texts.com/earth/za/za18.htm#page_54.

276.Samuel Rowbotham, Zetetic Astronomy, Earth Not a Globe!, at 55-56 (1881) http://www.sacred-texts.com/earth/za/za18.htm#page_54.

277.Mark Fonstad, William Pugatch, and Brandon Vogt, Kansas is Flatter Than a Pancake, Improbable Research, 2003, http://www.improbable.com/airchives/paperair/volume9/v9i3/kansas.html.

278.Mark Fonstad, William Pugatch, and Brandon Vogt, Kansas is Flatter Than a Pancake, Improbable Research, 2003, http://www.improbable.com/airchives/paperair/volume

9/v9i3/kansas.html.

279.Matt Sokiak, Is Kansas Really "Flatter than a Pancake"?, September 22, 2014, http://mentalfloss.com/article/58976/kansas-really-flatter-pancake.

280.Matt Sokiak, Is Kansas Really "Flatter than a Pancake"?, September 22, 2014, http://mentalfloss.com/article/58976/kansas-really-flatter-pancake.

281.Matt Sokiak, Is Kansas Really "Flatter than a Pancake"?, September 22, 2014, http://mentalfloss.com/article/58976/kansas-really-flatter-pancake.

282.Brendan Lynch, Research: If You Think Kansas Is the Flattest U.S. State, You're Plain Wrong, The University of Kansas, February 18, 2014, https://news.ku.edu/2014/02/06/research-if-you-think-kansas-flattest-us-state-youre-plain-wrong. Dobson, J. E., and J. S. Campbell. 2014. "The Flatness of U. S. States." Geographical Review 104(1):1-9.

283.Megan Garber, Science: Several U.S. States, Led by Florida, Are Flatter Than a Pancake, The Atlantic, March 11, 2014, http://www.theatlantic.com/technology/archive/2014/03/science-several-us-states-led-by-florida-are-flatter-than-a-pancake/284348/.

284.Kansans Flattered by Flatness vs Kansas Flatness Decriers, March 6, 2014, http://www.improbable.com/2014/03/05/kansans-flattered-by-flatness-vs-kansas-flatness-decriers/.

285.Jon McIntyre, Flat Earth ... A Mountain of Evidence, Sep 1, 2016, https://www.youtube.com/watch?v=DPDtMQqlprk.

286.Jon McIntyre, Flat Earth ... A Mountain of Evidence, at the 47:00 point in the video, Sep 1, 2016, https://www.youtube.com/watch?v=DPDtMQqlprk.

287.Jon McIntyre, Flat Earth ... A Mountain of Evidence, Sep 1, 2016, https://www.youtube.com/watch?v=DPDtMQqlprk.

288.Debunked: View of Blue Ridge Mountains Impossible on Spherical Earth, https://www.metabunk.org/debunked-view-of-blue-ridge-mountains-impossible-on-spherical-earth.t7941/page-2.

289.Debunked: View of Blue Ridge Mountains Impossible on Spherical Earth, https://www.metabunk.org/debunked-view-of-blue-ridge-mountains-impossible-on-spherical-earth.t7941/page-2.

290.December 17, 2017, 10:11 a.m. email From Dr. Jim Fetzer to Edward Hendrie, Alex Suder, and True Ott.

291.December 17, 2017, 11:47 a.m., email From Dr. Jim Fetzer to Edward Hendrie, Alex Suder, and True Ott.

292.Jesse Kozlowski, Jon McIntyre 's Mountain Of Evidence Evaluated, September 26, 2016, https://www.youtube.com/watch?v=0vQxLueDnRw&t=577s.

293.Rob Alphanostrum, Experiment which proves the flat earth or the round earth, May 6, 2016, Https://www.youtube.com/watch?v=iixntrq 8eu.

294.Matthew Hibbard, How to Take Transit to the Forest Park Balloon Glow, Balloon Race, September 8, 2017, https://www.metrostlouis.org/nextstop/how-to-take-transit-to-the-forest-park-balloon-glow-balloon-race/. Photo used by Hibbard was "Courtesy of Great Forest Park Balloon Race."

295.Rob Alphanostrum, Experiment which proves the flat earth or the round earth, May 6, 2016, Https://www.youtube.com/watch?v=iixntrq 8eu.

296.Malcolm Bowden, Flat Earth Stupidity Exposed, February 5, 2016, https://www.youtube.com/watch?v=nCh-tbTd Pc&t=1053s.

297.Malcolm Bowden, Flat Earth Stupidity Exposed, February 5, 2016, https://www.youtube.com/watch?v=nCh-tbTd Pc&t=1053s.

298.Malcolm Bowden, Flat Earth Stupidity Exposed, February 5, 2016, https://www.youtube.com/watch?v=nCh-tbTd Pc&t=1053s.

299.Nick Berry, Consequences of Living on a Globe, DataGenetics, http://datagenetics.com/blog/june32012/index.html (last visited on December 25, 2016).

300.Nick Berry, Consequences of Living on a Globe, DataGenetics, http://datagenetics.com/blog/june32012/index.html (last visited on December 25, 2016).

301.Facts & Figures About the Bridge, Golden Gate Bridge Highway & Transportation District, http://goldengate.org/exhibits/facts-and-figures-about-the-bridge.php (last visited on December 25, 2016).

302.Sarah L. Billington, David P. Billington, Jr., Engineering the Golden Gate, June 22, 2012, https://www.curee.org/projects/GGB/conference/agenda_files/presentations-day2/10_History_of_Research-Billington.pdf.

303.E.g., D. B. Steinman, A.M., C.E., Ph.D., A Practical Treatise on Suspension Bridges, 1922, https://archive.org/details/practicaltreatis00steiuoft.

304.Sumitomo Group Public Affairs Committee, Akashi Kaikyo Bridge and Advanced Technology, http://www.sumitomo.gr.jp/english/discoveries/special/74_02.html (last visited on December 25, 2016).

305.North Tower Blueprints, Table of World Trade Center Drawings, http://911research.wtc7.net/wtc/evidence/plans/table.html.

306.Ronald Hamburger, et al., FEMA 403, World Trade Center Building Performance Study (2002), https://www.fema.gov/pdf/library/fema403_ch2.pdf.

307.Final Reports from the NIST World Trade Center Disaster Investigation, Final Reports Released in November 2005, and Final Report on the Collapse of

World Trade Center Building 7, Released in November 2008, https://www.nist.gov/engineering-laboratory/final-reports-nist-world-trade-center-disaster-investigation.

308.Tower Blueprints, Surviving Evidence of the World Trade Center Attack, http://911research.wtc7.net/wtc/evidence/blueprints.html#ref1 (last visited on December 31, 2016).

309.Tower Blueprints, Surviving Evidence of the World Trade Center Attack, http://911research.wtc7.net/wtc/evidence/blueprints.html#ref1 (last visited on December 31, 2016).

310.Ronald Hamburger, et al., FEMA 403, World Trade Center Building Performance Study (2002), Chapter 2, WTC1 and WTC2, https://www.fema.gov/pdf/library/fema403_ch2.pdf.

311.Eric Dubay, The Flat Earth Conspiracy, at 46 (2014).

312.Brian Mullin, Official Launch! Force the Line Is On, July 16, 2016, https://www.youtube.com/watch?v=iu1t0jBBuTI.

313.Brian Mullin, Official Launch! Force the Line Is On, July 16, 2016, https://www.youtube.com/watch?v=iu1t0jBBuTI.

314.Brian Mullin, Official Launch! Force the Line Is On, July 16, 2016, https://www.youtube.com/watch?v=iu1t0jBBuTI. Comment by Yolanda Whittle.

315.Brian Mullin, Official Launch! Force the Line Is On, July 16, 2016, https://www.youtube.com/watch?v=iu1t0jBBuTI. Comment by Yolanda Whittle.

316.David Barnard Steiman, Levy Award, Engineering, The Design of the Mackinac Bridge, The Franklin Institute, 1957, https://www.fi.edu/laureates/david-barnard-steinman.

317.E.g., D. B. Steinman, A.M., C.E., Ph.D., A Practical Treatise on Suspension Bridges, 1922, https://archive.org/details/practicaltreatis00steiuoft.

318.Eric Dubay, The Flat Earth Conspiracy, at 46 (2014).

319.Target Generation Facility (TGF), Federal Aviation Administration, May 3, 2012, http://hf.tc.faa.gov/capabilities/tgf.htm.

320.Target Generation Facility (TGF), Federal Aviation Administration, May 3, 2012, http://hf.tc.faa.gov/capabilities/tgf.htm.

321.Mark Peters and Michael A. Konyak, The Engineering Analysis and Design of the Aircraft Dynamics Model For the FAA Target Generation Facility, Simulation Branch, Laboratory Services Division , Federal Aviation Administration, William J. Hughes Technical Center, October 2012, at 1, http://www.faa.gov/about/office_org/headquarters_offices/ang/offices/tc/about/campus/faa_host/labs/tgf/media/AircraftDynamicsModel.pdf.

322.Mark Peters and Michael A. Konyak, The Engineering Analysis and Design of the Aircraft Dynamics Model For the FAA Target Generation Facility, Simulation Branch, Laboratory Services

Division , Federal Aviation Administration, William J. Hughes Technical Center, October 2012, at 32, http://www.faa.gov/about/office_org/headquarters_offices/ang/offices/tc/about/campus/faa_host/labs/tgf/media/AircraftDynamicsModel.pdf.

323.Peters and Konyak, The Engineering Analysis and Design of the Aircraft Dynamics Model For the FAA Target Generation Facility, supra, at 1-2.

324.Peters and Konyak, The Engineering Analysis and Design of the Aircraft Dynamics Model For the FAA Target Generation Facility, supra, at 2.

325.Peters and Konyak, The Engineering Analysis and Design of the Aircraft Dynamics Model For the FAA Target Generation Facility, supra, at 234.

326.Eugene Duke, Robert Antoniewicz, and Keith Krambeer, NASA Reference Publication 1207, Derivation and Definition of a Linear Aircraft Model, NASA Scientific and Technical Information Division, Ames Research Center, Dryden Flight Research Facility, 1988, https://www.nasa.gov/centers/dryden/pdf/88104main_H-1391.pdf.

327.Atmospheric Refraction Errors for Optical Instrumentation, Preliminary Report, Technical Memorandum, No. 104, White Sands Proving Ground, Las Cruces, New Mexico, Systems Engineering Branch, Flight Determination Laboratory, United States Air Force, October 1953, http://www.dtic.mil/dtic/tr/fulltext/u2/020861.pdf.

328.Atmospheric Refraction Errors for Optical Instrumentation, Preliminary Report, Technical

Memorandum, No. 104, White Sands Proving Ground, Las Cruces, New Mexico, Systems Engineering Branch, Flight Determination Laboratory, United States Air Force, October 1953, http://www.dtic.mil/dtic/tr/fulltext/u2/020861.pdf.

329.Atmospheric Refraction Errors for Optical Instrumentation, Preliminary Report, Technical Memorandum, No. 104, White Sands Proving Ground, Las Cruces, New Mexico, Systems Engineering Branch, Flight Determination Laboratory, United States Air Force, October 1953, http://www.dtic.mil/dtic/tr/fulltext/u2/020861.pdf.

330.Atmospheric Refraction Errors for Optical Instrumentation, Preliminary Report, Technical Memorandum, No. 104, White Sands Proving Ground, Las Cruces, New Mexico, Systems Engineering Branch, Flight Determination Laboratory, United States Air Force, October 1953, http://www.dtic.mil/dtic/tr/fulltext/u2/020861.pdf.

331.Atmospheric Refraction Errors for Optical Instrumentation, Preliminary Report, Technical Memorandum, No. 104, White Sands Proving Ground, Las Cruces, New Mexico, Systems Engineering Branch, Flight Determination Laboratory, United States Air Force, October 1953, at 5, http://www.dtic.mil/dtic/tr/fulltext/u2/020861.pdf.

332.Gene R. Cooper, Adding Liquid Payloads Effects to the 6-DOF Trajectory of Spinning Projectiles, Army Research Laboratory, Weapons and Materials Research Directorate, ARL-TR-5118, at 1, March 2010, https://www.arl.army.mil/arlreports/2010/ARL-TR-5118.pdf.

333.Joseph R. Miletta, Propagation of Electromagnetic Fields Over Flat Earth, Army Research Laboratory, Sensors and Electron Devices Directorate, ARL-TR-2352, February 2001, https://www.arl.army.mil/arlreports/2001/ARL-TR-2352.pdf.

334.Joseph R. Miletta, Propagation of Electromagnetic Fields Over Flat Earth, Army Research Laboratory, Sensors and Electron Devices Directorate, ARL-TR-2352, February 2001, at 1, https://www.arl.army.mil/arlreports/2001/ARL-TR-2352.pdf.

335.A.J. Lineberger, H.D. Edwards, Atmospheric Oscillations, Prepared for NASA by Georgia Institute of Technology, April 1965, at 10, https://ntrs.nasa.gov/archive/nasa/casi.ntrs.nasa.gov/19650015408.pdf.

336.Brian E. Mihok, A Property-based System Design Method with Application to a Targeting System for Small UAVs, MIT, at 3, 2006, https://dspace.mit.edu/handle/1721.1/35571#files-area.

337.Brian E. Mihok, A Property-based System Design Method with Application to a Targeting System for Small UAVs, MIT, at 124, 2006, https://dspace.mit.edu/handle/1721.1/35571#files-area.

338.MQ-1B Predator, U.S. Air Force Fact Sheet, July 20, 2010, https://web.archive.org/web/20111216191110/http://www.af.mil/information/factsheets/factsheet.asp?fsID=122.

339.Brian E. Mihok, A Property-based System Design Method with Application to a Targeting System for Small UAVs, MIT, at 77, 2006, https://dspace.mit.edu/handle/1721.1/35571#files-area.

340.Brian E. Mihok, A Property-based System Design Method with Application to a Targeting System for Small UAVs, MIT, at 121-22, 2006, https://dspace.mit.edu/handle/1721.1/35571#files-area.

341.Barton J. Bacon and Irene M. Gregory, General Equations of Motion for a Damaged Asymmetric Aircraft, NASA Langley Research Center, https://ntrs.nasa.gov/archive/nasa/casi.ntrs.nasa.gov/20070030307.pdf (last visited on August 1, 2018).

342.Barton J. Bacon and Irene M. Gregory, General Equations of Motion for a Damaged Asymmetric Aircraft, NASA Langley Research Center, https://ntrs.nasa.gov/archive/nasa/casi.ntrs.nasa.gov/20070030307.pdf (last visited on August 1, 2018).

343.John S. Preisser, Determination of Angles of Attack and Sideslip from Radar Data and a Roll-Stabilized Platform, NASA Langley Research Center, March 1972, https://ntrs.nasa.gov/archive/nasa/casi.ntrs.nasa.gov/19720012071.pdf.

344.John S. Preisser, Determination of Angles of Attack and Sideslip from Radar Data and a Roll-Stabilized Platform, NASA Langley Research Center, March 1972, https://ntrs.nasa.gov/archive/nasa/casi.ntrs.nasa.gov/19720012071.pdf.

345.John S. Preisser, Determination of Angles of Attack and Sideslip from Radar Data and a Roll-Stabilized Platform, NASA Langley Research Center, March 1972, at 5, https://ntrs.nasa.gov/archive/nasa/casi.ntrs.nasa.gov/19720012071.pdf.

346.United States Navy Training Film, Gyroscopes. See also United States Navy Training Film, FN-8048e, Instrument Flight Control, Gyroscopic Instruments Part II, 1960.

347.odiupicku, THE FLAT EARTH - A GYRO EXPERIMENT, November 12, 2015, https://www.youtube.com/watch?v=hlHOSUpmgQ8.

348.Exalt Sets New World Record for Microwave Link Distance, February 08, 2016, http://www.marketwired.com/press-release/exalt-sets-new-world-record-for-microwave-link-distance-2094649.htm.

349.Microwave Link Networks, http://ethw.org/Microwave_Link_Networks (last visited on April 1, 2018).

350.Andy Singer, The World's Longest Microwave Radio Link, March 23, 2016, http://www.singerexecutivedevelopment.com/the-worlds-longest-microwave-radio-link/.

351.Dan Jones, Exalt Communications is Dead, Long Live Exalt Wireless!, October 29, 2014, https://www.lightreading.com/mobile/backhaul/exalt-communications-is-dead-long-live-exalt-wireless!/d/d-id/711725.

352.ExploreAir® LR - Highest capacity. Longest range. Smallest antenna, ExploreAir LR Spotlight, http://exaltcom.com/ExploreAir-LR-Spotlight.aspx (last visited on April 1, 2018).

353.Sherif Shaalan, Flat Earth Proof – Line-of-Sight Microwave Transmission (EDITED/ENHANCED), https://www.youtube.com/watch?v=HQ77t2kKXWw.

354.Radio Wave Basics, Yerkes Summer Institute, Center for Cosmological Physics, at ii, 2002, https://kicp.uchicago.edu/education/explorers/2002summer-YERKES/pdfs-sum02/background.pdf.

355.Propagation of Light Waves, Vigyan Prasar, http://vigyanprasar.gov.in/propagation-of-radio-waves/ (last visited on March 30, 2018)..

356.Propagation of Light Waves, Vigyan Prasar, http://vigyanprasar.gov.in/propagation-of-radio-waves/ (last visited on March 30, 2018)..

357.Sir Edward Appleton, The Discovery of the Properties of the Ionosphere, Nortown Amateur Radio Club of Toronto, http://www.ve3nar.org/meetings_files/Presentations/Appleton.pdf (last visited on June 11, 2018).

358.How to Use Amateur Radio Moonbounce, EME Propagation, https://www.electronics-notes.com/articles/ham_radio/amateur-propagation/moonbounce-propagation-eme.php (last visited on April 7, 2018).

359.Get ready for 2m MoonBounce Communications, https://www.dxmaps.com/jt65bintro.html (last visited on April 7, 2018). See also VE2ZAZ - First EME

(Moonbounce) Radio Contact with HB9Q - Higher Quality Vidéo, April 2, 2017, https://www.youtube.com/watch?v=nfrPwdmmky4.

360.Elizabeth Howell, What Are Cosmic Rays?, April 21, 2016, https://www.space.com/32644-cosmic-rays.html.

361.Cosmic Rays, http://astronomy.swin.edu.au/cosmos/C/Cosmic+Rays (last visited on April 7, 2018).

362.Electromagnetic Spectrum, October 30, 2012, https://www.nasa.gov/directorates/heo/scan/spectrum/txt_electromagnetic_spectrum.html.

363.Awe130 a Journey Towards the Truth Apollo Hoax, http://www.awe130.com/apollo-hoax-explained/178-apollo-12-astronaut-conrad-this-is-the-blackest-black-i-ever-saw (last visited on April 9, 2018).

364.Charlotte Lytton, Secrets of a Spaceman: Meet Eugene Cernan, the Last Man to Ever Walk on the Moon, The Telegraph, 17 January 2017, https://www.telegraph.co.uk/men/the-filter/secrets-of-a-spaceman-meet-eugene-cernan-the-last-man-to-ever-wa/.

365.The Strange Truth Project. Winner! Best Truth documentary of all time. TEAMYAHAWASHI #TheSTP, at 00:29:00, December 19, 2017, https://www.youtube.com/watch?v=kj-W5jB1Nv4.

366.The Strange Truth Project. Winner! Best Truth documentary of all time. TEAMYAHAWASHI #TheSTP, at 00:29:00, December 19, 2017, https://www.youtube.com/watch?v=kj-W5jB1Nv4.

367.Radio Waves to Gamma-rays, https://www.e-education.psu.edu/astro801/content/l3_p4.html (last visited on April 7, 2018).

368.The Electrical Experimenter, November, 1915, page 321, https://earlyradiohistory.us/1915ATT1.htm. See also Aaron Dover, Flat Earth-Impossible Radio Transmissions, Apr 20, 2016, https://www.youtube.com/watch?v=W08CFPSvzCg.

369.The Electrical Experimenter, November, 1915, page 321, https://earlyradiohistory.us/1915ATT1.htm. See also Aaron Dover, Flat Earth-Impossible Radio Transmissions, Apr 20, 2016, https://www.youtube.com/watch?v=W08CFPSvzCg.

370.Julia Zorthian, Speech Crossed The Atlantic for the First Time 100 Years Ago This Week, Time Magazine, October 22, 2015, http://time.com/4081211/transatlantic-speech-transmission-1915/.

371.The Arlington Naval Radio Station, Arlington Historical Society, https://arlingtonhistoricalsociety.org/arlington-naval-radio-station/ (last visited on April 2, 2018).

372.Loran, Radio Navigation, last updated on March 22, 2016, https://www.britannica.com/technology/loran.

373.Loran-c General Information, U.S. Coast Guard, https://www.navcen.uscg.gov/?pageName=loranMain (last visited on April 3, 2018).

374.Christopher Lagen, VIDEO: Demolition of America's largest LORAN Tower, April 29, 2010, http://coastguard.dodlive.mil/2010/04/video-demolition-of-americas-largest-loran-tower/.

375.Smithsonian National Air and Space Museum, How Loran Works, March 29, 2013, https://www.youtube.com/watch?v=PDtHulWGMGg.

376.MUST SEE!!! No satellites in space! Just balloons above the Flat Earth, October 12, 2016, https://www.youtube.com/watch?v=PxXGcc5OdmA.

377.NASA Awards Agency-wide Helium Contract, CONTRACT RELEASE C14-036, September 25, 2014, https://www.nasa.gov/press/2014/september/nasa-awards-agency-wide-helium-contract/. See also NASA Awards Helium Contract, CONTRACT RELEASE : C09-045, Sept. 8, 2009, https://www.nasa.gov/home/hqnews/2009/sep/HQ_C09-045_Helium_Contract.html.

378.NASA Awards Agency-wide Helium Contract, CONTRACT RELEASE C14-036, September 25, 2014, https://www.nasa.gov/press/2014/september/nasa-awards-agency-wide-helium-contract/. See also NASA Awards Helium Contract, CONTRACT RELEASE : C09-045, Sept. 8, 2009, https://www.nasa.gov/home/hqnews/2009/sep/HQ_C09-045_Helium_Contract.html.

379.Report to Congress on Items Required by Sec. 19 of the Helium Stewardship Act of 2013 Public Law 113-40, as Authorized under 50 U.s.c. 167 by the Secretary of the Interior, at 3, September 2015, https://www.blm.gov/sites/blm.gov/files/NM_Report%20to%20Congress%20Section%2019%20HSA_FINAL%20to%20DC%2006201 5%20Sec%20508%20Compliant.pdf.

380.Report to Congress on Items Required by Sec. 19 of the Helium Stewardship Act of 2013 Public Law 113-40, as Authorized under 50 U.s.c. 167 by the Secretary of the Interior, at 3, September 2015, https://www.blm.gov/sites/blm.gov/files/NM_Report%20to%20Congress%20Section%2019%20HSA_FINAL%20to%20DC%2006201 5%20Sec%20508%20Compliant.pdf.

381.NASA Awards Agency-wide Helium Contract, CONTRACT RELEASE C14-036, September 25, 2014, https://www.nasa.gov/press/2014/september/nasa-awards-agency-wide-helium-contract/. See also NASA Awards Helium Contract, CONTRACT RELEASE : C09-045, Sept. 8, 2009, https://www.nasa.gov/home/hqnews/2009/sep/HQ_C09-045_Helium_Contract.html.

382.Report to Congress on Items Required by Sec. 19 of the Helium Stewardship Act of 2013 Public Law 113-40, as Authorized under 50 U.s.c. 167 by the Secretary of the Interior, at 4, September 2015, https://www.blm.gov/sites/blm.gov/files/NM_Report%20to%20Congress%20Section%2019%20HSA_FINAL%20to%20DC%2006201 5%20Sec%20508%20Compliant.pdf.

383.Project Loon, Google, https://x.company/loon/ (last visited on April 15, 2018).

384.Project Loon, Google, https://x.company/loon/technology/ (last visited on April 15, 2018).

385.Project Loon, Google, https://x.company/loon/technology/ (last visited on April 15, 2018).

386.Project Loon, Google, https://x.company/loon/technology/ (last visited on April 15, 2018).

387.OneWeb Satellites, http://onewebsatellites.com/ (last visited on April 15, 2018).

388.Satellite 2018, March 12-18, 2018, http://2018.satshow.com/.

389.Eric Berger, There's Something Strange Going on amid the Satellite Internet Rush, 2/23/2108, https://arstechnica.com/science/2018/02/theres-something-strange-going-on-amid-the-satellite-internet-rush/.

390.Ron Nixon, Africa, Offline: Waiting for the Web, New York Times, JULY 22, 2007, https://www.nytimes.com/2007/07/22/business/yourmoney/22rwanda.html?_r=0.

391.Ron Nixon, Africa, Offline: Waiting for the Web, New York Times, JULY 22, 2007, https://www.nytimes.com/2007/07/22/business/yourmoney/22rwanda.html?_r=0.

392.Ron Nixon, Africa, Offline: Waiting for the Web, New York Times, JULY 22, 2007, https://www.nytimes.com/2007/07/22/business/yourmoney/22rwanda.html?_r=0.

393.Ron Nixon, Africa, Offline: Waiting for the Web, New York Times, JULY 22, 2007, https://www.nytimes.com/2007/07/22/business/yourmoney/22rwanda.html?_r=0.

394.Eric Dubay, The Flat Earth Conspiracy, at 35 (2014), quoting B. Chas. Brough, "The Zetetic" Volume 1 Number 1, July 1872.

395.Eric Dubay, The Always Horizontal Horizon Proves Earth Flat, January 14, 2015, http://www.atlanteanconspiracy.com/2015/01/flat-earth-horizon.html, quoting Dr. Samuel Rowbotham, "Zetetic Astronomy, Earth Not a Globe!"

396.Samuel Birley Rowbotham (Parallax), Zetetic Astronomy, Earth Not a Globe, at 24-25 (1881), http://www.sacred-texts.com/earth/za/za12.htm#page_25.

397.Samuel Birley Rowbotham (Parallax), Zetetic Astronomy, Earth Not a Globe, at 26-27 (1881), http://www.sacred-texts.com/earth/za/za13.htm#page_26.

398.Eric Dubay, The Always Horizontal Horizon Proves Earth Flat, January 14, 2015, http://www.atlanteanconspiracy.com/2015/01/flat-earth-horizon.html.

399.Samuel Birley Rowbotham, Zetetic Astronomy, Earth Not a Globe!, at 20 (1865).

400.Samuel Rowbotham, Zetetic Astronomy, Earth Not a Globe, at 40 (1881), http://www.sacred-texts.com/earth/za/za16.htm#page_40.

401.Davide James, A Level Horizon at 38,000 ft = A level Earth (Theodolite Calibration - iPhone 7), August 20, 2018, https://www.youtube.com/watch?v=QbaFThY6o7g.

402.David C James, A Level Horizon At 38,000 Feet Means A Level Earth, August 15, 2018, https://www.youtube.com/watch?v=J7IjvwKmNaU.

403.Eric Dubay, The Always Horizontal Horizon Proves Earth Flat, January 14, 2015, http://www.atlanteanconspiracy.com/2015/01/flat-earth-horizon.html.

404.NASA Unexplained Files 2015 Exposed Photo Expert Reveals PsyOp, October 23, 2015, https://www.youtube.com/watch?v=OeDhSbYsKXA. See also BUSTED! - 2015 NASA Photo Fakery Revealed, September 27, 2015, https://www.youtube.com/watch?v=18lGhRt37pM.

405.Flights of Fantasy, Mythbusters, Discovery Channel, https://www.discovery.com/tv-shows/mythbusters/full-episodes/flights-of-fantasy. Mythbusters 15x08 Flights of Fantasy Part 08.mkv, at 1:30, https://www.youtube.com/watch?v=c1df5x4-wVc, April 12, 2017. Video Revealing Mythbusters Lying About Curvature Suppressed, https://www.youtube.com/watch?v=JX-s_ZubBmk, June 1, 2018.

406.Tony Reichhardt, First Photo From Space, Air & Space Magazine, November 2006, http://www.airspacemag.com/space/the-first-photo-from-space-13721411/?no-ist.

407.First Photo From Space, NASA, http://solarsystem.nasa.gov/planets/earth (last visited on December 10, 2015).

408.Tony Reichhardt, First Photo From Space, Air & Space Magazine, November 2006, http://www.airspacemag.com/space/the-first-photo-from-space-13721411/?no-ist.

409.Flat Earth View from 317 000 Feet - 96 Km (NASA Footage), December 1, 2015, https://www.youtube.com/watch?v=l7BhcufC-5Y.

410.Gustave-Gaspard Coriolis, French physicist, Encyclopedia Britannica, http://www.britannica.com/biography/Gustave-Gaspard-Coriolis (last visited on September 26, 2015).

411.Gustave-Gaspard Coriolis, French physicist, Encyclopedia Britannica, supra.

412.Coriolis Effect, National Geographic, http://education.nationalgeographic.com/encyclopedia/coriolis-effect/ (last visited on September 26, 2015).

413.Mark Piggott, British Airways 777 Plane Blown Across Atlantic at Supersonic Speeds by Freak Jetstream, International Business Times, January 11, 2015, http://www.ibtimes.co.uk/british-airways-777-plane-blown-across-atlantic-supersonic-speeds-by-freak-jetstream-1482875.

414.David Wardlaw Scott, Terra Firma: The Earth Not a Planet Proved From Scripture, Reason, and Fact, 89-90 (1901).

415.Coriolis Effect, Coastal Practice Network, http://www.coastalpractice.net/glossary/coriolis%20effect%20%20.htm (last visited on October 26, 2015).

416.E.g., John J. McGrath, Fire for Effect: Field Artillery and Close Air Support in the U.S. Army, Combat Studies Institute Press, U.S. Army Combined Arms Center, Fort Leavenworth, Kansas.

417.E.g., Sniper Training, Field Manual, FM-2310, Headquarters, Department of the Army, 17 August 1994. See also U.S. Navy SEAL Sniper Training Program, Navy Special Warfare.

418. E.g., Colonel H.A. Bethel, Modern Artillery in the Field: A Description of the Artillery of the Field, A Discription of the Artillaery of the Field Army, and the Principles and Methods of Its Employment, 1911. See also Instruction for Field Artillery Prepared by a Board of Artillery Officers, 1860; Artillerist's Manual, Compiled from Various Sources and Adapted to The Service of the United States, 1863; Joseph Roberts, The Hand-Book of Artillery, For the Service of the United States, (Army and Militia.) With the Manual of Heavy Artillery, Including That of the New Iron Carriage, 1863.

419.Georgia Diebelius, Caught on Camera: Amazing Moment a Woman Gives Birth to a Premature Baby Girl at 30,000ft on a Bali to Los Angeles Flight with Crew and Passengers Helping, Daily Mail, 13 October 2015, http://www.dailymail.co.uk/travel/travel_news/article-3270382/Caught-camera-amazing-moment-woman-gives-birth-premature-baby-girl-30-000ft-Bali-Los-Angeles-flight-crew-passengers-helping-out.html.

420.Taiwanese Woman Gives Birth at 30,000 Feet on CAL Flight, October 12, 2015, http://www.chinapost.com.tw/taiwan/national/national-news/2015/10/12/448127/Taiwanese-woman.htm.

421.Taiwanese Woman Gives Birth at 30,000 Feet on CAL Flight, October 12, 2015, http://www.chinapost.com.tw/taiwan/national/national-news/2015/10/12/448127/Taiwanese-woman.htm.

422.UAL 7941, https://flightaware.com/live/flight/UAL7941/history/20171017/1620Z/KIAD/RJAA.

423.Flight Log of UAL 7941, https://flightaware.com/live/flight/UAL7941/history/20171017/1620Z/KIAD/RJAA/tracklog.

424.Bob Crilly, Jet Stream Blasts Ba Plane Across Atlantic in Record Time, The Telegraph, 10 January 2015, http://www.telegraph.co.uk/news/worldnews/northamerica/usa/11337617/Jet-stream-blasts-BA-plane-across-Atlantic-in-record-time.html.

425.Southern Flight Routes & The Jet Streams [Flat Earth], November 13, 2016, 7th Day Truth Seeker, https://www.youtube.com/watch?v=gAKMHT-ycE8.

426.Qantas Flight 63, Sydney to Johannesburg Flight Data, November 30, 2017, https://flightaware.com/live/flight/QFA63/history/20171130/0120Z/YSSY/FAOR/tracklog.

427.Qantas Flight 64, Johannesburg to Syndey Flight Data, December 7, 2017, https://flightaware.com/live/flight/QFA64/history/2017

1207/1650Z/FAOR/YSSY/tracklog.

428.December 10, 2017, Qantas Flight 63, https://flightaware.com/live/flight/QFA63/history/20171212/0055Z/YSSY/FAOR/tracklog.

429.Flat Earth, Banjo, USA, Japan and Brazil, Flat Earth: The AE map is correct and the FE A**hole (a.k.a Jake Gibson) is DEAD WRONG, December 14, 2017, https://www.youtube.com/watch?v=EO9FOXmyhkI.

430.How is this view of Earth's daily cycle similar to a clock?, Annenberg Learner, http://www.learner.org/jnorth/tm/mclass/jr/LongNorthPoleAnimationAns.html (last visited on December 10, 2015).

431.Now It's Your Turn: Self-Study Guide, University of California at Berkeley, http://cse.ssl.berkeley.edu/lessons/indiv/spin/selfstudy.html (December 10, 2015).

432.Spin the Globe, University of Chicago Department of Astronomy and Astrophysics, http://astro.uchicago.edu/cara/southpole.edu/spin.html (last visited on December 10, 2015).

433.Professor of Astronomy & Author Courtney Seligman, Rotation Period and Day Length, Online Astronomy eText: The Sky, March 3, 2015, http://cseligman.com/text/sky/rotationvsday.htm.

434.Professor of Astronomy & Author Courtney Seligman, Rotation Period and Day Length, Online Astronomy eText: The Sky, March 3, 2015, http://cseligman.com/text/sky/rotationvsday.htm.

435.Lesson 6.2 Start at the Source: Earth Rotating Around the Sun, Pennsylvania State University, The John A. Dutton e-Education Institute, https://www.e-education.psu.edu/meteo300/node/681 (last visited on November 2, 2015).

436.James I. Sammons, Navigating Around the World by Observing the Sun, Jamestown School Rhode Island, http://www.pbs.org/wgbh/nova/education/ideas/sammons/packet.html (last visited on December 10, 2015).

437.Sidereal Time, http://www.astro.cornell.edu/academics/courses/astro201/sidereal.htm (last visited on December 10, 2015).

438.Jerry Coffey, How Long is a Day on Earth?, June 3, 2008, http://www.universetoday.com/14700/how-long-is-a-day-on-earth/.

439.Sidereal Time, Cornell University Department of Astronomy, http://www.astro.cornell.edu/academics/courses/astro201/sidereal.htm (last visited on December 10, 2015).

440.Sidereal, Merriam-Webster Dictionary.

441.Justin Sink, Obama Mocks Skeptics of Climate Change as 'Flat-earth Society', The Hill, June 25, 2013, http://thehill.com/blogs/blog-briefing-room/news/307655-obama-we-don't-have-time-for-a-meeting-of-the-flat-earth-society.

442.North Norway, http://www.norwaves.com/north-norway.html (last visited on December 1, 2015).

443.Photographer, Amulf Husmo, describes the photograph as follows: “Time-lapse photography. Taken between July 21st, 1900 hrs and July 22nd, 1800 hrs. It is a 360 degrees panorama showing the sun's cycle North of the Arctic Circle. The location is the island of Loppa in Norway.” Norway Midnight Sun Montage, http://www.gettyimages.com/detail/photo/norway-midnight-sun-high-res-stock-photography/290186-001 (last visited on December 1, 2015).

444.My Perspective, Flat earth and the Maze, March 9, 2015, https://www.youtube.com/watch?v=dADEYcXvFhI.

445.Samuel Rowbotham, Zetetic Astronomy, at 91 (1865), citing "Astronomy and Astronomical Instruments," p. 105. By George G. Carey.

446.Samuel Rowbotham, Zetetic Astronomy, at 91 (1865), citing McCulloch's Geography, p. 85.

447.Samuel Rowbotham, Zetetic Astronomy, at 91 (1865), citing McCulloch's Geography, p. 85.

448.Samuel Rowbotham, Zetetic Astronomy, at 131 (1881). See "Daily Telegraph," July 16th, 1870. http://www.sacred-texts.com/earth/za/za29.htm#fr_23.

449.Joe Rao, Total Lunar Eclipse On Wednesday Will Be a Rare 'Selenelion', October 5, 2014, http://www.space.com/27338-total-lunar-eclipse-rare-sunrise-selenelion.html.

450.Joe Rao, Total Lunar Eclipse On Wednesday Will Be a Rare 'Selenelion', October 5, 2014, http://www.space.com/27338-total-lunar-eclipse-rare-sunrise-selenelion.html.

451.Impossible "Selenelion" eclipse, Dec 2011, https://www.youtube.com/watch?v=jIyw6xuEJxk, uploaded December 10, 2011. See also Daytime Lunar Eclipse Footage Debunks The Ball Earth - The Flat Earth Reality, https://www.youtube.com/watch?v=xYClIRXKiT4, published on July, 2016. Jenny Lea, Lunar Eclipse at Sunrise: Selenelion Viewed from Cahokia Mounds, December 14, 2011, https://www.youtube.com/watch?v=QUkjb4bbjpc.

452.John Kees, Partial Lunar Eclipse, Madison Wisconsin 12-10-2011, https://www.youtube.com/watch?v=mn6WrC_30IM.

453.Jenny Lea, Lunar Eclipse at Sunrise: Selenelion Viewed from Cahokia Mounds, December 14, 2011, https://www.youtube.com/watch?v=QUkjb4bbjpc.

454.Jenny Lea, Lunar Eclipse at Sunrise: Selenelion Viewed from Cahokia Mounds, December 14, 2011, https://www.youtube.com/watch?v=QUkjb4bbjpc.

455.Jenny Lea, Lunar Eclipse at Sunrise: Selenelion Viewed from Cahokia Mounds, December 14, 2011, https://www.youtube.com/watch?v=QUkjb4bbjpc.

456.Jenny Lea, Lunar Eclipse at Sunrise: Selenelion Viewed from Cahokia Mounds, December 14, 2011, https://www.youtube.com/watch?v=QUkjb4bbjpc.

457.Jenny Lea, Lunar Eclipse at Sunrise: Selenelion Viewed from Cahokia Mounds, December 14, 2011, https://www.youtube.com/watch?v=QUkjb4bbjpc.

458.Samuel Rowbotham, Zetetic Astronomy, at 148 (1881), http://www.sacred-texts.com/earth/za/za29.htm#fr_23.

459.Samuel Rowbotham, Zetetic Astronomy, at 149 (1881), quoting "Herschel's Astronomy," pp. 521 and 616, http://www.sacred-texts.com/earth/za/za29.htm#fr_23.

460.Samuel Rowbotham, Zetetic Astronomy, at 149 (1881), quoting "Philosophical Magazine" for 1848, p. 80.

461.Samuel Rowbotham, Zetetic Astronomy, at 149 (1881), quoting "Encyclopædia Londinensis." Art., "Fixed Stars."

462.Samuel Rowbotham, Zetetic Astronomy, at 150-51 (1881).

463.Samuel Rowbotham, Zetetic Astronomy, at 136 (1881), http://www.sacred-texts.com/earth/za/za29.htm#fr_23.

464.Samuel Rowbotham, Zetetic Astronomy, at 139-141 (1881), http://www.sacred-texts.com/earth/za/za29.htm#fr_23.

465.Samuel Rowbotham, Zetetic Astronomy, at 145 (1881), http://www.sacred-texts.com/earth/za/za29.htm#fr_23.

466."The Stevenson Effect" The Cold hard Truth Moonlight is Cold" Flat earth VS Mandela Effect, December 16, 2016, https://www.youtube.com/watch?v=DPi3n9br58M.

467.Moon Light Test Results - Full Moon & New Moon Compared, April 9, 2016, https://www.youtube.com/watch?v=rSHs2_116JQ.

468.Does Moonlight Make You Cold? CONFIRMED!, July 30, 2015, https://www.youtube.com/watch?v=6uZ9B_iMG5k.

469.AMAZING!! - Moonlight Experiment - Repeatable, July 2, 2015, https://www.youtube.com/watch?v=7sKe0POov-M&t=22s.

470.Enno Logic eT650D Dual Laser Infrared Thermometer.

471.R. Henry Noad, "Lectures on Chemistry," p. 334, quoted in Samuel Rowbotham, Zetetic Astronomy, at 145 (1881).

472.The Light of Day and The FLAT EARTH -- By My Perspective, February 5, 2017, https://www.youtube.com/watch?v=W4NBlfJ3pLk&t=412s.

473.Twilight, http://www.dictionary.com/browse/twilight?s=t (last visited on February 5, 2018).

474.Dawn, http://webstersdictionary1828.com/Dictionary/dawn.

475.https://en.oxforddictionaries.com/definition/us/dawn

476.Twilight, https://www.merriam-webster.com/dictionary/twilight (last visited on February 5, 2018).

477.Calla Cofield, What You'll See During the 2017 Total Solar Eclipse, April 27, 2017, https://www.space.com/36637-what-you-will-see-during-a-total-solar-eclipse.html.

478.Calla Cofield, What You'll See During the 2017 Total Solar Eclipse, April 27, 2017, https://www.space.com/36637-what-you-will-see-during-a-total-solar-eclipse.html.

479.Mike Killian, Totality Intercept in a Fighter Jet - 2017 Total Solar Eclipse, August 31, 2017, https://www.youtube.com/watch?v=9-LLVxAd_VU. Total Eclipse from atop Grand Teton!!! BEST Eclipse View EVER!, August 23, 2017, https://www.youtube.com/watch?v=GKFryKqXWFo. Guardian News, Live: total solar eclipse 2017, at 2:57:00, August 21, 2017, https://www.youtube.com/watch?v=Ozu5l0nnAKQ. Solar Eclipse 2017 Totality in Missouri - My Experience in Midwest America, August 21, 2017, https://www.youtube.com/watch?v=bdH2njpsACM&t=558s. Josh Weinstock, THE DARKEST DAY: Total Solar Eclipse Over Kansas City, August 21, 2017, https://www.youtube.com/watch?v=qRHAR-mAm3c.

480.Brent Butler, 2017 Total Solar Eclipse - St. Joseph, MO (4K), August 21, 217, https://www.youtube.com/watch?v=6norBo3H5uY.

481."The Flat Earth Debunked" (KJV Baptist Preaching), June 3, 2018, https://www.youtube.com/watch?v=Uwj6AQBUukE.

482."The Flat Earth Debunked" (KJV Baptist Preaching), June 3, 2018, https://www.youtube.com/watch?v=Uwj6AQBUukE.

483.PASTOR Steven Anderson attempts to destroy God's FIRMAMENT! (FLAT EARTH), https://www.youtube.com/watch?v=A2tkuTRGm8w (June 9, 2018).

484.Jay Earle Miller, $5,000 for Proving the Earth is a Globe, Modern Mechanix, October 1931, http://blog.modernmechanix.com/5000-for-proving-the-earth-is-a-globe/.

485.Hey Bill Nye, "How Do We Know the Earth Is Round?" #TuesdaysWithBill, May 24, 2016, https://www.youtube.com/watch?v=3oz7k7Wn_vo.

486.Hey Bill Nye, "How Do We Know the Earth Is Round?" #TuesdaysWithBill, May 24, 2016, https://www.youtube.com/watch?v=3oz7k7Wn_vo.

487.Hey Bill Nye, "How Do We Know the Earth Is Round?" #TuesdaysWithBill, May 24, 2016, https://www.youtube.com/watch?v=3oz7k7Wn_vo.

488.Hey Bill Nye, "How Do We Know the Earth Is Round?" #TuesdaysWithBill, May 24, 2016, https://www.youtube.com/watch?v=3oz7k7Wn_vo.

489.Earth From Space: 15 Amazing Things in 15 Years, NASA, December 18, 2014, https://www.nasa.gov/content/goddard/earth-from-space-15-amazing-things-in-15-years.

490.Elizabeth M. Jarrell, Lead Data Visualizer and Information Designer, Robert Simmon, Senior Program Analyst, Code 613, Climate and Radiation Branch, Earth Sciences Division, Sciences and Exploration Directorate, Conversations With Goddard, June 12, 2012, http://www.nasa.gov/centers/goddard/about/people/RSimmon.html (last visited on May 8, 2016).

491.NASA Admits They Dont Use Real Images from Space, Dec 12, 2014, https://www.youtube.com/watch?v=4p_0bnICtDc.

492.NASA Admits They Dont Use Real Images from Space, Dec 12, 2014, https://www.youtube.com/watch?v=4p_0bnICtDc.

493.Elizabeth M. Jarrell, Lead Data Visualizer and Information Designer, Robert Simmon, Senior Program Analyst, Code 613, Climate and Radiation Branch, Earth Sciences Division, Sciences and Exploration Directorate, Conversations With Goddard, June 12, 2012, http://www.nasa.gov/centers/goddard/about/people/RSimmon.html (last visited on May 8, 2016).

494.Robert Sungenis, Flat Earth Flat Wrong, at 336 (2018).

495.Robert Sungenis, Flat Earth Flat Wrong, at 338 (2018).

496.Robert Sungenis, Flat Earth Flat Wrong, at 338 (2018).

497.Earth - Apollo 16, Full Earth Showing North America and the Pacific Ocean, NASA, http://nssdc.gsfc.nasa.gov/imgcat/html/object_page/a16_h_118_18885.html (last visited on October 31, 2015).

498.Conscious Truth, NASA Fake Images of Earth - Smoking Gun [Part 3], March 29, 2015, https://www.youtube.com/watch?v=kyzRNv3KJqE.

499.Earth - Apollo 17, Full Earth Showing Africa and Antarctica, NASA, http://nssdc.gsfc.nasa.gov/imgcat/html/object_page/a17_h_148_22725.html (last visited on October 31, 2015).

500.Conscious Truth, NASA Fake Images of Earth - Smoking Gun [Part 1], January 31, 2015. https://www.youtube.com/watch?v=35n9FLAoiA8.

501.National Geospatial-Intelligence Agency, Mission, https://www.nga.mil/ProductsServices/Pages/default.aspx (last visited on October 31, 2015).

502.Earth - Apollo 17, Full Earth showing Africa and Antarctica, NASA, http://nssdc.gsfc.nasa.gov/imgcat/html/object_page/a17_h_148_22725.html (last visited on October 31, 2015).

503.Eric Hartwell's Infodabble, http://www.ehartwell.com/Apollo17/ (last visited on February 20, 2016).

504.Eric Hartwell's Infodabble, http://www.ehartwell.com/Apollo17/ (last visited on February 20, 2016).

505.Apollo 17 Image Library, last revised 14 December 2015, http://history.nasa.gov/alsj/a17/images17.html#MagNN.

506.The Blue Marble Explained, http://everything.explained.today/The_Blue_Marble/#Ref-7 (last visited on February 20, 2016).

507.Antonio Subirats, Flat Earth Slam Dunk! The Blue Marble's Missing Continents, November 18, 2015, https://www.youtube.com/watch?v=QtFvJtXu_jI.

508.Day and Night World Map, http://www.timeanddate.com/worldclock/sunearth.html?month=12&day=7&year=1972&hour=10&min=39&sec=0&n=&ntxt=&earth=0 (last visited on February 20. 2016).

509.William Brower, Epic Moon Transit Problems - Flat Earth NASA Lies, Aug 13, 2015, https://www.youtube.com/watch?v=YIuYfMz1Jl0.

510.NASA Captures "EPIC" Earth Image, July 20, 2015, https://www.nasa.gov/image-feature/nasa-captures-epic-earth-image.

511.NASA Captures "EPIC" Earth Image, July 20, 2015, https://www.nasa.gov/image-feature/nasa-captures-epic-earth-image.

512.Globe Skeptic, NASA Unexplained Files 2015 Exposed Photo Expert Reveals PsyOp, October 23, 2015, https://www.youtube.com/watch?v=OeDhSbYsKXA.

513.Globe Skeptic, NASA Unexplained Files 2015 Exposed Photo Expert Reveals PsyOp, October 23, 2015, https://www.youtube.com/watch?v=OeDhSbYsKXA.

514.Colgrove, W. G., The Planet Earth, Journal of the Royal Astronomical Society of Canada, Vol. 23, p.163,165, http://adsabs.harvard.edu/full/1929JRASC..23..163C.

515.NASA Captures "EPIC" Earth Image, July 20, 2015,

https://www.nasa.gov/image-feature/nasa-captures-epic-earth-image.

516.Blue Marble, 2012, January 25, 2012, NASA, http://www.nasa.gov/multimedia/imagegallery/image_feature_2159.html.

517.Jeranism, Flat Earth Obama and Unfalsifiable "Science", Published on Apr 6, 2016, https://www.youtube.com/watch?v=YqiiA4bOt_o&nohtml5=False.

518.WDTV Live 42 - Archive Footage, Apollo 10 "To Sort Out the Unknowns" - NASA Space Program & Moon Landings Documentary - WDTVLIVE42, uploaded to You Tube on November 8, 2011 ("This film has been made available thanks to NASA / courtesy of nasaimages.org"), https://www.youtube.com/watch?v=R4_30-0YTO0.

519.WDTV Live 42 - Archive Footage, Apollo 10 "To Sort Out the Unknowns" - NASA Space Program & Moon Landings Documentary - WDTVLIVE42, uploaded to You Tube on November 8, 2011 ("This film has been made available thanks to NASA / courtesy of nasaimages.org"), https://www.youtube.com/watch?v=R4_30-0YTO0.

520.Flat Earth Claims That Rocked My World (1 of 5), October 20, 2015, https://www.youtube.com/watch?v=FkjBx1f_OZ0.

521.Rob Skiba's (expanded) presentation from the 2017 FE International Conference - Part 1, November 13, 2017, https://www.youtube.com/watch?v=Ih1LPFqHSSM.

522.Robert Sungenis, Flat Earth Flat Wrong, at 340 (2018).

523.LROC, Looking Over The Limb, http://lroc.sese.asu.edu/posts/895 (last visited on July 24, 2018). Earthrise Reimagined, October 12, 2015, https://visibleearth.nasa.gov/view.php?id=87233.

524.Earthrise Reimagined, October 12, 2015, https://visibleearth.nasa.gov/view.php?id=87233.

525.Earthrise Reimagined, October 12, 2015, https://visibleearth.nasa.gov/view.php?id=87233.

526.Earthrise Reimagined, October 12, 2015, https://visibleearth.nasa.gov/view.php?id=87233.

527.Flat Earth - Fake Nasa Images, July 26, 2016, https://www.youtube.com/watch?v=ch8TRSD89-I. Paul On The Plane, FAKING SPACE - S1:E7 - DSCOVR Image Analysis [Part 4 of 5] FLAT EARTH, August 3, 2017, https://www.youtube.com/watch?v=7mOY7UTkKoM. Paul On The Plane, FAKING SPACE – S1:E9 - LRO Image Analysis [Part 4 of 4] FLAT EARTH, August 19, 2017, https://www.youtube.com/watch?v=o4ppZMltIQc.

528.FAKING SPACE - S1:E3 - LRO Image Analysis [Part 3 of 4] FLAT EARTH, July 28, 2017, https://www.youtube.com/watch?v=OZcjGM9zL7Q.

529.FAKING SPACE - S1:E3 - LRO Image Analysis [Part 3 of 4] FLAT EARTH, July 28, 2017, https://www.youtube.com/watch?v=OZcjGM9zL7Q.

530.FAKING SPACE - S1:E2 - LRO Image Analysis [Part 2 of 4] FLAT EARTH, July 26, 2017, https://www.youtube.com/watch?v=v4VeLIWt4sU.

531.Paul On The Plane, FAKING SPACE - S1:E1 - LRO Image Analysis [Part 1 of 4] FLAT EARTH, July 25, 2017, https://www.youtube.com/watch?v=IkqW5y4g4AU. FAKING SPACE - S1:E2 - LRO Image Analysis [Part 2 of 4] FLAT EARTH, July 26, 2017, https://www.youtube.com/watch?v=v4VeLIWt4sU.

532.Earth Polychromatic Imaging Camera, https://epic.gsfc.nasa.gov/, earth image at October 12, 2015.

533.Paul On The Plane, FAKING SPACE - S1:E7 - DSCOVR Image Analysis [Part 4 of 5] FLAT EARTH, August 3, 2017, https://www.youtube.com/watch?v=7mOY7UTkKoM. Paul On The Plane, FAKING SPACE – S1:E9 - LRO Image Analysis [Part 4 of 4] FLAT EARTH, August 19, 2017, https://www.youtube.com/watch?v=o4ppZMltIQc.

534.Paul On The Plane, FAKING SPACE - S1:E7 - DSCOVR Image Analysis [Part 4 of 5] FLAT EARTH, August 3, 2017, https://www.youtube.com/watch?v=7mOY7UTkKoM.

535.Gallery: Lunar Transit 2015, https://epic.gsfc.nasa.gov/galleries/2015/lunar_transit (last visited on July 28, 2018).

536.Valentin Vasilyevich, I am proud to be accused of having introduced Yury Gagarin to Orthodoxy, 12 April 2006, http://www.interfax-religion.com/?act=interview&div=24.

537.Valentin Vasilyevich, I am proud to be accused of having introduced Yury Gagarin to Orthodoxy, 12 April 2006, http://www.interfax-religion.com/?act=interview&div=24.

538.Leanna Garfield, The CIA's EarthViewer was basically the original Google Earth, Business Insider, December 30, 2015, http://www.businessinsider.com/the-cias-earthviewer-was-the-original-google-earth-2015-11.

539.Leanna Garfield, The CIA's EarthViewer was basically the original Google Earth, Business Insider, December 30, 2015, http://www.businessinsider.com/the-cias-earthviewer-was-the-original-google-earth-2015-11.

540. CIA's Impact on Technology, February 18, 2014, https://www.cia.gov/about-cia/cia-museum/experience-the-collection/text-version/stories/cias-impact-on-technology.html.

541. CIA's Impact on Technology, February 18, 2014, https://www.cia.gov/about-cia/cia-museum/experience-the-collection/text-version/stories/cias-impact-on-technology.html.

542.Google Earth Is Not a Globe!!! Recent discovery!, Zetetic Flat Earth, June 24, 2017, https://www.youtube.com/watch?v=Le99Vac7KL0.

543.Google Earth Is Not a Globe!!! Recent discovery!, Zetetic Flat Earth, June 24, 2017, https://www.youtube.com/watch?v=Le99Vac7KL0.

544.Mike Helmick, Is Google Earth really a Globe, June 27, 2017, https://www.youtube.com/watch?v=77B-UV9uiEM.

545.GLOBE FRAUD- RIGHT IN YOUR FACE - See Yourself on Google Earth, August 25, 2018, https://www.youtube.com/watch?v=jW36phP08DE.

546.Did We Go to the Moon? The NASA "Images", http://aplanetruth.info/2015/03/05/6-how-did-we-go-to-the-moon-the-nasa-pictures/ (last visited on December 27, 2015).

547.Did We Go to the Moon? The NASA "Images", http://aplanetruth.info/2015/03/05/6-how-did-we-go-to-the-moon-the-nasa-pictures/ (last visited on December 27, 2015).

548.Jack White's Apollo Studies – Index 1, http://www.aulis.com/jackstudies_index1.html (last visited on December 27, 2015).

549.Jack White's Studies – Apollo 15 File, http://www.aulis.com/jackstudies_11.html (last visited on December 27, 2015).

550.Jack White's Studies – Apollo 17 File, http://www.aulis.com/jackstudies_19.html (last visited on December 27, 2015).

551.Jack White's Mystery of the Apollo Moonrovers, http://www.aulis.com/jackstudies_23.html (December 27, 2015).

552.International Astronautical Federation, 1972: Apollo 16 (NASA), March 8, 2009, https://www.youtube.com/watch?v=vsuBd0nUKZw.

553.International Astronautical Federation, 1972: Apollo 16 (NASA), March 8, 2009, https://www.youtube.com/watch?v=vsuBd0nUKZw.

554.Apollo 16 Video Library, https://www.hq.nasa.gov/alsj/a16/video16.html#lrvdeploy (last visited on January 3, 2016).

555.Ralph Rene, NASA Mooned America, at page c, 1994, http://www.checktheevidence.com/pdf/Ralph%20Rene%20-%20NASA_mooned_america.pdf.

556.Ralph Rene, NASA Mooned America, at page d, 1994, http://www.checktheevidence.com/pdf/Ralph%20Rene%20-%20NASA_mooned_america.pdf.

557.Ralph Rene, NASA Mooned America, at page d, 1994, http://www.checktheevidence.com/pdf/Ralph%20Rene%20-%20NASA_mooned_america.pdf.

558.Ralph Rene, NASA Mooned America, at page d, 1994, http://www.checktheevidence.com/pdf/Ralph%20Rene%20-%20NASA_mooned_america.pdf.

559.Jarrah White, Carrying The Fire, 1975 Edition, December 17, 2007, https://www.youtube.com/watch?v=qIDYGsBmdeQ.

560.Jarrah White, Carrying The Fire, 1975 Edition, December 17, 2007, https://www.youtube.com/watch?v=qIDYGsBmdeQ. See also [Busted] NASA caught faking Gemini-10 Spacewalk Photos?, http://www.abovetopsecret.com/forum/thread482640/pg1 (last visited on January 9, 2016).

561.Jarrah White, Carrying The Fire, 1975 Edition, December 17, 2007, https://www.youtube.com/watch?v=qIDYGsBmdeQ.

562.Russ Brown, Secret Moon - NASA Fake Moon and Mars Hoax Proven, https://www.youtube.com/watch?v=iNxxzEIzDVU.

563.Russ Brown, Secret Moon - NASA Fake Moon and Mars Hoax Proven, https://www.youtube.com/watch?v=iNxxzEIzDVU.

564.Brian Mullin, Ball's Out Physics Episode 5.0: Propulsion in Space, April 1, 2016, https://www.youtube.com/watch?v=q8C0mStHdV0. Brian Mullin, Ball's Out Physics Episode 5.1: Propulsion in a Vacuum Chamber, April 4, 2016, https://www.youtube.com/watch?v=Mq4fGmYoiqs. Mirrored at https://www.youtube.com/watch?v=rrNeuOhjBh4 and https://www.youtube.com/watch?v=JyPb5DByIPY. See also Flat Earth - Physics 101 - Newton's 3rd Law, October 11, 2017, https://www.youtube.com/watch?v=1X86guBXWfo.

565.Scientist Shows Proof That Rockets Do Not Work In The Vacuum of Space, October 11, 2017, https://www.youtube.com/watch?v=Fnf_f4rogtg&index=2&list=WL.

566.Snorkels in Space: Nasa's 'MacGyver' Solution to Protect Spacewalking Astronauts, Fox News, December 19, 2013, http://www.foxnews.com/science/2013/12/19/snorkels-in-space-macgyver-solution-to-protect-spacewalking-astronauts.html.

567.Eric Dubay, ISS Hoax - The International Space Station Does Not Exist!, February 1, 2015, https://www.youtube.com/watch?v=5e-RnKAN9qY. See also Steve Blakey, ISS Space Station Hoax Yes the ISS and Pretty Much Everything Else to Do with NASA, May 12, 2014, https://www.youtube.com/watch?v=Kgt3V-2xPFU.

568.Lin Edwards, Moon Rock Turns Out To Be Fake, September 1, 2009, http://phys.org/news/2009-09-moon-fake.html.

569.Lin Edwards, Moon Rock Turns Out To Be Fake, September 1, 2009, http://phys.org/news/2009-09-moon-fake.html.

570.The fake Dutch moon rock and the Queen, October 9, 1969, Nieuwsblad van het Noorden, http://www.awe130.com/apollo-hoax-explained/304-the-fake-dutch-moon-rock-and-the-queen

571.The fake Dutch moon rock and the Prince, http://www.awe130.com/apollo-hoax-explained/305-the-fake-dutch-moon-rock-and-the-prince (last visited on December 28, 2015).

572.The fake Dutch moon rock and the Prince, http://www.awe130.com/apollo-hoax-explained/305-the-fake-dutch-moon-rock-and-the-prince (last visited on December 28, 2015).

573.Mary Bennett and David Percy, Dark Moon, Apollo and the Whistle-Blowers, at 268 (2001).

574.Documentary - Moon Landing Hoax - Conspiracy Theory - Part (1), https://www.youtube.com/watch?v=Y5MVVtFYTSo (last visited on January 4, 2016).

575.Astronaut Brian O'leary Is Not Entirely Convinced That the Apollo Program Succeeded, http://www.clavius.org/oleary.html (last visited on January 4, 2016). Quotations taken from a letter from Brian O'Leary, 29 August 2001, and from several letters in March 2009. All material used with his permission.

576.Astronaut Brian O'leary Is Not Entirely Convinced That the Apollo Program Succeeded, http://www.clavius.org/oleary.html (last visited on January 4, 2016). Quotations taken from a letter from Brian O'Leary, 29 August 2001, and from several letters in March 2009. All material used with his permission.

577.Astronaut Brian O'leary Is Not Entirely Convinced That the Apollo Program Succeeded, http://www.clavius.org/oleary.html (last visited on January 4, 2016). Quotations taken from a letter from Brian O'Leary, 29 August 2001, and from several letters in March 2009. All material used with his permission.

578.Astronaut Brian O'leary Is Not Entirely Convinced That the Apollo Program Succeeded, http://www.clavius.org/oleary.html (last visited on January 4, 2016). Quotations taken from a letter from Brian O'Leary, 29 August 2001, and from several letters in March 2009. All material used with his permission.

579.Astronaut Brian O'leary Is Not Entirely Convinced That the Apollo Program Succeeded, http://www.clavius.org/oleary.html (last visited on January 4, 2016). Quotations taken from a letter from Brian O'Leary, 29 August 2001, and from several letters in March 2009. All material used with his permission.

580.Christopher Ruddy, Apollo Astronaut Was Murdered, February 4, 2000, http://www.theforbiddenknowledge.com/hardtruth/astronaut_murdered.htm.

581.Why Did So Many Apollo Astronots Die Mysterious Deaths in Just Three Years Time?, http://aplanetruth.info/24-why-did-so-many-apollo-astronots-die-mysterious-deaths-in-just-three-years-time/ (last visited on January 6, 2016).

582.Bart Sibrel, Bill Kaysing Interview (Never Broadcast), April 14, 2013, https://www.youtube.com/watch?v=IJxHnpa90w4.

583.Ralph Rene, NASA Mooned America!, at 49 (1994).

584.Christopher Ruddy, Apollo Astronaut Was Murdered, February 4, 2000, http://www.theforbiddenknowledge.com/hardtruth/astronaut_murdered.htm.

585.Christopher Ruddy, Apollo Astronaut Was Murdered, February 4, 2000, http://www.theforbiddenknowledge.com/hardtruth/astronaut_murdered.htm.

586.Christopher Ruddy, Apollo Astronaut Was Murdered, February 4, 2000, http://www.theforbiddenknowledge.com/hardtruth/astronaut_murdered.htm.

587.Bart Sibrel, Bill Kaysing Interview (Never Broadcast), April 14, 2013, https://www.youtube.com/watch?v=IJxHnpa90w4.

588.Bart Sibrel, Bill Kaysing Interview (Never Broadcast), April 14, 2013, https://www.youtube.com/watch?v=IJxHnpa90w4.

589.Bart Sibrel, Bill Kaysing Interview (Never Broadcast), April 14, 2013, https://www.youtube.com/watch?v=IJxHnpa90w4.

590.Bart Sibrel, Bill Kaysing Interview (Never Broadcast), April 14, 2013, https://www.youtube.com/watch?v=IJxHnpa90w4.

591.Ralph Rene, NASA Mooned America!, at 46 (1994).

592.Ralph Rene, NASA Mooned America!, at 49 (1994).

593.Ralph Rene, NASA Mooned America!, at 44 (1994).

594.Ralph Rene, NASA Mooned America!, at 49 (1994).

595.Ralph Rene, NASA Mooned America!, at 49 (1994).

596.Ralph Rene, NASA Mooned America!, at 44 (1994).

597.Ralph Rene, NASA Mooned America!, at 47 (1994).

598.Ralph Rene, NASA Mooned America!, at 47 (1994).

599.Amy Shira Teitel, The Apollo 1 Conspiracy Theory, Discover News, February 4, 2013, http://news.discovery.com/space/history-of-space/the-apollo-1-conspiracy-theory-130204.htm.

600.Christopher Ruddy, Apollo Astronaut Was Murdered, February 4, 2000, http://www.theforbiddenknowledge.com/hardtruth/astronaut_murdered.htm.

601.Ralph Rene, NASA Mooned America!, at 50 (1994).

602.Bart Sibrel, Bill Kaysing Interview (Never Broadcast), April 14, 2013, https://www.youtube.com/watch?v=IJxHnpa90w4.

603.Bart Sibrel, Bill Kaysing Interview (Never Broadcast), April 14, 2013, https://www.youtube.com/watch?v=IJxHnpa90w4.

604.Why Did So Many Apollo Astronots Die Mysterious Deaths in Just Three Years Time?, http://aplanetruth.info/24-why-did-so-many-apollo-astronots-die-mysterious-deaths-in-just-three-years-time/ (last visited on January 6, 2016).

605.President Bush Offers New Vision For NASA, January 14, 2004, http://www.nasa.gov/missions/solarsystem/bush_vision.html.

606.President Bush Offers New Vision For NASA, January 14, 2004, http://www.nasa.gov/missions/solarsystem/bush_vision.html.

607.Ralph Rene, NASA Mooned America, at page 18, 1994, http://www.checktheevidence.com/pdf/Ralph%20Rene%20-%20NASA_mooned_america.pdf.

608.Hannah Osborn, Video by Sho Murakoshi, NASA Astronaut Don Pettit: Next Logical Step is to go Back to the Moon - Then Mars and Beyond, Updated September 21, 2016 12:13 BST, http://www.ibtimes.co.uk/nasa-astronaut-don-pettit-next-logical-step-go-back-moon-then-mars-beyond-1582401.

609.Ralph Rene, NASA Mooned America, at page 18, 1994, http://www.checktheevidence.com/pdf/Ralph%20Rene%20-%20NASA_mooned_america.pdf.

610.The Apollo 11 Telemetry Data Recordings: A Final Report, https://www.hq.nasa.gov/alsj/a11/Apollo_11_TV_Tapes_Report.pdf (last visited on January 11, 2016).

611.The Apollo 11 Telemetry Data Recordings: A Final Report, at 13, https://www.hq.nasa.gov/alsj/a11/Apollo_11_TV_Tapes_Report.pdf (last visited on January 11, 2016).

612.Brother Bart's Blog, Did We Walk on the Moon with 1960's Technology?, http://www.moontruth.org/bart/index.htm (last visited on February 12, 2016). See also, Bart Sibrel, CRAPPY

ANNIVERSARY: Did We Walk on the Moon 46 Years Ago With 1960's Technology? (VIDEOS), July 20, 2015, http://www.thesleuthjournal.com/land-moon-1960s-technology/.

613.Russianvids, Ron Howard Says The 1969 Apollo 11 Moon Landing Was Faked In A Studio, April 9, 2016, https://www.youtube.com/watch?v=NyivxOiKuBs.

614.Thomas W. Africa, Copernicus' relation to Aristarchus and Pythagoras, Isis, Vol. 52, No. 3. (Sep., 1961), pp. 403-409, http://www.faculty.umb.edu/gary_zabel/Courses/Spinoza/Texts/Copernicus'%20Relation%20to%20Aristarchus%20and%20Pythagoras.pdf. Stable URL: http://www.jstor.org/stable/228080?seq=1#page_scan_tab_contents.

615.Africa, at 403, citing J. Kepler, letter to Michael Mastlin, 11 June 1598, Gesammelte Werke, ed. Max Caspar (Munich: Beck'sche, 1955), XIII, 219.

616.Africa, at 404.

617.Africa, at 404.

618.Thomas W. Africa, Copernicus' relation to Aristarchus and Pythagoras, Isis, Vol. 52, No. 3. (Sep., 1961), pp. 403-409, http://www.faculty.umb.edu/gary_zabel/Courses/Spinoza/Texts/Copernicus'%20Relation%20to%20Aristarchus%20and%20Pythagoras.pdf. Stable URL: http://www.jstor.org/stable/228080?seq=1#page_scan_tab_contents.

619.Jose Wudka, The Pythagoreans, http://physics.ucr.edu/~wudka/Physics7/Notes_www/node32.html (last visited on December 8, 2016).

620.Jose Wudka, The Pythagoreans, http://physics.ucr.edu/~wudka/Physics7/Notes_www/node32.html (last visited on December 8, 2016).

621.Eric Dubay, The Flat Earth Conspiracy, at 149 (2015).

622.Eric Dubay, The Flat Earth Conspiracy, at 149 (2015).

623.Eric Dubay, The Flat Earth Conspiracy, at 149 (2015).

624.Albert Mackey, Encyclopedia of Freemasonry, http://www.phoenixmasonry.org/mackeys_encyclopedia/p.htm (last visited on January 1, 2016).

625.Jewish Encyclopedia, Cabala, at http://www.jewishencyclopedia.com/view.jsp?artid=1&letter=C#4 (last visited on April 18, 2010).

626.Jewish Encyclopedia, Cabala, at http://www.jewishencyclopedia.com/view.jsp?artid=1&letter=C#4 (last visited on April 18, 2010).

627. J. J. O'Connor and E. F. Robertson, Pythagoras of Samos, http://www-history.mcs.st-andrews.ac.uk/Biographies/Pythagoras.html (last visited on December 8, 2016).

628. J. J. O'Connor and E. F. Robertson, Pythagoras of Samos, http://www-history.mcs.st-andrews.ac.uk/Biographies/Pythagoras.html (last visited on December 8, 2016).

629.Jewish Encyclopedia, Cabala, at http://www.jewishencyclopedia.com/view.jsp?artid=1&letter=C#4 (last visited on April 18, 2010).

630.Jose Wudka, The Pythagoreans, http://physics.ucr.edu/~wudka/Physics7/Notes_www/node32.html (last visited on December 8, 2016).

631.The Divine Origin of the Kabbalah, Jewish Alchemy: the Kabbalah, http://www.alchemylab.com/jewish_alchemy.htm (last visited on January 5, 2016).

632.The Divine Origin of the Kabbalah, Jewish Alchemy: the Kabbalah, http://www.alchemylab.com/jewish_alchemy.htm (last visited on January 5, 2016).

633.The Divine Origin of the Kabbalah, Jewish Alchemy: the Kabbalah, http://www.alchemylab.com/jewish_alchemy.htm (last visited on January 5, 2016).

634.Johannes Reuchlin, The Wisdom of Kabbalah, http://www.kabbalah.info/eng/content/view/frame/80159?/eng/content/view/full/80159&main (last visited on January 5, 2016).

635.Johannes Reuchlin, The Wisdom of Kabbalah, http://www.kabbalah.info/eng/content/view/frame/80159?/eng/content/view/full/80159&main (last visited on January 5, 2016).

636.Johannes Reuchlin, The Wisdom of Kabbalah, http://www.kabbalah.info/eng/content/view/frame/80159?/eng/content/view/full/80159&main (last visited on January 5, 2016).

637.Albert Pike, Morals and Dogma of the Ancient and Accepted Scottish Rite of Freemasonry, P. 741 (1871).

638.Albert Mackey, Encyclopedia of Freemasonry, http://www.phoenixmasonry.org/mackeys_encyclopedia/c.htm (last visited on January 1, 2016).

639.Martin L. Wagner, Freemasonry: An Interpretation (1912), available at http://www.mindserpent.com/American_History/organization/mason/freemasonry/freemasonry.html.

640.Martin L. Wagner, Freemasonry: An Interpretation, at 182 (1912) available at http://www.mindserpent.com/American_History/organization/mason/freemasonry/freemasonry.html.

641.MICHAEL A. HOFFMAN, JUDAISM'S STRANGE GODS, at p. 88, (2000).

642.MICHAEL A. HOFFMAN, JUDAISM'S STRANGE GODS, at p. 88, (2000). See also Michael Hoffman, *Judaism Discovered*, at 779 (2008) (quoting Gershom Scholem, *Kabbalah* pp.183-84).

643.Protocols of the Learned Elders of Zion, Protocol 4, http://www.Biblebelievers.org.au/przion3.htm#PROTOCOL%20No.%204 (last visited on August 31, 2012). See also Des Griffin, Fourth Reich of the Rich, p. 216 (1993).

644.Texe Marrs, Masonic Jews Plot to Control World, Power of Prophecy, April 2003, http://www.texemarrs.com/masonic_jews_plot_world_control.htm (website address current as of April 4, 2003).

645.Texe Marrs, Masonic Jews Plot to Control World, Power of Prophecy, April 2003, http://www.texemarrs.com/masonic_jews_plot_world_control.htm (website address current as of April 4, 2003).

646.Protocols of the Learned Elders of Zion, Protocol 15, http://www.Biblebelievers.org.au/przion5.htm#protocol%20No.%2015 (last visited on August 25, 2012).

647.Protocols of the Learned Elders of Zion, Protocol 11, http://www.Biblebelievers.org.au/przion4.htm#protocol%20No.%2011 (last visited on August 25, 2012).

648.Martin L. Wagner, Freemasonry: An Interpretation, at 113 (1912), quoting Mystic Masonry, pp. 119, 130, 138, 139, 140, available at http://www.mindserpent.com/American_History/organization/mason/freemasonry/freemasonry.html.

649.Martin L. Wagner, Freemasonry: An Interpretation, at 182 (1912), available at http://www.mindserpent.com/American_History/organization/mason/freemasonry/freemasonry.html.

650.Texe Marrs, Masonic Jews Plot to Control World, Power of Prophecy, April 2003, http://www.texemarrs.com/masonic_jews_plot_world_control.htm (website address current as of April 4, 2003).

651.Texe Marrs, Masonic Jews Plot to Control World, Power of Prophecy, April 2003, http://www.texemarrs.com/masonic_jews_plot_world_control.htm (website address current as of April 4,

2003).

652.Michael Hoffman, *Judaism Discovered,* at 768 (2008) (citing "The Place of Kabbalah in the Doctrine of Russian Freemasons," in Aries: Journal for the Study Western Esotericism, vol. 4, no. 1 (Brill Academic 2004) and Heimbichner, *Blood on the Alter*).

653.Henry Makow, Illuminati 2, Deceit and Seduction, at 10 (2010).

654.Henry Makow, Illuminati 2, Deceit and Seduction, at 10 (2010).

655.Albert Pike, "Instructions to the 23 Supreme Councils of the World" (July 14, 1889), as recorded by Abel Clarin de La Rive, La Femme et l'Enfant dans la Franc-Maçonnerie Universelle (1894): 588, http://amazingdiscoveries.org/S-deception-Freemason_Lucifer_Albert_Pike#footnotevii. See also Occult Theocrasy, pp. 220-21.

656.Albert Pike, Morals and Dogma of the Ancient and Accepted Scottish Rite of Freemasonry Prepared for the Supreme Council of the Thirty-Third Degree, for the Southern Jurisdiction of the United States, and Published by Its Authority (Richmond, Virginia: L.H. Jenkins, 1871, New and Revised Edition 1950), 321.

657.Manly Palmer Hall, "The Fellow Craft," The Lost Keys of Freemasonry (Richmond, Virginia: Macoy Publishing, 1931), http://www.manlyphall.org/text/the-lost-keys-of-freemasonry/chapter-iv-the-fellow-craft/.

658.Blanche Barton, The Secret Life of a Satanist: The Authorized Biography of Anton Szandor LaVey, at 55 (1992).

659.Anton Szandor LaVey, Jew Age, http://www.jewage.org/wiki/en/Article:Anton_LaVey_-_Biography (last visited on August 31, 2018). See also Matt Lebovic, in Haunted Salem, a Jewish Church Founder Preaches the Art of 'Satanic' Social Change, The Times of Israel, October 20, 2016, https://www.timesofisrael.com/in-haunted-salem-a-jewish-church-founder-preaches-the-art-of-satanic-social-change/.

660.Anton Szandor LaVey, Jew Age, http://www.jewage.org/wiki/en/Article:Anton_LaVey_-_Biography (last visited on August 31, 2018). See also Jews Control U.S.A., Therefore the World – Is That a Good Thing?, at 8, http://www.solargeneral.org/wp-content/uploads/library/jews-control-usa-therefore-the-world-is-that-a-good-thing.pdf (last visited on August 31, 2018).

661.Matt Lebovic, in Haunted Salem, a Jewish Church Founder Preaches the Art of 'Satanic' Social Change, The Times of Israel, October 20, 2016, https://www.timesofisrael.com/in-haunted-salem-a-jewish-church-founder-preaches-the-art-of-satanic-social-change/.

662.Matt Lebovic, in Haunted Salem, a Jewish Church Founder Preaches the Art of 'Satanic' Social Change, The Times of Israel, October 20, 2016, https://www.timesofisrael.com/in-haunted-salem-a-jewish-church-founder-preaches-the-art-of-satanic-social-change/.

663.Matt Lebovic, in Haunted Salem, a Jewish Church Founder Preaches the Art of 'Satanic' Social Change, The Times of Israel, October 20, 2016, https://www.timesofisrael.com/in-haunted-salem-a-jewish-church-founder-preaches-the-art-of-satanic-social-change/.

664.Matt Lebovic, in Haunted Salem, a Jewish Church Founder Preaches the Art of 'Satanic' Social Change, The Times of Israel, October 20, 2016, https://www.timesofisrael.com/in-haunted-salem-a-jewish-church-founder-preaches-the-art-of-satanic-social-change/.

665.E.g., Lodge Copernicus No. 505, New Zealand; Lodge Copernicus No. 246, Australia.

666.The Freemasons' Quarterly Review, at 369 (1843).

667.Freemasonry: When Man Reaches New Worlds, Masonry Will Be There, July 19, 2009, http://www.conspiracyarchive.com/2013/12/01/freemasonry-when-man-reaches-new-worlds-masonry-will-be-there/.

668.The Story of Tranquility Lodge No. 2000, Chartered by The Grand Lodge of Texas, A.F. & A.M., http://tl2k.org/history.htm.

669.Buzz Aldrin Lies About His Masonic Activities on the Moon, Secret Societies, August 18, 2009, http://www.conspiracyarchive.com/2013/12/03/buzz-aldrin-lies-about-his-masonic-activities-on-the-moon/. See also, Jeranism, We're Not in Kansas Anymore - FRAUD KINGS: NASA, BOLDEN, KELLY & GIFFORDS, at 32:30, December 14, 2015, https://www.youtube.com/watch?v=H1AyU7jgKvE.

670.Zoey interviews Buzz Aldrin at the National Book Festival in Washington, DC, on September 5, 2015, https://www.youtube.com/watch?v=Y4UP6nRMuGs.

671.The Dark Side of NASA, May 28, 2008, http://primarysources.newsvine.com/_news/2008/05/28/1513441-the-dark-side-of-nasa.

672.1979 10th Anniversary Our Flags on the Moon Medallion, http://www.phoenixmasonry.org/masonicmuseum/sr_man_on_the_moon_medallion_1979.htm (last visited on December 31, 2015.

673.The Dark Side of NASA, May 28, 2008, http://primarysources.newsvine.com/_news/2008/05/28/1513441-the-dark-side-of-nasa.

674.Science Proves that NASA Faked the Moon Landings - Moon Landing Hoax, December 6, 2012, https://www.youtube.com/watch?v=q7pzg9xpAOE&index=6&list=RDXtW72nT7cYQ.

675.Science Proves that NASA Faked the Moon Landings - Moon Landing Hoax, December 6, 2012, https://www.youtube.com/watch?v=q7pzg9xpAOE&index=6&list=RDXtW72nT7cYQ.

676.Edinburgh Encyclopedia (ed. David Brewster) (J.&E. Parker, 1832) Vol. II at 230, http://www.jesuswordsonly.com/books/379-apollyon-Bible-study.html.

677.Wakeman Ryno, Amen: The God of the Amonians Or a Key to the Mansions in Heaven (Kessinger 2006), at 121, http://www.jesuswordsonly.com/books/379-apollyon-

Bible-study.html.

678.Texe Marrs, The Eagle Has Landed: Magic, Alchemy, and the Illuminati Conquest of Outer Space, http://www.texemarrs.com/032003/eagle_has_landed.htm (last visited on January 2, 2015).

679.Nasa, The KJV Old Testament Hebrew Lexicon, http://www.Biblestudytools.com/lexicons/hebrew/kjv/nasa.html (last visited on January 22, 2016).

680.Nasa, The KJV Old Testament Hebrew Lexicon, http://www.Biblestudytools.com/lexicons/hebrew/kjv/nasa.html (last visited on January 22, 2016).

681.Shira Schoenberg & Mitchell Bard, Israel Science & Technology: NASA-Israel Collaboration, October 2015, https://www.jewishvirtuallibrary.org/jsource/US-Israel/nasa.html.

682.Shira Schoenberg & Mitchell Bard, Israel Science & Technology: NASA-Israel Collaboration, October 2015, https://www.jewishvirtuallibrary.org/jsource/US-Israel/nasa.html.

683.Shira Schoenberg & Mitchell Bard, Israel Science & Technology: NASA-Israel Collaboration, October 2015, https://www.jewishvirtuallibrary.org/jsource/US-Israel/nasa.html.

684.Shira Schoenberg & Mitchell Bard, Israel Science & Technology: NASA-Israel Collaboration, October 2015, https://www.jewishvirtuallibrary.org/jsource/US-Israel/

nasa.html.

685.Bart Sibrel, Bill Kaysing Interview (Never Broadcast), April 14, 2013, https://www.youtube.com/watch?v=IJxHnpa90w4.

686.Bart Sibrel, Bill Kaysing Interview (Never Broadcast), April 14, 2013, https://www.youtube.com/watch?v=IJxHnpa90w4.

687.Bart Sibrel, Bill Kaysing Interview (Never Broadcast), April 14, 2013, https://www.youtube.com/watch?v=IJxHnpa90w4.

688.Website Terms of Service, ProBoards, https://proboards.com/tos (last visited on January 10, 2016).

689.Eric Dubay and IFERS Banned From the Internet, January 9, 2016, http://www.atlanteanconspiracy.com/.

690.Eric Dubay and IFERS Banned From the Internet, January 9, 2016, http://www.atlanteanconspiracy.com/.

691.Eric Dubay and IFERS Banned From the Internet, January 9, 2016, http://www.atlanteanconspiracy.com/.

692.John Whitley, *Seven Jewish Americans Control Most US Media*, Real News 24/7 (November 21, 2003), *available at* http://www.realnews247.com/seven_jewish_americans_control_media_rense.htm.

693.Id.

694.Id.

695.Id.

696.Guy Grimland, Facebook, Google Founders to Attend Jerusalem Conference in May, Ha´aretz 01/04/2008, cited in the Jewish Hand Behind Internet - Google, Facebook, Wikipedia, Yahoo!, MySpace, eBay, Freedom Research, June, 2009, http://www.fourwinds10.net/siterun_data/media/internet/news.php?q=1262024557.

697.Guy Rolnik, Google Founder Plans R&D Center in Israel, Ha´aretz 30/01/2006, cited in The Jewish Hand Behind Internet - Google, Facebook, Wikipedia, Yahoo!, MySpace, eBay, Freedom Research, June, 2009, http://www.fourwinds10.net/siterun_data/media/internet/news.php?q=1262024557.

698.The Jewish Hand Behind Internet - Google, Facebook, Wikipedia, Yahoo!, MySpace, eBay, Freedom Research, June, 2009, http://www.fourwinds10.net/siterun_data/media/internet/news.php?q=1262024557.

699.The Jewish Hand Behind Internet - Google, Facebook, Wikipedia, Yahoo!, MySpace, eBay, Freedom Research, June, 2009, http://www.fourwinds10.net/siterun_data/media/internet/news.php?q=1262024557.

700.IFERS Forum Terminated/TFESSboards.net Activates!, http://aplanetruth.info/2015/12/30/ifers-forum-terminatedtfessboards-net-activates/ (last visited on January 11, 2016). See The Flat Earth Skeptic's Society forum at http://tfess.proboards.com/forum (last visited on January 11, 2016).

701.The Flat Earth Skeptic's Society, http://tfess.proboards.com/forum (last visited on January 11, 2016).

702.Rob Skiba Folds, Yet Somehow this Only Pushes Me Further, Augst 17, 2015, https://weseeasthroughaglassdarkly.wordpress.com/2015/08/17/rob-skiba-folds-yet-somehow-this-only-pushes-me-further/.

703.Rob Skiba, The Flat Earth Controversy: A Biblical World View And One Man's Quest For Truth, http://testingtheglobe.com/ (last visited on January 10, 2016).

704.Rob Skiba, The Flat Earth Controversy: A Biblical World View And One Man's Quest For Truth, http://testingtheglobe.com/ (last visited on January 10, 2016).

705.Rob Skiba, The Flat Earth Controversy: A Biblical World View And One Man's Quest For Truth, http://testingtheglobe.com/ (last visited on January 10, 2016).

706.Rob Skiba, The Flat Earth Controversy: A Biblical World View And One Man's Quest For Truth, http://testingtheglobe.com/quest1.html (last visited on March 16, 2016).

707.Map Projections, United Steates Geological Survey, http://egsc.usgs.gov/isb//pubs/MapProjections/projections.html (last visited on November 2, 2015).

708.Map Projections, United Steates Geological Survey,

http://egsc.usgs.gov/isb//pubs/MapProjections/projections.html (last visited on November 2, 2015).

709.Eric Dubay, The South Pole Does Not Exist!, The Atlantean Conspiracy, June 13, 2015, citing Greely, General A. W. "Antarctica, or the Hypothetical Southern Continent." Cosmopolitan 17 (1894): p. 296. See also The Ice Wall, https://wiki.tfes.org/The_Ice_Wall, (Last visited on February 19, 2016), Photograph is used under a Creative Commons Attribution-Share Alike 4.0 International License.

710.The Ice Wall, https://wiki.tfes.org/The_Ice_Wall, (Last visited on February 19, 2016).

711.Polarstern, http://www.eurofleets.eu/np4/73.html (last visited on February 10, 2019).

712.Our World Heritage Travel Blog, December 7, 2006, https://ourworldheritagebe.wordpress.com/2013/06/15/antarctica-3-reaching-the-ice-and-neumayer-station/.

713.Samuel Birley Rowbotham (Parallax), Zetetic Astronomy, Earth Not a Globe (1881), at 258, quoting from a book published by Gordon & Gotch, 85, Collins Street West, Melbourne, and 121, Holborn Hill, London.

714.Samuel Birley Rowbotham (Parallax), Zetetic Astronomy, Earth Not a Globe (1881), at 258, quoting from a book published by Gordon & Gotch, 85, Collins Street West, Melbourne, and 121, Holborn Hill, London.

715.Samuel Birley Rowbotham (Parallax), Zetetic Astronomy, Earth Not a Globe (1881), at 258-59.

716.Samuel Birley Rowbotham (Parallax), Zetetic Astronomy, Earth Not a Globe (1881), at 259, quoting "South Sea Voyages." By Sir J. C. Ross, p. 96, vol. i.

717.Samuel Birley Rowbotham (Parallax), Zetetic Astronomy, Earth Not a Globe (1881), at 260, quoting from a book published by Gordon & Gotch, 85, Collins Street West, Melbourne, and 121, Holborn Hill, London.

718.Samuel Birley Rowbotham (Parallax), Zetetic Astronomy, Earth Not a Globe (1881), at 260, quoting "The Builder." Sept. 20th, 1862.

719.Samuel Birley Rowbotham (Parallax), Zetetic Astronomy, Earth Not a Globe (1881), at 260-61, citing "South Sea Voyages." By Sir J. C. Ross, p. 96, vol. i.

720.Thomas Jo, Earth Is Pear Shaped - Neil deGrasse Tyson, February 6, 2016, https://www.youtube.com/watch?v=nTOE4Ar0Dfo.

721.The White House, Inside the White House: The Situation Room, December 18, 2009, https://www.youtube.com/watch?v=T7ch13ZuMu8.

722.Eric Dubay, The Arctic and Antarctic Prove Flat Earth, Atlantean Conspiracy, March 18, 2016, http://www.atlanteanconspiracy.com/2016/03/arctic-antarctic-prove-flat-earth.html. See also The Fat Earth Conspiracy, at 64 (2014).

723.Thomas Winship, Zetetic Cosmogeny, at 8 (1899).

724.Eric Dubay, The Arctic and Antarctic Prove Flat Earth, Atlantean Conspiracy, March 18, 2016, http://www.atlanteanconspiracy.com/2016/03/arctic-antarctic-prove-flat-earth.html. See also The Fat Earth Conspiracy, at 64 (2014).

725.Thomas Winship, Zetetic Cosmogeny, at 9 (1899).

726.Eric Dubay, The Arctic and Antarctic Prove Flat Earth, Atlantean Conspiracy, March 18, 2016, http://www.atlanteanconspiracy.com/2016/03/arctic-antarctic-prove-flat-earth.html. See also The Fat Earth Conspiracy, at 64-65 (2014).

727.Eric Dubay, The Arctic and Antarctic Prove Flat Earth, Atlantean Conspiracy, March 18, 2016, http://www.atlanteanconspiracy.com/2016/03/arctic-antarctic-prove-flat-earth.html. See also The Fat Earth Conspiracy, at 64-65 (2014).

728.Eric Dubay, The Arctic and Antarctic Prove Flat Earth, Atlantean Conspiracy, March 18, 2016, http://www.atlanteanconspiracy.com/2016/03/arctic-antarctic-prove-flat-earth.html. See also The Fat Earth Conspiracy, at 65 (2014).

729.Eric Dubay, The Arctic and Antarctic Prove Flat Earth, Atlantean Conspiracy, March 18, 2016, http://www.atlanteanconspiracy.com/2016/03/arctic-antarctic-prove-flat-earth.html. See also The Fat Earth Conspiracy, at 65 (2014).

730.Eric Dubay, The Arctic and Antarctic Prove Flat Earth, Atlantean Conspiracy, March 18, 2016, http://www.atlanteanconspiracy.com/2016/03/arctic-antarctic-prove-flat-earth.html. See also The Fat Earth Conspiracy, at 65-66 (2014), quoting W. & R.

Chambers, Arctic Explorations.

731.Eric Dubay, The Arctic and Antarctic Prove Flat Earth, Atlantean Conspiracy, March 18, 2016, http://www.atlanteanconspiracy.com/2016/03/arctic-antarctic-prove-flat-earth.html. See also The Fat Earth Conspiracy, at 66-67 (2014), quoting Samuel Rowbotham, Zetetic Astronomy, Earth Not a Globe!, at 116-121 (1865).

732.William Carpenter, 100 Proofs the Earth is Not a Globe, proof 53, at page 16 (1885).

733.Time and Date Calculator for Sunrise and Sunset in Toronto, Canada on June 21, 2017, https://www.timeanddate.com/sun/canada/toronto?month=6&year=2017.

734.Time and Date Calculator for Sunrise and Sunset in Toronto, Canada on December 21, 2017, https://www.timeanddate.com/sun/canada/toronto?month=12&year=2017.

735.Time and Date Calculator for Sunrise and Sunset in Christchurch, New Zealand, on June 21, 2017, https://www.timeanddate.com/sun/new-zealand/christchurch?month=6&year=2017.

736.Time and Date Calculator for Sunrise and Sunset in Christchurch, New Zealand, on December 21, 2017, https://www.timeanddate.com/sun/new-zealand/christchurch?month=12&year=2017.

737.Samuel Rowbotham, Zetetic Astronomy, Earth Not a Globe!, at 119 (1881), http://www.sacred-texts.com/earth/za/za26.htm.

738.Samuel Rowbotham, Zetetic Astronomy, Earth Not a Globe!, at 119 (1881), http://www.sacred-texts.com/earth/za/za26.htm.

739.Samuel Rowbotham, Zetetic Astronomy, Earth Not a Globe!, at 122-23 (1881), http://www.sacred-texts.com/earth/za/za26.htm.

740.Samuel Rowbotham, Zetetic Astronomy, Earth Not a Globe!, at 104 (1881), http://www.sacred-texts.com/earth/za/za23.htm (sun is calculated to be approximately 700 statute miles above the earth). Thomas Winship, Zetetic Cosmogony, 118 (1899) (sun is calculated to be be approximately 2,700 nautical miles or approximately 3,000 statute miles above the earth). Dr. Zack, Flat Earth: How To Track The Sun (Subtitulado), February 18, 2016, https://www.youtube.com/watch?v=lozvuBI5Bmg (sun is calculated to be approximately 3,052 miles above the earth).

741.South Pole, National Geographic, https://www.nationalgeographic.org/encyclopedia/south-pole/ (last visited on June 25, 2017).

742.South Pole, https://en.wikipedia.org/wiki/South_Pole (last visited on June 11, 2018).

743.North Pole, https://en.wikipedia.org/wiki/North_Pole (last visited on June 11, 2018).

744.Jane J. Lee, There Are 3 South Poles, So Which One Did Prince Harry Reach?, National Geographic, December 14, 2013, http://news.nationalgeographic.com/news/2013/12/131

213-south-pole-where-is-it-geography-science/.

745.Antarctica A Year on Ice, 2013, https://www.youtube.com/watch?v=SAcNyZ1rzJs&t=347s.

746.Antarctica A Year on Ice, 2013, https://www.youtube.com/watch?v=SAcNyZ1rzJs&t=347s.

747.Antarctica A Year on Ice, 2013, https://www.youtube.com/watch?v=SAcNyZ1rzJs&t=347s.

748.Steven Anderson, Flat Earth Debunked by Antarctica Documentary, January 28, 2016, https://www.youtube.com/watch?v=gRCWjJPUXyE&t=113s.

749.David Weiss, Antarctica 24 hour sun DISPROVES the Flat Earth (UPDATED), DITRH, February 1, 2016, https://www.youtube.com/watch?v=8Vzjex7L8-E.

750.David Weiss, Antarctica 24 hour sun DISPROVES the Flat Earth (UPDATED), DITRH, February 1, 2016, https://www.youtube.com/watch?v=8Vzjex7L8-E.

751.Jeran Campanella, Antarctica Is NOT What You Think - No 24 Hour Sun!, Jeranism, October 1, 2016, https://www.youtube.com/watch?v=7Mdg5k-WBFw.

752.Jeran Campanella, Antarctica Is NOT What You Think - No 24 Hour Sun!, Jeranism, October 1, 2016, https://www.youtube.com/watch?v=7Mdg5k-WBFw, at 16:15 of the video.

753.Can Christianity and Science Coexist?, ABC News, June 11, 2009, https://www.youtube.com/watch?v=M5sMva2ydoU.

754.Compass, American Dictionary of the English Language, 1828, http://1828.mshaffer.com/d/word/compass.

755.Image: Flat Earth versus Spherical Earth (Matthew 4 - 8 and Luke 4 - 5), for RW.PNG, http://rationalwiki.org/wiki/File:Flat_Earth_versus_Spherical_Earth_%28Matthew_4_-_8_and_Luke_4_-_5%29,_for_RW.PNG. Copyright for image was released by its author into the public domain.

756.Corner, American Dictionary of the English Language, 1828, http://1828.mshaffer.com/d/word/corner (last visited on December 23, 2015).

757.Corner, American Dictionary of the English Language, 1828, http://1828.mshaffer.com/d/word/corner (last visited on December 23, 2015).

758.Corner, Webster's Revised Unabridged Dictionary (1913 + 1828), http://machaut.uchicago.edu/?resource=Webster%27s&word=corner&use1913=on (last visited on December 23, 2015).

759.Corner, MacMillan Dictionary, http://www.macmillandictionary.com/dictionary/american/corner_1#corner_20 (last visited on December 23, 2015).

760.Corner, Merriam Webster Dictionary, http://www.merriam-webster.com/dictionary/corner (last visited on December 23, 2015).

761.Corner, Merriam Webster Dictionary, http://www.merriam-webster.com/dictionary/corner (last visited on December 23, 2015).

762.Turn, Noah Webster American Dictionary of the English Language, 1828, http://webstersdictionary1828.com/Dictionary/turn.

763.Turn, Noah Webster American Dictionary of the English Language, 1828, http://webstersdictionary1828.com/Dictionary/turn.

764.Turn, Merriam Webster Dictionary, https://www.merriam-webster.com/dictionary/turn.

765.Turn, Noah Webster American Dictionary of the English Language, 1828, http://webstersdictionary1828.com/Dictionary/turn.

766.Turn, Merriam Webster Dictionary, https://www.merriam-webster.com/dictionary/turn.

767.SAMUEL C. GIPP, AN UNDERSTANDABLE HISTORY OF THE Bible, p. 70 (1987).

768.*Id.*

769.*Id.* at p. 71.

770.*Id.* at p. 70.

771.*Id.* at p. 71.

772.Samuel KJBB, Tares Among The Wheat (Sequel to "A Lamp in the Dark"), Published on April 29, 2014, https://www.youtube.com/watch?v=-aiHcghIdjM&spfreload=5.

773.Samuel KJBB, Tares Among The Wheat (Sequel to "A Lamp in the Dark"), Published on April 29, 2014, https://www.youtube.com/watch?v=-aiHcghIdjM&spfreload=5.

774.Samuel KJBB, Tares Among The Wheat (Sequel to "A Lamp in the Dark"), Published on April 29, 2014, https://www.youtube.com/watch?v=-aiHcghIdjM&spfreload=5.

775.Samuel KJBB, Tares Among The Wheat (Sequel to "A Lamp in the Dark"), Published on April 29, 2014, https://www.youtube.com/watch?v=-aiHcghIdjM&spfreload=5.

776.*Id.* at p. 71.

777.*Id.*

778.*Id.* at p. 72.

779.Les Garrett, Which Bible Can We Trust?, p. 151 (1982).

780.G.A. Riplinger, New Age Bible Versions, p. 435 (1993).

781.Samuel C. Gipp, An Understandable History of the Bible, p. 116-130 (1987).

782.*Id.*

783.*Id.* at 126-29.

784.*Id.* at 131-68.

785.*Id.*

786.*Id.*

787.*Id.*

788.*Id.*

789.*Id.*

790.*Id.* at p. 405.

791.*Id.* at p. 400.

792.*Id.*

793.*Id.* at p. 406.

794.*Id.*at p. 432.

795.G. A. Riplinger, The Language of the King James Bible, p. 66 (1998).

796.*Id.* at p. 132 (quoting *Carlo Martini, In the Thick of the Ministry,* p. 42, the Liturgical Press, Collegeville, Minn., 1990).

797.G.A. Riplinger, Blind Guides, p. 19.

798.G.A. Riplinger, Blind Guides, p. 19.

799.Company History: Zondervan Corporation, http://www.answers.com/topic/zondervan (last visited on April 5, 2010).

800.G. A. Riplinger, The Language of the King James Bible, p. 114 (1998).

801.G.A. Riplinger, New Age Bible Versions, p. 2 (1993).

802.G. A. Riplinger, The Language of the King James Bible, p. 128 (1998).

803.Company History: Zondervan Corporation, http://www.answers.com/topic/zondervan (last visited on April 5, 2010).

804.Anton La Vey, Satanic Bible, Harper Collins, *available at* http://www.harpercollins.com/book/index.aspx?isbn=9780380015399 (on sale 12/1/1976); NIV Bible, Harper Collins *available at* http://www.harpercollins.com/books/9780310949862/NIV_Bible/index.aspx.

805.G. A. Riplinger, The Language of the King James Bible, p. 128 (1998).

806.G. A. Riplinger, The Language of the King James Bible, p. 128 (1998).

807.Eric Dubay, The Flat Earth Conspiracy, at 114 (2014).

808.Eric Dubay, The Flat Earth Conspiracy, at 113-114 (2014).

809.David Wardlaw Scott, Terra Firma, at 8 (1901).

810.William Orem, Emerson College, https://www.emerson.edu/academics/faculty-guide/profile/william-orem/2558 (last visited on February 1, 2018).

811.https://www.google.com/search?q=Why+does+a+rock+sink+in+water%3F&rlz=1C1GCEA_enUS751US752&oq=Why+does+a+rock+sink+in+water%3F&aqs=chrome..69i57j0l5.1029j0j4&sourceid=chrome&ie=UTF-8. William Orem, The Archimedes' Principal And Rock Density, Moment of Science, April 1, 2011, https://indianapublicmedia.org/amomentofscience/archimedes-principal-rock-density/.

812.https://www.google.com/search?q=Why+does+a+rock+fall+to+the+earth%3F&rlz=1C1GCEA_enUS751US752&oq=Why+does+a+rock+fall+to+the+earth%3F&aqs=chrome..69i57.1112j0j9&sourceid=chrome&ie=UTF-8. Department of Physics, University of Illinois at Urbana-Champaign, https://van.physics.illinois.edu/qa/listing.php?id=164.

813.The Divine Origin of the Kabbalah, Jewish Alchemy: the Kabbalah, http://www.alchemylab.com/jewish_alchemy.htm (last visited on January 5, 2016).

814.The Divine Origin of the Kabbalah, Jewish Alchemy: the Kabbalah, http://www.alchemylab.com/jewish_alchemy.htm (last visited on January 5, 2016).

815.Sir Isaac Newton and the Jews, Jewish Currents, July 4, 2015, http://jewishcurrents.org/july-5-sir-isaac-newton-and-the-jews-37605.

816.Sir Isaac Newton and the Jews, Jewish Currents, July 4, 2015, http://jewishcurrents.org/july-5-sir-isaac-newton-and-the-jews-37605.

817.Sarah, Dry, Saving Isaac Newton: How a Jewish Collector Brought the Physicist's Papers to America, May 5, 2014, http://www.tabletmag.com/jewish-arts-and-culture/books/170960/saving-isaac-newton.

818.Sarah, Dry, Saving Isaac Newton: How a Jewish Collector Brought the Physicist's Papers to America, May 5, 2014, http://www.tabletmag.com/jewish-arts-and-culture/books/170960/saving-isaac-newton.

819.The Religious Background and Religious Beliefs of Albert Einstein, http://www.adherents.com/people/pe/Albert_Einstein.html, citing Dr. Arthur J. Deikman, who noted that Einstein echoed Isaac Newton's belief in the reality of the mystical (source: "A Functional Approach to Mysticism" in Journal of Consciousness Studies, Vol. 7, No. 11-12, November/December 2000; special issue: 'Cognitive Models and Spiritual Maps'; URL: http://www.deikman.com/functional.html).

820.Sarah, Dry, Saving Isaac Newton: How a Jewish Collector Brought the Physicist's Papers to America, May 5, 2014, http://www.tabletmag.com/jewish-arts-and-culture/books/170960/saving-isaac-newton.

821.Sarah, Dry, Saving Isaac Newton: How a Jewish Collector Brought the Physicist's Papers to America, May 5, 2014, http://www.tabletmag.com/jewish-arts-and-culture/books/170960/saving-isaac-newton.

822.Sarah, Dry, Saving Isaac Newton: How a Jewish Collector Brought the Physicist's Papers to America,

May 5, 2014, http://www.tabletmag.com/jewish-arts-and-culture/books/170960/saving-isaac-newton.

823.Marshall Hall, Sola Scriptura - IV, Kabbala Phariseeism or Bible Christianity?, http://www.fixedearth.com/sola-scriptura-4.html (last visited on December 8, 2015).

824.Marshall Hall, Sola Scriptura - IV, Kabbala Phariseeism or Bible Christianity?, http://www.fixedearth.com/sola-scriptura-4.html (last visited on December 8, 2015).

825.Gershom Scholem, The Kabbalah, Meridian Books, New York 1978, http://www.mishkan.com/kbltalk1.html.

826.Migene Gonzalez-Wippler, The Kabbalah & Magic of Angels, at 95-96 (2013).

827.Athol Bloomer, *The Eucharist and The Jewish Mystical Tradition • Part 1*, Association of Hebrew Catholics, *at* http://hebrewcatholic.org/PrayerandSpirituality/eucharistjewishm.html (originally published in The Hebrew Catholic #77, pp 15-18 (Summer-Fall 2002)).

828.Athol Bloomer, *The Eucharist and The Jewish Mystical Tradition • Part 1*, Association of Hebrew Catholics, *at* http://hebrewcatholic.org/PrayerandSpirituality/eucharistjewishm.html (originally published in The Hebrew Catholic #77, pp 15-18 (Summer-Fall 2002)).

829.Athol Bloomer, *The Eucharist and The Jewish Mystical Tradition • Part 1*, Association of Hebrew

Catholics, *at* http://hebrewcatholic.org/PrayerandSpirituality/eucharistjewishm.html (originally published in The Hebrew Catholic #77, pp 15-18 (Summer-Fall 2002)).

830.Rabbi Geoffrey W. Dennis, *The Encyclopedia of Jewish Myth, Magic, and Mysticism*, at 199 (2007), quoted by Michael Hoffman in *Judaism Discovered*, at 239-40 (2008).

831.Dennis, at 199.

832.Dan Cohn-Sherbok and Lavinia Cohn-Sherbok, Jewish and Christian Mysticism: An Introduction, at 167 (1994).

833.Rabbi Eli Mallon, M.Ed., LMSW, Shir ha-Shirim: The Song of Songs, April, 23, 2011, http://rabbielimallon.wordpress.com/2011/04/23/4-23-11-shir-ha-shirim-the-song-of-songs/, quoting, Schochet, J. Immanuel, transl. and ed.; Tzva'at ha-Rivash (The Ethical Will of the Baal Shem Tov); section 68; Kehot Publication Society.

834.Rabbi Eli Mallon, M.Ed., LMSW, Shir ha-Shirim: The Song of Songs, April, 23, 2011, http://rabbielimallon.wordpress.com/2011/04/23/4-23-11-shir-ha-shirim-the-song-of-songs/.

835.Francis and the Shekinah, Call Me Jorge, May 25, 2014, http://callmejorgebergoglio.blogspot.com/2014/05/francis-shekinah.html. See also, Brother Nathaniel Enlightens Us: Satan at the "Wailing Wall," http://givingpsychologyaway.com/?tag=davening and http://www.realjewnews.com/?p=798.

836.Brother Nathaniel Kapner, Satan at the Wailing Wall, February 13, 2013, http://www.realjewnews.com/?p=798.

837. Rabbi Michael Leo Samuel, Davening at Victoria's Secret, San Diego Jewish Herald, 27 December 2013, http://www.sdjewishworld.com/2013/12/27/davening-at-victorias-secret/, quoting Jacob I. Schochet, Tzavat HaRivash (Brooklyn, NY: Kehot Publication, 1998), pp.54-55.

838. Rabbi Michael Leo Samuel, Davening at Victoria's Secret, San Diego Jewish Herald, 27 December 2013, http://www.sdjewishworld.com/2013/12/27/davening-at-victorias-secret/, quoting Jacob I. Schochet, Tzavat HaRivash (Brooklyn, NY: Kehot Publication, 1998), pp.54-55.

839.Michael Hoffman, *Judaism Discovered*, at 239 (2008).

840.Michael Hoffman, *Judaism Discovered*, at 240 (2008).

841.Michael Hoffman, *Judaism Discovered*, at 240 (2008).

842.Moshe Idel, *Hasidism Between Ecstacy and Magic*, at 103.

843.Ithmar Gruenwald, *Israel Oriental Studies* 1 (1971); pp. 132-177 and *Temerin*, vol. 7 (Jerusalem, 1972) pp. 101-139. Gershom Scholem, *Jewish Gnosticism, Merkabah Mysticism and Talmudic Tradition* (Jewish Theological Seminary of America,

1965) cited by Michael Hoffman in *Judaism Discovered*, at 241 (2008).

844.Michael Hoffman, *Judaism Discovered*, at 241 (2008).

845.Tiferet, Jewish Virtual Library, *at* http://www.jewishvirtuallibrary.org/jsource/Judaism/Tiferet.html (last visited on March 2, 2010).

846.Tree of Life created by Friedhelm Wessel (25 October 2008), at http://commons.wikimedia.org/wiki/File:Tree_of_Life_%28Sephiroth%29.svg.

847. Laura Ellen Shulman, Judaism, Jewish *Mysticism Kabbalah and the Sefirot* (March 13, 2007), *at* http://www.nvcc.edu/home/lshulman/Rel232/resource/sefirot.htm.

848.MARSHAL HALL, THE EARTH IS NOT MOVING, p. 97 (1991)

849.GERARDUS BOUW, GEOCENTRICITY, p. 254-56 (1992).

850.Dr. Neville Thomas Jones, Ph.D., Airy's Experiment, http://www.geocentricuniverse.com/Airy.htm (last visited on August 24, 2012).

851.Malcolm Bowden, The Basic Scientific Arguments for Geocentricity, http://homepage.ntlworld.com/malcolmbowden/Geocexpl.htm.

852.Malcolm Bowden, The Basic Scientific Arguments for Geocentricity, http://homepage.ntlworld.com/malcolmbowden/Geocexpl.htm.

853.Gerrard Hickson, Kings Dethroned, at 60 (1922).

854.Herbert Dingle, Science at the Crossroads (1972), at 114.

855.Malcolm Bowden, Geocentricity is Scientific, Heliocentricity is a Lie!, December 28, 2014, https://www.youtube.com/watch?v=bxMSL9h2ziY.

856.Herbert Dingle, Science at the Crossroads (1972), at 108.

857.Robert Sungenis, Review of Neil deGrasse Tyson's Video on Special Relativity (last visited on July 16, 2018).

858.Robert Sungenis, Review of Neil deGrasse Tyson's Video on Special Relativity (last visited on July 16, 2018), quoting Äther und Relativitätstheorie. Rede gehalten am 5. Mai 1920 an der Reichs-Universität zu Leiden. Berlin: Springer, 1920 [Vol. 7, Doc. 38, 305–323; trans. 160–182]; http://en.wikisource.org/wiki/Ether_and_the_Theory_of_Relativity.

859.Nikola Tesla, http://www.brainyquote.com/quotes/quotes/n/nikolatesl401270.html (last visited on September 9, 2015).

860.New York Times, 11 July 1935, p23, c.8.

861.Herbert Dingle, Science at the Crossroads (1972), at 14, http://blog.hasslberger.com/Dingle_SCIENCE_at_the_Crossroads.pdf.

862.Herbert Dingle, Science at the Crossroads (1972), at 16, http://blog.hasslberger.com/Dingle_SCIENCE_at_the_Crossroads.pdf.

863.Herbert Dingle, Science at the Crossroads (1972), at 5, http://blog.hasslberger.com/Dingle_SCIENCE_at_the_Crossroads.pdf.

864.Herbert Dingle, Science at the Crossroads (1972), at 18, http://blog.hasslberger.com/Dingle_SCIENCE_at_the_Crossroads.pdf.

865.Herbert Dingle, Science at the Crossroads (1972), at 18, http://blog.hasslberger.com/Dingle_SCIENCE_at_the_Crossroads.pdf.

866.Herbert Dingle, Science at the Crossroads (1972), at 64, http://blog.hasslberger.com/Dingle_SCIENCE_at_the_Crossroads.pdf.

867.Marshal Hall, The Earth is Not Moving, at 117 (1991), (quoting Ronald W. Clark, Einstein: The Life and Times, at 87 (1971).

868.Herbert Dingle, Science at the Crossroads (1972), at 5, http://blog.hasslberger.com/Dingle_SCIENCE_at_the_Crossroads.pdf.

869.N. Martin Gwynne, Einstein and Modern Physics, at 11, http://www.alcazar.net/einstein.pdf (last visited on October 6, 2015).

870.Herbert Dingle, Science at the Crossroads (1972), at 6,
http://blog.hasslberger.com/Dingle_SCIENCE_at_the_Crossroads.pdf.

871.Herbert Dingle, Science at the Crossroads (1972), at 7,
http://blog.hasslberger.com/Dingle_SCIENCE_at_the_Crossroads.pdf.

872.Herbert Dingle, Science at the Crossroads (1972), at 33.

873.See Robert Sungenis, Review of Neil deGrasse Tyson's Video on Special Relativity (last visited on July 16, 2018), explaining Theory of relativity explained in 7 mins, November 1, 2014, https://www.youtube.com/watch?v=ttZCKAMpcAo.

874.Robert Sungenis, Review of Neil deGrasse Tyson's Video on Special Relativity, at 7-8 (last visited on July 16, 2018).

875.Robert Sungenis, Review of Neil deGrasse Tyson's Video on Special Relativity (last visited on July 16, 2018).

876.Michelson-Morley Experiment, World Heritage Encyclopedia,
http://self.gutenberg.org/articles/Michelson-Morley_experiment (last visited on October 28, 2016).

877.Gerrard Hickson, Kings Dethroned, at 65 (1922).

878.Gerrard Hickson, Kings Dethroned, at 65 (1922).

879.N. Martin Gwynne, Einstein and Modern Physics, at 3, http://www.alcazar.net/einstein.pdf (last visited on

October 6, 2015).

880.N. Martin Gwynne, Einstein and Modern Physics, at 31, http://www.alcazar.net/einstein.pdf (last visited on October 6, 2015).

881.N. Martin Gwynne, Einstein and Modern Physics, at 31-32, http://www.alcazar.net/einstein.pdf (last visited on October 6, 2015).

882.N. Martin Gwynne, Einstein and Modern Physics, at 32, http://www.alcazar.net/einstein.pdf (last visited on October 6, 2015).

883.N. Martin Gwynne, Einstein and Modern Physics, at 3-4, http://www.alcazar.net/einstein.pdf (last visited on October 6, 2015).

884.N. Martin Gwynne, Einstein and Modern Physics, at 5, http://www.alcazar.net/einstein.pdf (last visited on October 6, 2015).

885.N. Martin Gwynne, Einstein and Modern Physics, at 5-6, http://www.alcazar.net/einstein.pdf (last visited on October 6, 2015).

886.Malcolm Bowden, Geocentricity is Scientific, Heliocentricity is a Lie!, December 28, 2014, https://www.youtube.com/watch?v=bxMSL9h2ziY. See also, Jeroen van Dongen, infra.

887.Malcolm Bowden, Geocentricity is Scientific, Heliocentricity is a Lie!, December 28, 2014, https://www.youtube.com/watch?v=bxMSL9h2ziY, citing Ronald W. Clark, Einstein, The Life and Times, at 80 (1971).

888.Robert Sungenis, Review of Neil deGrasse Tyson's Video on Special Relativity (last visited on July 16, 2018), quoting Einstein: The Life and Times, pp. 109-110.

889.Robert Sungenis, Review of Neil deGrasse Tyson's Video on Special Relativity, http://www.galileowaswrong.com/wp-content/uploads/2015/06/Neil-Tyson-Review.pdf (last visited on July 16, 2018).

890.Robert Sungenis, Review of Neil deGrasse Tyson's Video on Special Relativity, at 9, http://www.galileowaswrong.com/wp-content/uploads/2015/06/Neil-Tyson-Review.pdf (last visited on July 16, 2018).

891.Jeroen van Dongen, On the role of the Michelson-Morley experiment: Einstein in Chicago, Institute for History and Foundations of Science & Descartes Centre, Utrecht University & Einstein Papers Project, Caltech, http://philsci-archive.pitt.edu/4778/1/Einstein_Chicago_Web2.pdf (last visited on October 6, 2015).

892.Jeroen van Dongen, supra (quoting Einstein to Michelson's biographer, Bernard Jaffe, on 17 March 1942; as in (Holton 1969/1995), p. 340.).

893.Jeroen van Dongen, supra.

894.Albert Einstein, How I Created the Theory of Relativity, Translated by Yoshimasa A. Ono, Physics Today, Vol. 35, No.8, pp. 45-47, August 1982, http://citeseerx.ist.psu.edu/viewdoc/download?doi=10.1.1.451.9501&rep=rep1&type=pdf.

895.Danian Hu, China and Albert Einstein, Harvard University Press, at 85 (2005).

896.Albert Einstein, How I Created the Theory of Relativity, Translated by Yoshimasa A. Ono, Physics Today, Vol. 35, No.8, pp. 45-47, August 1982, http://citeseerx.ist.psu.edu/viewdoc/download?doi=10.1.1.451.9501&rep=rep1&type=pdf.

897.Jeroen van Dongen, supra.

898.Jeroen van Dongen, supra.

899.Frederic Golden, Einstein, Time, December 31, 1999, http://content.time.com/time/magazine/article/0,9171,993017,00.html.

900.ANTHONY C. SUTTON, AMERICA'S SECRET ESTABLISHMENT, AN INTRODUCTION TO THE ORDER OF SKULL & BONES, at p. 212 (1986).

901.ANTHONY C. SUTTON, AMERICA'S SECRET ESTABLISHMENT, AN INTRODUCTION TO THE ORDER OF SKULL & BONES, at p. 212 (1986).

902.ANTHONY C. SUTTON, AMERICA'S SECRET ESTABLISHMENT, AN INTRODUCTION TO THE ORDER OF SKULL & BONES, at p. 7 (1986).

903.ANTHONY C. SUTTON, AMERICA'S SECRET ESTABLISHMENT, AN INTRODUCTION TO THE ORDER OF SKULL & BONES, at p. 200 (1986).

904.David Bay, Secret Societies Killed Jesus Christ, www.cuttingedge.org http://home.talkcity.com/InspirationAv/jforjesus/secret_societies.html (current as of October 1, 2001).

905.George H.W Bush, Address Before a Joint Session of the Congress on the State of the Union, January 29, 1991, The American Presidency Project, University of California,
http://www.presidency.ucsb.edu/ws/?pid=19253.

906.M.S. King, The Bad War, The Truth Never Taught About World War II, at 155 (2015). See also NEW WORLD ORDER PLEDGED TO JEWS; Arthur Greenwood of British War Cabinet Sends Message of Assurance Here RIGHTING OF WRONGS SEEN English Rabbi Delivers to Dr. S.S. Wise New Statement on Question After War, The New York Times, October 6, 1940,
http://query.nytimes.com/gst/abstract.html?res=9B03EFDC1639E23ABC4E53DFB667838B659EDE.

907.M.S. King, The Bad War, The Truth Never Taught About World War II, at 155 (2015). See also NEW WORLD ORDER PLEDGED TO JEWS; Arthur Greenwood of British War Cabinet Sends Message of Assurance Here RIGHTING OF WRONGS SEEN English Rabbi Delivers to Dr. S.S. Wise New Statement on Question After War, The New York Times, October 6, 1940,
http://query.nytimes.com/gst/abstract.html?res=9B03EFDC1639E23ABC4E53DFB667838B659EDE.

908.Herbert Dingle, Science at the Crossroads (1972), at 19.

909.Letter Offering Albert Einstein the Presidency of Israel, Embassy of Israel, November 17, 1952, Jewish Virtual Library,
https://www.jewishvirtuallibrary.org/jsource/Politics/ei

nsteinlet.html.

910.Protocols of the Learned Elders of Zion, Protocol 2, http://www.Biblebelievers.org.au/przion2.htm#PROTOCOL%20No.%202 (Last visited on August 22, 2012).

911.Ivan Fraser, The Protocols of the Learned Elders of Zion: Proofs of an Ancient Conspiracy, quoting The American Bulletin, May 5, 1935, http://www.bibliotecapleyades.net/sociopolitica/esp_sociopol_zion04.htm (last visited on October 17, 2012).

912.Ivan Fraser, The Protocols of the Learned Elders of Zion: Proofs of an Ancient Conspiracy, quoting The American Bulletin, May 5, 1935, http://www.bibliotecapleyades.net/sociopolitica/esp_sociopol_zion04.htm (last visited on October 17, 2012).

913.MICHAEL A. HOFFMAN II, JEWISH COMMUNISTS: THE DOCUMENTARY RECORD, http://www.hoffman-info.com/communist.html (website current as of March 5, 2003) (citing N.Y. Times, Oct. 7, 1999 and Newsweek, Oct. 18, 1999, p. 30).

914.New World Encyclopedia, http://www.newworldencyclopedia.org/entry/Comintern#cite_note-0 (last visited on September 19, 2012) (citing MI-5, History, The Inter-War Period. Retrieved December 19, 2007).

915.Henry Makow, Illuminati: The Cult That Hijacked the World, at 56 (2011).

916.MICHAEL A. HOFFMAN II, JEWISH COMMUNISTS: THE DOCUMENTARY RECORD, http://www.hoffman-info.com/communist.html (website current as of March 5, 2003).

917.MICHAEL A. HOFFMAN II, JEWISH COMMUNISTS: THE DOCUMENTARY RECORD, http://www.hoffman-info.com/communist.html (website current as of March 5, 2003) (quoting Dmitri Volkogonov, *Lenin: A New Biography,* p. 112).

918.MICHAEL A. HOFFMAN II, JEWISH COMMUNISTS: THE DOCUMENTARY RECORD, http://www.hoffman-info.com/communist.html (website current as of March 5, 2003).

919.ROBERT WILTON, THE LAST DAYS OF THE ROMANOVS (1920).

920.Gordon "Jack" Mohr, The Talmudic Effect on Judeo-Christianity, http://www.christianBiblestudy.org/OPS/JM/jm0027c.htm (current as of September 19, 2001).

921.ROBERT WILTON, THE LAST DAYS OF THE ROMANOVS, p. 148 (1920).

922.COLONEL GORDON "JACK" MOHR, THE TALMUDIC EFFECT ON JUDEO-CHRISTIANITY, http://www.christianBiblestudy.org/OPS/JM/jm0027e.htm (current as of September 17, 2001).

923.*Izvestia,* July 27, 1918.

924.HENRY FORD, THE INTERNATIONAL JEW, vol. 1, p. 225 (1920).

925.HENRY FORD, THE INTERNATIONAL JEW, vol. 1, p. 225 (1920).

926.MICHAEL A. HOFFMAN II, JEWISH COMMUNISTS: THE DOCUMENTARY RECORD, http://www.hoffman-info.com/communist.html (website current as of March 5, 2003) (quoting The Christian News, Jan. 8, 1996, p. 2).

927.MICHAEL A. HOFFMAN II, JEWISH COMMUNISTS: THE DOCUMENTARY RECORD, http://www.hoffman-info.com/communist.html (website current as of March 5, 2003).

928.Henry Makow, Communism - A Ruse for Illuminati Jewish Theft & Murder, April 28, 2008, http://www.savethemales.ca/communism_disguises_illuminati.html.

929.Slava Katamidze, Loyal Comrades, Ruthless Killers -The Secret Services of the USSR 1917-1991, at 25, quoted by Henry Makow, Communism - A Ruse for Illuminati Jewish Theft & Murder, April 28, 2008, http://www.savethemales.ca/communism_disguises_illuminati.html.

930.Henry Makow, Communism - A Ruse for Illuminati Jewish Theft & Murder, April 28, 2008, http://www.savethemales.ca/communism_disguises_illuminati.html.

931.Henry Makow, Communism - A Ruse for Illuminati Jewish Theft & Murder, April 28, 2008, http://www.savethemales.ca/communism_disguises_illuminati.html.

932.Sever Plocker, Stalin's Jews, December 26, 2006, http://www.ynetnews.com/articles/0,7340,L-3342999,00.html.

933.DES GRIFFIN, FOURTH REICH OF THE RICH, p. 62 (1976).

934.*Id.*

935.Protocols of the Learned Elders of Zion, Protocol 3, http://www.Biblebelievers.org.au/przion2.htm#PROTOCOL%20No.%203 (last visited on October 17, 2012).

936.DONN DE GRAND PRE, BARBARIANS INSIDE THE GATES, THE BLACK BOOK OF BOLSHEVISM, p. 209 (2000) (quoting BEJAMIN FREEDMAN, FACTS ARE FACTS (1954)).

937.Baruch Levy, Letter to Karl Marx, `La Revue de Paris', p. 574, June 1, 1928. http://www4.stormfront.org/posterity/ci/tjg.html (current as of September 9, 2001). *See also* DON DE GRAND PRE, BARBARIANS INSIDE THE GATES, p. 64 (2000).

938.DONN DE GRAND PRE, BARBARIANS INSIDE THE GATES, THE BLACK BOOK OF BOLSHEVISM, p. 313-14 (2000) (citing JACK BERNSTEIN, THE LIFE OF AN AMERICAN JEW IN RACIST MARXIST ISRAEL (1984)).

939.DONN DE GRAND PRE, BARBARIANS INSIDE THE GATES, THE BLACK BOOK OF BOLSHEVISM, p. 313-14 (2000) (citing JACK BERNSTEIN, THE LIFE OF AN AMERICAN JEW

IN RACIST MARXIST ISRAEL (1984)).

940.DONN DE GRAND PRE, BARBARIANS INSIDE THE GATES, THE BLACK BOOK OF BOLSHEVISM, p. 313-14 (2000) (citing JACK BERNSTEIN, THE LIFE OF AN AMERICAN JEW IN RACIST MARXIST ISRAEL (1984)).

941.JACK BERNSTEIN, MY FAREWELL TO ISRAEL THE THORN IN THE MIDEAST.

942.See John S. Torell, Showdown in Jerusalem, The Dove, winter 1995.

943.Walter White, Jr., The Hidden Tyranny, http://www.fourwinds10.com/corner/J224-ch4.pdf, http://www.antichristconspiracy.com/HTML%20Pages/Harold_Wallace_Rosenthal_Interview_1976.htm (web address current as of April 21, 2002).

944.Des Griffin, The Fourth Reich of the Rich, p. 70 (1993), citing Le Femme et l'enfant dans la France-Maconnerie Universelle, by A.C. De La Rive, P. 588, and Occult Theocrasy, pp. 220-21..

945.ALBERT PIKE, MORALS AND DOGMA OF THE ANCIENT AND ACCEPTED SCOTTISH RITE OF FREEMASONRY, p. 741 (1871).

946.RABBI DAVID A. COOPER, GOD IS A VERB, KABBALAH AND THE PRACTICE OF MYSTICAL JUDAISM, at 156 (1997) (emphasis added).

947.*Id.* at 160 (emphasis added).

948.Henry Makow, Illuminati 2, Deceit and Seduction, at 13 (2010).

949.Henry Makow, Illuminati 2, Deceit and Seduction, at 207 (2010).

950.See John S. Torell, Showdown in Jerusalem, The Dove, winter 1995.

951.MICHAEL A. HOFFMAN, JUDAISM'S STRANGE GODS, at p. 92, (2000).

952.Henry Makow, Illuminati, the Cult that Hijacked the World (2011), http://antimatrix.org/Convert/Books/Henry_Makow/Henry_Makow_Illuminati_The_Cult_that_Hijacked_the_World.html.

953.Henry Makow, Illuminati: The Cult That Hijacked the World, at 59 (2011).

954.Henry Makow, Illuminati: The Cult That Hijacked the World, at 59 (2011).

955.Texe Marrs, The Holy Serpent of the Jews, http://www.texemarrs.com/092012/holy_serpent_of_jews.htm (last visited on September 18, 2012).

956.Texe Marrs, The Holy Serpent of the Jews, http://www.texemarrs.com/092012/holy_serpent_of_jews.htm (last visited on September 18, 2012).

957.Chika Flint, Examining Judaism and the Origin of the Six Pointed Star, December 18, 2011.

958.Chika Flint, Examining Judaism and the Origin of the Six Pointed Star, December 18, 2011.

959.Protocols of the Learned Elders of Zion, Preface, http://www.Biblebelievers.org.au/przion1.htm#PREFACE (last visited on October 17, 2012).

960.See John S. Torell, Showdown in Jerusalem, The Dove, winter 1995.

961.Baruch Levy, Letter to Karl Marx, `La Revue de Paris', p. 574, June 1, 1928. http://www4.stormfront.org/posterity/ci/tjg.html (current as of September 9, 2001). *See also* DON DE GRAND PRE, BARBARIANS INSIDE THE GATES, p. 64 (2000).

962.Babylonian Talmud: *Tractate Sanhedrin, Folio 58b*, Sanhedrin Translated into English with Notes, Glossary and Indices Chapters I - VI by Jacob Shachter, Chapters VII - XI
by H. Freedman, B.A., Ph.D., Under the Editorship of Rabbi Dr I. Epstein B.A., Ph.D., D. Lit. (1961), *available at* http://www.come-and-hear.com/sanhedrin/sahedrin_58.html.

963.Michael Hoffman, *Judaism Discovered*, at 196 (2008) (quoting *Zohar Hadash, Bereshit* 18d-19a).

964.Henry Makow, Illuminati: The Cult That Hijacked the World, at 60-61 (2011).

965.The Kibbutz, Jewish Virtual Library, http://www.jewishvirtuallibrary.org/jsource/Society_&_Culture/kibbutz.html (last visited on September 19, 2012).

966.The Kibbutz, Jewish Virtual Library, http://www.jewishvirtuallibrary.org/jsource/Society_&_Culture/kibbutz.html (last visited on September 19, 2012).

967.Karl Marx, Critique of the Gotha Programme, http://www.marxists.org/archive/marx/works/1875/gotha/ch01.htm (last visited on September 19, 2012).

968.Henry Makow, Illuminati: The Cult That Hijacked the World, at 60-61 (2011).

969.Henry Makow, Illuminati: The Cult That Hijacked the World, at 60-61 (2011).

970.Harmony Daws, Why Evangelicals Want to be Jews and Not Christians,
EVANGELICALS: IN THE IMITATION OF JEWS,
April 28, 2010,
http://mysteryworshipers.wordpress.com/2011/05/04/why-evangelicals-want-to-be-jews-and-not-christians/.

971.Edith Starr Miller, Occult Theocrasy, pp. 77 (1933) (quoting Flavien Brenier. Source: Lt. Gen. A. Netchvolodow, Nicolas II et les Juifs, p. 139.).

972.Edith Starr Miller, Occult Theocrasy, pp. 78-79 (1933) (quoting Flavien Brenier. Source: Lt. Gen. A. Netchvolodow, Nicolas II et les Juifs, p. 139.).

973.Edith Starr Miller, Occult Theocrasy, p. 80 (1933).

974.Edith Starr Miller, Occult Theocrasy, p. 81 (1933).

975.Babylonian Talmud: *Tractate Sanhedrin, Folio 58b*, Sanhedrin Translated into English with Notes, Glossary and Indices Chapters I - VI by Jacob Shachter, Chapters VII - XI
by H. Freedman, B.A., Ph.D., Under the Editorship of Rabbi Dr I. Epstein B.A., Ph.D., D. Lit. (1961), *available at*
http://www.come-and-hear.com/sanhedrin/sahedrin_58.html.

976.*Sanhedrin, Folio 58b.*

977.Babylonian Talmud: *Tractate Yebamoth, Folio 98a,* Yebamoth Translated into English with Notes Glossary and Indices by Rev. Dr. Israel W. Slotki, M.A., Litt.D., Under the Editorship of Rabbi Dr I. Epstein B.A., Ph.D., D. Lit. (1961), *available at* http://www.come-and-hear.com/yebamoth/yebamoth_98.html .

978.The Curse Against Christians at Jamnia About 90 AD, http://www.defendingthebride.com/bb/curse.html#end (last visited on August 20, 2012).

979.The Curse Against Christians at Jamnia About 90 AD, http://www.defendingthebride.com/bb/curse.html#end (last visited on August 20, 2012).

980.Ariel Toaff, *Blood Passover, The Jews of Europe and Ritual Murder*, translated into English by Gian Marco Lucchese and Pietro Gianetti, 24 August 2007, English version available in HTML at http://www.bloodpassover.com/, English version available in PDF at http://www.israelshamir.net/BLOODPASSOVER.pdf, original Italian version available in PDF at http://www.laboratorio99.com/upload/Pasque%20di%20Sangue.pdf.

981.Bar Ilan University, Ramat-Gan, 52900 Israel, *at* http://www1.biu.ac.il/indexE.php (last visited on April 7, 2010).

982.Adi Schwartz, *The Wayward Son*, Haaretz, February 22, 2007, cited by Michael Hoffman in

Judaism Discovered, at 581 (2008).

983.Ariel Toaff, *Blood Passover*, at 10-11.

984.Ariel Toaff, *Blood Passover*, at 167.

985.Ariel Toaff, *Blood Passover*, at 167.

986.Ariel Toaff, *Blood Passover*, at 170.

987.Isabella of Spain, Chapter 25, The Trial of Benito Garcia— Expulsion of the Jews— Ritual Murder, http://www.jrbooksonline.com/PDF_Books/QueenIsabella.pdf.

988.Ariel Toaff, *Blood Passover*, at 181, http://www.israelshamir.net/BLOODPASSOVER.pdf.

989.Ariel Toaff, *Blood Passover*, at 182-83.

990.Ariel Toaff, *Blood Passover*, at 186.

991.Ariel Toaff, *Blood Passover*, at 187.

992.Talmud Unmasked , Appendix II, http://www.jrbooksonline.com/Talmud%20Unmasked/APPENDIX%20II.htm (last visited on January 12, 2016).

993.Talmud Unmasked , Appendix II, http://www.jrbooksonline.com/Talmud%20Unmasked/APPENDIX%20II.htm (last visited on January 12, 2016).

994.Royal Edict of the King and Queen of Spain,The Alhambra Decree, 31 March 1492, http://www.jrbooksonline.com/PDF_Books/QueenIsabella.pdf.

995.Matt Hale, Jewish Occult Murders How and Why Jews Have Sacrificed Non-Jewish Children as Part of Their Religion, 2002, http://antimatrix.org/Convert/Books/Matt_Hale/Jewish_Occult_Murders/Jewish_Occult_Murder.html.

996.Matt Hale, Jewish Occult Murders How and Why Jews Have Sacrificed Non-Jewish Children as Part of Their Religion, 2002, http://antimatrix.org/Convert/Books/Matt_Hale/Jewish_Occult_Murders/Jewish_Occult_Murder.html.

997.Matt Hale, Jewish Occult Murders How and Why Jews Have Sacrificed Non-Jewish Children as Part of Their Religion, 2002, http://antimatrix.org/Convert/Books/Matt_Hale/Jewish_Occult_Murders/Jewish_Occult_Murder.html.

998.Matt Hale, Jewish Occult Murders How and Why Jews Have Sacrificed Non-Jewish Children as Part of Their Religion, 2002, http://antimatrix.org/Convert/Books/Matt_Hale/Jewish_Occult_Murders/Jewish_Occult_Murder.html.

999.Matt Hale, Jewish Occult Murders How and Why Jews Have Sacrificed Non-Jewish Children as Part of Their Religion, 2002, http://antimatrix.org/Convert/Books/Matt_Hale/Jewish_Occult_Murders/Jewish_Occult_Murder.html.

1000.Matt Hale, Jewish Occult Murders How and Why Jews Have Sacrificed Non-Jewish Children as Part of Their Religion, 2002, http://antimatrix.org/Convert/Books/Matt_Hale/Jewish_Occult_Murders/Jewish_Occult_Murder.html.

1001.Matt Hale, Jewish Occult Murders How and Why Jews Have Sacrificed Non-Jewish Children as Part of Their Religion, 2002, http://antimatrix.org/Convert/Books/Matt_Hale/Jewish_Occult_Murders/Jewish_Occult_Murder.html.

1002.The Murder of Andrei Yushchinsky, 1917, JR Rare Books and Commentary, http://www.jrbooksonline.com/leese.htm. (last visited on January 12, 2016).

1003.The Murder of Andrei Yushchinsky, 1917, JR Rare Books and Commentary, http://www.jrbooksonline.com/leese.htm. (last visited on January 12, 2016).

1004.New York Herald, April 6, 1850, Matt Hale, Jewish Occult Murders How and Why Jews Have Sacrificed Non-Jewish Children as Part of Their Religion, 2002, http://antimatrix.org/Convert/Books/Matt_Hale/Jewish_Occult_Murders/Jewish_Occult_Murder.html.

1005.Jewish Ritual-Murder, A Historical Investigation, at 311 (1943), English translation 2007.

1006.Matt Hale, Jewish Occult Murders How and Why Jews Have Sacrificed Non-Jewish Children as Part of Their Religion, 2002, http://antimatrix.org/Convert/Books/Matt_Hale/Jewish_Occult_Murders/Jewish_Occult_Murder.html.

1007.Matt Hale, Jewish Occult Murders How and Why Jews Have Sacrificed Non-Jewish Children as Part of Their Religion, 2002, http://antimatrix.org/Convert/Books/Matt_Hale/Jewish_Occult_Murders/Jewish_Occult_Murder.html.

1008.Human Sacrifice Discussed on Oprah, October 7, 2011, https://www.youtube.com/watch?v=ImoZv1rprc0.

1009.Human Sacrifice Discussed on Oprah, October 7, 2011, https://www.youtube.com/watch?v=ImoZv1rprc0.

1010.Kitty Kelley, Oprah, at 201 (2010), http://www.lukeford.net/profiles/profiles/vicki_polin.htm (last visited on January 13, 2016).

1011.Kitty Kelley, Oprah, at 201 (2010), http://www.lukeford.net/profiles/profiles/vicki_polin.htm (last visited on January 13, 2016).

1012.Kitty Kelley, Oprah, at 201 (2010), http://www.lukeford.net/profiles/profiles/vicki_polin.htm (last visited on January 13, 2016).

1013.Kitty Kelley, Oprah, at 201 (2010), http://www.lukeford.net/profiles/profiles/vicki_polin.htm (last visited on January 13, 2016).

1014.Michael A. Hoffman II, *Judaism's Strange God's*, at 364 (2000) (quoting *Romemut Yisrael Ufarashat Hagalut*).

1015.Michael A. Hoffman II, *Judaism's Strange God's*, at 364 (2000) (quoting *Romemut Yisrael Ufarashat Hagalut*).

1016.Michael Hoffman, *Judaism Discovered*, at 529 (2008) (quoting Roman A. Foxbrunner, *Habad: The Hasidism of Shneur of Lyady*, at 108-09 (1993) (quoting Rabbi Shneur Zalman)).

1017.Babylonian Talmud: Tractate ‘Abodah Zarah, Folio 17a.

1018.Babylonian Talmud: Tractate Sanhedrin Folio 106a.

1019.Babylonian Talmud: Tractate Gittin Folio 57a.

1020.Michael Hoffman, *Judaism Discovered*, at 357 (2008).

1021.Chagigah 27a (a/k/a Hagigah 27a), *A Translation of the Treatise Chagigah from the Babylonian Talmud*, With Introduction, Notes, Glossary, and Indices by the Rev. A. W. Strean, M.A. Fellow, and Divinity and Hebrew Lecturer, of Corpus Christi College, Cambridge, and Formerly Tyrwhitt's Hebrew Scholar, at 145 (1891), *available at* http://www.archive.org/details/chagigahbabyloni00unknuoft.

1022.Tractate Baba Bathra Folio 75a at footnote 15.

1023.Michael Hoffman, *Judaism Discovered*, at 196 (2008) (quoting *Zohar Hadash, Bereshit* 18d-19a).

1024.Michael Hoffman, *Judaism Discovered*, at 196 (2008).

1025.Michael Hoffman, *Judaism Discovered*, at 203 (2008).

1026.William Wotton, *Miscellaneous Discourses Relating to the Traditions and Usages of the Scribes and Pharisees in Our Blessed Saviour Jesus Christ's Time* (London 1718) (quoted by Michael Hoffman, *Judaism Discovered*, at 193 (2008)).

1027.Michael Hoffman, *Judaism Discovered*, at 192 n.156 (2008).

1028.Babylonian Talmud, http://come-and-hear.com/sanhedrin/sanhedrin_57.html#57a_33.

1029.Babylonian Talmud, http://come-and-hear.com/sanhedrin/sanhedrin_57.html#57a_33.

1030.Michael Hoffman, *Judaism Discovered*, at 660 (2008).

1031.Michael Hoffman, *Judaism Discovered*, at 660 (2008).

1032.Michael Hoffman, *Judaism Discovered*, at 660 (2008).

1033.Michael Hoffman, *Judaism Discovered*, at 664 (2008) (quoting Israel Shahak, Jewish History, Jewish Religion, pp. 22-23).

1034.Michael Hoffman, *Judaism Discovered*, at 663 (2008).

1035.Jewish Encyclopedia, *Balaam*, *available at* http://www.jewishencyclopedia.com/view.jsp?artid=161&letter=B (last visited on March 10, 2010).

1036.Babylonian Talmud: *Tractate Sanhedrin, Folio 106b*, Sanhedrin Translated into English with Notes, Glossary and Indices Chapters I - VI by Jacob Shachter, Chapters VII - XI
by H. Freedman, B.A., Ph.D., Under the Editorship of Rabbi Dr I. Epstein B.A., Ph.D., D. Lit. (1961), *available at*

http://www.come-and-hear.com/sanhedrin/sanhedrin_106.html.

1037.Babylonian Talmud: *Tractate Sanhedrin, Folio 106b, at n.6,* Sanhedrin Translated into English with Notes, Glossary and Indices Chapters I - VI by Jacob Shachter, Chapters VII - XI
by H. Freedman, B.A., Ph.D., Under the Editorship of Rabbi Dr I. Epstein B.A., Ph.D., D. Lit. (1961), *available at* http://www.come-and-hear.com/sanhedrin/sanhedrin_106.html.

1038.Elizabeth Dilling, *The Jewish Religion, Its Influence Today* (1963), *at* http://www.come-and-hear.com/dilling/chapt03.html#T159.

1039.The scholars listed as contributing to the Jewish Encyclopedia entry for *Balaam* are: Morris Jastrow, Jr., Ph.D., Professor of Semitic Languages, University of Pennsylvania; Ira Maurice Price, Ph.D., LL.D., Professor of Semitic Languages and Literature, University of Chicago; Marcus Jastrow, Ph.D., Rabbi Emeritus of the Congregation Rodef Shalom, Philadelphia; H. M. Speaker, Gratz College; John Dyneley Prince, Ph.D., Professor of Semitic Languages, New York University; Dr. S. Mühsam, Chief Rabbi of Gratz, Austria; Kaufmann Kohler, Ph.D., Rabbi Emeritus of Temple Beth-El, New York, President of the Hebrew Union College.

1040.Jewish Encyclopedia, *Jesus of Nazareth, available at* http://www.jewishencyclopedia.com/view.jsp?artid=254&letter=J&search=jesus (last visited on March 10, 2010).

1041.Edith Starr Miller, Occult Theocrasy, p. 86 (1933).

1042.Edith Starr Miller, Occult Theocrasy, p. 87 (1933).

1043.Babylonian Talmud, http://come-and-hear.com/sanhedrin/sanhedrin_52.html. See also English Baylonian Talmud, http://halakhah.com (last visited on December 2, 2015).

1044.Talmud, *Tractate Sanhedrin, Folio 100b*, n.6 (1961), *available at* http://www.come-and-hear.com/sanhedrin/sanhedrin_100.html (emphasis added).

1045.Judaism vs. Christianity: The War The Lamb Wins, http://www.fixedearth.com/talmud.html (current as of September 11, 2001).

1046.DONN DE GRAND PRE, BARBARIANS INSIDE THE GATES, THE BLACK BOOK OF BOLSHEVISM, p. 209 (2000) (quoting BEJAMIN FREEDMAN, FACTS ARE FACTS (1954)).

1047.Babylonian Talmud: *Tractate Sanhedrin, Folio 106a*, Sanhedrin Translated into English with Notes, Glossary and Indices Chapters I - VI by Jacob Shachter, Chapters VII - XI by H. Freedman, B.A., Ph.D., Under the Editorship of Rabbi Dr I. Epstein B.A., Ph.D., D. Lit. (1961), *available at* http://www.come-and-hear.com/sanhedrin/sanhedrin_106.html.

1048.Babylonian Talmud: *Tractate Sanhedrin, Folio 106a, at n.42*, Sanhedrin Translated into English with Notes, Glossary and Indices Chapters I - VI by Jacob Shachter, Chapters VII - XI
by H. Freedman, B.A., Ph.D., Under the Editorship of Rabbi Dr I. Epstein B.A., Ph.D., D. Lit. (1961), *available at* http://www.come-and-hear.com/sanhedrin/sanhedrin_106.html.

1049.Babylonian Talmud: *Tractate Gittin, Folio 57a*, Gittin Translated into English with Notes, Glossary and Indices By Maurice Simon, M.A. Under the Editorship of Rabbi Dr. I. Epstein B.A., Ph.D., D. Lit. (1961), *available at* http://www.come-and-hear.com/gittin/gittin_57.html.

1050.Jewish Encyclopedia, *Balaam*, *available at* http://www.jewishencyclopedia.com/view.jsp?artid=161&letter=B (last visited on March 10, 2010).

1051.Babylonian Talmud: *Tractate Gittin, Folio 57a, at n.3,* Gittin Translated into English with Notes, Glossary and Indices By Maurice Simon, M.A. Under the Editorship of Rabbi Dr. I. Epstein B.A., Ph.D., D. Lit. (1961), *available at* http://www.come-and-hear.com/gittin/gittin_57.html.

1052.Soncino Press, http://www.soncino.com/about (last visited on December 2, 2015).

1053.Michael Hoffman, *Judaism Discovered*, at 785 (2008).

1054.Lawrence Fine, Chapter on Kabbalistic Texts, From: *Back to the Sources: Reading the Classic Jewish Texts* ("The First Complete Modern Guide to the Great

Books of the Jewish Tradition: What They Are and How to Read Them"), at p. 337 (2006) (bold emphasis added, italics in original).

1055.Lawrence Fine, Chapter on Kabbalistic Texts, From: *Back to the Sources: Reading the Classic Jewish Texts* ("The First Complete Modern Guide to the Great Books of the Jewish Tradition: What They Are and How to Read Them"), at p. 337 (2006) (quoting Zohar III, 152a).

1056.Blavatsky, Theosophical Glossary, p. 168 (quoted by Barbara Aho, Mystery, Babylon the Great Catholic or Jewish?, at http://watch.pair.com/mystery-babylon.html#cabala (last visited on April 17, 2010)).

1057.MICHAEL A. HOFFMAN, JUDAISM'S STRANGE GODS, at p. 88, (2000).

1058.MICHAEL A. HOFFMAN, JUDAISM'S STRANGE GODS, at p. 88, (2000). See also Michael Hoffman, *Judaism Discovered*, at 779 (2008) (quoting Gershom Scholem, *Kabbalah* pp.183-84).

1059.MICHAEL A. HOFFMAN, JUDAISM'S STRANGE GODS, at p. 91, (2000).

1060.Michael Hoffman, *Judaism Discovered*, at 780 (2008) (quoting Helen Jacaobus, *Eye Jinx*, Jewish Chronicle, May 7, 1999).

1061.Protocols of the Learned Elders of Zion, Protocol 16, http://www.Biblebelievers.org.au/przion5.htm#protocol%20No.%2016 (last visited on October 9, 2012).

1062.Protocols of the Learned Elders of Zion, Protocol 4, http://www.Biblebelievers.org.au/przion3.htm#PROTOCOL%20No.%204 (Last visited on August 23, 2012).

1063.George Orwell, 1984, at chapter 3, https://ebooks.adelaide.edu.au/o/orwell/george/o79n/chapter3.3.html.

1064.George Orwell, 1984, at part 3, chapter 3, https://ebooks.adelaide.edu.au/o/orwell/george/o79n/chapter3.3.html.

1065.George Orwell, 1984, at part 3, chapter 3, https://ebooks.adelaide.edu.au/o/orwell/george/o79n/chapter3.3.html.

1066.George Orwell, 1984, at part 1, chapter 3.

1067.George Orwell, 1984, at part 2, chapter 9.

1068.Eric Dubay Interview, http://www.atlanteanconspiracy.com/2015/04/eric-dubay-flat-earth-interviews.html (last visited on September 9, 2015).

1069.Eric Dubay Interview, http://www.atlanteanconspiracy.com/2015/04/eric-dubay-flat-earth-interviews.html (last visited on September 9, 2015).

1070.Trilateral Commission, http://trilateral.org/page/3/about-trilateral (last visited on June 20, 2017).

1071.Archibald E. Roberts, The Most Secret Science, at 116 (1984).

1072.Archibald E. Roberts, The Most Secret Science, at 116 (1984).

1073.Statement of James P. Warburg on Senate Resolution 56 at the Hearings before a Subcommittee of the Committee on Foreign Relations, United States Senate, 81st Congress, 2d Session on Resolutions relative to the United Nations charter, Atlantic Union, World Federation, Feb. 2, 3, 6, 8, 9, 13, 15, 17, and 20, 1950, Printed for the use of the Committee on Foreign Relations, U.S. GOVERNMENT PRINTING OFFICE, WASHINGTON, 1950: 64429
PP. 494-508,
https://en.wikisource.org/wiki/James_Warburg_before_the_Subcommittee_on_Revision_of_the_United_Nations_Charter. See also Duane L. Wilde, Government: A Christians Civil Responsibility, at 50 (2014).

1074.Congressional Record of June 7, 1949, pages 7356 and 7357. Corresponds to Senate Concurrent Resolution 56, the Tobey or 'World Federalist' Resolution. Final Warning: A History of the New World Order, Illuminism and the master plan for world domination, by: David Allen Rivera, 1994, source: darivera.com, Modern History Project, http://modernhistoryproject.org/mhp?Article=FinalWarning&C=8.1#Resolution.

1075.Final Warning: A History of the New World Order, Illuminism and the master plan for world domination, by: David Allen Rivera, 1994, source: darivera.com, Modern History Project, http://modernhistoryproject.org/mhp?Article=FinalWarning&C=8.1#Resolution.

1076.Statement of James P. Warburg on Senate Resolution 56 at the Hearings before a Subcommittee

of the Committee on Foreign Relations, United States Senate, 81st Congress, 2d Session on Resolutions relative to the United Nations charter, Atlantic Union, World Federation, Feb. 2, 3, 6, 8, 9, 13, 15, 17, and 20, 1950, Printed for the use of the Committee on Foreign Relations, U.S. GOVERNMENT PRINTING OFFICE, WASHINGTON, 1950: 64429
PP. 494-508,
https://en.wikisource.org/wiki/James_Warburg_before_the_Subcommittee_on_Revision_of_the_United_Nations_Charter. See also Duane L. Wilde, Government: A Christians Civil Responsibility, at 50 (2014).

1077.Final Warning: A History of the New World Order, Illuminism and the master plan for world domination, by: David Allen Rivera, 1994, source: darivera.com, Modern History Project, http://modernhistoryproject.org/mhp?Article=FinalWarning&C=8.1#Resolution.

1078.David Rockefeller, Memoirs, at 406 (2002).

1079.David Ben Gurion, Look Magazine, January 1962,
http://davidduke.com/jewish-supremacist-delusions-grandeur-exemplified-first-israeli-prime-minister/.

1080.Francis X. Clines, Reagan's Doubts on Dr. King Disclosed, The New York Times, October 23, 1983, http://www.nytimes.com/1983/10/22/us/reagan-s-doubts-on-dr-king-disclosed.html.

1081.Marcus Epstein, Myths of Martin Luther King, January 18, 2003,
https://www.lewrockwell.com/2003/01/marcus-epstein/myths-of-martin-luther-king/.

1082.Jimmy T. (Gunny) LaBaume, When Judged by the Content of His Character, January 18, 2016 (citing Starrett, Mary, How a Marxist Came to be an American Hero, NewsWithViews.com January 16, 2004. Quoted within that article: King, A Critical Biography by David Lewis, Martin Luther King, The Man Behind The Myth, by Des Griffin and the National Observer, 1963.)

1083.Labaume, supra.

1084.Jimmy T. (Gunny) LaBaume, When Judged by the Content of His Character, January 18, 2016 (citing Steele, Edgar J., The Perfect American Holiday; Horne, Cadrmeron L., HAPPY BIRTHDAY, ROBERT E. LEE or Happy Birthday, Martin Luther King, Jr.?; Hoffman, Michael, 1992, Holiday for a Cheater, Wiswell Ruffin House, Dresden, New York; Montgomery, Victor M. Letter to the Editor. Greenville, NC newspaper of record.

1085.Don Boys, Was MLK a Communist?, January 19, 2009, http://www.conservativetruth.org/article.php?id=928.

1086.Don Boys, Was MLK a Communist?, January 19, 2009, http://www.conservativetruth.org/article.php?id=928.

1087.Don Boys, Was MLK a Communist?, January 19, 2009, http://www.conservativetruth.org/article.php?id=928.

1088.Don Boys, Was MLK a Communist?, January 19, 2009, http://www.conservativetruth.org/article.php?id=928.

1089.Don Boys, Was MLK a Communist?, January 19, 2009, http://www.conservativetruth.org/article.php?id=928.

1090.Did You Say Dr. Martin Luther King Jr. Was a Phony?, http://frontline.org.za/index.php?option=com_content&id=1506:did-you-say-dr-martin-luther-king-jr-was-a-phony#sthash.9nFC0e2O.dpuf (last visited on May 15, 2016).

1091.Did You Say Dr. Martin Luther King Jr. Was a Phony?, http://frontline.org.za/index.php?option=com_content&id=1506:did-you-say-dr-martin-luther-king-jr-was-a-phony#sthash.9nFC0e2O.dpuf (last visited on May 15, 2016).

1092.Don Boys, Was MLK a Communist?, January 19, 2009, http://www.conservativetruth.org/article.php?id=928.

1093.Kevin Alfred Strom, The Beast as Saint: The Truth About Martin Luther King, Jr., http://www.jesus-is-savior.com/Wolves/king_jr-exposed.htm (last visited on May 21, 2016).

1094.Kevin Alfred Strom, The Beast as Saint: The Truth About Martin Luther King, Jr., http://www.jesus-is-savior.com/Wolves/king_jr-exposed.htm (last visited on May 21, 2016).

1095.Todd Starnes, Christian Bakers Fined $135,000 for Refusing to Make Wedding Cake for Lesbians, Fox News, July 3, 2015, http://www.foxnews.com/opinion/2015/07/03/christian-bakers-fined-135000-for-refusing-to-make-wedding-c ake-for-lesbians.html.

1096.Todd Starnes, supra.

1097.Did CIA Director William Casey Really Say 'We'll Know Our Disinformation Program Is Complete When Everything the American Public Believes Is False'?, Infinite Unknown, http://www.infiniteunknown.net (last visited on March 17, 2016).

1098.Did William Casey (CIA Director) really say, "We'll know our disinformation program is complete when everything the American public believes is false."?, https://www.quora.com/Did-William-Casey-CIA-Director-really-say-Well-know-our-disinformation-program-is-complete-when-everything-the-American-public-believes-is-false (last visited on April 28, 2016).

1099.MOCKINGBIRD: The Subversion Of The Free Press By The CIA, www.whatreallyhappened.com (last visited on April 24, 2016).

1100.MOCKINGBIRD: The Subversion Of The Free Press By The CIA, www.whatreallyhappened.com, citing Katherine The Great, by Deborah Davis (New York: Sheridan Square Press, 1991).

1101.Michael Collins Piper, Final Judgement, 6th Edition, at 260 (2007).

1102.David Wise, The American Police State: The Government Against the People. (New York: Random House, 1976), pp. 200-201, quoted in Michael Collins Piper, Final Judgement, 6th Edition, at 260 (2007).

1103.David Wise, supra, at 200-201.

1104.Alex Constantine, The Depraved Spies and Moguls of the CIA's Operation MOCKINGBIRD, www.whatreallyhappened.com (last visited on April 24, 2016), citing Deborah Davis, Katharine The Great, New York: Sheridan Square Press, at 119-132, 1991, and Carl Bernstein, The CIA and the Media, October 20, 1977.

1105.Carl Bernstein, The CIA and the Media, October 20, 1977, http://www.carlbernstein.com/magazine_cia_and_media.php.

1106.Carl Bernstein, supra.

1107.Carl Bernstein, supra.

1108.Carl Bernstein, supra.

1109.MOCKINGBIRD: The Subversion Of The Free Press By The CIA, www.whatreallyhappened.com (last visited on April 24, 2016).

1110.Edward Bernays, Propaganda, at 37 (1928).

1111.Bernays at 38.

1112.Cong. Rec. 2947-2948 (February 9, 1917) (speech of Rep. Callaway), http://www.iahf.com/media.html (current as of October 3, 2001).

1113.Granville Williams, *Bestriding The World, Campaign for Press and Broadcasting Freedom*, New Internationalist magazine, http://www.mediachannel.org/ownership/granville.shtml (last visited on May 25, 2008). Christopher Bollyn, *Arnon Milchan & Israeli Control of the Media*, http://www.bollyn.com/index/?id=10568 (September 1, 2007).

1114. Cong. Rec. 2947-2948 (February 9, 1917) (speech of Rep. Callaway), http://www.iahf.com/media.html (current as of October 3, 2001).

1115. ERIC JON PHELPS, VATICAN ASSASSINS: "WOUNDED IN THE HOUSE OF MY FRIENDS," p. 465 (2001) (quoting A U.S. Police Action: Operation Vampire Killer, pp. 18-19 (1992)).

1116. Building 7, the Secrecy Shrouded Building Holding Guiliani's Command Center, http://911research.wtc7.net/wtc/background/wtc7.html (web address current as of October 24, 2004).

1117. Rudolph Guiliani Interview, The Hall of Public Service, http://www.achievement.org/autodoc/page/giu0int-7 (web address current as of October 24, 2004).

1118. BBC Reported Building 7 Collapse 20 Minutes Before It Fell, http://www.youtube.com/watch?v=C7SwOT29gbc&mode=related&search= (last visited on June 2, 2008).

1119. Richard Porter, *Part of the conspiracy?*, BBC, February 27, 2007, http://www.bbc.co.uk/blogs/theeditors/2007/02/part_of_the_conspiracy.html.

1120. Richard Porter, *Part of the conspiracy?*, BBC, February 27, 2007, http://www.bbc.co.uk/blogs/theeditors/2007/02/part_of_the_conspiracy.html.

1121.BBC Anchor Agrees WTC-7 Collapse May Be a Conspiracy, May 2, 2008, http://noworldsystem.com/category/jane-standley/.

1122.Richard Porter, *Part of the conspiracy?*, BBC, February 27, 2007, http://www.bbc.co.uk/blogs/theeditors/2007/02/part_of_the_conspiracy.html.

1123.Statement of Principles, New American Century, January 3, 1997, http://www.newamericancentury.org/statementofprinciples.htm.

1124.Rebuilding America's Defenses, Strategy, Forces and Resources for a New Century, A Report of The Project for a New American Century, at 51, September 2000, http://www.newamericancentury.org/RebuildingAmericasDefenses.pdf.

1125.Vince Palamara, *The Secret Service: On the Job in Dallas*, contained at pg. 160 in Murder in Dealey Plaza, edited by James H. Fetzer, Ph.D.

1126.Vince Palamara, *The Secret Service: On the Job in Dallas*, contained at pg. 161 in Murder in Dealey Plaza, edited by James H. Fetzer, Ph.D.

1127.David W. Mantick, M.D., Ph.D., *The Paradoxes of the JFK Assassination, The Zabruder Film Controversy, The Chief Arguments Against Authenticity*, contained at pg. 341 in Murder in Dealey Plaza, edited by James H. Fetzer, Ph.D.

1128.Vince Palamara, *The Secret Service: On the Job in Dallas*, contained at pg. 169 in Murder in Dealey

Plaza, edited by James H. Fetzer, Ph.D.

1129.JIM GARRISON, ON THE TRAIL OF THE ASSASSINS, pg. 221 (1988).

1130.JIM GARRISON, ON THE TRAIL OF THE ASSASSINS, pg. 221 (1988).

1131.JIM GARRISON, ON THE TRAIL OF THE ASSASSINS, pg. 221 (1988).

1132.JIM GARRISON, ON THE TRAIL OF THE ASSASSINS, pg. 222-23 (1988).

1133.JIM GARRISON, ON THE TRAIL OF THE ASSASSINS, pg. 222-23 (1988).

1134.JIM GARRISON, ON THE TRAIL OF THE ASSASSINS, pg. xiii (1988).

1135.JIM GARRISON, ON THE TRAIL OF THE ASSASSINS, pg. xiii (1988).

1136.JIM GARRISON, ON THE TRAIL OF THE ASSASSINS, pg. xiii (1988).

1137.JIM GARRISON, ON THE TRAIL OF THE ASSASSINS, pg. 217 (1988).

1138.JIM GARRISON, ON THE TRAIL OF THE ASSASSINS, pg. xiii (1988).

1139.JIM GARRISON, ON THE TRAIL OF THE ASSASSINS, pg. 219 (1988).

1140.JIM GARRISON, ON THE TRAIL OF THE ASSASSINS, pg. 219 (1988).

1141.JIM GARRISON, ON THE TRAIL OF THE ASSASSINS, pg. 218 (1988).

1142.JIM GARRISON, ON THE TRAIL OF THE ASSASSINS, pg. xiii (1988).

1143.JIM GARRISON, ON THE TRAIL OF THE ASSASSINS, pg. xiii (1988).

1144.Morphology of Experimental Assault Rifle Skin Wounds, International Journal of Legal Medicine (January 22, 2002), *available at* http://www.springerlink.com/content/egtlulugcd8qh9hx/fulltext.html, *also available at* http://www.springerlink.com/content/egtlulugcd8qh9hx/fulltext.pdf.

1145.Patterns of Tissue Injury, Mercer University School of Medicine, http://library.med.utah.edu/WebPath/TUTORIAL/GUNS/GUNINJ.html (last visited on June 18, 2008).

1146.Department of Forensic Medicine, University of Dundee http://www.dundee.ac.uk/forensicmedicine/notes/gunshot.pdf (last visited on June 18, 2008).

1147.The Autopsy Pictures, JFK Murder Solved, http://www.jfkmurdersolved.com/autopsy.htm (last visited on June 17, 2008).

1148.Charles Crenshaw, http://www.spartacus.schoolnet.co.uk/JFKcrenshawC.htm (Citing Charles Crenshaw, Let's Set the Record Straight, Assassination Science (1998)) (last visited on June 17, 2008).

1149.Milicent Cranor, Vincent Bugliosi's Reclaiming History: A Crime Scene Between Two Hard Covers, Citizens for Truth about the Kennedy Assassination, http://www.ctka.net/bug_mc_2.html (last visited on June 20, 2008).

1150.Jacob Hornberger, The Mystery of Robert Knudsen, July 6, 2018, https://www.fff.org/2018/07/06/the-mystery-of-robert-knudsen/.

1151.Jacob Hornberger, The Mystery of Robert Knudsen, July 6, 2018, https://www.fff.org/2018/07/06/the-mystery-of-robert-knudsen/.

1152.Jacob Hornberger, The Mystery of Robert Knudsen, July 6, 2018, https://www.fff.org/2018/07/06/the-mystery-of-robert-knudsen/.

1153.Jacob Hornberger, The Mystery of Robert Knudsen, July 6, 2018, https://www.fff.org/2018/07/06/the-mystery-of-robert-knudsen/.

1154.Jacob Hornberger, The Mystery of Robert Knudsen, July 6, 2018, https://www.fff.org/2018/07/06/the-mystery-of-robert-knudsen/.

1155.James H. Fetzer, Murder in Dealey Plaza, 174 (2002) (citing Robert Groden, The Killing of a President, 86-88 (1963)).

1156.Testimony of Dr. William Kemp Clark taken at 11:50 a.m., on March 21, 1964, at Parkland Memorial

Hospital, Dallas, Tex., by Mr. Arlen Specter, assistant counsel of the President's Commission, http://mcadams.posc.mu.edu/russ/testimony/clark_w.htm (last visited on June 20, 2008).

1157.The Death Of John Kennedy, The Media Helped Sell the Lie of the Lone Assassin, http://www.whatreallyhappened.com/RANCHO/POLITICS/JFK/postphotos.html (last visited on May 31, 2008).

1158.Testimony of Billy Nolan Lovelady Before the Warren Commission, http://jfkassassination.net/russ/testimony/lovelady.htm (last visited on May 31, 2008) (The testimony of Billy Nolan Lovelady was taken at 3:50 p.m., on April 7, 1964, in the office of the U.S. attorney, 301 Post Office Building, Bryan and Ervay Streets, Dallas, Tex., by Messrs. Joseph A. Ball and Samuel A. Stern, assistant counsel of the President's Commission.).

1159.Id.

1160.Id.

1161.Id.

1162.JIM GARRISON, ON THE TRAIL OF THE ASSASSINS, pg. xiii (1988).

1163.*See* JIM GARRISON, ON THE TRAIL OF THE ASSASSINS, pg. xiii (1988).

1164.JIM GARRISON, ON THE TRAIL OF THE ASSASSINS, pg. 168 (1988).

1165.JIM GARRISON, ON THE TRAIL OF THE ASSASSINS, pg. 168 (1988).

1166.JIM GARRISON, ON THE TRAIL OF THE ASSASSINS, pg. 160-72 (1988).

1167.MICHAEL COLLINS PIPER, FINAL JUDGMENT, pg. 204-05 (2004).

1168.JIM GARRISON, ON THE TRAIL OF THE ASSASSINS, pg. 132-36 (1988).

1169.JIM GARRISON, ON THE TRAIL OF THE ASSASSINS, pg. 132-36 (1988).

1170.MICHAEL COLLINS PIPER, FINAL JUDGMENT, pg. 204-05 (2004).

1171.MICHAEL COLLINS PIPER, FINAL JUDGMENT, pg. 204-05 (2004).

1172.MICHAEL COLLINS PIPER, FINAL JUDGMENT, pg. 204-05 (2004).

1173.MICHAEL COLLINS PIPER, FINAL JUDGMENT, pg. 259 (2004).

1174.The JFK 100: Reprisals Against Jim Garrison, http://www.jfk-online.com/jfk100repris.html (web address current as of May 31, 2004)

1175.New Orleans Businessman Clay Shaw Interview 1967, Assassination of John F. Kennedy, Encyclopedia Britannica, https://www.youtube.com/watch?v=0VDgc3_Ci-w.

1176.CIA Document #1035-960, RE: Concerning Criticism of the Warren Report.

1177.CIA Document #1035-960, RE: Concerning Criticism of the Warren Report.

1178.Arieh O'Sullivan, *Vanunu: Israel behind JFK Assassination*, Jerusalem Post, Jul. 25, 2004.

1179.President John F. Kennedy Waldorf-Astoria Hotel, New York, NY, April 27, 1961.

1180.Marshall Hall, Kabbala,Part VII, The Concept of a Rotating Earth Orbiting the Sun Is Both the Pseudo-Scientific Foundation and The Achilles Heel of Modern Kabbala-based Cosmology, http://www.fixedearth.com/kabbala-7.html (last visited on September 17, 2015).

1181.The Secret Wisdom of Qabalah", by J.F.C. Fuller, http://216.109.125.130/search/cache?ei=UTF-8&p=Aristarchus+Copernicus+Kabbalism&fr... p. 19 of 56.

1182.Marshall Hall, Sola Scriptura - IV: Kabbala Phariseeism or Bible Christianity?, http://www.fixedearth.com/sola-scriptura-4.html (last visited on September 18, 2015).

1183.Chazal's View of the World, Torat Emet, http://www.aishdas.org/toratemet/en_shape.html (last visited on September 17, 2015).

1184.What the Sages Knew About the Shape of the Earth, http://www.daatemet.org/articles/article.cfm?article_id=16 (last visited on September 18, 2015).

1185.Talmud, Tractate Abodah Zarah at Folio 41a, http://www.come-and-hear.com/zarah/zarah_41.html (last visited on September 17, 2015).

1186. Jerusalem Tractate Abodah Zarah, Chapter 3, http://www.jewishvirtuallibrary.org/jsource/Talmud/av

odazara3.html (last visited on September 17, 2015).

1187.Chazal's View of the World, Torat Emet, http://www.aishdas.org/toratemet/en_shape.html (last visited on September 17, 2015). See also, Ronald H. Isaacs, Divination, Magic, and Healing: The Book of Jewish Folklore, at 33, 1998.

1188.Chazal's View of the World, Torat Emet, http://www.aishdas.org/toratemet/en_shape.html (last visited on September 17, 2015).

1189.George William Rutler, Nicolaus Copernicus, Catholic Education Resource Center, http://www.catholiceducation.org/en/science/faith-and-science/nicolaus-copernicus.html (last visited on September 17, 2015).

1190.Nicolaus Copernicus (1543), Dedication of the Revolutions of the Heavenly Bodies to Pope Paul III, http://www.bartleby.com/39/12.html.

1191.Nicolaus Copernicus (1543), Dedication of the Revolutions of the Heavenly Bodies to Pope Paul III, http://www.bartleby.com/39/12.html.

1192.Michael Hoffman, *The Truth About the Talmud, at* http://www.revisionisthistory.org/talmudtruth.html (last visited on March 10, 2010) (excerpted from MICHAEL A. HOFFMAN, JUDAISM'S STRANGE GODS (2000)).

1193.Robert Goldberg, *Talmud, Back to the Sources: Reading the Classic Jewish Texts* (New York: Simon and Schuster, 1984), p. 130 (quoted by Michael Hoffman, *Judaism Discovered*, at 187 (2008)).

1194.Michael Hoffman, *Judaism Discovered*, at 189 (2008).

1195.Rabbi Ben Zion Bokser, *Judaism and the Christian Predicament* (1966), pp. 59 and 159 (quoted by Michael Hoffman, *Judaism Discovered*, at 190 (2008)).

1196.Michael Hoffman, *The Truth About the Talmud, at* http://www.revisionisthistory.org/talmudtruth.html (excerpted from MICHAEL A. HOFFMAN, JUDAISM'S STRANGE GODS (2000)).

1197.Michael Hoffman, *The Truth About the Talmud, at* http://www.revisionisthistory.org/talmudtruth.html (excerpted from MICHAEL A. HOFFMAN, JUDAISM'S STRANGE GODS (2000)) (quoting Ari Goldman, New York Times, April 10, 1993, p. 38).

1198.Michael Hoffman, *The Truth About the Talmud, at* http://www.revisionisthistory.org/talmudtruth.html (excerpted from MICHAEL A. HOFFMAN, JUDAISM'S STRANGE GODS (2000)).

1199.Michael Hoffman, *The Truth About the Talmud, at* http://www.revisionisthistory.org/talmudtruth.html (excerpted from MICHAEL A. HOFFMAN, JUDAISM'S STRANGE GODS (2000)).

1200.Jewish Encyclopedia, Cabala, at http://www.jewishencyclopedia.com/view.jsp?artid=1&letter=C#4 (last visited on April 18, 2010).

1201.Marshall Hall, Kabbala, Part III, Judaism: Theological Anarchy or Kabbalists in Control All Along?, http://www.fixedearth.com/kabbala-3.html.

1202.Marshall Hall, Kabbala, Part IV, Some Insights to Ponder and Evaluate About the Hidden, Secretive Makeup of Kabbalism, http://www.fixedearth.com/kabbala-4.html, quoting Mail. Jewish Mailing List, vol. 20, #6, p. 5 of 6; http://www.shamash.org.

1203.Tracey R Rich, Judaism 101, Sages and Scholars, http://www.jewfaq.org/sages.htm (last visited on September 11, 2015).

1204.Nachmanides, http://www.creatia.org/timeline/nachmanides.htm (last visited on September 11, 2015).

1205.Yakov Leib haKohain, http://www.kheper.net/topics/Kabbalah/Yakov.htm (last visited on September 12, 2015).

1206.Yakov Leib HaKohain, Short essay on Lurianic Kabbalah, http://www.kheper.net/topics/Kabbalah/LurianicKabbalah.htm (last visited on September 12, 2015).

1207.Gerald Schroeder, The Age of the Universe, http://www.geraldschroeder.com/AgeUniverse.aspx (last visited on September 12, 2015).

1208.Schroeder, supra.

1209.Marshal Hall, Kabbala, Part III, Judaism: Theological Anarchy, Or Kabbalists in Control All Along?, http://www.fixedearth.com/kabbala-3.html (last visited on September 12, 2015).

1210.Edgar B. Herwick III, Big Bang Theory: A Roman Catholic Creation, March 20, 2014, http://wgbhnews.org/post/big-bang-theory-roman-cath

olic-creation.

1211.Mark Midbon, 'A Day Without Yesterday': Georges Lemaitre & the Big Bang, Commonweal Magazine Vol. 127 No. 6 (March 24, 2000) 18-19, republished by the Catholic Education Resource Center, http://www.catholiceducation.org/en/science/faith-and-science/a-day-without-yesterday-georges-lemaitre-amp-the-big-bang.html.

1212.Israel Modern History: Offering the Presidency of Israel to Albert Einstein (November 17, 1952), https://www.jewishvirtuallibrary.org/jsource/Politics/einsteinlet.html.

1213.Anti-Zion, Jews on the Jewish Question, http://www.diac.com/~bkennedy/az/A-E.html (current as of September 10, 2001).

1214.Edith Starr Miller (Lady Queenborough), Occult Theocrasy, p. 319 (1933).

1215.Ivan Fraser, Protocols of the Learned Elders of Zion, Proofs of an Ancient Conspiracy, http://www.vegan.swinternet.co.uk/articles/conspiracies/protocols_proof.html (current as of September 10, 2001).

1216.Barbara Aho, *Mystery, Babylon the Great - Catholic or Jewish?, Part III, The Jesuits*, Watch Unto Prayer, *at* http://watch.pair.com/mystery-babylon.html (last visited on February 8, 2010).

1217.The Abbate Leone, The Jesuit Conspiracy, The Secret Plan of the Order, at 183-84 (1848).

1218.John S. Torell, European-American Evangelical Association, July 1999, http://www.eaec.org/NL99jul.htm (current as of October 2, 2001).

1219.John S. Torell, European-American Evangelical Association, July 1999, http://www.eaec.org/NL99jul.htm (current as of October 2, 2001).

1220.John S. Torell, European-American Evangelical Association, July 1999, http://www.eaec.org/NL99jul.htm (current as of October 2, 2001).

1221.MANFRED BARTHEL, THE JESUITS, HISTORY AND LEGEND OF THE SOCIETY OF JESUS, p. 16 (1984).

1222.COLLIER'S ENCYCLOPEDIA, vol. 13, p. 550 (1992).

1223.Melinda Henneberger, *Vatican Says Jews' Wait for Messiah Is Validated by the Old Testament, New York Times,* January 18, 2002, http://www.hughhewitt.com/past_news_links_01.02/01.18.02.Vatican_Says_Wait_for_Messiah.html (Current as of February 10, 2002).

1224.Michael Hoffman II, Secret Societies and Psychological Warfare, at p. 75 (2001).

1225.Michael Hoffman II, Secret Societies and Psychological Warfare, at p. 75 (2001).

1226.Vatican Tells Catholics Not to Attempt to Convert Jews, The Guardian, December 10, 2015, http://www.theguardian.com/world/2015/dec/10/vatica

n-tells-catholics-not-to-attempt-to-convert-jews.

1227.Edward Hendrie, *The Anti-Gospel, The Perversion of Christ's Grace Gospel, available at* www.antichristconspiracy.com and www.lulu.com.

1228.G. Richard Fisher, The Other Gospel Of John Hagee: Christian Zionism And Ethnic Salvation, *Personal Freedom Outreach*, http://www.pfo.org/jonhagee.htm (last visited on May 15, 2012).

1229.Steven Paas, Christian Zionism Examined, A Review of Ideas on Israel , the Church, and the Kingdom, at 28 (2012).

1230.Steven Paas, Christian Zionism Examined, A Review of Ideas on Israel , the Church, and the Kingdom, at 98 (2012).

1231.Steven Paas, Christian Zionism Examined, A Review of Ideas on Israel , the Church, and the Kingdom, at 102 (2012).

1232.Babylonian Talmud: *Tractate Sanhedrin, Folio 90a*, Sanhedrin Translated into English with Notes, Glossary and Indices Chapters I - VI by Jacob Shachter, Chapters VII - XI
by H. Freedman, B.A., Ph.D., Under the Editorship of Rabbi Dr I. Epstein B.A., Ph.D., D. Lit. (1961), *available at* http://www.come-and-hear.com/sanhedrin/sanhedrin_90.html.

1233.Babylonian Talmud: *Tractate Sanhedrin, Folio 90a*, Sanhedrin Translated into English with Notes, Glossary and Indices Chapters I - VI by Jacob

Shachter, Chapters VII - XI
by H. Freedman, B.A., Ph.D., Under the Editorship of Rabbi Dr I. Epstein B.A., Ph.D., D. Lit. (1961), *available at* http://www.come-and-hear.com/sanhedrin/sanhedrin_90.html.

1234.Michael Hoffman, *Judaism Discovered*, at 534 (2008).

1235.Elizabeth Dilling, *The Jewish Religion, Its Influence Today* (1963), *at* http://www.come-and-hear.com/dilling/chapt03.html#The_18_Benedictions.

1236.Steven Paas, Christian Zionism Examined, A Review of Ideas on Israel , the Church, and the Kingdom, at 105 (2012).

1237.Steven Paas, Christian Zionism Examined, A Review of Ideas on Israel , the Church, and the Kingdom, at 113 (2012).

1238.ALBERTO RIVERA, DOUBLE CROSS, Chick Publications, p. 12, 1981. *See also,* EDWIN A. SHERMAN, THE ENGINEER CORPS OF HELL, Library of Congress catalog card # 66-43354, p. 118 (1883); Congressional Record, House Bill 1523, contested election case of Eugene C. Bonniwell against Thos. S. Butler, February 15, 1913, at pp. 3215-16; BURKE MCCARTY, THE SUPPRESSED TRUTH ABOUT THE ASSASSINATION OF ABRAHAM LINCOLN, at pp. 14-16.

1239.EDMOND PARIS, THE SECRET HISTORY OF THE JESUITS, p. 29 (1975).

1240.*Id.*

1241.*Id.*

1242.EDMOND PARIS, THE SECRET HISTORY OF THE JESUITS, p. 26 (1975).

1243.Barbara Aho, *Mystery, Babylon the Great - Catholic or Jewish?, Part III, The Jesuits*, Watch Unto Prayer, *at* http://watch.pair.com/mystery-babylon.html (last visited on February 8, 2010) (quoting Maurice Pinay, *The Plot Against the Church, Chapter Thirty-Nine, Jewish-Freemasonic Infiltration into the Jesuit Order*, http://www.catholicvoice.co.uk/pinay/part4d.htm (last visited February 10, 2010)).

1244.THE PROTOCOLS OF THE LEARNED ELDERS OF ZION, Paragraph 3, Protocol 17, http://www.thewinds.org/library/protocols_of_zion.html (current as of September 9, 2001).

1245.THE PROTOCOLS OF THE LEARNED ELDERS OF ZION, Paragraph 4, Protocol 17, http://www.thewinds.org/library/protocols_of_zion.html (current as of September 9, 2001).

1246.ALBERTO RIVERA, DOUBLE CROSS, Chick Publications, p. 12, 1981. *See also,* EDWIN A. SHERMAN, THE ENGINEER CORPS OF HELL, Library of Congress catalog card # 66-43354, p. 118 (1883); Congressional Record, House Bill 1523, contested election case of Eugene C. Bonniwell against Thos. S. Butler, February 15, 1913, at pp. 3215-16; BURKE MCCARTY, THE SUPPRESSED TRUTH ABOUT THE ASSASSINATION OF ABRAHAM LINCOLN, at pp. 14-16.

1247.EDMOND PARIS, THE SECRET HISTORY OF THE JESUITS, p. 70 (1975).

1248.COLLIER'S ENCYCLOPEDIA, volume 13, p. 550 (1991); *see also,* EDMOND PARIS, THE SECRET HISTORY OF THE JESUITS, p. 70 (1975).

1249.EDMOND PARIS, THE SECRET HISTORY OF THE JESUITS, p. 73 (1975).

1250.EDMOND PARIS, THE SECRET HISTORY OF THE JESUITS, p. 70 (1975).

1251.*Id.* at 70-71.

1252.*Id.* at 71.

1253.COLLIER'S ENCYCLOPEDIA, volume 12, p. 516 (1991).

1254.SIDNEY HUNTER, IS ALBERTO FOR REAL?, p. 21 (1991); *see also,* EDMOND PARIS, THE SECRET HISTORY OF THE JESUITS, p. 35 (1975).

1255.SIDNEY HUNTER, IS ALBERTO FOR REAL?, Chick Publications, p. 21-23 (1988).

1256.ERIC JON PHELPS, VATICAN ASSASSINS: "WOUNDED IN THE HOUSE OF MY FRIENDS," P. 206 (2001).

1257.WILLIAM STILL, NEW WORLD ORDER, The Ancient Plan of Secret Societies, p. 79 (1990).

1258.ERIC JON PHELPS, VATICAN ASSASSINS: "WOUNDED IN THE HOUSE OF MY FRIENDS," p. 205 (2001).

1259.ERIC JON PHELPS, VATICAN ASSASSINS: "WOUNDED IN THE HOUSE OF MY FRIENDS," p. 205 (2001).

1260.WILLIAM STILL, NEW WORLD ORDER, The Ancient Plan of Secret Societies, pp. 81-91 (1990).

1261.COLLIER'S ENCYCLOPEDIA, volume 13, p. 550 (1991).

1262.EDMOND PARIS, THE SECRET HISTORY OF THE JESUITS, p. 75 (1975).

1263.*Id.*

1264.SAMUEL FINLEY BREESE MORSE, IMMINENT DANGERS TO THE FREE INSTITUTIONS OF THE UNITED STATES THROUGH FOREIGN IMMIGRATION AND THE PRESENT STATE OF THE NATURALIZATION LAWS, p. 9-10 (1835).

1265.COLLIER'S ENCYCLOPEDIA, vol. 13, p. 550 (1991).

1266.*Id.*

1267.*Id. See also,* EDMOND PARIS, THE SECRET HISTORY OF THE JESUITS, p. 39 (1975).

1268.COLLIER'S ENCYCLOPEDIA, volume 13, p. 550 (1991).

1269.PARIS at p. 64.

1270.PARIS at p. 65.

1271.PARIS at p. 65.

1272.PARIS at p. 65.

1273.JOHN W. ROBBINS, ECCLESIASTICAL MEGALOMANIA, at p. 32 (1999) (quoting *Summa Theologiae*, ii-ii, 7th article).

1274. JOHN W. ROBBINS, ECCLESIASTICAL MEGALOMANIA, at p. 40 (1999) (quoting The Second Vatican Council, *Gaudium et Spes, Pastoral Constitution on the Church in the Modern World,* at p. 69 (1965)).

1275.Babylonian Talmud, *Tractate Baba Bathra, Folio 54b* (1961), *available at* http://www.come-and-hear.com/bababathra/bababathra_54.html.

1276.Babylonian Talmud, *Tractate Baba Kamma, Folio 113b*, Translated into English with Notes, Glossary and Indices by E. W. Kirzner, M.A., Ph.D., M.Sc, under the Editorship of Rabbi Dr I. Epstein, B.A., Ph.D., D. Lit. (1961), *available at* http://www.come-and-hear.com/babakamma/babakamma_113.html.

1277.J. E. C. SHEPHERD, THE BABINGTON PLOT, Wittenburg Publications, p.14, 1987.

1278.EDMOND PARIS, THE SECRET HISTORY OF THE JESUITS, p. 26 (1975).

1279.J. E. C. SHEPHERD, THE BABINGTON PLOT, Wittenburg Publications, p.16, 1987 (quoting Marianus de Luce, S.J., Professor Canon Law, Gregorian University of Rome, *Institutes of Public Ecclesiastical Law,* with personal commendation from Pope Leo XIII, 1901).

1280.EDMOND PARIS, THE SECRET HISTORY OF THE JESUITS, Chick Publications, p. 166-167, 1975.

1281.Babylonian Talmud: *Tractate Sanhedrin, Folio 57b*, Sanhedrin Translated into English with Notes, Glossary and Indices Chapters I - VI by Jacob Shachter, Chapters VII - XI
by H. Freedman, B.A., Ph.D., Under the Editorship of Rabbi Dr I. Epstein B.A., Ph.D., D. Lit. (1961), *available at* http://www.come-and-hear.com/sanhedrin/sanhedrin_57.html (the section actually uses the word "Cuthean," but an endnote explains: "'Cuthean' (Samaritan) was here substituted by the censor for the original goy (heathen).").

1282.Babylonian Talmud: *Tractate Sanhedrin, Folio 57b*, supra.

1283.Michael Hoffman, *Judaism Discovered*, at 441 (2008).

1284.Michael Hoffman, *Judaism Discovered*, at 441 (2008).

1285.Canadian Jewish News, p. 9, August 23, 1980, quoted by Michael Hoffman in *Judaism Discovered*, at 442 (2008).

1286.Israel Shahak, Jewish History, Jewish Religion: The Weight of Two Thousand Years (1994) (citing Maimonides, Mishneh Torah, 'Laws on Murderers' 2, 11; Talmudic Encyclopedia, 'Goy'), *available at* http://www.Biblebelievers.org.au/jewhis5.htm#The%20Laws%20Against%20Non-Jews.

1287.Shahak, Jewish History (citing R. Yo'el Sirkis, Bayit Hadash, commentary on Beyt Josef, 'Yoreh De'ah' 158. Shahak explains in an endnote: “The two rules just mentioned apply even if the Gentile victim is ger toshav, that is a 'resident alien' who has undertaken in front of three Jewish witnesses to keep the 'seven Noahide precepts' (seven biblical laws considered by the Talmud to be addressed to Gentiles”)).

1288.Shahak, Jewish History (citing R. David Halevi (Poland, 17th century), Turey Zahav" on Shulhan 'Arukh, 'Yoreh De'ah' 158).

1289.Shahak, Jewish History (citing Talmudic Encyclopedia, 'Ger' (= convert to Judaism)).

1290.Shahak, Jewish History (citing Shulhan 'Arukh, 'Hoshen Mishpat' 426).

1291.Shahak, Jewish History (Tractate 'Avodah Zarah', p. 26b).

1292.Shahak, Jewish History (Maimonides, op. cit., 'Murderer' 4, 11).

1293.Shahak, Jewish History (citing Leviticus, 19:16). Shahak also states in an endnote: “Concerning the rendering 'thy fellow', see note 14 to Chapter 3.”

1294.Shahak, Jewish History (citing Maimonides, op. cit., 'Idolatry' 10, 1-2).

1295.Shahak, Jewish History (Shahak states in an endnote: “In both cases in section 'Yoreh De'ah' 158. The Shulhan 'Arukh repeats the same doctrine in 'Hoshen Mishpat' 425").

1296.Shahak, Jewish History, *supra*.

1297.EDMOND PARIS, THE SECRET HISTORY OF THE JESUITS, p. 21 (1975) (quoting *H. Boehmer*, professor at the University of Bonn, *Les Jesuits* (1910)).

1298.J. E. C. SHEPHERD, THE BABINGTON PLOT, Wittenburg Publications, p.118, 1987.

1299.J. E. C. SHEPHERD, THE BABINGTON PLOT, Wittenburg Publications, 1987.

1300.*Id.* at p. 104-117. *See also* COLLIER'S ENCYCLOPEDIA, volume 9, p. 97 (1991). *See also*, LES GARRETT, WHICH Bible CAN WE TRUST?, p. 60 (1982).

1301.COLLIER'S ENCYCLOPEDIA, volume 9, p. 620 (1991).

1302.COLLIER'S ENCYCLOPEDIA, volume 11, p. 536 (1991).

1303.COLLIER'S ENCYCLOPEDIA, volume 12, p. 192 (1991).

1304.COLLIER'S ENCYCLOPEDIA, volume 13, p. 550 (1991).

1305.COLLIER'S ENCYCLOPEDIA, volume 13, p. 550 (1991).

1306.EDMOND PARIS, THE SECRET HISTORY OF THE JESUITS, p. 69 (1975).

1307.FOXE'S BOOK OF MARTYRS, edited by William Byron Forbush, http://www.ccel.org/foxe/martyrs/fox117.htm.

1308.Norman Hammond, Archaeology Correspondent, Expert Views Differ on Jesuit's Role in the Piltdown Man forgery, *London Times*, July 15, 1980, http://www.clarku.edu/~piltdown/map_prim_suspects/Teilhard_de_Chardin/Chardin_Prosecution/expertsdiffer.html (website address current as of February 28, 2005).

1309.Ken Ham, The Lie: Evolution (1987), *available at* http://www.creationists.org/evolutionism-is-a-religion.html.

1310.Samuel KJBB, Tares Among The Wheat (Sequel to "A Lamp in the Dark"), Published on April 29, 2014, at 39:20 of video, https://www.youtube.com/watch?v=-aiHcghIdjM&spfreload=5.

1311.Israel Shahak, Jewish History, Jewish Religion: The Weight of Two Thousand Years (1994), *available at* http://www.Biblebelievers.org.au/jewhis2.htm#Prejudice%20and%20Prevarication.

1312.Moses Maiminodes, The Guide of the Perplexed [*Moreh Nevuk'him*], translated by Schlomo Pines [Chicago: University of Chicago Press, 1963], vol. 2, pp 618-19 (quoted by Michael Hoffman in Judaism's Strange Gods, at 65 (2002).

1313.Bobby O'Connor, *The Racism of Evolution Theory,* CHARLESTON GAZETTE, June 25, 1998 at P18. (Quoting CHARLES DARWIN, THE DESCENT OF MAN (1874)). *See also* Benno C. Schmidt, *Principle and Prejudice: The Supreme Court and Race in the Progressive Era. Part 1: The Heyday of Jim Crow,* 82 COLUM. L. REV. 444, 453 (1982).

1314.Rabbi Shraga Simmons, Ask Rabbi Simmons, Evolution vs. Creation, *at* http://judaism.about.com/library/3_askrabbi_o/bl_simmons_evolution.htm.

1315.Jewish Encyclopedia, Evolution, at http://www.jewishencyclopedia.com/view.jsp?artid=534&letter=E (last visited on April 17, 2010).

1316.Dennis Bratcher, *The Five Articles of the Remonstrants (1610)*, http://www.crivoice.org/creedremonstrants.html (last visited on October 13, 2011).

1317.Stephen Tomkins, John Wesley, A Biography, at 168 (2003) (emphasis added).

1318.Kenneth Talbot, Gary W. Crampton, D. James Kennedy, *Calvinism, Hyper-Calvinism, & Arminianism: A Workbook*, at 38 (2007).

1319.Society of Evangelical Arminians, http://evangelicalarminians.org/ (last visited on November 28, 2011).

1320.Micael Horton, Evangelical Arminians, Option or Oxymoron?, November 28, 2011, http://www.reformationonline.com/arminians.htm.

1321.Franklin Graham, *Expect Suffering, But Not Forever*, April 27, 2011, Billy Graham Evangelistic Association, http://www.billygraham.org/articlepage.asp?articleid=1162.

1322.Ken Johnson, The Gnostic Origins of Calvinism, at 69-70 (2013).

1323.Ken Johnson, The Gnostic Origins of Calvinism, at 69-70 (2013).

1324.Joseph Flatley, Flat Earth, March 21, 2015, https://www.youtube.com/watch?v=KTjjW2L2png&list=PL8hYGBs2qN3tv0NpSS9y3gzFte8fchZg-

1325.Joseph Flatley, Flat Earth, March 21, 2015, supra.

1326.Joseph L. Flatley, The Earth Is Flat and 'They' Don't Want You to Know, The Kernel, September 20th, 2015, http://kernelmag.dailydot.com/issue-sections/features-issue-sections/14360/flat-earth-truthers-youtube/

1327.Shulamit Aloni: "It's a Trick We Always Use," http://wideeyecinema.com/?p=3804 (last visited on December 12, 2010).

1328.Shulamit Aloni, Jewish Virtual Library, http://www.jewishvirtuallibrary.org/jsource/biography/aloni.html (last visited on December 12, 2010).

1329.ALBERTO RIVERA, DOUBLE CROSS, Chick Publications, p. 12, 1981. *See also,* EDWIN A. SHERMAN, THE ENGINEER CORPS OF HELL, Library of Congress catalog card # 66-43354, p. 118 (1883); Congressional Record, House Bill 1523, contested election case of Eugene C. Bonniwell against Thos. S. Butler, February 15, 1913, at pp. 3215-16; BURKE MCCARTY, THE SUPPRESSED TRUTH ABOUT THE ASSASSINATION OF ABRAHAM LINCOLN, at pp. 14-16.

1330.Joseph L. Flatley, The Earth Is Flat and 'They' Don't Want You to Know, The Kernel, September 20th, 2015, http://kernelmag.dailydot.com/issue-sections/features-issue-sections/14360/flat-earth-truthers-youtube/

1331.Joseph L. Flatley, Propaganda 2.0: Why Israel and Hamas Are Fighting a War with Rockets and Tweets, The IDF Takes to Social Media in its War on Hamas, The Verge, November 15, 2012, http://www.theverge.com/2012/11/15/3649792/israel-hamas-social-networking-twitter-gaza.

1332.Joseph L. Flatley, Realtime War: Israeli Military Liveblogs, Tweets Attack on Hamas, The Verge, November 14, 2012, http://www.theverge.com/2012/11/14/3645426/israel-hamas-military-liveblog-tweet-warfare.

1333.George Washington, The World's First Social Media War: Israel v. Hamas, Zero Hedge, November 12, 2012, http://www.zerohedge.com/contributed/2012-11-21/israel-winning-war-%E2%80%A6-social-media.

1334.Jonathan Cook, Internet Users Paid to Spread Israeli Propaganda, 21 July 2009, The Electronic Intifada, https://electronicintifada.net/content/internet-users-paid-spread-israeli-propaganda/8355.

1335.George Washington, The World's First Social Media War: Israel v. Hamas, Zero Hedge, November 12, 2012, http://www.zerohedge.com/contributed/2012-11-21/israel-winning-war-%E2%80%A6-social-media.

1336.Ali Abunimah, Israeli Students to Get $2,000 to Spread State Propaganda on Facebook, 4 January 2012, The Electronic Intifada, https://electronicintifada.net/blogs/ali-abunimah/israeli-students-get-2000-spread-state-propaganda-facebook.

1337.Israel to Pay Students to Defend it Online, Associated Press, USA Today, August 14, 2013, http://www.usatoday.com/story/news/world/2013/08/14/israel-students-social-media/2651715/.

1338.Robert Carter & Jonathan Sarfati, Why the Universe Does Not Revolve Around the Earth, Refuting Absolute Geocentrism, Creation Ministries International, 12 February 2015, http://creation.com/refuting-absolute-geocentrism.

1339.Robert Carter & Jonathan Sarfati, supra.

1340.Dr Robert W. Carter, Creation Ministries International, http://creation.com/dr-robert-carter (last visited on October 4, 2015).

1341.What We Believe, Creation Ministries International, http://creation.com/about-us#what_we_believe.

1342.What We Believe, Creation Ministries International, http://creation.com/about-us#what_we_believe.

1343.Robert Carter and Jonathan Sarfati, Flat Earth, and Other Nonsense, Dealing with Ideas That Would Not Exist Were it Not for the Internet, Creation Ministries, February 13, 2016, http://creation.com/refuting-flat-earth.

1344.Robert Carter and Jonathan Sarfati, Flat Earth, and Other Nonsense, Dealing with Ideas That Would Not Exist Were it Not for the Internet, Creation Ministries, February 13, 2016, http://creation.com/refuting-flat-earth.

1345.Robert Carter and Jonathan Sarfati, supra.

1346.Robert Carter and Jonathan Sarfati, supra.

1347.Robert Carter and Jonathan Sarfati, supra.

1348.Robert Carter and Jonathan Sarfati, supra.

1349.Robert Carter and Jonathan Sarfati, supra.

1350.Robert Carter and Jonathan Sarfati, supra.

1351.Robert Carter and Jonathan Sarfati, supra.

1352.Robert Carter and Jonathan Sarfati, supra.

1353.Robert Carter and Jonathan Sarfati, supra.

1354. Louis Doré, Professor Brian Cox had the perfect response to Flat Earthers, January 2018, https://www.indy100.com/article/brian-cox-flat-earth-theory-response-scientists-flat-earthers-twitter-globe-nasa-8121836.

1355. Louis Doré, Professor Brian Cox had the perfect response to Flat Earthers, January 2018, https://www.indy100.com/article/brian-cox-flat-earth-theory-response-scientists-flat-earthers-twitter-globe-nasa-8121836.

1356.H. Paul Honsinger, If Your Kid's Teacher Is a Flat Earther and Keeps Trying to Convince Them the

Earth Is Flat, What Will You Do?, https://www.quora.com/If-your-kids-teacher-is-a-flat-earther-and-keeps-trying-to-convince-them-the-earth-is-flat-what-will-you-do. (last visited on January 19, 2018).

1357.Pastor Charles Lawson Sermons and Teaching from the KJB, http://pastorcharleslawson.org/ (last visited on June 10, 2018).

1358.Angry Christian Pastor Speaks Out Against Flat Earth!! Never Question Men, NASA & SpaceX, April 30, 2018, https://www.youtube.com/watch?v=Kk0p8hAKs6k&t=3s.

1359.Flat Earth Debunked by Christian Mathematician (Isaiah 40:22, Conspiracy, Geocentric, Heliocentric), June 3, 2013, https://www.youtube.com/watch?v=Hultq1apHaE.

1360."The Flat Earth Debunked" (KJV Baptist Preaching), June 3, 2018, https://www.youtube.com/watch?v=Uwj6AQBUukE.

1361.Creation Research Society, Statement of Belief, https://www.creationresearch.org/index.php/about-crs/statement-of-belief (last visited on March 21, 1956).

1362.Creation Research Society, Statement of Faith, supra.

1363.Danny Faulkner, Geocentrism and Creation, Answers in Genesis, August 1, 2001, https://answersingenesis.org/creationism/arguments-to-avoid/geocentrism-and-creation/.

1364.Glenn Elert, The Scriptural Basis for a Geocentric Cosmology, 25 April 1999, http://hypertextbook.com/eworld/geocentric.shtml.

1365.Glenn Elert, supra.

1366.Glenn Elert, supra.

1367.The Kent Hovind Flat Earth Challenge, August 2, 2015, https://www.youtube.com/watch?v=gqYqs2R4wDA.

1368.The Kent Hovind Flat Earth Challenge, August 2, 2015, https://www.youtube.com/watch?v=gqYqs2R4wDA.

1369.Hemant Mehta, Neil deGrasse Tyson Explains Why He Believes Faith and Reason Are Irreconcilable, January 21, 2014, http://www.patheos.com/blogs/friendlyatheist/2014/01/21/neil-degrasse-tyson-explains-why-he-believes-faith-and-reason-are-irreconcilable/#QywEmBMHV2F0fyjk.99.

Other books available from Great Mountain Publishing®

The Sphere of Influence: The Heliocentric Perversion of the Gospel
Edward Hendrie
ISBN: 978-1-943056-06-4

This book is a sequel to *The Greatest Lie on Earth (Expanded Edition): Proof That Our World Is Not a Moving Globe.* It will primarily focus on the infiltration into the church of the superstitious myth of heliocentrism and how that infiltration has served to undermine the gospel. The gospel is the entire Holy Bible, not just some of it. Matthew 4:4. Christian belief is an all or nothing proposition. "All scripture is given by inspiration of God, and is profitable for doctrine, for reproof, for correction, for instruction in righteousness." 2 Timothy 3:16. God's account of his creation is part and parcel of the gospel. A person with genuine faith believes what Jesus said about both heavenly and earthly things. "If I have told you earthly things, and ye believe not, how shall ye believe, if I tell you of heavenly things?" John 3:12. Jesus is God. Jesus created all things in heaven and on earth. See Colossians 1:16-18. God has revealed himself through his creation. "[T]hat which may be known of God is manifest in them; for God hath shewed it unto them. For the invisible things of him from the creation of the world are clearly seen, being understood by the things that are made, even his eternal power and Godhead; so that they are without excuse." Romans 1:19-20. If men have a misunderstanding of God's creation, they will also have a misunderstanding of who God is. If people believe in a creation that does not exist, they consequently also believe in a creator that does not exist. It is essential, therefore, to have an accurate understanding of God's creation. God did not make a movable, spherical earth. If men believe in a heliocentric creation, they will necessarily believe in a heliocentric creator. A heliocentric creation does not exist. So also, a heliocentric creator does not exist. A heliocentric creator is a false god. We have been warned to avoid the preaching of a false gospel, which presents a false Jesus. "For if he that cometh preacheth another Jesus, whom we have not preached, or if ye receive another spirit, which ye have not received, or another gospel, which ye have not accepted, ye might well bear with him." 2 Corinthians 11:4.

Antichrist: The Beast Revealed
Edward Hendrie
ISBN-13: 978-0-9832627-8-7

The antichrist is among us, here and now. This book proves it by comparing the biblical prophecies about the antichrist with the evidence that those prophecies have been fulfilled. This book documents the man of sin's esoteric confession that he is the antichrist. You will learn how the antichrist has changed times and laws as prophesied by Daniel, and how he is today sitting in the temple of God, "shewing himself that he is God," in fulfillment of Paul's prophecy in 2 Thessalonians 2:4. The beast of Revelation has come into the world, "after the working of Satan with all power and signs and lying wonders, and with all deceivableness of unrighteousness," as prophesied in 2 Thessalonians 2:10. The antichrist's adeptness as a hypocrite is the reason for his evil success. Indeed, to be the antichrist, his evil character must be concealed beneath a facade of piety. "And no marvel; for Satan himself is transformed into an angel of light. Therefore it is no great thing if his ministers also be transformed as the ministers of righteousness; whose end shall be according to their works." 2 Corinthians 11:14-15. The key to revealing the identity of the antichrist is to uncover his hypocrisy. Because the hypocrisy of the antichrist is so extreme, those who have been hoodwinked by his religious doctrines will be shocked to learn of it. This book exposes the concealed iniquity of the antichrist and juxtaposes it against his publicly proclaimed false persona of righteousness, thus bringing into clear relief that man of sin, the son of perdition, who is truly a ravening wolf in sheep's clothing, speaking lies in hypocrisy. See Matthew 7:15 and 1 Timothy 4:1-3.

9/11-Enemies Foreign and Domestic
Edward Hendrie
ISBN-13: 978-0983262732

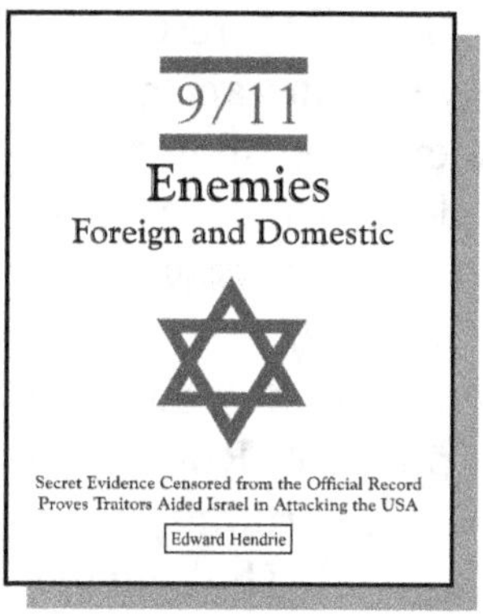

9/11-Enemies Foreign and Domestic proves beyond a reasonable doubt that the U.S. Government's conspiracy theory of the attacks on September 11, 2001, is a preposterous cover story. The evidence in 9/11-Enemies Foreign and Domestic has been suppressed from the official government reports and censored from the mass media. The evidence proves that powerful Zionists ordered the 9/11 attacks, which were perpetrated by Israel's Mossad, aided and abetted by treacherous high officials in the U.S. Government. 9/11-Enemies Foreign and Domestic identifies the traitors by name and details their subversive crimes. There is sufficient evidence in 9/11-Enemies Foreign and Domestic to indict important officials of the U.S. Government for high treason. The reader will understand how the U.S. Government really works and what Sir John Harrington (1561-1612) meant when he said: "Treason doth never prosper: what's the reason? Why if it prosper, none dare call it treason." There are millions of Americans who have taken an oath to defend the U.S. Constitution against all enemies foreign and domestic. The mass media, which is under the control of a disloyal cabal, keeps those patriotic Americans ignorant of the traitors among them. J. Edgar Hoover, former Director of the FBI, explained: "The individual is handicapped by coming face-to-face with a conspiracy so monstrous-he simply cannot believe it exists." 9/11-Enemies Foreign and Domestic erases any doubt about the existence of the monstrous conspiracy described by Hoover and arms the reader with the knowledge required to save our great nation. "My people are destroyed for lack of knowledge." Hosea 4:6.

Solving the Mystery of BABYLON THE GREAT
Edward Hendrie
ISBN-13: 978-0983262701

"Attorney and Christian researcher Edward Hendrie investigates and reveals one of the greatest exposés of all time. . . . a book you don't want to miss. Solving the Mystery of Babylon the Great is packed with documentation. Never before have the crypto-Jews who seized the reins of power in Rome been put under such intense scrutiny." Texe Marrs, Power of Prophecy. The evidence presented in this book leads to the ineluctable conclusion that the Roman Catholic Church was established by crypto-Jews as a false "Christian" front for a Judaic/Babylonian religion. That religion is the core of a world conspiracy against man and God. That is not a conspiracy theory based upon speculation, but rather the hard truth based upon authoritative evidence, which is documented in this book. Texe Marrs explains in his foreword to the book: "Who is Mystery Babylon? What is the meaning of the sinister symbols found in these passages? Which city is being described as the 'great city' so full of sin and decadence, and who are its citizens? Why do the woman and beast of Revelation seek the destruction of the holy people, the saints and martyrs of Jesus? What does it all mean for you and me today? Solving the Mystery of Babylon the Great answers these questions and more. Edward Hendrie's discoveries are not based on prejudice but on solid evidence aligned forthrightly with the 'whole counsel of God.' He does not condone nor will he be a part of any project in which Bible verses are taken out of context, or in which scriptures are twisted to mean what they do not say. Again and again you will find that Mr. Hendrie documents his assertions, backing up what he says with historical facts and proofs. Most important is that he buttresses his findings with scriptural understanding. The foundation for his research is sturdy because it is based on the bedrock of God's unshakeable Word."

The Anti-Gospel
Edward Hendrie
ISBN-13: 978-0983262749

Edward Hendrie uses God's word to strip the sheep's clothing from false Christian ministers and expose them as ravening wolves preaching an anti-gospel. The anti-gospel is based on a myth that all men have a will that is free from the bondage of sin to choose whether to believe in Jesus. The Holy Bible, however, states that all men are spiritually dead and cannot believe in Jesus unless they are born again of the Holy Spirit. Ephesians 2:1-7; John 3:3-8. God has chosen his elect to be saved by his grace through faith in Jesus Christ. Ephesians 1:3-9; 2:8-10. God imbues his elect with the faith needed to believe in Jesus. Hebrews 12:2; John 1:12-13. The devil's false gospel contradicts the word of God and reverses the order of things. Under the anti-gospel, instead of a sovereign God choosing his elect, sovereign man decides whether to choose God. The calling of the Lord Jesus Christ is effectual; all who are chosen for salvation will believe in Jesus. John 6:37-44. The anti-gospel has a false Jesus, who only offers the possibility of salvation, with no assurance. The anti-gospel blasphemously makes God out to be a liar by denying the total depravity of man and the sovereign election of God. All who preach that false gospel are under a curse from God. Galatians 1:6-9.

Bloody Zion
Edward Hendrie
ISBN-13: 978-0983262763

Jesus told Pontius Pilate: "My kingdom is not of this world." John 18:36. God has a spiritual Zion that is in a heavenly Jerusalem. Hebrews 12:22; Revelation 21:10. Jesus Christ is the chief corner stone laid by God in Zion. 1 Peter 2:6. Those who believe in Jesus Christ are living stones in the spiritual house of God. 1 Peter 2:5; Ephesians 2:20-22. Believers are in Jesus and Jesus is in believers. John 14:20; 17:20-23. All who are elected by God to believe in Jesus Christ are part of the heavenly Zion, without regard to whether they are Jews or Gentiles. Romans 10:12. Satan is a great adversary of God, who has created his own mystery religions. During the Babylonian captivity (2 Chronicles 36:20), an occult society of Jews replaced God's commands with Satan's Babylonian dogma. Their new religion became Judaism. Jesus explained the corruption of the Judaic religion: "Howbeit in vain do they worship me, teaching for doctrines the commandments of men." Mark 7:7. Jesus revealed the Satanic origin of Judaism when he stated: "Ye are of your father the devil, and the lusts of your father ye will do." John 8:44. Babylonian Judaism remains the religion of the Jews today. Satan has infected many nominal "Christian" denominations with his Babylonian occultism, which has given rise to "Christian" Zionism. "Christian" Zionism advocates a counterfeit, earthly Zion, within which fleshly Jews take primacy over the spiritual church of Jesus Christ. This book exposes "Christian" Zionism as a false gospel and subversive political movement that sustains Israel's war against God and man.

Murder, Rape, and Torture in a Catholic Nunnery
Edward Hendrie
ISBN-13: 978-1-943056-00-2

There has probably not been a person more maligned by the powerful forces of the Roman Catholic Church than Maria Monk. In 1836 she published the famous book, *Awful Disclosures of the Hotel Dieu Nunnery of Montreal*. In that book, she told of murder, rape, and torture behind the walls of the cloistered nunnery. Because the evidence was verifiably true, the Catholic hierarchy found it necessary to fabricate evidence and suborn perjury in an attempt to destroy the credibility of Maria Monk. The Catholic Church has kept up the character assassination of Maria Monk now for over 175 years. Even today, there can be found on the internet websites devoted to libeling Maria Monk. Edward Hendrie has examined the evidence and set it forth for the readers to decide for themselves whether Maria Monk was an impostor, as claimed by the Roman Catholic Church, or whether she was a brave victim. An objective view of the evidence leads to the ineluctable conclusion that Maria Monk told the truth about what happened behind the walls of the Hotel Dieu Nunnery of Montreal. The Roman Catholic Church, which is the most powerful religious and political organization in the world, has engaged in an unceasing campaign of vilification against Maria Monk. Their crusade against Maria Monk, however, can only affect the opinion of the uninformed. It cannot change the evidence. The evidence speaks clearly to those who will look at the case objectively. The evidence reveals that the much maligned Maria Monk was a reliable witness who made awful but accurate disclosures about life in a cloistered nunnery.

What Shall I Do to Inherit Eternal Life?
Edward Hendrie
ISBN-13: 978-0983262770

A certain ruler posed to Jesus the most important question ever asked: "Good Master, what shall I do to inherit eternal life?" (Luke 18:18) The man came to the right person. Jesus is God, and therefore his answer to that question is authoritative. This book examines Jesus' surprising answer and definitively explains how one inherits eternal life. This is a book about God's revelation to man. Except for the Holy Bible, this is the most important book you will ever read.

Rome's Responsibility for the Assassination of Abraham Lincoln, With an Appendix Containing Conversations Between Abraham Lincoln and Charles Chiniquy
Thomas M. Harris
ISBN-13: 978-0983262794

The author of this book, General Thomas Maley Harris, was a medical doctor, who recruited and served as commander of the Tenth West Virginia Volunteers during the Civil War. He rose in rank through meritorious service to become a brigadier general in the Union Army. General Harris established a reputation for faithfulness, industriousness, intelligence, and efficiency. He was noted for his leadership in preparing his troops and leading them in battle. He was brevetted a major general for "gallant conduct in the assault on Petersburg."

After the Civil War, General Harris served one term as a representative in the West Virginia legislature, and was West Virginia's Adjutant General from 1869 to 1870. General Harris was a member of the Military Commission that tried and convicted the conspirators who assassinated President Abraham Lincoln. He had first hand knowledge of the sworn testimony of the witnesses in that trial. This book summarizes the salient evidence brought out during the military trial and adds information from other sources to present before the public the ineluctable conclusion that the assassination of Abraham Lincoln was the work of the Roman Catholic Church. The Roman Catholic Church has been largely successful in suppressing the circulation of this book. This book has never been given a place on bookstore shelves, as it exposed too much for the Roman Catholic hierarchy to tolerate. Any display of this book would bring an instant boycott of the bookstore. It is only now, in the age of the internet, where the marketplace of ideas has been opened wide, that this book can be found by those searching for the truth of who was behind the assassination of Abraham Lincoln.

The above books can be ordered from bookstores and from internet sites, including, but not limited to:

www.antichristconspiracy.com
https://greatmountainpublishing.com
www.911enemies.com
www.mysterybabylonthegreat.net
www.antigospel.com
https://play.google.com
www.barnesandnoble.com
www.amazon.com

Edward Hendrie
edwardhendrie@gmail.com

www.ingramcontent.com/pod-product-compliance
Lightning Source LLC
LaVergne TN
LVHW052010200926
848481LV00005B/1459